Rock Mechanics and Engineering

Rock Mechanics and Engineering

Volume 2: Laboratory and Field Testing

Editor

Xia-Ting Feng

Institute of Rock and Soil Mechanics, Chinese Academy of Sciences,
State Key Laboratory of Geomechanics and Geotechnical Engineering,
Wuhan, China

CRC Press
Taylor & Francis Group
Boca Raton London New York

CRC Press is an imprint of the
Taylor & Francis Group, an **informa** business

A BALKEMA BOOK

First published in paperback 2024

First published 2017
by CRC Press/Balkema
4 Park Square, Milton Park, Abingdon, Oxon, OX14 4RN

and by CRC Press
2385 NW Executive Center Drive, Suite 320, Boca Raton FL 33431

CRC Press/Balkema is an imprint of the Taylor & Francis Group,
an informa business

Publisher's Note
The publisher has gone to great lengths to ensure the quality of this reprint
but points out that some imperfections in the original copies may be
apparent.

Library of Congress Cataloging-in-Publication Data

ISBN: 978-1-138-02760-2 (hbk)
ISBN: 978-1-03-291797-9 (pbk)
ISBN: 978-1-315-36425-4 (ebk)

DOI: 10.1201/9781315364254

Cover Photo: Rock Core Sample in Rock Mechanics Laboratory with Piece of
Crack Sample in Bucket.
Copyright: Namthip Muanthongthae
Courtesy of: www.shutterstock.com

Typeset by Integra Software Services Private Ltd

Visit the Taylor & Francis Web site at
http://www.taylorandfrancis.com

and the CRC Press Web site at
http://www.crcpress.com

Contents

Foreword   ix
Introduction   xi

Triaxial or True-triaxial Tests under Condition of Loading and Unloading   1

1 Introductory longer review for rock mechanics testing methods   3
   R. ULUSAY & H. GERCEK

2 Progress and applications of rock true triaxial testing systems   67
   X. LI, A. WANG & L. SHI

3 Rockburst concept and mechanism   97
   M.C. HE, G.L. ZHU & W.L. GONG

4 Laboratory acoustic emission study review   115
   X.L. LEI

5 Damage of rock joints using acoustic emissions   153
   Z. MORADIAN

Joint Tests   177

6 Morphological parameters of both surfaces of coupled joints   179
   C. PING, L. JIE & F. XIANG

7 Rock joints shearing testing system   217
   Y. JIANG

Dynamic and Creep Tests   251

8   Coupled static and dynamic test   253
    X.B. LI, M. TAO, L. WENG & Z.L. ZHOU

9   Dynamic behavior   285
    C. MENNA, D. ASPRONE & E. CADONI

10 Dynamic rock failure and its containment   305
    T.R. STACEY

11 Tests on creep characteristics of rocks   333
    Ö. AYDAN, T. ITO & F. RASSOULI

Physical Modeling Tests   365

12 Physical, empirical and numerical modeling of jointed rock
    mass strength   367
    P.H.S.W. KULATILAKE

13 Some recent progress in physical modeling of rock stability
    of tunnels or underground caverns in China   395
    W.S. ZHU, Q.Y. ZHANG, L.P. LI & Y. LI

Field Testing and URLs   431

14 Underground research laboratories   433
    J.S.Y. WANG, X-T. FENG & J.A. HUDSON

15 The Mont Terri rock laboratory   457
    P. BOSSART, F. BURRUS, D. JAEGGI & C. NUSSBAUM

16 Thermal properties and experiment at Äspö HRL   499
    J. SUNDBERG

17 Excavation response studies at AECL's underground research
    laboratory—1982 to 2010   531
    R.S. READ

18 URL and rock mechanics in Finland   567
    E. JOHANSSON

19 The Meuse/Haute-Marne underground research Laboratory:
    Mechanical behavior of the Callovo-Oxfordian claystone   593
    G. ARMAND, F. BUMBIELER, N. CONIL, S. CARARRETO, R. DE LA VAISSIÈRE,
    A. NOIRET, D. SEYEDI, J. TALANDIER, M.N. VU & J. ZGHONDI

    Series page   633

Foreword

Although engineering activities involving rock have been underway for millennia, we can mark the beginning of the modern era from the year 1962 when the International Society for Rock Mechanics (ISRM) was formally established in Salzburg, Austria. Since that time, both rock engineering itself and the associated rock mechanics research have increased in activity by leaps and bounds, so much so that it is difficult for an engineer or researcher to be aware of all the emerging developments, especially since the information is widely spread in reports, magazines, journals, books and the internet. It is appropriate, if not essential, therefore that periodically an easily accessible structured survey should be made of the currently available knowledge. Thus, we are most grateful to Professor Xia-Ting Feng and his team, and to the Taylor & Francis Group, for preparing this extensive 2017 "Rock Mechanics and Engineering" compendium outlining the state of the art—and which is a publication fitting well within the Taylor & Francis portfolio of ground engineering related titles.

There has previously only been one similar such survey, "Comprehensive Rock Engineering", which was also published as a five-volume set but by Pergamon Press in 1993. Given the exponential increase in rock engineering related activities and research since that year, we must also congratulate Professor Feng and the publisher on the production of this current five-volume survey. Volumes 1 and 2 are concerned with principles plus laboratory and field testing, *i.e.*, understanding the subject and obtaining the key rock property information. Volume 3 covers analysis, modelling and design, *i.e.*, the procedures by which one can predict the rock behaviour in engineering practice. Then, Volume 4 describes engineering procedures and Volume 5 presents a variety of case examples, both these volumes illustrating 'how things are done'. Hence, the volumes with their constituent chapters run through essentially the complete spectrum of rock mechanics and rock engineering knowledge and associated activities.

In looking through the contents of this compendium, I am particularly pleased that Professor Feng has placed emphasis on the strength of rock, modelling rock failure, field testing and Underground Research Laboratories (URLs), numerical modelling methods—which have revolutionised the approach to rock engineering design—and the progression of excavation, support and monitoring, together with supporting case histories. These subjects, enhanced by the other contributions, are the essence of our subject of rock mechanics and rock engineering. To read through the chapters is not only to understand the subject but also to comprehend the state of current knowledge.

I have worked with Professor Feng on a variety of rock mechanics and rock engineering projects and am delighted to say that his efforts in initiating, developing and seeing through the preparation of this encyclopaedic contribution once again demonstrate his flair for providing significant assistance to the rock mechanics and engineering subject and community. Each of the authors of the contributory chapters is also thanked: they are the virtuosos who have taken time out to write up their expertise within the structured framework of the "Rock Mechanics and Engineering" volumes. There is no doubt that this compendium not only will be of great assistance to all those working in the subject area, whether in research or practice, but it also marks just how far the subject has developed in the 50+ years since 1962 and especially in the 20+ years since the last such survey.

John A. Hudson, Emeritus Professor, Imperial College London, UK
President of the International Society for Rock Mechanics (ISRM) 2007–2011

Introduction

The five-volume book "Comprehensive Rock Engineering" (Editor-in-Chief, Professor John A. Hudson) which was published in 1993 had an important influence on the development of rock mechanics and rock engineering. Indeed the significant and extensive achievements in rock mechanics and engineering during the last 20 years now justify a second compilation. Thus, we are happy to publish 'ROCK MECHANICS AND ENGINEERING', a highly prestigious, multi-volume work, with the editorial advice of Professor John A. Hudson. This new compilation offers an extremely wide-ranging and comprehensive overview of the state-of-the-art in rock mechanics and rock engineering. Intended for an audience of geological, civil, mining and structural engineers, it is composed of reviewed, dedicated contributions by key authors worldwide. The aim has been to make this a leading publication in the field, one which will deserve a place in the library of every engineer involved with rock mechanics and engineering.

We have sought the best contributions from experts in the field to make these five volumes a success, and I really appreciate their hard work and contributions to this project. Also I am extremely grateful to staff at CRC Press / Balkema, Taylor and Francis Group, in particular Mr. Alistair Bright, for his excellent work and kind help. I would like to thank Prof. John A. Hudson for his great help in initiating this publication. I would also thank Dr. Yan Guo for her tireless work on this project.

Editor
Xia-Ting Feng
President of the International Society for Rock Mechanics (ISRM) 2011–2015
July 4, 2016

Triaxial or True-triaxial Tests under Condition of Loading and Unloading

Chapter 1

Introductory longer review for rock mechanics testing methods

R. Ulusay[1] *& H. Gercek*[2]
[1]*Department of Geological Engineering, Hacettepe University, Ankara, Turkey*
[2]*Department of Mining Engineering, Bulent Ecevit University, Zonguldak, Turkey*

Abstract: Rock mechanics involves characterizing the strength of rock material and the geometry and mechanical properties of the natural discontinuities of the rock mass. Rock engineering is concerned with specific engineering circumstances, for example, how much load will the rock support and whether reinforcement is necessary. Since the establishment of the International Society for Rock Mechanics (ISRM) in the 1960s, there have been important scientific developments and technological advances both in rock mechanics and rock engineering. Particularly, modeling of rock behavior, design methodologies for rock structures and rock testing methods are the main issues in these developments and advances. The models developed depend considerably on the input parameters such as boundary conditions and material, discontinuity and rock mass properties. For this reason, establishing how to obtain these input parameters for a particular site, rock mass and project is important. In this chapter, first, a brief historical account of material testing is given with a special emphasis on the evolution of testing of rocks. Next, rock mechanics testing methods including those for rock material, discontinuities and rock masses are critically reviewed in terms of laboratory and *in-situ* tests. Then, standardization of rock testing methods is mentioned within the context of the Suggested Methods (SMs) by the ISRM. Finally, current developments and future trends in rock testing methods are briefly discussed.

1 INTRODUCTION

Rocks have been used as a construction material since the dawn of civilization and different structures have been built in or on rock. The term "rock mechanics" refers to the basic science of mechanics applied to rocks. While the term "rock engineering" refers to any engineering activity involving rocks, in other words, or the use of rock mechanics in rock engineering within the context of civil, mining and petroleum engineering such as dams, rock slopes, tunnels, caverns, hydroelectric schemes, mines, building foundations etc. (Hudson & Harrison, 2000). Table 1 shows main areas of application of rock engineering.

The application of mechanics on a large scale to a pre-stressed, naturally occurring material is the main factor distinguishing rock mechanics from other engineering disciplines. Although, as early as 1773, Coulomb included results of tests on rocks collected from France in his paper (Coulomb, 1776; Heyman, 1972), the subject of rock mechanics started in the 1950s from a rock physics base and gradually became a discipline in its own

Table 1 Main areas of applications of rock engineering.

Eng'g	Underground	Surface
Mining	• Design and support of long-term (galleries, shafts, etc.) and short-term (gate roads, etc.) service openings • Design and support of production excavations (*e.g.* longwalls, stopes, room-and-pillar panels, etc.) • Design of pillars for room-and-pillar works, long-wall panels, shafts, etc. • Surface effects (*i.e.* subsidence) due to underground excavations • Rock or coal bursts, acoustic emission	Open-pit planning and design • Stability of rock slopes • Bench design • Road design
	Fragmentation (*i.e.* breaking, crushing, grinding) of rocks for mineral processing	
	Drilling, blasting, fracturing, cutting, digging, ripping, etc.	
Civil	Design and support of tunnels for • Transportation (road, railway, subway, navigational) • Conveyance (water, drainage, sewer) • Utility (water, electricity, cable, gas) • Power plants (access, intake, pressure, tailrace, etc.) Design and support of caverns for • Energy and science – Hydroelectric power plants – Nuclear power plants – Research facilities (*e.g.* CERN, neutrino detector) • Storage – Oil, water, natural gas, compressed air – Waste (chemical, nuclear) – Others (grain, food, etc.) • Public – Dwellings, train or subway stations, parking garages – Shopping, cultural, and sports centers – Offices, factories • Defense – Public (shelters, storage) – Military (arms, ammunition, vehicles, planes) – Nuclear (ICBM silos, defense command centers)	Stability of rock slopes (natural or man-made) for • highways or railways • canals • etc. Rock foundations for • buildings, • dams, • bridges, • etc.
Petroleum & Natural Gas	• Mechanical properties and behavior of cap and reservoir rocks • Drilling wells • Design and stability of wellbores • Hydro fracturing	

right during the 1960s. Rock mechanics was born as a new discipline in 1962 in Salzburg, Austria, mainly by the efforts of Professor Leopold Müller and he officially endorsed it at the first congress of the International Society for Rock Mechanics (ISRM) in 1966.

Since the formation of the ISRM, there have been many developments and technological advances in both rock mechanics and rock engineering. Nevertheless, the subject remains essentially concerned with rock modeling behavior, whether as a research subject or to support the design of structures to be built on or in rock masses. The models developed depend critically on the input parameters, such as boundary conditions (*e.g.*, *in-situ* stresses), intact rock (rock material) and rock mass properties. Laboratory and *in-situ* (field) tests provide important inputs for rock modeling and rock engineering design approaches. Therefore, for proper consideration given to most economical and safe performance of rock engineering project, adequate information on the properties of surface and subsurface rocks must be available. Correct evaluation of the properties of rock material and rock mass frequently requires laboratory and *in-situ* tests, supplemented with a high degree of experience and judgment.

Mechanical testing of materials has been carried out since about 1500 and testing machines have been in existence since the early 18th century (Timoshenko, 1953; Gray, 1988). In the 1920s, Josef Stini was probably the first to emphasize the importance of structural discontinuities as related to the engineering behavior of rock masses. Other notable scientists and engineers from a variety of disciplines, such as Von Karman (1911), King (1912), Griggs (1936), Ide (1936), and Terzaghi (1946) worked on the failure of rock materials (Hoek, 2007). In 1921 and 1931, Griffith proposed his theory of brittle material failure (Griffith, 1921) and Bucky started using a centrifuge to study the failure of mine models under simulated gravity loading, respectively. However, after the formal development of rock mechanics as an engineering discipline in the early 1960s, better understanding of the importance of rock mechanics in engineering practice, increasing demands from rock engineering studies and rapid advances in technology resulted in development of a number of laboratory and *in-situ* rock testing methods.

In this chapter, first, a brief historical account of material testing is given with a special emphasis on the evolution of testing of rocks. Next, rock mechanics testing methods are reviewed in terms of laboratory and *in-situ* tests. Then, standardization of rock testing methods is mentioned within the context of the Suggested Methods (SMs) by the ISRM. Finally, current developments and future trends in rock testing methods are briefly discussed.

2 HISTORICAL BACKGROUND

Interest in materials had begun and mechanical testing procedures possibly have been developed thousands of years ago during one of the eras when large-scale wood and stone structures were being built. Mankind has been utilizing rocks in different forms since early times. The earlier uses involved the natural caves and cliffs for accommodation and protecting people against their enemies. They also utilized rocks as excavation, cutting, fighting tools and creating fires through friction of rock. Although some of them were initially accidental findings, they later improved their knowledge and knew what type of rocks could be used. The positive science, which constitutes the basics of

(a)    (b) 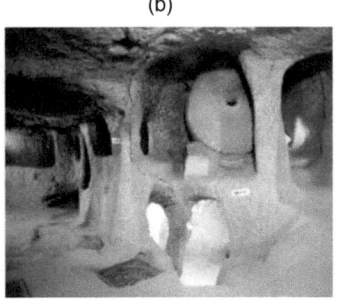   (c)

Figure 1 Examples of man-made historical cliff settlements and underground structures caved in rocks: (a) a cliff settlement and (b) an underground city in Cappadocia (Turkey), and (c) a rock-hewn settlement in Bezelik (East Turkmenistan) (Aydan, 2012).

rock mechanics and rock engineering of the modern time, is said to have been started following the Renaissance period. However, it is quite arguable who were the pioneers of mechanical laws governing solids and fluids and their testing and monitoring techniques in view of huge engineered structures related to rock built in the lands of Turan, China, India, Middle East, Egypt, Central America, Peru as well as Roman and old Greek lands and some of which were built more than thousands years ago with a high precision of modern days (Figure 1). These achievements could not be simply intuitive and an experience only, and there is no doubt that there were some mechanics and mathematics behind in their achievements, which need further through investigations to understand our achievements in rock mechanics and rock engineering. All these earlier civilizations had precise unit systems for measuring physical quantities, angles and time, which were the most fundamental elements of testing and monitoring in the past.

Da Vinci (ca. 1500) tested the tensile strength of wire, and his note "Testing the Strength of Iron Wires of Various Lengths" was the first record of mechanical testing (Figure 2a). He also studied the strength of columns and the influence of the width and length on the strength of a beam. During the 16th and 17th centuries, some experiments on mechanical properties of materials were carried out with simple testing apparatus. Galileo (1638) presented the first serious mathematical treatment of the elastic strength of a material in a structure subjected to bending (Loveday *et al.*, 2004). He also considered the strength of stone columns and this is illustrated in the well-known drawing that appeared in his *Discorsi e Dimostrazioni Matematiche* published in Leiden (Figure 3b), as discussed by Todhunter & Pearson (1886). Young (1773–1829) is associated with the measurement of the modulus of elasticity of materials. One of the earliest machines used for the systematic measurement of tensile strength was developed by a Dutch physicist Van Musschenbroek (1729) at the University of Leiden.

The first rock mechanics experimental studies were performed by Gauthey, who built a testing machine using the lever system and measured the compressive strength of cubic specimens (Figure 3c), in about 1770 for the design of the pillars for the Sainte Genevieve Church in Paris. Gauthey noted that the compressive strength of longer specimens was lower than the cube strength (Hudson *et al.*, 1972). The systematic assessment of the strength of materials at high temperatures was an important contribution by Fairbairn

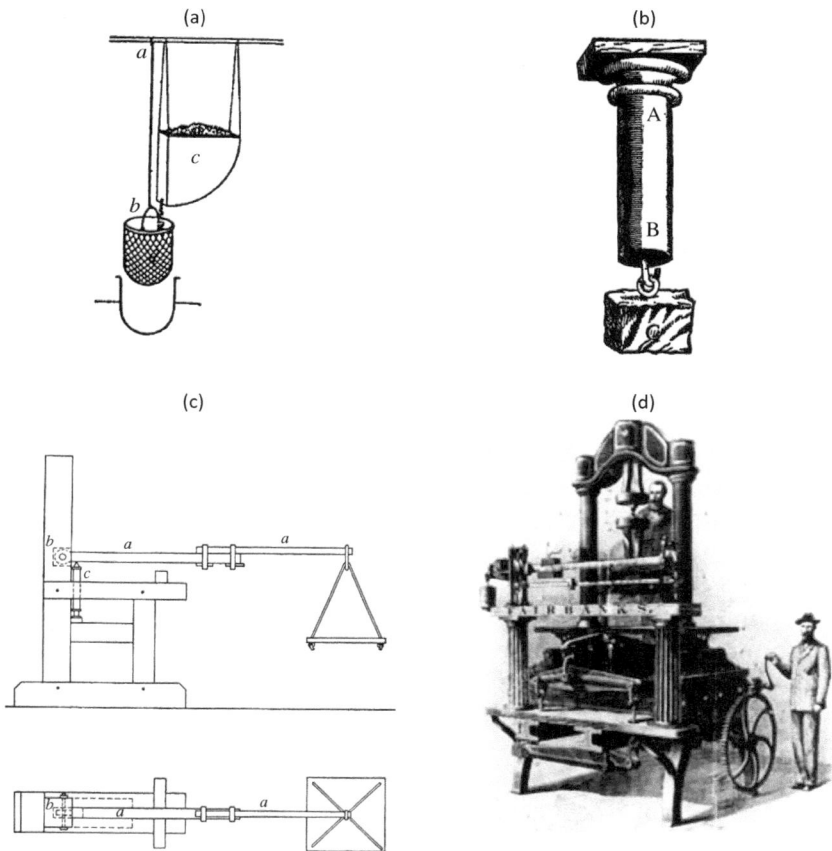

Figure 2 (a) Da Vinci's testing setup for the tensile strength testing of iron wires (a: wire, b: basket, c: hopper with sand) (after Lund & Byrne, 2001), (b) Galileo's illustration of tensile test (after Timoshenko, 1953), (c) Gauthey's testing machine (after Timoshenko, 1953), (d) a testing machine of the 1880s (after Abbott, 1884).

(1856). David Kirkaldy also made an important contribution to the determination of the strength of materials by designing and building a large horizontal hydraulic testing machine in order to undertake testing to uniform standards (Smith, 1982) and it was used in the first commercial testing laboratory of Kirkaldy in London. A typical testing machine of the 1880s is shown in Figure 3d.

During the early part of the 20th century, interesting work on the failure of rock materials was conducted by Von Karman (1911) and King (1912) in Europe and Griggs (1936) and Handin (1953) in the US, playing pioneering roles in the development of high pressure loading testing machines. In experimental rock mechanics, important developments were performed between 1945 and 1960, based on laboratory large-scaled experimental works by Mogi (1959), the studies on friction of discontinuities by Jaeger (1959, 1960) and large-scale triaxial tests performed by

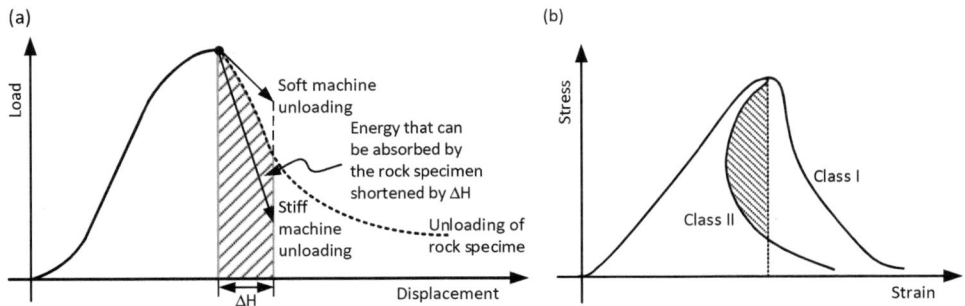

Figure 3 (a) Comparison of load-displacement curves obtained from stiff an soft machines (arranged from Hudson, 1989), (b) Complete load-displacement curves for rock samples under uniaxial compression obtained by Wawersik (1968) using a stiff testing machine to identify the so-called 'kick-back' or Class II post-peak response of very brittle rock.

Blanks & McHenry (1945), and Golder & Akroyd (1954). In addition, the studies by Rocha *et al.* (1955) and John (1962) motivated a more common use of large scale field shear testing of rock discontinuities in many parts of the world. In the absence of modern fracture mechanics theory and scaling laws, Professor Fernando L.L.B. Carneiro from Brazil, tried to establish a correlation between compressive strength and flexural tensile strength. In 1943, a challenging engineering problem inspired Carneiro to develop a new test method that is known as the Brazilian test (Fairbairn & Ulm, 2002).

Another important advance in rock testing was the development of stiff and servo-controlled testing machines. In 1966, it was recognized that the stiffness of the testing machine (relative to the slope of the post-peak load-displacement curve) determined whether failure of the specimen is stable or unstable. A soft machine causes sudden failure by the violent release of stored strain energy, *i.e.* by the testing system itself (Figure 3a). In their state of the art review, Hudson *et al.* (1972) indicated that the advantage of developing stiff testing machines was first suggested by Spaeth (1935). In 1969, complete load-displacement curves for rock samples under uniaxial compression was obtained by Wolfgang Wawersik using the stiff testing machine which was first developed by Cook at the Chamber of Mines (South Africa) and then modified by Wawersik to increase its stiffness. By adding simple 'post-peak control jacks' between the lower and basal crossheads to oppose the rapid release of crosshead energy during the post-peak unloading regime, Wawersik was able to obtain the response illustrated in Figure 3b and to identify the so-called Class II post-peak response of very brittle rock (Wawersik, 1968). Then, laboratory tests on machine stiffness and rock failure and the development of such machines were continued by several investigators (*e.g.*, Cook, 1965; Bieniawski, 1966; Waversik & Fairhurst, 1970; Hudson *et al.*, 1971; Martin, 1997).

After the establishment of the ISRM Commission on Testing Methods in 1966, a number of laboratory and field testing methods to be used in rock engineering were developed and/or improved with the efforts of the Commission, its Working Groups and cooperation among other ISRM Commissions (ISRM, 1981, 2007, 2014), based on the previous experiences and new developments in technology. These methods are briefly given in Section 3. In this period, the use of computerized methods of test control and automatic test data collection and analysis also became popular and some experimental contributions were made on the determination of shear strength (e.g., Barla et al., 2007) and deformability characteristics.

For small scale excavations in rock, the data obtained for the intact rock from laboratory testing might be sufficient to carry out an adequate design. However, rock is usually intersected by many geological planes of weakness and, if a significant number of these is involved in the excavation, intact rock data alone will not be sufficient. The properties and number of these weaknesses will modify the behavior of the rock mass to such an extent that the behavior of the intact rock may become almost irrelevant. This situation suggested that laboratory tests can quantify the behavior of intact rock and the extension of this approach to quantify rock mass behavior is to carry out large-scale in-situ (field) tests. Therefore, in addition to rock mechanics laboratory methods, particularly after the establishment of the ISRM, in-situ (field) tests were considered to also have vital importance in rock engineering applications and they gained an increasing popularity both in research and practice. This was a result of inevitable appreciation of the differences between the mechanical behaviors of intact rock and rock mass, as well as the realization of scale effect.

From the second half of the 20th century to the present, important contributions were made to the development and improvement of the in-situ testing methods. One of the categories considered in in-situ tests includes the tests used for determining in-situ deformability of rock masses, such as plate loading, flat jack and dilatometer tests which are discussed in Section 3.2.

The other category of in-situ methods commonly applied in rock engineering practice is geophysical techniques. The main emphasis of geophysical surveys in the formative years was for petroleum and mineral exploration. From the 1950s until the present time, the geophysical methods have enjoyed an increasing role in geotechnical projects, and now are used in an almost routine manner to provide information on site parameters, such as in-situ dynamic properties, depth to and condition of rock that in some instances are not obtainable by other methods, degree of saturation, chemistry, and thermal properties of rocks, etc. Since 1981, a number of geophysical methods were accepted as ISRM SMs (ISRM, 1981, 2007) and now are being commonly used in practice. For the last two decades, seismic imaging has had an increasing popularity particularly as it relates to rock-burst investigations (Young, 1993).

The behavior of rocks is significantly influenced by the in-situ stress field and other factors such as water, which are also usually subject to significant local and regional variability. The need for understanding of in-situ stresses in rocks has been recognized by engineers and geologists for a long time, and many methods to measure these stresses have been proposed since the early 1930s. One of the earliest measurements of in-situ stresses using surface relief methods was reported by Lieurance (1933, 1939) from the US Bureau of Reclamation in Denver. Pierre Habib was involved in the development and application of the flat jack method as early as 1950 (Habib, 1950; Mayer et al., 1951; Habib & Marchand, 1952), and this method was also used to measure the in-situ moduli of rock masses (Habib, 1950), as were dynamic methods (Brown & Robertshaw, 1953; Evison, 1953). After the 1960s a wide range of methods of rock stress measurement had been developed. These methods, such as hydraulic fracturing, the CCBO (Compact

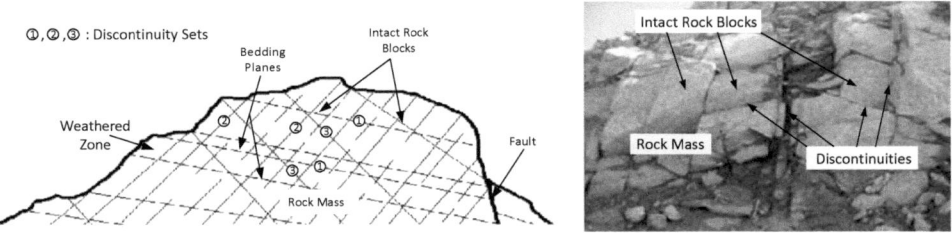

Figure 4 Rock mass and its two elements (intact rock and discontinuities).

Conical-ended Borehole Overcoring) technique, other overcoring methods, the flat jack method and other issues considered *in-situ* stress measurements were also accepted as ISRM SMs and published by the ISRM (ISRM, 2007, 2014; Sugawara & Obara, 1999; Hudson *et al.*, 2003; Sjöberg *et al.*, 2003; Haimson & Cornet, 2003; Christiansson & Hudson, 2003; Stephansson & Zang, 2012).

3 ROCK TESTING METHODS

The rock mass is composed of intact blocks of rock separated by discontinuities such as bedding planes, joints, schistosity planes, faults and sheared zones (Figure 4). These rock blocks may vary from fresh and unaltered rock to highly decomposed and disintegrated rock. Under applied stresses, the rock mass behavior is generally governed by the interaction of the intact rock blocks with the discontinuities. By considering the rock mass itself and its two elements, rock testing methods can mainly be categorized into three groups, such as tests on intact rock, discontinuities, and rock masses. Intact rock properties and those of smooth and slickensided discontinuities can be determined with the aid of laboratory tests. But, since the strength of rock masses depends on the nature of both intact rock material and discontinuities, and due to the difficulties in sampling from rock masses and scale effect, determination of geomechanical properties of rock masses is one of the main problems in rock engineering. In other words, it is often difficult to explain the behavior of a rock structure designed on the basis of intact rock properties determined in the laboratory. This requires field measurements for rock masses.

In addition, knowledge of *in-situ* stresses has a vital importance in most rock engineering studies. Although initial estimates can be made based on simple guidelines, field measurements of *in-situ* stresses are the only true guide for critical structures. Therefore, in terms of testing environment, rock mechanics tests can also be categorized as "laboratory tests" and "*in-situ* (field) tests". These methods are known as direct methods. However, there are also indirect methods, such as empirical correlations and estimations from some rock mass classifications, combination of rock material and discontinuity properties using analytical and numerical methods and back-analyzing using field observations.

In the following sub-sections, mainly based on the available ISRM SMs, direct tests and the typical applications of the intact rock, discontinuities and rock mass properties in rock engineering are briefly reviewed.

3.1 Laboratory tests

Rock mechanics laboratory tests are performed to determine a physical and/or mechanical property of intact rock or discontinuity. The property determined by the test is generally used for

(a)   classification and characterization of intact rock and
(b)   rock engineering design by analytical, numerical and empirical (*e.g.* rock mass classification systems) methods.

3.1.1 Laboratory tests on intact rock

The laboratory test methods for intact rock are mainly divided into two categories:

(a)   Classification and characterization tests:
    (a.1)   Unit weight, porosity, water content, absorption (physical properties) tests
    (a.2)   Hardness tests (Schmidt rebound hardness, Shore hardness, indentation hardness index)
    (a.3)   Strength index tests (point load strength index, block punch strength index, needle penetration index)
    (a.4)   Resistance to abrasion (Cerchar abrasivity index, Los Angeles abrasion)
    (a.5)   Uniaxial compressive strength (UCS) and deformability tests
    (a.6)   Other index tests (slake durability index, swelling index)
    (a.7)   Sound velocity tests
    (a.8)   Permeability

(b)   Fundamental tests to determine intact rock properties to be used in rock engineering design:
    (b.1)   UCS and deformability (Young's modulus and Poisson's ratio) tests
    (b.2)   Triaxial compressive strength test (shear strength of rock material)
    (b.3)   Tensile strength tests (direct and Brazilian tests)
    (b.4)   Creep tests (time dependent properties)

Typical applications of the intact rock properties are given in Table 2, where only those properties determined by the ISRM SMs are considered.

Index properties most closely relate to the behavior of intact rock, but are of lesser importance and require caution when used in the prediction of rock mass behavior. Index properties of intact rock are generally used; (i) to further aid in geo-engineering classification and as indicators of rock mass behavior, (ii) to provide a measure of the "quality" of the rock, and (iii) to indirectly estimate fundamental rock properties by empirical relationships.

Although index tests are cheap and can be performed quickly, they do not determine an intrinsic rock property and are not considered in rock engineering design. Determination of the engineering properties of rocks is an important part of rock engineering studies and is conducted with the aid of fundamental laboratory tests listed above. Rock engineer should consider whether emphasis is to be placed on index tests, fundamental tests or combination of the two. Some standards (such as of ASTM) and suggested methods (such as of ISRM) provide guidance related to the specific procedures for performing these laboratory tests.

Table 2 Typical applications of intact rock properties based on the current ISRM SMs.

Property				Symbol	Typical Applications
Physical	Density			ρ	General, 1, 2, 6
	Porosity			n	General, 1, 2
	Water Content			w	General, 2
	Permeability			k	Flow of fluids through rock
Mechanical	Strength	Static	Uniaxial Compressive Strength (UCS)	σ_c	1, 2, 3, 4, 5, 6, 9
			Tensile Strength — Direct	σ_t	1, 2, 4, 6
			Brazilian	σ_{tB}	
		Dyn.	UCS and Brazilian Tensile Strengths by Split-Hopkinson Pressure Bar	$(\sigma_c)_{dyn}$ $(\sigma_{tB})_{dyn}$	2, 7
			Triaxial Compressive Strength	$\sigma = f(p)$	Determination of cohesion & angle of internal friction, 4
			Direct Shear Strength	$\tau = f(\sigma)$	
	Deformability	Stat.	Complete Stress-Strain Curve in Uniaxial Compression	$\sigma = f(\varepsilon_a)$ $\sigma = f(\varepsilon_d)$	Studies on intact rock behavior
			Young's Modulus	E	1, 6
			Poisson's Ratio	ν	1, 6, 8
		Dyn.	Propagation Velocities of Elastic (P- & S-) Waves	v_P v_S	Determination of E_{dyn} & v_{dyn}, 3, 7
		Creep	Creep Characteristics in Uniaxial, Triaxial Comp. & Brazilian Tests	$\varepsilon = f(t)$	Time-dependent stress analysis
	Fracture Toughness	Mode I	Chevron Bend Specimen	K_{CB}	Rock fracture mechanics studies
			Short Rod Specimen	K_{SR}	
			Cracked Chevron Notched Brazilian Disk Specimen	K_{IC}	
			Notched Semi-Circular Bend Specimen	K_{ISCB} & $K_{IC}(t)$	
		Mode II	Punch-Through Shear with Confining Pressure	K_{IIC}	

1. Classification of intact rock
2. Empirical estimation of other material properties
3. Empirical estimation of rock mass properties
4. Parameter for failure criteria
5. Input for rock mass classification systems
6. Input for static stress analysis
7. Input for dynamic stress analysis
8. In-situ stress estimation & measurement
9. Mechanical excavation studies
10. Weathering properties

p: confining pressure
σ: normal stress
τ: shear stress
t: time
ε: strain
ε_a: axial strain
ε_d: diametric strain

Index	Strength	Point Load Strength Index	$I_{S(50)}$	1, 2, 5
		Block Punch Strength Index	BPI	1, 2
		Needle Penetration Index	NPI	2, 10
	Hardness	Schmidt Rebound Hardness	SRH	1, 2, 9
		Shore Hardness	SH	1, 2, 9
		Indentation Hardness Index	IHI	1, 2, 9
	Others	Slake Durability Index	I_{d2}	1, 2, 10
		Swelling Index	I_{sp} I_{ss}	Determination of swelling pressure & strain, 2, 10
		Los Angeles Abrasion	—	Determination of aggregate resistance to abrasion
		Cerchar Abrasivity Index	CAI	1, 9

As one of the physical properties of intact rock, water content is an indirect indication of porosity of intact rock or clay content of sedimentary rock. Unit weight, which is a measure of mass per unit of volume, is an indirect indication of weathering and soundness, and it depends on the mineralogical composition, porosity and the material filling the voids. Porosity is an indirect indicator of weathering and soundness and governs permeability, and it varies with grain size distribution, grain shape, depth and pressure.

As an index test to indirectly estimate the UCS of intact rock and to be used as input in rock mass classification (Bieniawski, 1989), the point load strength index test (ISRM, 1981, 2007) has been widely used in practice due to its testing ease, simplicity of specimen preparation and field applications. It gives the standard point load strength index, $I_{S(50)}$, calculated from the point load at failure and the size of the specimen, with size correction to an equivalent core diameter of 50 mm. It is customary to convert the $I_{S(50)}$ to an equivalent UCS by multiplying by a factor of k. A wide variety of k values have, therefore, been recommended by various investigators following theoretical considerations and experimental studies. Point load tests by Reed et al. (1980) have shown that the factor k varies with both rock type and weathering grade. It was also reported by Norbury (1986) that wide scatter of values of k ranging from 13 to 50 had appeared in the literature although there was an accumulation of values between about 16 and 24. Therefore, in rock engineering community, it is now agreed that $I_{S(50)}$ should be used carefully as an index in its own right. The failure mode for point load test is primarily by tensile fracturing. As the point load failure is due to Mode I fracturing, it is therefore anticipated some correlations between Mode I Fracture Toughness (K_{IC}) and $I_{S(50)}$.

When rock cores are only divided into small discs, due to the presence of thin weakness planes, the core length may be too short to allow preparation of the specimens long enough even for the point load strength index test. In addition, the degradation of the surrounding rock due to various causes may increase and sampling for laboratory testing becomes difficult. By considering these and some limitations associated with the estimation of UCS from $I_{S(50)}$ mentioned above, two alternative strength index tests, block punch strength index (BPI) test (Ulusay et al., 2001) and needle penetration (NP) test (Ulusay et al., 2014) were developed and also accepted as ISRM SMs (ISRM, 2007, 2014). The BPI test can be performed using a thin rock disk with a portable apparatus fitted to the columns of the point load test frame, and so may be conducted in the laboratory and field (Figure 5a). It is mainly used to predict the UCS and tensile strength of rock material with a lower estimation error and strength classification. The NP test determines the needle penetration index (NPI) with the aid of a portable light-weight non-destructive device (Figure 5b) and is used for the estimation of UCS and some other properties of soft and weak rocks both in the field and laboratory (Table 2).

Hardness is the characteristic of a solid material expressing its resistance to permanent deformation. Hardness of an intact rock mainly depends on mineral composition and density. Typical measures are the Schmidt rebound hardness (SRH) number, Shore hardness and indentation hardness index. SRH is a measure of the hardness of the intact rock by count the rebound degree and can be determined using a portable equipment both in the field and laboratory (ISRM, 1981, 2007). At the same time, the SRH can be used to estimate the UCS of the intact rock and as an input of the failure

(a)   (b)

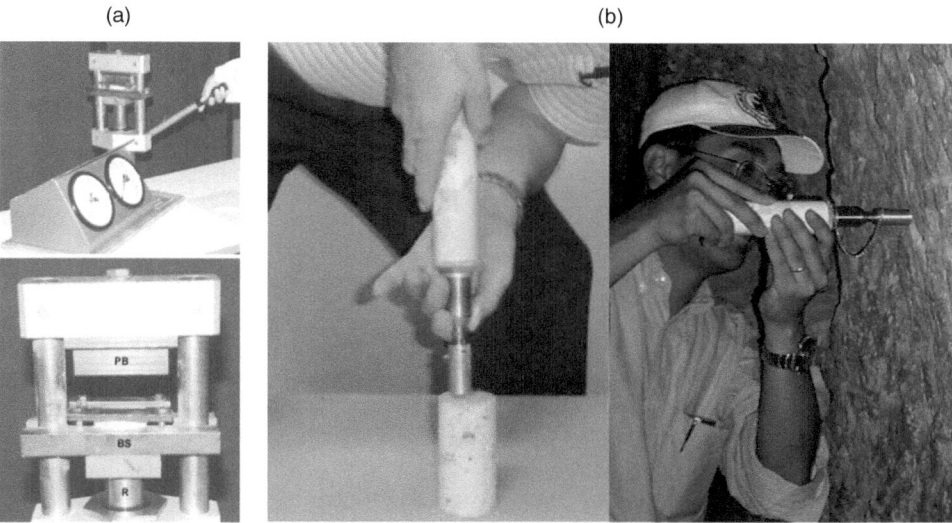

Figure 5 (a) BPI test device fitted into a point load testing frame (PB: punching block; BS: base support; R: ram) (Ulusay *et al.*, 2001), (b) the needle penetration test in laboratory and field (Ulusay *et al.*, 2014).

criterion developed by Barton (1976) for the estimation of the shear strength of rough and undulated discontinuity surfaces. Aydin (2009) proposed a revised SM, which supersedes the portion of earlier ISRM document, for determining the SRH of rock surfaces both in laboratory conditions and *in situ* with an emphasis on the use of this hardness value as an index of the UCS and E of intact rock. However, the method has some limitations: (i) highly fractured and closely jointed rocks are difficult to test, (ii) the method is not applicable to extremely weak rocks, and (iii) non-homogenous rocks are difficult to test. Shore hardness (SH) test is a convenient and non-destructive method in measuring the hardness of intact rock and used in rock mechanics since it can be used a predictor of other mechanical properties of rocks, especially the UCS. It was also accepted as an SM by the ISRM in 2006 (Altindag & Guney, 2006; ISRM, 2007). Indentation hardness index (IHI) test is another method for characterization of hardness of rock material (ISRM, 1981, 2007). IHI is used to assess some strength properties (UCS and tensile strength) with which it can be correlated, and may also be used in the prediction of drillability and cuttability of rocks.

The velocity measurements provide correlation to physical properties of rocks (mainly with porosity, strength and static modulus) in terms of compaction degree of the material and/or are used in determining dynamic elastic constants and as an index in their own right indicating anisotropy and/or inhomogeneity. Wave velocities are also commonly used to assess the degree of rock mass fracturing at large scale. Measurements of wave velocity are often done by using P (compression)-wave and, sometimes, S (shear)-wave. A well compacted rock has generally high velocity as the grains are all in good contact and waves are traveling through the solid. Wave velocity in rock cores is easily determined by measuring the travel time of vibrational waves introduced by piezoelectric crystals. Wave velocity can be determined in laboratory using one of the three non-destructive methods including the high and low frequency ultrasonic pulse techniques and the resonant method as suggested by ISRM (1981, 2007). However, most recently, the first two methods were upgraded to unify the two ultrasonic methods by a generalized scheme applicable to any specimen shape/size at any frequency within the ultrasonic range (> 20 kHz) and to suggest possible modifications in test procedures and specimen preparation to account for the special microfractures encountered in common rock types (Aydin, 2014; ISRM, 2014).

Changes in rock properties due to processes of chemical and mechanical breakdown can be very important in engineering applications. The ability of a material to resist abrasion, wear and breakdown with time is known as durability. Durability is particularly important for soft/weak rocks, such as shales, marls, mudstones, claystones and tuffs. The durability of such rocks, as a measurement of their deterioration over time, strongly depends on the interaction between the rock and water. This interaction is referred to as slaking and it often results in dissolution of particles, creation of fractures and flaking of surface layers (Santi, 1996). This non-durable behavior of rocks may be responsible for slope stability problems due to rapid slope degradation by loss of strength of the surface material, embankment failures and long-term loss of intact strength affecting the stability of underground openings. For example, a tunnel excavated in shale, which is one of the materials which degrades, may initially be stable, but it may collapse a few days later. Because of the physical interdependence between durability and slaking, durability of rocks is mainly measured by an index test called Slake Durability Test. This test is intended to assess the resistance offered by a rock sample to weakening and disintegration when subjected to two standard cycles of wetting and drying. The loss of sample weight is a measure of the susceptibility of the rock to the combined action of slaking and mechanical erosion and the test provides the calculation of the slake durability index (I_{d2}), as described by both ISRM (1981, 2007) and ASTM (2008c). The experimental studies on clay-bearing weak and soft rocks conducted by Gokceoglu *et al.* (2000) suggest that a series of repeated wetting and drying processes contribute to an increase in the amount of disintegrated clay minerals from the sample, and therefore, it seems a better way to apply more than two cycles in the slake durability test in order to obtain I_d values that better represent the slaking behavior of such rocks. Similar conclusions were also drawn by some other researchers (Taylor, 1988; Moon & Beattie, 1995; Ulusay *et al.*, 1995; Bell *et al.*, 1997) who studied soft and clay-bearing rocks.

Abrasivity measures the abrasiveness of rock materials against other materials, *e.g.* steel. Rock abrasivity plays an important role in characterizing a rock material for excavation purpose. Abrasivity is highly influenced by the amount of quartz mineral in the rock material. The higher quartz content gives higher abrasivity. Abrasivity is measured by several tests, however, the Cerchar abrasivity test and Los Angeles abrasion test are the most commonly used methods. The Cerchar test was proposed by the Laboratoire du Centre d'Etudes et Recherches des Charbonnages (Cerchar) in France (Valantin, 1973) and is intended as an index test for classifying the abrasivity of an intact rock. The test measures the wear on the tip of a steel stylus having a Rockwell Hardness of HRC 55 and determines an index called Cerchar Abrasivity Index (CAI) for the rock's abrasivity. The Los Angeles test developed for highway aggregates subjects a graded sample to attrition due to wear between rock pieces and also to impact forces produced by an abrasive charge of steel spheres. The procedures for both tests are available in ISRM (2007, 2014; Alber *et al.*, 2014) and ASTM (2010, 2014a).

Permeability is a measure of the ability of a material to transmit fluids. It is a critical property in defining the flow capacity of a rock sample. Most rocks, including igneous, metamorphic and chemical sedimentary rocks, generally have very low permeability. The permeability of intact rock is governed by porosity. Porous rocks such as sandstones usually have high permeability while granites have low permeability. Generally, the permeability of intact rock, which is known as primary permeability and is the rate of fluid flow through pore spaces, is very low, mostly less than 10^{-7}–10^{-6} cm/s. Since there are many fractures and karstic cavities in rock mass, the permeability of rock mass or secondary permeability, which is the rate of flow through secondary pores, cavities and fractures, is far higher than that of intact rock. Rocks such as granite and massive limestone with low primary permeability may be very permeable if they have been intersected by a network of discontinuities. Therefore, permeability of intact rock and rock mass is separately considered, and in most rock engineering applications, the secondary permeability dominates the design and construction. Determination of secondary permeability for rock mass is discussed in Section 3.2. The permeability of rock material is generally used for the purposes of classification, characterization, and in the studies related to the flow of fluids through rock. There are some laboratory methods proposed for determination of the permeability of the intact rock. A standard method for determining the coefficient of specific permeability for the flow of air through intact rock was recommended by ASTM (2013a).

Rock strengths are very different depending on the stress field applied to the rock. All rocks are very much stronger in compression than in tension. In terms of compression, the two common laboratory tests to determine the compressive strength of rock are UCS test and triaxial confined compression test. The other type of rock strength is the tensile strength. In addition to its use in strength classification (*e.g.*, Deere & Miller, 1966) and rock mass classification (*e.g.*, RMR system of Bieniawski, 1989), characterization of the intact rock, indirect estimation of the tensile strength of the rock material and as an input for the rock mass strength criterion (Hoek *et al.*, 2002), the UCS is one of the most important mechanical properties of intact rock. It is used in design, analysis and modeling and also most useful as a means for comparing rocks and classifying their likely behavior (Table 2). The method for determination of the UCS has been standardized by ASTM (2014b) and suggested by ISRM (1981, 2007). The method is simple, but it is time consuming, expensive and requires test specimens of particular sizes in order to fulfill testing standards, which is particularly difficult for weak and soft rocks. Therefore, indirect tests such as point load strength index, block punch strength index, Schmidt hammer and needle penetration tests are often performed to indirectly estimate the UCS by using empirical relationships between these index properties and UCS.

Shear strength is used to describe the strength of intact rock, to resist deformation due to shear stress. Rock resists against shear stress by two internal mechanisms: cohesion and internal friction. Cohesion is a measure of internal bonding of the rock material and internal friction is caused by contact between particles. Shear strength of rock material can be determined by direct shear test and triaxial compression test. In practice, the latter method is widely used and accepted. With a series of triaxial tests conducted at different confining pressures, peak axial stresses (σ_1) are obtained at

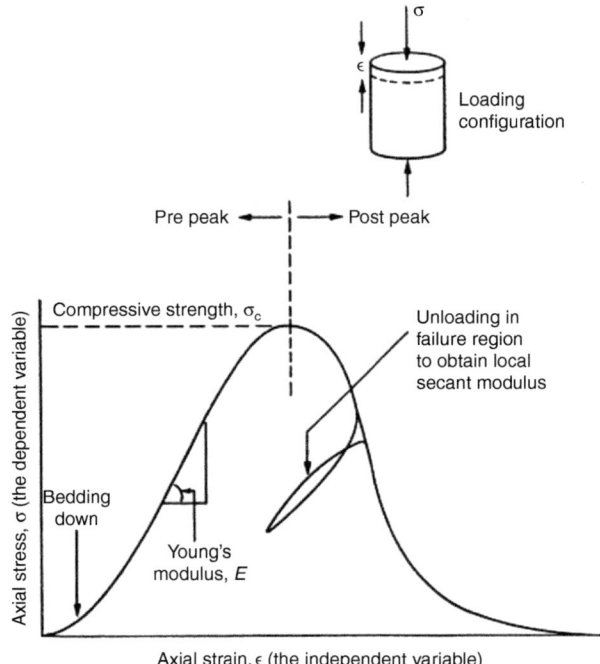

Figure 6 Complete stress-strain curve for an intact rock specimen showing the pre-peak Young's modulus, UCS and post-peak Young's modulus (Fairhurst & Hudson, 1999; ISRM, 2007).

various lateral stresses ($\sigma_2 = \sigma_3$). By plotting Mohr circles, the shear strength envelope is defined, which gives the cohesion and angle of internal friction. Compilation of some of these relationships can be found in Zhang (2005).

When an intact rock is subjected to compression, whether tested in uniaxial compression or in a confined state, "complete stress-strain curve" is very useful in order to understand the total process of specimen deformation, cracking and eventual disintegration and to provide insight into potential *in-situ* rock mass behavior (Fairhurst & Hudson, 1999; ISRM, 2007). This term refers to the displacement of the specimen ends from initial loading, through the linear elastic pre-peak region, through the onset of significant cracking, through the UCS, into the post-peak failure locus, and through to the residual strength (Figure 6). This method is intended for the characterization of intact rock (Table 2).

Tensile strength of intact rock is the maximum tensile stress to which the intact rock can withstand. Intact rock generally has a low tensile strength due to the existence of microcracks in the rock. The microcracks may also be the cause of rock failing suddenly in tension with a small strain. There are a variety of tests to determine the tensile strength of rock, such as direct pull test, Brazilian test and beam flexure test, etc. (Figure 7). For direct tension test, sometimes the rock specimen is to be prepared in dog-bone shape with a thin middle. The specimen is then loaded in tension by pulling from the two ends. Direct tension test on intact rock is not common, due to the difficulty in specimen preparation and proper axial loading. Therefore, the most

Name	Specimen Geometry and Loading	Strength and Notes
Direct Tensile Test	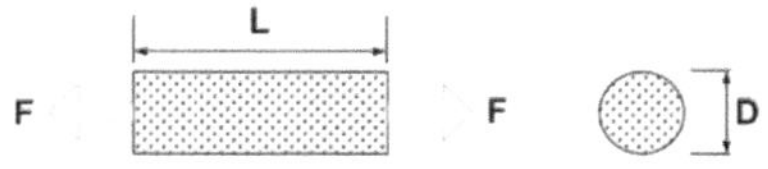	$\sigma_t = 4F/(\pi D^2)$ ISRM (1981, 2007): $L/D = 2.5 - 3.0$; $D_{min} = 54$ mm ASTM D 2936-08 (ASTM, 2008a): $L/D = 2.0 - 2.5$; $D_{min} = 47$ mm
Brazilian or Splitting Tensile Test	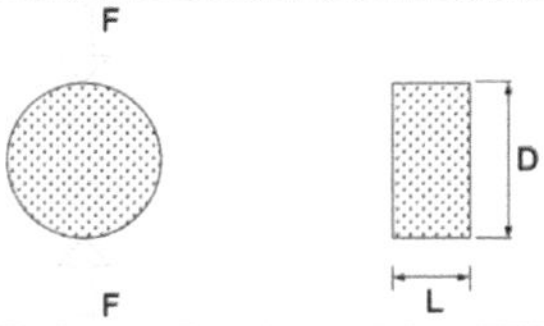	$\sigma_{tB} = 2F/(\pi DL)$ ISRM (1981, 2007): $L/D \approx 1.0$; $D_{min} = 54$ mm ASTM D 3967-08 (ASTM, 2008b): $L/D = 0.2 - 0.75$; $D_{min} = 54$ mm
3-Point Flexural Strength Test	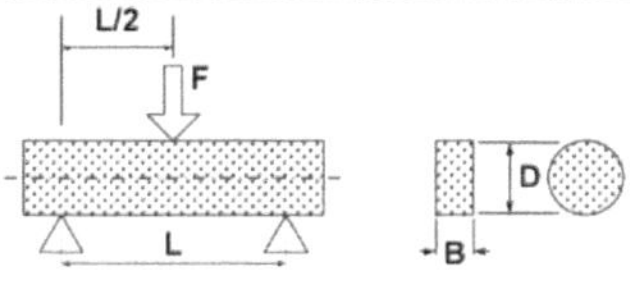	Rectangular section: $(\sigma_r)_{3P} = 3FL/(2 BD^2)$ Circular section: $(\sigma_r)_{3P} = 8FL/(\pi D^3)$
4 - Point Flexural Strength Test	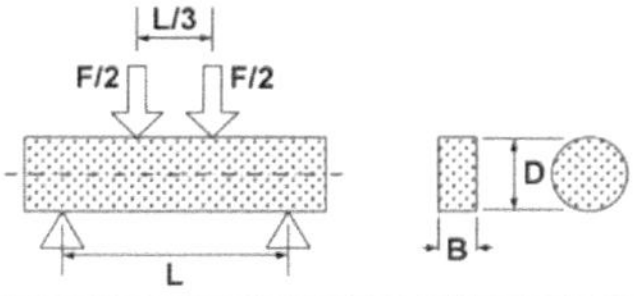	Rectangular section: $(\sigma_r)_{4P} = FL/(BD^2)$ Circular section $(\sigma_r)_{4P} = 16FL/(3\pi D^3)$
Diametric Loading of Circular Disc with Hole	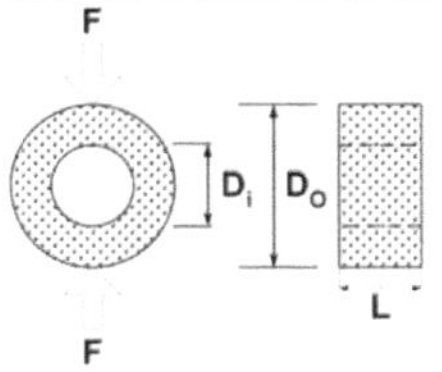	Hobbs (1965): $\sigma_t = 2F(6 + 38\rho^2)/(\pi Do\, L)$ $\rho = Di/Do \approx 0.1 - 0.2$ $Do \geq 45$ mm; $L/D \approx 0.5$
Hollow Cylinder with Internal Pressure	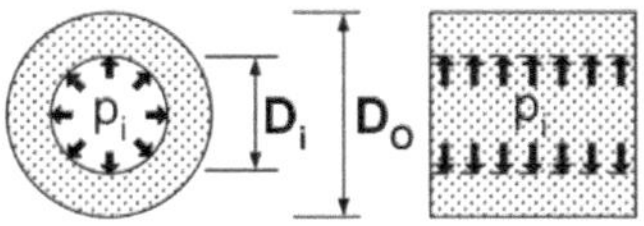	$\sigma_t = P_i(Do^2 + Di^2)/(Do^2 - Di^2)$
Luong Test	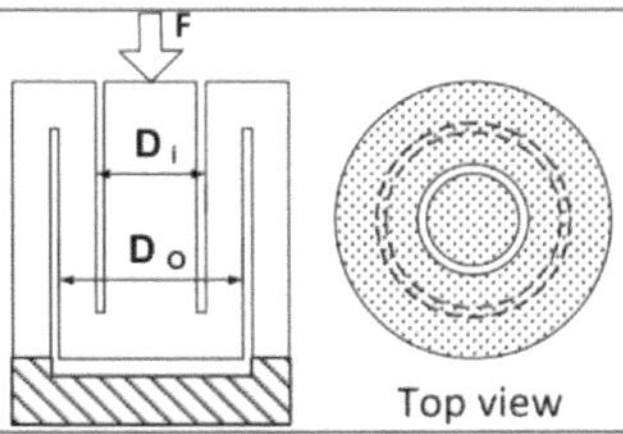	Luong (1986, 1988): $\sigma_t = 4F/[\pi(Do^2 - Di^2)]$
Triaxial Compression -Tensile Test	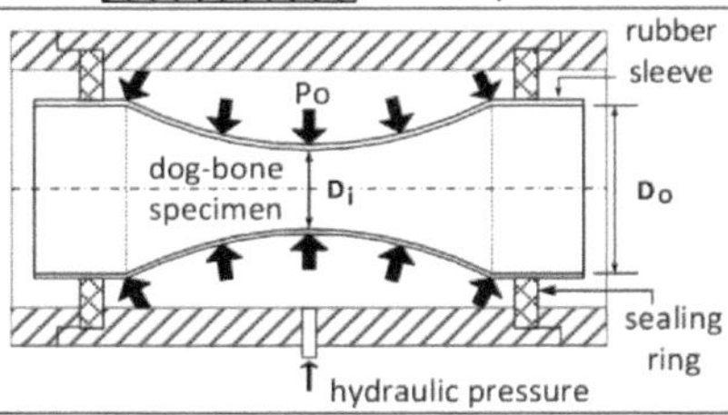	Hoek and Brown (1980): $\sigma_t = Po(Do^2 - Di^2)/Di^2$

Note: F = breaking load, L = length, D = diameter (cylinder) or height (prism), B = width (prism), P = failure pressure

Figure 7 Some tests for determining tensile strength of intact rock (Gercek & Ozarslan, 2011).

common tensile strength determination in practice is by the Brazilian test. As shown by the Griffith criterion (Griffith, 1921), theoretical tensile strength of a brittle material is 1/8 of its UCS. Typically, tensile strength of intact rocks is about 1/20 to 1/10 of the UCS. It is important to be aware of the fact that compressive strength is significantly greater than tensile strength for rocks. Since the failure mode for point load test is primarily by tensile fracturing, the correlation between tensile strength and $I_{s(50)}$ is more consistent; however, it can vary with a significant margin. In addition to the static strength tests, most recently, new ISRM SMs for determining dynamic UCS and indirect tensile strength by the Brazilian test were also developed (Zhou *et al.*, 2012; ISRM, 2014). These two methods are mainly intended for dynamic strength classification and characterization of intact rock. The tensile strength is used in strength classification of intact rock, design (analysis of rock structures subjected to tensile stresses, such as wide roof spans) and numerical analyses, such as flexural toppling analyses, continuum and discontinuous models.

The two main deformability properties of intact rock are Young's modulus and Poisson's ratio. Young's modulus is the modulus of elasticity measuring of the stiffness of an intact rock. It is defined as the ratio, for small strains, of the rate of change of stress with strain. The usual method to determine the Young's modulus of intact rock is to conduct UCS test on pieces of rock core and it can be experimentally determined from the slope of a stress-strain curve obtained during compression or tensile tests conducted on a rock sample (ISRM, 1981, 2007; ASTM, 2014b). Similar to strength, Young's modulus of intact rocks varies widely with rock type. For extremely hard and strong rocks, Young's modulus can be as high as 100 GPa. Since specimen preparation for this test is time consuming and expensive and especially extremely difficult or not possible for weak and soft rocks, indirect tests particularly wave velocity tests are used to estimate the Young's modulus by using empirical relationships. Young's modulus determined from the wave velocity measurements is the dynamic elastic property of the intact rock and is usually larger than the static modulus determined from the UCS test. This property is commonly used for classification of intact rock (Table 3) (*e.g.*, Deere & Miller, 1966) and indirect estimation of rock mass deformation modulus, and in design (such as estimation of deformations in various rock engineering structures, settlement for foundations in homogeneous, isotropic rock conditions) and in numerical analyses.

Poisson's ratio measures the negative ratio of lateral strain to axial strain of a specimen under uniaxial stress at linear-elastic region. For most rocks, Poisson's ratio is between 0.15 and 0.4. Generally, Poisson's ratio of intact rock can be determined in the laboratory either indirectly by dynamic methods (wave velocity tests) or directly (UCS test) (ISRM, 1981, 2007; ASTM, 2014b). Poisson's ratio is no less significant than some of the intact rock properties for which classifications have been proposed. However, by considering that a Poisson's ratio classification could be useful for a qualitative assessment of laboratory test results and since the theoretical upper limit is 0.5 and there seems to be an observed lower limit of zero, Gercek (2007) suggested two practical classification alternatives for this mechanical property (Table 3). The classifications recommended for Poisson's ratio of rocks are simple and easy to remember, and they can be utilized for qualitative grouping of quantitative test data. Since both deformability properties and mechanical properties, that play a role in the deformation of elastic materials, they are utilized in rock engineering problems

Table 3 Some of the classifications based on the intact rock material properties determined by the ISRM SMs.

Property	Symbol (unit)	Class							References
		Extremely Low	Very Low	Low	Moderate or Medium	High	Very High	Extremely High	
Density	ρ (Mg/m³)		< 1.8	1.8 – 2.2	2.2 – 2.55	2.55 – 2.75	> 2.75		IAEG (1979)
Porosity	n (%)		< 1	1 – 5	5 – 15	15 – 30	> 30		IAEG (1979)
Uniaxial Comp. Strength	σ_c (MPa)	0.25 – 1.0	1 – 5	5 – 25	25 – 50	50 – 100	100 – 250	> 250	ISRM (1981, 2007)
Tensile Strength	σ_t (MPa)	< 0.1	0.1 – 0.5	0.5 – 2.5	2.5 – 5.0	5 – 10	10 – 25	> 25	Backstrom et al., (2009)
Young's Modulus	E (GPa)		< 5	5 – 15	15 – 30	30 – 60	> 60		IAEG (1979)
Poisson's Ratio	ν (–)		0 – 0.1	0.1 – 0.2 1 – (1/6)	0.2 – 0.3 (1/6) – (1/3)	0.3 – 0.4 (1/3) – (1/2)	0.4 – 0.5		Gercek (2007)
Sonic Velocity	v_p (km/s)		< 2.5	2.5 – 3.5	3.5 – 4.0	4.0 – 5.0	> 5.0		IAEG (1979)
Point Load Index	$I_{s(50)}$ (MPa)		< 1	1 – 2	2 – 4	4 – 10	> 10		Bieniawski (1989)
Block Punch Strength Index	BPI (MPa)		< 1	1 – 5	5 – 10 (Moderate) 10 – 20 (Medium)	20 – 50	> 50		Sulukcu & Ulusay (2001)
CERCHAR Abrasivity Index	CAI	0.1 – 0.4	0.5 – 0.9	1.0 – 1.9	2.0 – 2.9	3.0 – 3.9	4.0 – 4.9	> 5	ISRM (2014)
Slake Durability Index	I_{d2} (%)		< 30	30 – 60	60 – 85	85 – 95 (Med. High) 95 – 98 (High)	> 98		Gamble (1971)
Modulus Ratio	E / σ_c (–)		< 50	50 – 100	100 – 200	200 – 500	> 500		Ramamurthy & Arora (1991)
Strength Ratio	σ_c / σ_t (–)		< 5	5 – 10	10 – 20	20 – 40	> 40		Gercek & Ozarslan (2011)

associated with the elastic deformation of rocks, *e.g.* they are required computational inputs for the numerical stress analyses and also as the inputs of intact rock classifications (Table 2), and the value of Poisson's ratio of intact rock is required for evaluation and interpretation of overcoring *in-situ* stress measurement methods.

Fracture is a failure mechanism of brittle materials that has great importance to the performance of structures. Large scale rapid failures in rocks cause significant hazards and damage to rock engineering structures. Ability to recognize pre-failure rock mass behavior may result in predicting or averting the potential for geotechnical and geological failure (Szwedzicki, 2003). Rock fracture mechanics is an approach to resolve this task. Particularly in the last three decades, more attention has been paid to the application of fracture mechanics principles to the field of geo-engineering, such as hydraulic fracturing, rock cutting, drilling and blasting, slope stability, etc. The resistance of rock to the initiation and propagation of fractures is described in terms of fracture toughness. The fracture toughness is the limit of local stress increase due to an existing fracture before its critical extension takes place. In other words, fracture toughness of intact rocks measures the effectiveness of rock fracturing. It is typically measured by a toughness test. There are three fracture modes: Mode I (the crack tip is subjected to displacements perpendicular to the crack plane), Mode II (the crack faces move relatively to each other in the crack plane) and Mode III (shear displacement is acting parallel to the front in the crack plane) as shown in Figure 8a, and correspondingly, there are three fracture toughness: K_{IC}, K_{IIC}, and K_{IIIC}.

There are three fracture modes: Mode I (the crack tip is subjected to displacements perpendicular to the crack plane), Mode II (the crack faces move relatively to each other in the crack plane) and Mode III (shear displacement is acting parallel to the front in the crack plane) as shown in Figure 8a, and correspondingly, there are three fracture toughness: K_{IC}, K_{IIC}, and K_{IIIC}. For determination of Mode I fracture toughness, there exist three ISRM SMs, such as tests using Chevron Bend (CB), Short Rod (SR) and Cracked Chevron Notched Brazilian Disk (CCNBD) specimens (ISRM, 2007).

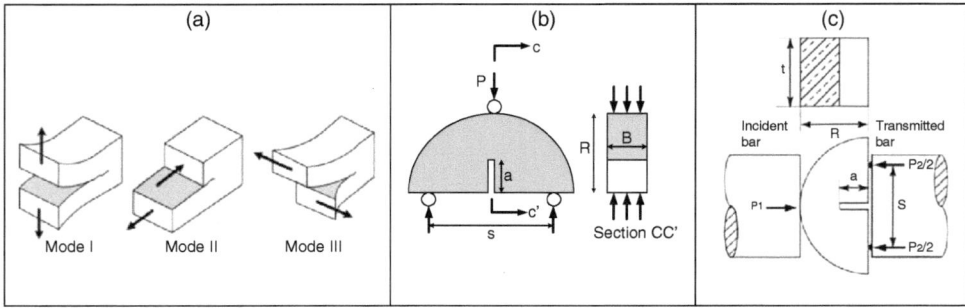

Figure 8 (a) Modes of fracture toughness (Backers, 2004), (b) SCB specimen geometry and schematic loading arrangement (R: radius of the specimen, B: thickness, a: notch length, s: distance between the two supporting cylindrical rollers, P: monotonically increasing compressive load applied at the central loading roller of the three-point bend loading; Kuruppu *et al.*, 2014) (c) the notched semi-circular bend (NSCB) specimen in the split Hopkinson pressure bar (SHPB) system (R: radius of the specimen, t: thickness of the sample, a: notch length, S: distance between the two supporting pins, P1 and P2 are the dynamic forces on both ends of the sample; Zhou *et al.*, 2013).

However, most recently, two new methods to determine the static and dynamic Mode I fracture toughness were also developed. The static test method (Kuruppu *et al.*, 2014; ISRM, 2014), which uses semi-circular bend (SCB) specimens (Figure 8b), is intended to determine the Mode I static fracture toughness (K_{ISCB}) under slow and steady loading where dynamic effects are negligible. In this method, the advantages of using the SCB specimens are: (a) material requirement per specimen is small, (b) machining is relatively simple, and (c) only the maximum compressive load is required to determine the fracture toughness. The new dynamic method (Zhou *et al.*, 2012; ISRM, 2014) is intended to determine the dynamic Mode I fracture toughness, $K_I(t)$of a rock material using the notched semicircular bend specimen by placing it in the split Hopkinson bar system (Figure 8c). This test intends for the classification and characterization of the intact rock with respect to its resistance to the crack propagation, and $K_I(t)$ serves as an index for rock fragmentation processes involving drilling, crushing and tunnel boring. Most recently, an ISRM Suggested Method called "Punch-through shear with confining pressure" was also developed for the direct determination of Mode II fracture toughness (K_{IIC}) of rock material (Backers & Stephansson, 2012; ISRM, 2014). The experimental setup of this method allows the determination of K_{IIC} at different levels of confining pressure. Although some experimental methods have been proposed (*e.g.*, Yacoub-Tokatly *et al.*, 1989), at present, no ISRM SM for the determination of Mode III fracture toughness exists.

Ground often responds to excavation operations (particularly in tunnels) with a considerable delay due to time-dependent behavior of rocks. Time-dependency is a general term encompassing concept like creep, rate dependent behavior, delayed fracturing and long-term strength (Malan *et al.*, 1997). Creep in rock mechanics is an irreversible deformation under constant or sustaining stress without fracturing and is observed mainly in soft rocks and less in all other kinds of rocks within long enough time intervals (Cristescu & Hunsche, 1996) (Figure 9a). The time-dependent creep deformation process consists of three stages: primary, secondary, and tertiary creep phases (Figure 9b). The initial (time- independent or elastic) deformation can roughly be predicted by its stress-strain modulus. A material, capable of creep, will continue to deform slowly with time indefinitely. It is possible if applied stresses are

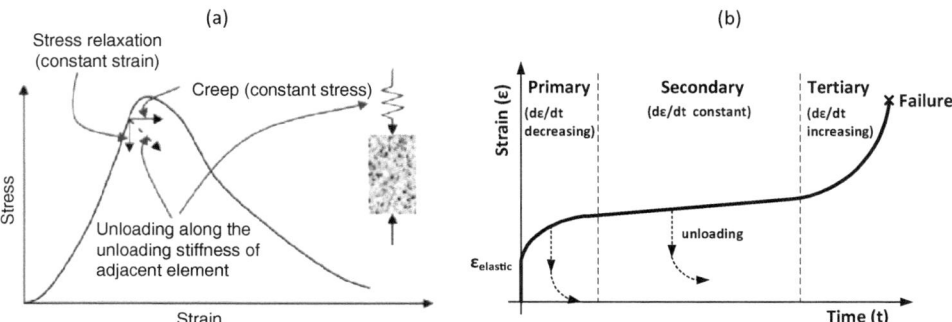

Figure 9 (a) Definitions of creep, stress relaxation and time dependent unloading along the stiffness of the adjacent element (Hagros *et al.*, 2008), (b) theoretical strain-time behavior of rock under sustained load.

low or until rupture causes failure if the stressing is high enough. The primary region is the early stage of loading when the strain rate decreases rapidly over time. Then, it reaches a steady state which is called the secondary creep stage followed by a rapid increase in strain rate (tertiary stage) and fracture, if the applied stresses are high. The tertiary stage is always connected with the phenomenon of time-dependent failure (Jeremic, 1994).

The creep characteristics are often used to determine the time-dependent strength and/or time dependent modulus of rocks and in the assessment of long-term stability of rock engineering structures (Aydan et al., 2014). Since the early part of 1900s, studies on mechanical creep behavior of rocks have been conducted (Phillips, 1932; Griggs, 1939). Since the rocks show significant creep under stress and temperature conditions, which are easily applied in the laboratory, extensive laboratory experiments focused on the softer rocks such as coal, rock salt and shale. However, hard rocks, such as granite, were also investigated with the aid of servo-controlled hydraulic systems, conventional compression test machines and mechanical loading equipment. Detailed summary of creep studies can be found in the literature (e.g., Lama & Vutukuri, 1978; Dusseault & Fordham, 1993). Although a standard test method was developed by ASTM (2008d), until 2014, no ISRM SM for laboratory creep tests was available. In 2014, ISRM SMs for determining the creep characteristics of intact rock (Aydan et al., 2014; ISRM, 2014) were developed. These separate three methods concern the creep characteristics of intact rock under the indirect tensile stress regime of the Brazilian test, and the uniaxial and triaxial compression tests in the light of available creep testing techniques used in the field of rock mechanics under laboratory conditions.

Some rocks, such as clay-bearing rocks (e.g., mudstone and claystone) and anhydrite-bearing rocks, are swelling rocks, and their volumes increase if water is allowed to infiltrate. Swelling deformations reduce with the logarithm of stress, and the swelling deformations can be completely surpassed by sufficiently high pressure, resulting in damages to the structures. Rock swelling is measured in confined and unconfined conditions. A number of recommended test methods to determinate swelling behavior have been developed. However, majority of these methods were developed particularly for soils and available in ASTM Standards.

The ISRM SMs developed for swelling rocks include only the determination of the axial swelling stress, axial and radial free swelling strain, and axial stress as a function of axial swelling strain (Madsen, 1999; ISRM, 2007). The third method is practicable only on purely argillaceous rocks. In these tests, two conditions, i.e. unconfined and confined swelling, are considered. Unconfined swelling is measured by the percentage increase of length in three perpendicular directions, when a rock specimen is placed in water. The confined swelling index measures swelling in one direction, while deformations in other two directions are constrained.

3.1.2 Laboratory tests on discontinuities

Rocks are heterogeneous and quite often discontinuous. These discontinuities may be in the form of bedding planes, joints, faults or other recurrent planar or undulating fractures. The appropriate modeling of mechanical behavior of discontinuities and the quantitative determination of their geomechanical characteristics have an important role on evaluation of stability and deformation behavior of structures in the discontinuous rock mass. Discontinuities usually have negligible tensile strength and their shear strength, in generally, is smaller than that of the surrounding rock material. Various parameters such as discontinuity roughness, scale (size of discontinuity), stiffness of the surrounding rock mass, shear rate, condition of the discontinuity (e.g. presence of infill material, its type, thickness and drainage condition), which should be measured and/or described in the field (ISRM, 1981, 2007), influence the shear behavior of rock discontinuity. The main discontinuity properties considered in rock engineering practice are shear strength (cohesion and friction angle) and stiffness (normal and shear).

The portable field shear box apparatus is used for rapid determination of discontinuity strength properties at various normal stresses, along planes of discontinuity or weakness both in the field and laboratory. However, it is rather insensitive and difficult to use at the relatively low normal stresses associated with relatively shallow rock engineering excavations and structures such as slopes and dams. In addition, higher normal loads produced by this apparatus would unquestionably have resulted in more severe damage to the sheared weakness planes, particularly in weak and clay-bearing rocks, and control of the normal and shear displacements is also extremely difficult (Ulusay & Yoleri, 1993). Therefore, the shear behavior and shear strength properties of planar discontinuities are commonly determined in the laboratory by using a conventional direct shear apparatus.

For the determination of the shear behavior of planar discontinuities using shear box device, the normal load is kept constant (CNL) during the shearing process. This mode of shearing is suitable if the surrounding rock freely allows the discontinuity to shear without restricting the dilation (Figure 10 a). This boundary condition is applicable to rock engineering problems such as the sliding block near the ground surface, as seen in Figure 10a, where the shear plane is subjected to a constant normal load generated by the weight of the blocks in a slope. But in case of rough discontinuities, shearing results in dilation as one asperity overrides another, and if the surrounding rock mass is unable to deform sufficiently, then an increase in the normal stress occurs during shearing. In such cases, shearing of rough discontinuities no longer takes place under CNL, but rather under variable normal load where stiffness of the surrounding rock mass plays an important role in the shear behavior. This particular mode of shearing is called as shearing under constant normal stiffness (CNS) boundary condition (Figure 10b). The case, where shear is subjected to constant stiffness due to the constraints of lateral displacement in a tunnel, is given in Figure 10b as an example for CNS condition. Typical plots from a shear test for CNL and CNS conditions are also shown in Figures 10c and 10d, respectively. Although there exists an ISRM SM (ISRM, 1981, 2007) and a standard method (ASTM, 2008e), the ISRM SM for laboratory determination of the shear strength of discontinuities was upgraded with consideration on the technological advances since its initial publication. This upgraded version intends to cover the requirements and laboratory procedures for performing direct shear test by considering CNL and CNS conditions to determine peak and residual cohesion and friction angle (Muralha et al., 2014; ISRM, 2014).

Normal and shear stiffness of discontinuities relate discontinuity stress to relative displacement between opposite points on the two surfaces of discontinuities. Joint

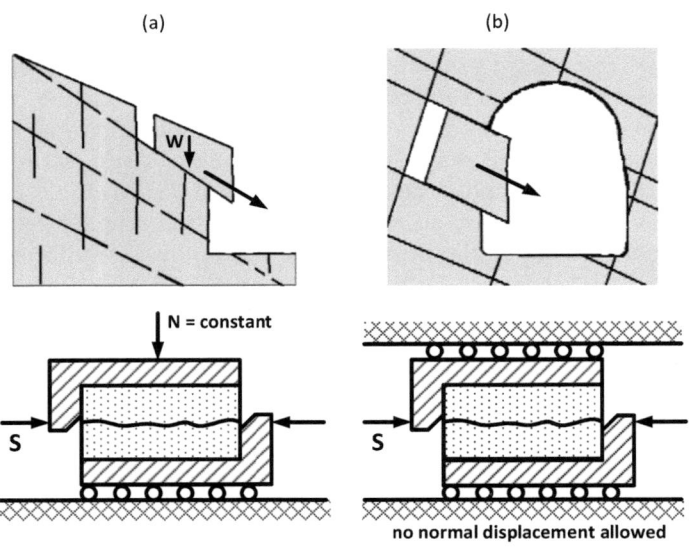

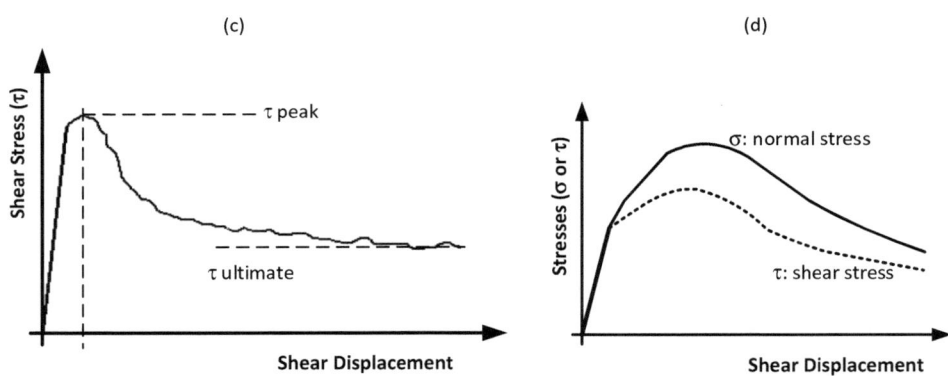

Figure 10 (a) Controlled normal load (CNL) and (b) controlled normal stiffness (CNS) shearing modes and tests (rearranged from Brady & Brown, 1993), and shear stress-shear displacement plots for shear test under (c) CNL and (d) CNS conditions (Muralha *et al.*, 2014).

shear stiffness (k_s) is defined as the ratio of shear stress corresponding shear displacement prior to reaching the peak shear strength. Similarly, the joint normal stiffness (k_n) is the normal stress per unit closure of the joint before reaching the peak strength. In short, the stiffness parameters of a joint describe the stress-deformation characteristics of the discontinuity. Normal and shear joint stiffness values are used in numerical stress analyses that allow modeling of joint behavior (*e.g.*, discontinuum models such as UDEC). Since the shear and normal displacement measurements are available as a result of tests involving shear strength of

(a)   (b)

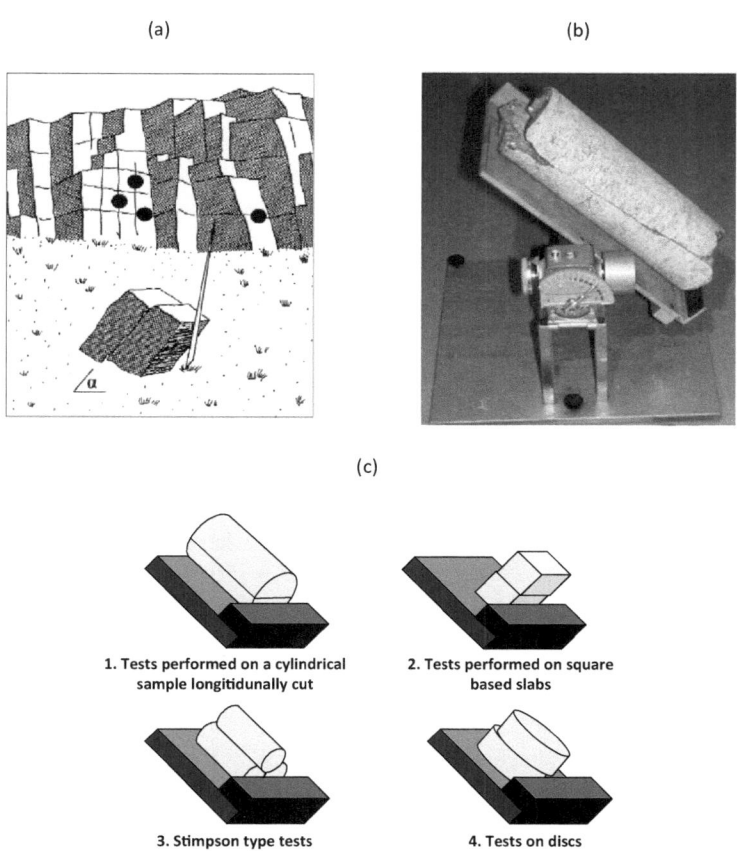

(c)

1. Tests performed on a cylindrical   2. Tests performed on square
   sample longitidunally cut   based slabs

3. Stimpson type tests   4. Tests on discs

Figure 11 Tilt test: (a) at field (self-weight gravity shear test; Barton & Bandis, 1980) and (b) in laboratory (Barton, 2013), and (c) different setups for the tilt tests (rearranged from Alejano *et al.*, 2012).

joints (Muralha *et al.*, 2014; ISRM, 2014), the normal and shear stiffness values of the samples may also be derived from the tests.

Although the triaxial compression test is primarily used to measure the shear strength and in some cases the elastic properties of intact rock, by orienting planes of weakness the strength of discontinuities can also be measured. The oriented plane variation is particularly useful for obtaining strength information on thinly filled discontinuities containing soft material. The primary disadvantage of the triaxial test is that stresses normal to the failure plane cannot be directly controlled.

One of the parameters used by the Barton's empirical criterion (Barton, 1976) in the estimation of shear strength of discontinuities with rough surfaces is the basic friction angle (ϕ_b). This parameter is usually derived from different types of tilt tests. The test can be either conducted simply by hand in the field as illustrated in Figure 11a or a machine-operated tilt test equipment in laboratory (Figure 11b). At present there is no any ISRM SM or a standard method recommended by any institution for the tilt test.

However, there exist some methods described for the tilt test in the literature. The method proposed by Stimpson (1981) uses rock cores to perform tilt tests with a cylinder-shaped sample placed over other two equal-dimension cylinder-shaped samples. The other proposals (Horn & Deere, 1962; Cruden & Hu, 1988; Bruce *et al.*, 1989) do not provide full indications for normalizing tilt testing. Recently, Alejano *et al.* (2012), who reported that the current methods produce varying ϕ_b values, investigated these differences on results by conducting an experimental study with four types of rocks submitted to different types of inclination tests illustrated in Figure 11c. Based on the test results, these researchers concluded that the mechanisms of sliding along cylinder generatrixes (Stimpson's method, Figure 11c.3) and planar surfaces (Figure 11c.1-2-4) are quite different, and that tests based on sliding along generatrixes are not appropriate for determining reliable ϕ_b values. Tests on small specimens are also not recommended by Alejano *et al.* (2012) for geometry reasons and since ensuring reliable stress condition is difficult. Alejano *et al.* (2012) also recommend that a fourth supplementary repetition of the test on a specimen should be performed when the maximum deviation between one of the results and the median is larger than $\pm 3°$. This study suggests that further efforts are needed to develop a suggested method or a standard for the tilt test.

3.2 *In-situ* tests

The properties of a rock mass are significantly different from the properties of the same rock material. The strength and mechanical behavior of the rock mass are commonly dominated more by the nature of its mass properties than by its material properties. A rock mass comprised of even the strongest intact rock material is greatly weakened by the occurrence of closely spaced discontinuities. Material properties, however, tend to control the strength of the rock mass if discontinuities are widely spaced or if the intact rock material is inherently weak or altered. Discontinuities within a rock mass, therefore, reduce its strength, stability and the energy required to excavate or erode it. In addition to the strength and deformability characteristics of rock masses, *in-situ* stresses and permeability are the other two classes of property which are vitally important on their own right and as they influence strength and deformability. Laboratory tests can quantify the behavior of intact rock. But, first of all, for laboratory tests, sampling from large volumes of rock mass representing its two components, intact rock and discontinuities, is not possible. In addition, there are three principal influences which determine the relevancy of laboratory testing methods:

(i)   Although great care could be taken during sampling some degree of structural disturbance and stress change occur and a further disturbance during the transportation of the sample to laboratory may also be possible.

(ii) Scale effect: The performance of rock engineering structures is governed by the whole characteristics of the rock mass. However, in attempting to model the rock mass behavior in laboratory tests, the size of samples used will be limited for practical reasons.

(iii)   The original *in-situ* stress conditions cannot be reimposed prior to laboratory testing.

The extension of laboratory approach to quantify rock mass behavior is to carry out large-scale *in-situ* or field tests. The determination of rock mass properties can be approached in two ways (Hudson & Harrison, 2000):

(i)   using some empirical relations via the properties of the intact rock and the properties of the discontinuities, which together make up the rock mass properties; or

(ii)   via the properties of the rock mass as measured or estimated directly with the aid of *in-situ* tests.

In this section, only *in-situ* tests, which involve subjecting a large volume of rock to load and monitoring the deformation, are briefly reviewed. *In-situ* tests can be categorized based on the following main purposes: (i) characterization of rock mass, (ii) estimation of *in-situ* stresses, (iii) determination of the rock mass and discontinuity properties (deformability tests and *in-situ* shear strength of discontinuity), which are used in rock engineering design assessments.

3.2.1 In-situ tests for rock mass characterization

In this category, geophysical methods and rock mass permeability determinations are the common methods. Geotechnical geophysics is the application of geophysics to geotechnical engineering problems; such investigations normally extend to a total depth of less than a hundred meters but can be extended to several hundred meters in some instances. A geophysical survey is often the most cost-effective and rapid means of obtaining subsurface information, especially over large study areas. Geotechnical geophysics can be used to select borehole locations and can provide reliable information about the nature and variability of the subsurface between existing boreholes. Other advantages of geotechnical geophysics are related to site accessibility, portability, non-invasiveness, and operator safety (Anderson & Croxton, 2008). Geophysical methods can be used for establishing soil and rock stratification, but also for determining engineering properties of rock masses by direct or indirect methods. During the past decades, geophysical methods have become highly scientific tools, especially as a result of the powerful electronic measuring and data acquisition systems, sophisticated data interpretation and presentation methods.

Geophysical surveys in rock engineering are performed on the ground surface, within and between boreholes. These methods are mainly seismic testing (reflection and refraction), electrical resistivity, electromagnetic, gravity, radiometric method, ground penetrating radar (GPR) and seismic tomography. These methods, with the exception of seismic tomography, are primarily surface-based techniques. The methods for geophysical logging of boreholes were published by ISRM (ISRM, 1981, 2007). Then new suggested methods including all the above mentioned methods were developed and compiled in the ISRM Suggested Methods book (ISRM, 2007). The properties determined with the aid of different geophysical methods in geotechnical applications can be found in the literature (*e.g.*, Massarch, 2000; Anderson & Croxton, 2008) and the geophysical testing procedures are described in the ISRM SMs (ISRM, 1981, 2007).

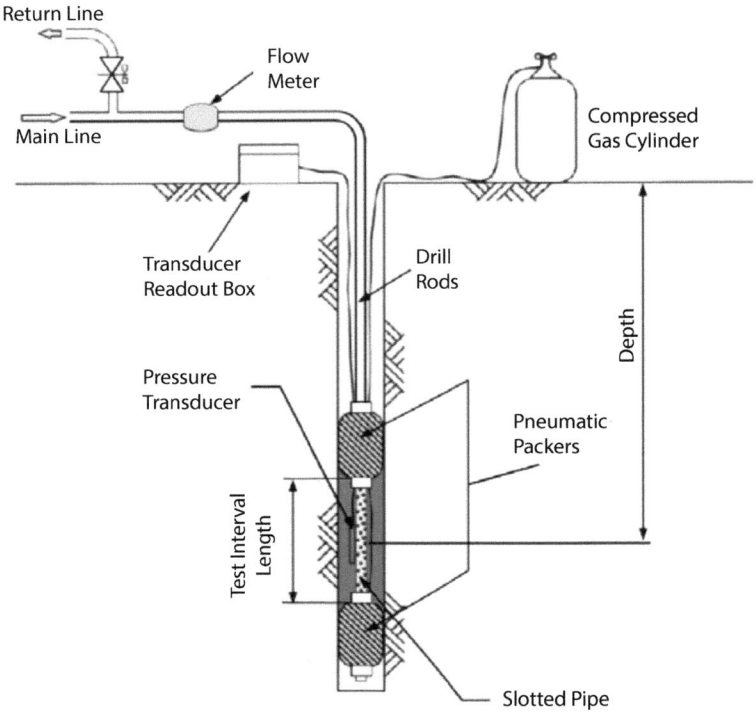

Figure 12 Lugeon test configuration (after Camilo Quiñones-Rozo, 2010).

Hydrogeological characteristics of rock masses, in other words, understanding water or gas flow through a rock mass and its influence on overall permeability of rock mass, are very important in various fields of rock engineering, such as slopes, dam foundations, underground excavations, oil recovery, geothermal reservoirs and underground nuclear waste disposal. Unlike soils, where seepage takes place through a series of small closely spaced, interconnected pore spaces, and seepage through rock masses occurs mostly along discrete discontinuities. Therefore, in rock masses the permeability depends on the aperture, spacing and infilling characteristics of its discontinuities (Goodman, 1989) and accurate estimates of the permeability of a rock mass can only be determined with the aim of *in-situ* tests. While the falling head and constant head test methods are used for soil materials, the "Lugeon test" or based on the name of testing equipment, "Packer test", or pressurized water test is the most commonly used *in-situ* test to estimate permeability of rock masses. The Lugeon test, which was developed in France by Maurice Lugeon (Lugeon, 1933), is a constant head type test that takes place in an isolated portion of a borehole. Water at constant pressure is injected into the rock mass through a slotted pipe bounded by pneumatic packers. A pneumatic packer is an inflatable rubber sleeve that expands radially to seal the annulus space between the drill rods and the boring walls (Figure 12). There is no any standard method for this test. The current Lugeon interpretation practice is mainly derived from the work performed by Houlsby (1976). Most recently, an ISRM SM for the Lugeon test is under preparation.

3.2.2 In-situ stress measurements

Stresses in rock can be divided into two groups, such as *in-situ* or virgin stresses and induced stresses. The virgin or *in-situ* stresses are the natural stresses that exist in the ground prior to any disturbance, while induced stresses are associated with man-made disturbance (excavation, drilling, pumping, loading etc.) or are induced by changes in natural conditions (drying, swelling, consolidation etc.) (Amadei & Stephansson, 1997). Their magnitudes and orientation are determined by the weight of the overlying strata and the geological history of the rock mass. The principal stress directions are often vertical and horizontal. They are likely to be similar in orientation and relative magnitude to those that caused the most recent deformations.

Since *in-situ* stresses determine the boundary conditions for stress analyses and affect stresses and deformations that develop when an opening is created, knowledge of *in-situ* stresses in rock engineering applications (civil and mining engineering applications, such as stability of underground excavations, drilling and blasting, design of pillars and support systems, prediction of rock burst, dams and stability of slopes; in energy development, such as borehole stability, fracturing and fracture propagation, fluid flow and geothermal problems, reservoir production management, energy extraction and storage, etc.). Some of the simplest clues to stress orientation can be estimated from knowledge of a region's structural geology and its recent geologic history (*e.g.*, Johnson, 1970; Ramberg, 1981; Price & Cosgrove, 1990). Although initial estimates can be made based on simple guidelines, field measurements of *in-situ* stresses are the only true guide for critical structures.

When compared to other rock mass properties, it is difficult to quantitatively measure *in-situ* rock stress. The stress estimation methods can mainly be divided into two groups as the direct (field measurements) methods and other methods (indicator methods and indirect methods). These methods with their brief descriptions and features are given in Table 4.

In addition to the *in-situ* stress estimation methods given in Table 4, there also exist three useful guides (as ISRM SMs) to be used in stress measurements. These SMs involve:

(i) the recommended strategy of approach for estimating the state of stress in a rock mass within the context of rock mechanics modeling and rock engineering design (Hudson *et al.*, 2003; ISRM, 2007),

(ii) quality control (technical auditing issues) of rock stress estimation (Christiansson & Hudson, 2003; ISRM, 2007), and

(iii) how a model for the *in-situ* stress at a given site is established (Stephansson & Zang, 2012; ISRM, 2014).

3.2.3 In-situ shear strength, deformability, and creep tests

Laboratory testing of rock masses still plays a disproportionately large role in the determination of strength and deformability of rock masses. Hoek (2007) suggests that only 10 to 20 percent of a balanced rock mechanics investigation should be

Table 4 Methods of *in-situ* stress measurements (compiled mainly from Amadei & Stephansson, 1997; ISRM, 2007; and the references cited in the table).

Method	Description	Feature
Stress Relief Methods:	A rock sample is isolated from the stress field in the surrounding mass and its response is monitored.	
(a) Surface relief method (Obert & Duvall, 1967)	On the rock surfaces instrumented by gages or pins, the response of rock to stress relief by drilling or cutting is obtained by recording the gages/pins before and after the relief process.	Performance of gages/pins can be affected by humidity and dust, strains are measured on weathered or damaged rock faces.
(b) Overcoring methods using • USBM-type drillhole deformation gage (ISRM, 2007) • CSIR or CSIRO-type cell (ISRM, 2007) • Compact conical-ended borehole overcoring (CCBO) technique (Sugawara & Obara, 1999; ISRM, 2007) • Borre (SSBP) probe (Sjöberg et al., 2003; ISRM, 2007)	A borehole is drilled to a desired depth and then a probe is installed in the hole. The stress tensor in rock is determined by measuring the displacements or strains occurring in the walls of the drill hole when the stresses are relieved by overcoring.	3-D stress state can be estimated. No assumption needs to be made regarding the in situ stress field. Most commonly used stress relief methods. They are expensive and time consuming.
Hydraulic Methods:	A borehole is drilled from the surface or a roadway; hydraulic pressure is applied along a section of the borehole isolated by packers, and it is increased until existing fractures are opened or new fractures are created. In-situ stress is estimated from hydraulic fracturing data.	
Hydraulic fracturing methods: • Hydraulic fracturing (HF) method (ISRM, 2007)	The most popular method, which was first recommended by Fairhurst (1964). The orientation of the resulting fracture is obtained using televiewers or impression packers.	Only horizontal stresses usually estimated. It can be applied up to several kilometers deep.
• Hydraulic testing of pre-existing fracture (HTPF method) (Haimson & Cornet, 2003; ISRM, 2007)	The only in-situ stress determination method at great depth. It consists of reopening an existing fracture of known orientation previously been isolated in between two packers.	Works well in homogeneous rocks, but does not work well in heterogeneous (stratified) rocks.
• Sleeve fracturing (Stephansson, 1983)	Similar to HF method, except that it has the major advantage	Compared with HF method, the breakdown pressure is not well

Table 4 (Cont.)

Method	Description	Feature
	that no fluid penetrates the rock upon fracturing. No standard or SM for this method exists yet.	defined, thus complicating the interpretation of the field test results. In addition, the induced fractures do not propagate far from the borehole wall.
Stress Compensating (Jacking) Method: Flat jack method (ISRM, 2007)	It determines the rock stress parallel to and near the exposed surface in an excavation. Stress is relieved measuring displacement or strain. Stress is applied until the displacement or strain recovers to the values before the stress relief. When using flat jack, the cancellation pressure is used as a direct estimate of the stress normal to the jack.	Elastic constants are not required to estimate rocks stress. However, since each flat jack test yields one component of the in-situ stress field, minimum six tests, at different orientations, have to be carried out at six different locations to obtain the complete in-situ stress field.
Strain Recovery Methods:	Based on monitoring the response of core samples following drilling. No standard or SM exists for these methods yet.	
(a) Anelastic strain recovery (ANS) method (Teufel, 1982)	It consists of instrumenting an oriented core sample following its removal from a borehole and monitoring its strain response as it continues to recover from the in situ state of stress.	These methods are well suited for stress measurements in deep to very deep wells for which many of the other methods do not work and for which only small core samples are available.
(b) Differential strain curve analysis (DSCA) method (Strickland & Ren, 1980)	It consists of applying a hydrostatic pressure to a cubic sample cut from an oriented drill core following its removal	
Borehole Breakout Method:	It is used as an indicator method of stress estimation. The rock around a circular opening may not be able to sustain the compressive stress concentration induced drilling process and breakage of the rock results in zones of enlargement called breakout. The main idea in this method is that the breakout occurs in the direction parallel to the minimum in-situ stress component.	The method can be used in boreholes several kilometers deep and all rock types. However, the theory of the method has limited value if the rock is anisotropic or time dependent or borehole wall yields. No standard or SM exists for this method yet.
Other Methods: **(a) Indicator Methods:**		
• Fault-slip data analysis	Measurement of striations on faults can be used to determination of the orientation	Advanced knowledge of the rock deformability properties is not required. In case of the

Table 4 (Cont.)

Method	Description	Feature
	as well as the magnitude of in-situ stress field. Simple methods along a single discontinuity under axi-symmetric loading (Jaeger, 1960; Bray, 1967) and more complex methods under 2- and 3-D stress fields (Jaeger & Cook, 1976; Amadei & Savage, 1989; Morris et al., 1996) were proposed. The most recent method, proposed by Aydan (2000), is capable of inferring the stress state from a single fault.	estimation of the current is-situ stress field, there must be sufficient evidence that the striations used are related to that stress field.
• Earthquake focal mechanism	By analyzing the earthquake fault-plane solution, a best fit regional stress tensor can be determined by means of inversion technique. Estimates relative magnitudes of the three in-situ principal stress components and their orientation.	Provides data about in-situ stresses at greater depths (5–20 km) and is most effective for large earthquakes.
• Core disking (e.g., Dyke, 1989; Natau et al., 1989; Haimson & Lee, 1995)	Geometry of stress-induced core fracturing due to high horizontal stresses indicates stress components.	The morphology of the disks can be used as an indicator of the direction and approximate ratio of the horizontal stresses.

(b) Indirect Methods:

- Acoustic method (Rivkin et al., 1956)
- Seismic and microseismic methods (Swolfs & Handin, 1976; Talebi & Young, 1989; Martin et al., 1990)
- Blasthole-damage method (Aydan, 2013)
- Sonic and ultrasonic methods (Aggson, 1978; Pitt & Klosterman, 1984; Sun & Peng, 1989)
- Radioisotope method (Riznichanko, 1967)
- Atomic magnetic resonance method (Cook, 1972)
- Electromagnetic methods (Petukhov et al., 1961)
- Holographic methods (Smither et al., 1988; Smither & Ahrens, 1991; Schmitt & Li, 1993)

Kaiser effect method (e.g. Kanagawa et al., 1976; Holcomb & Martin, 1985; Momayez & Hassani, 1992; Seto et al., 1992, 1997)

Indirect methods measure stresses by looking at changes in some physical, mechanical or other rock properties as a result of a change in stress.

Except for the Kaiser effect method which is based on AE measurements, these methods have not gained much popularity in practice and there exists no standard or SM for them.

allocated toward laboratory testing. Laboratory tests can usually only be carried out on intact rocks of small sample sizes due to the limited size and loading capacity of the testing equipment. Therefore, the laboratory test specimens will be much smaller than the scale of interest for a typical engineering project. The results will then be representative of the extreme end of the strength and deformability values for a jointed rock mass and provide very little consideration of the influence of the discontinuity network on the strength and deformability of the rock mass.

(a) *In-situ* direct shear test for discontinuities:

Because of scale effects, there is no simple method of predicting the *in-situ* shear strength of a rock discontinuity from the results of laboratory tests on small specimens, and therefore, *in-situ* tests on large specimens are the most reliable means.

The advantages with the *in-situ* shear test, in addition to minimize possible scale effects, are that the joint can be tested in undistributed conditions. By doing so, the effect on the peak friction angle from infilling materials such as silt and loose pieces of weathered rock are considered. For this purpose, *in-situ* direct shear test is conducted in the field. Since it is an expensive test, it is generally used for critically located discontinuities filled with a very weak material, such as soil-like material or gouge. A relatively large surface area is tested to address unknown scale effects and the method covers the measurement of peak and residual shear strength of *in-situ* rock discontinuities as a function of normal stress to the sheared plane (Figure 13). A further important advantage of direct shear tests is that generally they also permit large shearing displacements. The testing procedure is given in ISRM (1981, 2007) and ASTM (2012).

(b) *In-situ* large triaxial tests for rock masses:

In many countries, the number of construction works for large scale structures has increased recently, and in addition to stability-based designs, displacement-based designs also became important. Therefore, measuring stress-strain relationship of rock masses is very essential. With the aid of recent developments in laboratory testing methods, evaluation of the stress-strain relationships and shear strength of rocks much more accurately than it was before. Yet, due to previously mentioned issues such as scale effect and sampling quality, the sampling and testing of undisturbed samples that are sufficiently large to represent rock mass properties is extremely difficult. Therefore, the direct measurement of rock mass properties is the most accurate and reliable method. For studying the deformation properties of rock masses, large-scale field tests such as plate load test, flat jack test, etc. are very common in rock engineering applications.

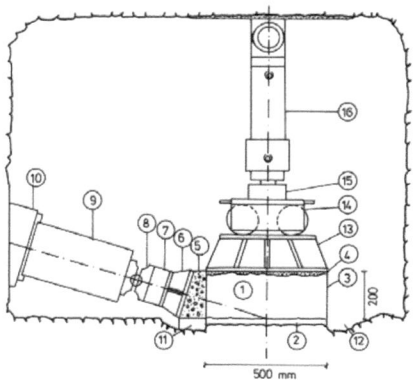

Figure 13 Typical arrangement of equipment for *in situ* direct shear test and a view from the test at the bottom of a shaft (GIF, 2004a).

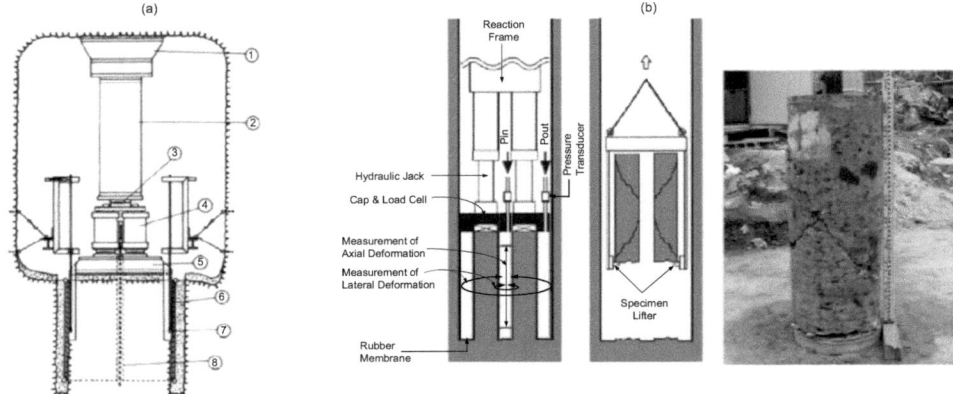

Figure 14 (a) *In-situ* test on a large triaxial specimen (1: abutment in the ridge of the test gallery, 2: reaction beam, 3: spherical seat, 4: three pressure cylinders, 5: load distribution plate, 6: pressure pad, 7: horizontal displacement gauge, 8: vertical displacement gauge; GIF, 2004b), (b) loading and lifting stages of the test procedure of in-situ triaxial test for rock masses suggested by Tani et al. (2003) and a specimen retrieved after *in-situ* triaxial test (Okada et al., 2006).

But, stress-strain relationships cannot be measured directly, and they are mostly suffering from loosening and disturbance effects. Apart from deformation characteristics, the strength characteristics of rock masses have been investigated separately in general, by *in-situ* rock shear tests.

Since Leopold Müller and his team (Müller, 1961) conducted *in-situ* triaxial tests during construction of the Kurobe Dam (Japan), these tests have belonged to the standard repertoire of *in-situ* tests. Either the test specimen is removed from the rock and then subjected to a genuine triaxial test with hydraulic cylinders or, as a variation, the specimen is loaded triaxially with hydraulic presses and the mean and smallest main orthogonal stress is applied by pressure pad (GIF, 2004b) (Figure 14a). After the development of a sampling and testing technique for 60 cm diameter specimens of jointed hard rock, it was possible to bring rock mass into the laboratory and to test it under well-defined conditions (Natau et al., 1983). Thus, a method to obtain large scale samples from the field and to determine the shear strength of jointed rock using a large scale apparatus in laboratory under radial symmetric triaxial states of stress was developed and approved by the ISRM as a SM (Natau et al., 1989; ISRM, 2007).

In this method, because of its high cost, a multi-stage technique using a single sample is recommended. This approach can be very useful for the case of non-homogeneous rocks or when the number of specimens is limited. However, the shear strength properties are prone to be underestimated by this method (Kim & Ko, 1979), and this technique results in sample disturbance under different loadings during testing. It is also noted that this method is a laboratory test, and large scale sampling is an expensive and time-consuming procedure. In order to solve the above mentioned problems, in recent years, some attempts have been made to directly measuring average stress-strain relationships and to investigate strength and deformation characteristics of rock masses using *in-situ* triaxial testing methods. For example, a down-hole large triaxial testing equipment was developed by Tani *et al.* (2003) in Japan. In 2006, a new trial series of tests using this method was carried out at the site of rhyolitic-tuffacious rock and rudaceous rock (Okada *et al.*, 2006). The results were found acceptable and it was generally agreeable with previous studies results in the same site. The test is conducted on a hollow cylindrical specimen prepared at the bottom of a drill-hole (Figure 14b). Average axial as well as lateral strains can be measured in a center hole and an outer slit by a novel technique of instrumentation for cavity deformation (Figure 14b). Another effort was made by Zhang *et al.* (2011), who developed a triaxial test system to test rock mass samples with a size of 50 × 50 × 100 cm, and the system provides new means for the estimation of deformation, strength and breaking properties of deep and complicated rock masses. However, no standard or a SM for the *in-situ* triaxial test is available yet and such studies are rather limited. Therefore, further experiments are needed.

In-situ measurements of deformation modulus:
There are two main geotechnical considerations in the design of any rock engineering structure: (i) the maximum load that can be supported by the rock mass without catastrophic failure, and (ii) the relative movement of the rock mass under the application (or removal) of loads. In terms of strength, the maximum load, that a rock mass can support, is called failure strength. The movement or deformation of a rock mass under a given stress can be estimated from the deformation modulus. Estimating both the failure strength and deformation modulus is important for designers. The static deformation modulus is among the parameters that best represent the mechanical behavior of a rock and of a rock mass, in particular, when it comes to underground excavations. This is why most numerical (*e.g.* finite element, boundary element) analyses for studies of the stress and displacement distribution around underground excavations are based on this parameter (Palmström & Singh, 2001). For its determination, there are two alternative ways:

(i)   via the indirect estimates based on the properties of the intact rock and discontinuities which together make up the rock mass properties (empirical equations), and
(ii) via the properties of the rock mass as measured or estimated directly in the field.

A number of empirical equations have been developed that correlate various rock properties or rock mass classification systems, such as RMR (Bieniawski, 1989) and Q (Barton *et al.*, 1974) systems, to *in-situ* modulus. A compilation of some empirical relationships is listed in Palmström & Singh (2001) and most recently majority of them have been compiled by Aydan *et al.* (2013). The indirect procedures to estimate the deformation modulus are simple and cost-effective, when compared with the *in-situ* tests. The rock mass classification schemes have been originally developed to determine support systems for tunnels, based on practical experience, a database of geological properties and performance of the support systems used in previous underground engineering projects. But, since

Table 5 In-situ deformability tests covered by the ISRM SMs and ASTM.

Test Method		Description
Plate Loading Test[1, 2]	Superficial Loading	Two areas (each ≈ 1 m in diam.) opposite to each other in small tunnel or adit are simultaneously loaded by flat jacks, and rock mass deformations are measured in boreholes behind each loaded area.
	Down a Borehole	Load is applied to the flattened end of a large borehole (min. diameter 0.5 m), and displacements of the bottom of the hole from a reference level are measured.
Radial Jacking Test[3]		Certain length of circular tunnel is loaded radially using a test chamber with circular cross section. The loading is provided either internally pressurizing the chamber or by hydraulic jacks located circumferentially around it. The radial displacements occurring in the rock mass surrounding the chamber area are measured by the extensometers located in the radial boreholes.
Large Flat Jack Test[4]		The rock mass is loaded using large flat jacks located in slots created by cutting rock with large disk saw or by line drilling a series of boreholes. The displacements of slot walls are measured at several points by deformeters built in the jacks.
Borehole Expansion Tests[5]	Flexible Dilatometer	An expanding probe (dilatometer) is used to apply pressure on the walls of a borehole, and either the volume change or radial displacement caused by the expansion is measured.
	Stiff Dilatometer	Expansion pressure is applied against opposite walls of a borehole by the stiff loading platens of a dilatometer (i.e. Borehole or Goodman Jack) and the change in borehole diameter is measured.

Notes: (1) ASTM D4394-08 (ASTM, 2008f), (2) ASTM D4395-08 (ASTM, 2008g), (3) ASTM D4506-13 (ASTM, 2013b), (4) ASTM D4729-08 (ASTM, 2008h), (5) ASTM D4971-08 (ASTM, 2008i)

the application of classification methods is mostly limited to the preliminary stages of engineering projects, detailed *in-situ* testing methods should not be fully replaced by them. There are different *in-situ* test methods available to estimate deformation modulus of rock masses. However, only the followings have ISRM suggested methods (ISRM, 2007): plate load tests, radial jacking test, flat jack test, and the dilatometer tests (Table 5). Other test methods, that are not standardized, are described in the literature.

(d) *In-situ* creep test:

Dusseault & Fordham (1993) indicate that in cases where creep is along a large planar feature, such as a pillar roof line in stratified deposits, it is impossible to obtain representative laboratory data, thus extrapolation from laboratory tests is not feasible, and *in-situ* creep test becomes the only source of design information. *In-situ* creep testing gives bulk parameters and they may also be used to verify and refine numerical model predictions. Despite some disadvantages (such as rarely homogenous stress and strain states, stresses seldom known accurately, concentration of creep in a single material or interface, and obtaining the test data

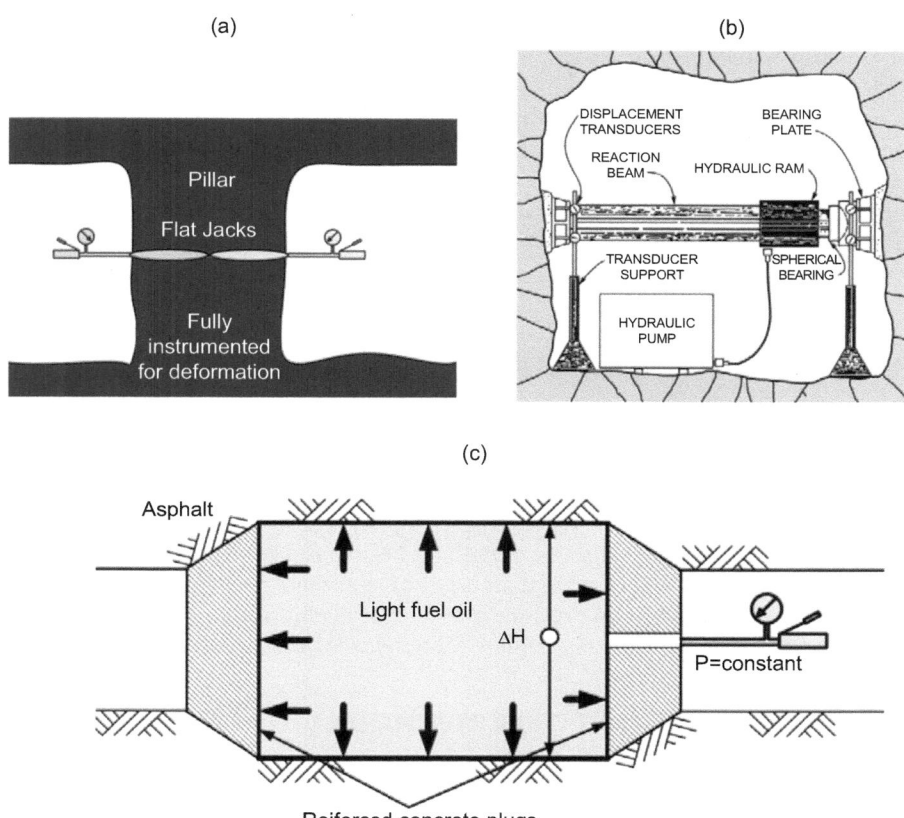

Figure 15 In-situ creep measurements: (a) flat jacks loading a pillar (redrawn from Dusseault & Fordham, 1993), (b) *in-situ* creep testing of a room by rigid plate bearing (ASTM, 2008j), (c) isolation of a tunnel section in salt rock for internal pressure creep testing (redrawn from Dusseault & Fordham, 1993).

reflecting behavior of damaged or non-representative material; Dusseault & Fordham, 1993), *in-situ* creep test is useful particularly for a large mine or civil engineering structure susceptible to creep, perhaps extended for several years into the construction phase. *In-situ* creep behavior of rock mass can be performed on a joint plane or a pillar by introducing flat jacks (Figure 15a) or loading an entire structural component such as a room or a section of a tunnel (Figure 15b,c), and testing of isolated elements such as a series of clay-filled, steeply inclined faults, is a vital method for *in-situ* creep assessment (Dusseault & Fordham, 1993). Due to its expensive and time-consuming nature, application of *in-situ* creep test is rather limited.

However, there exists a standard method for *in-situ* creep measurements recommended by ASTM (2008j), *i.e.*, ASTM D4553-08: Standard Test Method for Determining In Situ Creep Characteristics of Rock.

3.2.4 Most recently developed in-situ test methods

Most recently, some new *in-situ* test methods, which were also approved by the ISRM as SMs (ISRM, 2014), were developed for rock fracture observations, for measuring rock mass displacements, and for estimating elastic stiffness, strength and hydraulic properties of the fractures. One of these methods (Li *et al.*, 2013a) is intended to directly observe fractures in a rock mass using a digital optical borehole camera through pre-drilled boreholes, with characteristics of the fractures being surveyed in both air and clear fluid-filled borehole. In this method, by comparing the fractures observed at different times, the fracture initiation, propagation and closure occurring in the rock mass can be evaluated. In addition, with the aid of an adoption to detect possible stress induced damages in the borehole, the method can also be helpful to estimate *in-situ* stress orientation. The second method (Li *et al.*, 2013b) aims to measure rock mass displacement occurring as the result of surface and underground excavations, movement of artificial slopes, and foundation loads using a sliding micrometer with a precision up to ± 0.002 mm/m. The method can also be applied to measure settlement in earth or rock fill dams and dam abutments. The third method, called "Step-rate injection method for fracture *in-situ* properties (SIMFIP)", is suggested to estimate normal and shear stiffness, strength (cohesion and friction angle) and hydraulic properties (hydraulic aperture and storage) of the fractures using a step-rate injection of a given water volume to produce micro-scale elastic and inelastic deformations of a localized fractured rock mass volume (Guglielmi *et al.*, 2014).

4 STANDARDIZATION OF TEST METHODS

"Test method" is a definitive procedure for the identification, measurement and evaluation of one or more qualities, characteristics or properties of a material. Numerous tests methods have been developed for direct or indirect determination of a certain physical or mechanical property of rock materials; however, only a few of them have become widely-used or recognized by the people who need or use the certain property in their field of application. For example, some of the test methods employed to determine the tensile strength of intact rock are depicted in Figure 7. As it is well known, only one method (*i.e.* the Brazilian or splitting tensile strength test) has become the most widely-used one in rock engineering. Although the direct tensile test is the other one of the two recommended test methods to determine the tensile strength, it has not been as popular as the Brazilian test due to the difficulties involved. Yet, the easiness or practicality of a test method may bring important pitfalls. As a matter of fact, erroneous measurements or interpretations are inevitable if one ignores the theoretical facts and assumptions involved in a particular testing method. Again, considering the Brazilian test as an example, some of the precautions are as follows:

(i)   The rock specimen must be homogeneous.
(i)   The compressive strength of the rock must be larger than the three times of its uniaxial strength (generally this requirement is satisfied).
(ii) The Young's moduli of the rock in compression and in tension should be equal. If the rock is bimodular, a correction must be made (Chen & Stimpson, 1993).

(iii) The failure of the specimen should start with a crack at the center of the specimen and the failure load should be taken as the value at crack initiation level (Diederichs, 1999).

(iv) The rock specimen should not demonstrate transversal anisotropy; otherwise, a correction is required for the tensile stress occurring at the center of the disk (Claesson & Bohloli, 2002).

Furthermore, as the people involved in rock mechanics testing well know, repeated execution of the same test method on the same rock material, whether by the same operator in the same laboratory using the same equipment or by different operators in different laboratories using equipment of similar design, will not always yield comparable results (Lau, 2009). In this respect, one should consider the "repeatability" and "reproducibility" of a particular testing method, which generally are not readily available. The repeatability of a rock mechanics experiment is the precision determined when the same methods and equipment, used by the same operator, under identical conditions are used to make multiple measurements on identical rock specimens; on the other hand, reproducibility refers to the precision determined when the same methods, but different equipment and operators are used to make measurements on identical rock specimens (Glassel, 2014). In short, the aforementioned and other considerations have necessitated the use of standard or suggested test methods.

"Standard" is a document that has been developed and established within the consensus principles of a society and that meets the approval requirements of that society's principles and regulations. Standards become legally binding only when a government body references them in regulations or when they are cited in a contract. There is also the practical aspect that it may be wished to specify something about the rock conditions in contracts, then it is useful to use standardized methods within contractual procedures. Hudson & Harrison (2000) indicate that although the strategy of rock characterization is a function of the engineering objectives, the tactical approach to individual tests can be standardized and describe the advantages of the standardization of rock testing methods as follow:

(i) the standardization guidance is helpful to anyone conducting the test;

(ii) the results obtained by different organizations on rocks at different sites can be compared in the knowledge that 'like is being compared with like'; and

(iii) there is a source of recommended procedures for use in contracts, if required.

There are national bodies which produce standards for their own countries. In particular, American Society for Testing and Materials (ASTM) in the United States has produced an extensive series of methods for rock testing via Committee D18.12. Many other countries also have their own wide range of standards, such as British Standards (BS) in the UK and Deutsche Industrie Normen (DIN) in Germany, and the methods suggested by Japanese Geotechnical Society (JGS), etc.

After the formation of the ISRM in 1962 in Salzburg, some Commissions on different aspects of rock mechanics and rock engineering were established by the ISRM. One of these Commissions was the "Commission on Standardization of Laboratory and Field Tests" which was established in 1966 at the time of the 1st ISRM Congress. In 1979, its name was changed to "Commission on Testing Methods" at the 4th ISRM Congress held in Switzerland. The objectives of the ISRM Commission on Testing Methods are:

(i)   to generate and publish SMs for testing or measuring properties of rocks and rock masses, as well as for monitoring the performance of rock engineering structures,

(ii)   to raise or upgrade the existing SMs based on recent developments and publish them in book form,

(iii)   to solicit and invite researchers to develop new methods, procedures or equipment for tests, measurements and the monitoring required for rock mechanics and laboratory or field studies, and

(iv)   to encourage collaboration of those who practice in rock mechanics testing.

The Commission also cooperates with other ISRM Commissions for the development of new SMs as most recently successfully done with the ISRM Commission on Rock Dynamics and Commission on Petroleum Geomechanics.

The ISRM Commission on Testing Methods has been producing SMs for rock covering a wide range of subjects since 1978, and these are widely used by engineers, scientists, government agencies, and companies. The term 'Suggested Method' has been carefully chosen: these are not standards *per se*; they are explanations of recommended procedures to follow in the various aspects of rock characterization, testing and monitoring. An "ISRM SM" is a document that has been developed and established within the consensus principles of the ISRM and that meets the approval requirements of the ISRM procedures and regulations. If someone has not been involved with a particular subject before and if this subject is part of a SM, they will find the guidance to be most helpful. For example, rock stress estimation is not an easy task and anyone involved in measuring rock stresses should not take on the task lightly. The five SMs concerning rock stress estimation cover the understanding of rock stress, overcoring, hydraulic fracturing, quality assurance and establishing a model for the *in-situ* stress at a given site. In other words, the two main stress measurement methods of overcoring and hydraulic fracturing are bracketed, firstly by ensuring that the reader is aware of the rock stress pitfalls, and secondly by ensuring that the necessary quality checks have been highlighted. The SMs can be used as standards on a particular project if required for contractual reasons, but they are intended more as guidance. The purpose of the ISRM SMs is, therefore, to offer guidance for rock characterization procedures, laboratory and field testing and monitoring in rock engineering. These methods provide a definitive procedure for the identification, measurement and evaluation of one or more qualities, characteristics or properties of rocks or rock systems that produce a test result.

The SMs are developed voluntarily by the Working Groups established by the ISRM Commission on Testing Methods. An ISRM SM is subject to revision at any time by the responsible technical committee. From 1974 to the present the ISRM has generated 62 SMs. They are classified into four groups, namely: Site Characterization, Laboratory Testing, Field Testing and Monitoring. Although some index tests, such as the Point Load Test, Schmidt Hammer Test and Needle Penetration Test can be performed either in the laboratory or in the field using portable laboratory equipment, all index and mechanical tests, along with the petrographic description of rocks, are considered in the "Laboratory Testing" group. Note that the 1975 version of the SM for shear strength of rock joints, and 1978 versions of the SMs concerning triaxial compressive strength testing, the measurement of Shore hardness, Schmidt hammer test and sound velocity test were revised in 2014, 1983, 2006, 2009 and 2014, respectively. In the

"Field Testing" group, the tests are divided into five sub-groups: Deformability Tests, *In-situ* Stress Measurements, Geophysical Testing, Other Tests, and Bolting and Anchoring Tests. The ISRM SMs for laboratory and field tests are listed in Table 6 in chronological order. In addition, the ISRM SMs books: the Yellow Book (ISRM, 1981), the Blue Book (ISRM, 2007), and the Orange Book (ISRM, 2014), which include fuller descriptions of the tests on intact rock and rock mass, are also mentioned in this table.

Table 6 List of the ISRM SMs for laboratory and *in-situ* (field) tests published between 1974 and 2014 (in chronological order).

SM for Determining Shear Strength[a,b] – 1974
SM for Rockbolt Testing[a,b] – 1974
SM for Determining Water Content – Porosity – Density – Absorption and Related Properties and Swelling and Slake-Durability Index Properties[a,b] – 1977
SM for Determining Sound Velocity[a,b] – 1978
SM for Determining Tensile Strength of Rock Materials[a,b] – 1978
SM for Determining Hardness and Abrasiveness of Rocks[a,b] – 1978
SM for Determining the Strength of Rock Materials in Triaxial Compression[a,b] – 1978
SM for Petrographic Description of Rocks[a,b] – 1978
SM for Determining *In-situ* Deformability of Rock[a,b] – 1979
SM for Determining the Uniaxial Compressive Strength and Deformability of Rock Materials[a,b] – 1979
SM for Geophysical Logging of Boreholes[a,b] – 1981
SM for Determining the Strength of Rock Materials in Triaxial Compression: Revised Version[b] – 1983
SM for Determining Point Load Strength[b] – 1985
SM for Rock Anchorage Testing[b] – 1985
SM for Deformability Determination Using a Large Flat Jack Technique[b] – 1986
SM for Deformability Determination Using a Flexible Dilatometer[b] – 1987
SM for Rock Stress Determination[b] – 1987
SM for Determining the Fracture Toughness of Rock[b] – 1988
SM for Seismic Testing Within and Between Boreholes[b] – 1988
SM for Laboratory Testing of Argillaceous Swelling Rocks[b] – 1989
SM for Large Scale Sampling and Triaxial Testing of Jointed Rock[b] – 1989
SM for Determining Mode I Fracture Toughness Using Cracked Chevron Notched Brazilian Disk[b] – 1995
SM for Deformability Determination Using a Stiff Dilatometer[b] – 1996
SM for Determining the Indentation Hardness Index of Rock Materials[b] – 1998
SM for Complete Stress-Strain Curve for Intact Rock in Uniaxial Compression[b] – 1999
SM for in Situ Stress Measurement Using the Compact Conical-Ended Borehole Overcoring Technique[b] – 1999
SM for Laboratory Testing of Swelling Rocks[b] – 1999
SM for Determining Block Punch Strength Index[b] – 2001
SM for Rock Stress Estimation – Part 1: Strategy for Rock Stress Estimation[b] – 2003
SM for Rock Stress Estimation – Part 2: Overcoring Methods[b] – 2003
SM for Rock Stress Estimation – Part 3: Hydraulic Fracturing (HF) and/or hydraulic testing of pre-existing fractures (HTPF)[b] – 2003
SM for Rock Stress Estimation – Part 4: Quality Control of Rock Stress Estimation[b] – 2003
SM for Land Geophysics in Rock Engineering[b] – 2004
SM for Determining the Shore Hardness Value for Rock[b] – 2006 (updated version)
SM for Determination of the Schmidt Hammer Rebound Hardness: Revised version [c] – 2009

Table 6 (Cont.)

SMs for Determining the Dynamic Strength Parameters and Mode I Fracture Toughness of Rock Materials[c] – 2012
SM for the Determination of Mode II Fracture Toughness[c]– 2012
SM for Rock Stress Estimation – Part 5: Establishing a Model for the In-situ Stress at a Given Site[c] – 2012
SM for Measuring Rock Mass Displacement Using a Sliding Micrometer[c] – 2013
SM for Rock Fractures Observations Using a Borehole Digital Optical Televiewer[c] – 2013
SM for Determining the Mode-I Static Fracture Toughness Using Semi-Circular Bend Specimen[c] – 2014
SM for Reporting Rock Laboratory Test Data in Electronic Format[c] – 2014
SM for Determining Sound Velocity by Ultrasonic Pulse: Upgraded Version[c] – 2014
SM for Determining the Creep Characteristics of Rock Materials[c] – 2014
SM for Laboratory Determination of the Shear Strength of Rock Joints: Revised Version[c] – 2014
SM for Determining the Abrasivity of Rock by the Cerchar Abrasivity Test[c] – 2014
SM for Step-Rate Injection Method for Fracture In-situ Properties (SIMFIP): Using a 3-Components Borehole Deformation[c] – 2014
SM for the Needle Penetration Test[c] – 2014

Notes: [a] Published in the Yellow Book (ISRM, 1981); [b] Published in the Blue Book (ISRM, 2007), [c] Published in the Orange Book (ISRM, 2014)

Rock mechanics test data are very important in the design, construction and research of rock engineering. Successful strategy of using rock mechanics test data largely depends on the efficient data integration, sharing and management among different departments or organizations. In the past decades, significant development has been achieved in the engineering test databases and information systems. As a consequence, people can operate the data more efficiently in many engineering and research projects by these tools (Li *et al.*, 2012). Since the laboratory testing methods have different contents, their reports are individually somewhat different. In addition, the output format of the test data obtained from different testing devices also varies.

Some researchers (Toll & Cubitt, 2003; Toll, 2007, 2008) indicate that, since usually the reporting of testing results only retained by the tester or in publications, it is difficult to use and compare them for the same rock types from different sites or different rock types. This situation created a widely shared concern among rock engineers. In order to eliminate this concern, some attempts have been made to develop and approach leading to a digital standardized format for the same rock type and different rock types worldwide (AGS, 1999, 2005; Swift *et al.*, 2004; Exadaktylos *et al.*, 2007; Zheng *et al.*, 2010; Li *et al.*, 2012). One of these efforts were transformed into an ISRM SM by Zheng *et al.* (2014; ISRM 2014) for reporting the results of the ISRM SMs for rock laboratory tests in electronic format. With this standard electronic format, users in different locations can upload the information and can store their own data on the web file and they can be shared worldwide.

5 CURRENT DEVELOPMENTS AND FUTURE NEEDS IN ROCK TESTING METHODS

Considering the current and new areas of application for rock engineering, the level of sophistication reached in electronic measurement and control systems, the advances in data acquisition and processing methods, and the developments in the testing of other materials, etc., rock testing methods covered by the ISRM SMs are far from complete. Experimental rock mechanics has a very wide scope ranging from laboratory to field, and there exist some issues requiring further researches and a need for further developments in experimental methods which may lead to new ISRM SMs. As a matter of fact, there are already new working groups occupied in developing new ISRM SMs. These methods, which are under preparation and/or in review, are as follows:

(i)   Thermal properties of rock (rock material)
(ii) Lugeon test to determine the permeability of rock mass
(iii) *In-situ* microseismicity monitoring of the fracturing process in rock masses
(iv) Laboratory Acoustic Emission (AE) monitoring
(v)   Uniaxial-strain compressibility testing for reservoir geomechanics

Near future trends and needs on experimental rock mechanics are briefly given in the following paragraphs.

Behavior of rocks under high temperature is more complex and hydro-mechanical properties are particularly important for the projects involving repositories for spent nuclear fuel, radioactive nuclear waste disposal, natural gas storage, gas production from deep coal seams, carbon dioxide sequestration in deep underground, and geothermal energy extraction. Particularly, the nuclear waste disposal is one of hot topics in countries utilizing nuclear energy and/or having nuclear weaponry. In this issue, the design time frame ranges from 10,000 to 1,000,000 years. The constitutive law parameters among coupling of diffusion (C), heat flow (T) and seepage (p) are generally unknown and further experimental studies are required to obtain the actual values of Dufour & Sorret coefficients for a meaningful assessment of fully coupled thermo-hydro-diffusion phenomena (Aydan, 2008).

As a branch of rock mechanics, rock dynamics deals with the responses of rock under dynamic stress fields, where an increased rate of loading (or impulsive loading) induces a change in the mechanical behavior of the rock materials and rock masses. Due to the additional 4th dimension of time, dynamics has been a more challenging topic to understand and to apply. Rock dynamics remains, at least in the discipline of rock mechanics, a relatively virgin territory where research and knowledge are limited. When compared to other aspects of rock mechanics, except a dynamic laboratory test method suggested by the ISRM (Zhou *et al.*, 2012; ISRM, 2014), guidance and standards/SMs for rock dynamics testing are generally lacking. Much of the research works done on rock dynamics is for military. Therefore, there are many issues in rock dynamics testing requiring further investigations, as summarized by Zhao (2011) in a most recently published book, such as shear strength of rock joints under dynamic loads in order to understand the rate effects on shear strength and dilation, and assessment of mechanical and physical causes of the rate effects on the rock strength and failure pattern, etc.

Since stress is a tensorial quantity requiring six independent components, estimation of rock stress is one of the most important and problematic issues in rock engineering due to the considerable variation in the rock stress at all scales (caused *inter alia* by various types of fracturing). As emphasized by Hudson (2008, 2011) and Bieniawski (2008), although there are some rock stress measurement techniques recommended, the development of a method of rapidly and reliably estimating the six components of the rock stress tensor at a given location is an important need. Since the boreholes drilled for *in-situ* stress measurements starts to fail as depth increases, some *in-situ* stress influence methods using the experiments in laboratory have also been developed. As mentioned in Section 3.2.2, AE technique (Kaiser Effect Method) is one of these potential methods providing simpler measurements. Although the results obtained by several researchers (*e.g.*, Kanagawa *et al.*, 1976; Holcomb & Martin, 1985; Momayez & Hassani, 1992; Seto *et al.*, 1992, 1997; Daido *et al.*, 2003) showed a fairly good correlation between stresses determined with the Kaiser Effect and with some *in-situ* methods, Holcomb (1993) indicated that using the AE emitted during uniaxial compression laboratory tests to infer *in-situ* stresses could not be justified. Further studies to compare stresses inferred from this method using AE measurement applied suitably on oriented samples under uniaxial loading and those of well-known *in-situ* stress determination methods together with an SM are still a need.

Determination of the strength and deformability for "difficult rocks" is another important issue in terms of experimental rock mechanics. This term mainly includes soft rocks and block-in matrix rocks. Rocks whose UCS falls approximately in the range 0.5–25 MPa as suggested by ISRM is considered as soft rocks. He (2014) also uses another term "engineering soft rock", which refers to the rocks that can produce significant plastic deformation under engineering forces. These rocks are usually sediments in the process of consolidation and solidification or can also be weathered rocks. Soft rocks are critical geomaterials since they present several types of problems. They may present undesirable behaviors, such as low strength, disaggregation, crumbling, high plasticity, slaking, fast weathering and many other characteristics. They have intermediate strength between soils and hard rocks; therefore, in some cases, they are too soft to be tested in rock mechanics equipment and too hard for soil mechanics equipment, and their mechanical properties are highly sensitive to variations in their water content (Kanji, 2014). Sampling from soft rocks, their site characterization and classification under the usual systems such as RMR and Q, which are generally applicable to discontinuous media made of hard rocks, are other difficulties. In his paper on the critical issues in testing of soft rocks, Kanji (2014) indicates that they may lead to false results: (i) when square prisms are used as an alternative to cylindrical samples, which are difficult to obtain from soft rocks, they yield an ultimate strength 30% higher than that for cylindrical ones, and (ii) soft rock samples significantly deform before failure, even in case of point load strength index and Brazilian tests as an alternative to the UCS test. He (2014), based on the latest progresses in China on soft rock mechanics, reported that the large deformation mechanism of engineering soft rocks using sophisticated equipment is to be understood through numerous experiments. However, soft rocks are still difficult to characterize, sample, test, and predict. Therefore, there seems to be a consensus that the soft rocks are still not fully understood in engineering practice and there is a need for further investigations to develop new laboratory and *in-situ* testing methods in conjunction with the adaptation of some existing methods for soft rocks.

Block-in matrix rocks (Bimrocks), which are the mixture of rocks composed of geotechnically significant blocks within bonded matrix of finer texture such as melanges, faulted/fractured rocks and other complex geological mixtures (Figure 16a) (*e.g.*, Medley, 1994), are considered to be another group of "difficult rocks". Due to their complex heterogeneity and mechanical variability, the correct geomechanical characterization and determination of their strength and deformability are quite

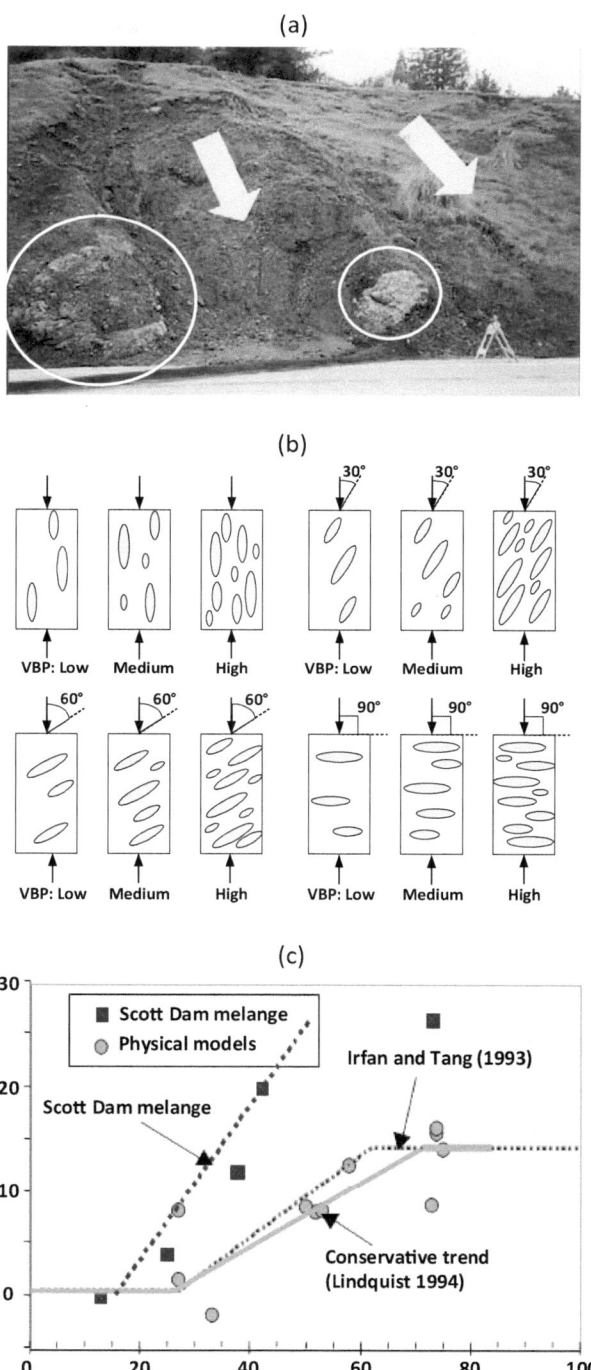

Figure 16 (a) A typical bimrock consisting of blocks in a sheared shale matrix (Franciscan Complex melange; Medley, 2007), (b) different block orientations and volumetric block proportions (rearranged from Lindquist, 1994), (c) strength of bimrocks increasing with volumetric block proportion (after Medley, 2008: from the data of Lindquist, 1994 and Irfan & Tang, 1993).

challenging issues, and in such cases, it is necessary to reduce expensive and inconvenient surprises in rock engineering applications. Determination of strength of bimrocks is one of the difficult issues in rock engineering. Mechanical properties of the matrix, the volumetric block proportion, shape and size distribution of blocks, and their orientation relative to failure surfaces are the main factors affecting the overall mechanical properties of bimrocks. Figure 16b shows different physical model bimrocks prepared by Lindquist (1994).

The arrows in Figure 16b indicate the axial loading and the angle indicated (0°, 30°, 60° and 90°) is the angle between the axial direction and the orientation in which the blocks are aligned. Each specimen had volumetric block proportions of about 30% (low), 50% (medium), or 75% (high). These samples were tested in a Hoek triaxial cell by Lindquist (1994), and it was found that as volumetric block proportion increased, frictional strength increased and cohesion decreased. Neglecting the contributions of blocks to overall bimrock strength, choosing instead to design on the basis of the strength of the weak matrix may be too conservative for many bimrocks in terms of slope design (Medley, 2008). Based on the study on a physical model melange by Lindquist (1994), when the block proportions are between about 25% and 70%, the increase in the overall mechanical properties of bimrocks are mainly related to the volumetric block proportion (VBP) in the rock mass (Figure 16c). Some efforts have been performed to assess the strength of bimrocks or faulted/fractured zones based on physical models and empirical approaches (e.g., Lindquist, 1994; Medley, 1997; Aydan et al., 1997; Sonmez et al., 2009; Pilgerstorfer, 2014), in-situ tests (e.g., Li et al., 2004; Xu et al., 2007; Coli et al., 2011), and equivalent material techniques (e.g., Aydan et al., 1995). Although in case of small blocks floating in a soft matrix, there is a chance to correlate VBP and bimrock friction angle (Coli et al., 2011) by in-situ large shear box tests; however, when the size of huge blocks exceed the dimension of the large shear box, in-situ testing for bimrocks becomes insufficient. Since there is still no consensus on the available methods to determine strength and deformability properties of bimrocks, further studies and combination of their results with existing experiences to develop more efficient methods are needed.

The preparation of smaller samples from weak and soft rocks even for some index tests is extremely difficult. In addition, sampling from historical sites, monuments and buildings for determination of geomechanical properties of rocks is generally discouraged. Therefore, in order to overcome these difficulties, the use of non-destructive techniques has been receiving great attention in recent years. The needle penetration test mentioned in Section 3.1.1 is one of the non-destructive testing methods. Although the use of X-ray imaging in experimental geomechanics dates back to the 1960s and has been mostly considered in soil mechanics (Viggiani & Hall, 2012), X-ray computed tomography (CT) scanning technique has becoming widely used in rock engineering and a quite promising non-destructive method. With the aid of this technique, it is possible to visualize and to investigate various conditions and processes in porous and fractured rocks without any disturbance to samples and quantitative evaluations are possible (Figure 17a) (e.g., Otani, 2004; Ito et al., 2004; Sato & Aydan, 2014). Several scientific studies have been carried out in recent years on the infrared radiation in the process of rock deformation leading to fracturing and failure (e.g. Aydan et al., 2003; Liu et al., 2006; Prendes-Gero et al., 2013; Luong & Emami, 2014). The thermal response of geo-materials would be observed as mechanical energy which is

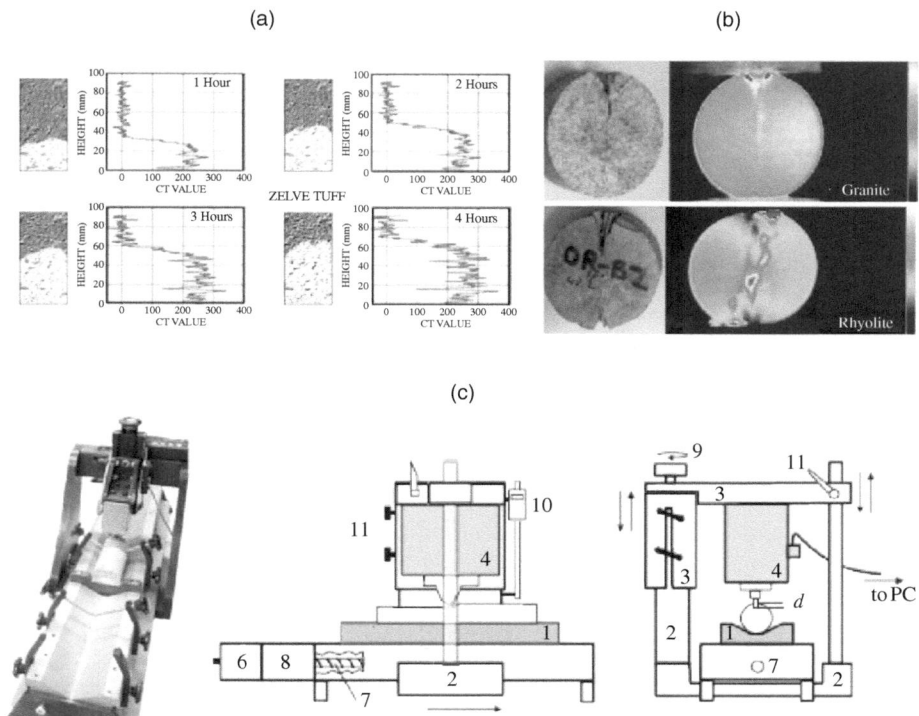

Figure 17 Examples of some promising non-destructive test methods: (a) X-Ray CT scan images and CT value distribution with height at different time intervals for a tuff sample (Sato & Aydan, 2014), (b) infrared thermography images of samples in Brazilian experiments (Aydan, 2014), (c) rock strength device used in scratch test (www.cefor.umn.edu; Richard *et al.*, 2012).

transformed into heat during deformation and fracturing. Infrared thermography technique allows imaging and measuring temperature from radiation in the infrared spectral band. The procedure makes it possible to precisely determine the potential for localized detachments in rock masses based on objective criteria, taking the thermal variations obtained using thermography techniques as data to eliminate their inherent risk (Prendes-Gero *et al.*, 2013). Figure 17b shows an example of the infrared thermography images of samples in Brazilian compression experiments associated with fracturing. As noted from the infrared thermograph images, high temperature bands appear along some zones before rupture and these high temperature bands eventually constitute the major fracture zones. The application and use of this technique to detect and evaluate quantitatively the extent of damage in brittle geomaterials owing to the non-linear coupled thermo-mechanical effects is quite promising.

A new non-destructive test method, called "the scratch test", is based on the effort initiated at the University of Minnesota in the mid-90s (Detournay *et al.*, 1997) to build a scientific apparatus to study the cutting action of a single cutter in order to assess the dependence of the cutting force on the rock mechanical properties and on the UCS. The UCS of rocks can be estimated from the scratch test performed under controlled conditions, namely with a sharp cutter and at depth of cut small enough to guarantee that cutting takes place in the ductile regime (Figure 17c). This method offers several advantages over conventional tests. These are: (i) the results are not affected by the sample dimension, and thus only a small volume of intact rock is required to assess its strength, (ii) the sample preparation is limited as it only requires performing a pre-cut on the sample surface to obtain a flat reference surface that ensures a constant depth of cut along the groove, (iii) the test offers a high degree of repeatability, and (iv) the semi-destructive nature of the test is an interesting asset, as it allows additional group of tests to be conducted on the same sample (porosity, permeability, sound velocity, uniaxial compression, triaxial test, etc.) (Richard *et al.*, 2012). The scratch test seems an attractive alternative to conventional UCS test and to indirect methods such as the point load test and the Schmidt hammer test.

In conventional triaxial tests, the cylindrical rock specimen is subject to axisymmetric state of stresses in which two of the principal stresses are equal to each other. For example, in the conventional triaxial compression (CTC) tests, the axial stress is larger than the confining stresses (*i.e.*, the intermediate and minimum principal stresses are equal or $\sigma_1 > \sigma_2 = \sigma_3$). The failure characteristics as well as the strength parameters obtained in such tests are used in failure criteria that ignore the effect of intermediate principal stress on the failure of rocks. Similarly, the conventional triaxial extension (CTE) tests, in which the confining stresses are larger than the axial stress (*i.e.*, the intermediate and maximum principal stresses are equal or $\sigma_1 = \sigma_2 > \sigma_3$), suffer from the same shortcoming. Yet, in the true triaxial test (TTT), all principal stresses acting on the sample are unequal ($\sigma_1 \neq \sigma_2 \neq \sigma_3$). The results of the TTTs on rocks, which started in the late 1960s and early 1970s by Mogi (1969, 1971, 1972), showed that the intermediate principal stress indeed affects the strength and failure of rock material. As a matter of fact, there exists a failure criterion which is based on true triaxial testing of rocks (Chang & Haimson, 2012; ISRM, 2014). Presentations of typical test results obtained from such tests are shown in Figure 18. Further information on the TTT of rocks can be

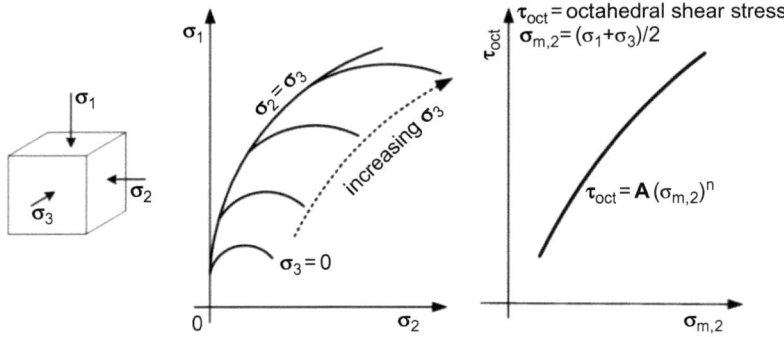

Figure 18 Typical presentation of results from the true triaxial tests (σ_1, σ_2, and σ_3 are the maximum, intermediate, and minimum principal stresses, respectively) (arranged from Chang & Haimson, 2012; ISRM, 2014).

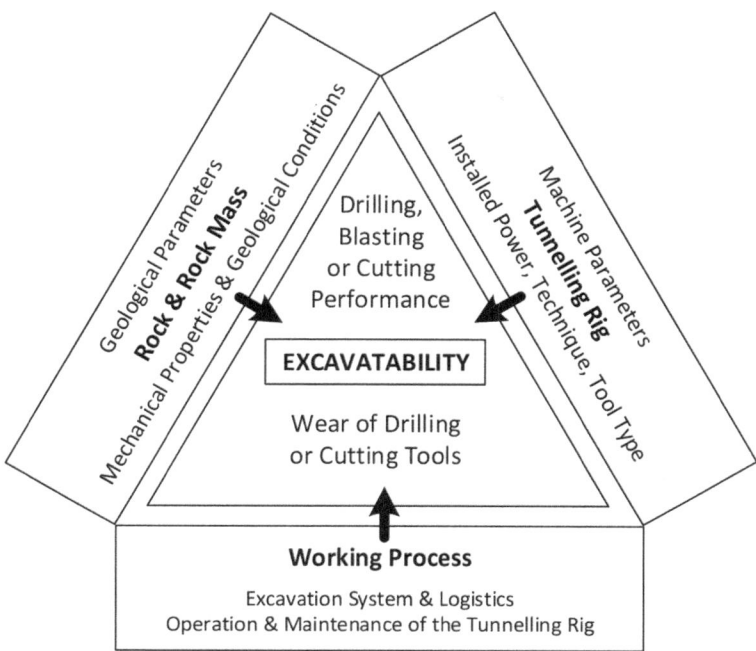

Figure 19 Conceptual overview of the three main parameters influencing excavatability (arranged from Thuro & Plinninger, 2003).

found in Mogi (2006), Kwasniewski *et al.* (2012), etc. An important limitation of the criterion is that it requires the use of a true triaxial testing apparatus; also, there is no any accepted standard or SM for the TTT.

Cutting and drilling performances as well as the wear of tools and equipment are decisive for the progress of excavation and drilling works. Excavatability is a term used in underground construction to describe the influence of a number of parameters on the drilling, blasting or cutting rate (excavation performance) and the tool wear of a drilling rig, road header or tunnel boring machine (TBM) (Thuro & Plinninger, 2003). As can be seen from Figure 19, which illustrates the interactions of the main factors involved in excavatability, mechanical properties of rock and rock mass play an important role in excavation performance. There are a number of laboratory test methods to determine the properties of rocks in terms of excavatability for the proper selection, performance prediction of mechanical miners, tool wear, etc., and they are given in the literature in necessary detail (*e.g.*, Bruland, 1998; Bilgin *et al.*, 2014). However, some of the methods have still no standard or suggested method or under improvement. By considering the increasing interest in TBMs and deep borings, some improvements on determination of excavatability and drillability parameters and the preparation of associated ISRM SMs are also some of the near future expectations which may assist considerably in the effort of predicting excavatability and in the assessment of drilling performance.

A number of geophysical methods are available to be used in rock engineering. However, newer sophisticated instrumentation with increased measurement sensitivities will permit geophysical techniques to play an increasingly important role in rock engineering. There is a need to obtain more rock property information, particularly on the geometry and mechanical properties of rock fractures. More emphasis will be given on geophysical methods in site investigation through rapidly developing seismic techniques, especially tomography and associated 3D visualization methods.

As emphasized by the ISRM Commission on Geophysics (Matsuoka, 2011), because carbon capture and storage (CCS) is becoming one of the key technologies for the reduction of CO_2 emission in the atmosphere, rock mechanics is expected to contribute to the procedures. Geophysics is also expected to play a central role for monitoring and verifying CO_2 movement in the ground. Although geophysics has been applied already to several CCS fields, there still remain many challenges to be solved in the future.

As a result of extracting oil from deeper and more difficult geological settings, the use of rock mechanics in petroleum engineering has become increasingly important since the 1970s (*e.g.*, Roegiers, 1999). In terms of rock testing, the factors are mainly the measurement of *in-situ* stresses, particularly shale and sandstone characterization, and petroleum engineering related laboratory tests such as the thermo-hydro-mechanical behavior of shales (ARMA, 2012). Boring and testing issues including coring guidelines and best practices, minimizing core damage, identifying core damage, sample preparation and handling, "best-practice" testing protocols, index testing, non-standard tests (*e.g.* creep, high temperature, high pressure, reactive fluids and fractured rock) and the use of analogue materials will be the important developments expected in this area in the near future.

When the stresses at the excavation boundary reach the rock mass strength, a brittle failure occurs that is often called "spalling". The spalling phenomenon takes place as a high compressive stress induces crack growth behind excavated surface, and buckling of thin rock slabs occurs (Figure 20a, b) (Siren *et al.*, 2011). Rock spalling is an important aspect in rock engineering, particularly in underground studies and in the preservation of man-made historical underground openings. As emphasized by the ISRM Commission on Rock Spalling (Diederichs, 2008), the focus is mainly on spalling in hard and low porosity rocks. In terms of experimental rock mechanics, the near future primary tasks are providing guidelines for laboratory procedures to detect damage thresholds and suggesting field observations using the televiewer, core disking (Figure 20c), etc., which can be used during investigations to assess the spalling potential. The exact mechanism of spalling in foliated rocks also needs clarification.

(a)   (b)   (c)

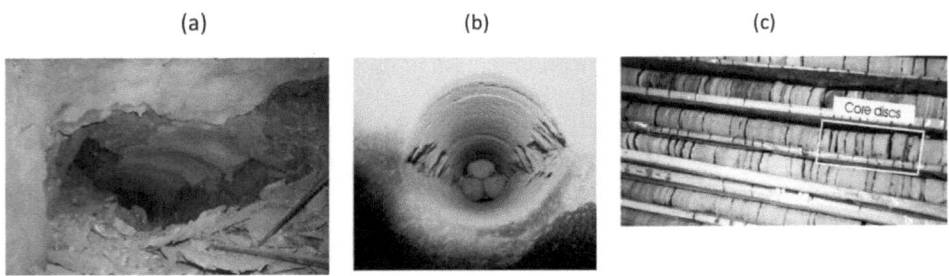

Figure 20 (a) and (b) examples of spalling in underground openings (Kaiser, 2010), (c) core disking.

The ISRM SM for compressive testing of intact rock samples does not discuss the measurement of crack damage thresholds in a compressive test. The ISRM Commission on Rock Spalling developed a guideline (Ghazvinian *et al.*, 2012) for estimation of onset of crack damage thresholds for brittle rock in laboratory, and it is expected that the updated and comprehensive version of this guideline would be a candidate ISRM SM in near future.

Long-term maintenance and preservation of man-made historical and modern rock structures as well as waste disposal sites become important issues in geo-engineering. Although they are well-known issues, quantitative evaluation methods are still lacking. Important issues are how to evaluate the weathering and degradation rates and effect of variations in water content on rocks with minerals or particles susceptible to water, and to incorporate these in the stability assessments (*e.g.*, Aydan, 2003; Ulusay & Aydan, 2011). Available methods such as slake durability, drying and wetting, freezing and thawing, and swelling tests can be used for the purpose. However, disintegration of rocks during wetting-drying and freezing-thawing laboratory tests, in which weather conditions are simulated, occurs faster than the natural processes *in situ*, and they are also insufficient to provide experimental data for constitutive and mechanical modeling. Therefore, the development of new experimental techniques and/or modification of the existing methods to solve this problem is urgently needed.

In summary, although a considerable progress has been achieved during the 50 years of the ISRM Commission on Testing Methods, the future promises many important and exciting developments in rock mechanics testing.

REFERENCES

Abbot, A.V. (1884) *Testing Machines: Their History, Construction and Use*. New York, Van Nostrand.

Aggson, J.R. (1978) *The potential application of ultrasonic spectroscopy to underground site characterization*. Presented at the 48th Annual International Meeting of Society of Exploration Geophysicists, San Francisco.

AGS (1999) *Electronic transfer of geotechnical and geoenvironmental data*. 3rd edn. [Online] Association of Geotechnical and Geoenvironmental Specialists, Beckenham, Kent, Available from: www.ags.org.uk/ [Accessed 15th January 2015].

AGS (2005) *Electronic transfer of geotechnical and geoenvironmental data using XML data format*. [Online] Association of Geotechnical and Geoenvironmental Specialists, Beckenham, Kent, Available from: http://www.ags.org.uk/agsml/AGSMLAugust2005Report.pdf [Accessed 15th January 2015].

Alber, M., Yarali, O., Dahl, F., Bruland, A., Kasling, H., Michalakopoulos, T.N., Cardu, M., Hagan, P., Aydin, H. & Ozarslan, A. (2014) ISRM suggested method for determining the abrasivity of rock by the CERCHAR Abrasivity Test. *Rock Mechanics and Rock Engineering*, 47, 261–266.

Alejano, L.R., Gonzalez, J. & Muralha, J. (2012) Comparison of different techniques of tilt testing and basic friction angle variability assessment. *Rock Mechanics and Rock Engineering*, 45, 1023–1035.

Altindag, R. & Guney, A. (2006) ISRM suggested method for determining the Shore Hardness value for rock. *International Journal of Rock Mechanics and Mining Sciences*, 43, 19–22.

Amadei, B. & Savage, W.Z. (1989) Anisotropic nature of jointed rock mass strength. *Journal of Engineering Mechanics ASCE*, 115, 525–542.

Amadei, B. & Stephansson, O. (1997) *Rock Stress and Its Measurement*. London, Chapman & Hall.

Anderson, N. & Croxton, N. (2008) Introduction to geotechnical geophysics. In: *Geophysical Methods Commonly Employed for Geotechnical Site Characterization*, Transportation Research Circular E-C130, Washington, pp. 2–12.

ARMA (2012) *Workshop on petroleum geomechanics testing*. [Online] Available from: http://www.arma.org/conference/2012/Chicago.aspx [Accessed 15th January 2015].

ASTM (2008a) ASTM D2936-08. *Standard test method for direct tensile strength of intact rock core specimens*. West Conshohocken, PA, ASTM International.

ASTM (2008b) ASTM D3967-08. *Standard test method for splitting tensile strength of intact rock core specimens*. West Conshohocken, PA, ASTM International.

ASTM (2008c) ASTM D4644-08. *Standard test method for slake durability of shales and similar weak rocks*. West Conshohocken, PA, ASTM International.

ASTM (2008d) ASTM D7070-08. *Standard test methods for creep of rock core under constant stress and temperature*. West Conshohocken, PA, ASTM International.

ASTM (2008e) ASTM D5607-08. *Standard test method for performing laboratory direct shear strength tests of rock specimens under constant normal force*. West Conshohocken, PA, ASTM International.

ASTM (2008f) ASTM D4394-08. *Standard test method for determining in situ modulus of deformation of rock mass using rigid plate loading method*. West Conshohocken, PA, ASTM International.

ASTM (2008g) ASTM D4395-08. *Standard test method for determining in situ modulus of deformation of rock mass using flexible plate loading method*. West Conshohocken, PA, ASTM International.

ASTM (2008h) ASTM D4729-08. *Standard test method for in situ stress and modulus of deformation using flatjack method*. West Conshohocken, PA, ASTM International.

ASTM (2008i) ASTM D4971-08. *Standard test method for determining in situ modulus of deformation of rock using diametrically loaded 76-mm (3-in.) borehole jack*. West Conshohocken, PA, ASTM International.

ASTM (2008j) ASTM D4553-08. *Standard test method for determining in situ creep characteristics of rock*. West Conshohocken, PA, ASTM International.

ASTM (2010) ASTM D7625-10. *Standard test method for laboratory determination of abrasiveness of rock using the CERCHAR method*. West Conshohocken, PA, ASTM International.

ASTM (2012) STM D4554-12. *Standard test method for in situ determination of direct shear strength of rock discontinuities*. West Conshohocken, PA, ASTM International.

ASTM (2013a) ASTM D4525-13. *Standard test method for permeability of rocks by flowing air*. West Conshohocken, PA, ASTM International.

ASTM (2013b) ASTM D4506-13. *Standard test method for determining in situ modulus of deformation of rock mass using radial jacking test*. West Conshohocken, PA, ASTM International.

ASTM (2014a) ASTM C131/131M-14. *Standard test method for resistance to degradation of small-size coarse aggregate by abrasion and impact in the Los Angeles machine*. West Conshohocken, PA, ASTM International.

ASTM (2014b) ASTM D7012-14. *Standard test methods for compressive strength and elastic moduli of intact rock core specimens under varying states of stress and temperatures*. West Conshohocken, PA, ASTM International.

Aydan, Ö. (2000) A new stress inference method for the stress state of earth's crust and its application. *Yerbilimleri / Earth Sciences*, 22, 223–236 [in Turkish].

Aydan, Ö. (2003) The moisture migration characteristics of clay-bearing geo-materials and the variations of their physical and mechanical properties with water content. In: Proceedings of the 2nd Asian Conference on Saturated Soils (UNSAT-ASIA 2003), *Osaka*, pp. 383–388.

Aydan, Ö. (2008) New directions of rock mechanics and rock engineering: Geomechanics and geoengineering. In: Majdi, A. & Ghazvinian, A. (eds.), *ARMS5: Proceedings of the Asian Rock Mechanics Symposium, Tehran*, Vol. 1, pp. 3–21.

Aydan, Ö. (2012) *Historical Rock Mechanics and Rock Engineering*. Tokai University, Japan [Unpublished Notes].

Aydan, Ö. (2013) In situ stress inference from damage around blasted holes. *Journal of Geosystem Engineering*, 16 (1), 83–91.

Aydan, Ö. (2014) Future advancement of rock mechanics and rock engineering (RMRE). In: Sariisik, A., Ozkan, E., & Sariisik, G. (eds.), ROCKMEC'2014: Proceedings of the XIth Regional Rock Mechanics Symposium, *Afyon, Turkey*, pp, 27–50.

Aydan, Ö., Seiki, T., Jeong, G.C. & Akagi, T. (1995) A comparative study on various approaches to model discontinuous rock mass as equivalent continuum. In: Proceedings of 2nd International Conference on Mechanics of Jointed and Fractured Rocks, *Vienna*, pp. 560–574.

Aydan, Ö., Shimizu, Y., Akagi, T. & Kawamoto, T. (1997) Tests for mechanical properties of model fracture zones. In: ARMS'96: Proceedings of the 1st Asian Rock Mechanics Symposium, *Seoul, Korea*, pp. 643–648.

Aydan, Ö., Tokashiki, N., Ito, T., Akagi, T., Ulusay, R. & Bilgin, H.A. (2003) An experimental study on the electrical potential of non-piezoelectric geomaterials during fracturing and sliding. In: Proceedings of the 9th ISRM Congress, *South Africa*, pp. 73–78.

Aydan, Ö., Ulusay, R. & Tokashiki, N. (2013) A new rock mass quality rating system: Rock Mass Quality Rating (RMQR) and its application to the estimation of geomechanical characteristics of rock masses. *Rock Mechanics and Rock Engineering*, 47, 1255–1276.

Aydan, Ö., Ito, T., Ozbay, U., Kwasniewski, M., Shariar, K., Okuno, T., Ozgenoglu, A., Malan, D.F. & Okada, T. (2014) ISRM suggested methods for determining the creep characteristics of rock. *Rock Mechanics and Rock Engineering*, 47, 275–290.

Aydin, A. (2009) ISRM suggested method for determination of the Schmidt hammer rebound hardness: Revised version. *International Journal of Rock Mechanics and Mining Sciences*, 46, 627–634.

Aydin, A. (2014) Upgraded ISRM suggested method for determining sound velocity by ultrasonic pulse transmission technique. *Rock Mechanics and Rock Engineering*, 47, 255–259.

Backers, T. (2004) Fracture toughness determination and micromechanics of rock under Mode I and Mode II loading. PhD Thesis, Institut für Geowissenschaften Sektion 3.2 Deformation und Rheologie des GFZ Potsdam, Germany.

Backers, T. & Stephansson, O. (2012) ISRM suggested method for the determination of Mode II fracture toughness. *Rock Mechanics and Rock Engineering*, 45, 1011–1022.

Backstrom, A.L., Metcalf, J.G. & McKelvie, S. (2009) What happens in Las Vegas: The Apex Tunnel geologic investigation. In: Almeraris, G. & Mariucci, B. (eds), Proceedings of 2009 Rapid Excavation and Tunneling Conference, SME, *Littleton, CO*, pp. 534–547.

Barla, G., Barla, M., Camusso, M. & Martinotti, M.E. (2007) Setting up a new direct shear testing apparatus. In: Ribeiro e Sousa, L., Olalla, C. & Grossmann, N.F. (eds.), Proceedings of 11th Congress of International Society for Rock Mechanics, Lisbon, Taylor & Francis, pp. 415–418.

Barton, N. (1976) Shear strength of rock and rock joints. *International Journal of Rock Mechanics and Mining Sciences & Geomechanics Abstracts*, 13 (9), 255–279.

Barton, N. (2013) Shear strength criteria for rock, rock joints, rockfill and rock masses: Problems and some solutions. *Journal of Rock Mechanics and Geotechnical Engineering*, 5, 249–261.

Barton, N. & Bandis, S. (1980) Some effects of scale on the shear strength of joints. *International Journal of Rock Mechanics, Mining Sciences & Geomechanics Abstracts*, 17, 69–73.

Barton, N., Lien, R. & Lunde, J. (1974) Engineering classification of rock masses for the design of tunnel support. *Rock Mechanics*, 6, 189–236.

Bell, F.G., Entwisle, D.C. & Culshaw, M.G. (1997) A geotechnnical survey of some British Coal Measures mudstones, with particular emphasis on durability. *Engineering Geology*, 46, 115–129.

Bieniawski, Z.T. (1966) *Mechanism of rock fracture in compression*. Report of the Rock Mechanics Division, National Mechanical Engineering Institute, South African Council of Scientific and Industrial Research, No. MEG 459.

Bieniawski, Z.T. (1975) Point load test in geotechnical practice. *Engineering Geology*, 9, 1–11.

Bieniawski, Z.T. (1989) *Engineering Rock Mass Classifications*. New York, John Wiley & Sons.

Bieniawski, Z.T. (2008) Reflections on new horizons in rock mechanics design: Theory, education and practice. In: Majdi, A. & Ghazvinian, A. (eds.), *ARMS5: Proceedings of the 5th Asian Rock Mechanics Symposium, Tehran*, Vol. 1, pp. 37–50.

Bilgin, N., Copur, H. & Balci, C. (2014) *Mechanical Excavation in Mining and Civil Industries*. London, CRC Press, Taylor & Francis Group.

Blanks, R.F. & McHenry, D. (1945) Large triaxial testing machine built by Bureau of Reclamation. *Engineering News Record*, 135 (6), 171–172.

Brady, B.H.G. & Brown, E.T. (1993) *Rock Mechanics for Underground Mining*. 2nd edn., London, Chapman & Hall.

Bray, J.W. (1967) A study of jointed and fractured rock-Part 1: Fracture patterns and their characteristics. *Rock Mechanics and Engineering Geology*, 5–6 (2–3), 117–136.

Brown, P.D. & Robertshaw, J. (1953) The in-situ measurement of Young's modulus for rock by a dynamic method. *Géotechnique*, 3 (7), 283–286.

Bruce, I.G., Cruden, D.M. & Eaton, T.M. (1989) Use of a tilting table to determine the basic friction angle of hard rock samples. *Canadian Geotechnical Journal*, 26, 474–479.

Bruland, A. (1998) *Drillability test methods-hard rock tunnel boring*. [Online] Available from NTNU: www.drillability.com/13A-98eng.pdf [Accessed 19th December 2014].

Camilo Quiñones-Rozo, P.E. (2010) Lugeon test interpretation, revisited. In: Collaborative Management of Integrated Watersheds 30th Annual USSD Conference, *Sacramento, California*, pp. 405–414.

Claesson, J. & Bohloli, B. (2002) Brazilian test: Stress field and tensile strength of anisotropic rocks using an analytical solution. *International Journal of Rock Mechanics and Mining Sciences*, 39, 991–1004.

Chang, C. & Haimson, B. (2012) A failure criterion for rocks based on true triaxial testing. *Rock Mechanics and Rock Engineering*, 45, 1007–1010.

Chen, R. & Stimpson, B. (1993) Interpretation of indirect tensile strength when moduli of deformation in compression and in tension are different. *Rock Mechanics and Rock Engineering*, 26 (2), 183–189.

Christiansson, R. & Hudson, J.A. (2003) Suggested method for rock stress estimation-Part 4: Quality control of rock stress estimation. *International Journal of Rock Mechanics and Mining Sciences*, 40, 1021–1025.

Coli, N., Berry, P. & Boldini, D. (2011) In situ non-conventional shear tests for the mechanical characterisation of a bimrock. *International Journal of Rock Mechanics and Mining Sciences*, 48, 95–102.

Cook, J.C. (1972) *Semi-annual report on electronic measurements of rock stress*. US Bureau of Mines Technical Report No. 72–10.

Cook, N.G.W. (1965) The failure of rock. *International Journal of Rock Mechanics and Mining Sciences*, 2 (4), 389–403.

Coulomb, C.A. (1776) Essai sur une application des regles de maximis et minimis a quelques problemes de statique, relatifs a l'architecture. *Memoires de Mathematique & de Physique*, 7, 343–382.

Cristescu, N. & Hunsche, U. (1996) A comprehensive constitutive equation for rock salt determination and application. In: Proceedings of the Third Conference on the Mechanical Behavior of Salt, *Clausthal-Zellerfeld, Trans Tech Publications*, pp. 191–205.

Cruden, D.M. & Hu, X.Q. (1988) Basic friction angles of carbonate rocks from Kananaskis country, Canada. *Bulletin of the International Association of Engineering Geology*, 38, 55–59.

Daido, M., Aydan, Ö., Kuwae, H. & Sakoda, S. (2003) An experimental study on the validity of Kaiser effect for in-situ stress measurements by Acoustic Emission Method (AEM) in rocks subjected to cyclic loads. *Journal of the School of Marine Science and Technology*, 1 (1), 17–22.

Deere, D.U. & Miller, R.P. (1966) *Classification and index properties of intact rock*. Technical Report AFWL-TR-65-116, AF Special Weapons Center, Kirtland Air Force Base, New Mexico.

Detournay, E., Drescher, A. & Hultman, D.A., (1997) Portable rock strength evaluation device. United States Patent 5670711.

Diederichs, M.S. (1999) Instability of hard rock masses: The role of tensile damage and relaxation. PhD Thesis, University of Waterloo, Canada.

Diederichs, M.S. (2008) ISRM Rock Spalling Commission: Report for 2008. *ISRM News Journal*, 11, 50–51.

Dusseault, M.B. & Fordham, J.C. (1993) Time dependent behaviour of rocks. In: Hudson, J.A. (ed.), *Comprehensive Rock Engineering – Principles, Practice & Projects*, Vol. 3, pp. 119–149.

Dyke, C.G. (1989) Core disking: Its potential as an indicator of principal in situ stress directions. In: Maury, V. & Fourmaintraux, D. (eds.), *Proceedings ISRM-SPE International Symposium on Rock at Great Depth, Pau*, Rotterdam, Balkema, Vol. 2, pp. 1057–1064.

Evison, F.F. (1953) The seismic determination of Young's modulus and Poisson's ratio for rocks in situ. *Géotechnique*, 6 (3), 118–123.

Exadaktylos, G., Liolios, P. & Barakos, G. (2007) Some new developments on the representation and standardization of rock mechanics data: From the laboratory to the full scale project. [Online] In: Toll, D.G. & Chen, Z. (eds.), *Proceedings of Specialized Session S02 of the 11th ISRM Congress, Lisbon, Portugal*, pp. 1–8. Available from: http://community.dur.ac.uk/geoengineering/jtc2/ISRM2007/ISRM_Specialised_Session_S02.pdf [Accessed 15th January 2015].

Fairbairn, W. (1856) On the tensile strength of wrought iron at various temperatures. In: *British Association for the Advancement of Science Annual Report*, pp. 405–422.

Fairbairn, E.M.R. & Ulm, F.J. (2002) A tribute to Fernando LLB Carneiro (1913–2001), engineer and scientist who invented the Brazilian test. *Materials and Structures*, 35, 195–196.

Fairhurst, C.E. (1964) Measurement of in-situ rock stress with particular reference to hydraulic fracturing. *Rock Mechanics and Engineering Geology*, 2, 129–147.

Fairhurst, C.E. & Hudson, J.A. (1999) ISRM suggested method for the complete stress-strain curve for intact rock in uniaxial compression. *International Journal of Rock Mechanics and Mining Sciences*, 36, 279–298.

Galileo, G. (1638) *Two New Sciences*. Crew, H. & de Salvio, A. (trans.), New York, Macmillan (1914).

Gamble, J.C. (1971) Durability-plasticity classification of shales and other argillaceous rocks. PhD Thesis, University of Illinois at Urbana-Champaign.

Gercek, H. (2007) Poisson's ratio values for rocks. *International Journal of Rock Mechanics and Mining Sciences*, 44 (1), 1–13.

Gercek, H. & Ozarslan, A. (2011) Tensile strength classification of rock material. In: Proceedings of ROCMEC'2011: Xth Regional Rock Mechanics Symposium, *Ankara, Turkey*, pp. 105–116 [in Turkish].

Ghazvinian, E., Diederichs, M., Martin, D., Christiansson, R., Hakala, M., Gorski, B., Perras, M. & Jacobsson, L. (2012) *Prediction thresholds for crack initiation and propagation in crystalline rocks*. ISRM Commission on Spall Prediction Report on Testing Procedures 2012.

GIF (2004a) *In situ shear tests*. [Online] Report of Geotechnisches Ingenieurbüro Prof. Fecker & Partner GmbH Chapter 17, Germany, Available from: www.gif-ettlingen.de/engl/pdf/engl/Kap._17.pdf [Accessed 1st February 2015].

GIF (2004b) *In situ triaxial tests*. [Online] Report of Geotechnisches Ingenieurbüro Prof. Fecker & Partner GmbH Chapter 15, Germany, Available from: www.gif-ettlingen.de/engl/pdf/engl/Kap._15.pdf [Accessed 1st February 2015].

Glassel, P.R. (2014) *Engineering 101: Repeatability and reproducibility*. [Online] P.R. Glassel and Associates, Inc., Available from: www.prga.com:8080/eng101Repeatability.cshtml [Accessed 15th December 2014].

Gokceoglu, C., Ulusay, R. & Sonmez, H. (2000) Factors affecting the durability of selected weak and clay-bearing rock from Turkey, with particular emphasis on the influence of the number of drying and wetting cycles. *Engineering Geology*, 57, 215–237.

Golder, H.Q. & Akroyd, T.N.W. (1954) An apparatus for triaxial compression tests at high pressures. *Géotechnique*, 4 (4), 131–136.

Goodman, R.E. (1989) *Introduction to Rock Mechanics*. 2nd edn., Chichester, John Wiley.

Gray, T.G.F. (1988) Tensile testing: Chp. 1. In: *Mechanical Testing*, Book 445, London, Pub Inst. of Metals, pp. 1–42.

Griffith, A.A. (1921) The phenomena of rupture and flow in solids. *Philosophical Transactions of the Royal Society, Series A: Mathematical, Physical and Engineering Sciences*, 221, 163–198.

Griggs, D.T. (1936) Deformation of rocks under high confining pressures. *Journal of Geology*, 44, 541–577.

Griggs, D.T. (1939) Creep of rocks. *Journal of Geology*, 47 (3), 225–251.

Guglielmi, Y., Cappa, F., Lancon, H., Janowczyk, J.B., Rutqvist, J., Tsang, C.F. & Wang, J.S.Y. (2014) ISRM suggested method for step-rate injection method for fracture in-situ properties (SIMFIP): Using a 3-components borehole deformation sensor. *Rock Mechanics and Rock Engineering*, 47, 303–311.

Habib, P. (1950) Détermination du module d'élasticité des roches en place. *Annales de l'Institut Technique du Bâtiment et des Travaux Publics*, 145, 27–35.

Habib, P. & Marchand, R. (1952) Mesures des pressions de terrains par l'essai de vérin plat. Suppléments aux *Annales de l'Institut Technique du Bâtiment et des Travaux Publics, Série Sols et Foundations*, 58, 967–971.

Hagros, A., Johanson, E. & Hudson, J.A. (2008) *Time dependency in the mechanical properties of crystalline rocks: A literature survey*. [Online] Posiva working Report 2008-68, Finland, Posiva Oy, Available from: http://www.posiva.fi/files/818/WR2008-68web.pdf [Accessed 21th January 2015].

Haimson, B.C. & Lee, M.Y. (1995) Estimating in situ stress conditions from borehole breakouts and core disking-experimental results in granite. In: *Proceedings of International Workshop on Rock Stress Measurement at Great Depth – 8th ISRM Congress, Tokyo*, pp. 19–24.

Haimson, B.C. & Cornet, F.H. (2003) ISRM suggested methods for rock stress estimation-Part 3: Hydraulic fracturing (FH) and/or hydraulic testing of pre-existing fractures (HTPF). *International Journal of Rock Mechanics and Mining Sciences*, 40, 1011–1020.

Handin, J. (1953) An application of high pressure geophysics: Experimental rock mechanics. *Transactions American Society of Mechanical Engineers*, 75, 315–324.

He, M. (2014) Latest progress of soft rock mechanics and engineering in China. *Journal of Rock Mechanics and Geotechnical Engineering*, 6, 165–179.

Heyman, J. (1972) *Coulomb's Memoir on Statics: An Essay in the History of Civil Engineering*. Cambridge, Cambridge University Press.

Hobbs, D.W. (1965) An assessment of a technique for determining the tensile strength of rock. *British Journal of Applied Physics*, 16, 259–268.

Hoek, E. (2007) *Practical rock engineering.* [Online] Available from: https://www.rocscience. com/education/hoeks_corner [Accessed 15th January 2015].

Hoek, E. & Brown, E.T. (1980) *Underground Excavations in Rock.* London, Institution of Mining and Metallurgy.

Hoek, E., Caranza-Torres, C.T. & Corkum, B. (2002) Hoek–Brown failure criterion – 2002 edition. In: Proceedings of the 5th North American Rock Mechanics Symposium, *Toronto, Canada*, Vol. 1, pp. 267–273.

Holcomb, D.J. (1993) Observations of the Kaiser effect under multiaxial stress state: Implications for its use in determining in situ stress. *Geophysical Research Letter*, 20, 2119–2122.

Holcomb, D.J. & Martin, R.J. (1985) Determining peak stress history using acoustic emissions. In: Asworth, E. (ed.), Proceedings of 26th US Symposium on Rock Mechanics, Rapid City, Rotterdam, Balkema, pp. 715–722.

Horn, H.M. & Deere, D.U. (1962) Frictional characteristics of minerals. *Géotechnique*, 12, 319–335.

Houlsby, A.C. (1976) Routine interpretation of the Lugeon water-test. *Quarterly Journal of Engineering Geology and Hydrogeology*, 9, 303–313.

Hudson, J.A. (1989) *Rock Mechanics Principles in Engineering Practice.* London, CIRIA Butterworths.

Hudson, J.A. (2008) The future for rock mechanics and the ISRM. In: Majdi, A. & Ghazvinian, A. (eds.), *ARMS5: Proceedings of the Asian Rock Mechanics Symposium, Tehran*, Vol. 1, pp. 105–118.

Hudson, J.A. (2011) The next 50 years of the ISRM and anticipated future progress in rock mechanics. In: Qian, Q. & Zhou, Y. (eds.), *Proceedings of the 12th International Congress on Rock Mechanics, Beijing, CRC Press*, pp. 47–55.

Hudson, J.A. & Harrison, J.P. (2000) *Engineering Rock Mechanics-An Introduction to the Principles.* 2nd edn., Amsterdam, Pergamon.

Hudson, J.A., Brown, E.T. & Fairhurst, C. (1971) Optimizing the control of rock failure in servo-controlled laboratory tests. *Rock Mechanics*, 3, 217–224.

Hudson, J.A., Crouch, S.L. & Fairhurst, C. (1972) Soft, stiff and servo-controlled testing machines: A review with reference to rock failure. *Engineering Geology*, 6, 155–189.

Hudson, J.A., Cornet, F.H. & Christiansson, R. (2003) ISRM suggested methods for rock stress estimation-Part 1: Strategy for rock stress estimation. *International Journal of Rock Mechanics and Mining Sciences*, 40, 991–998.

IAEG (1979) Classification of rocks and soils for engineering geological mapping part I: Rock and soil materials. *Bulletin of Engineering Geology*, 19 (1), 364–371.

Ide, J.M. (1936) Comparison of statically and dynamically determined Young's modulus of rock. *Proceedings of National Academy of Sciences*, 22, 81–92.

Irfan, T.Y. & Tang, K.Y. (1993) *Effect of the coarse fraction on the shear strength of colluvium in Hong Kong.* Hong Kong Geotechnical Engineering Office, TN 4/92.

ISRM (1981) *Rock Characterization, Testing and Monitoring, ISRM Suggested Methods.* Brown, E.T. (ed.), Oxford, Pergamon Press.

ISRM (2007) *The Complete ISRM Suggested Methods for Rock Characterization, Testing and Monitoring: 1974–2006.* Ulusay, R. & Hudson, J.A. (eds.), Suggested Methods Prepared by the Commission on Testing Methods, International Society for Rock Mechanics, Compilation Arranged by the ISRM Turkish National Group, Ankara, Turkey.

ISRM (2014) *The ISRM Suggested Methods for Rock Characterization, Testing and Monitoring: 2007–2014.* Ulusay, R. (ed.), Suggested Methods Prepared by the Commission on Testing Methods, International Society for Rock Mechanics, Heidelberg, Springer.

Ito, F., Aoki, T. & Obara, Y. (2004) Visualization of bond failure in a pull-out test of rock bolts and cable bolts using X-ray CT. In: Otani, J. & Obara, Y. (eds.), *GEOX2003: Proceedings of the International Workshop on X-ray CT for Geomaterials, Kumamoto, Japan*, pp. 305–314.

Jaeger, J.C. (1959) The frictional properties of joints in rock. *Geofisica Pura e Applicata*, 43 (Part 2), 148–158.

Jaeger, J.C. (1960) Shear fracture of anisotropic rocks. *Geological Magazine*, 97, 65–72.

Jaeger, J.C. & Cook, N.G.W. (1976) *Fundamentals of Rock Mechanics*. 2nd edn., London, Chapman & Hall.

Jeremic, M.L. (1994) *Rock Mechanics in Salt Mining*. Rotterdam, Balkema.

John, K.W. (1962) An approach to rock mechanics. *Journal of Soil Mechanics & Foundation Division, ASCE*, 88 (SM4), 1–30.

Johnson, A.M. (1970) *Physical Processes in Geology*. San Francisco, Freeman Cooper & Co.

Kaiser, P.K. (2010) *Practical implication of brittle failure on hard rock tunnelling construction*. [Online] Presentation at Universitat Politéchnicade Catalunya Barcelona, Spain, Available from: www.etcg.upc.edu/estudis/aula-paymacotas/granit/ponencies/kaiser.pdf [Accessed 7th February 2015].

Kanagawa, T., Hayashi, M. & Nakasa, H. (1976) *Estimation of spatial geostress components in rock samples using the Kaiser effect of acoustic emission*. Central Research Institute of Electrical Power Industry, Abiko, Report No. 375017, Japan.

Kanji, M. (2014) Critical issues in soft rocks. *Journal of Rock Mechanics and Geotechnical Engineering*, 6, 186–195.

Kim, M.M. & Ko, H.Y. (1979) Multistage triaxial testing of rocks. *Geotechnical Testing Journal*, 2, 98–105.

King, L.V. (1912) On the limiting strength of rocks under conditions of stress existing in the earth's interior. *Journal of Geology*, 20, 119–138.

Kuruppu, M.D., Obara, Y., Ayatollahi, M.R., Chong, K.P. & Funatsu, T. (2014) ISRM-Suggested Method for Determining the Mode I Static Fracture Toughness Using Semi-Circular Bend Specimen. *Rock Mechanics and Rock Engineering*, 47, 267–274.

Kwasniewski, M., Li, X. & Takahashi, M. (eds.) (2012) True triaxial testing of rocks. Proceedings of the TTT Workshop, London, CRC Press.

Lama, R.D. & Vutukuri, V.S. (1978) *Handbook on Mechanical Properties of Rocks-Testing Techniques and Results*, Vol. III. Clausthal, Trans Tech Publications, pp. 209–320

Lau, A. (2009) *What are repeatability and reproducibility?* [Online] Available from: http://www.astm.org/SNEWS/MA_2009/datapoints_ma09.html [Accessed 17th January 2015].

Li, S.J., Feng, X.T., Wang, C.Y. & Hudson, J.A. (2013a) ISRM suggested method for rock fractures observations using a borehole digital optical televiewer. *Rock Mechanics and Rock Engineering*, 46, 635–644.

Li, S.J., Feng, X.T. & Hudson, J.A. (2013b) ISRM suggested method for measuring rock mass displacement using a sliding micrometer. *Rock Mechanics and Rock Engineering*, 46, 645–653.

Li, X., Lia, Q.I. & He, J.M. (2004) In situ tests and stochastic structural model of rock and soil aggregate in the three Gorges Reservoir area. China. *International Journal of Rock Mechanics and Mining Sciences*, 41 (3), 702–707.

Li, X., Wang, G. & Zhu, H. (2012) A data model for exchanging and sharing ISRM rock mechanics test data. *Electronic Journal of Geotechnical Engineering (EJGE)*, 17 (Bund. D), 377–401.

Lieurance, R.S. (1933) *Stresses in foundation at Boulder (Hoover) dam*. US Bureau of Reclamation Technical Memorandum No. 346.

Lieurance, R.S. (1939) *Boulder canyon project final report: Part V-Technical investigation*. Bulletin 4, pp. 265–268.

Lindquist, E.S. (1994) The strength and deformation properties of mélange. PhD Dissertation, Departmet of Civil Engineering, University of California at Berkeley, California.

Liu, S., Wu, L. & Wu, Y. (2006) Infrared radiation of rock at failure. *International Journal of Rock Mechanics and Mining Sciences*, 43 (6), 972–979.

Loveday, M.S., Gray, T. & Aegerter, J. (2004) *Tensile Testing of Metallic Materials: A Review*. [Online] Tenstand – Work Package 1 – Final Report, Available from: http://resource.npl.co.uk/docs/science_technology/materials/measurement_techniques/tenstand/test_method_review.pdf [Accessed 14th December 2014].

Lugeon, M. (1933) *Barrages et Géologie*. Paris, Dunod.

Lund, J.R. & Byrne, J.P. (2001) Leonardo Da Vinci's tensile strength tests: Implications for the discovery of engineering mechanics. *Civil Engineering & Environmental Systems*, 18 (3), 243–250.

Luong, M.P. (1986) Un nouvel essai pour la mesure de la résistance á la traction. *Revue Française de Géotechnique*, 34, 69–74.

Luong, M.P. (1988) Direct tensile and direct shear strengths of Fontainebleau sandstone. In: Cundall, P.A., Starfield, A.M. & Sterling, R.L. (eds.), *Key Questions in Rock Mechanics: Proceedings of the 29th US Symposium on Rock Mechanics, Minneapolis, Minnesota*, Rotterdam, Balkema, pp. 237–246.

Luong, M.P. & Emami, M. (2014) Characterization of mechanical damage in granite. *Frattura ed Integrità Strutturale*, 27, 38–42.

Madsen, F.T. (1999) Suggested methods for laboratory testing on swelling rocks. *International Journal of Rock Mechanics and Mining Sciences*, 36 (3), 291–306.

Malan, D.F., Vogler, U.W. & Drescher, K. (1997) Time-dependent behaviour of hard rock in deep level gold mines. *Journal of the South African Inst. of Mining and Metallurgy, May/June* 1997, 135–147.

Martin, C.D. (1997) The effect of cohesion loss and stress path on brittle rock strength. *Canadian Geotechnical Journal*, 34 (5), 698–725.

Martin, C.D., Read, R.S. & Lang, P.A. (1990) Seven years of in situ stress measurements: Some findings at the Underground Research Laboratory. In: *Proceedings of the 1st International Workshop on Scale Effects in Rock Masses, Loen, Norway*, Rotterdam, Balkema, pp. 307–316.

Massarch, K.R. (2000) Geophysical methods for geotechnical, geoenvironmental and geo-dynamic site characterisation – An overwiev. In: *Proceedings of the 3rd International Workshop on the Application of Geophysics to Rock and Soil Engineering, University of Melbourne, Melbourne*, pp. 1–5.

Matsuoka, T. (2011) Annual report of the ISRM Geophysics Commission. *ISRM News Journal*, 14, 54.

Mayer, A., Habib, P. & Marchand, R. (1951) Mesure en place des pressions de terrains. In: *Proc. Conf. Int. sur les Pressions de Terrains et le Soutènement dans les Chantiers d'Exploration, Liège*, pp 217–221.

Medley, E.W. (1994) The engineering characterization of melanges and similar block-in-matrix rocks (bimrocks). PhD dissertation, Department of Civil Engineering, University of California at Berkeley, California.

Medley, E.W. (1997) Uncertainty in estimates of block volumetric proportion in mélange bimrocks. In: Marinos, P.G., Koukis, G.C., Tsiambaos, G.C. & Stournaras, G.C. (eds.), *Engineering Geology and the Environment, Proceedings of International Symposium, Athens, Greece*, Rotterdam, Balkema, Part A, pp. 267–272.

Medley, E.W. (2007) Bimrocks-Part 1: Introduction. *Newsletter of HSSMGE*, 7, 17–21

Medley, E.W. (2008) Engineering the geological chaos of Franciscan and other bimrocks. In: *Proceedings of the 42nd US Rock Mechanics and 2nd US-Canada Rock Mechanics Symposium, San Francisco*, Paper No. ARMA08-316.

Mogi, K. (1959) Experimental study of deformation and fracture of marble (1): On the fluctuation of compressive strength of marble and relation to the rate of stress application. *Bulletin of Earthquake Research Institute, University of Tokyo*, 37, 155–170.

Mogi, K. (1969) On a new triaxial compression test of rocks. In: Abstr. 1969 Meeting Seismol. Soc. *Japan*, 3.

Mogi, K. (1971) Fracture and flow of rocks under high triaxial compression. *Journal of Geophysical Research*, 76 (5), 1255–1269.

Mogi, K. (1972) Effect of the triaxial stress system on fracture and flow of rocks. *Physics of the Earth and Planetary Interiors*, 5, 318–324.

Mogi, K. (2006) *Experimental Rock Mechanics*. Leiden, Taylor & Francis / Balkema.

Momayez, M. & Hassani, F.R. (1992) Application of Kaiser effect to measure in-situ stresses in underground mines. In: Tillerson, J.R. & Wawersik, W.R. (eds.), Proceedings of the 33rd US Symposium on Rock Mechanics, *Santa Fe, New Mexico*, Balkema, Rotterdam, pp. 979–987.

Moon, V.G. & Beattie, A.G. (1995) Textural and microstructural influences on the durability of Waikato coal measures mudrocks. *Quarterly Journal of Engineering Geology*, 28, 303–312.

Morris, A., Ferrill, D.A. & Henderson, D.B. (1996) Slip-tendency analysis and fault reactivation. *Geology*, 24 (3), 275–278.

Muralha, J., Grasselli, G., Tatone, B., Blumel, M., Chryssanthakis, P. & Yujing, J. (2014) ISRM suggested method for laboratory determination of the shear strength of rock joints: Revised version. *Rock Mechanics and Rock Engineering*, 47, 291–302.

Müller, L. (1961) Grundsatzliches über gebrigstechnologische Großversuche. *Geologie und Bauwesen*, 27 (H.1), 3–8.

Natau, O., Borm, G. & Rockel, T. (1989) Influnce of lithology and geological structure on the stability of KTB pilot hole. In: Maury, V. & Fourmaintraux, D. (eds.), Proceedings ISRM-SPE International Symposium on Rock at Great Depth, *Pau*, Rotterdam, Balkema, pp. 1487–1490.

Natau, O., Fröhlich, B. & Mutscher, Th. (1983) Recent developments of the large scale triaxial test. In: Proceedings of 5th International Congress of ISRM, *Melbourne*, pp. A65–A74.

Norbury, D.R. (1986) The point load test. In: Hawkins, A.B. (ed.), *Site Investigation Practice: Assessing BS 5930*, Geological Society of Engineering Geology Special Publication, No. 2, pp. 325–329.

Obert, L. & Duvall, W.I. (1967) *Rock Mechanics and the Design of Structures in Rock*. New York, John Wiley & Sons.

Okada, T., Tani, K., Ootsu, H., Toyooka, Y., Hosono, T., Tsujino, T., Kimura, H., Naya, T. & Kaneko, S. (2006) Development of in-situ triaxial test for rock masses. *International Journal of JCRM*, 2 (1), 7–12.

Otani, J. (2004) State of the art report on geotechnical X-ray CT research at Kumamoto University. In: Otani, J. & Obara, Y. (eds.), *GEOX2003: Proceedings of the International Workshop on X-ray CT for Geomaterials, Kumamoto, Japan*, pp. 43–78.

Palmström, A. & Singh, R. (2001) The deformation modulus of rock masses – comparisons between in situ tests and indirect estimates. *Tunnelling and Underground Space Technology*, 16 (3), 115–131.

Pethukov, L.M., Marmorshteyn, L.M. & Morozov, G.I. (1961) Use of changes in electrical conductivity of rock to study the stress state in the rock mass and its aquifer properties. *Trudy VNIMI*, 42, 110–118.

Phillips, D.W. (1932) Further investigation of the physical properties of coal-measure rocks and experimental work on development of fractures. *Transactions Institution of Mining Engineering*, 82, 432–450.

Pilgerstorfer, T. (2014) Mechanical characterization of fault zones. PhD Thesis, Graz University of Technology, Graz, Austria.

Pitt, J.M. & Klosterman, L.A. (1984) In-situ stress by pulse velocity monitoring of induced fractures. In: Dowding, C.H. & Singh M.M. (eds.), *Rock Mechanics in Productivity and Protection: Proceedings of 25th US Rock Mechanics Symposium, Evanston, Illinois*, pp. 186–193.

Prendes-Gero, M.B., Suárez-Domínguez, F.J., González-Nicieza, C. & Álvarez-Fernández, M. I. (2013) Infrared thermography methodology applied to detect localized rock falls in self-supporting underground mines. In: Kwasniewski, M. & Łydzba, D. (eds.), *EUROCK2013: Rock Mechanics for Resources, Energy and Environment, Wroclaw, Poland*, London, Taylor & Francis Group, pp. 825–829.

Price, N.J. & Cosgrove, J.W. (1990) *Analysis of Geological Structures*. Cambridge, Cambridge University Press.

Ramamurthy, T. & Arora, V.K. (1991) A simple stress–strain model for jointed rocks. In: Proceedings of 7th International Congress on Rock Mechanics, *Aachen, Germany*, Vol. 1, pp. 323–326.

Ramberg, H. (1981) *Gravity, Deformation and the Earth's Crust*. 2nd edn., London, Academic Press.

Reed, J.R.L., Thornton, P.N. & Regan, W.M. (1980) A rational approach to the point load test. In: *Proceedings of the Australian–New Zealand Conference on Geomechanixs, Wellington, New Zealand*, Vol. 1, pp. 35–39.

Richard, T., Dagrain, F., Poyol, E. & Detournay, E. (2012) Rock strength determination from scratch tests. *Engineering Geology*, 147–148, 91–100.

Rivkin, I.D., Zapolskiy, V.P. & Bogdanov, P.A. (1956) *Sonometric Method for the Observation of Rock Pressure Effects*. Moscow, Ketallrgizdat Press.

Riznichanko, Y.V. (1967) *Study of Rock Stress by Geophysical Methods*. Moscow, Nauka Press.

Rocha, M., Serafim, J.L., Silveira, A. & Neto, J.R. (1955) Deformability of foundation rocks. In: *Proceedings of 5th Congress on Large Dams, Paris*, R75, 3, pp. 531–559.

Roegiers, J.C. (1999) The importance of rock mechanics to the petroleum industry. In: Vouille, G. & Berest, P. (eds.), *Proceedings of 9th Congress of International Society for Rock Mechanics, Paris*, Rotterdam, Balkema, Vol. 3, pp. 1525–1549.

Santi, P.M. (1996) Improving the jar slake, slake index and slake durability index tests for shales. *Environmental and Engineering Geoscience*, IV (3), 385–396.

Sato, A. & Aydan, Ö. (2014) An X-ray CT imaging of water absorption process of soft rocks. In: Khalili, N., Russell, A. & Khoshghalb A. (eds.), *Proceedings of International Symposium on Unsaturated Soils: Research and Applications*, CRC Press/Balkema, Leiden, The Netherlands, pp. 675–678.

Schmitt, D.R. & Li, Y. (1993) Influence of a stress relief hole's depth on induced displacements: application in interferometric stress determination. *International Journal of Rock Mechanics and Mining Sciences & Geomechanics Abstracts*, 30, 985–988.

Seto, M., Utagawa, M. & Katsuyama, K. (1992) The estimation of pre-stress from AE in cyclic loading of pre-stressed rock. In: *Proceedings of 11th International Symposium on Acoustic Emission, The Japanese Society for NDI, Fukuoka, Japan*, pp. 159–166.

Seto, M., Utagawal, K., Katsuyama, K., Nag, D.K. & Vutukuri, V.S. (1997) In situ stress determination by acoustic emission technique. *International Journal of Rock Mechanics and Mining Sciences*, 34 (3–4), Paper No. 281.

Siren, T., Martinelli, D. & Uotinen, L. (2011) *Assessment of the potential for rock spalling in the technical rooms of the ONKALO*. Working Report 2011-35, Posiva Oy, Finland.

Sjöberg, J., Christiansson, R. & Hudson, J.A. (2003) ISRM suggested method for rock stress estimation-Part 2: Overcoring methods. *International Journal of Rock Mechanics and Mining Sciences*, 40, 999–1010.

Smith, D. (1982) David Kirkaldy (1820–1897) and engineering materials testing. *Newcomen Society of Engineering & Technology Transactions*, 52, 49–65.

Smither, C.L. & Ahrens, T.J. (1991) Displacements from relief by in situ stress by a cylindrical hole. *International Journal of Rock Mechanics and Mining Sciences & Geomechanics Abstracts*, 28, 175–186.

Smither, C.L., Schmitt, D.R. & Ahrens, T.J. (1988) Analysis and modelling of holographic measurements of in-situ stress. *International Journal of Rock Mechanics and Mining Sciences & Geomechanics Abstracts*, 25, 353–362.

Sonmez, H., Kasapoglu, K.E., Coskun, A., Tunusluoglu, C., Medley, E.W. & Zimmerman, R.W. (2009) A Conceptual empirical approach for the overall strength of unwelded bimrocks. In: Vrkljan, I. (ed.), *Rock Engineering in Difficult Ground Conditions, Soft Rock and Karst: Proceedings of the ISRM Regional Symposium, Dubrovnik, Croatia*, pp. 357–360.

Spaeth, W. (1935) Einfluss der federung der Zerreissmaschine auf das spannungs-Denhungs-Schaubild. *Archiv für das Eisenhüttenwesen*, 6, 277–283.

Stephansson, O. (1983) Rock stress measurement by sleeve fracturing. In: *Proceedings of the 5th ISRM Congress, Melbourne*, Rotterdam, Balkema, pp. 129–137.

Stephansson, O. & Zang, A. (2012) ISRM suggested methods for rock stress estimation – Part 5: Establishing a model for in-situ stress at a given site. *Rock Mechanics and Rock Engineering*, 45, 955–969.

Stimpson, B. (1981) A suggested technique for determining the basic friction angle of rock surfaces using core. *International Journal of Rock Mechanics and Mining Sciences & Geomechanics Abstracts*, 18, 63–65.

Strickland, F.G. & Ren, N.K. (1980) Use of differential strain curve analysis in predicting the in situ stress state for deep wells. In: *Proceedings of the 21st US Symposium on Rock Mechanics, Rolla, University of Missouri Publication*, pp. 523–532.

Sugawara, K. & Obara, Y. (1999) ISRM suggested method for in-situ stress measurement using the compact conical-ended borehole overcoring (CCBO) technique. *International Journal of Rock Mechanics and Mining Sciences*, 36, 307–322.

Sulukcu, S. & Ulusay, R. (2001) Evaluation of the block punch index test with particular reference to the size effect, failure mechanism and its effectiveness in predicting rock strength. *International Journal of Rock Mechanics and Mining Sciences*, 38, 1091–1111.

Sun, Y.L. & Peng, S.S. (1989) Development of in-situ stress measurement technique using ultrasonic wave attenuation method: A progress report. In: *Proceedings of 30th US Rock Mechanics Symposium, Morgantown*, Rotterdam, Balkema, pp. 477–484.

Swift, J., Bobbitt, J., Roblee, C., Futrelle, J., Tiwana, S., Peters, A., Castro, J., Ali, M., Nasir, F., Javed, A., Khan, Y. & Stepp, C. (2004) *Cosmos/peer lifelines geotechnical virtual data center.* [Online] COSMOS Workshop 1, 15 October 2004. Available from: www.cosmoseq.org/Projects/GSMA/Presentation/GSMA2004 [Accessed 15th January 2015].

Swolfs, H.S. & Handin, J. (1976) Dependence of sonic velocity on size and in situ stress in a rock mass. In: *Proceedings of the ISRM Symposium on Investigation of Stress in Rock: Advances in Stress Measurement, Sydney*, The Institution of Engineers, Australia, pp. 41–43.

Szwedzicki, T. (2003) Quality assurance in mine ground control management. *International Journal of Rock Mechanics and Mining Sciences*, 40 (4), 565–572.

Talebi, S. & Young, R.P. (1989) Failure mechanism of crack propagation induced by shaft excavation at the Underground Research Laboratory. In: Maury, V. & Fourmaintraux, D. (eds.), *Proceedings ISRM-SPE International Symposium Rock at Great Depth, Pau*, Rotterdam, Balkema, Vol. 3, pp. 1455–1461.

Tani, K., Nozaki, T., Kaneko, S., Toyo-oka, Y. & Tachikawa, H. (2003) Down-hole triaxial test to measure average stress-strain relationship of rock mass. *Soils and Foundations*, 43 (5), 53–62.

Taylor, R.K. (1988) Coal measures mudrocks: Composition, classification and weathering processes. *Quarterly Journal of Engineering Geology*, 21, 85–99.

Terzaghi, K. (1946) Rock defects and loads on tunnel supports. In: Proctor, R.V. & White, T.L. (eds.), *Rock Tunnelling with Steel Supports*, Youngstown, OH, Commercial Shearing and Stamping Co., pp. 17–99.

Teufel, L.W. (1982) Prediction of hydraulic fracture azimuth from anelastic strain recovery measurements of oriented core. In: *Proceedings of the 23rd US Symposium on Rock Mechanics, Berkeley, SME/AIME*, pp. 238–245.

Thuro, K. & Plinninger, R.J. (2003) Hard rock tunnel boring, cutting, drilling and blasting: Rock parameters for excavatability. In: *Technology Roadmap for Rock Mechanics: Proceedings of the 9th ISRM Congress, South Africa, SAIMM*, pp. 1227–1233.

Timeshenko, S.P. (1953) *History of Strength of Materials*. New York, McGraw-Hill.

Todhunter, I. & Pearson, K. (1886) *A History of the Theory of Elasticity and the Strength of Materials from Galilei to the Present Time*. Cambridge, Cambridge University Press, Vol. 1, pp. 1–6.

Toll, D.G. (2007) Geo-engineering data: Representation and standardisation. [Online] *Electronic Journal of Geotechnical Engineering*, Available from: www.ejge.com/2007/Ppr0699/Ppr0699.htm [Accessed 15th January 2015].

Toll, D.G. (2008) International data exchange: The future for geoengineering. [Online] *The 12th International Conference of International Association for Computer Methods and Advances in Geomechanics (IACMAG), Goa, India*, Available from: http://community.dur.ac.uk/geo-engineering/jtc2/documents/2008-Toll-IACMAG.pdf [Accessed 15th January 2015].

Toll, D.G. & Cubitt, A.C. (2003) Representing geotechnical entities on the World Wide Web. *Advanced Engineering Software*, 34 (11–12), 729–736.

Ulusay, R. & Yoleri, M.F. (1993) Shear strength characteristics of discontinuities in weak, stratified, clay-bearing coal measures encountered in Turkish surface coal mining. *Bulletin of the International Association of Engineering Geology*, 48, 63–71.

Ulusay, R., Arikan, F., Yoleri, M.F. & Caglan, D. (1995). Engineering geological characterization of coal mine waste material and an evaluation in the context of back-analysis of spoil pile instabilities in a strip mine SW Turkey. *Engineering Geology*, 40, 77–101.

Ulusay, R., Gokceoglu, C. & Sulukcu, S. (2001) ISRM suggested method for determining block punch strength index (BPI). *International Journal of Rock Mechanics and Mining Sciences*, 38, 1113–1119.

Ulusay, R. & Aydan, Ö. (2011) Issues on short- and long-term stability of historical and modern man-made cavities in the Cappadocia Region of Turkey. In: *Proceedings of the 1st Asian and 9th Iranian Tunnelling Symposium, Tehran* [on CD].

Ulusay, R., Aydan, Ö., Erguler, Z.A., Ngan-Tillard, D.J.M., Seiki, T., Verwaal, W., Sasaki, Y. & Sato, A. (2014) ISRM Suggested Method for the needle penetration test. *Rock Mechanics and Rock Engineering*, 47, 1073–1085.

Valentin, A. (1973) Test Cerchar pour la mesure de la dureté et de l'abrasivité des roches. Annex of: Examen des Differents Procédes Classiques de Determination de la Nocivité des Roches vis-a-vis de l'Appatage Mechanique. *Proc. Tech. Creusement*, Luxembourg. pp. 133–154.

Van Musschenbroek, P. (1729) *Introductio ad cohaerentiam firmorum in Physicae Experimentales et Geometricae Dissertationes*. Lugduni (Leiden), Referenced in Materials Testing Machines (CH Gibbons), Pittsburgh, Instrument Publishing Company.

Viggiani, G. & Hall, S.A. (2012) Full-field measurements in experimental geomechanics: Historical perspective, current trends and recent results. In: Viggiani, A., Hall, S.A. & Romero, E. (eds.), *ALERT Doctoral School 2012: Advanced Experimental Techniques in Geomechanics, Dresden*, pp. 3–67.

Von Karman, T. (1911) Festigkeitsversuche unter allseitigem Druck. *Zeitscrift der Vereines Deutscher Ingenieur*, 55, pp. 1749–1757.

Wawersik, W.R. (1968) Detailed analysis of failure in laboratory compression tests. PhD Thesis, University of Minnesota.

Wawersik, W.R. & Fairhurst, C. (1970) A study of brittle rock fracture in laboratory compression experiments. *International Journal of Rock Mechanics and Mining Sciences*, 7 (5), 561–575.

Xu, W., Hu, R. & Ta, R. (2007) Some geomechanical properties of soil-rock mixture in the Hutiao Gorge area, China. *Géotechnique*, 3, 255–264.

Yacoub-Tokatly, Z., Barr, B. & Norris, P. (1989) Mode III fracture – a tentative test geometry. In: Shah, S.P., Swartz, S.E. & Barr, B. (eds.), *Fracture of Concrete and Rock – Recent Developments*, London, Elsevier Applied Science, pp. 596–604.

Young, R.P. (1993) Seismic methods applied to rock mechanics. *ISRM News Journal*, 1 (3), 4–18.

Zhang, L. (2005) *Engineering Properties of Rocks*. Elsevier Geo-Engineering Book Series, Hudson, J.A. (Series Editor), Vol. 4, Amsterdam, Elsevier,

Zhang, Y., Zhou, H., Zhong, Z., Xiong, S. & Hao, S. (2011) In situ rock masses triaxial test system YXSW–12 and its application. *Chinese Journal of Rock Mechanics and Engineering*, 11, 2312–2320 [in Chinese].

Zhao, J. (2011) An overview of some recent progress in rock dynamics research: Chp. 2. In: Zhou, Y. & Zhao, J. (eds.), *Advances in Rock Dynamics and Applications*, CRC Press/Balkema, Leiden, The Netherlands, pp. 5–33.

Zheng, H., Feng, X-T. & Chen, Z. (2010) Standardization and digitization for ISRM suggested methods of rock mechanics laboratory tests. *Chinese Journal of Rock Mechanics and Engineering*, 29 (12), 2456–2468.

Zheng, H., Feng, X-T., Chen, Z., Hudson, J.A. & Wang, Y. (2014) ISRM Suggested method for reporting rock laboratory test data in electronic format. *Rock Mechanics and Rock Engineering*, 47, 221–254.

Zhou, Y.X., Xia, K., Li, X.B., Li, H.B., Ma, G.W., Zhao, J., Zhou, Z.L. & Dai, F. (2012) Suggested methods for determining the dynamic strength parameters and mode-I fracture toughness of rock materials. *International Journal of Rock Mechanics and Mining Sciences*, 49, 105–112.

Chapter 2

Progress and applications of rock true triaxial testing systems

X. Li, A. Wang & L. Shi
State Key Laboratory of Geomechanics and Geotechnical Engineering, Institute of Rock and Soil Mechanics, Chinese Academy of Sciences, Wuhan, Hubei, China

Abstract: This paper intends to make a global and historical review of the development and application of rock true triaxial testing (TTT) systems. The TTT systems can generally be classified into three types, *i.e.*, rigid loading type (Type-I), flexible loading type (Type-II) and mixed loading type (Type-III).Each type has their respective advantages and disadvantages in terms of loading capacity, the end friction effect, blank corner effect, experimental accuracy as well as the functional extendibility. Type-III has higher level of overall performance. Type-I appeared first and dominated in number at the early time. After 1980s, the number of Type-III true triaxial machine has steadily increased. Now, Type-I and Type-III machines are approximately the same in number. They are both widely been used in rock mechanics. The paper also focuses on the results obtained from Type-III regarding the influence of intermediate principal stress on strength, deformation characteristics, permeability change features, etc. In the end, comments and recommendations are made for the further development of TTT techniques.

I INTRODUCTION

The three principal stresses in the earth's crust, generally speaking, are not equal to one another ($\sigma_1 > \sigma_2 > \sigma_3$). Their differences are more obvious in the shallow part of the earth's crust (Brown *et al.*, 1978; McGarr *et al.*, 1978; Jing *et al.*, 2007) and may further enlarge and change dynamically in the surrounding rocks of caverns or underground chamber. It is these differences and dynamic changes that determine whether the earth's crust and engineering surrounding rocks will be damaged and their damage and failure modes. Conventional triaxial testing systems ($\sigma_1 > \sigma_2 = \sigma_3$ or $\sigma_1 = \sigma_2 > \sigma_3$) can be used to partly apply and control the above-mentioned stress states, thus they may play a great role in understanding the mechanical behaviors of surrounding rocks. However, they can only independently control confining pressure and axial pressure and limit the loading stress path in two fixed planes (σ_3 and σ_1) in the stress space, which is not enough to support us to completely understand the mechanical properties of rocks beyond these two planes (σ_2 is between σ_3 and σ_1). Since the 1960s, more and more researchers have paid great attention to the experimental methods used to apply the general stress state, and developed various devices based on different principles and structures.

Previous experiments have shown that the intermediate principal stress has important effects on the mechanical and seepage characteristics of rocks, and plays a unique role in understanding deep engineering destruction and fluid migration process.

The intermediate principal stress effect is crucial for understanding and analyzing rock burst disasters in deep mining. Deep rocks generally are subject to high stress and high deviation stress environments, excavation and unloading will result in the redistribution of stresses on cavern walls, causing the rock body to transfer from its initial 3-D stress state to the 2-D stress state. For the wall rock of hard and brittle nature, then the elastic strain energy stored in the rock body is suddenly released, causing rock burst. Considering that the role of the intermediate principal stress could well explain the "Onion-skin-type fracture" failure phenomena (Cai, 2008) as well as rock outburst phenomena (Gong *et al.*, 2012) observed in the field, as shown in Figure 1, most true triaxial testing machines developed in recent years are used to study rock bursts (He, 2007 # 283; Zhang, 2012 # 333; Du, 2013 # 328).

CO_2 capture and storage (CCS) is one of the effective measures to reduce CO_2 emission. CCS is to sequester physically and chemically captured CO_2 into underground at the supercritical fluid state. The long-term stability of sequester strata is the key for safe storage of CO_2. Li *et al.* (2003) pointed out that making sure that CO_2 staying within the predefined zone in a certain period of time is the key to ensure safe CO_2 sequestration. However, this depends on the sealability or tightness of the caprock subject to reservoir pressure build-up due to CO_2 injection. This is closely related to the mechanical processes of strata, such as the permeability change or fracturing of intact caprock and activation of existing fractures. Considering the economy and safety factors, the depth of CO_2 storage is generally in the range of 1000~3500 m (Li *et al.*, 2006; Li *et al.*, 2003) and the difference between the intermediate principal stress and the minimum principal stress will be great.

In addition to the above-mentioned two kinds of engineering projects, similar engineering problems also exist in high level radioactive nuclear waste depository, underground power house, and the like.

The above engineering and scientific needs have driven true triaxial testing techniques to advance steadily and diversely.

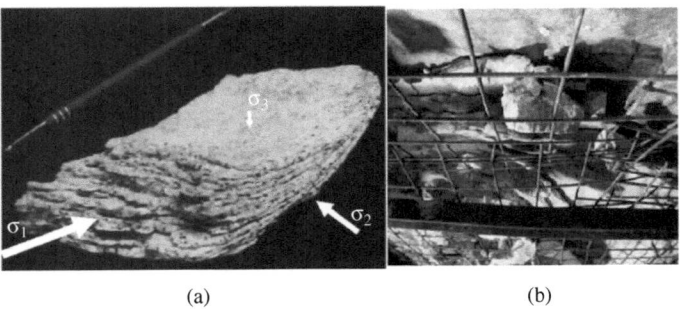

(a)   (b)

Figure 1 (a) Granite slab of layered fractures that occurred at the Mine-by tunnel (depth 420 m) at URL (Cai, 2008). (b) Rock burst in drainage tunnel (depth 454 m) in Jinping II Hydropower Station Project (Gong *et al.*, 2012).

2 DISCOVERY OF THE INTERMEDIATE PRINCIPAL STRESS EFFECT

The intermediate principal stress has an important effect on the strength of rocks. Based on the commonly applied Mohr-Coulomb criterion in rock mechanics, the strength of a material has nothing to do with the intermediate principal stress. However, it is not always true. Kármán (1911) conducted the conventional triaxial compression (CTC) test on Carrara marble ($\sigma_1 > \sigma_2 = \sigma_3$), that is, the cylindrical sample was first compressed by hydrostatic pressure, and then subjected to the constant confining pressure condition under axial pressure until it failed. During the test, the direction of the maximum principal stress σ_1 was consistent with the axis of the sample, and the intermediate principal stress was equal to the minimum principal stress and to the confining pressure. Thus, he obtained the lower curve shown in Figure 2 (a). Böker (1915) performed a conventional triaxial extension (CTE) test on Carrara marble ($\sigma_1 < \sigma_2 = \sigma_3$). Different from Kármán's test, the axial pressure kept invariant, increasing the confining pressure until the sample was failed, and thus he obtained the upper curve shown in Figure 2(a). The differences between these two curves revealed the impact of the intermediate principal stress on the strength of rocks: that is, the strength of CTE test was higher than that of the CTC test. Thereafter, Murrell (1965), Mogi (1967), and

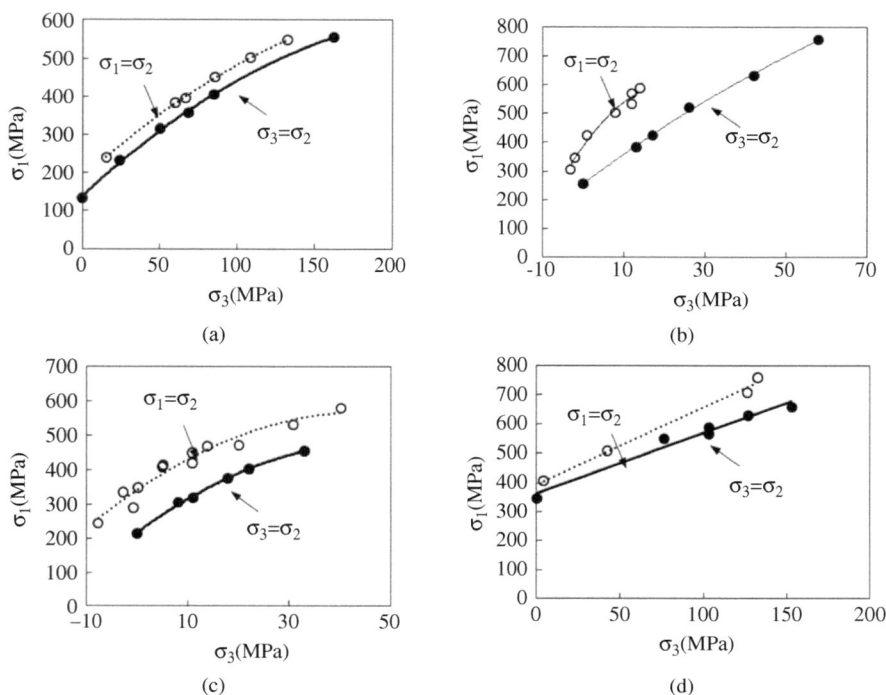

Figure 2 Intermediate principal stress effect found in triaxial compression tests and triaxial tensile tests. (a) Carrara marble: Tests carried out by Kármán (1911) and Böker (1915); (b) Westerly Granite: by Mogi (1967); (c) Dunham Dolomite: by Mogi (1967); (d) Solnhofenlime stone: by Handin (1967).

Handin *et al.* (1967) just to name a few, confirmed the conclusion, and further found that the impacting degree of the intermediate principal stress on the strength of different rocks were different, as shown in Figure 2.

The rock strength obtained in CTC tests is always lower than that in CTE tests. Therefore, the Mohr-Coulomb criterion based on CTC tests in rock engineering may underestimate the strength of rocks under the common stress state, leading to conservative engineering design and increased project costs. Furthermore, since the stresses used in CTC and CTE tests are all axisymmetric, they are not suitable to study the mechanical properties of rocks in general stress state ($\sigma_1 > \sigma_2 > \sigma_3$). To extend the study to the common stress state, that is, increasing the intermediate principle stress σ_2 from $\sigma_1 > \sigma_2 = \sigma_3$ to $\sigma_1 = \sigma_2 > \sigma_3$, it is necessary to explore and develop various true triaxial test (TTT) systems.

3 TTT DEVICES

Before further discussion, let us give a not very strict definition of a TTT device. The so-called TTT device refers specifically to those devices that can be used to directly apply the compressive stress or tensile stress on the surface of a sample to realize a uniform and general stress state in the sample.

3.1 Primary investigation

Before the advent of a TTT machine, to study the mechanical properties of rock in the common stress state, researchers mainly took the following measures to realize the common stress state: 1) applying the combination of confining pressure, axial pressure, and torsion on solid cylinders (Handin *et al.*, 1967; HANDIN *et al.*, 1960), 2) applying the combination of internal and external pressure, confining pressure, and torsion on hollow cylinders (da Gama, 2012; Hoskins, 1969; Jaeger *et al.*, 1966; Mazanti *et al.*, 1966; Robertson, 1955), and 3) applying punching shear with confining pressure and Brazilian splitting with confining pressure on disks (Jaeger & Hoskins, 1966; Robertson, 1955). Table 1 shows the primary studying methods for realization of general stress state.

It is, to some extent, feasible through a combination of loads upon the cylindrical and hollow cylindrical specimens to achieve the general stress state, but there are some drawbacks. With a hollow cylinder as an example, it is known from the derivation of elasticity that any stress state in the sample can be achieved through the combination of control loads, either axial pressure, external confining pressure, and internal confining pressure or axial pressure, external confining pressure, and torque. Thus, the contributions of these tests to researches on the strength of rocks in the general stress state are great. However, these tests had some problems. 1) Applying the formula of stress distribution derived based on the linear elastic theory in rocks is still controversial, especially as the rock is closer to failure, the dispute is fiercer. 2) The distribution of stresses in the sample is not uniform with a greater stress gradient, and the thicker the cylinder walls, the greater the stress gradient is. To solve these problems, Handin *et al.* (1967) conducted experiments on hollow cylinders of thin walls (thickness of 0.7 mm). Mogi (2007) pointed out that the data obtained by Handin *et al.* on thin cylinders were

Table 1 Primary research methods for the realization of general stress state.

Specimen style	Axial compression	Torsion	External confining pressure	Internal confining pressure	Punching	Brazilian loading	Researcher
Hollow cylindrical sample	1		1				Hobbs 1962, Jaeger & Hoskins 1966, Mazanti & Sowers 1966, Hoskins 1969
	1		1	1			Mazanti & Sowers 1966, Hoskins 1969, Dinis da Gama & Menezes 1974
	2		2				Robertson 1955
		1					Handin et al., 1967
		1	1	1			Handin et al., 1967
	1	1	1	1			Handin et al., 1967
Solid cylindrical sample		1					Handin et al., 1967
	1	1					Handin et al., 1967
		1	1				Handin et al., 1967
	1	1	1				Handin et al., 1960, 1967
Disk	2		2		1		Robertson 1955
	2		2			1	Jaeger & Hoskins 1966

Note: 1. Mainly from Kwaśniewski (2012); 2) In Table, "1" denotes the load can be independently controlled, "2" denotes the corresponding load kept consistent in the test.

very discrete because the rock sample had a lot of cracks produced during the preparation of these cylinders.

Still, these tests had some other disadvantages, such as the stress-strain curve and the experimental curve after rock failure could not be measured. All of these shortcomings limit their wide application in studies on rock mechanics and require researchers to find new ways.

3.2 Research and development of TTT devices

To study the mechanical properties of rocks in general stress state, since the 1960s, many researchers have designed a wide variety of TTT machines to reconstruct the mechanical response of rocks to the stress environment produced by TTT devices. Table 2 lists some TTT testing machines in literature.

3.3 Classification

According to different loading media, the loading ways are classified into rigid and flexible loading ways, both of which have their advantages and disadvantages. Rigid loading directly applies the pressure produced by the jack through a platen on sample

Table 2 Some TTT testing machines since the 1860s (incomplete statistics).

No.	Times	Designer	Loading way
1	1961	Weigler & Becker (1961)	Type-I
2	1964	Bertacchi (1964)	Type-I
3	1967	Niwa et al. (1967)	Type-I
4	1968	Hojem & Cook (1968)	Type-II
5	1970	Mills & Zimmerman (1970)	Type-I
6	1971	Mogi (Mogi, 1971b)	Type-III
7	1972	Launay & Gachon (1972)	Type-I
8	1973	Atkinson & Ko (1973)	Type-II
9	1976	Zhang (1976)	Type-I
10	1977	Andenaes et al. (1977)	Type-I
11	1981	Spetzler et al. (1981)	Type-III
12	1983	Takahashi & Koide (1983)	Type-III
13	1985	Michelis (1985b)	Type-II
14	1986	Furuzumi & Sugimoto (1986)	Type-I
15	1986	Li et al. (1986)	Type-I
16	1986	Li et al. (1986)	Type-III
17	1989	Esaki et al. (1989)	Type-I
18	1990	Xu et al. (1990)	Type-III
19	1995	King et al. (1995)	Type-I
20	1995	Skoczylas & Henry (1995)	Type-I
21	1995	Smart (1995)	Type-II
22	1997	Sibai et al. (1997)	Type-I
23	1997	Ohnaka (1997)	Type-III
24	1997	Wawersik et al. (1997)	Type-III
25	2000	Haimson & Chang (2000)	Type-III
26	2004	Alexeev et al. (2004)	Type-I
27	2004	Tiwari & Rao (2004)	Type-I
28	2006	Cheon et al. (2006)	Type-III
29	2007	Marone et al. (2007)	Type-III
30	2007	He et al. (2007)	Type-I
31	2009	Walsri (2009)	Type-I
32	2011	Bésuelle & Hall (2011)	Type-III
33	2011	Li et al. (Li et al., 2012)	Type-III
34	2012	Zhang et al. (Zhang et al., 2012a; Zhang et al., 2012b)	Type-III
35	2013	Du (Du, 2013)	Type-I

Note: The machine type will be introduced in the next subsection.

surface, which is kept to be flat in the loading process. The flexible loading directly applies fluid pressure provided generally by the hydraulic pump or pressure intensifier or directly loading through the flat jack on the wrapped sample surface. Thus, the loading medium flexibly contacts with the sample surface in the loading process and the surface can freely change its shape under the pressure. Although rigid loading has higher loading capacity, its rigid contact will result in a greater friction on the sample surface and the surface under the constraint of the rigid plane cannot freely deform. All these disadvantages are generally referred to as the end effect and result in an uneven stress distribution on the sample surface. In contrast, flexible loading has no end effect and the stress distribution on the sample surface is more uniform, but it has a smaller loading capacity.

Since a TTT machine can load independently in three directions, the loading ways in each direction can be either rigid loading or flexible loading. Thus, it can load in ways of four combinations: rigid loading in all three directions (3R), rigid loading in two directions plus flexible loading in the third direction (2R1F), rigid loading in one direction plus flexible loading in other two directions (1R2F), and flexible loading in all three directions (3F). Considering that 1R2F and 3F have no significant difference in the end effects, they are usually combined into one type (Kwasniewski, 2012; Li *et al.*, 2012; Mogi, 2007; Takahashi *et al.*, 2001). Therefore, TTT machines are often classified as rigid loading type (Type-I), flexible loading type (Type-II), and mixed loading type (Type-III).

3.3.1 Type-I: Rigid loading type

This type of testers is developed based on concrete biaxial testers developed in the 1960s (Weigler & Becker, 1961; Weigler *et al.*, 1963). According to its loading balance ways, it is further divided into two types: three jacks as shown in Figure 3 (a), and three pairs of jacks, as shown in Figure 3(b). Figure 4 shows their representatives (Furuzumi & Sugimoto, 1986; King *et al.*, 1995).

The tester developed by Furuzumi & Sugimoto uses three hydraulic pistons to separately load on the sample through rigid blocks. Correspondingly, it uses the loads produced by three fixed supports to balance the forces from the three pistons. Because the supports are fixed, the sample suffers from unidirectional compression toward the supports in the three stress axes, leading to an increase in end friction and sample eccentric compression, and eventually resulting in uneven stress distribution on the sample surface. As shown in Figure 4, the tester developed by King *et al.* uses three pairs of jacks to load. The jacks in the σ_2 and σ_3 directions are fixed on the reaction ring, the pressures provided by each pair of pistons are the same. If the difference between the frictions of two jacks in the same direction is less than the friction between the sample and the other two pairs of pistons, theoretically it is possible to ensure the sample to avoid eccentric compression during its deformation. However, three additional hydraulic pistons also increase the complexity and cost of the device.

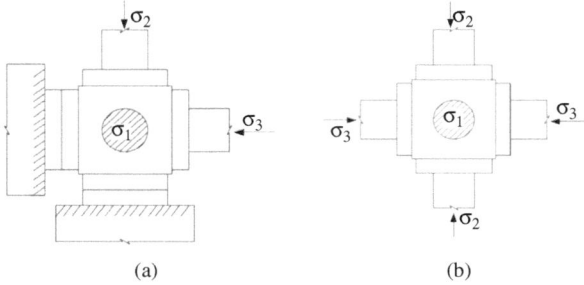

(a)   (b)

Figure 3 Rigid platen type of TTT devices. (a) Three jacks; (b) Three pairs of jacks.

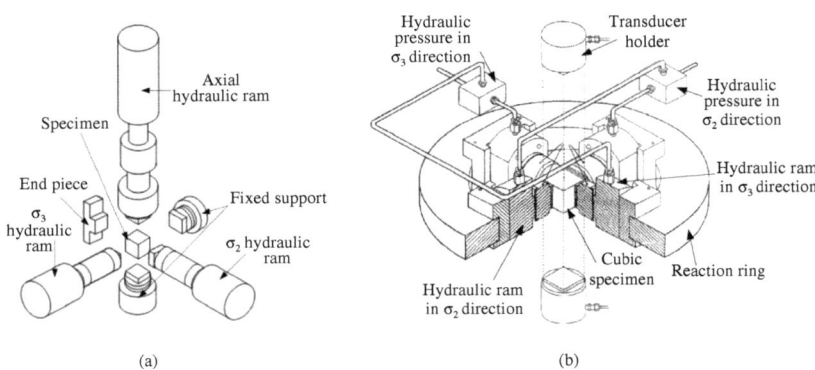

(a)   (b)

Figure 4 Schematic of TTT devices. (a) Furuzumi & Sugimoto Type (1986); (b) King *et al.* Type (1995).

Due to its rigid loading, this type of testers has a larger loading capacity. Thus they are widely applied in rock mechanics tests and have prominent advantages for samples of larger size. Of course, these testers also have some inherent issues. First, to avoid collision between the end platens, usually the platen size is slightly smaller than sample size. This will result in blank corner and subsequent stress concentration at the edges of the sample, impacting the test accuracy. Second, it is difficult to measure the permeability, especially in the case of a high pore pressure. In addition, to solve the sealing issue, the experimental procedure is too complex and the boundary flow between the platen and the sample surface remains unsolved. Finally, the acoustic emission (AE) sensor is difficult to install on the surface of the sample to locating AE events. These drawbacks to some extent limit the promotion of this type of testers.

3.3.2 Type-II: Flexible loading type

To solve the above mentioned problems, researchers developed TTT systems with flexible loading in more than two directions. This type of TTT could ensure even stress distribution on the sample surface and free sample deformation by transferring loading through flexible medium. Most flexible loading TTT use flat jack to load stress on the sample surface. The flexible loading TTT was first developed by Koyfma *et al.* (1964), followed by Hojem & Cook (1968), Atkinson & Ko (1973), Desai *et al.* (1982), Michelis (1985a), and Smart (1995). The materials of flat jack are rubber (Koyfma *et al.*, 1964), wrought copper sheet (Hojem & Cook, 1968), or polyvinyl chloride (Michelis, 1985a). Among these devices, the tester developed by Smart (1995) is more novel. As shown in Figure 5, the device uses a steel piston to load σ_1, while uses the hydraulic pressure from peripheral 24 PVC pipes to load σ_2 and σ_3, reconstructing the true triaxial stress state in the cylinder sample. Thus, its applications on cylindrical samples can be used to compare with those of conventional triaxial tests. Michelis (1985a) developed a tester with PVC hydraulic pack and rigid prism. Hojem & Cook (1968) developed a tester with a flat jack as the loading medium. Both of them have their own advantages.

This test machine via its flexible medium loading solves the problems such as the blank corner effect and the uneven stress distribution on the sample surface, thus it is an

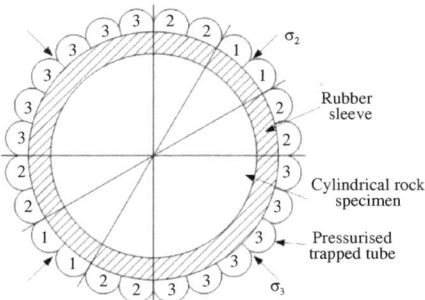

Figure 5 Plane view of a true triaxial loading system developed by Smart (1995). In the figure, 1, 2 and 3 stand for three independently servo-controlled hydraulic circuits.

ideal TTT devices. And if a rigid platen is installed in one direction, the water permeable plate could be installed in this direction to measure seepage. However, because the flat jack made of copper plate and PVC, etc., has limited loading capacity typically from a few MPa to 100 MPa, this type of devices is ideal for soil mass and soft rock tests, but not suitable for the hard rock tests in the high stress state.

3.3.3 Type-III: Mixed loading type

This type of TTT devices uses rigid loading in its two directions and flexible loading (oil or flat jack) in the third direction. Its representative device is the Karman triaxial tester modified by Mogi (1971a). As shown in Figure 6, it consists of one pressure chamber and two servo loading frames in vertical and horizontal directions. The pressure chamber is connected with the pressure intensifier. Each servo frame is installed with a jack. σ_3 is loaded by hydraulic pressure provided by the pressure intensifier, and σ_1 and σ_2 are loaded by corresponding jacks. The force needed to move the two servo frames corresponding to the jacks is smaller than the friction force between the tightly

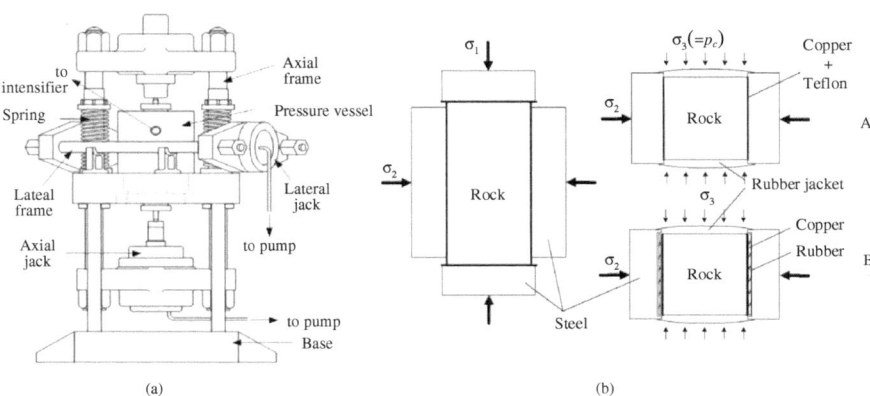

Figure 6 Schematic of TTT devices modified by Mogi (1970). (a) Front view; (b) Assembly of specimen and platens.

clamped sample and steel platen, thus ensuring the frame to move while the specimen center remains still in the testing process. Meanwhile, because the device uses only one jack in each direction, the equipment cost is reduced.

To compare the effect of using rigid loading and flexible loading in σ_2 direction, Mogi (2007, 2012) compared the situations A and B in Figure 6(b) and found no significant difference. Thus, he believed that the use of rigid loading in 2th principle stress directions does not bring substantial impact.

Compared with rigid loading TTT machine, mixed loading TTT has the following advantages: 1) It directly uses oil hydraulic loading on the direction with maximum impact on the sample strength, the σ_3 direction, thus ensuring even stress distribution on the sample surface; 2) It uses the movable frames in the σ_1 and σ_2 directions to make sure that the specimen center is still in the testing process; 3) It can accurately determine σ_3 through directly measuring the displacements between two surfaces in the three directions; 4) It is cost-effective; 5) It can be used to measure the permeability of the rock specimen. For example, the fractures obtained by using Mogi-type TTT machine are always in the σ_3 direction, making it more easy to form one dimensional flow, thus the measured permeability is more meaningful (Li, 2001).

Of course, mixed loading TTT devices also have some shortcomings, such as the blank corner effect and the end friction. But compared with Type-I devices, these issues have been greatly improved.

It is because Type-III has more advantages that they have obtained more rapid development. Table 2 lists various types of these devices. Among them, the most representative ones were developed by Takahashi & Koide (Takahashi *et al.*, 1983), Haimson & Chang (Haimson & Chang, 2000), and Li & Shi (Li *et al.*, 2012). The device designed by Haimson & Chang can provide a high stress on the sample of 19 mm × 19 mm × 38 mm, reaching 1600 MPa, 1600 MPa and 400 MPa in σ_1, σ_2 and σ_3, respectively. The device developed by Li & Shi uses two horizontal servo loading frames to make sure that the sample is not eccentric in the loading process and a liftable positioning device to ensure the initial position of the specimen center to coincide with the load center.

Figure 7 summarizes the advantages and disadvantages of various types of TTT devices concluded by Li *et al.* (2012).

3.3.4 Development path

Figure 8 shows the comparison of increments in the numbers of three TTT devices in various decades. It can be seen from the figure that the total number of TTT devices has been increased steadily (the decrease in the 2010s is due to statistics only from 2010 to 2015). Type-I devices were the mainstream of TTT devices in the 1960s-1970s and had a wide application after the 1980s, indicating that this type of testers has a large loading capacity and is really preferred for specimens of larger size. Type-II devices have been fewer in number, and no new tester appears since the 2000s. This possibly is due to their lower loading capacity, which limits the scope of their application. Type-III devices appear latest. But since the 1980s, its number has been increasing steadily. Figure 9 shows the active period of each Type-III devices. From the figure it can be seen that the number of operational test machines shows a rising trend overall within the same period, and 8 test machines are reported in the last five years, revealing that this type of testers has obtained a broad recognition and application.

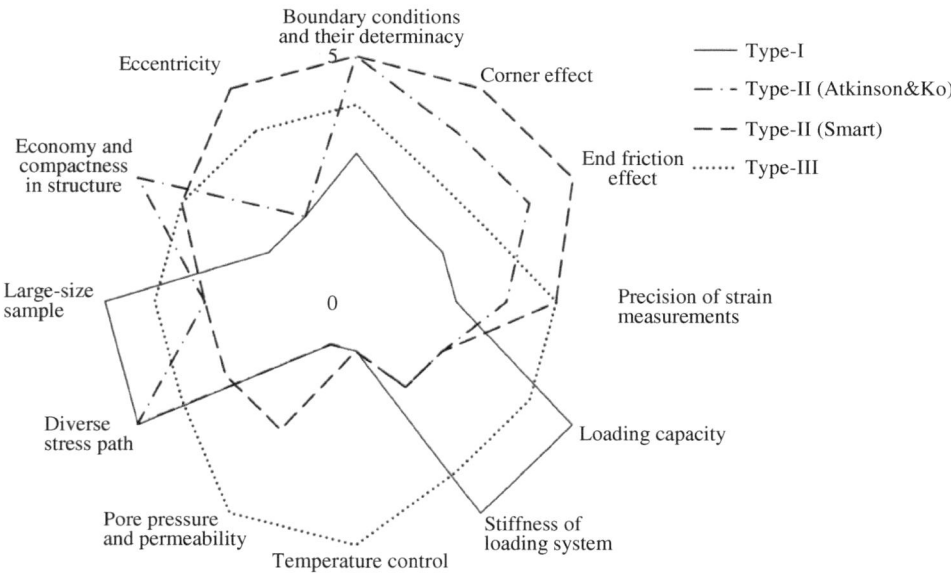

Figure 7 Comparison in performance among various TTT devices summarized by Li *et al.* (Li *et al.*, 2012). Evaluation is divided into five levels denoted by different radius. The meanings of various levels are: 1- impossible; 2- comparatively poor; 3- usual; 4- good; 5- excellent.

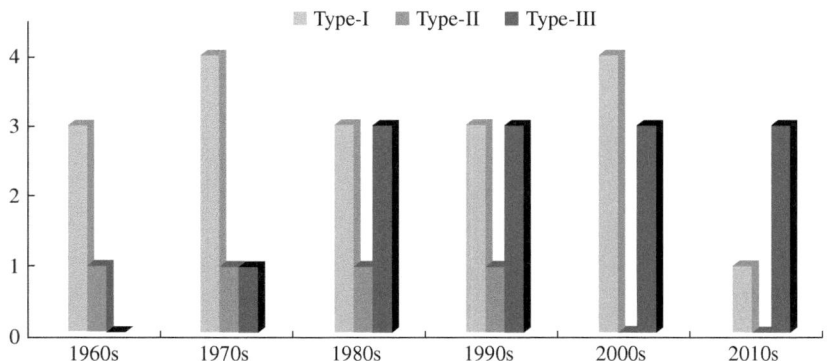

Figure 8 Comparison of increments in the numbers of three TTT devices. Statistical time from 1960 to 2015 with 10 years as a statistical unit. And in the 2010s: from 2010 to 2015.

Figure 10 shows the statistical results of relevant literature on Type-III TTT machines, which are similar to those of Li *et al.* (2012). Therefore, we discuss the developmental path by using the statistic results of Type-III device as a representative. As shown in Figure 10, research basing on Type-III devices gradually increased from scratch, reaching a peak in the 1980s, declining in the 1990s and then increased rapidly after 2000. It is worth noting the results from Figure 9 and Figure 10 are consistent. In

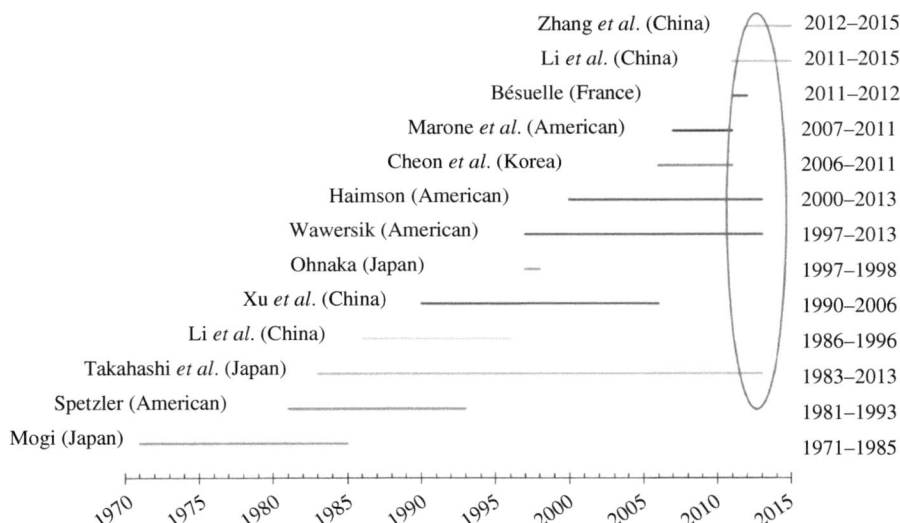

Figure 9 Operational state of Type-III TTT test machines. The statistic standard is the publication time of the first paper to the publication time of the latest paper. Zhang's device is considered as in 2015 based on that we clearly know it is operating.

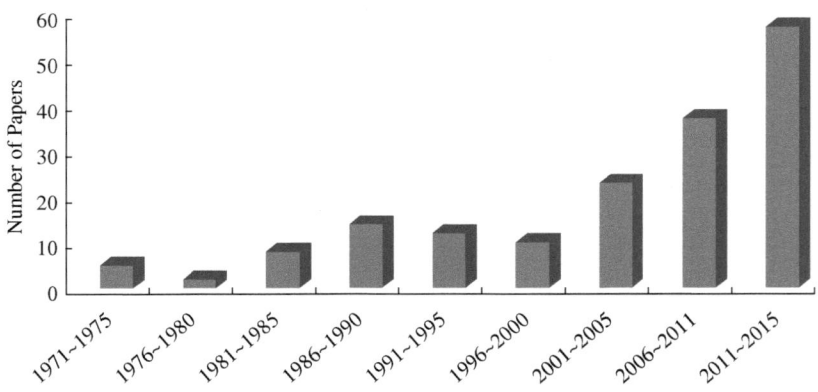

Figure 10 Change in the number of papers related to Type-III TTT machines.

addition, it is noted that quite a few papers analyzed the experimental results of other researchers, indicating the urgent need for TTT machines.

Figure 11 shows the progress of TTT technology and makes it possible to study the mechanical response of rocks in the TTT stress state from multiple aspects.

1) From pre-peak mechanical behaviors to post-peak mechanical behaviors: From the 1960s to the early 1980s, one of the important issues on TTT machines is the impact of the intermediate principal stress on rock strength (due to the lithologic

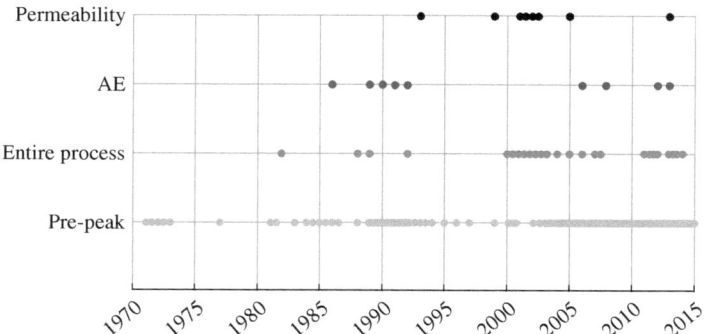

Figure 11 Charges in studying content and means of Type-III TTT machines.

complexity, researches on this issue continue in the whole TTT research history) and the mechanisms of deformation and failure of rocks before the peak. Therefore, the design of testing machines does not fully consider the rigidity of testing machine. The measured test curves are mainly pre-peak curves. Until 1982, Xu (1982) obtained the full process curve of soft and weak sandstone of high porosity through Mogi test machine, but he didn't analyze the post-failure rock zone. Xu *et al.* (1991, 1992) systematically studied the impact of TTT device rigidity on rock deformation and failure process. Haimson & Chang (Chang, 2001; Haimson & Chang, 2000) independently showed in experimental curves that their devices were able to control the whole process of hard rock deformation and failure, but they also didn't analyze the characteristics of the post-failure rock zone.

Although the post-failure rock zone with TTT machines haven't basically been systematically studied, the complete TTT stress-strain process curves obtained by some researchers provide us with the possibility to further study the mechanical behavior of post-peak rock.

2) Acquiring diversified information of whole testing process: Through primary test machines, one only could observe rock deformation during loading through displacement sensors, strain gauges, etc., and deduce the mechanism of sample deformation and failure from the stress-strain curves and the morphology of failure sample. Thus, the information is simpler and not be verified. In the 1980s, after Li *et al.* (1986) introduced acoustic emission (AE) testing devices into TTT tester, a few researchers applied AE testing devices to study the properties of development and propagation of cracks in the rock in the TTT stress state and obtained the location of microcracks in the rock samples compressed by TTT machines (Xu *et al.*, 1991; Xu *et al.*, 1992). The most representative research on this was by Ingraham (2012, 2013b), he used AE positioning technology and revealed the law of strain localization of deformation of rocks in the TTT state.

However, the application of AE in the TTT machines is still in its infancy and need to be further studied.

3) From pure solid mechanical behaviors to fluid-solid coupling mechanical behaviors: Based on the necessities of energy development, waste storage, and the like, since the 1990s, researchers began to study permeability change of rocks under the

TTT state (Takahashi *et al.*, 1993; Zhang *et al.*, 1999). Later, some researchers systematically studied permeability change of sandstone subject to complex stresses (Li, 2001; Li *et al.*, 2001; Takahashi *et al.*, 2005; Takahashi *et al.*, 2013). Recently, Zeng (2015) investigated the variation in permeability of mudstone in its deformation process. All these studies obtained many useful conclusions.

In short, for the study of the mechanical response of rock to general stresses, since the 1960s, a variety of TTT machines have been developed. They are mainly divided into three loading types, rigid loading, flexible loading, and mixed loading types. All these devices have their own advantages and disadvantages. Among the three, rigid loading and mixed loading have become mainstream TTT machines. Development of mixed loading TTT machines experiences a whole process from using them to obtain pre-peak mechanical behaviors to post-peak mechanical behaviors, to using them to acquire information diversification, and to using them to conduct pure mechanics to conduct fluid-solid coupling researches.

4 REVIEW OF TESTING RESULTS

In the process of TTT devices, researchers have obtained numerous experimental results and gradually revealed the mechanical properties of rock in the general stress state.

Deformation of rocks subject to TTT stresses or in the TTT stress state mainly affects the following mechanical properties: strength, deformation characteristics, failure mode, fracture angle, permeability, and acoustic emission characteristics. All of these are briefly concluded as follows and described in details in related papers.

For ease of discussion later, the following basic concepts defined by Mogi (1971a, 1972a, 1972b), as shown in Figure 12, are used:

1) Strength: differential stress in failure.
2) Yield stress: differential stress at the first occurrence of permanent deformation. For most rocks without obvious turning point, it can be defined as the differential stress corresponding to some small permanent strain, such as 0.2% of the differential stress.
3) Stress drop: amplitude in stress incline in failure.

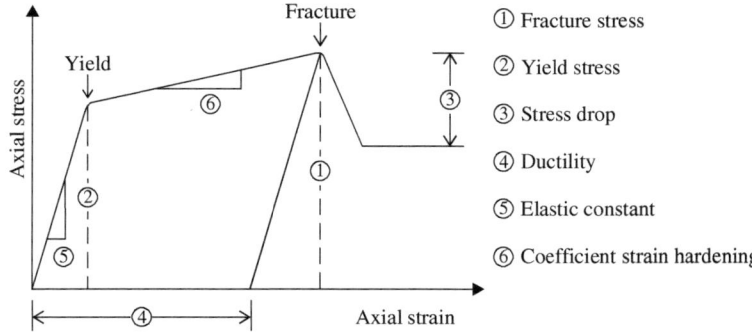

Figure 12 Schematic of basic concept definitions following Mogi (1971).

4)   Ductility: the ability of the test material to resist plastic deformation without failure. According to the definition by Heard (1960), when the ductility is less than 3%, the rock is brittle; when the ductility is in the range of 3–5%, the rock is in the brittle-ductile transition stage; and when the ductility is greater than 5%, the rock is ductile.

5)   Elastic modulus: generally considered to be the slope of the initial straight segment on the loaded stress – strain curve, and to be precise, the slope of the unloading segment.

6)   Strain hardening coefficient: the slope of the stress – strain curve after yield.

7)   Fracture angle: the angle between the normal of the failure surface and the maximum principal stress, which is complementary to that defined by Mogi.

8)   Dilatancy: Dilatancy reflects the volumetric deformation, mostly volume expansion of samples due to compression-induced deviation from the straight segment and is usually expressed as the volumetric strain $\Delta v / v = \varepsilon_1 + \varepsilon_2 + \varepsilon_3$.

(9)   Microfracture events and the number of microfracture events: the number of elastic impacts accompanied with microfractures.

4.1 Impact on strength

In general stress state, when σ_3 keeps constant, the strength of rock shows a trend of first increase followed by decline with σ_2 increasing. Mogi (1971a, 1971b, 1972a, 1972b, 1981, 2007, 2012) systematically conducted several TTT experiments to study the impacts of the intermediate principal stress on the strength of various rocks including dolomite, limestone, marble, trachyte, andesite, granite, schist, etc. and found that the intermediate principal stress σ_2 could significantly affect their strength which shows a trend of increase followed by decrease with σ_2 increasing, but this impact is obviously less than that of σ_3. With dolomite as an example, as shown in Figure 13, Mogi (1971a) plotted the contour lines of the rock strength vs both σ_2 and σ_3

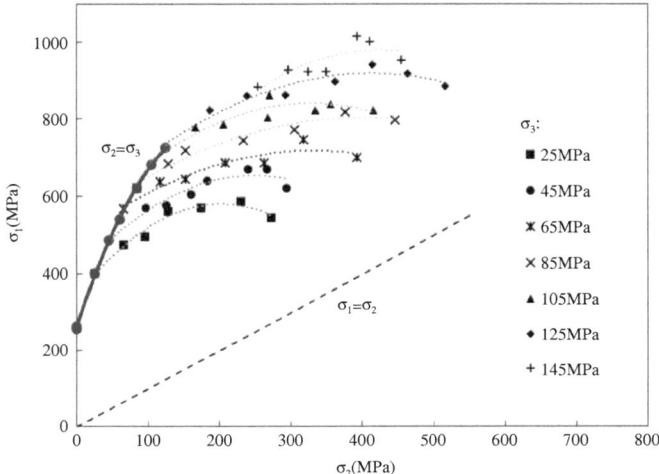

Figure 13 Stress at fracture as the function of σ_2 in Dunham dolomite. (Data from Mogi (2007)).

and estimated that the intermediate principal stress effect was about 0.2 times of that of the minimum principal stress. His conclusion was verified later by many other researchers (Chang, 2004; Chang et al., 2000; Haimson, 2006; Haimson, 2011; Haimson & Chang, 2000; Haimson et al., 2010a; Koide et al., 1991; Lee et al., 2011; Takahashi et al., 1989; Li et al., 1991; Li et al., 1994; Xu, 1982; Xu et al., 1984; Xu et al., 1985; Xu et al., 1986). Xu & Geng (1985) further found that the denser and harder the rock is, the greater the confining pressure and the intermediate principal stress effect on rock strength. In addition, rocks with a large confining pressure effect also have greater intermediate principal stress effect.

The minimum principal stress effect of rocks is less than that of the confining pressure effect. In the comparison between the intermediate principal stress effect and the minimum principal stress effect, the latter is often obtained through conventional triaxial compression tests, at this time, $\sigma_2 = \sigma_3$. What was obtained in this way is in fact the confining pressure effect, in which contains the intermediate principal stress effect, rather than solely the minimum principal stress effect. Kwaśniewski et al. (2006, 2013) first experimentally studied the minimum principal stress effect by exerting the same σ_2 but different σ_3 on different Rozbark sandstone samples to measure the strength of rocks and revealed that the minimum principal stress effect was significantly less than the confining pressure effect, but slightly greater than the intermediate principal stress effect, as shown in Figure 14.

The intermediate principal stress effect of rock strength is different for different rocks. For some rocks, rock strength is insensitive to the intermediate principal stress. Chang and Haimson (Chang et al., 2005; Haimson, 2006) found that hornfels and metapelites are not sensitive to the intermediate principal stress, as shown in Figure 15. Compared with that volumetric expansion occurs in all rocks with the intermediate principal stress effect, they believed that lack of the intermediate principal stress effect in these rocks is related to the fact that these rocks have no dilatancy before fracture.

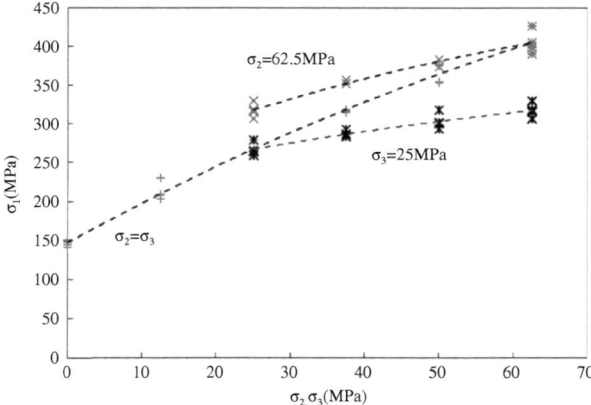

Figure 14 Effect of confining pressure, the intermediate principal stress, and minimum principal stress on the maximum principal stress at strength failure of the Rozbark sandstone (Kwaśniewski, 2013; Kwaśniewski & Takahashi, 2006).

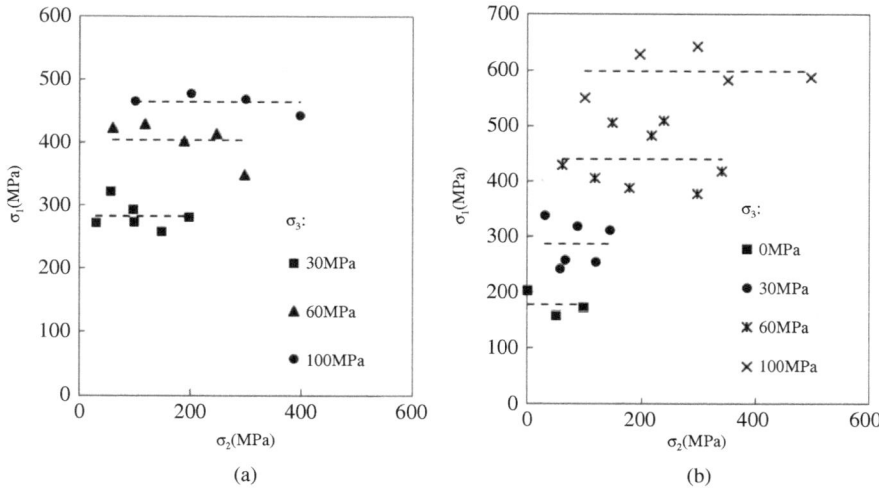

Figure 15 True triaxial strength (σ_1) as a function of σ_2 for different σ_3 levels (Chang & Haimson, 2005). (a) hornfelsand (b) metapelite.

Mogi (1972a, 2007) pointed out that the yield stress increased with the intermediate principal stress the increase of the intermediate principal stress, but didn't depend on the change in the minimum principal stress. Chang & Haimson (2007) carried out true triaxial experiments with impermeable Pohang rhyolite at the jacketed and unjacketed states, and found that the strength of unjacketed samples was close to the yield strength of the jacketed samples, and the yield strength showed a trend of first increase which inclines with σ_2 increasing.

The stress path has a certain influence on the strength of rocks. Xu & Geng (Geng *et al.*, 1985; Xu & Geng, 1984, 1986) studied the intermediate principal stress effects of granite and marble in several stress paths and found that the stress path of a high differential stress had a certain impact on the strength of rocks, while other paths had no obvious impacts. Ingraham *et al.* (2013a, 2011) analyzed those experimental results of different stress paths and found that the strength criterion of Castlegate sandstone is correlative to the third invariant of the stress, especially in the high average stress case. Xiang *et al.* (2008) simulated the failure process of rock body in the cutting and unloading stress path using the true triaxial stress state, They found that the strength of samples was more affected by unloading and believed that with the interval change of the intermediate principal stress. Samples mainly suffered from brittle fracturing. In addition, Mogi (1981) found that in some particular directions, the strength of anisotropic rocks decreased with the increase of σ_2.

About the true triaxial strength criterion, it is generally believed that the yield stress obeys Nadai's formula (Nadai, 1950), $\tau_{oct} = f_1(\sigma_{oct})$, while the failure stress is consistent with the Mogi's criterion (Mogi, 1971b), $\tau_{oct} = f_2(\sigma_{m,2})$. In many researches, the functions, f_1 and f_2, are linear functions, power functions, and so on.

4.2 Impact on deformation features

The effects of the intermediate principal stress on deformation characteristics mainly include the elastic modulus, strain hardening coefficient, ductility, dilatancy, failure mode, and fracture angle.

Dependence of the elastic modulus on the stress state is related to rock types. Mogi (1972b) analyzed conventional triaxial experimental data and found that the elastic module of monzonite and diabase were independent on the confining pressure, but those of granite and quartzite evidently increased with the confining pressure increasing. By comparing the curves of the strength and elastic modulus of marble, granite, quartzite, monzonite and diabase with the confining pressure increasing, Mogi also found that at a low confining pressure, for the rocks whose elastic modulus was more sensitive to the confining pressure, their strength was also more sensitive to the confining pressure and more deviant from the Coulomb strength criterion. Oku $et\ al.$ (2007) investigated the relationship of elastic modulus of siltstone to σ_2, and found that its elastic modulus showed a trend of first rise then decline with the intermediate principal stress increasing, which is related to the closure state of natural pores and cracks inside rocks. The varying characteristics of the elastic modulus of a rock are similar to those of its strength, which might imply some intrinsic relation between them.

It is commonly accepted that the strain hardening coefficient increases with the increase of σ_2, but doesn't rely on the minimum principal stress σ_3 (Mogi, 1972a). However, Xu (1982) experimentally studied the soft sandstone with high porosity and found that the strain hardening coefficient increased with the confining pressure increasing, but might decrease with the intermediate principal stress increasing.

Mogi (1971a, 1971b, 1972a) found that the ductility of a rock increased with the increase of σ_3 and decreased with the increase of σ_2, and can be expressed as $\varepsilon_n = f_3(\sigma_1 - \alpha\sigma_2)$, moreover, in the form of a logarithmic function as $log\varepsilon_n = K_1(\sigma_1 - \alpha\sigma_2) + K_2$. The result was further verified by Kwaśniewski & Takahashi (2006).

In 1977, Mogi (1977) first found that in the elastic deformation, the strains, ε_2 and ε_3, in the directions, σ_2 and σ_3, were roughly equal. However, in the non-elastic deformation, if σ_2 was much larger than σ_3, ε_3 is obviously larger than ε_2, shear dilatancy began to appear. Dilatancy marks the initiation of plastic deformation and was correspondent to the yield stress. Moreover, Mogi found that the higher the σ_2, the higher the corresponding ε_1 or σ_1 at the onset of dilatancy, the higher the ratio, $\varepsilon_3 / \varepsilon_2$. The anisotropic dilatancy could be reasonably explained by the fact that cracks always opened in the direction perpendicular to the minimum principal stress. Haimson & Chang (2000) also obtained similar conclusion, and believed that for a fixed σ_3, an increase in σ_2 could enlarge the range of quasi-linear elasticity. Kwasniewski and Takahashi studied the experiments of Śląsk sandstone and Rozbark sandstone (Kwaśniewski & Takahashi, 2006; Kwasniewski et al., 2003) and found that both the intermediate principal stress and confining pressure played an inhibitory effect on dilatancy. In other words, a larger ε_1 or σ_1 was needed in order to start dilatancy as increasing intermediate principle stress. When σ_2 was lower, rock failure occurred only after a larger dilatancy, while when σ_2 was higher, rock failure occurred without much dilatancy, which resulted in a sudden occurrence of rock failure, also was a demonstration of ductility decline. Chang & Haimson (2005) revealed that the strength of crystalline rocks had a stronger dependence on σ_2, which was relevant to the initial micro-cracks inside the rock, and an increase in σ_2 delayed the course of dilatancy damage to hornfels and metapelites. The fact that dilatancy did not occur in hornfels and metapelites also revealed that their strength was not associated with the intermediate principal stress. Chang's true triaxial tests with impermeable rhyolite in both the jacketed and unjacketed states (Chang & Haimson, 2007) showed that in the unjacketed state, the strength of Pohang rhyolite was equal to or slightly higher than the stress at the initiation of dilatancy, which provided an evidence for the initiation of dilatancy at the rock yield.

The fracture angle is affected by the stress state. According to the Mohr-Coulomb criterion, it is generally considered that the fracture angle of rocks is $45° + \phi/2$, and in fact, even in the conventional three-axis machines, the fracture angle inclines to some degree with the confining pressure increasing (Haimson & Chang, 2000). In true triaxial machines, the fracture angle increases with the intermediate principal stress increasing (Haimson, 2007; Haimson, 2011; Haimson & Chang, 2000; Haimson et al., 2005; Haimson et al., 2012; Haimson & Rudnicki, 2010a; Haimson et al., 2010b; Ingraham et al., 2013a; Ingraham et al., 2013b; Issen et al., 2011; Ma et al., 2013; Ma et al., 2014; Mogi, 1972a; Xu & Geng, 1984), see Figure 16. Some researchers recommended to use the Rudnicki and Rice formula (Rudnicki, 1975) to treat with the relationship between the failure angle and triaxial stresses. The formula is a little complex, and can be found in relevant literature (Haimson, 2007; Haimson, 2011; Haimson & Rudnicki, 2010a; Haimson & Rudnicki, 2010b; Ingraham et al., 2013a; Ma et al., 2014; Oku et al., 2007).

The failure pattern of rocks is divided into splitting failure, tension-shear failure, shear failure, compression-shear failure, compression failure, among which both tensile-shear failure and compression-shear failure are transitional types from splitting to shear failure and from shear to compression failure, respectively. In a uniaxial case, rock often suffers from splitting failure. At a low confining pressure, it shows compression-shear failure, while at a high confining pressure, it shows compression failure. At this time, the splitting failure surface is roughly parallel to the maximum principal stress direction, and the shear failure surface obliquely crosses the maximum principal

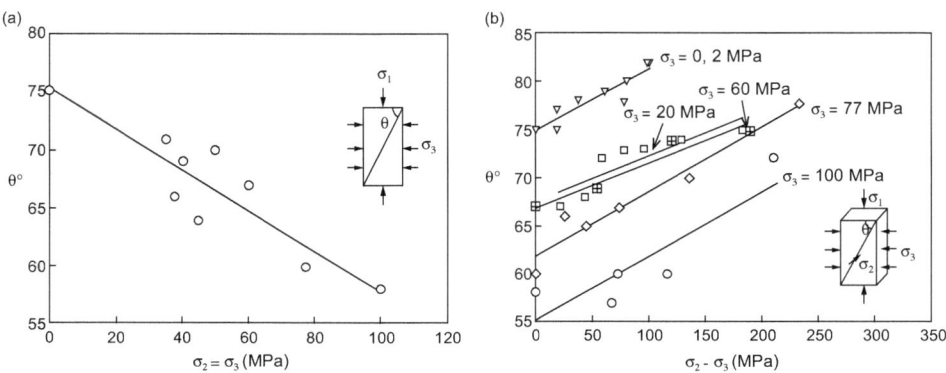

Figure 16 Failure angle as a function of (a) the confining pressure and (b) the intermediate principal stress. From Haimson & Chang (2000).

stress. However, the compression failure shows no obvious failure surface. In summary, in a conventional triaxial machine, increase in confining pressure corresponds to a decline in brittleness and an increase in ductility, and with the ductility increasing, rock has a transitional trend from splitting failure to compression failure. The intermediate principal stress σ_2 serves to reduce the ductility of rock. Therefore, with the increase of σ_2, rock has a trend from compression failure to splitting failure, which has been verified by many researchers (Ingraham *et al.*, 2013b; Li *et al.*, 1994; Xu *et al.*, 2000).

Xu *et al.* (1992) and Ingraham *et al.* (2013b) utilized AE monitoring methods to locate microfractures in granite and porous sandstone in the loading process, respectively, and obtained many beneficial conclusions. Xu showed that the acoustic emission from rock subject to true triaxial stresses undergoes three stages. At the first stage, there are quite a few microfractures inside the rock sample. At the second stage, microfractures distribute in the whole rock sample, but fewer than the first stage. At the third one, microfractures further grow in quantity and to the rock form some non-microfractures zones. At the time close to failure, they show a zonal distribution, indicating that only when it is close to failure, the major rupture surface forms. Ingraham performed a series of tests on stress Lode angle from -30° (corresponding to triaxial tension test) through 0° (corresponding to pure shear) to 30° (corresponding to triaxial compression test) under different average principal stresses. The results showed that with the mean stress increasing, the shear zone is difficult to occur. In this process, the angle of the shear zone continues to decrease until no significant shear zone occurs. The zonal distribution of AE (*i.e.* strain localization) forms earlier than the peak point.

4.3 Impact on permeability features

The impact of true triaxial loading on permeability is in nature the impact of rock deformation on permeability under true triaxial stress states. As what mentioned above, the confining pressure increases the ductility of rocks. Thus, under the action of axial stress, it can cause rocks to transit from splitting failure to compression-shear failure to compression failure, while the intermediate principal stress acts just opposite to the confining pressure. It makes cracks within the rock sample arrange along the $\sigma_1 \sim \sigma_2$ plane and finally the failure surface parallel to the direction of σ_2, causing plastic deformation and dilatancy to mainly occur in the direction of σ_3.

Takahashi *et al.* (1993) first studied the impact of the intermediate principal stress on rock permeability. Later, Li (2001) and Takahashi *et al.* (2013) systematically studied the relationships of the permeability of Shirahama sandstone to three principal stresses. Overall, at the initial stage of the maximum principal stress loading, the permeability decreases with load increasing, but the decreasing rate ($\partial k/\partial\varepsilon_1$ or $\partial k/\partial\sigma_1$) continues to decrease with strain or stress increasing until reaches its minimum, which corresponds to the closure of original cracks within rock samples. The minimum permeability corresponds to the starting of rock dilatancy, after which, the strains in three directions start to be dominant. From dilatancy initiation to prior to rock failure, the permeability increases slowly. In the process of σ_1 further growth, several processes occur in the rock. 1) At a lower confining pressure, brittle failure occurs and the permeability increases sharply. 2) At a higher confining pressure, no brittle failure occurs in the rock in the process from compression to shear failure, and at this time, the permeability restored later is smaller and slightly lower than the initial permeability. Zeng (2015) also obtained the similar conclusion from her tests on mudstone, and further demonstrated that with the intermediate principal stress increasing, the increase in permeability becomes more obvious after the dilatancy point.

Li *et al.* (2001, 2002) studied the effect of the stress path on the permeability of Shirahama sandstone and defined the impact coefficient of the effective stress σ_{je} to the permeability k_i, namely, $a_{ij} = \partial log\, k_i/\partial\sigma_{je}$. Their experimental results showed that a_{ij} was independent on the stress path. In addition, King (2002) utilized his full rigid loading tester to study the characteristics of elastic wave propagation and percolation in the rock mass with multiple parallel fractures.

4.4 Other aspects

In addition to above-mentioned studies, researchers also investigated the characteristics of elastic wave propagation (Takahashi, 2012; Takahashi, 1991; Luo *et al.*, 1992; Zhao *et al.*, 1996), and the characteristics of AE signals including amplitude, frequency, etc. (Mou *et al.*, 1990; Xu *et al.*, 1989) in the true triaxial loading process. These studies briefly revealed the characteristics of rock deformation in the true triaxial loading process.

He *et al.* (2010) conducted the unilateral dynamic loading TTT experiments to simulate the process of limestone rock burst and used AE monitoring technique during tests. Analysis of measured AE signal data indicated that there were two principal frequency ranges in AE signal. AE signal displayed high frequency and low amplitude at lower stress, while moved toward high amplitude with stress increasing, and showed low frequency and high amplitude as close to rock burst. All of these data provide some prediction basis for monitoring rock bursts.

Gong *et al.* (2012) and Du *et al.* (2015) reconstructed the slabbing failure phenomenon of rock along the maximum principal stress direction, in consistence with the excavation site failure pattern. Du *et al.* according to their AE monitored results and infrared radiation characteristics, further concluded that with the intermediate principal stress increasing, the failure mode transited from shear failure to slabbing failure.

Overall, with the intermediate principal stress he increasing 1) the strength of ordinary hard rock shows a trend of increase followed by decrease, but the CTE strength is always higher than the CTC one; 2) the ductility of hard rock decreases, 3) the brittleness, elastic modulus, strain hardening coefficient, and failure angle all increase accordingly, while 4) the dilatancy of hard rock is suppressed. Some soft rocks with high porosity and few hard rock are slightly different under the influence of the intermediate principal stress. Studies on the permeability show that the effect of the intermediate principal stress on the permeability of rock samples is in agreement with that of the intermediate principal stress on deformation. Studies on rock burst are one of the research and application hot spots of TTT machines in recent years. Experimental results showed that the intermediate principal stress is the cause for rock slabbing failure along the maximum principal stress direction, consistent with the failure phenomenon in actual engineering excavations.

5 PROSPECT FOR TTT MACHINES

Since the invention of TTT machine in the 1960s, the characteristics of strength, deformation and failure of rock in the general stress state have been well understood. With the introduction of acoustic emission monitoring technology, scanning electron microscopy, and other measurement means, the microscopic mechanisms of the deformation and failure processes also have been further understood. At the same time, with the help of both true triaxial force and flow field in the testers, the effect of three directional stresses (strains) on the seepage is further studied.

However, the TTT machine now is still at the laboratory stage and has not been accepted in the engineering field. This may be due to several reasons: 1) true triaxial test boundary conditions, such as the blank corner effect, end friction effect, and loading eccentricity are complex, resulting in poor repeatability and reliability of tests; 2) high test machine costs make it difficult for promotion; and 3) insufficient rock mechanics constitutive research in the true triaxial stress state results in application difficulty in engineering. Thus, studies need to be conducted to further improve the true triaxial testers and better understand the mechanism of rock failure with focuses on the following aspects:

1) Developing new flexible loading materials. Flexible loading is prominent over rigid loading in control of blank corner, end friction, and loading eccentricity. However, its lower loading ability makes its promotion more difficult. With the development of materials science and related technologies, flexible materials of high strength will be developed.
2) Developing new loading ways. Using conventional true triaxial loading way is difficult to avoid the corner effect and the end friction effect. Therefore, the non-contact loading ways using a pair of deformable sheets to achieve flexible loading in one direction such as the EMF-driven loading way will be more popular in the future.
3) Strengthening international cooperation. Because rock true triaxial testers are complex in structure and with high cost, previous explorations and attempts for the development of new equipments are very valuable. Therefore, it is very important to strengthen international exchange and cooperation, promote knowledge sharing and technology transfer, and seeking more financial support.
4) Establishing the database of rock true triaxial tests. In addition to rock TTT technology, the database of TTT experimental results is also very important.
5) Further strengthening the monitoring means of TTT machines. Although there are few testers being introduced AE monitoring means, experimental data are still relatively limited. Thus, more researches should be conducted. In addition, CT scanning monitoring means should also be introduced into future TTT systems to monitor the whole rock failure process and reveal its failure mechanism.
6) Realizing multi-field coupling. Current TTT systems have basically realized the fluid-solid coupling; but further studies need to be performed for more field coupling such as temperature, chemistry, etc.

We hope that TTT systems could be more wide applied in geotechnical engineering in the near future.

REFERENCES

Alexeev, A. D., Revva, V. N., Alyshev, N. A. & Zhitlyonok, D. M. (2004) True triaxial loading apparatus and its application to coal outburst prediction. International Journal of Coal Geology, 58 (4), 245–250.

Andenaes, E., Ko, H.-Y. & Gerstle, K. H. (1977) Response of mortar and concrete to biaxial compression. Journal of the Engineering Mechanics Division, 103 (4), 515–526.

Atkinson, R. & Ko, H.-Y. (1973) A fluid cushion, multiaxial cell for testing cubical rock specimens. International Journal of Rock Mechanics and Mining Sciences & Geomechanics Abstracts, 10, 351–361.

Bertacchi, P. (1964) Behavior of concrete under combined loads: A comparison of the concrete shearing strength values obtained from direct tests with the values determined from triaxial test. In: 8th International Congress on Large Dames, May 4–8, 1964, Edinburgh, UK, p. 279.

Bésuelle, P. & Hall, S. (2011) Characterization of the strain localization in a porous rock in plane strain condition using a new truetriaxial apparatus. In: Bonelli, S., Dascalu, C. & Nicot, F. (eds.) Advances in bifurcation and degradation in geomaterials. Springer, pp. 345–352.

Böker, R. (1915) Die Mechanik der bleibenden Formänderung in Kristallinisch aufgebauten Körpern. Ver. dtsch. Ing. Mitt. Forsch., 175, 1–51.

Brown, E. & Hoek, E. (1978) Trends in relationships between measured in-situ stresses and depth. International Journal of Rock Mechanics and Mining Sciences & Geomechanics Abstracts, 15, 211–215.

Cai, M. (2008) Influence of intermediate principal stress on rock fracturing and strength near excavation boundaries-Insight from numerical modeling. International Journal of Rock Mechanics and Mining Sciences, 45 (5), 763–772.

Chang, C. (2001) True triaxial strength and deformability of crystalline rocks. Doctor thesis, The University of Wisconsin – Madison.

Chang, C. (2004) Effect of intermediate principal stress on rock fractures. Journal of the Korean Earth Science Society, 25 (1), 22–31.

Chang, C. & Haimson, B. (2000) True triaxial strength and deformability of the German Continental Deep Drilling Program (KTB) deep hole amphibolite. Journal of Geophysical Research: Solid Earth (1978–2012), 105 (B8), 18999–19013.

Chang, C. & Haimson, B. (2005) Non-dilatant deformation and failure mechanism in two Long Valley Caldera rocks under true triaxial compression. International Journal of Rock Mechanics and Mining Sciences, 42 (3), 402–414.

Chang, C. & Haimson, B. (2007) Effect of fluid pressure on rock compressive failure in a nearly impermeable crystalline rock: Implication on mechanism of borehole breakouts. Engineering Geology, 89 (3), 230–242.

Cheon, D., Keon, S., Park, C. & Ryu, C. (2006) An experimental study on the brittle failure under true triaxial conditions. Tunnelling and Underground Space Technology, 21, 448–449.

Crawford, B., Smart, B., Main, I. & Liakopoulou-Morris, F. (1995) Strength characteristics and shear acoustic anisotropy of rock core subjected to true triaxial compression. International Journal of Rock Mechanics and Mining Sciences & Geomechanics Abstracts, 32 (3), 189–200.

Da Gama, C. D. (2012) The hollow cylinder test as an alternative to true triaxial loading of prismatic rock specimens. Kwasniewski, M., Li, X. & Takahashi, M. (eds.) True triaxial testing of rocks: Proceedings of the 12th ISRM International Congress on Rock Mechanics in 2011, 12th ISRM International Congress on Rock Mechanics in 2011, 16–21 October 2011, Beijing, China. London New York & Leiden, Taylor & Francis, vol. 4, pp. 73–82.

Desai, C., Janardahanam, R. & Sture, S. (1982) High capacity multiaxial testing device. ASTM Geotechnical Testing Journal, 5 (1–2), 26–33.

Du, K. (2013) Study on the failure characteristics of deep rock and the mechanisn of strainburst under true triaxial unloading condition. Central South University.

Du, K., Li, X.-B., Li, D.-Y. & Lei, W. (2015) Failure properties of rocks in true triaxial unloading compressive test. Transactions of Nonferrous Metals Society of China, 25 (2), 571–581.

Esaki, T. & Kimura, T. (1989) Mechanical behaviour of rocks under generalized high stress conditions. In: ISRM International Symposium on Rock at Great Depth, Pau, France, August 30–September 2, 123–130.

Furuzumi, M. & Sugimoto, F. (1986) Effect of intermediate principal stress on failure of rocks and failure condition of rocks under multiaxial stress. Journal of the Japan Society of Engineering Geology, 27 (1), 13–20.

Geng, N. & Xu, D. (1985) Rock rupture caused by decreasing the minimum principal stress. ACTA Geophysica Sinica (02), 191–197.

Gong, Q. M., Yin, L. J., Wu, S. Y., Zhao, J. & Ting, Y. (2012) Rock burst and slabbing failure and its influence on TBM excavation at headrace tunnels in Jinping II hydropower station. Engineering Geology, 124 (1), 98–108.

Haimson, B. (2006) True triaxial stresses and the brittle fracture of rock. True triaxial stresses and the brittle fracture of rock. Pure and Applied Geophysics, 163, 1101–1130.

Haimson, B. (2007) The effect of the intermediate principal stress on the brittle fracture of rocks. Geophysical Research Abstracts, 9, 02100.

Haimson, B. (2011) Consistent trends in the true triaxial strength and deformability of cores extracted from ICDP deep scientific holes on three continents. Tectonophysics, 503 (1), 45–51.

Haimson, B. & Chang, C. (2000) A new true triaxial cell for testing mechanical properties of rock, and its use to determine rock strength and deformability of Westerly granite. International Journal of Rock Mechanics and Mining Sciences, 37 (1), 285–296.

Haimson, B. & Chang, C. (2005) Brittle fracture in two crystalline rocks under true triaxial compressive stresses. Geological Society, London, Special Publications, 240 (1), 47–59.

Haimson, B. & Ma, X. (2012) From KTB amphibolite to Bentheim sandstone: The diminishing effect of the intermediate principal stress on faulting and fault angle. EGU General Assembly Conference Abstracts, 14, 5994.

Haimson, B. & Rudnicki, J. W. (2010a) The intermediate principal stress effect on faulting and fault orientation. EGU General Assembly Conference Abstracts, 12, 2070.

Haimson, B. & Rudnicki, J. W. (2010b) The effect of the intermediate principal stress on fault formation and fault angle in siltstone. Journal of Structural Geology, 32 (11), 1701–1711.

Handin, J., Heard, H. C. & And Magouirk, J. N. (1967) Effects of the intermediate principal stress on the failure of limestone, dolomite, and glass at different temperatures and strain rates. Journal of Geophysical Research, 72, 611–640.

Handin, J., Higgs, D. V. & O'brien, J. K. (1960) Torsion of Yule marble under confining pressure. Geological Society of America Memoirs, 79, 245–274.

He, M. C., Miao, J. L. & Feng, J. L. (2010) Rock burst process of limestone and its acoustic emission characteristics under true-triaxial unloading conditions. International Journal of Rock Mechanics and Mining Sciences, 47 (2), 286–298.

He, M., Miao, J., Li, D. & Wang, C. (2007) Experimental study on rockburst processes of granite specimen at great depth. Chinese Journal of Rock Mechanics and Engineering, 26 (5), 865–876.

Heard, H. C. (1960) Transition from brittle fracture to ductile flow in Solenhofen limestone as a function of temperature, confining pressure, and interstitial fluid pressure. Geological Society of America Memoirs, 79, 193–226.

Hojem, J. & Cook, N. (1968) The design and construction of a triaxial and polyaxial cell for testing rock specimens. South African Mechanical Engineer, 18 (2), 57–61.

Hoskins, E. R. (1969) The failure of thick-walled hollow cylinders of isotropic rock. International Journal of Rock Mechanics and Mining Sciences, 6, 99–125.

Ingraham, M. D. (2012) Investigation of localization and failure behavior of Castlegate sandstone using true triaxial testing. Clarkson University.

Ingraham, M. D., Issen, K. A. & Holcomb, D. J. (2011) Failure of Castlegate sandstone under true triaxial loading. In: Bonelli, S., Dascalu, C. & Nicot, F. (eds.), *Advances in Bifurcation and Degradation in Geomaterials*. Dordrecht, Netherlands, Springer, pp. 321–326.

Ingraham, M. D., Issen, K. A. & Holcomb, D. (2013a) Response of Castlegate sandstone to true triaxial states of stress. Journal of Geophysical Research: Solid Earth, 118, 536–552.

Ingraham, M. D., Issen, K. A. & Holcomb, D. J. (2013b) Use of acoustic emissions to investigate localization in high-porosity sandstone subjected to true triaxial stresses. Acta Geotechnica, 8 (6), 645–663.

Issen, K. A., Ingraham, M. D. & Dewers, T. A. (2011) Strain localization conditions under true triaxial stress states. In: *Advances in bifurcation and degradation in geomaterials*. Springer, pp. 309–314.

Jaeger, J. C. & Hoskins, E. R. (1966) Rock failure under the confined Brazilian test. Journal of Geophysical Research, 71, 2651–2659.

Jing, F., Sheng, Q., Zhang, Y., Luo, C. & Liu, Y. Research on distribution rule of shallow crustal geostress in china mainland. Chinese Journal of Rock Mechanics and Engineering, 26 (10), 2056–2062.

Kármán, T. V. (1911) Festigkeitsversuche unter allseitigem Druck. Ver. dtsch. Ing, 55 (42), 1749–1757.

King, M. (2002) Elastic wave propagation in and permeability for rocks with multiple parallel fractures. International Journal of Rock Mechanics and Mining Sciences, 39 (8), 1033–1043.

King, M., Chaudhry, N. & Shakeel, A. (1995) Experimental ultrasonic velocities and permeability for sandstones with aligned cracks. International Journal of Rock Mechanics and Mining Sciences & Geomechanics Abstracts, 32 (2), 155–163.

Koide, H. & Takahashi, M. (1991) Effect of anisotropic stress and fluid pressure on the stability of rock around an underground radioactive waste repository. In: The 11th international conference on structural mechanics in reactor technology, SMiRT 11, August 18–33, 1991, Tokyo, Japan, pp. 441–446.

Koyfma, M. P., Ilnitskaya, E. I. & Karpov, V. I. (1964) Strength of rocks under bulk stress state conditions. Moskva: Nauka.

Kwasniewski, M. (2012) Mechanical behavior of rocks under true triaxial compression conditions – A review. Kwasniewski, M., Li, X. & Takahashi, M. (eds.) True triaxial testing of rocks: Proceedings of the 12th ISRM International Congress on Rock Mechanics in 2011, 12th ISRM International Congress on Rock Mechanics in 2011, October 16–21, 2011, Beijing, China. London, New York & Leiden, Taylor & Francis, vol. 4, pp. 99–138.

Kwaśniewski, M. (2013) Recent Advances in Studies of the Strength of Rocks Under True Triaxial Compression Conditions/Postępy W Badaniach Nad Wytrzymałością Skał W Warunkach Prawdziwego Trójosiowego Ściskania. Archives of Mining Sciences, 58 (4), 1177–1200.

Kwaśniewski, M. & Takahashi, M. (2006) Behavior of a sandstone under axi- and asymmetric compressive stress conditions. In Leung, C.F. & Zhou, Y.X. (eds), Rock mechanics in underground construction (Proc. 4th Asian Rock Mech. Symp., Singapore, November 8–10, 2006), p. 320 + CD-ROM. Singapore: World Scientific Publishing Co. Pte. Ltd.

Kwaśniewski, M., Takahashi, M. & Li, X. (2003) Volume changes in sandstone under true triaxial compression conditions. In: Technology Roadmap for Rock Mechanics: Proceedings of the 10th Congress of the ISRM, Sandton, September 8–12, 2003, vol. 1, pp. 683–688.

Launay, P. & Gachon, H. (1972) Strain and ultimate strength of concrete under triaxial stress. Special Publication, 34, 268–282.

Lee, H. & Haimson, B. C. (2011) True triaxial strength, deformability, and brittle failure of granodiorite from the San Andreas Fault Observatory at Depth. International Journal of Rock Mechanics and Mining Sciences, 48 (7), 1199–1207.

Li, H., Xu, J., Wan, L., Yin, G. & Li, X. (1986) The mechanism of failure strength of brittle rocks under polyaxial stress states. Journal of Chongqing University (Natural Science Edition), 3, 87–97.

Li P., Hua P. & Xu Z. (1986) Development study of zsy-83 triaxial pressure-adjustable apparatus. Journal of Seismological Research, 11 (2), 209–197.

Li, X. (2001) Permeability change in sandstones under compressive stress conditions. Doctor thesis: Ibaraki University.

Li, X., Hitoshi, K. & Takashi, O. (2003) CO_2 aquifer storage and the related rock mechanics issues. Chinese Journal of Rock Mechanics and Engineering, 22 (06), 989–994.

Li, X., Liu, Y., Bai, B. & Fang, Z. (2006) Ranking and screening of CO_2 saline aquifer storage zones in china. Chinese Journal of Rock Mechanics and Engineering, 25 (5), 963–968.

Li, X., Shi, L., Bai, B., Li, Q. & Feng, X. (2012) True-tiraxial testing techniques for rocks-State of the atr and future perspectives. Kwasniewski, M., Li, X. & Takahashi, M. (eds.) True triaxial testing of rocks: Proceedings of the 12th ISRM International Congress on Rock Mechanics in 2011, 12th ISRM International Congress on Rock Mechanics in 2011, October 16–21, 2011, Beijing, China. London, New York & Leiden, vol. 4, pp. 3–18.

Li, X., Wu, Z., Takahashi, M. & Yasuhara, K. (2001) Hydrostatic and non-hydrostatic compressive stresses-induced permeability change in Kimachi sandstone. Frontiers of Rock Mechanics and Sustainable Development in the 21st Century, 201–204

Li, X., Wu, Z., Takahashi, M. & Yasuhara, K. (2002) Permeability anisotropy of Shirahama sandstone under true triaxial stresses. Journal of Geotechnical Engineering, JSCE, 708, 1–11.

Li, X., & Xu, D. (1991) Law and degree of effect the intermediate principle stress on strength of rock. Rock and Soil Mechanics, 12, 9–16.

Li, X., Xu, D. & Liu, S. (1994) The experimental research of the strength, deformation and failure properties of laxiwa granite under the status of true triaxial stress. Proceedings of the third Conference of Chinese Society of Rock Mechanics and Engineering, 153–159.

Luo, J., Cai, Z., Liu, K. & Li, X. (1992) Determination of the various stages in the process of rock failure under loading by the acoustic parameters. Rock and Soil Mechanics, 13 (1), 51–56.

Ma, X. & Haimson, B. (2013) From dilatancy to contraction: Stress-dependent failure mode progression in two porous sandstones subjected to true triaxial testing. In: EGU General Assembly Conference Abstracts, 15, 6665.

Ma, X., Rudnicki, J. & Haimson, B. (2014) Failure-plane angle in Bentheim sandstone subjected to true triaxial stresses: Experimental results and theoretical prediction. In: EGU General Assembly Conference Abstracts, 16, 1800.

Marone, C., Carperter, B., Elsworth, D., Faoro, I., Ikari, M., Knuth, M., Niemeijer, A., Saffer, D. & Samuelson, J. (2007) A pressure vessel for true-triaxial deformation and fluid flow during frictional shear. In: 29th Course of the International School of Geophysics, Euro-conference of Rock Physics and Geomechanics on Natural Hazards: Thermo-hydro-mechanical coupling processes in rocks, September 25–30, 2007, Erice, Italy. Oral presentation in session 1.

Mazanti, B. & Sowers, G. (1966) Laboratory testing of rock strength. Testing Techniques for Rock Mechanics, ASTM STP, 402, 207–231.

Mcgarr, A. & Gay, N. (1978) State of stress in the earth's crust. Annual Review of Earth and Planetary Sciences, 6, 405–436.

Michelis, P. (1985a) Polyaxial yielding of granular rock. Journal of Engineering Mechanics, 111 (8), 1049–1066.

Michelis, P. (1985b) A true triaxial cell for low and high pressure experiments. International Journal of Rock Mechanics and Mining Sciences & Geomechanics Abstracts, 22 (3), 183–188.

Mills, L. L. & Zimmerman, R. M. (1970) Compressive strength of plain concrete under multiaxial loading conditions. ACI Journal Proceedings, 66 (10), 802–807.

Mogi, K. (1967) Effect of the intermediate principal stress on rock failure. Journal of Geophysical Research, 72, 5117–5131.

Mogi, K. (1971a) Effect of the triaxial stress system on the failure of dolomite and limestone. Tectonophysics, 11 (2), 111–127.

Mogi, K. (1971b) Fracture and flow of rocks under high triaxial compression. Journal of Geophysical Research, 76 (5), 1255–1269.

Mogi, K. (1972a) Effect of the triaxial stress system on fracture and flow of rocks. Physics of the Earth and Planetary Interiors, 5, 318–324.

Mogi, K. (1972b) Fracture and flow of rocks. Tectonophysics, 13 (1), 541–568.

Mogi, K. (1977) Dilatancy of rocks under general triaxial stress states with special reference to earthquake precursors. Journal of Physics of the Earth, 25 (Supplement), S203–S217.

Mogi, K. (1981) Flow and fracture of rocks under general triaxial compression. Applied Mathematics and Mechanics, 2 (6), 635–651.

Mogi, K. (2007) Experimental rock mechanics. Leiden: Taylor & Francis/Balkema.

Mogi, K. (2012) How I developed a true triaxial rock testing machine. Kwasniewski, M., Li, X. & Takahashi, M. (eds.) True triaxial testing of rocks: Proceedings of the 12th ISRM International Congress on Rock Mechanics in 2011, 12th ISRM International Congress on Rock Mechanics in 2011, October 16–21, 2011, Beijing, China. London, New York & Leiden, Taylor & Francis, vol. 4, pp. 139–157.

Mou, J., Xu, Z. & Han, M. (1990) P-wave Q-value of acoustic emission in fracture process of granite specimens under true triaxial compression. Seismology and Geology, 12 (4), 277–281.

Murrell, S. A. F. (1965) The effect of triaxial stress systems on the strength of rocks at atmospheric temperatures. Geophysical Journal International, 10, 231–281.

Nadai, A. (1950) Theory of flow and fracture of solids. New York: McGraW-Hill.

Niwa, Y., Koyanagi, W. & Kobayashi, S. (1967) Failure criterion of light weight concrete subjected to triaxial compression. Proceedings, Japan Society of Civil Engineers, 143, 28–35.

Ohnaka, M., Akatsu, M., Mochizuki, H., Odedra, A., Tagashira, F. & Yamamoto, Y. (1997) A constitutive law for the shear failure of rock under lithospheric conditions. Tectonophysics, 277 (1), 1–27.

Oku, H., Haimson, B. & Song, S. R. (2007) True triaxial strength and deformability of the siltstone overlying the Chelungpu fault (Chi-Chi earthquake), Taiwan. Geophysical research letters, 34, L09306,

Robertson, E. C. (1955) Experimental study of the strength of rocks. Geological Society of America Bulletin, 66 (10), 1275–1314.

Sibai, M., Henry, J. P. & Gros, J. C. (1997) Hydraulic fracturing stress measurement using a true triaxial apparatus. International Journal of Rock Mechanics and Mining Sciences, 34 (3–4), 289.e281–289.e210.

Skoczylas, F. & Henry, J. (1995) A study of the intrinsic permeability of granite to gas. International Journal of Rock Mechanics and Mining Sciences & Geomechanics Abstracts, 32, 171–179.

Smart, B. (1995) A true triaxial cell for testing cylindrical rock specimens. International Journal of Rock Mechanics and Mining Sciences & Geomechanics Abstracts, 32, 269–275.

Spetzler, H. A., Sobolev, G. A., Sondergeld, C. H., Salov, B. G., Getting, I. C. & Koltsov, A. (1981) Surface deformation, crack formation, and acoustic velocity changes in pyrophyllite under polyaxial loading. Journal of Geophysical Research: Solid Earth (1978–2012), 86 (B2), 1070–1080.

Takahashi, M. (1991) Water infiltration into rock observed by P wave velocity and amplitude ratio. Journal of the Japan Society of Engineering Geology, 32 (5), 232–239.

Takahashi, M. (2012) Seismic wave velocity anisotropy in Westerly granite under a true triaxial compression test. Kwasniewski, M., Li, X. & Takahashi, M. (eds.) True triaxial testing of rocks: Proceedings of the 12th ISRM International Congress on Rock Mechanics in 2011, 12th ISRM International Congress on Rock Mechanics in 2011, October 16–21, 2011, Beijing, China. London, New York & Leiden, Taylor & Francis, vol. 4, pp. 193–202.

Takahashi, M. & Koide, H. (1989) Effect of the intermediate principal stress on strength and deformation behavior of sedimentary rocks at the depth shallower than 2000 m. Rock at Great Depth, 1, 19–26.

Takahashi, M., Koide, H. & Kinoshita, S. (1983) Characteristics of strength in sedimentary rocks under true triaxial compressional stress state and the increase of brittleness on the intermediate stress. Journal of the Japan Society of Engineering Geology, 24 (4), 150–157.

Takahashi, M., Lin, W., Li, X. & Kwasniewski, M. (2005) Mechanical and hydraulic behaviors in Shirahama sandstone under true triaxial compression stress. Proc. Int. Symp. ISRM, EUROCK, 2005. Brno. May 18–20, 2005. Brno: UCN, 2005, pp. 236–248.

Takahashi, M., Narita, T., Tomishima, Y. & Arai, R. (2001) Various loading systems for rock true triaxial compression test. Journal of the Japan Society of Engineering Geology, 42 (4), 242–247.

Takahashi, M., Park, H., Takahashi, N. & Fujii, Y. (2013) True triaxial tests–using permeability and extensional stress parameters to simulate geological history in rocks. Geosystem Engineering, 16 (1), 75–82.

Takahashi, M., Sugita, Y., Xue, Z., Oonishi, Y. & Tsukuba, H. (1993) Three principal stress effects on permeability of the shirahama sandstone – In case of stress state prior to dilatancy. Shigen – to – Sozai,109 (10), 803–809.

Tiwari, R. & Rao, K. (2004) Physical modeling of a rock mass under a true triaxial stress state. International Journal of Rock Mechanics and Mining Sciences, 41, 396–401.

Walsri, C. 2009. Compressive and tensile strengths of sandstones under true triaxial stresses. Master thesis: Suranaree University of Technology.

Wawersik, W. R., Carlson, L. W., Holcomb, D. J. & Williams, R. J. (1997) New method for true-triaxial rock testing. International Journal of Rock Mechanics and Mining Sciences, 34, 285–296.

Weigler, H. & Becker, G. (1961) Über das Bruch-und Verformungsverhalten von Beton bei mehrachsiger Beanspruchung. Der Bauingeneieur, 36, 390–396.

Weigler, H. & Becker, G. (1963) Untersuchungen über das Bruch- und Verformungsverhalten von Beton bei zweiachsiger Beanspruchung. Deutscher Ausschuss für Stahlbeton, Heft 157.

Xiang, T., Feng, X., Chen, B. & Jiang, Q. (2008) True triaxial and acoustic emission experimental study of failure process of hard rock under excavating and supporting stress paths. Rock and Soil Mechanics, 29 (Supp.), 500–507.

Xu, D. (1982) Mechanical properties of weak porous sandstone under general triaxial stress states. Rock and Soil Mechanics, 3 (01), 13–25.

Xu, D. & Geng, N. (1984) Rock rupture caused by change of the intermediate principal stress and earthquake. ACTA Seismologica Sinica, 6 (02), 159–166.

Xu, D. & Geng, N. (1985) The variation law of rock strength with increase of intermediate principal stress. ACTA Mechanica Solida Sinica, 1 (1), 72–80.

Xu, D. & Geng, N. (1986) The various stress paths causing deformation and failure in rocks. Rock and Soil Mechanics, 7 (02), 17–25.

Xu, D., Xing, Z., Li, X., Zhang, G. & Wei, M. (1990) Development of RT3 type rock high pressure true triaxial machine. Rock and Soil Mechanics, 11 (02), 1–14.

Xu, D., Zhang, G., Li, T., Tang, G., Li, Q. & Xu, Y. (2000) On the stress state in rock burst. Chinese Journal of Rock Mechanics and Engineering, 19 (2), 169–172.

Xu, Z., Bao, Y. & Li, Q. (1989) Variations of the amplitude and the predominant frequency of the S-wave coda trains during the fracture-preparing process of rock samples under true triaxial compressions. Journal of Seismological Research, 12 (2), 015.

Xu, Z., Mei, S., Zhuang, C. & Li P. (1991) The influence of stiffness of the true triaxial testing machine on the rupturing and acoustic emission of rocks and its correlation with seismic activities. ACTA Seismologica Sinica, 13 (2), 223–232.

Xu, Z., Mei, S., Zhuang, C. & Li, P. (1992) The influence of stiffness of the true triaxial testing machine on the rupturing and acoustic emission of rocks and its correlation with seismic activities. Acta Seismologica Sinica, 5 (2), 367–379.

Xu, Z., Mei, S., Zhuang, C., Yang, H. & Bao, Y. (1992) Preliminary location results of microfracture under true triaxial compression for several rock samples. ACTA Seismologica Sinica, 14 (Supp.), 702–709.

Zeng, Z. (2015) Study on permeability change of mudstone during deformation process. Doctor thesis: The University of Chinese Academy of Sciences.

Zhang, J. (1976) The measurement of three-dimensional strength of tunnel or rock slope. Metal Mine, 4, 18–20.

Zhang, M. Takahashi, M. & Esaki, T. (1999) State of the art: Integrated laboratory measurement of shear and flow properties of geologic materials. Jour. Journal of the Japan Society of Engineering Geology, 40 (4), 240–246.

Zhang, X., Feng, X. & Hao, Q. (2014) A rigid follow-up loading frame structure[P]. Liaoning: CN103822831A,2014-05-28

Zhang, X., Feng, X., Li, Y., Yang, C., Tian, J. & Ren, J. (2012) A true triaxial pressure chamber which can transfer the principal stress indirections in unloading process[P]. Liaoning: CN102735532A, 2012-10-17.

Zhao, J., Hu, Y., Wang, B. & Xu, Z. (1996) Variation characteristics of elastic wave during the fracture-preparing process of gneiss under true triaxial compression. Seismology and Geology, 18 (3), 277–281.

Chapter 3

Rockburst concept and mechanism

M.C. He, G.L. Zhu & W.L. Gong

State Key Laboratory for Geomechanics & Deep Underground Engineering, China University of Mining & Technology (Beijing), Beijing, China

Abstract: This chapter presents a variety of rockbursts in five sections: (i) the rockburst introduction, (ii) classification according to triggering mechanisms and modeling approaches, (iii) the study of the above classifications, (iv) the criterion of rockburst, and (v) a brief summary of this chapter. Innovative work has been done in developing the 'strainburst testing machine' and 'impact-induced rockburst testing machine' that create a new era in which varied rockburst phenomena can be produced in the laboratory. New concepts are proposed regarding the stress paths for artificially produced rockbursts, the strain-burst criteria, and impact-induced rockburst criteria that take into account both the static and dynamic stress states analogous to that at excavation boundaries. This research provides us with a basis for rockburst prediction and control in the field.

I INTRODUCTION

With fast development of economies and industrial construction, world-wide exploitation of the natural resources will inevitably proceed into deep ground. As countries across the world recognize their underground resources as a growth point for the national economy, geotechnical engineering goes into deep ground as well. The scale and depth of underground engineering for the nuclear industry, national defense industry, transportation industry, and hydraulic engineering are growing at particularly high speed. The frequencies of various disasters have increased significantly as a result of the increasing depths for underground projects. Among them the most serious problem is rockburst hazards induced by high ground stresses (Qian, 2010).

Rockbursts, as a catastrophic phenomenon in underground engineering, are characterized by a sudden, explosive expulsion or ejection of the cracked country rocks. In underground coal mining, the rockburst phenomenon is also known as coal bursts or coal bumps. Rockbursts or coal bumps cause destruction of the supporting equipment in a mining stope; distortion and destruction of the stope and drifts; casualties among the mine personnel; and even collapse of the ground surface accompanied by local seismicity. Therefore, mechanisms, forecasting and prediction, as well as control of rockbursts, have become key scientific problems to be tackled. The problems facing the community of rock mechanics are technically complex and difficult. Numerous research projects on rockburst mechanisms have been carried out across the world and notable discoveries made (He, 2006; He *et al.*, 2010, 2012a, 2012b, 2015; Tang &

Kaiser, 1998; Kaiser *et al.*, 1995; Haimson, 2006; Mogi, 2007; Jaeger *et al.*, 2007; Brady & Brown, 2005; Singh *et al.*, 2001; Zhou *et al.*, 2008). Most of the findings seen in the publications and international conferences have been on the basis of qualitative explanations. As for the forecasting and prediction of rockbursts, although quite a few successful experiences were achieved, no systematic theoretical results have been built up. Some experts even assert that there is no possibility for rockburst forecasting.

However, the degree of difficulty and scale of rock engineering in its current state in China is so large that it has been the focus of world attention. Scientists and researchers in our country should undertake overall and in-depth studies of rockburst mechanisms, and provide the development of rock engineering in deep ground with strong support in practical techniques and scientific theory (Qian, 2010). This paper presents the authors' achievements in researching the rockburst mechanisms and their control measures, involving development of laboratory rockburst testing machines; theoretical and experimental investigations; and development of the constant resistance large deformation (CRLD) bolt, as well as its practical use in roadway support under rockburst conditions (He, 2006; He *et al.*, 2007, 2009, 2010, 2012).

2 ROCKBURST CATEGORIZATION

A rockburst is generally defined as a sudden rock failure characterized by the breaking up and expulsion of rock from its surroundings, accompanied by a violent release of energy (Blake, 1972). Although the definition of rockburst differs from one author to another, the common ground of these definitions is the sudden release of energy in the form of a violent expulsion of rock (McMahon, 1988). Brown (1984) suggests that a rockburst should be considered as a particular manifestation of seismic activity that is induced by mining activities. In fact, the sudden failure that characterizes a rockburst can be, in itself, the source of the seismic event, or may have been triggered by a distant seismic event, or from a load transfer due to the latter (Gill & Aubertin, 1993).

It is well understood that classification and categorization of the naturally occurring phenomena in question is one of the major approaches, or the preliminary step, for further investigation in scientific research practice. Since Morrison (1942) and Blake (1972) first reported on an Indian gold mine, rockbursts have been categorized according to the phenomena (Kaiser *et al.*, 1995); the mechanisms (Simon, 1999); the energy (Corbett, 1996); the scale and location (Ghose & Rae, 1988); the cause and effect (Jiang *et al.*, 2007; Qi *et al.*, 2003), as well as the detonation source mechanism of rockbursts; the relationship between the parameters of the detonation source; the first motion signs of a seismic wave; the location where the rockburst occurred and the magnitude of released energy; the stress paths that induced a rockburst; and the causes for the formation of a rockburst event (Cook, 1965; Brady & Brown, 2005; Ryder, 1988; Henry *et al.*, 1989; Kuhnt, 1989; Corbett, 1996).

The authors categories the rockburst according to triggering mechanisms and the related laboratory physical modeling approaches. That is, a rockburst may happen in the following two conditions: (i) in a highly stressed rock mass storing a large amount of the strain energy during roadway excavations or a face stopping phase, and (ii) in less stressed and deformed rock storing a lesser amount of the strain energy after the excavations phase, but induced by the external disturbances in the far-field region

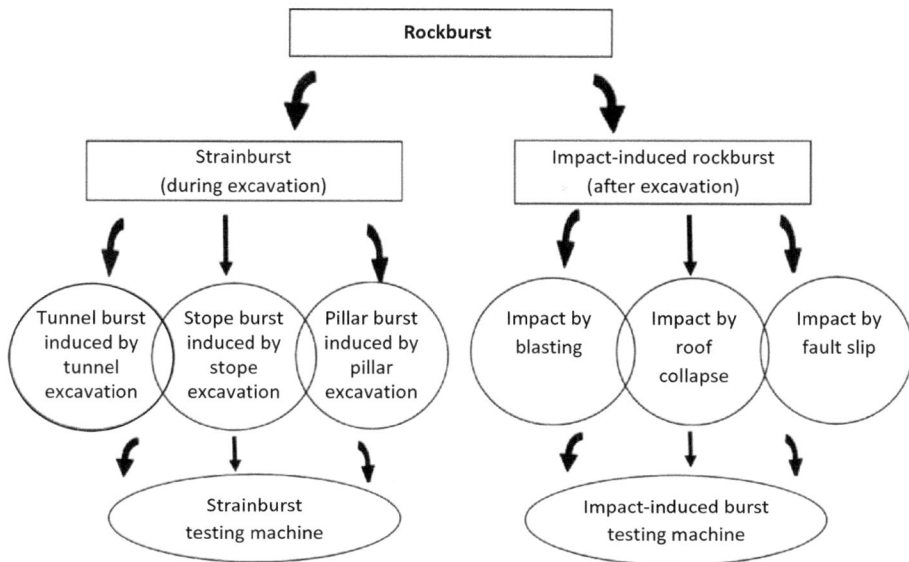

Figure 1 Rockburst classification based the triggering mechanism and experimental methods (He et al., 2012b).

such as blasting, caving, and adjacent tunneling, etc. Based on this awareness, rockburst phenomena under two different conditions can be categorized into two major classes (depicted in Figure 1):

Class I: the strainbursts or excavation-induced rockbursts occurred during process of the excavation

Class II: the impact-induced rockbursts occurred after the excavation.

Strainbursts (or stress-induced rockbursts) occur during the excavation phase and are induced by the redistribution of excavation-induced stress. They include *tunnel burst* (induced by tunnel excavation), *stope burst* (induced by stope excavation) and *pillar burst* (induced by pillar excavation). The seismic site and source for strainbursts are the same: at, or near, the excavation boundaries (for tunnel burst and stope burst) or at the pillar face (for pillar burst). Impact-induced rockbursts take place at the excavation boundaries and are initiated by far-field seismic waves, and could be further classified into three types: rockbursts induced by blasting or excavation; rockbursts induced by roof collapse; and rockbursts induced by fault slip. The dynamic nature of this kind of event will be dealt with specifically in the Section 5 of this paper.

3 ROCKBURST STUDY

It was known to all that indoor rockburst experiments play an important role in understanding its formation mechanisms, calibration of numerical models, evaluation of mechanical parameters, and identification of the stress state where a dynamic event may be initiated. The primary goal for researchers who were engaged in the physical modeling of rockburst phenomena is to reproduce the stress states in the laboratory in a faithful manner, to simulate the conditions under which a dynamic event may occur. Great efforts were made in developing tri-axial or true-tri-axial devices to achieve this end, involving work done by Mogi (1967), Crawford & Wylie (1987), Chang & Haimson (2000), Chen & Feng (2006), Haimson (2006), Lee & Haimson (2011), for example. Although some degree of success was achieved with these studies, they did not, however, reproduce conditions for testing rock specimen analogous to the *in situ* event, *i.e.* from a stationary or equilibrium state, transferred to the critical state and then attaining the final chaotic state where a large volume of the fractured rock is ejected. This Section presents the research achievements to accomplish these goals carried out in the State Key Laboratory for Geomechanics and Deep Underground Engineering (LGDUE) at the China University of Mining & Technology Beijing (CUMTB) involving the development of the strainburst testing machine and the impact-induced testing machine based on the above rockburst classifications (Class I and Class II).

3.1 Studies on strainburst

3.1.1 Strainburst testing machine

The strainburst testing machine, developed by He (2006), is shown in Figure 2; its main task is to simulate the stress-induced rockburst phenomena (Class I rockburst in Figure 1, Patent No. ZL 2007 1 0099297.1). Figure 2a is the schematic of the main unit of the testing machine, a modified true-tri-axial apparatus (MTTA). The MTTA can provide dynamic loading/unloading independently in three principal directions (He *et al.*, 2010). Load in the minimum principal stress direction (horizontal) can be unloaded abruptly on one face and displacement constraint condition is maintained on the opposite face, simulating the stress condition for rock mass at the excavation surface. Its major technical specifications are also shown in Figure 2(a).

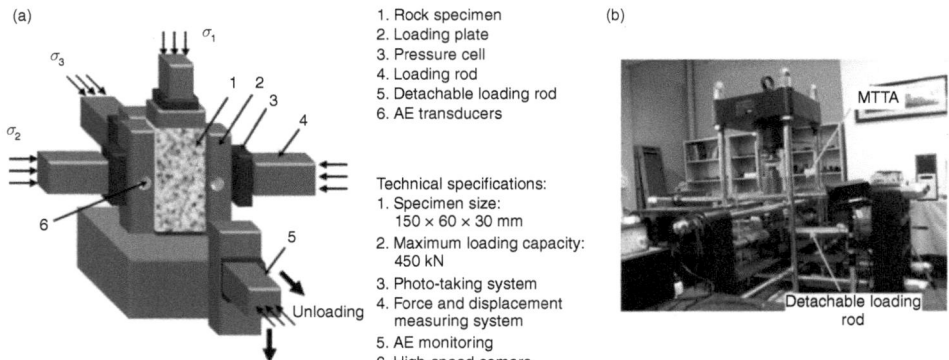

Figure 2 Strainburst testing machine; (a) schematic of the testing machine, and (b) photograph taken in the laboratory for an overview of the testing machine and the peripheral monitoring instruments.

Figure 2b is a photograph showing on overview of the strainburst testing machine. In addition to the MTTA, the testing machine is equipped with a hydraulic control system and monitoring instruments such as those for force monitoring, acoustic emission (AE) monitoring, and high-speed digital camera recording. The AE monitoring system manufactured by PAC (Physical Acoustics Corporation, USA) was used with two polarity transducers mounted on the two sides of the specimen (the intermediate principal direction, in Figure 2a). The resonance frequency is 150 kHz, the fairly flat response ranges from 0–400 kHz; the pre-amplification is 40 dB, gain amplification is 10 times, and the total amplification is 1000 times; the data acquisition rate was set to 1 MHz. AE events were monitored over 0–512 kHz of the frequency specimen. The recording speed of the high-speed camera was 1000 frames per second under full resolution.

3.1.2 Stress paths for instantaneous and delayed rockbursts

A strainburst is self-initiated due to the stresses in the near-face region. That is, the self-initiated rockbursts occur when the stresses near the boundary of an excavation exceed the rock mass strength, and failures proceed in an unstable or violent manner (when the stored strain energy in the rock mass is not dissipated during the fracturing process). The design of the MTTA can perform loading/unloading independently in three principal directions, which provide flexibilities for realized varied stress paths analogous to those at, or near, the surface of underground excavations or mine stopings. Typically, stress paths for the three types of the self-initiated rockbursts (Figure 1) can be designed.

Figure 3a shows the stress path for instantaneous rockbursts. The stress path was designed to resemble static loading and dynamic loading due to excavation at the excavation boundaries. The load is first increased proportionally and slowly to attain the hydrostatic stress state ($\sigma_1 = \sigma_2 = \sigma_3$) marked by letter A, under the convention of $\sigma_1 > \sigma_2 > \sigma_3$ for the principal stresses. Secondly, σ_1 and σ_2 were increased step by step until approaching the virginal stress state marked by B. After that, this stress state was maintained for a certain time to allow the equilibrium state to be attained inside the rock specimen, and then σ_3 was suddenly removed from one surface of the specimen (Figure 2a) while keeping σ_1 and σ_2 at constant. Thus the sample is at the possible

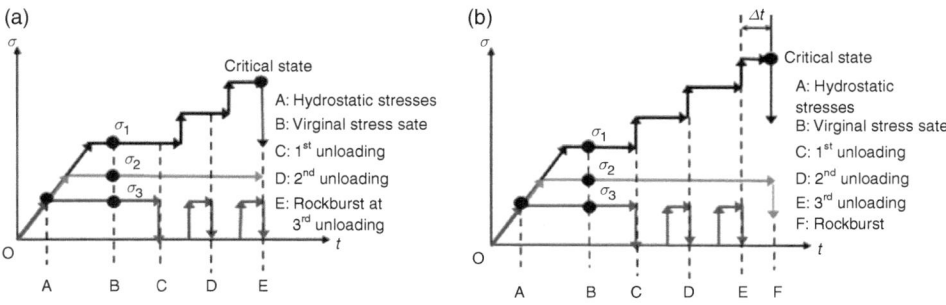

Figure 3 Stress paths; (a) stress path for instantaneous rockburst, and (b) stress path for delayed rockburst.

occurrence state of rockburst (the first unloading marked by C). If the rock sample does not fail after around 30 minutes of the first unloading (He *et al.*, 2010), the loading process is resumed. Typically, the stress paths in Figure 3 have three cycles of the loading/unloading, assuming that rockburst occurs at the third cycle.

Figure 3b shows the stress for delayed rockbursts. The major difference between 3a and 3b lies in the relationship between σ_1 and σ_3 at unloading. In the stress path for instantaneous rockbursts (Figure 3a), σ_1 is kept constant when unloading σ_3, and rockburst occurs after the unloading with a very small time lag (delayed rockburst mechanism). In the stress path for delayed rockbursts (Figure 3b), σ_1 increases at unloading and rockburst occurs undergoing a period of time, Δt, after unloading σ_3. For the instantaneous rockbursts at the critical stress state undergoing the stress path, σ_1, is equal to, or larger than, the unconfined compressive strength (UCS) of the rock mass. When σ_3 approaches zero instantaneously, the excessive potential energy stored in the highly stressed rock mass will be released and converted into kinematic energy, carrying the rock fractures into the excavation space. For the delayed rockburst, no failure occurs when σ_3 approaches zero instantaneously and σ_1 at a low level (less than the UCS). It is the excavation that causes the dramatic increase in σ_1 at unloading and rockburst will happen when σ_1 is sufficiently high. For a reflection of the stress concentration, σ_1 is increased to the level of 1.2–1.3 times the UCS.

3.1.3 Stress paths for pillar bursts

Figure 4 shows the stress path for pillar bursts. This stress path simulates the stress redistribution existing in a coal pillar during the mining phase. The load is first increased proportionally and slowly to attain the hydrostatic stress state ($\sigma_1 = \sigma_2 = \sigma_3$) marked by letter A. Secondly, σ_1 and σ_2 were increased step by step until approaching the virginal stress state marked by B. After that, this stress state was maintained for a certain time to allow the equilibrium state to be attained inside the rock specimen. Then the major principal stress was increased, simulating the stress concentration due to the decreasing of the cross-sectional section perpendicular to the vertical direction. At the same time, the minimum principal stress was decreased with a small stress increment,

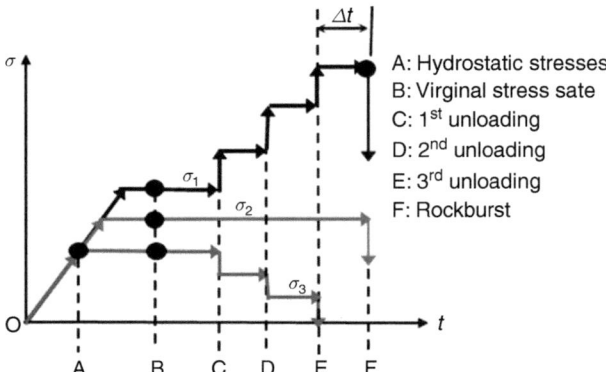

Figure 4 Stress paths for pillar bursts.

$-\Delta\sigma_3$, simulating the mining-induced unloading on the pillar's lateral direction (marked by C). It was assumed that the pillar-like specimen undergoes three cycles of the loading/unloading (denoted by C, D, and E) and the pillar burst occurs at the third cycle (marked by E, the critical state). The pillar burst also exhibits a delayed burst mechanism with a time lag by Δt.

Pillars were designed and remained when using the pillar mining method in underground coal mines. Progressively reduction of the cross-sectional area during face stoping can result in the concentration of stress in the vertical direction while setting the lateral face free, leading to the rockburst event. Pillar design plays a significant role in these events, *i.e.* increase in pillar size will effectively limit the transfer of abutment stresses to the longwall face and decrease the rockburst potential (Barton *et al.*, 1992).

Actually, the stress paths for pillar bursts are closely related to the mine design and more complex than that for the former two types of rockbursts. The methodology for the design of a stress path for the pillar burst will be dealt with in detail late in the document, along with the pillar burst criterion.

3.1.4 Experimental results for strainburst

Up to now, more than 300 rockburst experiments have been conducted employing the strainburst testing machine with rock samples of different rock types cored from different countries, including Italy, Canada, Iran, Singapore, and China. As an example, rockburst experiments on the granite from Laizhou (Shandong Province, eastern China) are presented. The cored rock blocks were machined into cubical specimens with a dimension of 150×60×30 mm. X-ray diffraction analysis of the granite was carried out. The major minerals are quartz (27%), feldspar (68%), and clay minerals (5%) with mica as most of the content. During the test, the stress path for the instantaneous rockburst (Figure 3a) was followed.

Figure 5a shows the measured stress path for the Laizhou granite sample. The virginal *in situ* stresses (σ_1, σ_2, σ_3) are 101, 60 and 30 MPa, respectively. The sample underwent two cycles of the loading/unloading and rockburst occurred at the second unloading with the critical state of stress at 129, 59, 0 MPa, respectively. That is, the polyaxial strength for rockburst is $\sigma_1 = 129$ MPa. Figure 5b shows the AE energy rate and accumulated AE energy rate monitored during the experiment. The accumulative

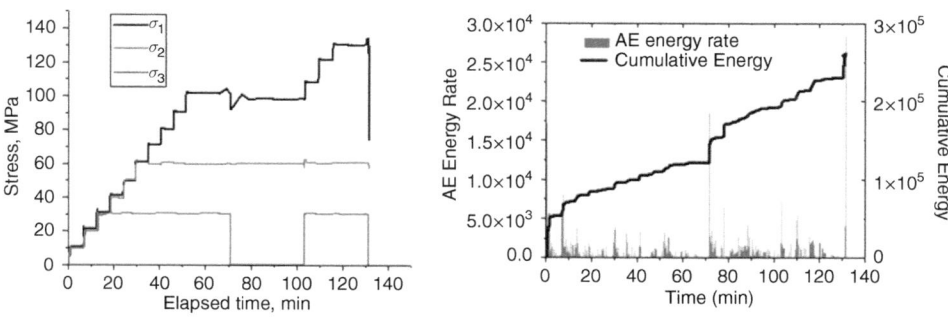

Figure 5 Pillar bursts; (a) the measured stress path, and (b) the accumulated AE energy paths.

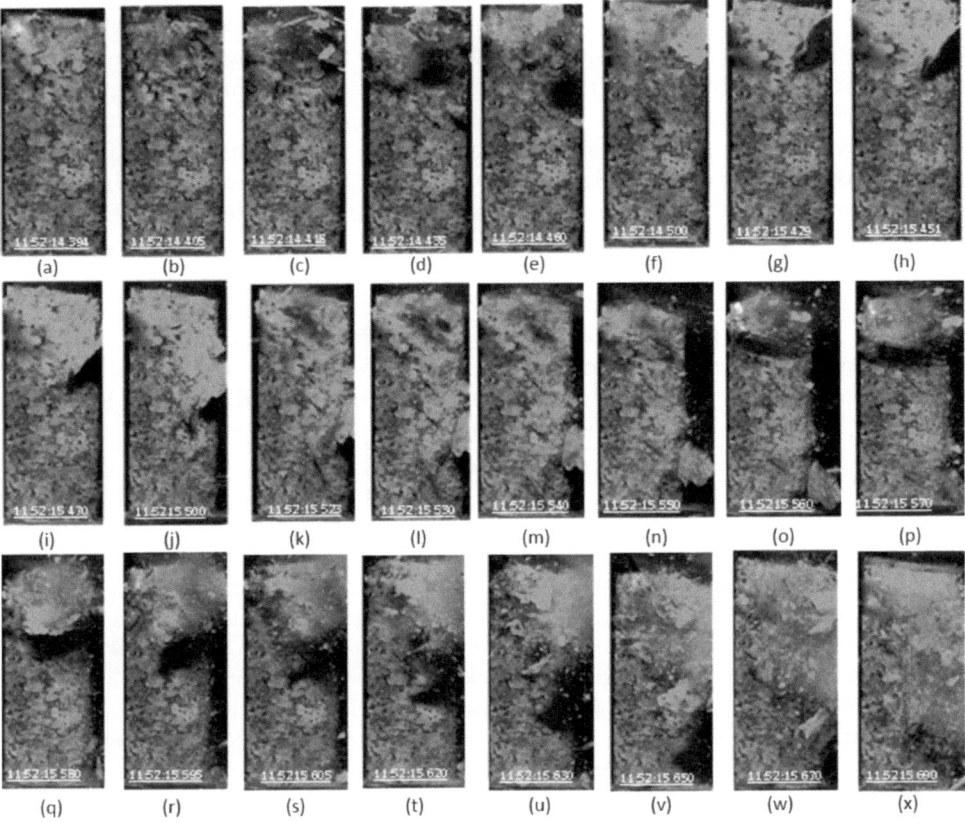

Figure 6 A selection of photographs showing the process of rockbursting in laboratory; the digits in the lower part of the photographs indicate the time scale: hour:minute:second:millisecond.

AE rate increases during the test and increases suddenly at unloading. Similarly, the AE energy rate has a high level at every unloading. Both the cumulation and distribution of AE energy have higher levels at the critical stress state, manifested by the dynamic nature of the event.

Figure 6 shows a selection of photographs of the unloading surface of the test rock specimen recorded by the high-speed camera during the rockburst experiment. Figure 6a shows the surface at the intact state. Figures 6b–f show that there were rock grains and fragments ejected from the upper area of the surface for the first time 11 microseconds after the unloading. Figures 6g–m show a large rock fracture split from the upper left area of the surface one second later. Figures 6n–x show the man-made rockburst event during the test. From these photographs we can observe the ejection and expulsion of rock fractures on the upper region of the exposed surface. The photographs in Figure 6 demonstrate that the strainburst event at the excavation boundaries can be reproduced very well in the laboratory using our innovative rockburst testing machine and designed stress path.

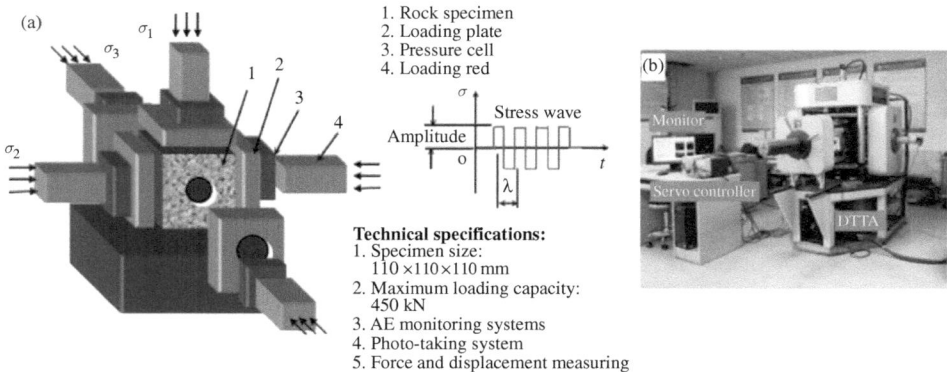

Figure 7 The impact-induced rockburst testing machine; (a) schematic of the dynamic true-tri-axial apparatus (DTTA), and (b) photograph of the testing machine and peripheral instruments.

3.2 Study on impact-induced burst

3.2.1 Impact-induced rockburst testing machine

Figure 7a shows the schematic of the main unit for the impact-induced rockburst testing machine (Patent No. ZL 200610113003.1). The main unit is actually a dynamic true-tri-axial apparatus (DTTA) which can accommodate a cubical specimen of 110×110×110 mm with a tunnel-like hole inside. Each of the loading devices for DTTA consists of a loading plate, a pressure cell, and a loading rod, and all the loading devices can produce the wave-formed dynamic loads independently in the three perpendicular directions. As reviewed above, a rockburst event at the excavation surface can also be triggered by a remote seismic source such as blasting, a roof fall, and adjacent caving, etc. The DTTA, developed by Professor He, was designed for creating the dynamic stress states analogous to those in real situations. Figure 7b shows a photograph of the overview of the impact-induced testing machine including the main unit and the peripheral controlling and monitoring instruments, which are the servo-controlled stress wave loading device (beyond the figure), the data monitoring system, and the imaging system.

3.2.2 Stress paths for impact-induced bursts

The DTTA is capable of producing the dynamic stress wave in any, or all, of the principal directions, providing flexibilities for the investigator to design different the static–dynamic combinations of the principal stresses (σ_1, σ_2, σ_3). Figure 8 shows a typical stress path for the impact-induced rockbursts in which σ_1 is designed as a squared-formed stress wave and σ_2 and σ_3 are stationary.

This stress path is designed to simulate caving or blasting-induced impact on the country rock mass. With the programmed computer code, the servo-controlled dynamic loading device of the DTTA can simulate stress waves generated by different sources such as by site excavation, blasting, caving, earthquakes, mechanical

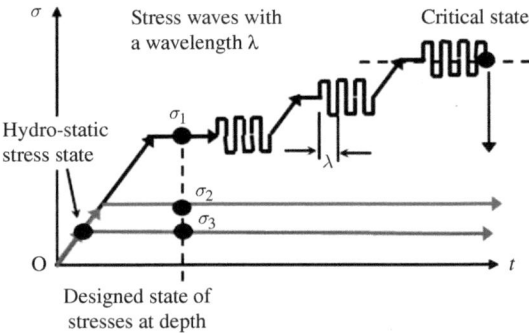

Figure 8 Stress paths for impact-induced rockbursts.

tunneling, etc. At present, sixteen kinds of the stress wave forms can be implemented in the testing machine. These are the ramp wave; sine wave; triangle wave; sawtooth wave; square wave; white noise; Gaussian noise; cycle random noise; ramp and circular wave; ramp and noise wave; circular and noise wave; ramp and circular and noise wave; loading single pulse; uninstall single pulse; Laplace pulse; and uninstall Laplace pulse.

3.2.3 Results

Employing the impact-induced rockburst testing machine, more than thirty experiments on the tunnel-like model were accomplished including the sandstone model, the limestone model, mudstone model, artificial-analogue model, and granite model. The experimental results regarding the granite model are presented here as an illustrative case. The tested granite tunnel-like model has a dimension of 110×110×110 mm with a tunnel-like hole inside. The routine analysis and tests are also performed on the granite (description is omitted) and a UCS of 68 MPa is used as a strength reference here for the model. The measured stress states and the loading path are shown in Figure 9. At the stationary loading stage, the stresses (σ_1, σ_2, σ_3) were increased proportionally and slowly to attain the hydrostatic stress state firstly, and then attain the virginal stress state of the simulated prototype of 20.7, 4.3 and 2.5 MPa, respectively, as was done in the strainburst experiment.

During the dynamic loading process (Figure 9), σ_1 (vertical direction) was modulated as a stress wave with an amplitude of 0.1 mm and frequency of 0.5 Hz, and σ_2, σ_3 are stationary. During the first two episodes of the applied stress waves, no rock fracturing or ejection occurred. In the third episode of the stress wave (at an averaged stress level of 26.6 MPa), ejection of rock grains took place. During the fourth episode of the wave loading (at an average stress level of 29.2 MPa, the critical stress for σ_1), rockburst occurred. Figure 10 shows photographs showing the rockburst process. Figures 10a and 10b show rock splitting on the left side wall and floor. Figures 10c and 10d shows the violent ejection of the rock fractures from the tunnel face during the fourth wave loading and Figures 10e and 10f show the final state of the face after the event.

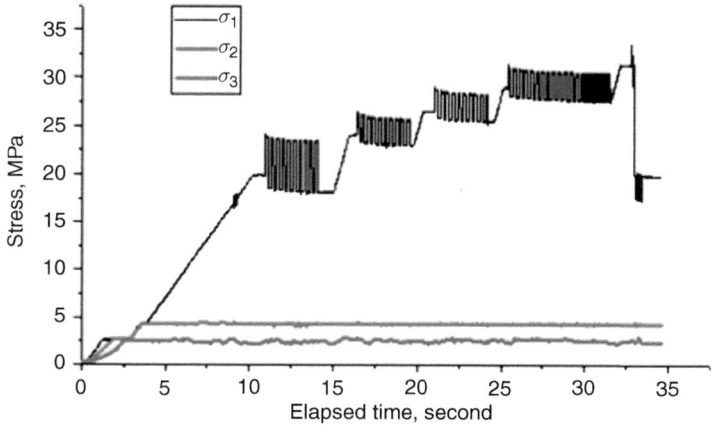

Figure 9 The measured stress paths in the impact-induced rockburst experiment.

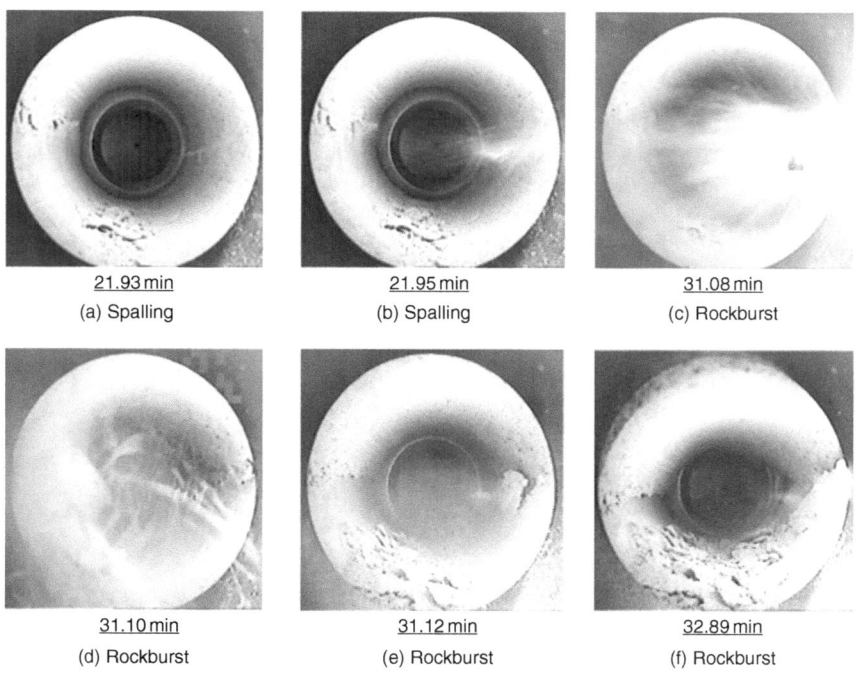

21.93 min
(a) Spalling

21.95 min
(b) Spalling

31.08 min
(c) Rockburst

31.10 min
(d) Rockburst

31.12 min
(e) Rockburst

32.89 min
(f) Rockburst

Figure 10 Photographs showing the process of impact-induced rockburst.

4 ROCKBURST CRITERION

Research on empirical equations for the determination of stress-strength behavior of a rock mass, including the discontinuity pattern, has been a very attractive topic in rock engineering, and numerous empirical equations have been proposed in the literature. Hoek and Brown introduced the most popular empirical approach to determine the strength of rock materials and rock masses (Hoek & Brown, 1980). The proposed approach by Hoek and Brown, known as the Hoek–Brown (H–B) empirical failure criterion, has been in the center of rock mechanics practice since 1980 (Sonmez & Gokceoglu, 2006). This section presents the empirical rockburst criteria based on insight into the rockburst mechanisms obtained in the rockburst experiments introduced above.

4.1 Study on impact-induced burst

The rockburst criteria for the three types of the strainbursts (see Figure 1), viz. the instantaneous rockbursts, the delayed rockbusts and the pillar bursts, are presented in this subsection with references to the H–B criterion.

4.1.1 Instantaneous rockburst criterion

Figure 11a shows the instantaneous rockburst criterion and the bursting path in the $\sigma_1-\sigma_3$ space. Point A represents the virginal stress state $(\sigma_1, \sigma_2, \sigma_3)$. σ_c and σ_c' are the UCS and the long-term peak strength for the country rocks, respectively. Instantaneous rockburst occurs under the condition where rapid release of the minimum principal stress, σ_3, and the maximum principal stress, σ_{1c}, is larger than the UCS, σ_c. These theoretical analyses were verified by the rockburst experiments (He et al., 2007, 2010, 2012a, 2012b). Any point falling into the area between the H–B curve and the UCS (marked by Zone 1) represents the stress state that may initiate a rockburst event.

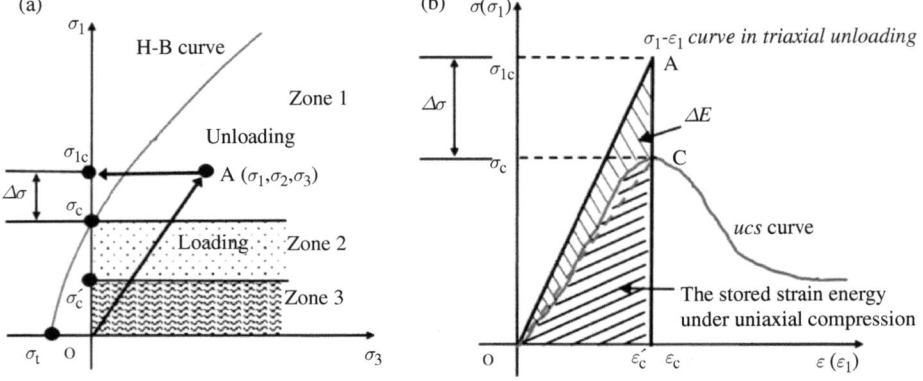

Figure 11 (a) The instantaneous rockburst criterion, and (b) energy relation between the uni-axial compression and true-tri-axial compression by suddenly unloading the confining pressure.

The experimentally obtained knowledge on the strainbursts leads to the following theoretical formulation of the energy relationship between the uni-axial compressive tests and the true-tri-axial unloading tests (Figure 11b). The blue line is the complete stress and strain curve of $\sigma-\varepsilon$ typically found in the UCS tests. The right-angled triangle drawn in a black line is an idealized curve of $\sigma_1-\varepsilon_1$ found in true-tri-axial unloading tests. The shaded area below the $\sigma-\varepsilon$ curve represents the energy consumed during the uni-axial loading process in the pre-peak region. The area of the triangle minus this area represents the energy stored in the country rocks (shaded area between the $\sigma_1-\varepsilon_1$ curve and $\sigma-\varepsilon$ curve) which is releasable during the dynamic event. The releasable energy is achieved due to the high level of σ_1 under the confinement at depth where the polyaxial strength σ1c is much higher than the UCS. The amount of the releasable energy (or seismic energy) could be estimated by the area encompassed by the triangle $\triangle$OAC in Figure 11b, and be expressed mathematically by:

$$\Delta E = \int_0^{\varepsilon_{1c}} \sigma_1 d\varepsilon_1 - \int_0^{\varepsilon_c} \sigma d\varepsilon \approx area(\triangle OAC) \tag{1}$$

Where ΔE is the releasable energy; σ_{1c} the polyaxial strength obtained in the rockburst tests; σ_c the UCS; ε_c the peak strain at the peak stress in the UCS test; and ε_{1c} the peak strain at σ_{1c} in the rockburst test. Thus, the generalized rockburst criterion can be written as:

$$\Delta E > 0 \tag{2}$$

4.1.2 Delayed rockburst criterion

Figure 12a shows the delayed rockburst criterion and the bursting path in the $\sigma_1-\sigma_3$ space. Point B represents the virginal stress state (σ_1, σ_2, σ_3). Under this condition, the rockburst may not occur at the unloading, σ_3; as a result σ1 is lower than the UCS, σ_c.

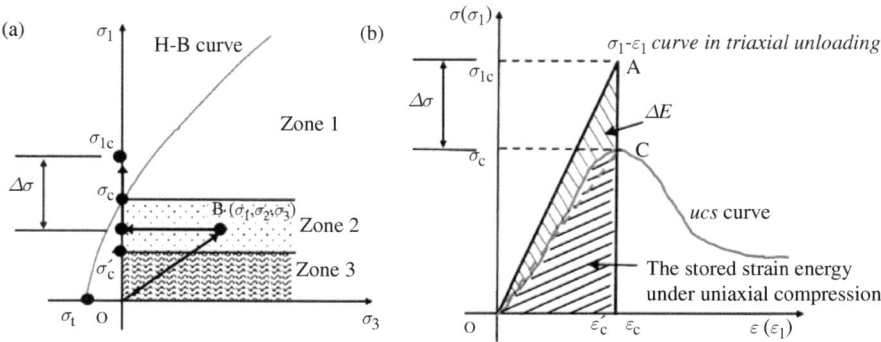

Figure 12 (a) The delayed rockburst criterion, and (b) energy relation between the uni-axial compression and true-tri-axial compression by suddenly unloading the confining pressure.

At or near the excavation surface, rockbursts may be initiated as σ_1 increases (the tangential stress concentration due to the excavation) to a level larger than UCS and the rockburst criterion in Equation 2 is met. This kind of stress path is manifested by the rock response, such that rockburst occurs usually in a period of time after the free surface was created. The theoretical formulation of the energy relationship between the uni-axial compressive tests and the true-tri-axial unloading tests is given in Figure 12b, and the releasable energy and rockburst criterion are expressed by Equations 1 and 2.

4.2 Impact-induced rockburst criteria

The impact-induced rockburst, as classified in Section 1, consists of three types of rockbursts, viz. induced by blasting, by roof collapse, and by fault slip. Besides the rock strength, wavelength was also considered in their criteria as the dynamic nature.

4.2.1 Physical model for the impact-induced rockbursts

Figure 13 shows a tunnel of diameter D in the path of a plane seismic wave of wave length. When a seismic wave propagates in an elastic continuum, a point or particle of this continuum oscillates around its stationary or equilibrium position. A particle is prevented from flying off by the atomic bond strength which increases as the particle moves away from its equilibrium position. On the other hand, a particle on the free boundary of an opening may fly off into the opening if the atomic bond is broken due to excessive acceleration of this particle. Similarly, rock blocks on the free boundary of an opening which are separated from the surrounding rock mass by joints and fractures can be carried or ejected into the opening by seismic waves. This seismic wave is assumed to be sinusoidal with positive (compressive) and negative (tensile) pulse. The tunnel will undergo a dynamic stress state under the condition that the tunnel diameter is much less than the wave length (McGarr *et al.*, 1981; Roberts & Brummer, 1988), that is:

$$D\lambda^{-1} \ll 1 \tag{3}$$

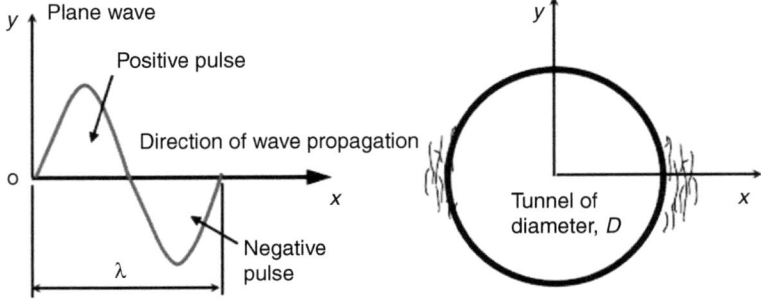

Figure 13 Cylindrical tunnel in the path of a propagating plane seismic wave (Xiaoping Yi, 1993).

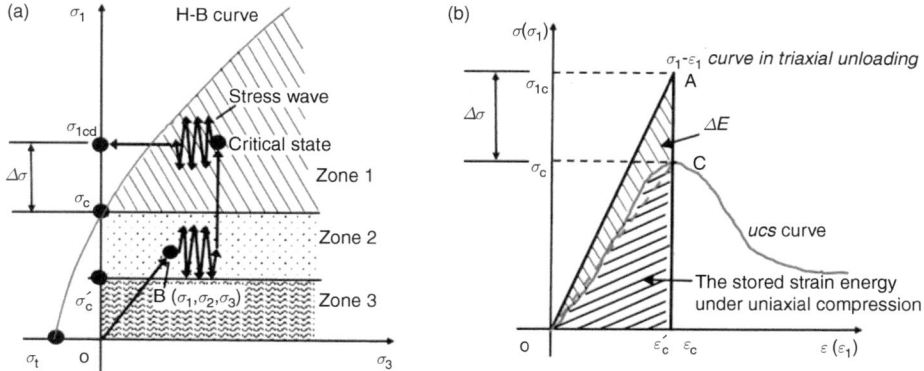

Figure 14 (a) Failure criterion for rockburst induced by blasting or excavation, and (b) energy relation between the uni-axial compression and true-tri-axial compression by suddenly unloading the confining pressure.

4.2.2 Delayed rockburst criterion

Figure 14a shows the criterion for impact-induced rockbursts and the bursting path in the σ_1–σ_3 space. Point B represents the virginal stress state (σ_1, σ_2, σ_3) at which the stress wave oscillates. Different from the static loading illustrated in Figures 11 and 14, the dynamic loading in Figure 15 is a rate-dependent processes and can also result in an increase in the tangential stress at the excavation surface, *i.e.* the increase in the major principal stress. The dynamic rockburst strength is hereby termed as σ_{1d} in comparison with the static rockburst strength σ_1 in the above section. The released energy ΔE can also be estimated using the method illustrated in Figure 14b and Equation 1 by substituting σ_1 with σ_{1d}. The necessary and sufficient condition for initiating the induced-types of rockbusts should follow the energy criterion expressed by Equation 2 and the dynamical condition ($D\lambda^{-1} \ll 1$) expressed by Equation 3 all together:

$$\Delta E > 0 \quad \text{and} \quad D\lambda^{-1} \ll 1 \tag{4}$$

As to the specific rockburst criterion for induced rockbursts, the differences lie in the configuration of the dynamical stress waves. That is, no matter what the type of seismic source – blasting, caving, roof collapse, and fault slip – it can be represented by adjusting the magnitude, the wave length, and the wave type. Therefore, a specific rockburst criterion could be formulated.

5 SUMMARY

A new concept was introduced into the classification of rockbursts according to their initiation mechanisms and results of indoor experimental methods. This classification of rockbursts into two general categories, viz. stress-induced rockbursts and impact-induced rockbursts, contributes a great deal to the development of state-of-the-art rockburst testing machines, and the establishment of a framework for the systematic rockburst studies undertaken by the author's research team.

The innovative work in developing the 'strainburst testing machine' and 'impact-induced rockburst testing machine' creates a new era at which in-depth and comprehensive investigations into the rockburst phenomena can be made in the laboratory. The artificially produced rockburst phenomena of different types in the novel testing machines can provide us with a convenient way of observing the rockburst process with various advanced monitoring techniques, which was originally unlikely in field.

The stress paths proposed in this paper, corresponding to the rockbursts in the new classification systems, are closely analogous to those at, or near, the surface of underground excavations. The corresponding real rockburst phenomena can be reproduced in the laboratory in the new testing machines using these stress paths and the related testing procedures.

Rockburst criteria for the stress-induced and impact-induced rockbursts are proposed based on experimental investigations both in the laboratory and in the field. The rockburst criteria take into account both the static and dynamic stress states, and are have general applications in rockburst prediction and rockburst control design.

REFERENCES

Blacke W., 1972. Rockburst mechanics. Quarterly of the Colorado School of Mines. 67: 1–64.

Brady B. H. G., Brown E. T., 2005. Rock mechanics for underground mining. Kluwer Academic Publishers, New York, Boston, Dordrecht, London, Moscow.

Barton T. M, Campoli A., Guana M., 1992. Rock mechanics research decreases longwall bump potential at a Southern Appalachian coal mine. Min. Eng. April: 347–352.

Cook N. G. W., 1965. A note on rockburst considered as a problem of stability. J. South Afr. Int. Min. Metall. 65: 437–446.

Crawford A. M., Wylie D. A., 1987. A modified multiple failure state triaxial testing method. In: Proceedings of 28th US symposium on rock mechanics, Tucson, 133–140.

Chang C., Haimson B. C., 2000. True triaxial strength and deformability of the German Continental Deep Drilling Program (KTB) deep hole amphibolite. J. Geophys. Res. 105 (B8): 1 8999–9013.

Chen J. T., Feng X. T., 2006. True triaxial experimental study on rock with high geostress. Chin. J. Rock Mech. Eng. 25: 1537–1543.

Cho S. H., Ogata Y., Kaneko K., 2005. A method for estimating the strength properties of a granitic rock subjected to dynamic loading. Int. J. Rock Mech. Min. Sci. 42(4): 561–568.

Corbett G. R., 1996. The Development of a Coal Mine Portable Microseismic Monitoring System for the Study of Rock Gas Outbursts in the Sydney Coal Field, Nova Scotia. Ph.D. Dissertation, McGill University, Montreal, March.

Ghose K, Rae H. Q., 1988. Theory of rock bursts and their classification, Chapter 1, I. M. Petukhov, Lenningrad USSR Rock bursts: Global Experiences. Papers Presented at The Fifth Plenary Scientific Session of Working Group on Rock Bursts of International Bureau of Strata Mechanics, February. Pp. 3–10, Edited by Assay.

Gill D. E., Aubertin M., Simon R., 1993. A practical engineering approach to the evaluation of rockburst potential. Rockbursts and Seismicity in Mines, Young (ed.), Balkema, 63–68.

Hasegawa H. S., Wetmiller R. J., Gendzwill D. J., 1989. Induced seismicity in mines in Canada-An overview. PAGEOPH, Vol. 129. Nos. 3/4.

Haimson B. C., 2006. True triaxial stresses and the brittle fracture of rock. Pure Appl. Geophy. 163: 1101–1130.

Hoek E., Brown E. T., 1980. Underground excavations in rock. The Institution of Mining Metallurgy, London, 527.

He M. C., Miao J. L., Li D. J., Wang C. G., 2007. Experimental study on rockburst processes of granite specimen at great depth. Chin. J. Rock Mech. Eng. 26(5): 865–876.

He Manchao, Yang Guoxing, Miao Jinli, et al., 2009. Classification and research methods of rockburst experimental fragments. Chin. J Rock Mech. Eng. 29(8): 1521–1529.

He M.C., Miao J. L., Feng J. L., 2010. Rock burst process of limestone and its acoustic emission characteristics under true-triaxial unloading conditions. Int. J. Rock Mech. Min. Sci. 47: 286–298.

He M.C., Jia X. N., Coli M., Livi E., Sousa L., 2012a. Experimental study of rockbursts in underground quarrying of Carrara marble. Int. J. Rock Mech. Min. Sci. 52: 1–8.

He M. C., Xia H., Jia X., Gong W., Zhao F., Liang K., 2012b. Studies on classification, criteria and control of rockbursts. J. Rock Mech. Geotech. Eng. 4(2): 97–114.

He Manchao, L. Ribeiro e Sousa, Tiago Miranda, Zhu Gualong. Rockburst laboratory tests database – Application of data mining techniques. Eng. Geol. 185(5) (February): 116–130

Jaeger J. C., Cook N. G. W., Zimmerman R. W., 2007. Fundamentals of rock mechanics. MA02148-5020 USA: Blackwell Publishing, 30–40.

Jiang Yaodong, Zhao Yixin, He Manchao, Peng Suping, 2007. Investigation on mechanism of coal mine bumps based on mesoscopic experiments. Chin. J. Rock Mech. Eng. 26(5): 901–907.

Kuhnt W., Knoll P., Grosser H., Behrens H. -J., 1989. Seismological models for mining-induced seismic events. PAGEOPH, Vol. 129, Nos. 3/4.

Li X., Ma C., 2004. Experimental study of dynamic response and failure behavior of rock under coupled static-dynamic loading// Aoki O. In: Proceedings of the ISRM international symposium 3rd ARMS. Rotterdam: Mill Press, 891–895.

Lee H., Haimson B. C., 2011. True triaxial strength, deformability, and brittle failure of granodiorite from the San Andreas Fault Observatory at Depth. Int. J. Rock Mech. Min. Sci. 48: 1199–1207.

Mogi K., 2007. Experimental rock mechanics. London, UK: Taylor & Francis Group.

McMahon T., 1988. Rockburst research and the Coeur d'Alene District. U. S. B. M., IC 9186.

Morrison R. G. K., 1942. Report on the rockburst situation in Ontario mines. Trans Can. Inst. Min. and Met. 45: 225–272.

Mogi K., 1967. Effect of intermediate principal stress on rock failure. J. Geophys. Res. 72: 5117–5131.

McGarr A., Green R. W. E., Spottiswood S. M., 1981. Strong ground motion of mine tremors: some implications for near-source motion parameters. Bull. Seismol. Soc. Am. 71(1): 295–319.

Pettitt W. S, King M. S., 2004. Acoustic emission and velocities associated with the formation of sets of parallel fractures in sandstones. Int. J. Rock Mech. Min. Sci. 41(sup.1): 1–6.

P. K. Kaiser et al., 1995. Rockburst research handbook, Chapter 2 Ground support. Canadian rockburst research program 1990–1995.

Qian Q. H., 2011. Preface in the Exploring into the rockburst mechanisms. In: he 51st Proceedings of the new concepts and ideas. Beijing: China Science Press, 1–3.

Qi Q., Chen, S., Wang H., Mao D., Wang Y., 2003. Study on the relations among coal bump, rockburst and mining tremor with numerical simulation. Chin. J. Rock Mech. Eng. 22(11): 1852–1858.

Roberts M. K. C., Brummer R. K., 1988. Support requirements in rockburst conditions. J. South Afr. Int. Min. Metall. 88(3): 97–104.

Ryder J. A. 1988. Excess shear stresses in the assessment of geologically hazardous situations. J. South Afr. Int. Min. Metall. 88(1): 27–39.

Richard Simon. 1999. Analysis of Fault-slip Mechanisms in Hard Rock Mining. Ph.D Dissertation, McGill University, Montreal Canada, January.

Stillborg B., Rockbolt and cablebolt tensile testing across a joint. Unpublished report, Mining and Geotechnical Consultants, Lulea, Sweden, 19.

Salamon M. D. G., 1970. Stability, instability and design of pillar workings. Int. J. Rock Mech. Min. Sci. Geomech. 7(6): 613–631.

Singh S. P., 1989. Classification of mine workings according to their rockburst proneness. Min. Sci. Tech, 8(3): 253–262.

Singh M., Raj A., Singh B., 2011. Modified Mohr–Coulomb criterion for non-linear triaxial and polyaxial strength of intact rock. Int. J. Rock Mech. Min. Sci. 48: 546–555.

Sonmez H., Gokceoglu C., 2006. Discussion of the paper by E. Hoek and M. S. Diederichs "Empirical estimation of rock mass modulus". Int. J. Rock Mech. Min. Sci. 43: 671–676.

Tang C. A., Kaiser P. K., 1998. Numerical simulation of cumulative damage and seismic energy release during brittle rock failure-Part I: Fundamentals. Int. J. Rock Mech. Min. Sci. 35(2): 113–121.

Xiaoping Yi, 1993. Dynamic response and design of support elements in rockburst conditions. Ph.D. thesis, Queen's University, Kingston, Ontario, Canada, February.

Zhou X. P., Qian Q. H., Yang H. P., 2008. Strength criteria of deep rock mass. Chin. J. Rock Mech. Eng. 27(1): 117–123.

Chapter 4

Laboratory acoustic emission study review

X.L. Lei

Geological Survey of Japan, National Institute of Advanced Industrial Science and Technology (AIST), Tsukuba, Ibaraki, Japan

Abstract: Since the discovery of the Kaiser effect in 1950 and the similarity in size distribution of earthquakes and acoustic emissions (AE) in the 1960s, many laboratory studies have been motivated by the need to provide tools for the estimation of in-situ stress, the understanding of pre-failure damaging and fracturing mechanism in brittle rocks, and the prediction of mining failures and natural earthquakes. This chapter aims to draw an outline of progress that has been made in AE technology and laboratory AE study and to highlight some significant and extensive achievements in rock mechanics and engineering during the last 20–30 years, such as aspects related to the pre-failure damage evolution, fault nucleation and growth in rocks and discuss factors governing these processes.

1 INTRODUCTION

Acoustic emission (AE) is an elastic wave of ultrasonic to sonic frequencies radiated by rapid cracking and friction in solid materials and rocks. AE technology, in which AE signal is monitored, has three major applications. As a tool of non-destructive inspection, it has a long history and has been applied in numerous areas including material sciences, medical sciences, and engineering fields. In rock mechanics, AE studies in laboratory scales and microseismicity in mine scales were motivated by a desire to estimate in-situ stress, to predict rock bursts, to monitor hydrofracturing, to examine damage around tunnels and boreholes, to develop new mining techniques, and so on. In earthquake seismology, AE is treated as an analogue of earthquake and has been intensively studied in laboratory (see recent review in (Lei & Ma, 2014)).

This chapter aims to draw an outline of laboratory AE studies that address rock mechanic problems with a focus on significant and extensive achievements in rock mechanics and engineering during the last 20–30 years. First, a brief outline of progress that has been made in AE technology and laboratory AE study is given. Then I describe several commonly accepted laws, in which the scale-independent similarities are involved. In sections 4 and 5, I summarize aspects relating to pre-failure damage evolution, fault nucleation and growth in brittle rocks, and discuss factors governing these processes. Finally, I discuss questions that remain poorly understood but might be interest issues for future works.

2 PROGRESS IN AE TECHNOLOGY

2.1 Advances in AE monitoring and data processing

During the last five decades, AE technology, including AE monitoring technology and data processing methodology, has been developed greatly. The following is a brief outline of some key points.

Fifty years ago, only the hitting time (time at which the signal amplitude reaches a pre-defined threshold) of an AE could be recorded with a single sensor or a small number of sensors. At the same time, the rock fracture test was performed under simple loading conditions such as uniaxial compression/extension and bending test. Even so, two discoveries have greatly promoted laboratory AE studies. The first one is the Kaiser effect discovered in 1950 (Kaiser, 1950). The second one is finding of the similarity in size distribution of earthquakes and acoustic emissions in the 1960s (Mogi, 1962; Scholz, 1968c).

Source location has been an important issue from the very beginning of laboratory AE study (Mogi, 1968; Scholz, 1968a). However, limited by AE monitoring technique, only a small number of events could be located. Later, a six-channel system to detect the arrival time, amplitude, and vibration direction of AE wave from PZT sensors mounted on a sample was developed in USGS laboratory (Byerlee & Lockner, 1977). With this system, the hypocenter of a large number of AEs could be determined (Byerlee & Lockner, 1977; Lockner et al., 1991a; Lockner et al., 1992). Developments in transient memory technique in the 1970s through to the 1980s led to the ability to make a digital recording of the full waveform of an AE at multi channels (Nishizawa et al., 1984; Sondergeld & Estey, 1981; Yanagidani et al., 1985). Hypocenter location was then improved greatly by the use of more precise arrival times pick-up using AR model to represent AE signal and background noise and AIC (Akaike Information Criteria, Akaike, 1973, see reprint (Akaike, 1998)) to find the optimal models (see 2.2 for details). With detailed AE hypocenter data, the localization of AE hypocenters in "homogeneous" rock samples, under various loading conditions particularly creeping, was reported by a number of groups (Hirata et al., 1987; Lei et al., 1992; Lockner & Byerlee, 1980; Nishizawa et al., 1982; Yanagidani et al., 1985). The term "homogeneous" indicates fine grained rocks free of mesoscale heterogeneities. Localization means the changing from random distribution to spatial clustering.

Since the 2000s, AE are usually monitored by more than 8 sensors (up to 32 sensors in some laboratories) with digital waveform recording at up to a 200 MHz sampling rate and up to a 16 bit A/D resolution. The dead time of a recording in event mode is sufficiently short and continuous recording is also possible by use of very large amounts of on-board memory. The waveform of most events can be captured with multiple channels, even for AE burst, in which the AE rate may reach several thousand a second, commonly occurred during the period of fracturing nucleation and dynamic failure. Rock fracture experiments can be performed under triaxial compression conditions with controlled fluid injection and pore pressure (Lei et al., 2011; Mayr et al., 2011; Stanchits et al., 2011). AE hypocenters can be determined with a location error of a few mms (Benson et al., 2007; Lei et al., 2000b, Lei et al., 2000c; Stanchits et al., 2006). Multi-channel waveform data allowed the determination of focal mechanisms and sources parameters of AE from the polarization of the first motions for individual events or a group of events *e.g.* (Lei *et al.*, 2000b). Moment tensor inversion has been used in laboratory and in-site fracturing tests, and stick-slip of fault (Aker *et al.*, 2014; Chang & Lee, 2004; Enoki & Kishi, 1988, Kwiatek *et al.*, 2014, Manthei *et al.*, 2001; Ohtsu, 2008; Yuyama *et al.*, 1995).

Recent systems were utilized with large amounts of memory, allowing continuous waveform recording on multiple channels and more detailed analysis of faulting nucleation, dynamic fracturing, and aftershock processes (Benson *et al.*, 2007; Lei, 2012; Stanchits *et al.*, 2006; Thompson, 2005; Thompson *et al.*, 2006). Benefitting from the fast speed of PCIe bus, cheaper waveform recorders are developed, by which AE waveforms can be digitized at a sample rate up to 20 MHz and streamed to hard disk arrays directly for continuous recording (Yoshimitsu *et al.*, 2014).

2.2 Pickup of P-wave first motion and determination of AE hypocenter

The same method used for the automatic measurement of the arrival times of seismic signals (Yokota *et al.*, 1981) has been used in the pickup of P-wave first motion of AE signals (Satoh *et al.*, 1987). In the method, autoregressive (AR) model is used to represent a time series x_i (i = 1,..N):

$$x_i = \sum_{j=1}^{M} a_j x_{i-j} + u_i \qquad (i = 1, ..., N) \tag{1}$$

where, M and a_j (*j* = 1, ..., *M)* are the order and coefficients of the AR model, respectively, and u_i is the residual error assuming to be white noise. Parameters associated with each model are estimated using the maximum likelihood method. As a practical application of the method, the Akaike's Information Criterion (AIC), (Akaike, 1973, see reprint (Akaike, 1998)) is introduced:

$$AIC = -2\ln(l_{\max}) + 2N_p \tag{2}$$

where, l_{max} is the maximum likelihood and N_p is the number of parameters.

There are several algorithms for the first arrival determination based on AR model and AIC (Yokota *et al.*, 1981). For AE signals, two algorithm termed MUPEO and MEPET are found being practically useful (Satoh *et al.*, 1987). In MUPEO, the so called F-model (of an order of L_f), which is obtained for a small section at the head of the noise section, is applied to the two sections separated at time-*k* of the whole record. The AIC at a given point *k* is given by:

$$AIC_k = k\ln(\widehat{\sigma}_F{}^2) + (N - k)\ln(\widehat{\sigma}_{FS}{}^2) + N\ln(2\pi) + N + 2(L_F + 2) \tag{3}$$

Where $\widehat{\sigma}_F{}^2$ and $\widehat{\sigma}_{FS}{}^2$ are the variances of prediction error for the first and second sections, respectively. In MUPET, the so called S-model (of and order of L_s), which is obtained for a small section at the tail of the signal section, is applied to the second section. The AIC at a given point *k* is then given by:

$$AIC_k = k\ln(\widehat{\sigma}_F{}^2) + (N - k)\ln(\widehat{\sigma}_S{}^2) + N\ln(2\pi) + N + 2(L_F + L_S + 2) \tag{4}$$

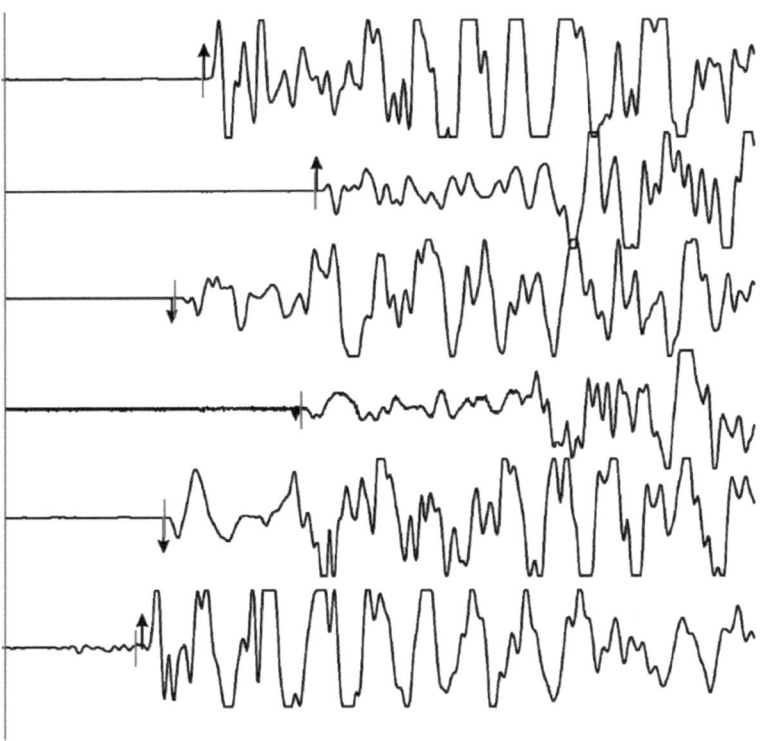

Figure 1 Advanced P wave first motion determination utilizing autoregressive (AR) model and Akaike information Cretia (AIC). Automatically determined arrival time and theoretic time indicated as black arrow and red line, respectively.

The first arrival time is identified by the minimum *AIC*. A multi-step approach is found especially effective. The procedure consists of two steps. At first, the MUPEO method is applied to the entire record for a preliminary estimation of arrival time. Next, both MUPEO and MUPET are applied to a shorter time window centered at the arrival time estimation by the first step, and the earlier arrival time is used as the final estimation.

AE hypocenters were determined automatically by using the first arrival time data of P-waves and measured P velocities during the test. For most rocks anisotropic velocities measured at different stages of deformation are required for better location precision. The first arrivals of a signal to noise ratio lower than given threshold are ruled out. As a practical approach, a loop between pickup of arrival times and source location is very powerful. Once a primary hypocenter is determined, pick up of P arrival time for channels of larger residual and channels been ruled out can be reprocessed within a relatively short window centered at the theoretical arrival times. Then the hypocenter is refined with the renewed arrival times. This procedure can be iterated for several times until the best resolution is found.

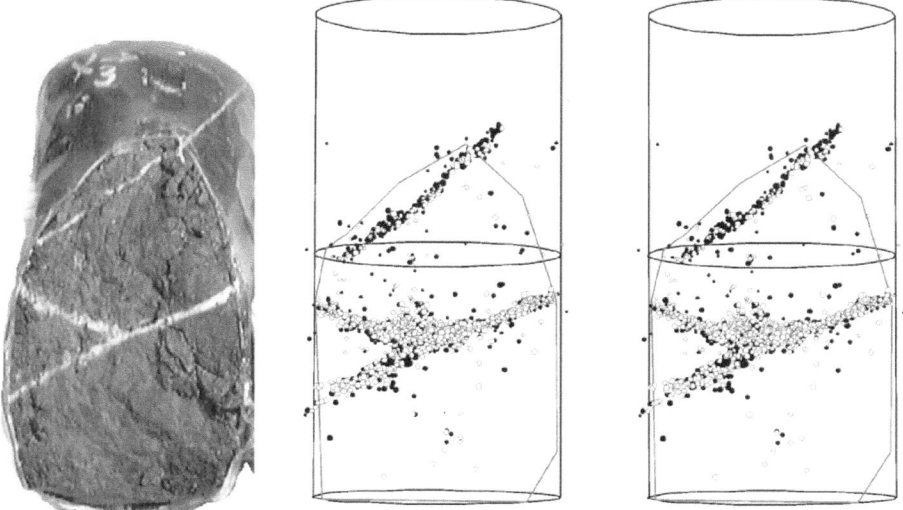

Figure 2 A comparison of photograph of a fractured rock sample and stereo plots of AE hypocenters, which are determined with an error less than 2–3 mm. White and black circles indicate shear (or shear-dominant) and tensile sources, respectively. The test sample is a mudstone containing 3 quartz veins. The mudstone demonstrates ductile fracturing behaviors, while the quartz veins show brittle fracturing, thus AE hypocenters are mainly distributed along the intersections of the veins and the final shear fracture plane (Modified from Lei *et al.*, 2000c).

2.3 Determination of source parameters and mechanism solutions of AE event

The observed AE magnitude is somewhat relative due to many unknown factors such as coupling effect and the sensitivity of AE sensor used. However, the relative magnitudes can be roughly shifted to equivalent earthquake magnitudes based on calibration using Laser Doppler velocitometer and corner frequency analysis (Lei *et al.*, 2003). Magnitude of large events with saturated waveform record can be estimated from their wave continuation time based on scaling law between magnitude and continuation time similar with that used for small earthquakes (Lei *et al.*, 2003). The magnitude of a stick-slip event as well as the main event which has split the whole sample can be estimated following the procedure presented by McGarr & Fletcher (2003). This method was used by Thompson *et al.* (2009) and Lei (2012) in determining magnitude of stick-slip event and rock failure in laboratory. Seismic moment of a fault embedded in an elastic medium is given by a function of shear modulus (G) of the host medium, displacement (u), and fault surface area (A)

$$M_O = GuA \tag{5}$$

For the case of rock sample having limited dimension under a stiff loading frame, an equivalent radius (r) for a circular asperity with elastic unloading stiffness k given by (Eshelby, 1957; Thompson *et al.*, 2009; Walsh, 1971) can be used for calculating the fault surface area.

$$r = \frac{7\pi}{16}\frac{G}{k} \tag{6}$$

Where, k can be estimated from stress-drop and displacement. At last, a moment magnitude (Mw) for an equivalent earthquake can be then calculated using the scaling relation between Mo (in dyne·cm) and Mw (Hanks & Kanamori, 1979).

$$M_W = \frac{2}{3} \log M_O - 10.07 \tag{7}$$

At the same time, it is important to note that McGarr method is basically a measure of elastic energy release in ultimate failure, which is pretty much controlled by stiffness of the loading apparatus used. It can become however large if soft apparatus is used (*i.e.* residual slip after the breakdown continues and achieves a big displacement). So, the obtained moment magnitude should be treated as an upper-bound. Smaller events are stopped by local heterogeneity within the sample, but the ultimate failure, which has split the whole sample, is unstoppable by this mechanism and continues until using up the excess stress stored in apparatus (hence moment can be however big). To make a fair comparison, the only way is to take the rupture dimension of the ultimate event as a lower-bound estimate of its size and to estimate its magnitude using the scaling relation between magnitude and fault rupture dimension:

$$M_W = a + b \log A \tag{8}$$

where a has a typical value of 4~4.5 and b is close to 1 for earthquakes (Dowrick & Rhoades, 2004).

The spectrum of AE signal recorded by PZT sensor can be corrected somehow based on calibration using laser doppler velocitometer, and thus the corner frequency and source dimension can be roughly estimated (Lei *et al.*, 2003). Recently, by use of wide band AE sensors embedded within the end-pieces, stress drop, source dimension were roughly estimated (Yoshimitsu *et al.*, 2014).

The focal mechanisms of AE can be determined from the polarization of the first motions recorded at multi sensors. In laboratory, five types of mechanism solution were distinguished and were categorized them as Type-C, Type-T, Type-S, Type-TS, and Type-TTS. Figure 3 is a schematic illustration showing the distribution of the first motions and the likely corresponding crack modes. Type-C, resulted from sudden closure of pores, has a pulling first motion at all directions. Type-C AEs are well observed in andesite and porous sandstones. In porous rocks, compaction and cataclastic flow is significant (Menéndez *et al.*, 1996), which can lead to Type-C AE events (Fortin *et al.*, 2009). Type-T, which corresponds to tensile cracking, has a pushing first motion at all directions. Type-S having quadrant distribution of P first motion is generated by a shear cracking. Type-TS event is assumed to be a slip along the crack with a tensile cracking at its end (it is also called wing crack (Ashby & Sammis, 1990)). Type-T, Type-S, and Type-TS have been observed in various rocks (*e.g.* Lei *et al.*, 2000b; Lei *et al.*, 1992; Stanchits *et al.*, 2006; Zang *et al.*, 1998). Type-TTS event shows distribution of polarization cannot be assigned to either Type-S or Type-TS cracks and can be modeled by a slip along a crack with two tensile cracks of different orientations at its ends. In other words, the combination of two Type-TS cracks (Lei *et al.*, 2000b). Under triaxial compression

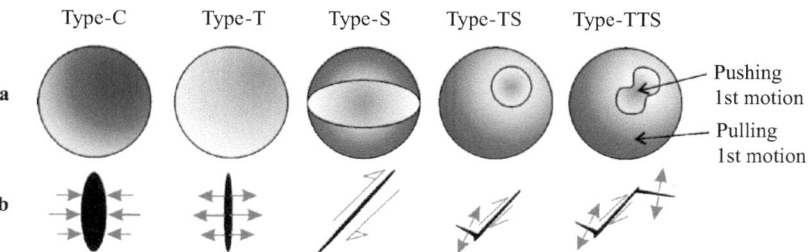

Figure 3 Five types of focal mechanism solution derived from P wave first motion direction. a) Distribution of the first P motions on an equal-area projection of the lower hemisphere of the focal sphere. b) Cracking modes corresponding to a). (modified from (Lei *et al.*, 2000b)).

conditions, both shear and tensile fracture are major microscopic fracture mechanism, and shear became more dominant as the confining pressure increased (Chang & Lee, 2004).

By using amplitude data, a full moment tensor (MT) can be estimated under the assumption of point source and a simple sources function. In laboratory condition, the recorded amplitudes should be carefully corrected for the sensor directivity (a cosine sensitivity function is normally used for P-wave type sensors) (Aker *et al.*, 2014). The moment tensor is represented as a real and symmetric 3×3 matrix, which can be decomposed into three components: 1) a double coupled (DC) shear source; 2) an isotropic (ISO) source (explosion or implosion); and 3) a compensated linear vector dipole (CLVD) source. This eigenvector-based approach was used in earthquake seismology (*e.g.* Hudson *et al.*, 1989). It is worth noting that the obtained isotropic tensor is unique, but the decomposition of the remaining deviatoric tensor into DC and CLVD terms is not. Consequently, there are several decomposition methods in the literature (Chapman & Godbee, 2012).

2.4 Estimation of in-situ stress

Following the discovery of the Kaiser effect: the production of AE in stressed rocks was somehow stress-history dependent (Kaiser, 1950), there are many laboratory studies were done for examining the Kaiser effect. The Kaiser effect is resulted from irreversible damages (micro-cracks) accumulated in the rock and is affected by many factors, such as rock types, loading rate and duration, interval, stress relaxation and strain recovery, stress level, temperature-history, water saturation, interactions between principal stresses, scales, and so on. The uniaxial loading method (ULM) was commonly used for determining in situ stress using the Kaiser effect and there are some successful cases. However, it was shown that the ULM cannot determine the previous stress imposed on a sample in the laboratory. It is thus suggested mechanism for the *in situ* Kaiser effect is AE produced by reclosure of relaxation cracks that opened as a result of stored strain when the *in situ* stress was removed by coring (Holcomb, 1993).

A detailed review has been presented in (Lavrov, 2003). In following, I will focus some new developments in recent years. In the last ten years, the Western Australian School of Mines (WASM) has researched, developed and successfully applied an in situ

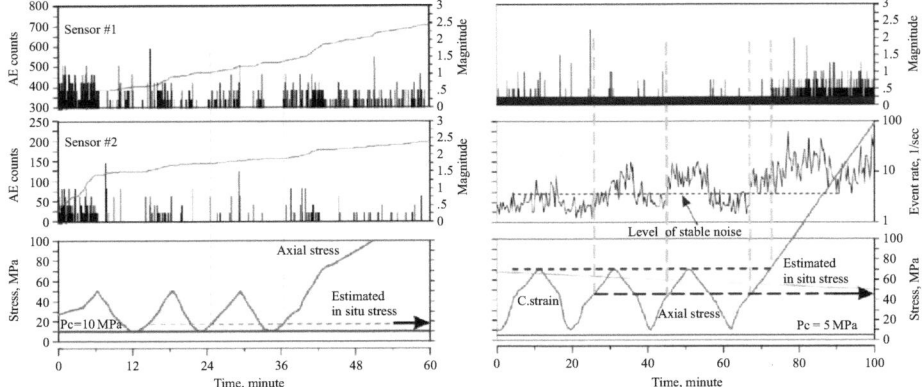

Figure 4 An example estimating in situ stress using AE data during cycled loading under triaxial compression. Left plot shows results of a metamorphic rock sample retrieved via coring at a depth of ~856 m at the Bendigo gold mine, Victoria Australia (Modified from Villaescusa *et al.*, 2009 with new data). Right plot shows results of a case, in which the AE activity is very low and the AE data contains stable noise as a result of lowered triggering threshold.

stress measurement technique known as the WASMAE method using oriented core from exploration, (Villaescusa *et al.*, 2002; Villaescusa *et al.*, 2003). The method is based upon the monitoring of intact rock specimen micro-structure mobilization under uniaxial loading (Seto *et al.*, 1997; Villaescusa *et al.*, 2002) or triaxial loading (Villaescusa *et al.*, 2009). It is possible to estimate the previous stress by using the AE signature in the second and/or third reloadings, even when the Kaiser effect is not clear in the first reloading due to noise associated with crack closure or compaction. Figure 4 shows 2 examples. In the right plot, the AE record system used a triggering threshold lower than the background noise, because AE activity at low stress level is very low and insufficient to observe the Kaiser effect if a greater trigger threshold was used. Fortunately, the background noise is time-stable and thus it is possible to explore the AE rate changes, which result in a good estimation of the *in situ* stress.

3 SCALE-INDEPENDENT AND FRACTAL PROPERTIES OF AE

AE shows scale-independent properties in many aspects through the full scale range, which are involved in several power laws. The common accepted laws are the Gutenberg and Richter (GR) relation in size distribution, modified Omori's law in aftershock decay, the accelerated moment release (AMR) preceding failure, and fractal distribution of AE hypocenter.

3.1 Gutenberg and Richter relation in frequency-magnitude distribution and seismic *b*-value

The first commonly accepted power law should be the Gutenberg and Richter (GR) relation frequency-magnitude distribution (Gutenberg & Richter, 1944):

$$\log_{10} N = a - bM \tag{9}$$

where N is the number of events of magnitude M or greater, and a and b are constants. The GR relation works well for AEs in rocks in laboratory scales (*e.g.* Lei, 2003; Liakopoulou-Morris *et al.*, 1994; Mogi, 1962; Scholz, 1968b) and mining scales (*e.g.* Kwiatek *et al.*, 2010; Yoshimitsu *et al.*, 2014). The seismic moment Mo is linked with moment magnitude Mw by equation (7), therefore, the GR relation is also a power law of moment:

$$N(M_O) = AM_O^{-2b/3} \tag{10}$$

The parameter b, commonly referred to as the "b-value", has a global mean value of 1.0. The GR relationship works for all magnitudes above a lower end cut-off magnitude due to the detecting ability of seismic stations or the existing of a fault break-down zone.

Beginning with Mogi's work published in 1962 (Mogi, 1962), a lot of laboratory works have been motivated by the expectation that precursory changes in b-value could have resulted from stress change and thus could be used for earthquake prediction and failure prediction in mines. Indeed, laboratory studies of AE events have consistently shown a decrease in b-value with increasing stress during the deformation of intact samples containing pre-existing microcracks (*e.g.* Lei, 2006; Lockner *et al.*, 1991b; Meredith *et al.*, 1990).

In order to link the variations in b-value to established physical mechanisms, some damage models were proposed which employ the constitutive laws of subcritical crack growth of crack populations with a fractal size distribution (Lei, 2006; Main *et al.*, 1989; Main *et al.*, 1993). As an example, three kinds of typical granite of different grain size distributions, in all nine samples, were loaded to failure at different stress rates. The b-value decreased with increasing stress and could be well modeled with the damage laws by assuming a grain size-dependent initial mean crack length (Lei, 2006).

Beside stress, b-value is strongly dependent on rock properties. The following lists the most important factors (in order of their relative importance, most important first).

1) **The distribution of pre-existing crack lengths.** This feature can be directly derived from the aforementioned damage laws. According to the damage laws, b-value ranges from 0.5 to 1.5 and are inversely correlated with the mean crack length. In most cases, the initial distribution of crack lengths within a given rock is related to the grain size distribution of the rock. Thus, fine-grained granite shows a b-value larger than coarse-grained granite. Under stress, a growing mean crack length leads to a decreasing b-value (see (Lei, 2006) for details).

2) **Macroscopic structures such as joints, veins, and foliations.** In these cases, the orientation and the healing strength of the structure planes are key factors. A number of metamorphic basalt samples cored along different directions from a deep mine has been used for fracture tests under triaxial compression. The results show that samples of unfavorably oriented foliation and optimally oriented foliation shows a primary b-value of ~1.5 and ~1.0, respectively.

3) **Homogeneity and grain size.** Homogeneous and fine-grained sedimentary rocks generally show large primary b-value. For example, Berea sandstone shows a b-value of ~1.5 (Lei *et al.*, 2011).

4) **Foliation orientation.** Results of some strongly foliated metamorphic rocks shows that the b-vale (prior to the peak stress) depends on the orientation of the foliation structure: it is close to unit for samples having favorably oriented foliations (has the minimum shear strength) and 1.3~1.5 for samples having unfavorably oriented foliations.

5) **Inelastic strain rate.** By keeping the inelastic strain rate constant, (Sano *et al.*, 1982) found that b-value is a function of inelastic strain rate rather than stress level. The effects of inelastic strain rate could be enhanced by increasing pressure and using a wet sample (Masuda *et al.*, 1988).

Recent results from jointed or foliated rocks have demonstrated a typical pattern in the change of b-value: a precursory drop (from 1.0~1.5 to ~0.5) in foreshocks and a consequent recovery (to 1 to ~1.2) in aftershocks (Lei *et al.*, 2013a).

Foreshocks underlie the fault nucleation phase that again strongly depends on rock properties and loading conditions. Due to the fact that b-value depends on several factors, a change of b-value could not be simply linked to stress change in both mining and natural fields. The usefulness of b-value as an earthquake predictor remains an area of continued debate. An integrated analysis of b-value with other AE statistics may lead to better indicator of failure (see 4.7).

3.2 Omori's law

It is well known that a large earthquake is generally followed by many aftershocks. In fact, both AE (Hirata *et al.*, 1987; Lei, 2003; Ojala, 2004) and earthquake of any size can produce its own aftershocks depending on its magnitude. Event rate of aftershocks (R_a) follows the modified version of Omori's law (presented in Utsu, 1961). It is an empirical power law of the time from the main event $(t - t_m)$, for aftershock rate:

$$R_a = K_a(c_a + t - t_m)^{-p} \tag{11}$$

where c_a, K_a, and p are constants. The constant c_a is a characteristic time between consecutive events. The p-value modifies the decay rate and typically falls in the range from 0.6 to 2.5 with a global mean of ~1 for earthquake aftershock sequences (Utsu & Ogata, 1995). In rock sample of laboratory scales and rock mass of mining scales, the foreshock rate R_f, before a main event obeys a similar power law of the time-to-failure rate $(t_m - t)$:

$$R_f = K_f(c_f + t_m - t)^{-p'} \tag{12}$$

AE bursts during creep tests (Hirata *et al.*, 1987; Lockner & Byerlee, 1977) and AE aftershock rates for slip events in sandstone show p-values in the range of 1 to 2 (Goebel *et al.*, 2012). Both foreshocks and aftershocks associated with the fracture of major asperities in a jointed rock can be roughly modeled with p and p' of 1.0 (Lei, 2003).

3.3 Critical point behavior and accelerated moment release model

Similar with earthquakes, chaotic behavior in AE activity can be resulted from the fractal or hierarchical complexities in cracks, joints, fractures and other heterogeneities in rocks, together with the non-linear interaction between cracks. Therefore, the concept of critical point behavior has been applied to AEs using time-to-failure analysis (Moura *et al.*, 2005; Moura *et al.*, 2006; Wang *et al.*, 2008); where the catastrophic event is considered a critical phenomenon occurring at a second-order phase transition in an analogy with percolation phenomena (Moura *et al.*, 2005). In the vicinity of the critical point, the variations in the energy release can be characterized by a power law of time-to-failure interspersed with log-periodic oscillations (Laherrere & Sornette, 1998). Mathematically, such oscillations correspond to adding an imaginary part to the exponent of the power law (Moura *et al.*, 2005; Moura *et al.*, 2006; Yukalov *et al.*, 2004):

$$\sum E(t) = A + B(t_f - t)^{a+i\omega} \tag{13}$$

where E may be any kind of energy release rate, t_f is the failure time, A is cumulative energy release at $t = t_f$, B is negative, a and ω are constant. The log-periodic oscillations correspond to an accelerating frequency modulation as the critical time is approached. By fitting experimental data of granites having various grain sizes, (Moura *et al.*, 2006) suggested that the imaginary part of the complex exponent ω has good correlation with grain size and loading rate. Larger grain size and faster loading rate result in a greater ω, which indicates a longer interaction range in space. By ignoring the oscillations in equation (6) we get the power law of the accelerated moment release (AMR).

$$\sum E(t) = A + B(t_f - t)^m \tag{14}$$

As expected from equation 14, progressively increasing AE activity before the catastrophic failure of brittle rocks is commonly observed (Lei, 2006; Main, 1991; Main *et al.*, 1993) and can be represented by the AMR model with an m of 0.2 to 0.3, which is in agreement with the typical value for natural earthquakes (Lei & Satoh, 2007; Tyupkin & Di Giovambattista, 2005; Wang *et al.*, 2008; Yin *et al.*, 2004).

AMR can be also derived from damage laws of crack population with a fractal size distribution. Under loading conditions of constant stress-rate w, the energy release rate $\dot{E}$ evaluated by the measured AE magnitude can be derived from laboratory-derived constitutive laws of the stress-induced subcritical growth of crack populations with a fractal size distribution (Lei, 2006):

$$\dot{E}(t)/\dot{E}(t = 0) = (1 - t/t_f)^{l+2-2l'/2-l}(1 + wt)^l \tag{15}$$

where l is referred to as the stress corrosion index (the exponent of the power law between the mean quasi-static rupture velocity of crack populations and the stress intensity factor (Main *et al.*, 1993)), l' is the exponent of the power law between the mean quasi-static rupture velocity of crack populations and AE rate, the failure time t_f is defined so that c approaches infinity. Under constant stress (creep) condition ($w=0$), equation (15) reduces to:

$$\dot{E}(t)/\dot{E}(0) = \left(1 - t/t_f\right)^{-m'}, \quad m' = l + 2 - 2l'/2 - l \tag{16}$$

The cumulative energy release is then obtained by integrating (16) with time:

$$\sum E = \int \dot{E}dt = A + B(t_f - t)^m \quad \left(B = -\frac{1}{(1-m')t_f}, m = 1 - m'\right) \tag{17}$$

Equation (17) is the same as equation (14). The exponent in the power law of AMR is thus linked with the exponents of another two power laws relating AE rate and mean crack length. In such a consideration, both event rate and moment release increase as a power of time-to-failure. The AMR is thus resulted from a cascade of small events progressively releasing stress before a large event.

If a system approaches a critical point, the spatial correlation length (hereafter abbreviated as SCL) is expected to grow according to a power law (Bruce & Wallace, 1989)

$$\xi(t) \propto (t_f - t)^{-k} \tag{18}$$

where k is positive. Assuming a scaling relation between moment release E and SCL, growing SCLs (GSCLs) can be derived from the AMR model in (14). The SCL of a set of N consecutive events can be estimated using single-link cluster analysis (Frohlich & Davis, 1990). Initially, each individual hypocenter is linked with its nearest neighbor hypocenter to form a set of clusters. Then, every cluster is linked with its nearest cluster. This process is repeated until N events are connected with N-1 links. Throughout the process, the distance between any two clusters is calculated based on their geometric centers. The SCL is here defined as the median of the length distribution of the N-1 links (Tyupkin & Di Giovambattista, 2005; Zöller et al., 2001). In order to reduce the dependence of SCL on event number and sample dimension, the dimensionless value is used:

$$\xi_R(t) = \xi(t)/l_R$$
$$l_R = l_\Omega/n^{1/3} \tag{19}$$

where l_Ω is the characteristic linear dimension of the rock sample. A decrease in SCL preceding precursory growth before large events (Lei & Satoh, 2007; Li et al., 2010; Tyupkin & Di Giovambattista, 2005).

3.4 Fractal and hierarchical property of AE hypocenters

Similar to natural earthquake, AE shows fractal self-similarity in both time (e.g. Feng & Seto, 1999) and space (e.g. Hirata et al., 1987; Lei et al., 1992). The generalized correlation-integral (Kurths & Herzel, 1987), was usually applied for multi-fractal analysis of the spatial clustering of AEs and earthquakes:

$$C_q(r) = \frac{1}{N}\left[\sum_{j=1}^{N}\left(\frac{N_j(R \le r)}{N-1}\right)^{q-1}\right]^{1/(q-1)} \quad (q = -\infty, ..., -1, 0, 1, ..., \infty) \tag{20}$$

where $N_j(R < r)$ is the number of hypocenter pairs separated by a distance equal to or less than r, q is an integer, and N is the total number of AE events analyzed. If the hypocenter distribution exhibits a power law for any q, $C_q(r) \propto r^{D_q}$, the hypocenter population can be considered to be multi-fractal, and D_q defines the fractal dimension that can be determined by the least-squares fit on a log-log plot. In the case of a homogeneous fractal, D_q does not vary with q. It can be easily proved that D_0, D_1 and D_2 coincide with the information, capacity and correlation dimensions, respectively.

Localization of AE activity leads to a decrease of fractal dimension, thus it was expected to be a possible indication of approaching failure. Gradual decreasing D was observed in some cases but not in others cases (Hirata *et al.*, 1987; Lei *et al.*, 1992; Lei & Satoh, 2007). By looking up results obtained so far, especially recent ones utilizing high-speed recording, we can draw some clearer conclusions now (see section 4 for details). In general, there is no gradual localization as ever expected. The final fracturing nucleated in a site with no clear correlation with the previous damage. The chance to observe decreasing D depends on the number of foreshocks which is somewhat determined by the critical size of the fault nucleation zone (over which the rock fails dynamically) as well as the growing velocity of the fault. In cases of a small nucleation zone or fast growing velocity, a smaller number of foreshocks are insufficient to cause a notable change in D.

3.5 Dragon-Kings in AEs

Similar with many geological systems of the Earth, AEs do not show perfect power law behaviors. There are extreme events of a magnitude which do not follow the Gutenberg-Richter power law governing the size distribution of other events. In laboratory, a test sample is loaded by a very rigid frame, thus the final fracture likely release almost strains in the sample. Therefore, it is dangerous to conclude that the final fracture is a dragon-king. Beside the final fracture, the fracture of major asperities in the shear plane also demonstrate some extreme events as the examples shown in Fig. 5. These

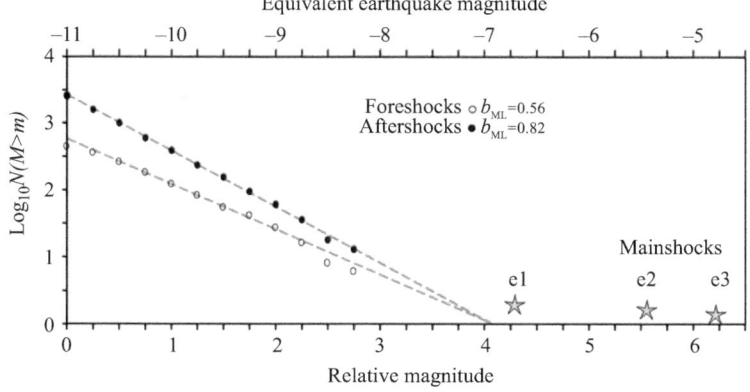

Figure 5 Magnitude-frequency distribution of fore- and after-shocks associated with fracture of a major asperity on a fault going to failure. The e1~3 mark the main events which are termed as "dragon-kings," extreme events departed from the power law (after Lei, 2012).

events have a magnitude significantly greater than that expected by the GR power-law relation between the magnitude-frequency distribution for either foreshocks or aftershocks. There are at least two mechanisms that may lead to dragon-kings: 1) the power-law event rate and moment release increasing and 2) hierarchical fracturing behavior resulting from hierarchical inhomogeneities in the sample (Lei, 2012). In the first mechanism, the final failure corresponds to the end-point of the progressive occurrence of events, and thus the resulting dragon-king event can be interpreted as a superposition of many small events. In the second mechanism, an event of extreme size is the result of fracture growth stepping from a lower hierarchy into a higher hierarchy on a fault surface having asperities characterized by hierarchical distribution (of size or strength) rather than simple fractal distribution. In both mechanisms, the underlying physics is that fracture in rocks is hard to stop beyond certain threshold.

4   AE IN INTACT ROCKS UNDER DIFFERENTIAL COMPRESSION

4.1   Lithology and confining pressure

At first, different rock type may demonstrate quite different pre-failure AE activity. In general, polycrystal rocks such as granitic rocks, andesite, and metamorphic rocks show very high AE productivity. AE activity is initiated at relatively lower stress and then increased progressively until failure showing fairly well AMR behaviors. In sedimentary rocks, quartz-rich and strongly cemented sandstones show high level AE activity. While clay-rich shale and mudstone, and homogeneous dolomite demonstrate low level pre-failure AEs but shale and dolomite may show strong AE energy release during the quasi-static to dynamic fracturing stages (Lei *et al.*, 2014). Coal shows high level of AE activity under uniaxial compression (Ranjith *et al.*, 2010), triaxial compression (Yin *et al.*, 2012), and during gas adsorbing/desorbing and swelling/shrinking (*e.g.* Ma *et al.*, 2012, Majewska *et al.*, 2009). CO_2 gas can be adsorbed onto the surface of the micro-, meso-and macropore and fracture systems within the coal. It was found that CO_2 saturation has a significant effect on crack initiation stress, which can be monitored by AEs (Ranjith *et al.*, 2010). In any lithology, high level AE activity resulted primarily from grain-size scale heterogeneities. Porous rocks such as sand stone generally shows high level of AE activity, but the AE energy might be very weak as a result of attenuation. In rock salt, AE activity is responsible for dilatancy strain under uniaxial and triaxial compression conditions (Alkan *et al.*, 2007; Xie *et al.*, 2011).

In general, AE activity is relatively higher under lower confining pressures. Ductile limestone may demonstrate some AE activity under low confining pressure or true-triaxial unloading conditions (He *et al.*, 2010).

4.2   Pre-existing microcrack density or damage index

In granitic rocks, the density of pre-existing microcracks is found to be the most important factor governing pre-failure AE activity. A microcrack-free rock demonstrates a fracture behavior similar to glass with very few AE events preceding the final failure, while rock of higher micro-crack density shows earlier AE initiation and progressively increasing AE activity approaching the failure (Lei *et al.*, 2000a). The damage index can be used to quantitatively define the pre-existing microcrack density. Damage index can be estimated use P wave velocity:

$$DI = 1 - \frac{M}{M_0} = 1 - \left(\frac{V}{V_0}\right)^2 \tag{21}$$

where, M and V represent the P wave modulus and velocity, respectively. The subscript 0 indicates measurement of damage free sample, which can be obtained by measuring the P wave velocity at sufficiently high hydrostatic pressure, saying 100 MPa is practically applicable for many rocks.

4.3 Role of mesoscale/macroscopic heterogeneities

For guaranteeing reproducibility of experimental results, samples as homogeneous as possible are required. A homogeneous sample means the heterogeneities of grain size scale are randomly distributed within the sample. However, results from carefully selected samples containing macroscopic heterogeneities and artificial structures in mesoscale or core scale are also meaningful for investigating their role (*e.g.* Pappas *et al.*, 1998; Satoh *et al.*, 1996). Faulting nucleation processes are strongly governed by heterogeneities. As an extreme case, it was found that brittle veins in ductile mudstones have an important role on pre-failure damaging. Fault segments along the bedding plane showed slip behavior with large compressive deformation before the peak stress, while the vein asperities showed large precursor dilatancy prior to dynamic rupture accompanied with bursts of AEs (Lei *et al.*, 2000c). A long-term decreasing trend and short-term fluctuation of the *b*-value in the phase immediately preceding dynamic fracture are identified as characteristic features of the failure of heterogeneous faults (Lei *et al.*, 2004, Lei *et al.*, 2000c).

4.4 Orientation of foliation and bedding structures

It is well known that brittle rocks under triaxial compression failure by shear fracture. In foliated metamorphic rocks and in sedimentary rocks containing clear bedding structures, the angle of the structures to the maximum principal stress is a key factor. In a recent work, a series of samples of foliated metamorphic basalt and metasediment from deep mines, cored along different directions with respect to the foliations, were tested under triaxial compression to examine the role of foliation in rock fractures (Lei *et al.*, 2013a; Villaescusa *et al.*, 2009). Fig. 6 shows schematic plots for a comparison of AE hypocenters patterns of various foliation orientations. Samples of favorably oriented foliations demonstrate a fracturing process that is somewhat similar to natural earthquakes, showing: 1) a small number of foreshocks, 2) a large number of aftershocks, and 3) a fault nucleation zone having a dimension of a small fraction of the fault dimension. Samples with unfavorably oriented foliations demonstrate more complicated fracturing processes and the final fracture is created through the linkage of damage clusters along the shear zone. Every individual cluster are controlled or affected by the foliations. Major clusters are aligned along the shear zone of the final fault plane. The vertical AE clusters are clearly related with the vertical foliations. All aspects concerning the fracture including the geometry of the final fracture planes, the failure

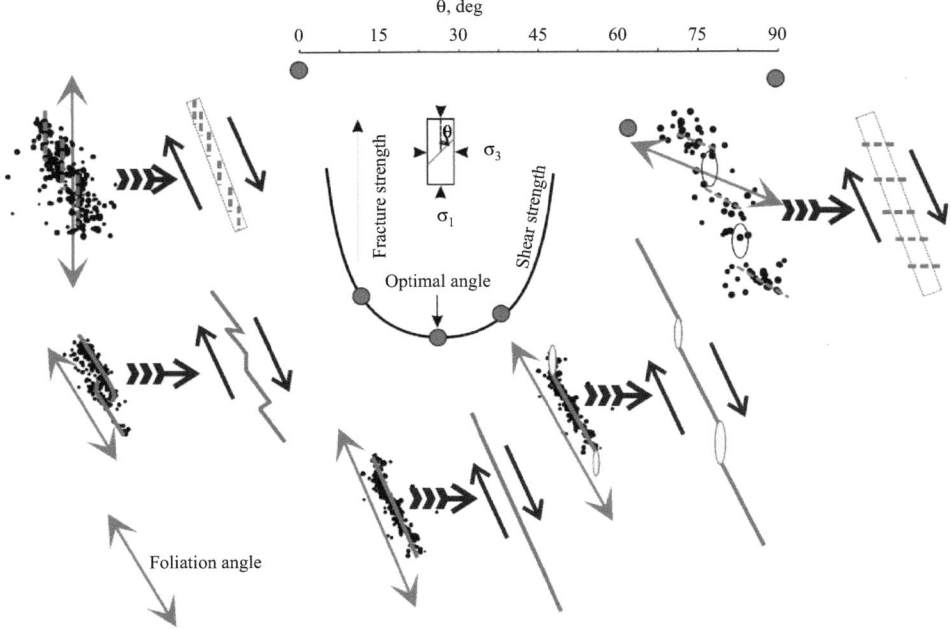

Figure 6 Schematic plots show a comparison of mesoscale patterns of AE hypocenters of various foliation orientations (modified from Lei *et al.*, 2013, 2015).

strength, pattern of pre-nucleation AEs, and nucleation processes, are systematically affected by the foliation angle.

4.5 Asperities on favorably oriented shear fracture plane

The aforementioned samples with favorably oriented planar structures are a better simulation model of earthquakes. Such works are of considerable interest and importance, because they could provide rules useful for understanding rock fracturing in mining sites and earthquake processes in the crust. Since favorably oriented faults have the minimum reaction strength (in term of $\sigma_1-\sigma_2$), the critical crust stress is somewhat controlled by such faults. AE activity during the fracture of a shear fault is controlled by the fault geometry and asperities on the fault plane. Rough fault surface, irregular fault geometries (bend, step overs) lead to more pre-failure AE events. A detailed study by Lei and colleagues (Lei, 2003; Lei *et al.*, 2003) on in a granitic porphyry reveals that a quasi-static nucleation of the shear faulting corresponds to the fracture of coupled asperities on the fault plane. Acoustic emissions caused by the fracture of individual asperities exhibit similar characteristics to the sequence for natural earthquakes, including foreshock, mainshock, and aftershock events. Foreshocks, initiated at the edge of the asperity, occur with an event rate that increases according to a power law of the temporal distance to the mainshock, and with a decreasing *b*-value (decrease from ~1.1 to ~0.5). One or a few mainshocks then initiate at the edge of the asperity or in the front of the foreshocks. The aftershock period is characterized by a remarkable increase and subsequent gradual decrease in *b*-value and a decreasing event rate obeying the modified Omori law. The fracture of neighboring asperities is then initiated after the mainshock of a particular asperity. The progressive fracturing of multiple coupled asperities during the nucleation of shear faulting results in short-term precursory fluctuations in both *b*-value and event rate, which may prove useful information in the prediction of failure of the main fault plane of earthquakes.

In a very recent study, by using the X-ray computer tomography (CT) technique and an AE monitoring approach, it was found that geometric asperities identified in CT scan images were connected to regions of low *b*-values, increased event densities and moment release over multiple stick-slip cycles (Goebel *et al.*, 2012). This result is consistent with the aforementioned fracture of unbroken asperities on a naturally healed fault.

4.6 Role of water saturation, pore pressure and hydrofracturing

Water saturation in polycrystalline porous rocks is an important factor in both short-term and long-term and time dependent mechanical behaviors. Studies have shown that time-dependent weakening is much more important for a saturated rock than for a dry one. Further, it has been shown that the weakening effect of water is more significant in long-term experiments than in short-term ones (*e.g.* Lajtai *et al.*, 1987). Uniaxial creep tests of iron under partially saturated conditions show that water saturation induces a strong increase in AE activity and dilatant inelastic volumetric strain and a decrease in Young's modulus and in the seismic *b*-value as the rock approaches failure, indicating that water saturation accelerates static fatigue through hydro-mechanical coupling and subcritical stress corrosion cracking (Grgic & Amitrano, 2009).

In the laboratory, the governing role of pore pressure in pre-failure damage and final failure in either low-porosity rocks, such as granites (Byerlee & Lockner, 1977; Kranz *et al.*, 1990; Masuda *et al.*, 1990), or high-porosity rocks, such as porous sandstones (Lei *et al.*, 2011; Mayr *et al.*, 2011; Schubnel *et al.*, 2007; Stanchits *et al.*, 2011), has been investigated. Basically, AEs could be triggered by a pore pressure increase over a critical pore pressure level in rocks under differential stress. In low-porosity rocks, a positive feedback between AE activity and pore pressure diffusion is especially significant because local permeability of the rock can be greatly improved by microcracks which cause AEs (Masuda *et al.*, 1990). It was found that water-pressure induced AEs involve a greater rupture velocity or a more equidimensional fault geometry than stress-induced ones (Masuda, 2013), suggesting that that pulse width analysis of P waveforms can be used to distinguish fluid induced events from those induced by regional stress.

Mayr *et al.* (2011) presented some interesting results in porous sandstones of a permeability of 10^{-17} to 10^{-16} m^2. In their experiments, AEs could be triggered by a pore pressure increase over a critical pore pressure level. The critical level was controlled by the applied pore pressure of the previous cycle according to an apparent Kaiser effect in terms of pore pressure. This memory effect of the rock vanished if additional axial stress was applied to the sample before the next injection cycle. In addition, they found that in a highly fractured rock the nucleation of the final failure was more likely to be controlled by the propagation of the fracture than by the diffusion of pore pressure.

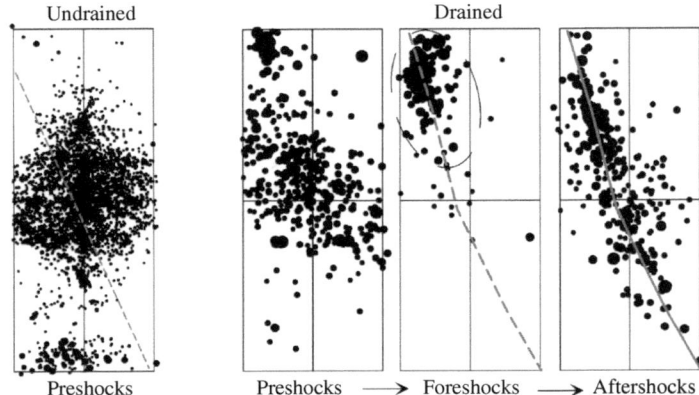

Figure 7 AE hypocenters in two Berea sandstone samples supplied to triaxial compression test under different drainage conditions. Under undrained condition (no fluid flow into and flow out from the sample), a large number of AEs, which formed several compaction bands, were observed before the dynamic fracture. Under drained condition (pore pressure at the end faces of the sample was kept constant), relatively small number of pre- and fore-shocks were observed (modified from Lei *et al.*, 2011).

Dilatancy and dilatancy hardening are generally observed to occur prior to brittle faulting. Pore pressure decreases resulting from dilatancy are the most important mechanism of dilatancy hardening in a wet sample. Therefore, fluid flowing and *in situ* drainage conditions are dominant factors that govern rock fracture behaviors. According to field evidence, an increase in fluid pressure may produce not only significant seismicity (*e.g.* Lei *et al.*, 2013b; Lei *et al.*, 2008) but also stable or aseismic slips along pre-existing faults (Cornet *et al.*, 1998). In a Berea sandstone, which has a porosity of ~20% and a permeability of ~10^{-13} m^2, the drainage conditions play a governing role in the deformation and fracturing processes (Fig. 7). The dilatancy-hardening effect can be greatly suppressed by dilatancy-driven fluid flowing under good drainage conditions. Fast diffusion of pore pressure leads to a significant reduction in rock strength and stabilization of the dynamic rupture process. Furthermore, good drainage conditions have the potential to enlarge the nucleation dimension and duration, thereby improving the predictability of the final catastrophic failure (Lei *et al.*, 2011).

In basalt, rapid post-failure decompression of the water-filled pore volume and damage zone triggered low-frequency events, which exhibited a weak component of shear (double-couple) slip, consistent with fluid-driven events occurring beneath active volcanoes (Benson *et al.*, 2008).

In these studies, the speed of pore pressure diffusion is a key factor. Since pore pressure diffusion is inversely proportional to the viscosity of the fluid, fluids of lower viscosity are thus expected to be more efficient to suppress the dilatancy-hardening effect. Evidence was recently obtained through hydraulic fracturing laboratory experiments with supercritical (SC-) and liquid (L-) CO_2, which have viscosity one order lower than water at low temperature (Ishida *et al.*, 2012). Their results show that AE hypocenters with the SC- and L-CO_2 injections tend to distribute in a larger area than those with water injection, and furthermore, SC-CO_2 tended to generate cracks extending more three dimensionally (rather than along a flat plane) than L-CO$_2$.

4.7 Typical phases of AEs during rock deformation and fracture

It is noted by many authors that the fracturing of rocks in laboratory experiments demonstrates some typical stages with different underlying physics corresponding to the stress level or deformation history. For example, creep tests normally show three deformation stages: primary, transient, and accelerated. Under uniaxial compression, AE activity in rocks shows five typical phases: 1) microcrack and pore closure; 2) linear elastic deformation; 3) microcrack growth; 4) fracture propagation and chipping; and 5) post-failure relaxation (Ohnaka & Mogi, 1982). Since the number of AEs is well correlated with volumetric strains (*e.g.* Lei *et al.*, 1992), AE activity directly reflects rock deformation and thus show corresponding phases. Detailed analysis of AEs is helpful for investigating the physics underlying each phase of deformation.

Through an integrated analysis of several AE statistics obtained from AE data collected with the high-speed AE waveform recording system, a three-phase pre-failure-damage model has been proposed in (Lei *et al.*, 2004) and further enforced with new data (Lei & Satoh, 2007; Lei *et al.*, 2006; Rao *et al.*, 2011). The lithology of the test samples covers granitic, sedimentary, and metamorphic rocks whilst the structure of the samples covers fine- and coarse-grained, jointed, and foliated. Time-dependent statistics include the energy release rate, b-value of the magnitude–frequency distribution, and fractal dimension and/or spatial correlation length (SCL) of AE hypocenters. The data from these studies indicate that the pre-failure damage process is characterized by three major phases of microcracking activity, termed the primary, secondary, and nucleation phases, respectively. Figure 4 shows a typical example in which all phases with a large number of AEs and the features of every phase are very clear (Lei *et al.*, 2006). In some cases, AEs during the dynamic fracturing process and following aftershock period could be recorded (*e.g.* Lei *et al.*, 2013a; Main *et al.*, 1992; Thompson *et al.*, 2006). A general summary of the phases follows.

Primary phase: The primary phase reflects the initial opening or ruptures of pre-existing microcracks, and it is characterized by an increase, with increasing stress, of both the event rate and the b-value. The rate of AE depends on the density of pre-existing cracks. The observed fact of initially low and subsequently increasing b-value suggests that relatively longer cracks likely rupture at relatively lower stress.

Secondary phase: The secondary phase involves the sub-critical growth of a crack population, revealed by an increase, with increasing stress, in the energy release rate and a decrease in the *b*-value.

Nucleation phase: The nucleation phase corresponds to the quasi-static fault nucleation processes. During the nucleation phase, the *b*-value decreases rapidly down to the global minimum value, in well monitored cases, it is around 0.5. See section 4.8 for further aspects on fault nucleation.

Dynamic phase: For most intact rocks it is impossible to distinguish individual AE events during the dynamic fracturing period since AE rate shows progressively increasing rate preceding the dynamic fracturing, resulting in noisy background (Benson *et al.*, 2007; Lei, 2012; Schubnel *et al.*, 2007; Thompson *et al.*, 2006;

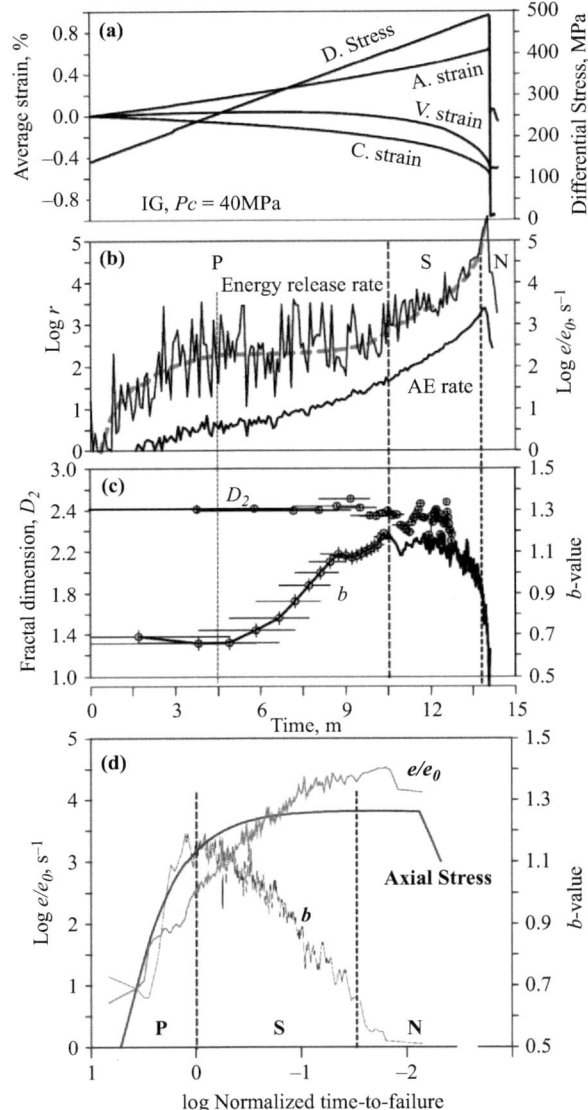

Figure 8 An example of AE data obtained from a typical experiment on a coarse-grained Inada granite sample under constant stress rate (27.5 MPa/min). (a) Differential stress, average axial/circumferential/volumetric strains. (b) AE energy release rate and event rate calculated consecutively every 10 s. (c) *b*-value and fractal dimension D_2. (d) Stress, energy rate and *b*-value against normalized time-to-failure in logarithm coordinate. P, S, and N denote the primary, secondary and nucleation phases, respectively. (Modified from Lei, 2006).

Thompson *et al.*, 2009). In some cases, such as that in samples of optimally oriented foliations, there is fewer foreshocks immediately before the dynamic fracturing, the main shock could be recorded and located by chance (Lei, 2012).

Aftershock phase: In well-jacked samples and samples of optimally oriented weak surfaces (foliated-, joint-, healed-fault, and so on), aftershocks could be recorded. Similar to earthquakes, aftershocks are concentrated in the fault plane and show a gradual recovering of *b*-value.

Stick-slip phase: In well-jacked samples, a few numbers of stick-slip events could be monitored with AE data (Thompson *et al.*, 2009).

For the secondary and the nucleation phases, *b*-value and release rate can be modeled as functions of the time-to-failure (section 3.3) and thus can be treated as an indicator of the critical point (Lei & Satoh, 2007). Both the failure of a major asperity on the fault plane and the catastrophic failures of the rock samples are generally preceded by 1) accelerated energy release, 2) a decrease in fractal dimension and SCL with a subsequent precursory increase indicating growth of fault nucleation, and 3) a decrease in *b*-value from ~1.5 to ~0.5 for fine-grained rocks, and from ~1.1 to ~0.8 for coarse-grained or weak rocks such as S-C cataclasite. However, each parameter also reveals more complicated temporal evolution due to either the heterogeneity of the rock (see section 4.3) or the micro-mechanics of shear fracturing (see section 4.8). AE statistics in the secondary and nucleation phases confirm the potential importance of integrated analysis of two or more parameters for successfully predicting the critical point. The decreasing b-value and increasing energy release may prove meaningful for intermediate-term prediction, while the precursory increase in fractal dimension and SCL are possible indications of approaching failure and can facilitate short-term prediction.

As reviewed in 4.1–4.6, lithology, density and size distribution of pre-existing cracks, meso-scale, macro-scale heterogeneities, pore fluid, and drainage condition all have an overall role in AEs. There are some cases in which some phases are not clear. In general, homogeneous (both fine-grained and coarse-grained) rocks with pre-existing cracks likely show all phases. Heterogeneous or weak rocks such S-C cataclasite normally show a lack of the primary phase. Samples with few pre-existing cracks and samples containing optimally oriented weak structures, likely show an unpredictable fracturing behavior as well as a lack of primary and secondary phases, in addition the nucleation phase has a small number of AEs.

4.8 Fracture nucleation and process zone

Earlier AE waveform recording systems could record a few tens of events per second, insufficient for exploring the details of fault nucleation which corresponds to an AE rate on the order up to several thousand events per second. Indeed, the final fracture plane could be mapped with pre-failure AEs in inhomogeneous samples having weak structures (Lockner & Byerlee, 1980; Lockner *et al.*, 1992; Satoh *et al.*, 1996). If the test sample contains optimally oriented weak interfaces, and if the environmental rigidity is sufficient high to sustain the quasi-static growth of fault nucleation, the fault could be nucleated at quite a low stress level relative to the peak stress. On the other hand, in homogeneous and brittle rocks, fault nucleation is likely created at a stress close to the peak stress, while the environmental rigidity is lowered by damage throughout the sample volume the rock fails rapidly with a large stress drop (Lockner, 1993; Lockner *et al.*, 1992; Satoh *et al.*, 1996). To overcome this difficulty, different approaches were applied among research groups.

In the first approach, the AE event rate was used to control the axial loading. With such a technique the rapid faulting nucleation process (in seconds) could be stabilized (in hours) and thus make it possible to observe the complete nucleation and quasi-static growth process of a fault through the analysis of AE hypocenters (Lockner *et al.*, 1991a; Lockner *et al.*, 1992).

In the second approach, a non-standard "asymmetric" loading was used to force shear faulting under uniaxial compression and earlier (and accordingly slowed down) faulting nucleation (Zang *et al.*, 1998; Zang *et al.*, 2000).

In the third approach, a rapid AE monitoring system was developed in the late 1990s (Lei *et al.*, 2000b), which can record AE waveforms without major loss of events, even for AE event rates on the order of several thousand events per second, typical for fault nucleation in brittle rocks. Therefore, it is possible to study cracking activity during the spontaneous and quasi-static growth of a fault in intact rocks. The main advantage of this approach over the AE feedback technique is that the process of fault growth is in a condition of constant stress rate or constant stress (creep) loading, which allows both quasi-static and dynamic crack growth to occur. As we know the constitutive relation of friction (*e.g.* Dieterich, 1992) and the mechanics of the interaction between cracks are indeed time-dependent so the fault growth under artificially slowed down conditions with time-varying strain rate may change the nature of fracturing. Recent systems were utilized with large amounts of memory, allowing continuous waveform recording on multiple channels and more detailed analysis of faulting nucleation, dynamic fracturing, and aftershock processes (Benson *et al.*, 2007; Lei, 2012; Stanchits *et al.*, 2006; Thompson, 2005; Thompson *et al.*, 2006). Benefited with the fast speed of PCIe bus, cheaper waveform recorders are developed, by which AE waveforms can be digitized at a sample rate up to 20 MHz and streamed to hard disk arrays directly for continuous recording (Yoshimitsu *et al.*, 2014).

By using the AE feedback approach, the nucleation and quasi-static growth of faults in Westerly granite and sandstone were observed (Lockner *et al.*, 1991a; Lockner *et al.*, 1992). These results are comprehensive. In granite samples, prior to the peak stress, AE activity was distributed evenly throughout the samples. The fault plane nucleated abruptly at a point on the sample surface, and then grew across the sample, accompanied by a gradual drop in axial stress. AE locations showed that the fault propagation was guided by a fracture front, termed process zone. After the process zone, a damage zone was created with AE activity of a low level. In sandstone samples, a diffuse damage zone appeared prior to peak strength and gradually localized into an incipient fault plane. It is interesting that after passing through peak stress, the shear fault plane grew in a way similar to that in the granite samples. Migration of AE hypocenters shows that the fault growth is not smooth but that there are periods of acceleration and deceleration. Such a kind of fault nucleation process was also observed using X-ray CT to view the damage created in rock samples which have been exposed to triaxial compression under constant strain (in a circumferential direction) rate and released at different stages of post peak stress (Kawakata *et al.*, 1997). By using either displacement or the rate of acoustic emissions to control the applied axial force, the propagation velocity of the process zone is varied from 2 mm/s to 2 μm/s (Zang *et al.*, 2000).

Comprehensive results are 1) the width of the process zone is about 9 times the grain diameter inferred from acoustic data but is only 2 times the grain size from optical crack inspection; 2) the process zone of fast propagating fractures is wider than for slow ones; 3) the density of microcracks and AE emissions increases approaching the main fracture; 4) shear displacement scales linearly with fracture length; and 5) the ratio of the process zone width to the fault length in Aue granite ranges from 0.01 to 0.1 inferred from crack data and acoustic emissions, respectively. With a combination of AE feedback and continuous waveform recording, accelerating propagation speed of fault nucleation from ~1 mm/s to ~100 m/s could be fingered out through 3D AE locations (Thompson et al., 2006).

The use of a high-speed waveform recording system is successfully in monitoring AE hypocenters during spontaneous fault nucleation and even the unstable failure process in intact brittle rocks under constant stress (rate) loading conditions. A number of different kinds of samples including intact rocks (Lei et al., 2000a; Lei et al., 2000b), jointed rocks with unbroken asperities (Lei et al., 2003), shale with thin quartz veins (Lei et al., 2000c), and foliated rocks (Lei et al., 2013a), were investigated and more detailed structures of faulting nucleation and process zone were obtained.

It has been observed that, "unlike tensile cracks, shear fractures do not grow by simple propagation in their planes but show a complex breakdown process" (Scholz et al., 1993). The process zone is defined as the damage zone at the tip of a propagating fault. It is a very important aspect of fault mechanics and has been studied by laboratory works (Cox & Scholz, 1988a; Cox & Scholz, 1988b; Lei et al., 2000b; Zang et al., 2000). The process zone geometry and size, relative to the fault contained within it, can be used to evaluate fault growth models. The process zone may also play an important role in the permeability structure of fault zones. Understanding process zones as a whole, as well as understanding the size, structure, and orientation of elements within process zones, will lead to a better understanding of the role faults play in fluid migration within the Earth's crust.

It was observed that a fault initiated at a site with slight preceding damage and then propagated into un-faulted rock with a process zone of intense cracking (Reches & Lockner, 1994). Reches & Lockner (1994) proposed a model based on the mutual enhancement of cracking due to stress induction and illustrated the propagation of a fault through the interaction of tensile cracks. However, based on knowledge of the focal mechanism type, it was observed that tensile cracking was dominant only in the front area of the fault, i.e., within the process zone (Lei et al., 2000b). When the density of the damage zone increased or when the fault growth was accelerated, shear mode or other modes containing a shear component became the major/dominant modes of cracking; this is in agreement with a suggestion by Cox & Scholz (1988a) based on microstructural examination.

Fig. 9 shows a renewed fault model with major features illustrated. As shown in the figure, shear fault growth is guided by the progressive occurrence of tensile cracks in the process zone at the fault tip. By noting the optimal orientations of cracks in the process zone, it is easy to understand that a shear fault is likely to bend. At the same time the entire shear fault is driven to propagate along the optimal direction as defined by the Mohr-Coulomb failure criterion. As a result of such micro-mechanics and macro-mechanics, a shear fault is likely to show complicated processes such as temporal fluctuations (acceleration and deceleration) and spatial step-overs, as derived from

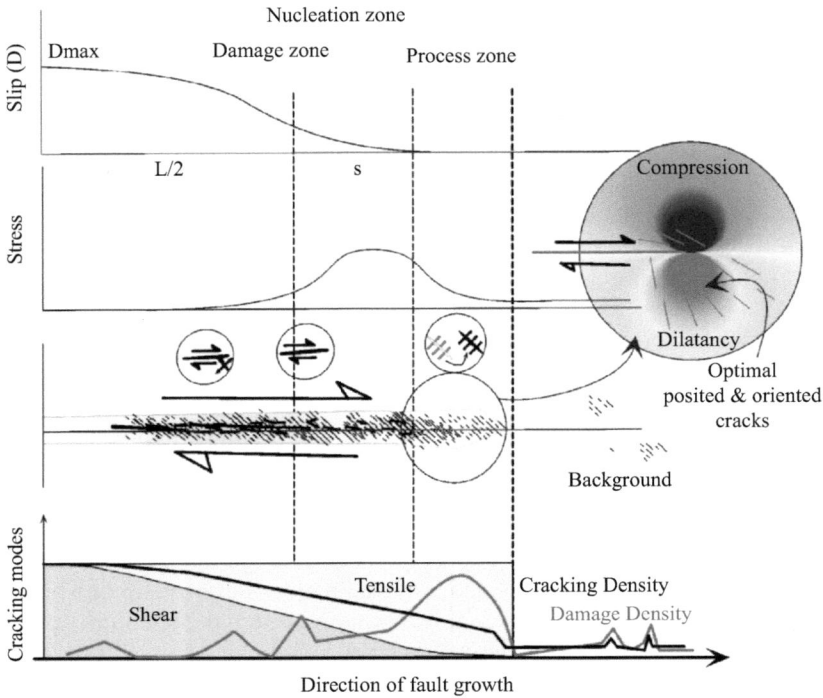

Figure 9 Schematic section of a quasi-statically growing shear fault (modified from Lei *et al.*, 2000b). The fault is growing toward the right. The fault consists of two portions: the process zone at the fault tip and the damage zone behind it. The AE activity in the process zone is dominated by tensile cracking, a *b*-value lower than the background, fewer larger events, and strong self-excitation. The damage zone is characterized by shear cracking, low *b*-value, larger events, and weak self-excitation.

AE monitoring (Lei *et al.*, 2013a; Lei *et al.*, 2000b) and microstructural observations (Kawakata *et al.*, 2000). The core of the shear fault is the cataclasite zone, which may form gouge layers after repeated slips and thus may reduce the permeability along the direction perpendicular to the fault. While at step-overs and fault ends, the remaining damage zone of tensile cracks may greatly enhance the permeability.

5 AE IN PRE-CUT SAMPLES

Experimental studies on fault stick-slip behavior are motivated by the need to provide fault frictional models which are required for faulting simulation or earthquake cycling simulation. AE events have been observed during frictional sliding on saw-cut surfaces and naturally fractured surfaces in laboratory rock samples. New insights could be gained from detailed AE monitoring. Beside the lithology of the blocks, the roughness of surface, geometries (bend, orientation) of the fault, gouge, and loading speed are major factors have been investigated. A summary of the effects of these characteristics follows.

5.1 Surface roughness

The sliding surfaces were normally ground with abrasive of desired particle size. The characteristic scale of asperities on the sliding surfaces corresponds to this particle size (*cf.* Brown & Scholz, 1985). A drop in the *b*-value of foreshocks before a stick slip event was observed in a large granite sample containing a saw cut (Weeks *et al.*, 1978). The AE sources were located on the pre-cut fault, and their focal mechanism solution was consistent with that expected theoretically for macroscopic sliding. It is not surprising that the roughness of the fault surface is a first order factor governing the fractional behavior and AE activity. Frictional sliding experiments on coarse-grained Inada granite blocks in double shear show that: (1) smooth-ground (600#, 10 µm) surfaces produce the maximum number of AE events and (2) *b*-values are related to the surface topographic fractal dimensions, *i.e.*, smooth surfaces exhibit lower *b*-values than rough-ground (#60, 300 µm) surfaces (Sammonds & Ohnaka, 1998). Thus, the change of *b*-value (decreasing from 1.5 to 1.0, and from ~1.2 to 0.5 for rough and smooth surfaces, respectively) with accumulative amount of slip can be interpreted in terms of evolving fractal crack damage during frictional sliding of the fault surface. It was also suggested that that the grain-scale topography determines the AE source dimension, while the fractal-domain asperities control the magnitude of the stress drop (Yabe, 2008). The source radii of large AE events may reach the order of 10 mm (as derived from the widths of the first P-wave pulses). High frequency AE signals during the passing rupture of a stick-slip event, observed by near fault AE monitoring, show similar source dimensions (Kato *et al.*, 1994). A single AE event cannot be due to the rupture of a single asperity and is instead caused by coherent rupture of many asperities which have a spot size of contact on the order 10–100 µm (Yabe *et al.*, 2003).

Detailed AE data from recent studies show that off-fault AE density decay with distance from the slip'surface following a power-law, which exponents are sensitive to both fault roughness and normal stress variations. Larger normal stresses and increased roughness lead to slower AE density decay with fault-normal distance. This emphasizes that both roughness and stress have to be considered when trying to understand microseismic event distributions in the proximity of fault zones (Goebel *et al.*, 2014; Goebel *et al.*, 2013b). The friction experiments on macroscopically non-homogeneous faults indicate that there exist two types of nucleation phase for stick-slip instability of non-homogeneous faults, which coincide with the preslip model and the cascade model, respectively (Ma *et al.*, 2002). A very recent study of stick-slip test of saw cut fault (hand lapped with 600 grit abrasive) shows that each stick slip begins as an AE event that rapidly (~20µs) grows about 2 orders of magnitude in linear dimension and ruptures the entire of the simulated fault (150mm length) in aseismic slip which weakens the fault and produces AEs that will eventually cascade-up to initiate the larger dynamic rupture (McLaskey & Lockner, 2014). Sliding on naturally fractured surfaces, which can be considered as a proxy for a very young fault (or a very complex fault zone), showed, in addition to double-couple components, significant volumetric contributions, especially during the inter-slip periods and immediately after stick-slip events indicating substantial shear-enhanced compaction within a relatively broad damage zone. The obtained results fault roughness controls the kinematics of microseismicity during different periods of the seismic cycle (Kwiatek *et al.*, 2014).

5.2 Strain rate- and slip-dependence, gouge

Strain rate- and slip-dependent AE activity was confirmed by a number of experiments during the stable sliding of bare fault surfaces (*e.g.* Yabe *et al.*, 2003). Acoustic emission per unit slip was found to decrease with increasing strain rate.

The results of AE activity obtained during shear of granular layers show that, for a given loading strain rate, the AE rate decreased with accumulating slip. Step increases in strain rate led to immediate and sustained increases in AE rate; however, AE per unit slip decreased with increasing strain rate (Mair *et al.*, 2007). The slip-dependent behaviors are consistent with those of bare surface experiments (Yabe *et al.*, 2003). Single AE events generated in sheared granular materials require the coherent rupture of many grain contacts.

5.3 Fault geometry

For a slightly bent (5 degree) fault (600#, ground 10 μm), AEs are likely to cluster in the bend; this is caused by cracks in the host rock due to stress concentration rather than the rupture of asperities on the surface (Kato *et al.*, 1999).

5.4 AEs during stick-slip of unfavorably oriented fault

Very recent results on stick-slip tests of unfavorable oriented faults indicates that there are two competing mechanisms involved in fault stick-slips (Lei *et al.*, 2015). On one hand, the fault plane is smoothed by fault slip as a result of failure of asperities on the fault plane. Thus we can see a decreasing tendency in AE activity and friction coefficient with increasing number of stick-slip events. The friction coefficient of pre-cut fault depends only on the history of stick-slip, independent on fault angle. In all cases, the fault friction drops from ~0.75 to 0.6 after a few numbers of stick-slip. On the other hand, the fault plane is roughened by new damages and leads to fluctuations in AE activity and frictional behaviors. In D45 (the number indicates angle between the maximum principal stress axis and fault plane) samples smoothing mechanism plays a dominant role. While, D50 samples roughening mechanism and formation of sub-fault along optimal directions are important (Fig. 10).

6 A GENERAL SUMMARY AND CHALLENGES

In summary, the pre-failure AE activity in intact rocks, primarily depend on several factors including rock lithology, mesoscale/macroscopic heterogeneities, pre-existing microcrack density, foliation and bedding structures. In rocks of a favorably oriented fault, AE activity during the fracture of the fault is controlled by the fault geometry and asperities on the fault plane. If the fault is as strong as the host rock then the fracture makes no difference and the rock remains intact. Furthermore, a homogeneous fault or rock mass appears to fracture in unpredictable ways without a consistent trend in precursory statistics, while inhomogeneous faults fracture with clear precursors related to the nature of the heterogeneity. Fig. 11 schematically shows major feature of AE activity and the key factors governing it. Finally, it is worth noting that in a specified

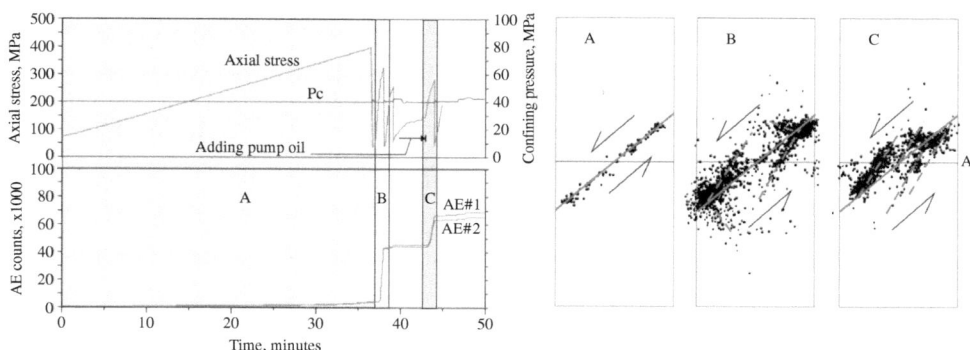

Figure 10 Results for sample with 50° pre-cut showing axial stress, confining pressure (Pc), and AE counts recorded at 2 selected sensors and AE hypocenters on a vertical profile perpendicular to the pre-cut fault.

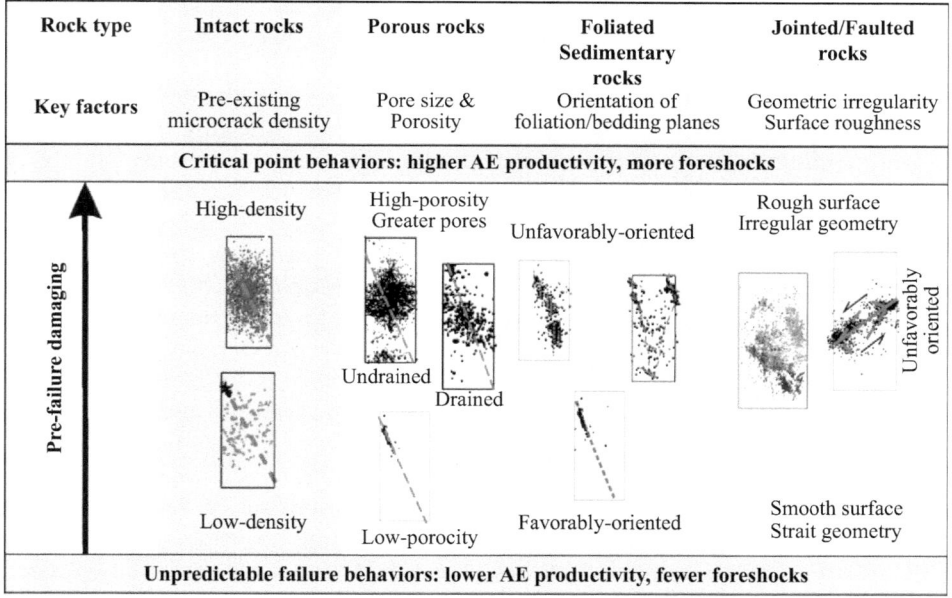

Figure 11 Schematic plot show a summary of microcrack activity preceding rupture of different rock types and the key factors governing the pre-failure damaging processes.

sample, all factors are normally mixed with each other and can thus play different role in different stages of deformation.

Since the similarity between the size distribution of earthquakes and AEs has been documented (Mogi, 1962; Scholz, 1968b), a half century has passed. Considerable numbers of studies have been carried out on AE activity before, during, and after the fracture of rocks. Together with other experimental studies, particularly those concerning the frictional behavior of a fault, laboratory studies have shed light onto rock failure and earthquake seismology. Rules obtained at the laboratory scale are helpful for understanding rock fracture on significantly larger scales. However, we cannot simply bridge laboratory scale to a scale several orders larger. At every step up from a smaller scale to a larger scale, we encountered something different. The difference could be small for each step but, after many steps, we could see something quite different. Studies on all scales are important. With such considerations in mind, we can list some issues worth being addressed in future studies.

Thus far, it has been observed that a weak strength (relative to the host rock) optimally oriented planar structure shows fracturing behavior underlined with several parallels with natural earthquakes. For example, asperity regions in lab and field studies are connected to spatial b-value anomalies, and such regions appear to play an important role in controlling the nucleation spots of dynamic slip events (Goebel et al., 2012; Lei et al., 2003; McLaskey & Lockner, 2014). It is suggested that naturally fractured surface resembled natural fault structure more closely (Goebel et al., 2013a).

The process zone at the front of a propagating fault involves many important scaling laws (cf. Scholz et al., 1993). It is a very interesting work to systematically examine these laws using the newest AE monitoring technology and better analogue models. An increase in process zone size with increasing fault length is also expected but is not yet proved.

Through indentation test, AE data can be used to predict drillability of rock (Jung et al., 1994). Impact test is used for the investigation of damaging in rocks during dynamic loading. In such a case, since thousands of AEs occurred in a very short time, generally in mini seconds, AE signals overlap with each other and are recorded as continuous waveforms (Chmel & Shcherbakov, 2013; Chmel & Shcherbakov, 2014). It is impossible to distinguish individual events, and thus how to explore useful information from the waveforms is an important issue required further study.

AE measurement under very high temperature (up to more than 1000 °C) and pressure, as analogy of volcano seismicity/tremor and deep focused earthquakes, was applied in laboratory experiment reproducing magma migration in fractures. It was found that opening fractures emit high-frequency acoustic events, while the flow of the melt in the fractures accompanies low frequency and harmonic tremor (Burlini et al., 2007). It is technical challenge to be able observe AE signals at multi-sensors under such high temperature. By using a D-DIA cell, phase transformations of metastable olivine were recently investigated in laboratory using an AE monitor. Several laboratory deformation experiments on germanium olivine (Mg_2GeO_4) under differential stress at high pressure (P = 2 to 5 GPa) and within a temperature over 1000 °C. Tense AEs were observed during the dynamical propagation of fractures nucleated at the onset of the olivine-to-spinel transition. Moment tensor inversion shows that most acoustic emissions arise from pure shear sources, similar to deep-focus earthquakes (Schubnel et al., 2013).

In nature, fault healing is an important process. It is difficulty to simulate such process in laboratory using rock samples. Since glassy polymer poly (methyl methacrylate) (PMMA) has a low hardness and melting temperature (160 °C), the behavior of PMMA–PMMA interfaces at room temperature and modest stress levels (100 kPa) may be somewhat representative of the behavior of rocks at depth and thus commonly used as a model material for fault rupture and friction (McLaskey & Glaser, 2011). With such a model material, it was found that fault healing promotes high-frequency acoustic emissions in laboratory experiments (McLaskey *et al.*, 2012). To represent the healing process in rocks is a challenge of wide interests.

The role of fluids in rock fracturing is an interesting issue not only in earthquake seismology but also in industrial applications in which fluids are injected into the Earth. Pore fluid is a very important factor that must be properly addressed in order to understand the faulting behaviors of crustal rocks, particularly the nucleation process. In fault mechanics, the scenario of stability, dilatancy hardening and poor drainage has been argued since Rice in the 1980s (Rice, 1983). This fault zone dilatancy theory is the result of the negative feedback between dilatancy-hardening and slip-weakening during the mainshock fault movement. Lei *et al.* (2011) proposed a different scenario of stability with no dilatancy hardening and good drainage for the failure of intact porous rocks. Drained intact rock comes to be stable via a lowered strength that has been realized before the fault starts moving. The competing mechanism between dilatancy hardening and pore pressure diffusion is strongly dependent on local hydraulic conditions and thus may result in different fracturing behaviors. Thus, systematic experiments under various drainage conditions using various rocks of different hydraulic properties are required. Quantitative investigation of rock fracture through such experiments by means of AE techniques is an interesting issue for the future.

Hydrofracturing is a fundamental technology in applications including: enhanced geothermal systems (EGS), fracking shale gas, tight gas and corebed gas, and CO_2 geological storage. It is also an effective method to control blasting by increasing permeability (*e.g.* Huang *et al.*, 2011). In such applications, fracture networks and fault reactivation (unnecessary but cannot be avoided) are directly driven by fluid pressure. Laboratory AE study may provide a fundamental technical background promoting these applications and thus has shown increasing attentions. At the same time, it is also very important to address problems raised by fault reactivation, such as induced earthquake and fluid seepage and leakage. Water-pressure induced AEs might involve different source process such as rupture velocity as compared with stress-induced ones, thus study on source process of AE event may be helpful for distinguishing fluid induced events from those induced by regional stress in fields of water injection.

REFERENCES

Akaike, H. (1998). Information theory and an extension of the maximum likelihood principle. *Selected Papers of Hirotugu Akaike*: Springer, 199–213.

Aker, E., Kühn, D., Vavryčuk, V., Soldal, M., & Oye, V. (2014). Experimental investigation of acoustic emissions and their moment tensors in rock during failure. *International Journal of Rock Mechanics and Mining Sciences* 70, 286–295.

Alkan, H., Cinar, Y., & Pusch, G. (2007). Rock salt dilatancy boundary from combined acoustic emission and triaxial compression tests. *International Journal of Rock Mechanics and Mining Sciences* 44, 108–119.

Ashby, M. & Sammis, C. (1990). The damage mechanics of brittle solids in compression. *Pure and Applied Geophysics* 133, 489–521.

Benson, P. M., Thompson, B. D., Meredith, P. G., Vinciguerra, S., & Young, R. P. (2007). Imaging slow failure in triaxially deformed Etna basalt using 3D acoustic-emission location and X-ray computed tomography. *Geophysical Research Letters* 34, L03303, doi:10.1029/2006GL028721.

Benson, P. M., Vinciguerra, S., Meredith, P. G., & Young, R. P. (2008). Laboratory simulation of volcano seismicity. *Science* **322**, 249–252.

Brown, S. R. & Scholz, C. H. (1985). Closure of random elastic surfaces in contact. *Journal of Geophysical Research* **90**, 5531.

Bruce, A. & Wallace, D. (1989). Critical point phenomena: Universal physics at large length scales. *The new physics*, 236–267.

Burlini, L., Vinciguerra, S., Di Toro, G., De Natale, G., Meredith, P., & Burg, J.-P. (2007). Seismicity preceding volcanic eruptions: New experimental insights. *Geology* **35**, 183.

Byerlee, J. D. & Lockner, D. (1977). Acoustic emission during fluid injection into rock. *Proceedings of 1st Conference on Acoustic Emission/Microseismic Activity in Geological Structures and Materials*: Trans Tech Publications, 87–98.

Chang, S.-H. & Lee, C.-I. (2004). Estimation of cracking and damage mechanisms in rock under triaxial compression by moment tensor analysis of acoustic emission. *International Journal of Rock Mechanics and Mining Sciences* **41**, 1069–1086.

Chapman, M. C. & Godbee, R. W. (2012). Modeling geometrical spreading and the relative amplitudes of vertical and horizontal high-frequency ground motions in Eastern North America. *Bulletin of the Seismological Society of America* **102**, 1957–1975.

Chmel, A. & Shcherbakov, I. (2013). A comparative acoustic emission study of compression and impact fracture in granite. *International Journal of Rock Mechanics and Mining Sciences* **64**, 56–59.

Chmel, A. & Shcherbakov, I. (2014). Temperature dependence of acoustic emission from impact fractured granites. *Tectonophysics* **632**, 218–213.

Cornet, F., Helm, J., Poitrenaud, H., & Etchecopar, A. (1998). Seismic and aseismic slips induced by large-scale fluid injections. *Seismicity Associated with Mines, Reservoirs and Fluid Injections*: Springer, 563–583.

Cox, S. J. D. & Scholz, C. H. (1988a). On the formation and growth of faults: An experimental study. *Journal of Structural Geology* **10**, 413–430.

Cox, S. J. D. & Scholz, C. H. (1988b). Rupture initiation in shear fracture of rocks: An experimental study. *Journal of Geophysical Research* **93**, 3307.

Dieterich, J. H. (1992). Earthquake nucleation on faults with rate-and state-dependent strength. *Tectonophysics* **211**, 115–134.

Dowrick, D. J. & Rhoades, D. A. (2004). Relations between earthquake magnitude and fault rupture dimensions: How regionally variable are they? *Bulletin of the Seismological Society of America* **94**, 776–788.

Enoki, M. & Kishi, T. (1988). Theory and analysis of deformation moment tensor due to microcracking. *International Journal of Fracture* **38**, 295–310.

Eshelby, J. D. (1957). The determination of the elastic field of an ellipsoidal inclusion, and related problems. *Proceedings of the Royal Society of London A: Mathematical, Physical and Engineering Sciences*: The Royal Society, 376–396.

Feng, X.-T. & Seto, M. (1999). Fractal structure of the time distribution of microfracturing in rocks. *Geophysical Journal International* **136**, 275–285.

Fortin, J., Stanchits, S., Dresen, G., & Gueguen, Y. (2009). Acoustic emissions monitoring during inelastic deformation of porous sandstone: Comparison of three modes of deformation. *Pure and Applied Geophysics* **166**, 823–841.

Frohlich, C. & Davis, S. D. (1990). Single-link cluster analysis as a method to evaluate spatial and temporal properties of earthquake catalogues. *Geophysical Journal International* **100**, 19–32.

Goebel, T., Sammis, C., Becker, T., Dresen, G., & Schorlemmer, D. (2013a). A comparison of seismicity characteristics and fault structure between stick–slip experiments and nature. *Pure and Applied Geophysics*, 1–18.

Goebel, T. H. W., Becker, T. W., Sammis, C. G., Dresen, G., & Schorlemmer, D. (2014). Off-fault damage and acoustic emission distributions during the evolution of structurally complex faults over series of stick-slip events. *Geophysical Journal International* **197**, 1705–1718.

Goebel, T. H. W., Becker, T. W., Schorlemmer, D., Stanchits, S., Sammis, C., Rybacki, E., Dresen, G. (2012). Identifying fault heterogeneity through mapping spatial anomalies in acoustic emission statistics. *Journal of Geophysical Research* **117**.

Goebel, T. H. W., Candela, T., Sammis, C. G., Becker, T. W., Dresen, G., Schorlemmer, D. (2013b). Seismic event distributions and off-fault damage during frictional sliding of saw-cut surfaces with pre-defined roughness. *Geophysical Journal International* **196**, 612–625.

Grgic, D. & Amitrano, D. (2009). Creep of a porous rock and associated acoustic emission under different hydrous conditions. *Journal of Geophysical Research: Solid Earth* **114**, B10201, doi:10.1029/2006JB004881.

Gutenberg, B. & Richter, C. F. (1944). Frequency of earthquakes in California. *Bulletin of the Seismological Society of America* **34**, 185–188.

Hanks, T. C. & Kanamori, H. (1979). A moment magnitude scale. *Journal of Geophysical Research* **84**, 2348.

He, M. C., Miao, J. L., & Feng, J. L. (2010). Rock burst process of limestone and its acoustic emission characteristics under true-triaxial unloading conditions. *International Journal of Rock Mechanics and Mining Sciences* **47**, 286–298.

Hirata, T., Satoh, T., & Ito, K. (1987). Fractal structure of spatial distribution of microfracturing in rock. *Geophysical Journal International* **90**, 369–374.

Holcomb, D. J. (1993). General theory of the Kaiser effect. *International Journal of Rock Mechanics and Mining Sciences & Geomechanics Abstracts*: Elsevier, 929–935.

Huang, B., Liu, C., Fu, J., & Guan, H. (2011). Hydraulic fracturing after water pressure control blasting for increased fracturing. *International Journal of Rock Mechanics and Mining Sciences* **48**, 976–983.

Hudson, J., Pearce, R., & Rogers, R. (1989). Source type plot for inversion of the moment tensor. *Journal of Geophysical Research: Solid Earth (1978–2012)* **94**, 765–774.

Ishida, T., Aoyagi, K., Niwa, T., Chen, Y., Murata, S., Chen, Q., & Nakayama, Y. (2012). Acoustic emission monitoring of hydraulic fracturing laboratory experiment with super-critical and liquid CO_2. *Geophysical Research Letters* **39**, n/a–n/a.

Jung, S., Prisbrey, K., & Wu, G. (1994). Prediction of rock hardness and drillability using acoustic emission signatures during indentation. *International Journal of Rock Mechanics and Mining Sciences & Geomechanics Abstracts*: Elsevier, 561–567.

Kaiser, J. (1950). An investigation into the occurrence of noises in tensile tests, or a study of acoustic phenomena in tensile tests. Ph.D Thesis, Tech. Hosch. Munchen, Munich, Germany.

Kanamori, H. (1977). The energy release in great earthquakes. *Journal of Geophysical Research* **82**, 2981–2987.

Kato, N., Satoh, T., Lei, X., Yamamoto, K., & Hirasawa, T. (1999). Effect of fault bend on the rupture propagation process of stick-slip. *Tectonophysics* **310**, 81–99.

Kato, N., Yamamoto, K., & Hirasawa, T. (1994). Microfracture processes in the breakdown zone during dynamic shear rupture inferred from laboratory observation of near-fault high-frequency strong motion. *Pure and Applied Geophysics* **142**, 713–734.

Kawakata, H., Cho, A., Yanagidani, T., & Shimada, M. (1997). The observations of faulting in Westerly granite under triaxial compression by X-ray CT scan. *International Journal of Rock Mechanics and Mining Sciences* **34**, 151.e151–151.e112.

Kawakata, H., Cho, A., Yanagidani, T., & Shimada, M. (2000). Gross structure of a fault during its formation process in Westerly granite. *Tectonophysics* **323**, 61–76.

Kranz, R. L., Satoh, T., Nishizawa, O., Kusunose, K., Takahashi, M., Masuda, K., & Hirata, A. (1990). Laboratory study of fluid pressure diffusion in rock using acoustic emissions. *Journal of Geophysical Research* **95**, 21593.

Kurths, J. & Herzel, H. (1987). An attractor in a solar time series. *Physica D: Nonlinear Phenomena* 25, 165–172.

Kwiatek, G., Goebel, T., & Dresen, G. (2014). Seismic moment tensor and b value variations over successive seismic cycles in laboratory stick-slip experiments. *Geophysical Research Letters* 41, 5838–5846.

Kwiatek, G., Plenkers, K., Nakatani, M., Yabe, Y., & Dresen, G. (2010). Frequency-magnitude characteristics down to magnitude -4.4 for induced seismicity recorded at Mponeng Gold Mine, South Africa. *Bulletin of the Seismological Society of America* 100, 1165–1173.

Laherrere, J. & Sornette, D. (1998). Stretched exponential distributions in nature and economy: "fat tails" with characteristic scales. *The European Physical Journal B-Condensed Matter and Complex Systems* 2, 525–539.

Lajtai, E., Schmidtke, R., & Bielus, L. (1987). The effect of water on the time-dependent deformation and fracture of a granite. *International Journal of Rock Mechanics and Mining Sciences & Geomechanics Abstracts*: Elsevier, 247–255.

Lavrov, A. (2003). The Kaiser effect in rocks: Principles and stress estimation techniques. *International Journal of Rock Mechanics and Mining Sciences* 40, 151–171.

Lei, X. (2003). How do asperities fracture? An experimental study of unbroken asperities. *Earth and Planetary Science Letters* 213, 347–359.

Lei, X. (2006). Typical phases of pre-failure damage in granitic rocks under differential compression. *Geological Society, London, Special Publications* 261, 11–29.

Lei, X. (2012). Dragon-Kings in rock fracturing: Insights gained from rock fracture tests in the laboratory. *The European Physical Journal Special Topics* 205, 217–230.

Lei, X., Funatsu, T., & Villaescusa, E. (2013a). Fault formation in foliated rock – insights gained from a laboratory study. *Proceedings of the 8th International Symposium on Rockbursts and Seismicity in Mines (RaSim8) (ed. Malovichko A, Malovichko D)*: Russia Saint-Peterburg-Moscow, 41–49.

Lei, X., Kusunose, K., Nishizawa, O., Cho, A., & Satoh, T. (2000a). On the spatio-temporal distribution of acoustic emissions in two granitic rocks under triaxial compression: The role of pre-existing cracks. *Geophysical Research Letters* 27, 1997–2000.

Lei, X., Kusunose, K., Rao, M. V. M. S., Nishizawa, & O., Satoh, T. (2000b). Quasi-static fault growth and cracking in homogeneous brittle rock under triaxial compression using acoustic emission monitoring. *Journal of Geophysical Research* 105, 6127.

Lei, X., Kusunose, K., Satoh, T., & Nishizawa, O. (2003). The hierarchical rupture process of a fault: An experimental study. *Physics of the Earth and Planetary Interiors* 137, 213–228.

Lei, X., Li, X., & Li, Q. (2014). Insights on injection-induced seismicity gained from laboratory AE study—fracture behavior of sedimentary rocks. In: Shimizu, N., Kaneko, K., Kodama, J. (eds.) *8th Asian Rock Mechanics Symposium*. Sapporo, Japan: Japanese Committee for Rock Mechanics, 947–953.

Lei, X., Li, S., & Liu, L. (2015). Fracturing behaviors of unfavorably oriented faults investigated using an acoustic emission monitor. *World Conference on Acoustic Emission Technology*: Hawaii, USA.

Lei, X. & Ma, S. (2014). Laboratory acoustic emission study for earthquake generation process. *Earthquake Science* 27, 627–646.

Lei, X., Ma, S., Su, J., & Wang, X. (2013b). Inelastic triggering of the 2013 Mw6.6 Lushan earthquake by the 2008 Mw7.9 Wenchuan earthquake. *Seismology and Geology* 35, 411–422.

Lei, X., Masuda, K., Nishizawa, O., Jouniaux, L., Liu, L., Ma, W., Satoh, T., & Kusunose, K. (2004). Detailed analysis of acoustic emission activity during catastrophic fracture of faults in rock. *Journal of Structural Geology* 26, 247–258.

Lei, X., Nishizawa, O., Kusunose, K., & Satoh, T. (1992). Fractal structure of the hypocenter distributions and focal mechanism solutions of acoustic emission in two granites of different grain sizes. *Journal of Physics of the Earth* 40, 617–634.

Lei, X. & Satoh, T. (2007). Indicators of critical point behavior prior to rock failure inferred from pre-failure damage. *Tectonophysics* **431**, 97–111.

Lei, X., Satoh, T., & Nishizawa, O. (2006). Experimental study on stress-induced pre-failure damage in rocks and its applications to earthquake source-process research based on AE. *Geological Society*, London, Special Publications **261**, 11–29.

Lei, X., Tamagawa, T., Tezuka, K., & Takahashi, M. (2011) Role of drainage conditions in deformation and fracture of porous rocks under triaxial compression in the laboratory. *Geophysical Research Letters* **38**, L24310, doi:10.1029/2011GL049888.

Lei, X., Yu, G., Ma, S., Wen, X., & Wang, Q. (2008). Earthquakes induced by water injection at ~3 km depth within the Rongchang gas field, Chongqing, China. *Journal of Geophysical Research* **113**, B10310, doi:10.1029/2008JB005604.

Lei, X. L., Nishizawa, O., Kusunose, K., Cho, A., Satoh, T., & Nishizawa, O. (2000c). Compressive failure of mudstone samples containing quartz veins using rapid AE monitoring: The role of asperities. *Tectonophysics* **328**, 329–340.

Li, Y. H., Liu, J. P., Zhao, X. D., & Yang, Y. J. (2010). Experimental studies of the change of spatial correlation length of acoustic emission events during rock fracture process. *International Journal of Rock Mechanics and Mining Sciences* **47**, 1254–1262.

Liakopoulou-Morris, F., Main, I. G., Crawford, B. R., & Smart, B. G. (1994). Microseismic properties of a homogeneous sandstone during fault nucleation and frictional sliding. *Geophysical Journal International* **119**, 219–230.

Lockner, D. (1993). The role of acoustic emission in the study of rock fracture. *International Journal of Rock Mechanics and Mining Sciences & Geomechanics Abstracts* **30**, 883–899.

Lockner, D. & Byerlee, J. (1977). Acoustic emission and creep in rock at high confining pressure and differential stress. *Bulletin of the Seismological Society of America* **67**, 247–258.

Lockner, D. & Byerlee, J. (1980). Development of fracture planes during creep in granite. *Proceedings of the 2nd Conference on Acoustic Emission/Microseismic Activity in Geological Structures and Materials* (Edited by H. R. Hardy and W. F. Leighton), pp. 11–25. Trans-Tech Publications, Clausthal-Zellerfeld, Germany, 11–25.

Lockner, D. A., Byerlee, J., Kuksenko, V., Ponomarev, A., & Sidorin, A. (1991a). Quasi-static fault growth and shear fracture energy in granite. *Nature* **350**, 39–42.

Lockner, D. A., Byerlee, J. D., Kuksenko, V., Ponomarev, A., & Sidorin, A. (1991b). Quasi-static fault growth and shear fracture energy in granite. *Nature* **350**, 39–42.

Lockner, D. A., Byerlee, J., Kuksenko, V., Ponomarev, A., & Sidorin, A. (1992). Observations of quasistatic fault growth from acoustic emissions. *International Geophysics* **51**, 3–31.

Ma, S., Ma, J., & Liu, L. (2002). Experimental evidence for seismic nucleation phase. *Chinese Science Bulletin* **47**, 769–773.

Ma, Y., Wang, E., Xiao, D., Li, Z., Liu, J., & Gan, L. (2012). Acoustic emission generated during the gas sorption–desorption process in coal. *International Journal of Mining Science and Technology* **22**, 391–397.

Main, I. G. (1991). A modified Griffith criterion for the evolution of damage with a fractal distribution of crack lengths: Application to seismic event rates and b-values. *Geophysical Journal International* **107**, 353–362.

Main, I. G., Meredith, P. G., & Jones, C. (1989). A reinterpretation of the precursory seismic b-value anomaly from fracture mechanics. *Geophysical Journal International* **96**, 131–138.

Main, I. G., Meredith, P. G., & Sammonds, P. R. (1992). Temporal variations in seismic event rate and b-values from stress corrosion constitutive laws. *Tectonophysics* **211**, 233–246.

Main, I. G., Sammonds, P. R., & Meredith, P. G. (1993). Application of a modified Griffith criterion to the evolution of fractal damage during compressive rock failure. *Geophysical Journal International* **115**, 367–380.

Mair, K., Marone, C., & Young, R. P. (2007). Rate Dependence of Acoustic Emissions Generated during Shear of Simulated Fault Gouge. *Bulletin of the Seismological Society of America* **97**, 1841–1849.

Majewska, Z., Ceglarska-Stefańska, G., Majewski, S., & Ziętek, J. (2009). Binary gas sorption/desorption experiments on a bituminous coal: Simultaneous measurements on sorption kinetics, volumetric strain and acoustic emission. *International Journal of Coal Geology* 77, 90–102.

Manthei, G., Eisenblätter, J., & Dahm, T. (2001). Moment tensor evaluation of acoustic emission sources in salt rock. *Construction and Building Materials* 15, 297–309.

Masuda, K. (2013). Source duration of stress and water-pressure induced seismicity derived from experimental analysis of P wave pulse width in granite. *Geophysical Research Letters* 40, 3567–3571.

Masuda, K., Mizutani, H., Yamada, I., & Imanishi, Y. (1988). Effects of water on time-dependent behavior of granite. *Journal of Physics of the Earth* 36, 291–313.

Masuda, K., Nishizawa, O., Kusunose, K., Satoh, T., Takahashi, M., & Kranz, R. L. (1990). Positive feedback fracture process induced by nonuniform high-pressure water flow in dilatant granite. *Journal of Geophysical Research: Solid Earth (1978–2012)* 95, 21583–21592.

Mayr, S. I., Stanchits, S., Langenbruch, C., Dresen, G., & Shapiro, S. A. (2011). Acoustic emission induced by pore-pressure changes in sandstone samples. *Geophysics* 76, MA21–MA32.

McGarr, A. & Fletcher, J. B. (2003). Maximum slip in earthquake fault zones, apparent stress, and stick-slip friction. *Bulletin of the Seismological Society of America* 93, 2355–2362.

McLaskey, G. C. & Glaser, S. D. (2011). Micromechanics of asperity rupture during laboratory stick slip experiments. *Geophysical Research Letters* 38, L12302, doi:10.1029/2011GL047507.

McLaskey, G. C. & Lockner, D. A. (2014). Preslip and cascade processes initiating laboratory stick slip. *Journal of Geophysical Research: Solid Earth*.

McLaskey, G. C., Thomas, A. M., Glaser, S. D., & Nadeau, R. M. (2012). Fault healing promotes high-frequency earthquakes in laboratory experiments and on natural faults. *Nature* 491, 101–104.

Menéndez, B., Zhu, W., & Wong, T.-F. (1996). Micromechanics of brittle faulting and cataclastic flow in Berea sandstone. *Journal of Structural Geology* 18, 1–16.

Meredith, P. G., Main, I. G., & Jones, C. (1990). Temporal variations in seismicity during quasi-static and dynamic rock failure. *Tectonophysics* 175, 249–268.

Mogi, K. (1962). Study of elastic shocks caused by the fracture of heterogeneous materials and its relations to earthquake phenomena. Bulletin of the Earthquake Research Institute, Tokyo University, 40, 126–173.

Mogi, K. (1968). Source locations of elastic shocks in the fracturing process in rocks (1). Bulletin of the Earthquake Research Institute, Tokyo University, 46, 1103–1125.

Moura, A., Lei, X., & Nishisawa, O. (2005). Prediction scheme for the catastrophic failure of highly loaded brittle materials or rocks. *Journal of the Mechanics and Physics of Solids* 53, 2435–2455.

Moura, A., Lei, X., & Nishisawa, O. (2006). Self-similarity in rock cracking and related complex critical exponents. *Journal of the Mechanics and Physics of Solids* 54, 2544–2553.

Nishizawa, O., Kusunose K Yanagidani, T., Oguchi, F., & Ehara, S. (1982). Stochastic process of the occurrence of AE events and hypocenter distributions during creep in Ohshima granite, Zisin. *Journal of the Seismological Society of Japan* 35, 117–132.

Nishizawa, O., Onai, K., & Kusunose, K. (1984). Hypocenter distribution and focal mechanism of AE events during two stress stage creep in Yugawara andesite. *Pure and Applied Geophysics* 122, 36–52.

Ohnaka, M. & Mogi, K. (1982). Frequency characteristics of acoustic emission in rocks under uniaxial compression and its relation to the fracturing process to failure. *Journal of Geophysical Research: Solid Earth (1978–2012)* 87, 3873–3884.

Ohtsu, M. (2008). Moment tensor analysis. *Acoustic Emission Testing*: Springer, 175–200.

Ojala, I. O. (2004). Strain rate and temperature dependence of Omori law scaling constants of AE data: Implications for earthquake foreshock-aftershock sequences. *Geophysical Research Letters* 31, L24617, doi:10.1029/2004GL020781.

Pappas, Y., Markopoulos, Y., & Kostopoulos, V. (1998). Failure mechanisms analysis of 2D carbon/carbon using acoustic emission monitoring. *NDT & E International* **31**, 157–163.

Ranjith, P., Jasinge, D., Choi, S.-K., Mehic, M., & Shannon, B. (2010). The effect of CO_2 saturation on mechanical properties of Australian black coal using acoustic emission. *Fuel* **89**, 2110–2117.

Rao, M., Lakshmi, K. P., Rao, G. N., Vijayakumar, R., & Udayakumar, S. (2011). Precursory microcracking and brittle failure of Latur basalt and migmatite gneiss under compressive loading. *Current Science(Bangalore)* **101**, 1053–1059.

Reches, Z. e. & Lockner, D. A. (1994). Nucleation and growth of faults in brittle rocks. *Journal of Geophysical Research: Solid Earth (1978–2012)* **99**, 18159–18173.

Rice, J. R. (1983). Constitutive relations for fault slip and earthquake instabilities. *Pure and Applied Geophysics* **121**, 443–475.

Sammonds, P. & Ohnaka, M. (1998). Evolution of microseismicity during frictional sliding. *Geophysical Research Letters* **25**, 699–702.

Sano, O., Terada, M., & Ehara, S. (1982). A study on the time-dependent microfracturing and strength of Oshima granite. *Tectonophysics* **84**, 343–362.

Satoh, T., Kusunose, K., & Nishizawa, O. (1987). A minicomputer system for measuring and processing AE waveform -High speed digital recording and automatic hypocenter determination-. *Bulletin of the Geological Survey Japan (in Japanese)* **38**, 295–303.

Satoh, T., Shivakumar, K., Nishizawa, O., & Kusunose, K. (1996). Precursory localization and development of microfractures along the ultimate fracture plane in amphibolite under triaxial creep. *Geophysical Research Letters* **23**, 865–868.

Scholz, C. (1968a). Experimental study of the fracturing process in brittle rock. *Journal of Geophysical Research* **73**, 1447–1454.

Scholz, C. (1968b). The frequency-magnitude relation of microfracturing in rock and its relation to earthquakes. *Bulletin of the Seismological Society of America* **58**, 399–415.

Scholz, C. (1968c). Microfracturing and the inelastic deformation of rock in compression. *Journal of Geophysical Research* **73**, 1417–1432.

Scholz, C., Dawers, N., Yu, J. Z., Anders, M., & Cowie, P. (1993). Fault growth and fault scaling laws: Preliminary results. *Journal of Geophysical Research: Solid Earth (1978–2012)* **98**, 21951–21961.

Schubnel, A., Brunet, F., Hilairet, N., Gasc, J., Wang, Y., & Green, H. W. (2013). Deep-focus earthquake analogs recorded at high pressure and temperature in the laboratory. *Science* **341**, 1377–1380.

Schubnel, A., Thompson, B. D., Fortin, J., Guéguen, Y., & Young, R. P. (2007). Fluid-induced rupture experiment on Fontainebleau sandstone: Premonitory activity, rupture propagation, and aftershocks. *Geophysical Research Letters* **34**, L19307, doi:10.1029/2007GL031076.

Seto, M., Utagawa, M., Katsuyama, K., Nag, D., & Vutukuri, V. (1997). In situ stress determination by acoustic emission technique. *International Journal of Rock Mechanics and Mining Sciences* **34**, 281. e281–281. e216.

Sondergeld, C. H. & Estey, L. H. (1981). Acoustic emission study of microfracturing during the cyclic loading of westerly granite. *Journal of Geophysical Research* **86**, 2915.

Stanchits, S., Mayr, S., Shapiro, & S., Dresen, G. (2011). Fracturing of porous rock induced by fluid injection. *Tectonophysics* **503**, 129–145.

Stanchits, S., Vinciguerra, S., & Dresen, G. (2006). Ultrasonic velocities, acoustic emission characteristics and crack damage of basalt and granite. *Pure and Applied Geophysics* **163**, 975–994.

Thompson, B. D. (2005). Observations of premonitory acoustic emission and slip nucleation during a stick slip experiment in smooth faulted Westerly granite. *Geophysical Research Letters* **32**, L10304, doi:10.1029/2005GL022750.

Thompson, B. D., Young, R. P., & Lockner, D. A. (2006). Fracture in Westerly granite under AE feedback and constant strain rate loading: Nucleation, quasi-static propagation, and the transition to unstable fracture propagation. *Pure and Applied Geophysics* **163**, 995–1019.

Thompson, B. D., Young, R. P., & Lockner, D. A. (2009). Premonitory acoustic emissions and stick-slip in natural and smooth-faulted Westerly granite. *Journal of Geophysical Research* **114**.

Tyupkin, Y. S. & Di Giovambattista, R. (2005). Correlation length as an indicator of critical point behavior prior to a large earthquake. *Earth and Planetary Science Letters* **230**, 85–96.

Utsu, T. (1961). A statistical study on the occurrence of aftershocks. *Geophysical Magazine* **30**, 521–605.

Utsu, T. & Ogata, Y. (1995). The centenary of the Omori formula for a decay law of aftershock activity. *Journal of Physics of the Earth* **43**, 1–33.

Villaescusa, E., Lei, X., Nishizawa, O., & Funatsu, T. (2009). Laboratory testing of brittle intact rock–implications for in situ stress measurements and rock mass failure. *Australian Mining* **226**.

Villaescusa, E., Seto, M., & Baird, G. (2002). Stress measurements from oriented core. *International Journal of Rock Mechanics and Mining Sciences* **39**, 603–615.

Walsh, J. (1971). Stiffness in faulting and in friction experiments. *Journal of Geophysical Research* **76**, 8597–8598.

Wang, L., Ma, S., & Ma, L. (2008). Accelerating moment release of acoustic emission during rock deformation in the laboratory. *Pure and Applied Geophysics* **165**, 181–199.

Weeks, J., Lockner, D., & Byerlee, J. (1978). Change in b-values during movement on cut surfaces in granite. *Bulletin of the Seismological Society of America* **68**, 333–341.

Xie, H. P., Liu, J. F., Ju, Y., Li, J., & Xie, L. Z. (2011). Fractal property of spatial distribution of acoustic emissions during the failure process of bedded rock salt. *International Journal of Rock Mechanics and Mining Sciences* **48**, 1344–1351.

Yabe, Y. (2008). Evolution of source characteristics of AE events during frictional sliding. *Earth, Planets and Space* **60**, e5–e8.

Yabe, Y., Kato, N., Yamamoto, K., & Hirasawa, T. (2003). Effect of sliding rate on the activity of acoustic emission during stable sliding. *Pure and Applied Geophysics* **160**, 1163–1189.

Yanagidani, T., Ehara, S., Nishizawa, O., & Kusunose, K., & Terada, M. (1985). Localization of dilatancy in Ohshima granite under constant uniaxial stress. *Journal of Geophysical Research: Solid Earth (1978–2012)* **90**, 6840–6858.

Yin, G., Qin, H., Huang, G., Lv, Y., & Dai, Z. (2012). Acoustic emission from gas-filled coal under triaxial compression. *International Journal of Mining Science and Technology* **22**, 775–778.

Yin, X.-c., Yu, H.-z., Kukshenko, V., Xu, Z.-Y., Wu, Z., Li, M., Peng, K., Elizarov, S., & Li, Q. (2004). Load-unload response ratio (LURR), accelerating moment/energy release (AM/ER) and state vector saltation as precursors to failure of rock specimens. *Computational Earthquake Science Part II*: Springer, 2405–2416.

Yokota, T., Zhou, S., Mizoue, M., & Nakamura, I. (1981). An automatic-measurement of arrival-time of seismic-waves and its application to an online processing system. *Bulletintin of the Earthquake Research Institute-University of Tokyo* **56**, 449–484.

Yoshimitsu, N., Kawakata, H., & Takahashi, N. (2014). Magnitude– 7 level earthquakes: A new lower limit of self-similarity in seismic scaling relationships. *Geophysical Research Letters* **41**, 4495–4502.

Yukalov, V. I., Moura, A., & Nechad, H. (2004). Self-similar law of energy release before materials fracture. *Journal of the Mechanics and Physics of Solids* **52**, 453–465.

Yuyama, S., Okamoto, T., Shigeishi, M., & Ohtsu, M. (1995). Quantitative evaluation and visualization of cracking process in reinforced concrete by a moment tensor analysis of acoustic emission. *Materials evaluation* **53**, 751–752.

Zöller, G., Hainzl, S., & Kurths, J. (2001). Observation of growing correlation length as an indicator for critical point behavior prior to large earthquakes. *Journal of Geophysical Research* **106**, 2167.

Zang, A., Christian Wagner, F., Stanchits, S., Dresen, G., Andresen, R., & Haidekker, M. A. (1998). Source analysis of acoustic emissions in Aue granite cores under symmetric and asymmetric compressive loads. *Geophysical Journal International* **135**, 1113–1130.

Zang, A., Wagner, F. C., Stanchits, S., Janssen, C., & Dresen, G. (2000). Fracture process zone in granite. *Journal of Geophysical Research: Solid Earth (1978–2012)* **105**, 23651–23661.

Chapter 5

Damage of rock joints using acoustic emissions

Z. Moradian

Department of Civil and Environmental Engineering and Earth Resources Lab (ERL), Massachusetts Institute of Technology (MIT), Cambridge, MA, USA

Abstract: This chapter focuses on using acoustic emission (AE) technique for investigating shear mechanisms of rock discontinuities from the initial movement up to the residual state. Direct shear tests were done on rock joints and AE technique along with image analysis and 3D topography of the surfaces, scanned by a laser profilometer, were used to obtain insights into several stages in the shear failure process of rock discontinuities. These stages are: I: pre-peak linear period, II: pre-peak non-linear period, III: post-peak period and IV: residual shear strength period. The rate and cumulative graphs of the AE parameters such as number of hits, and energy were correlated to the shear stress-shear displacement of the tested rock joints. These correlations revealed that AE has a high competency as a precursor prior to shear failure of discontinuities and therefore it can be used as an indicator of instability in slopes and structures suffering from sliding along discontinuities. These observations can also open a door for a better understanding of the mechanisms of faulting and finally for earthquake prediction. Locations of the AE event sources were determined from propagation velocity of acoustic waves and the traveling time from the event source to the AE sensor. These sources correspond to damage zones that are caused by active asperities in shearing process, mostly asperities facing the shear direction. The distribution of the source locations and their associated AE energy can provide an estimation of contact areas between upper and lower surfaces of the rock joint. It also helps to detect location, size, and failure intensity of the damaged zones during each stage in the shearing process. Presence of the AE events with low energy before shear stress peak revealed that slipping/shearing process may start from zones with less frictional resistance and then it will be controlled by rough asperities with higher frictional resistance.

1 INTRODUCTION

Discontinuities have an important role in controlling behavior of rock mass under normal and shear loading conditions. They reduce strength and increase deformability in the rock mass. Thus, the safe management of projects on or in the rock requires a precise evaluation of the rock mass stability in terms of the shear strength of the discontinuities (Patton, 1966; Ladanyi & Archambault, 1970; Barton, 1973; Kulatilake *et al.*, 1995; Seidel & Haberfield, 2002; Moradian *et al.*, 2013; Gravel *et al.*, 2015). Observations of experimental shear tests on rock discontinuities have shown valuable information in improving our knowledge about earthquakes, since the mechanism of stick-slip of rock joints in the laboratory is similar to fault mechanisms and earthquakes (Mogi, 1962, Brace & Byerlee, 1966; Scholz, 1998). Thus, understanding earthquake processes relies on studying shear fracturing in rocks (Scholz, 2002). Studying shear mechanism of rock joints also shed light on understanding mechanisms of induced earthquakes by fluid injection in several industrial applications including enhanced geothermal systems (EGS), hydraulic fracturing, geological storage of $CO2$ and waste water disposal (Johnston *et al.*, 1987; Thomason *et al.*, 2009; Maxwell *et al.*, 2009; Warpinski *et al.*, 2012; Zoback & Gorelick 2012; Ellsworth 2013).

Shear strength of rock joints depends on the applied normal stress, the roughness of the joint surfaces, the contact area between joint surfaces (degree of matching), the compression and tensile strength of the rock, the loading rate, the size of the joint (scale effect) and the environmental conditions, *e.g.* weathering, presence of water and pore pressure (Ladanyi & Archambault, 1970; Barton, 1977; Byerlee, 1978). Size, shape, and distribution of contact areas are mainly related to the geometry of the asperities, loading conditions, mechanical parameters of the rock, and shear displacement (Zhao, 1997; Gentier *et al.*, 2000; Grasselli, 2006; Park *et al.*, 2013; Moradian *et al.*, 2010; Fathi *et al.*, 2016a).

Grasselli (2006), Gentier *et al.* (2000), Moradian *et al.* (2012b) and Fathi *et al.* (2016a) stated that degradation mostly occur in steeper asperities. Therefore, instead of considering the whole contact area between surfaces, an effective contact area should be considered in the shearing process. They explained that effective contact areas consist of asperities that are facing the shear direction (Figure 1). Thus the identification of the potential damaged areas only requires the determination of the areas which face the shear direction and which, among them, are steep enough to be involved (Grasselli, 2006). Some other researchers believe that the shear strength along a joint surface may not be uniform because of the existence of zones with low

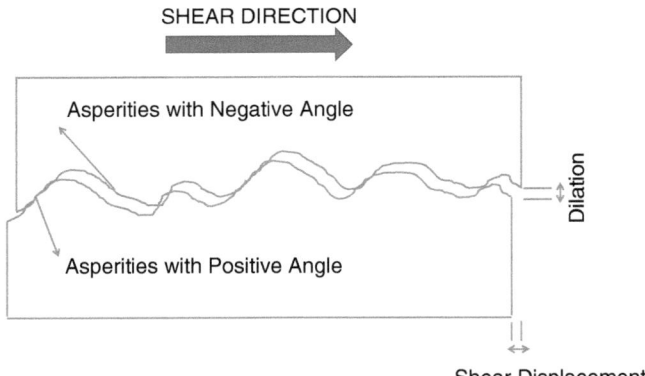

Figure 1 A schematic cartoon showing asperities with positive and negative angles toward the shear direction. Asperities with negative angles open right after slip initiation leaving asperities with positive angle to participate in shearing process. The real contact area is the sum of the in-contact asperities, mostly those with positive angle towards the shear direction.

and high frictional properties. As a result, slip may initiate from a region of low frictional resistance and spread asymmetrically to regions with higher resistance resulting in non-uniform normal and shear stress distributions along the discontinuity (Mutlu & Bobet, 2006; Hedayat *et al.*, 2014). Variation of frictional resistance along a joint, non-uniform stress field, inelastic deformations near joint boundaries, and variation of elastic modulus of the intact rock may cause non-uniform shear stress distributions (Comninou & Dundurs, 1983; Burgmann *et al.*, 1994; Gorbatikh *et al.*, 2001; Mutlu & Bobet, 2006; Malanchuk, 2011; Hedayat *et al.*, 2014).

Topography (roughness) of the rock discontinuities can be divided into two categories: first order asperities (waviness) and second order asperities (unwaviness). Second order asperities are considered as small dents on the surface of the first order asperities or on the flat areas over the joint surface. It is believed that in small shear displacements, second order asperities govern the shearing process and for large shear displacements the first other asperities take the role. It has also been shown that the effect of surface roughness on shear strength of rock joints is more pronounced for relatively low values of normal stress and it decreases with increasing normal stress (Byerlee, 1970; Huang *et al.*, 2002; Grasselli, 2006). It can be stated that at very high normal stresses, roughness of the joint surface loses its effect and the friction between rock grains replaces it.

Scale effect is the most important factor affecting roughness. Recent paper published by Tatone & Grasselli (2013) showed that with a constant measurement resolution, the scale effect is positive (increase in roughness with increasing scale). As a conclusion of their work, Tatone & Grasselli (2013) suggested that the common thought of "decrease in roughness with increasing sample size" may be resulted from inconsistent measurement resolution.

The shear strength of the rock discontinuities consists of two components: cohesion and friction. Cohesion consists of cohesive bond between the upper and the lower joint surfaces as well as the internal cohesive bond between rocks' minerals in the intact asperities. On the other hand friction of a rock discontinuity consists of basic friction angle of the rock (ϕ_b), (ϕ_r) for unweathered surfaces, and friction of the rough asperities (i). Open joints under low normal stress don't have cohesion. They may show some cohesion under high normal stress due to the breakage of intact asperities. Barton (2013) has strongly mentioned, *"Even rough open joints do not have any cohesion, but instead have very high friction angles at low stress, due to strong dilation"*. As a result, he states that defining cohesion for open joints, even as an apparent cohesion overestimates the shear strength of the rock discontinuities.

Contrarily to open joints, cohesion is a very important parameter affecting the shear strength of the closed (bonded) joints, even under low normal stress. Saiang *et al.* (2005) conducted laboratory tests on shotcrete–rock joints in direct shear test. They showed that for low values of normal load, the shear strength is determined by the bond strength for genuinely bonded shotcrete–rock interfaces. Moradian *et al.* (2011, 2012a) also showed that the cohesive bond between concrete and rock has the most important effect on shear mechanism of concrete–rock interfaces.

Mobilization of the cohesion and friction in the shear process has been under controversy. Along with Hajiabdolmajid *et al.* (2002), Barton (2013) believes that cohesion is broken at small strain, while friction is mobilized at larger strain and it remains to the end of the shear process. He proposes that the criterion 'c then σntan Φ' should replace 'c plus σntanΦ' for improved fit to reality. In other words Barton declares that the traditional thought of "cohesion and friction", used in Mohr-Coulomb and Hoek-Brown models must be replaced by "cohesion then friction" (Barton, 2013).

Damage of rock joints can be divided into two mechanisms: 1) sliding which causes overriding (dilation) of the asperities without breaking the intact asperities 2) shear damage which partially or completely breaks the intact asperities. It is believed that under low normal stress, sliding (dilation) process occurs more while under high normal stresses dilation is entirely replaced by shear damage. Gentier *et al.* (2000) and Moradian *et al.* (2010a) found that a little damage can occur before shear stress peak and most of the asperity damages occur during post-peak softening and residual period.

Barbosa (2009) categorized the shear mechanism of the pre-peak shear strength into elastic and plastic stages. In the elastic stage there is neither degradation nor dilation, thus, there is no decrement in the asperities angles. After the elastic stage, the joint starts to slide over the asperities. At this stage, degradation and dilation are initiated. Hutson and Dowding (1990) stated that under high normal stresses, asperity degradation could occur during small shear displacements. Conversely, under low normal stresses, asperity degradation can only arise if the shear displacement is large enough. In other words, if the applied normal stress is low, the amount of gouge materials is not significant. Nevertheless, at higher shear displacements where the contact area is reduced due to dilation, the local shear and normal stresses increase significantly and due to stress concentration asperity degradation happens and some gouge materials may be produced. Although increasing normal load normally increases the shear strength of the rock joints, Grasselli (2006) has shown that after a certain point (sigma n/sigma c = 0.2), the effect of normal load on shear strength will be neutral.

Researchers found out that tensile failure, rather than compressive failure, plays a major role in the failure of individual asperities (Fishman, 1990; Handanyan *et al.*, 1990; Kutter & Otto, 1990; Pereira & De Freitas, 1993; Huang *et al.*, 2002; Grasselli, 2006).

Shear strength of rock joints is shown to decrease with increasing f/a ratio where f is infilling thickness and a is mean roughness amplitude, and approaches a minimum when f/a is between 1.25 and 1.5, depending on the amount of normal stress (Papaliangas *et al.*, 1993). Residual strength decreases less markedly with increasing f/a ratio and tends towards a constant value when f/a > 1.0 (Papaliangas *et al.*, 1993). For rock joints with infilling thicker than the mean roughness amplitude, failure planes may develop through the infilling rather than along the joint interface.

The effect of fluid flow and pore pressure on the shear strength of rock discontinuities is another important factor that must be taken into account when dealing with shear strength of rock discontinuities. Generally fluid flow in the rock discontinuities decreases the shear strength in two ways: 1) the fluid facilities the sliding through the discontinuities by decreasing the friction angle especially in the existence of clay fillings and 2) by increasing the pore pressure, it decreases the effective normal stress and consequently the shear resistance of the rock discontinuity.

Predictions of rock slope failures and earthquakes triggered by stick-slip events have been an interesting but challenging field of research. Premonitory phenomena such as changes in dilatancy, creep, electrical resistivity, gas emission, ratio of seismic velocities, and seismic wave attenuation have been used to monitor and even predict slip or stick-slip events along joints and faults (Aggarwal *et al.*, 1973; Byerlee, 1978; Cicerone *et al.*, 2009; Hedayat *et al.*, 2014). The direct shear tests and stick-slip experiments are considered to be a laboratory analog of rock slope instabilities and earthquakes, which generate ultrasonic signals that are recorded as a microseismic event (Shiotani 2006; Ishida *et al.*, 2010; McLaskey *et al.*, 2014).

Acoustic emission (AE) is a transient elastic wave that is generated by the rapid release of energy within a material (Koerner *et al.*, 1981; Lockner, 1993). Besides conventional monitoring techniques such as stress and strain measurement instruments, the AE monitoring has been experimented in several civil engineering structures. However the scope of this chapter will be limited to a narrow field of study in which AE is used to study the damage of rock discontinuities.

A few researchers have addressed the application of the AE for monitoring the shear behavior of the joints including the author's works (Moradian *et al.*, 2008–2013). Li & Nordlund (1990) characterized AE during shearing of rock joints using artificial and natural joints. Their test results indicated that the AE rate peaks coincide with the stress drops caused by fracturing of asperities during joint shear. Ishida *et al.* (2010) performed in situ direct shear tests on a large block to provide some insights on analog models of seismogenic faulting. In their study, AE sources were located with an accuracy expected to fall within 50 mm. Hong & Jeon (2006) performed a series of direct shear tests to investigate the influence of shear load on AE characteristics of rock–concrete interface under constant normal load. They showed that the location of the AE sources distributed over the entire shear zone before the shear stress reach converged residual value. They believed that after the residual shear stress is attained, the sources are localized. Finally they showed that the maximum rates of count and energy were observed when the stress dropped after peak shear stress. Several studies have also been done on rock core specimens containing smooth and ground saw-cut faults under high confining pressures to study stick-slip nucleation process (*e.g.* Thompson *et al.*, 2009; Goebel *et al.*, 2014; McLaskey & Lockner, 2014). In these studies AE signals have been detected to explore how earthquakes begin.

Although the previous researches are useful to improve our understanding of rock joint shear mechanism, they are not sufficient for investigating several stages in degradation of asperities in the shear failure process of rock joints. In the present research, laboratory direct shear tests in constant normal load condition (CNL) are conducted on rock joints with different characteristics and AE and shear graphs are correlated. The rate and the cumulative graphs of the AE hits and their energies are analyzed for monitoring different stages in shear stress–shear displacement graphs of joints. The capability of the AE as a precursor for predicting slip/shear initiation of rock discontinuities and faults is evaluated too. Then the source locations of the AE signals are correlated to the locations of the damaged zones observed by images in order to study the progressive degradation of the asperities based on their geometry.

2 DIRECT SHEAR TESTS ON ROCK JOINTS

Rock joint samples were prepared by tension splitting of the 150 mm diameter rock cores drilled on a medium-grained Barre granite block from Vermont, USA. Table 1

Table 1 Physical and mechanical properties of the Barre granite.

Specific gravity	P-wave Velocity (m/s)	Young's Modulus (GPa)	Poisson Ratio	Uniaxial Compressive Strength (MPa)
2.63	4675	58.10	0.30	179

contains physical and mechanical properties of the Baree granite. The author and his colleagues have extensively investigated fracture mechanics and AE properties of this rock (Li *et al.*, 2015; Goncalvez Da Silva *et al.*, 2015; Moradian *et al.*, 2016). A dark blue color was sprayed on joint surfaces to localize the damaged zones caused by shear loading. After putting the specimens in the shear test mold and surrounding them by Sika 212 cement grout, the joint surfaces were scanned using the scanner profilometer. The laser profilometer model Kreon was used in this study. The laser emits a red, luminous plane, with a wavelength of 670 nm and a maximum output power of 4 mW. The sensor has wavelength of 670 nm, number of points per second of 30 000, and depth and width of field of 90 and 25 mm, respectively. Profiles parallel and perpendicular to the direction of shear loading were drawn with a 0.5 mm interval over the scanned surfaces (Fathi *et al.*, 2016a,b).

Direct shear tests were performed on the joint specimens in constant normal load condition using a direct shear apparatus mounted inside a rigid loading frame of a rock and concrete testing machine fabricated by Materials Testing Systems (MTS). The rate of horizontal displacement in all tests was 0.15 mm/min, and the test was considered finished when the horizontal displacement reached 10 mm. Shear stress and shear and normal displacements from applied shear loads were recorded during each test (Moradian *et al.*, 2010a). Pictures of the joint surfaces after the shear tests were taken (Figure 2), and the joint surfaces were scanned after the test.

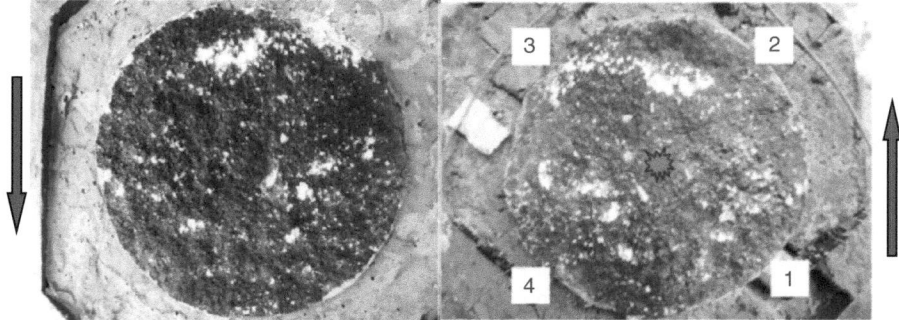

Figure 2 Photos of the joint specimen after testing. Left: fixed surface, right: mobile surface with attached AE sensors. Arrows show the shear direction. Numbers show the position and the order of the AE sensors.

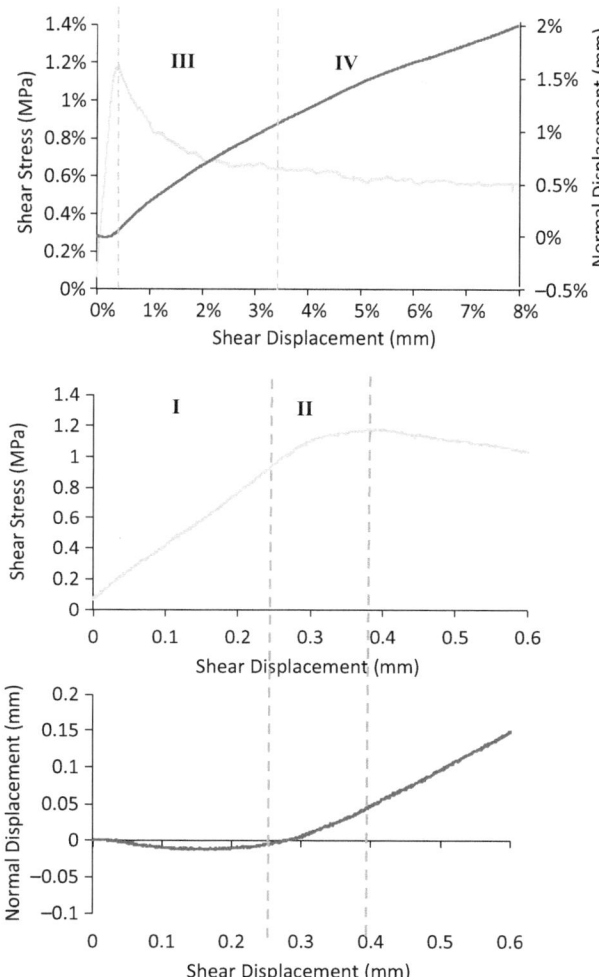

Figure 3 a) shear stress vs shear displacement, b) normal displacement vs shear displacement, c) a close-up of shear stress vs shear displacement d) a close-up of normal displacement vs shear displacement. The shear mechanism of the joint has been divided into four periods: I: pre-peak linear period, II: pre-peak non-linear period, III: post-peak period and IV: residual shear strength period.

The shear behavior of the rock joints can be divided into four periods (Figure 3):

1) Pre-peak linear period: By applying normal and shear load on the joint specimen, the upper and lower surfaces are settled and interlocked in this period. The stiffness and contact area increase and dilation normally shows a negative trend due to the contraction of the joint surfaces. There is neither degradation nor positive dilation in this period.

2) Pre-peak non-linear period (plastic period): Dilatancy is generated and increases along this period because of the sliding or damaging of the secondary asperities. In

this period, contact areas decreases and steep asperities facing the shear direction mobilize in shearing process while asperities with negative angle to the shear direction lose their contact producing hydraulic aperture. This period is ended by peak shear stress where steepest primary asperities start shearing and dilatancy shows its maximum slope.

3) Post-peak period: All secondary and primary asperities facing the shearing direction are either slipped or sheared in this period (depending on the amount of normal load) and the shear stress–shear displacement curve shows a stress drop and a progressive softening behavior. The magnitude of the stress drop (difference between maximum shear stress and the residual shear strength) represents the magnitude of the released energy or the magnitude of the generated seismic/acoustic event. Depending on the amount of normal stress and shear displacement, some gouge materials are produced from damage of asperities.

4) Residual period: Shear stress is stable in a residual stress and asperities degradation continues in a lower severity than post-peak period. Due to existence of the gouge materials on the joint surface, shear stress decreases during the residual period and the slope of the dilation graph start showing a decreasing trend. Due to dilation, the contact area between the joint surfaces is small therefore, a high stress concentration happens at those areas that may cause some asperity damage.

3   ACOUSTIC EMISSION MONITORING

The AE data acquisition system in this study was a μSAMOS from Physical Acoustic Corporation (PAC) with 16 channels. PAC R15a sensors with operating frequency range of 25–70 kHz and resonant frequency of 29 kHz were used. AE hardware was set up with a threshold of 35 dB, preamplification of 40 dB, sampling rate of 3 MSPS, sample length of 3K and a band-pass filtration of 20–400 KHz. PDT, HDT and HLT were set as 300, 800 and 1000 μsec respectively.

In addition to recording the number of hits and the signal waveforms, the AE system records certain properties of the AE signals. Common parametric features of the waveforms employed for evaluating AE characteristics are hits, amplitude, counts, duration, energy and rise time. Frequency domain features such as peak frequency and frequency centroid are also determined from the Fast Fourier Transform (FFT) of the recorded waveforms, though these parameters are very sensitive to the resonance frequency of the sensors. Figures 4 and 5 show the common AE features in the time and frequency domains.

Figure 6 shows waveforms and spectrums of a microseismic event located at the center of the specimen (star in the Figure 2) and detected by all four sensors. Table 2 lists the AE parameters of the waveforms displayed in Figure 6.

4   DETECTING DAMAGE STAGES IN SHEARING PROCESS OF ROCK JOINTS

A combination of the rate and cumulative graphs of AE hits and energy along with shear stress- shear displacement graphs have been used to correlate the shear behavior

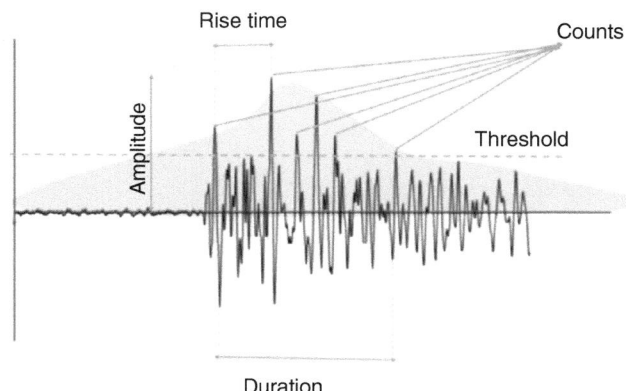

Figure 4 Common parameters of an AE waveform in time domain. Amplitude is the highest peak voltage of the signal, counts are the number of the times that the signal crosses the threshold, duration is the time interval between the first and the last threshold crossing, rise time is the time interval between first threshold crossing and the signal peak and energy is the area under the envelope of the signal (colored area) (Moradian *et al.*, 2016).

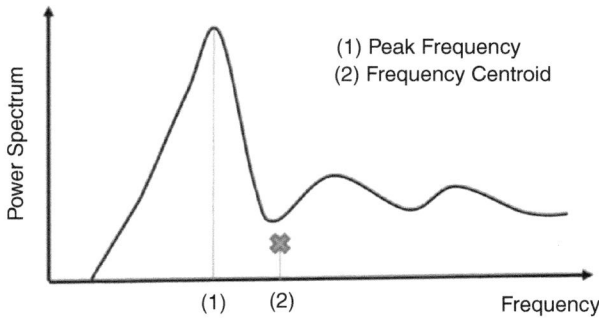

Figure 5 Peak frequency is the point where the power spectrum is greatest and frequency centroid is the center of mass of the power spectrum graph (Moradian *et al.*, 2016).

of the joints with generated AE signals. It is believed that the number of AE events is proportional to the number of damaged asperities and the amount of AE energy is proportional to the magnitude of the asperity damage. Shear behavior of joints has been divided into four periods based on their shear stress-shear displacement graphs and their AE characteristics. These periods are: I: pre-peak linear period, II: pre-peak non-linear period, III: post-peak period and IV: residual shear strength period. In this section, the evolution of the asperity damages during these periods will be discussed for rock joints. Figures 7–10 display shear stress, hits and energy vs. shear displacement for the four mentioned periods of rock joints. The graphs have been drawn based on shear displacement rather than on time to eliminate the effect of loading rate. As it can be seen in Figures 7–10, there is a distinct maximum shear stress occurring at a shear

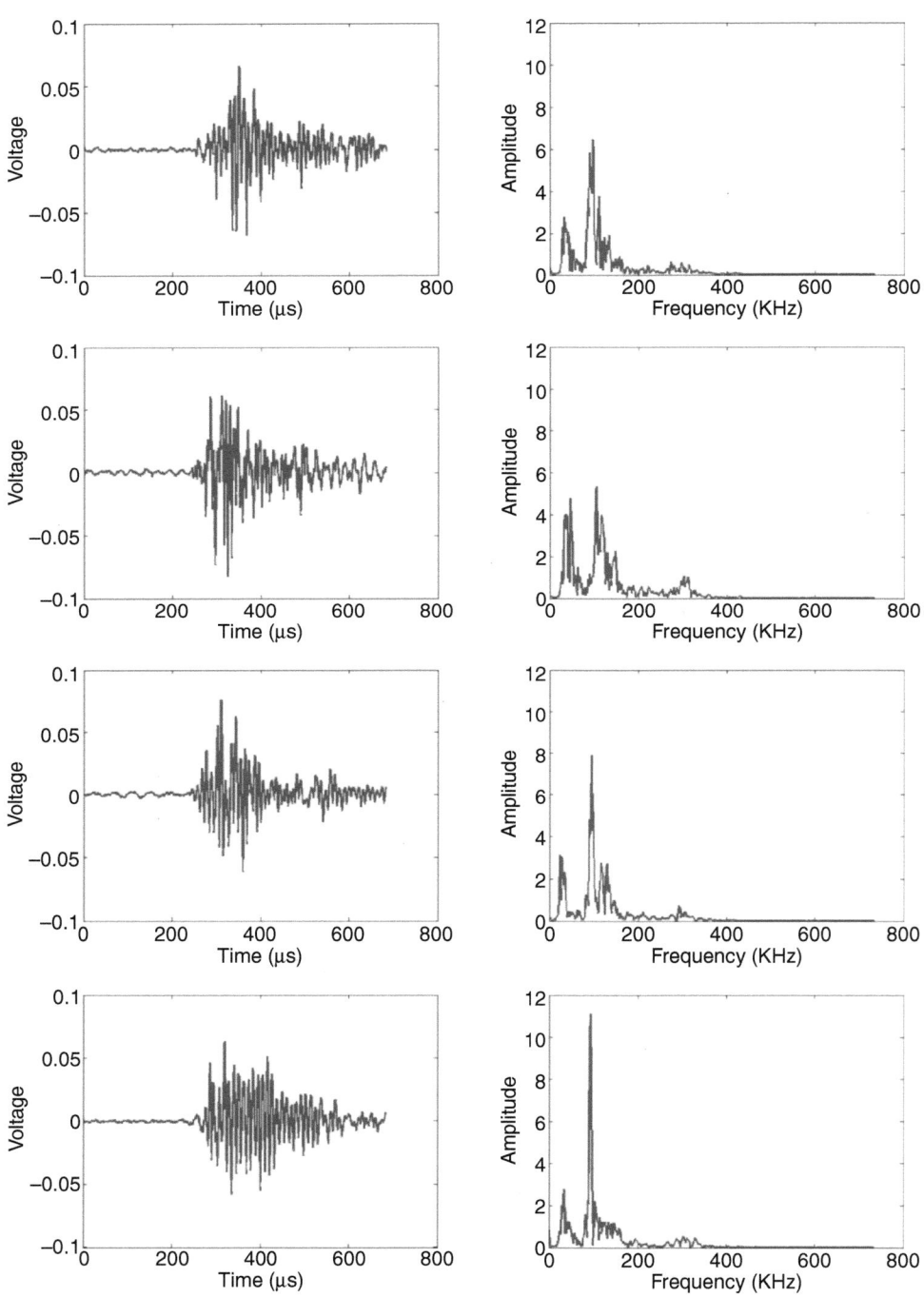

Figure 6 Waveforms and spectrums of an event located at the center of the specimen (star in the Figure 2) and detected by all four sensors.

Table 2 AE parameters of the waveforms displayed in Figure 6.

Arrival time (s)	Channels (ordered based on arrival time)	Rise time (μs)	Counts	Energy (10μvolt-sec/ count)	Duration (μs)	Amplitude (dB)	Threshold (dB)	Frequency centroid (KHz)	Peak frequency (KHz)
415.4792	4	109	39	8	694	57	35	98	96
415.4792	2	66	40	10	807	58	35	97	91
415.4792	1	52	37	8	733	58	35	98	93
415.4792	3	61	36	10	884	56	35	100	90

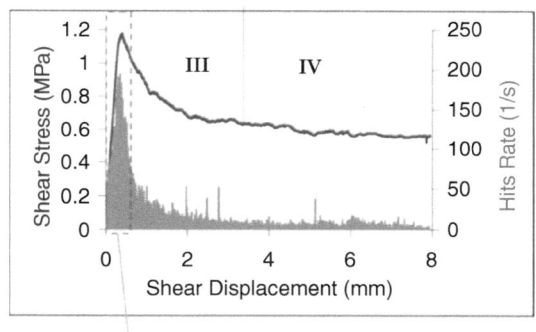

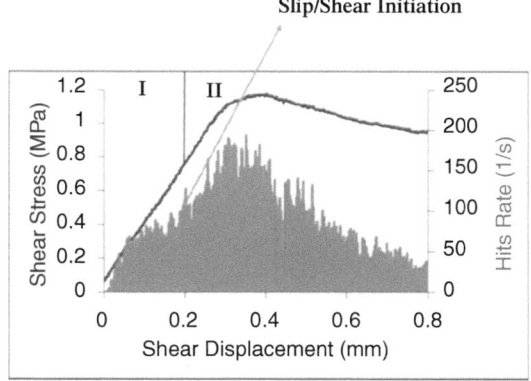

Figure 7 Shear stress and hits rate vs shear displacement for the rock joint. Slip/shear initiation point was observed at the beginning of the pre-peak non-linear period. AE hits rates show low AE activity in residual period except for some instant slipping/shearing of the remained in-contact asperities.

displacement of less than 1 mm. This maximum shear stress coincided well with the maximum peak of AE hits and energy.

AE activities generate right after applying shear load and starting of the shear displacement (Figures 7–10), therefore in pre-peak linear period (Period I), joints

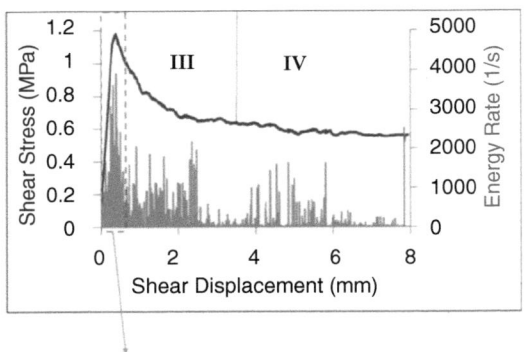

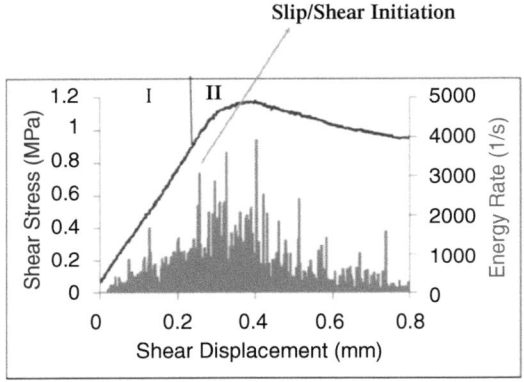

Figure 8 Shear stress and energy rate vs shear displacement for the rock joint. Slip/shear initiation point was observed at the beginning of the pre-peak non-linear period. AE energy rates may locally show high AE activity in residual period. Although instant slipping/shearing of the remained in-contact asperities in residual period produce a few AE hits, their associated energy is high due to high stress concentration.

show some AE activity for AE rate graphs and increasing with concavity in cumulative graphs. It is believed that these initial activities in pre-peak linear period come from sitting and locking of the upper and lower joint halves rather than asperity degradation. In pre-peak non-linear period (Period II), joints show an increase in values of AE parameters proportionally to the loading before maximum shear stress. These activities are generated from breaking and sliding of the secondary asperities. The slip/shear initiation of the rock joints was observed at the beginning of the pre-peak non-linear period for both rate and cumulative graphs though it was better observed for the latter. The failure process in slip/shear initiation point produces many small AE signals, as a result of degradation of secondary asperities, however the rate of these AE signals may not be very well distinguishable from the background noise, while the cumulative graph (as the sum of all these small AE signals) distinctly show the boost in AE activity in slip/shear initiation point. During post-peak period (Period III), all asperities (secondary and primary) are either slipped or sheared off (depending on the amount of the normal load) and joints show a sudden increase of AE parameters after shear stress peak, so

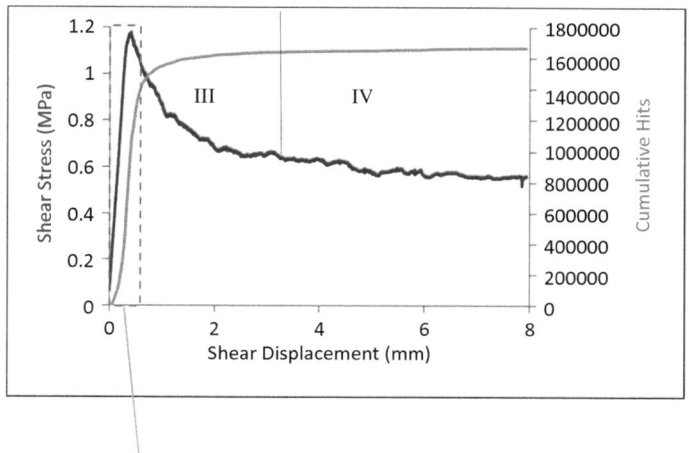

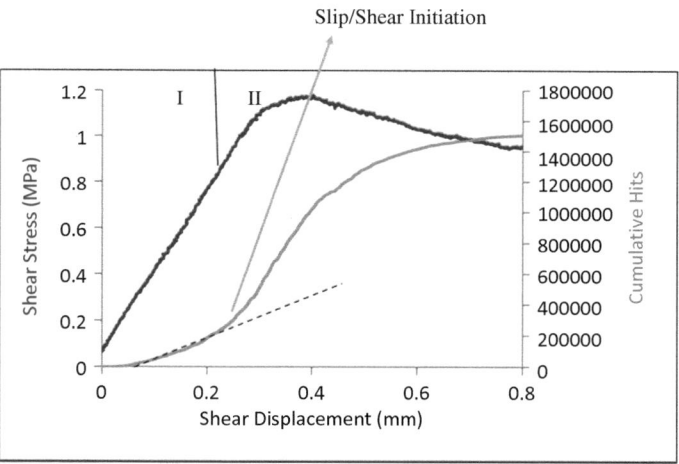

Figure 9 Shear stress and cumulative hits vs shear displacement for the rock joint. The slip/shear initiation point was observed at the beginning of the pre-peak non-linear period.

that the maximum value of the AE rates is observed in this period. At the end of post-peak period joints show a gradual decrease in AE activity. They show their minimum AE values in residual period (Period IV) indicating of small shearing process caused by sliding of the remained in-contact asperities and gouge materials. While in residual period, rock joints don't show a lot of AE activity for hits rate (period IV, Figure 7), they may show some activity for energy rate (period IV, Figure 8). Local breaking of the first order asperities due to stress concentration on in-contact areas produces this energy. After checking the joint surfaces at the end of the test, it was found that the amount of the gouge material wasn't remarkable. It is in good agreement with

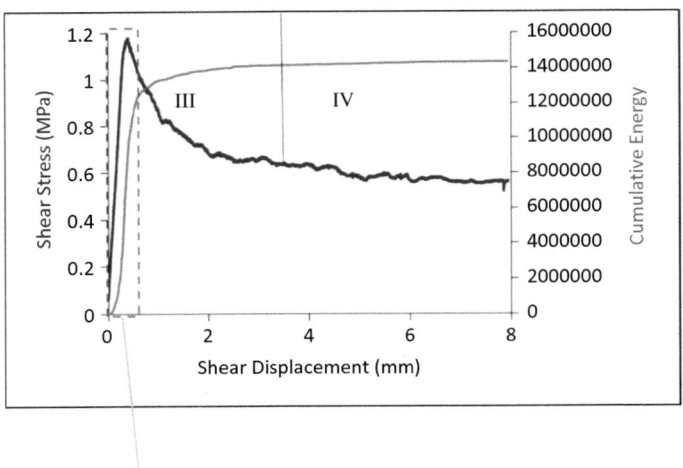

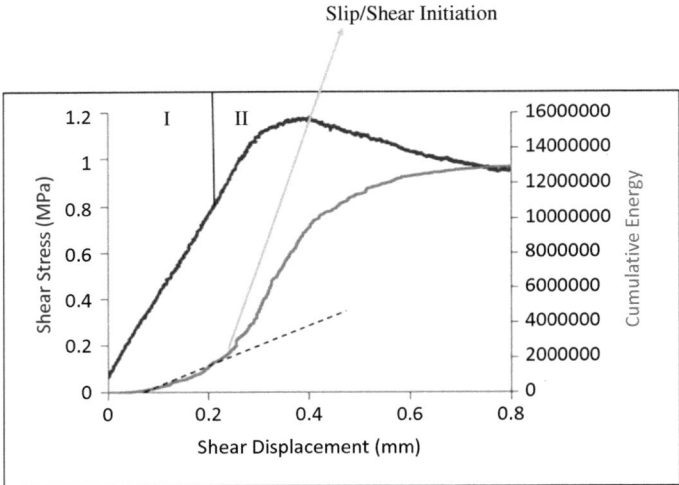

Figure 10 Shear stress and cumulative energy vs shear displacement for the rock joint. The slip/shear initiation point was observed at the beginning of the pre-peak non-linear period.

Barton & De Quadros (1997) who believed that under high JCS/normal stress, the amount of the gouge material is negligible.

Figures 7–10 for the tested rock joints in this study demonstrated a fracturing process similar to natural earthquakes showing: 1) small number of hits before peak (a few foreshocks), 2) a sudden release of AE energy (main shock), 3) large number of hits after peak (a lot of aftershocks) and 4) finally decreasing hits rate obeying the Omori's law (Lei & Ma, 2014).

Table 3 summaries the behavior of the rate and cumulative graphs of AE hits and energy in each shear mechanism period for rock joints.

Table 3 Different behaviors in shearing process of rock joints monitored by AE.

Periods	Behavior according to hits and energy rate	Behavior according to cumulative hits and energy
Pre-peak linear Period	Increasing from background in low values, maybe some instant peaks	Increasing with concavity
Pre-peak non-linear period	Increasing in a constant rate with some instant peaks	Linear increasing
Post-peak Period	Increasing dramatically and decreasing gradually	Sudden increasing with convexity
Residual Period	Attaining minimum values with some instant peaks	Increasing, with a very low rate

5 DETECTING LOCATION, SIZE, AND INTENSITY OF THE ASPERITY DAMAGED ZONES

Locations of the AE event sources were determined from propagation velocity of acoustic waves and the traveling time from the event source to the AE sensor. Using AE localization, scanned surfaces and image of shear surfaces simultaneously provide the possibility to investigate the degradation sequences of the asperities (Moradian *et al.*, 2012b). In order to identify AE activity at different periods of shear test, the Y positions of the AE events is drawn versus their X positions. To verify the accuracy of the source location technique pencil lead breaks (PLB) were conducted before test at the center of the rock joint (known point) and the source positions were measured. The error of the technique was found to be ± 2 mm.

Figure 11 displays the 3D roughness as well as the horizontal and vertical roughness profile of the mobile surface in which AE sensors have been attached for the rock joint. For measuring roughness of the joint surfaces, 0.5 mm interval was chosen and the average $Z2$ parameter was measured for the whole surface as 0.355, then JRC was calculated as 18 from JRC = 32.2 + 32.47 log ($Z2$) (Tse & Cruden, 1979). $Z2$ represents the root mean square of the first height derivative in the 2D profile and JRC is joint roughness coefficient.

In Figure 12, the 2D location of the AE events for four detected periods as well as the photo of the surface are shown. Each figure represents the top view of the joint. For each event in the 2D location graph, its associated energy has been shown in different colors. As it was stated earlier, in order to compare the AE source locations with asperity damaged zones, the joint surface was colored by spraying a dark blue paint before performing direct shear test. This procedure allows pointing out the damaged zones quite easily. If an asperity is crushed, the blue color is removed and the remaining light color exhibits damaged zone. Since the AE sensors have been attached to the mobile (upper) replica, the top view photo of the mobile replica is shown in Figure 12. It can be seen that there is a good correlation and similarity between damaged zones and zones with the cluster of the AE events.

From Figure 12, it can be seen that the locations of the AE sources are distributed over the entire surface of the joint in pre-peak linear and pre-peak non-linear periods. After the maximum shear stress, the sources became localized. These results could be

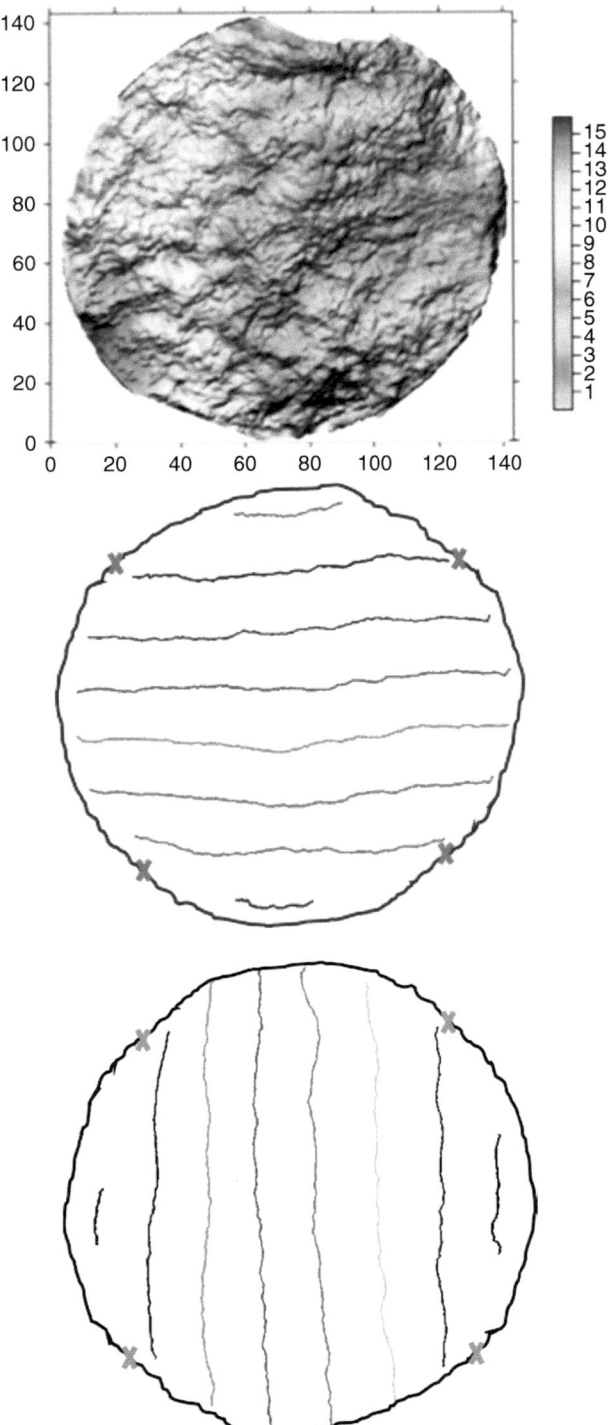

Figure 11 3D surface roughness along with horizontal and vertical roughness profiles of the monitored surface (mobile surface) for the rock joint. The rough asperities facing to the shear direction are located at the top and bottom of the joint surface. Z2=0.335 and JRC= 18.

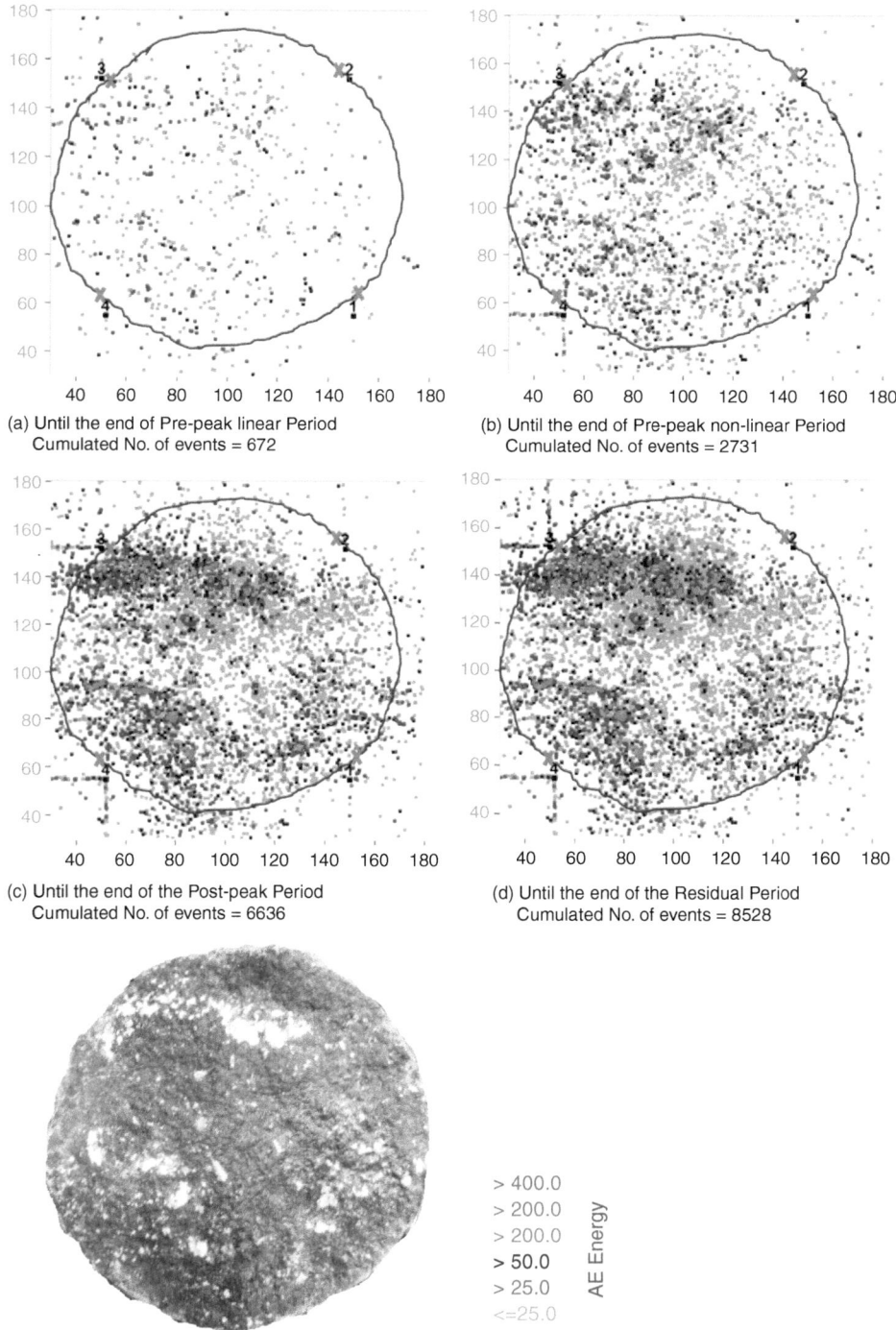

(a) Until the end of Pre-peak linear Period
Cumulated No. of events = 672

(b) Until the end of Pre-peak non-linear Period
Cumulated No. of events = 2731

(c) Until the end of the Post-peak Period
Cumulated No. of events = 6636

(d) Until the end of the Residual Period
Cumulated No. of events = 8528

> 400.0
> 200.0
> 200.0
> 50.0
> 25.0
<=25.0

AE Energy

Figure 12 AE source locations for rock joint and their associated energy. Energy of the first detected hit (closest sensor) has been considered as the energy of the event, to reduce the wave propagation effect. The event was defined when all of four sensors detected it (four hits for each event).

explained by the concept of first and second order asperities. It seems that secondary asperities, which are distributed in entire surface of joint, are sheared until peak shear stress. In other words, before peak shear stress, AE sources are distributed as a result of failure of the secondary asperities and they are localized after maximum shear strength as a result of failure of the primary asperities. This demonstrates that shearing process along a rock joint doesn't occur simultaneously. In fact it occurs first along the smooth surfaces and later along the rough surfaces that is in a good agreement with other researches (Comninou & Dundurs, 1983; Gorbatikh *et al.*, 2001; Malanchuk, 2011).

Generally the asperity degradation in residual shear strength period decreases. In some specimens, depending on the amount of the normal load, the associated energy of the AE events is high due to the failure of the remained in-contact asperities. As a conclusion, in the residual period, the number of AE events decreases but AE may show some local activity in the remained in-contact asperities due to high stress concentration.

It is believed that second order asperities determine the number of AE events while the first order asperities control the magnitude of the AE events or the shear strength of the joint (stress drop). In order to have a better insight about damage intensity of the asperities, energy of each event (magnitude of damage) has been shown by different colors. It can be seen that AE energy increases from low values in pre-peak linear period to pre-peak non-linear period. AEs reached their maximum energy in post-peak period and then they decreased in residual period.

Comparing Figure 11 with Figure 12, one can say that clustering point of AE events as well as damaged zones are related to the rough zones of the joint surfaces. Figure 11 shows that asperities facing the shear direction are located in upper and lower sides of the joint surface, while in the center of the specimen asperities are facing opposite to the shear direction. This kind of asperity distribution has caused damaged zones to occur in upper and lower sides of the joint surface, whereas in the center the lack of damaged zones is clear. It is worth noting that slip nucleates at a certain point(s) rather than occurring simultaneously along the entire fault (Martel & Pollard, 1989). Burgmann *et al.* (1994) stated that although slip distribution is commonly assumed to be symmetrical with respect to a central slip patch; however, it is often asymmetric due to non-uniformities in both normal and shear stress.

The joint surfaces have been scanned by an interval of 0.5 mm * 0.5 mm. As mentioned earlier in this section, the accuracy of the AE source location for known sources (pencil lead breaks) was measured as ±2 mm that is worse than the resolution of the scanner (0.5 mm), therefore it would be difficult to correlate the locations of the AE events to the roughness precisely. What AE source location technique can show are the zones of damaged asperities but not necessarily the individual damaged asperities. In other words, the rupture of several asperities can cause an AE event. This is in agreement with Yabe *et al.* (2003). They believed that a single AE event cannot be due to the rupture of a single asperity and is instead caused by rupture of many asperities that have a spot size of contact on the order 10–100 μm.

The best source locations result in a 2D geometry is achieved when an event is located between at least three sensors. This concept, which is called "triangulation", provides a good coverage of the event by the sensors. If an event doesn't fall in a triangle of the sensors, it may not be localized properly. This has happened in Figure 12 where some of the AE sources have been detected outside of the sensor array. Increasing the number of sensors and an improved sensor array will definitely increase the accuracy of the localization.

6 SUMMARY

AE was monitored during constant normal load (CNL) direct shear testing of rock joints. Shear behavior of joints was divided into four periods based on their shear stress-shear displacement graphs and their AE characteristics: I: pre-peak linear period, II: pre-peak non-linear period, III: post-peak period and IV: residual shear strength period. A distinct maximum shear stress occurred at a shear displacement of less than 1 mm. This maximum shear stress coincided well with the maximum peak of AE hits and energy.

The results showed that the rate and particularly cumulative graphs of the AE parameters could be used as a precursor to the shear failure of the rock discontinuities by showing a notable increasing at pre-peak non-linear period. These observations can also open a door for a better understanding of the mechanisms of faulting and finally for earthquake prediction. However it should be mentioned that while AE technique is very applicable for brittle failures, it is less sensitive to ductile deformation that doesn't produce considerable AE. As a result, smooth rock discontinuities under low normal stress or slow sliding (creep shearing) may not produce remarkable AE. This is what is called aseismic faulting. It is believed that following the development of AE technology, it will be possible to differentiate small events from background noise with sufficient precision and detect the precursors in a better way.

A comparison between damaged zones, AE source locations, asperities roughness profiles revealed that rock joint damage is resulted from breaking of the asperities that are facing to the shear direction. However presence of the AE events with low energy in pre-peak linear and pre-peak non-linear periods (periods I and II) revealed that slipping/shearing process may start from zones with less frictional resistance and then it will be controlled by rough asperities facing the shearing direction with higher frictional resistance.

The experimental tests presented in this chapter were done under low normal stresses by using only 4 sensors for detecting the AE signals. Further experimental tests under higher normal stresses (up to 100 MPa) and using higher number of AE sensors (up to 16) have been planned to be done using a servo-controlled triaxial loading machine.

ACKNOWLEDGMENT

The author would like to thank Professor Gérard Ballivy and Professor Patrice Rivard at Université de Sherbrooke for their support, advice, and suggestions throughout the duration of this research. Dr. Clermont Gravel and Dr. Serge Kodjo have assisted the author with many interesting discussions and comments. They are greatly acknowledged. Gratitude is also expressed to those working at Laboratory of Rock Mechanics and Applied Geology, Civil Engineering Department, Université de Sherbrooke especially, Georges Lalonde and Danick Charbonneau who have helped the author to carry out laboratory tests.

REFERENCES

Aggarwal, Y. P., Sykes, L. R., Armbruster, J., Sbar, M. L. (1973). Premonitory changes in seismic velocities and prediction of earthquakes. Nature, 241, 101–104.

Barbosa, R. E. (2009). Constitutive model for small rock joint samples in the lab and large rock joint surfaces in the field. In: Diederichs, M., Grasselli, G., (eds), Proceedings of the 3rd CANUS Rock Mechanics Symposium, Toronto.

Barton, N. (1973). Review of a new shear strength criterion for rock joints. Q. J. Eng. Geol. 7, 287–332.

Barton, N., De Quadros E. F. (1997). Joint aperture and roughness in the prediction of flow and groutability of rock masses. Int. J. Rock Mech. Min. Sci. 34(3-d), ISSN 0148–9062.

Barton, N. (2013). Shear strength criteria for rock, rock joints, rockfill and rock masses: Problems and some solutions. J. Rock. Mech. Geotech. Eng. 5(4), 249–261.

Brace, W. F., Byerlee, J. D. (1966). Stick-slip as a mechanism for earthquakes. Science. 153, 990–992.

Burgmann, R., Pollard, D. D., Martel, S. J. (1994). Slip distributions on faults: Effects of stress gradients, inelastic deformation, heterogeneous host-rock stiffness, and fault interaction. J. Struct. Geol. 16, 1675–1690.

Byerlee, J. D. (1970). The mechanics of stick-slip. Tectonophysics, 9(5), 475–486.

Byerlee, J. (1978). A review of rock mechanics studies in the United States pertinent to earthquake prediction. Pure Appl. Geophys. 116, 586–602.

Cicerone, R. D., Ebel, J. E., Britton, J. (2009). A systematic compilation of earthquake precursors. Tectonophysics, 476, 371–396.

Comninou, M., Dundurs, J. (1983). Spreading of slip from a region of low friction. Acta Mech. 47, 65–71.

Ellsworth, W. L. (2013). Injection-induced earthquakes. Science, 341(6142), 142.

Fathi, A., Moradian, Z., Rivard, P., Ballivy, G., Boyd, A. (2016a). Geometric effect of asperities on shear mechanism of rock joints. Rock Mech. Rock Eng. doi:10.1007/s00603-015-0799-6.

Fathi, A., Moradian, Z., Rivard, P., Ballivy, G. (2016b). Shear mechanism of rock joints under pre-peak cyclic loading condition. Int. J. Rock Mech. Min. Sci. 83(March), 197–210.

Fishman, Y. A. (1990). Failure mechanism and shear strength of joint wall asperities. In: Barton, N., Stephansson, O. (eds.), Rock joints. Balkema, Rotterdam, pp. 627–631.

Gentier, S., Riss, J., Archambault, G., Flamand, R., Hopkins, D. (2000). Influence of fracture geometry on shear behavior. Int. J. Rock. Mech. Min. Sci. & Geomech. Abstr. 37(1–2), 161–174.

Goebel, T. H. W., Candela, T., Sammis, C. G., Becker, T. W., Dresen, G., Schorlemmer, D. (2014). Seismic event distributions and off-fault damage during frictional sliding of saw-cut surfaces with pre-defined roughness. Geophys. J. Int. 196(1), 612–625, doi:10.1093/gji/ggt401.

Gorbatikh, L. B., Nuller, D., Kachanov, M. (2001). Sliding on cracks with non-uniform frictional characteristics. Int. J. Solids Struct. 38, 7501–7524.

Goncalves da Silva, B., Li, B., Moradian, Z., Germaine, J., Einstein, H. (2015). Development of a test setup capable of producing hydraulic fracturing in the laboratory with image and acoustic emission monitoring. 49th U.S. Rock Mechanics/Geomechanics Symposium, June 28–July 1, San Francisco, California.

Grasselli, G. (2006). Shear strength of rock joints based on quantified surface Description. Rock Mech. Rock Eng. 39(4), 295–314

Gravel, C., Moradian, Z., Fathi, A., Ballivy, G., Rivard, P. (2015). In situ shear testing of simulated dam concrete-rock interfaces. ISRM May 10–15, Montreal, Canada.

Hajiabdolmajid, V., Kaiser, P. K., Martin, C. D. (2002). Modeling brittle failure of rock. Int. J. Rock Mech. Min. Sci. 39, 731–741.

Handanyan, J. M., Danek, E. R., Dandrea, R. A., Sage, J. D. (1990). The role of tension in failure of jointed rock. In: Barton, N., Stephansson, O. (eds.), Rock joints. Balkema, Rotterdam, pp. 195–202.

Hedayat, A., Pyrak-Nolte, L. J., Bobet, A. (2014). Seismic precursors to the shear failure of rock discontinuities. Geophys. Res. Lett. http://dx.doi.org/10.1002/2014GL060848.

Hong, C., Jeon, S. (2006). Influence of shear load on the characteristics of acoustic emission of rock-concrete interface, Key Eng. Mater. 270–273, 1598–1603.

Huang, T. H., Chang, C. S., Chao, C. Y. (2002). Experimental and mathematical modeling for fracture of rock joint with regular asperities. Eng. Fract. Mech. 69(17), 1977–1996.

Hutson, R. W., Dowding, C. H. (1990). Joint asperity degradation during cyclic shear. Int. J. Rock Mech. Min. Sci. & Geomech. Abstr. 27(2), 109–119.

Ishida, T., Tadashi Kanagawa, Yuji Kanaori. (2010). Source distribution of acoustic emissions during an in-situ direct shear test: Implications for an analog model of seismogenic faulting in an inhomogeneous rock mass. Eng. Geol. 110(3–4), 66–76.

Johnston, M. J. S., Linde, A. T., Gladwin, M. T., Borcherdt, R. D. (1987). Fault failure with moderate earthquakes. Tectonophysics, 144, 189–206.

Koerner, R. M., McCabe, W. M., Lord, A. E. (1981). Overview of acoustic emission monitoring of rock structures. Rock Mech. 14, 27–35.

Kulatilake, P. H. S. W., Shou, G., Huang, T. H., Morgan, R. M. (1995). New peak shear strength criteria for anisotropic rock joints. Int. J. Rock Mech. Min. Sci. & Geomech. Abstr. 32, 673–697.

Kutter, H. K., Otto, F. (1990). Influence of parallel and cross joints on shear behaviour of rock discontinuities. In: Barton, N., Stephansson, O. (eds.), Rock joints. Balkema, Rotterdam, pp. 243–250.

Ladanyi, B., Archambault, G. (1970). Simulation of shear behavior of a jointed rock mass. 11th Symposium on Rock Mechanics, Berkeley, pp. 105–125.

Lei, X., Ma, S. (2014). Laboratory acoustic emission study for earthquake generation process. Earthquake Sci. 27(6), 627–646. http://doi.org/10.1007/s11589-014-0103-y

Li, B., Moradian, Z., Goncalves da Silva, B., Germaine, J. (2015). Observations of acoustic emissions in a hydraulically loaded granite specimen. 49th U.S. Rock Mechanics/Geomechanics Symposium, June 28– July 1, San Francisco, California.

Li, C., Nordlund, E. (1990). Characteristics of acoustic emissions during shearing of rock joints. In: Barton, N., Stephansson, O. (eds.), Proceedings of first international symposium on rock joints. Rotterdam, Balkema, pp. 251–258.

Lockner D. (1993). The role of acoustic emission in the study of rock fracture. Int. J. Rock Mech. Min. Sci. & Geomech. Abstr. 30(7), 883–899.

Malanchuk, N. I. (2011). Local slip of bodies caused by the inhomogeneous friction coefficient. Mater. Sci. 46(4), 543–552.

Martel, S. J., Pollard, D. D. (1989). Mechanics of slip and fracture along small faults and simple strike-slip fault zones in granitic rock. J. Geophys. Res. 94, 9417–9428.

Maxwell, S. C., Jones, M., Parker, R., Miong, S., Leany, S., Dorval, D., D'Amico, D., Logel, J., Anderson, E., Mammermaster, K. (2009). Fault activation during hydraulic fracturing, SEG Annual Meeting, October 25–30, Houston, Texas, 1552–1556.

McLaskey, G. C., Lockner, D. A. (2014). Preslip and cascade processes initiating laboratory stick-slip, J. Geophys. Res. 119, 6323–6336.

McLaskey, G. C., Kilgore, B. D., Lockner, D. A., Beeler, N. M. (2014). Laboratory generated M6 earthquakes. Pure Appl. Geophys, 171, 2601–2615.

Mogi, K. (1962). Magnitude-frequency relation for elastic shocks accompanying fractures of various materials and some related problems in earthquakes, Bull. Earthq. Res. Inst. 40, 831–853.

Moradian, Z. A., Ballivy, G., Gravel, C., Saleh, K. (2008). Analysis of the shear strength of the active joints using results of the constant normal load shear test. 5th Asian Rock Mechanics Conference, November 24–26, Tehran, Iran.

Moradian, Z. A, Ballivy, G., Rivard, P., Gravel, C., Rousseau, B. (2010a). Evaluating damage during shear tests of rock joints using acoustic emission. Int. J. Rock Mech. Min. Sci. 47(4), 590–598.

Moradian, Z. A., Ballivy, G., Rivard, P., André, C. (2010b), Effect of normal load on shear behavior and acoustic emissions of rock joints under direct shear loading. European Rock Mechanics Symposium (Eurock 2010), 15–17 June 2010, Lausanne, Switzerland.

Moradian, Z., Ballivy, G., Rivard, P. (2011). Role of adhesive bond on shear mechanism of bonded concrete-rock joints under direct shear test. 45th U.S. Rock Mechanics/Geomechanics Symposium, June 26–29, San Francisco.

Moradian, Z. A, Ballivy, G., Rivard, P. (2012a). Application of acoustic emission for monitoring shear behavior of bonded concrete-rock joints. Can. J. Civ. Eng. 39, 887–896.

Moradian, Z. A., Ballivy, G., Rivard, P. (2012b). Correlation acoustic emission source locations with damage zones of rock joints under direct shear test. Canadian Geotechnical Journal, 49(6), 710–718.

Moradian, Z., Gravel, C., Fathi, A., Ballivy, G., Rivard, P., Quirion, M. (2013). Developing a high capacity direct shear apparatus for large scale laboratory testing of rock joints. ISRM International Symposium EUROCK 2013, September 21–26, Wroclaw, Poland.

Moradian, Z., Einstein, H. H., Ballivy, G. (2016). Detection of cracking levels in brittle rocks by parametric analysis of the acoustic emission signals. Rock Mech. Rock Eng. 49(3), 785–800.

Mutlu, O., Bobet, A. (2006). Slip propagation along frictional discontinuities. Int. J. Rock Mech. Min. Sci., 43, 860–876.

Papaliangas, T., Hencher, S. R., Lumsden, A. C., Manolopoulou, S. (1993). The effect of frictional fill thickness on the shear strength of rock discontinuities. Int. J. Rock Mech. Min. Sci. & Geomech. Abstr. 30, 81–91.

Park, J. W., Song, J. J. (2013). Numerical method for the determination of contact areas of a rock joint under normal and shear loads. Int J Rock Mech. Min. Sci. 58, 8–22.

Patton, F. D. (1966). Multiple modes of shear failure in rocks. In: Proceedings of First Congress of International Society of Rock Mechanics, Portugal, Vol.1, pp. 509–513.

Pereira, J. P., De Freitas, M. H. (1993). Mechanism of shear failure in artificial fractures of sandstone and their implication for models of hydromechanical coupling. Rock Mech. Rock Eng. 10(1–2), 1–54.

Saiang, D., Malmgren, L., Nordlund, E. (2005). Laboratory tests on shotcrete-rock joints in direct shear, tension and compression. Rock Mech. Rock Eng. 38(4), 275–297.

Scholz, C. H. (1998). Earthquakes and friction laws. Nature. 391, 37–42.

Scholz, C. H. (2002). The mechanics of earthquakes and faulting. Cambridge, Cambridge University Press, 2nd edition.

Seidel, J. P., Haberfield, C. M. (2002). Laboratory testing of concrete-rock joints in constant normal stiffness direct shear. Geotech. Test. J. 25(4), 391–404.

Shiotani, T. (2006). Evaluation of long-term stability for rock slope by means of acoustic emission technique. NDT and E Int. 39(3), 217–228.

Thompson, B. D., Young, R. P., Lockner, D. A. (2009). Premonitory acoustic emissions and stick-slip in natural and smooth-faulted Westerly granite. J. Geophys. Res. 114, B02205. doi:10.1029/2008JB005753.

Tse, R., Cruden, D. M. (1979). Estimating joint roughness coefficients. Int. J. Rock Mech. Min. Sci. & Geomech. Abstr. 16(5), 303–307. doi:10. 1016/0148-9062(79)90241-9.

Warpinski, N. R., Mayerhofer, M. J., Agarwal, K., Du, J. (2012). Hydraulic fracture geomechanics and microseismic source mechanisms. SPE158935 In: SPE Annual Technical Conference & Exhibition, October, San Antonio, Texas., 8–10.

Yabe, Y., Kato, N., Yamamoto, K., Hirosawa, T. (2003). Effect of sliding rate on the activity of acoustic emission during stable sliding. Pure appl. Geophys. 160, 1163–1189.

Zhao, J. (1997). Joint surface matching and shear strength – Part A: Joint Matching Coefficient (JMC). Int. J. Rock Mech. Min. Sci. 34(2), 173–178

Zoback, M. D. and Gorelick, S. M. (2012). Earthquake triggering and large – Scale geologic storage of carbon dioxide. Proc. Natl. Acad. Sci. U.S.A. 109(26), 10164–10168.

Joint Tests

Chapter 6

Morphological parameters of both surfaces of coupled joints

Cao Ping, Liu Jie & Fan Xiang
Central South University, Changsha, Hunan, P.R. China

1 INTRODUCTION

Joints which are widely distributed in rock masses concern engineers when conducting underground constructions, such as the underground oil depots and power stations. It is well known that the roughness of discontinuities which are clean and unfilled will have great impacts on both hydraulic and strength characteristics of discontinuous rock masses (Tatone & Grasselli, 2010; Jang *et al.*, 2006). Therefore, extensive investigations have been conducted on the mechanical properties of joints and the morphology characteristics of joint surfaces.

Firstly, the accurate measurement of joint surfaces is prerequisite to investigate morphological parameters and to set up corresponding models. To data, non-contact and contact techniques are often used to gauge joint surfaces. The non-contact laser morphology instruments are widely used due to the high accuracy.

Secondly, to investigate the morphologies of the joints obtained, the statistical and fractal geometry theories are mostly applied. Initially, research interests were focused on 2D parameters of the joint surfaces. However, the 2D parameter descriptions are limited by the descriptions of the surface in 3 dimensions which are more realistic. Therefore, 3D parameters were developed. Xia (1996) quantified the height characterization parameters of joint surface topography with the mathematical method and identified the waviness and unevenness components for joint surface profiles. Zhao (1997a,b) pointed out that the mechanical properties of a joint, which are related to the coupling degree of two joint surfaces, were poor when the coupling degree was limited, and the joint coupling parameter which varies from 0 to 1 was put forward to describe the degree of coupling between the two halves. The JRC-JMC shear strength criterion was also brought up, based on the Barton JRC-JCS model (Barton & Choubey, 1977). To investigate the morphological evolution of joint surfaces under cyclic shear loads, Homand *et al.* (2001) developed several parameters, including θ_s, K_a and R_s, to quantify morphological characterization of joint surfaces, and proposed morph-mechanical model of direct shear tests. Grasselli and others (Grasselli, 2006; Grasselli & Egger, 2003; Grasselli *et al.*, 2002) proposed the 3D shear model based on 3D parameters of joint surfaces. Other researches on morphological parameters of rock joint surface have also been done in recent years (Lee *et al.*, 2001; Jiang *et al.*, 2006; Cao *et al.*, 2011; Chen *et al.*, 2010; Belem *et al.*, 2000). Barton & Choubey

(1977) proposed ten typical roughness profiles and the corresponding JRC values, and then one shear criterion containing JRC was further proposed. However, the estimation of JRC requires a great deal of experiences. Hence, lots of methods were proposed to estimate JRC values (Tse & Cruden, 1979; Andrade & Saraiva, 2008; Beer et al., 2002; Du et al., 2009; Yang et al., 2001). Tatone & Grasselli (2010) tried to establish empirical relationship between the new 2D roughness parameters and JRC which enabled shear strength estimation according to the Barton-Bandis shear strength criterion. Fractal geometry set up by Mandelbrot (1983) is a useful method to investigate irregularity in the nature. Irregular profiles of joint surfaces have self-similarity in statistics. Carr & Warriner (1989) introduced fractal theory to study morphology of joint surfaces firstly. The Fractal dimension and amplitude were used as parameters to describe morphology characteristics of joint surfaces. At first, the fractal dimension of profile was computed, later, that of joint surface was computed by different methods (Odling, 1994; Kulatilake et al., 1995). Xie et al. (1997) computed fractal dimension of profiles which was between 1 and 2, and that of joint surface was between 2 and 3.

Thirdly, coexistence of joints and water is frequently encountered in rock engineering. Colback & Wild (1965) found that the influence of water-rock interaction on rock mass strength is prominent. According to the experiment of Rebinder et al. (1994), at the prophase of the water–rock circle interaction, rock was under a more serious damage effects both physically and chemically, in which the cohesive strength and internal friction angle were influenced most. The physical and chemical damage effects on rock were reduced as the actuation duration was prolonged and times were added (Melo et al., 2008, 2007). Indratna et al. (1999) indicated that the influence on cohesive strength and internal friction angle was reduced as well and the changes became gentle. So, it can be inferred that the properties of the discontinuities, as special rock masses, will also be affected by the water-rock interaction, and rock strengths even engineering stabilities might be influenced by the water-rock interaction in further. Therefore, there is an urgent need to determine more accurately the relationship between rock and water–rock interaction (Zhao et al., 2006; Sari & Karpuz, 2006; Jiang et al., 2004; Pyrak-Nolte & Morris, 2000; Hoek & Diederichs, 2006). The surface properties with water–rock interaction have a major influence on the hydro-mechanical behavior of the rock masses and rock joints. Many parameters (Zhang et al., 2002; Jae-Joon, 2006; Zimmerman et al., 2004), for instance hydraulic conductivity, frictional resistance and resistance to shearing along discontinuities, will change with the effect of water–rock interaction. There are strong links between roughness and these parameters. Although previous studies have provided important information regarding accurate characterization of rock surface roughness, most studies concentrate on characterizing surface roughness of rock using small samples and no systematic study has been made to investigate the effect of water–rock interaction using large 3D rock samples.

To investigate the morphology characteristics of coupled joints, typical parameters used to characterize joint morphologies are introduced firstly. And a feasible method to generate the coupled joints has been proposed too, analysis based on above parameters has been conducted to investigate the characteristics of these coupled joints. And then, further investigation on the influence of water content on the joint morphologies obtained by conducting Brazil tensile tests has been conducted. Finally, the effects of water-rock interaction on rock and morphologies of the coupled joints by shear tests have been investigated.

2 MORPHOLOGY PARAMETERS

Morphology parameters which characterize surface features are mainly composed of two groups of parameters (Chen *et al.*, 2010). The first group parameters are the statistical parameters which quantify the Z-axis perpendicular to the scanning surface. The other group is the textural parameters. While the third group of parameters are density of peaks, bearing index and valley fluid retention index, *et al.*

2.1 Statistical parameters

1) Maximum height of joint surface, S_p, which denotes the height between the highest peak and the mean plane can be written as:

$$S_p = \max (S_{p1}, S_{p2}, \ldots S_{pn})$$

2) The maximum depth of valleys, S_m, which denotes the depth between the mean plane and the deepest valley can be written as:

$$S_m = \max (S_{m1}, S_{m2}, \ldots S_{mn})$$

3) The maximum height, S_h, which indicates the height between the highest peak and the deepest valley can be written as:

$$S_h = S_p + S_m$$

4) Arithmetical mean height of joint surface, S_a, which can be used to characterize the volatility and discreteness of the height distribution can be written as:

$$S_a = \frac{1}{A} \iint_A |Z(x,y)| dxdy$$

Where $Z(x,y)$ is a height function of joint surface, and A is the horizontal area of the sample surface.

5) Root mean square (RMS) height of joint surface, S_q, which is used to describe the standard deviation of the height distribution and to characterize the discreteness of the height distribution of joint surfaces can be written as:

$$S_q = \sqrt{\frac{1}{A} \iint_A Z(x,y)^2 dxdy}$$

6) Skewness of height distribution of joint surface, S_{sk}, which is used to quantify the symmetry of the height distribution of the joint surface can be written as:

$$S_{sk} = \frac{1}{S_q^3} \left[\frac{1}{A} \iint_A Z(x,y)^3 dxdy \right]$$

A negative S_{sk} indicates that the surface is composed of principally one plateau and deep and fine valleys. While, a positive S_{sk} indicates a joint surface is composed of lots of peaks on a plane. Due to the big exponent in the equation, this parameter is very sensitive to the sampling of the measurement.

7)   Kurtosis of the height distribution of joint surface, S_{ku}, which is expressed without any unit can be written as:

$$S_{ku} = \frac{1}{S_q^4} \left[\frac{1}{A} \iint_A Z(x,y)^4 dxdy \right]$$

This parameter is used to quantify the flatness and the concentration degree of a surface. If this parameter is higher than 3, the height distribution is leptokurtosis and concentrated. As this parameter equals to 3, the surface height is in a normal distribution. However, a S_{ku} which is less than 3 indicates that the scattered surface height distribution is in a platy kurtosis form.

2.2 Textural parameters

Textural parameters are composed of spatial parameters and hybrid parameters. Spatial parameters describe topographic characteristics based on spectral analysis. They are used to quantify the lateral information present on the x- and y-axes of the surface. While, hybrid parameters are a class of surface finish parameters which quantify the information present on the x-, y- and z-axes of the surface, *i.e.*, those criteria hats are both on the amplitude and the spacing, such as slopes, curvature, etc.

2.2.1 Spatial parameters

1)   The autocorrelation length, S_{al}
      The autocorrelation which represents the similarity of surface compared to itself when being translated. It helps to distinguish isotropic surfaces from anisotropic surfaces. Horizontal distance of the autocorrelation function (f_{ACE}) has the fastest decay to a specified value s, with $0 < s < 1$. The default value for s in the analysis software is 0.2. This parameter expresses the content in wavelength of the surface. A high value indicates that the surface has mainly high wavelengths but low frequencies. The autocorrelation length, S_{al}, can be written as:

$$S_{al} = \sqrt[min]{x^2 + y^2}$$

Where $R = \{(x,y) : f_{ACE}(x,y) \leq s\}$.

2)   Texture aspect ratio, S_{tr}
      This is the ratio of the shortest decrease length at 0.2 from the autocorrelation on the greatest length. This parameter ranges from 0 to 1. If the value is near 1, one can say that the surface is isotropic, *i.e.* it has the same characteristics in all directions. If the value is near 0, the surface is anisotropic, *i.e.* it has an oriented and/or periodical structure.

2.2.2 Hybrid parameters

1) Developed interfacial area ratio, S_{dr}
   This parameter is the ratio of the increment of the interfacial area of the scale limited surface within the specific area to the specific area. It is used to indicate the complexity of joint surface. The developed surface indicates the complexity of the surface due to the comparison of the curvilinear surface and the support surface. A completely flat surface will have an S_{dr} which nearly equals to 0, while a complex surface will have an S_{dr} of some percents.

$$S_{dr} = \frac{1}{A} \left\{ \iint_A \left[\sqrt{1 + \left(\frac{\partial z(x,y)}{\partial x} \right)^2 + \left(\frac{\partial z(x,y)}{\partial y} \right)^2} - 1 \right] dxdy \right\}$$

2) Root mean square gradient of joint surface, S_{dq}.
   This parameter is used to quantify the statistical property of morphology evolution, it can be written as:

$$S_{dq} = \sqrt{\frac{1}{A} \iint_A \left[\left(\frac{\partial z(x,y)}{\partial x} \right)^2 + \left(\frac{\partial z(x,y)}{\partial y} \right)^2 \right] dxdy}$$

2.3 Other parameters

1) Density of peaks, S_{pd}
   This parameter is expressed in peaks/mm^2. A point is considered as a peak if it is higher than its eight neighbors. The S_{pd} is only calculated through those significant peaks that remain after discrimination by segmentations.
2) Arithmetic mean summit curvature, S_{pc}.
   This parameter enables us to know the mean form of the peaks according to the mean value of the curvature of the surface at these points. While laser scanned every peak point, the final statistics of the average of all peak-point curves were recorded. It reflects the peak point of the specimen surface and the overall curvature of the situation.
3) Bearing index, S_{bi}.
   This parameter is the ratio of the RMS deviation over the surface height at 5% bearing area. The higher the S_{bi} index, the larger the number of wear shelves exist on the surface.
4) Core fluid retention index, S_{ci}
   This parameter is the ratio of the void volume at the core zone (5% to 80% bearing area) over the RMS deviation. A larger S_{ci} index indicates a good degree of fluid retention on the joint surface. It should be mentioned that the fracture aperture inside rock is not considered in present chapter.
5) Valley fluid retention index, S_{vi}.
   This parameter is the ratio of the void volume at the valley zone (80% to 100%) over the RMS deviation. A large S_{vi} indicates a good fluid retention in the valley zone on the rock surface.

3 MORPHOLOGICAL INVESTIGATIONS ON COUPLED JOINTS BY TENSILE TESTS

Owning to the influences exerted by human's underground mining activities, such as blast and drill, numerous artificial joints and cracks are generated, which result in a big menace to underground workers. So, the research conducted on coupled joints in a laboratory scale may help us to get a better understanding on the corresponding characteristics. Joints which are unfilled, clean, well-coupled and unweathered are difficult to be obtained from nature. Therefore, it is more difficult to process standard rock specimens with joints for rock mechanic tests. So, Brazilian split tests and shearing tests which have been adopted in present chapter are good methods to make artificial joints.

3.1 Morphological investigations on coupled joints by tensile tests

3.1.1 Test design and procedures

This series of morphological investigations are mainly composed of three procedures. The first procedure is the generation of artificial coupled joints, the second procedure is the scanning of morphologies, and the last is the morphological analysis on obtained results.

1) Generation of coupled joints
    Electro-hydraulic servo-controlled universal testing machine which is composed of host machine, oil pump, control system, a computer and data-processing system has been adopted in Brazilian splitting tests. The loading can be force- or displacement controlled. The loading velocity adopted is 200N/s in this set of Brazilian tests. The machine stops loading when the cylindrical specimen cracks, which could guarantee the specimen would not be crushed.
    As shown in Fig. 1 and Fig. 2 that 21 coupled joints (a total of 42 joint surfaces, see Fig. 2) were made by means of Brazilian split tests. The specimens were obtained from several kinds of hard rocks, such as limestone, sandstone, etc. The artificial joints made by Brazilian split test have good coupled conditions, which can be found from the following pictures and morphological parameters.

2) Morphology scanning
    As shown in Fig. 3 that the morphologies of the 42 surfaces are scanned by the three-dimensional high-accuracy non-contact laser morphology instrument. The dimension of the instrument is 800mm×600mm×970mm with the mass of 280kg. The scanning range in z-direction is defined by the range of the gauge. The maximum allowed weight of the sample is about 150N. As shown in Fig. 4 that the highest measurement accuracy can reach 0.5μm. The technique used for the laser triangulation gauge deduces the height of a surface point by sensing the position of a laser spot on the surface using a detector placed at a certain angle away from the incoming laser beam. A focused laser beam projects a spot on a rough surface. This spot is detected by a CCD (charge coupled device) sensor

Figure 1 Brazilian split tests.

Figure 2 Samples of coupled joint surfaces.

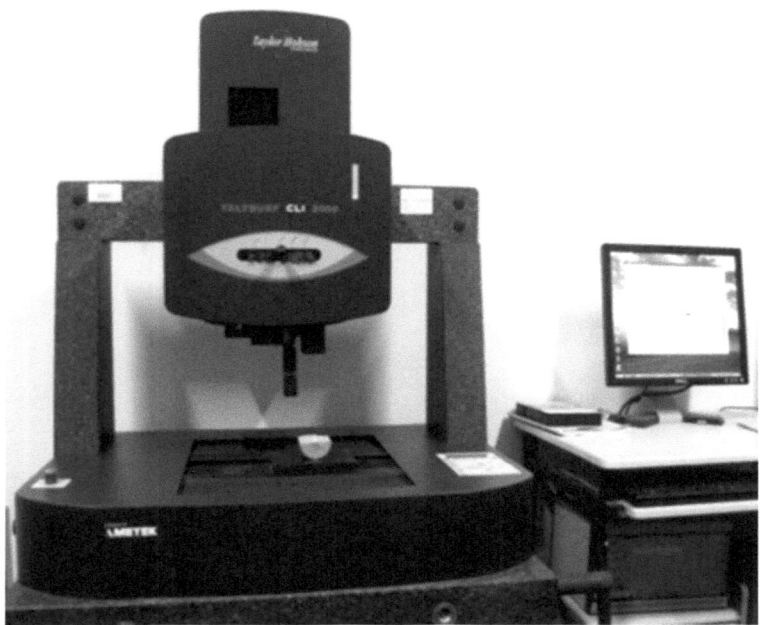

Figure 3 Talysurf CLI 2000.

placed at a different angle. The image of the laser spot on the CCD is focused on a position depending on the vertical position of the spot.

In the scanning process, specimens are placed on the slide, and then after the origin setting and scanning area definition, the morphologies of the joints are obtain by the scanning in the path shown in Fig. 5.

3)   Morphological analysis

Finally, the morphology parameters are computed with the aid of TalyMap5.0 which is the powerful software for morphology analysis. Details can be found in the Talysurf user's manual and the formula and definition can refer to the ISO25178 and EUR15178N.

3.1.2 Results and discussions

1)   Morphologies of joint surfaces

The typical morphologies of the joint surfaces obtained are shown in Fig. 6. The joint surface looks like rolling hills and has some similarity with topography after magnifying the height in z-direction by ten times. The valleys have significant impact on both seepage path and seepage speed of underground water in rock masses.

The undulation of joint surface can be judged from different colors of images. The positions of the highest peak and the deepest valley are known from 3D images. If a line located between Fig. 7(c) and Fig. 7(d), the two 3D images can be

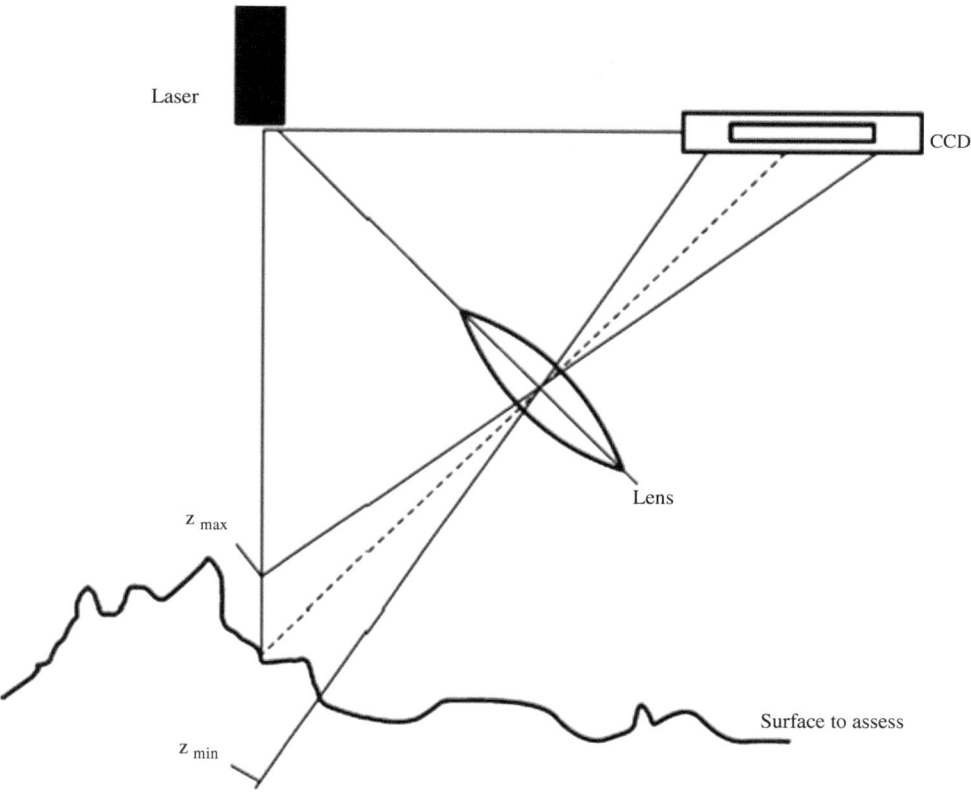

Figure 4 Working principle of Talysurf CLI 2000.

folded to a well-coupled joint along the line. Visualized, clear 3D images of joint surface can be made and parameters that are needed can be computed by the TalyMap5.0, which is the powerful software for analyzing surface.

2) Results and discussion on statistical parameters

The obtained results of statistical parameters are listed in Table 1. It can be known from Table 1 that both surfaces of the coupled joints have similar values of S_h, which vary significantly with different joint surfaces and reflect the fluctuation of the joint surfaces in certain degree. The shear strength generated from the interlocking effects of the peaks that stagger each other. Tiny distinctions of S_h were induced by a spot of rock debris dropped under the high normal stress. It can be seen from the data in Table 1 that the S_h and S_q values between two halves of a joint are very close. The relatively small error is generated from the process of making artificial joints. From the data, we can see that S_a is strongly related to S_h. Similar values of S_q explain the joint surfaces' consistent discreteness of height distribution. The surfaces of different joints have large difference with the S_q values, which can reflect that there are big differences in the aspect of height distributions, and S_q is obviously related to both S_h and S_a.

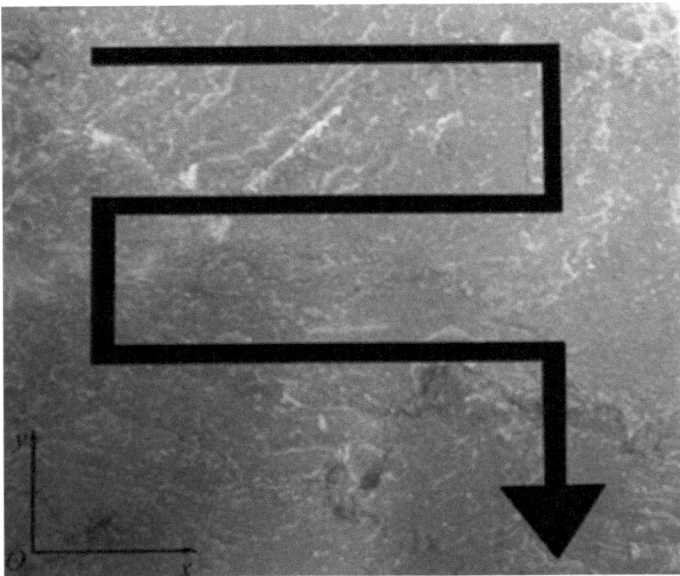

Figure 5 Scanning path.

S_{sk} is a characteristic of the deviation of $Z(x, y)$ that is the density function of height distribution from the origin. It can be found by comparing and analyzing that the S_{sk} of the coupled joint surfaces that the S_{sk} value of one side is positive and its height distribution is positive-skewed distribution, while the S_{sk} value of the other side is negative and its height distribution is negative-skewed distribution. However, their absolute values are approximate, which shows that the distances of $Z(x, y)$ from the origin are close. Two surfaces' height distributions of one sample among all samples are both negative skewness.

S_{ku} values of two halves are approximately equal, which means the concentration ratio of the upper and lower surface's height distribution is similar. S_{ku} value of joint surface varies from one other. Statistically, for example, high kurtosis distribution and low kurtosis distribution are common. These are in accordance with statistical laws of height distribution of joint surface morphology done by other scholars. Among all of the samples, there is one sample has near normal distribution because its upper and lower surfaces' S_{ku} values are closed to 3.

3)   Results and discussion on textural parameters

The corresponding textural parameters listed in Table 2 indicate that the S_{al} values of the upper and lower surfaces are very close, with the biggest gap which is less than 1mm, which means that both surfaces of coupled joints consist of waves have equal wavelengths and frequencies. The lengths of the auto-correlation are related to sectional dimension, and all S_{al} values are larger than 8mm. The larger the S_{al} values, the higher the relevance of points on the contour line will be. As shown in Fig. 8 that the upper and lower surfaces of coupled joints are exactly alike after

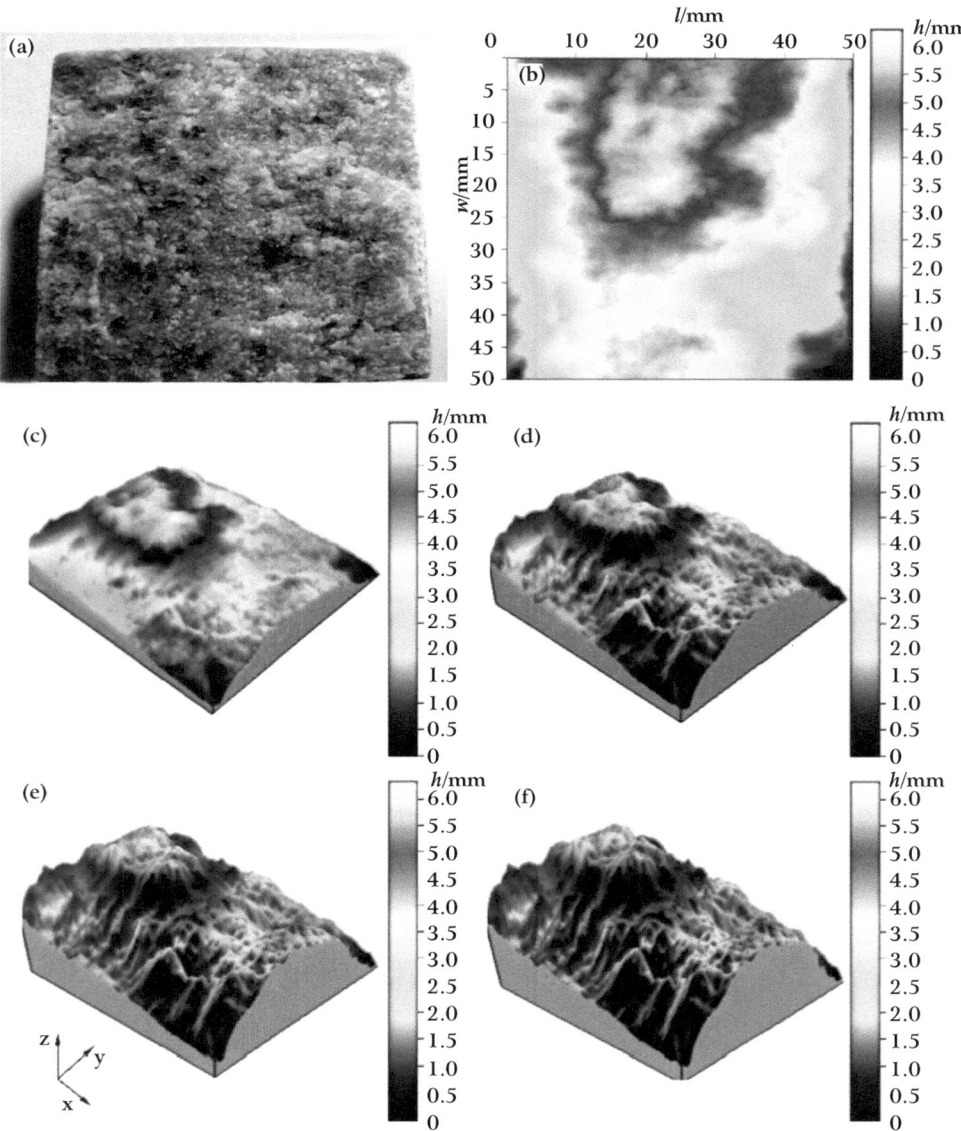

Figure 6   Images of joint surface: (a) Real joint surface; (b) 2D image of joint surface; (c) 3D image (height is magnified 2.5 times); (d) 3D image (height is magnified 5 times); (e) 3D image (height is magnified 7.5 times); (f) 3D image (height is magnified 10 times).

autocorrelation transformation, which indicates that they have the same wavelengths and frequencies. It can be seen from Fig. 8 that joint surfaces are anisotropic, because if they were isotropic, joint surfaces should have the same color in all direction. Any S_{tr} values of both surfaces of the coupled joints are less

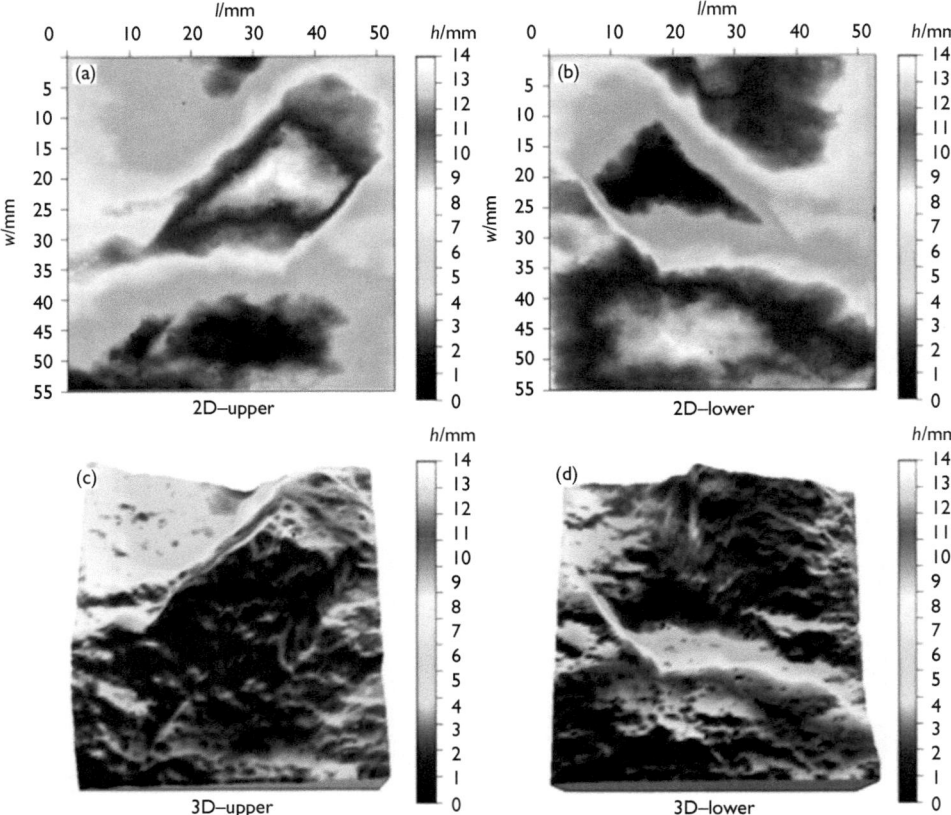

Figure 7 Morphologies of the coupled joints: (a) 2D morphologies of upper surface; (b) 2D morphologies of lower surface; (c) 3D morphologies of upper surface; (d) 3D morphologies of lower surface.

than 1 and approximately equal, which indicates both surfaces of the coupled joints are anisotropic again. All of the S_{tr} values of joint surfaces are in the range from 0.3 to 0.5 and the dimensions of joint surfaces are involved.

It can be found from the comparisons from the coupled joints that the S_{dr} values of two halves are very similar, which indicates that both sides of the coupled joints have the same complexity and near actual surface areas. After autocorrelation transformation, all joint surfaces' S_{dq} values are small and almost the same; however, different joint surfaces have different S_{dq} values, because different joint surfaces own different profiles.

4)   Envelop profile and envelop area
       A scan image of joint surface actually is formed by plenty of scan spots locating in different heights. Each spot has corresponding coordinate value and z is the vertical height above the reference plane. The reference plane is the least squares plane minimizing the sum of squares of the basic distances $Z(x, y)$ at the point

Table 1 Statistical parameters of joint surfaces.

Sample	Surface	S_h/mm	S_a/mm	S_q	S_{sk}	S_{ku}
1	Upper	8.18	1.217	1.431	0.0986	2.050
	Lower	8.40	1.292	1.529	−0.1517	2.103
2	Upper	4.93	0.668	0.813	0.6005	2.764
	Lower	4.66	0.634	0.779	−0.6424	2.927
3	Upper	7.56	0.894	1.120	0.4587	3.030
	Lower	8.51	0.908	1.126	−0.34019	2.937
4	Upper	11.73	2.426	2.753	0.3687	1.887
	Lower	11.85	2.414	2.762	−0.4164	1.989
5	Upper	9.99	1.604	1.844	0.2477	2.011
	Lower	9.85	1.489	1.715	−0.2655	2.063
6	Upper	11.93	2.013	2.406	0.4187	2.280
	Lower	12.27	1.888	2.405	−0.1432	2.540
7	Upper	5.33	0.709	0.960	0.2339	3.311
	Lower	5.54	0.716	0.968	−0.2138	3.353
8	Upper	13.65	1.771	2.161	0.6707	2.551
	Lower	14.15	1.764	2.140	−0.6733	2.570
9	Upper	12.36	1.504	1.835	0.2862	2.554
	Lower	12.11	1.425	1.739	−0.2252	2.461
10	Upper	14.14	2.335	2.851	0.4530	2.507
	Lower	13.76	2.284	2.772	−0.3777	2.464

Table 2 Textural parameters obtained.

Sample	Surface	Spatial parameter		Hybrid parameter	
		S_{al}/mm	S_{tr}	S_{dr}	S_{dq}
1	Upper	10.86	0.4117	0.1420	0.05331
	Lower	10.73	0.4026	0.1543	0.05557
2	Upper	9.63	0.3778	0.1439	0.05368
	Lower	9.76	0.3884	0.1323	0.05147
3	Upper	11.43	0.4203	0.1554	0.05577
	Lower	10.96	0.4125	0.1643	0.05908
4	Upper	10.67	0.4062	0.1993	0.06157
	Lower	10.51	0.4083	0.2008	0.06341
5	Upper	11.08	0.7199	0.1637	0.05724
	Lower	11.16	0.4174	0.1695	0.05826
6	Upper	9.52	0.4142	0.1248	0.04998
	Lower	9.53	0.4172	10.1345	0.05090
7	Upper	8.37	0.4822	0.1003	0.05107
	Lower	8.64	0.4821	0.1067	0.04922
8	Upper	10.41	0.4143	0.2095	0.06578
	Lower	10.64	0.4042	0.1944	0.06239
9	Upper	10.93	0.4098	0.2207	0.06672
	Lower	10.46	0.4107	0.2237	0.06694
10	Upper	11.36	0.4206	0.1513	0.05504
	Lower	11.63	0.4269	0.1526	0.05527

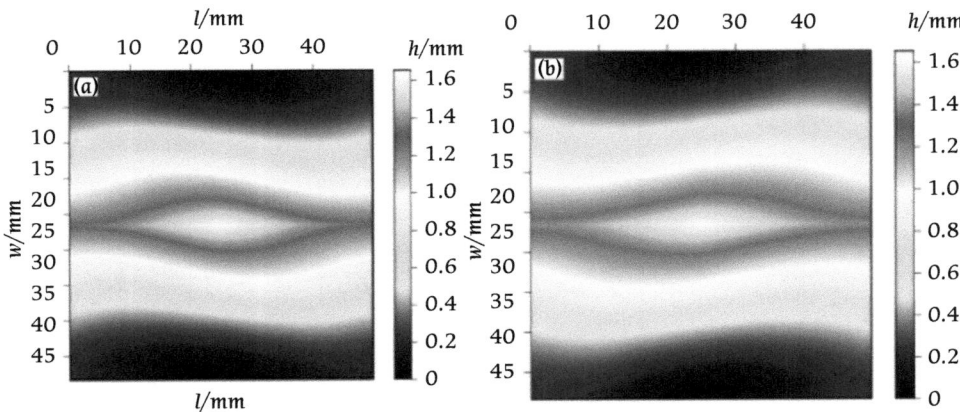

Figure 8 2D images of joint surface after being translated in Autocorrelation method: (a) Upper surface; (b) Lower surface.

(x, y, z), $z(x, y)$ is the distance between the point (x, y, z) of the surface and the point (x, y, z) of the plane, respectively. These spots are linked sequentially along the x-y direction and then it will form into profile along the x-y direction. Every one square millimeter has about 400 scan spots and each one millimeter span has about 20 profiles. If all the profiles in the x-direction are projected to the x-z plane, the projection area with approximate 1000 profiles will be too crowded to be clearly identified. Thus 20 profiles which are one in fifty of the total profiles of joint surface are projected on the x-z plane (Fig. 9). The upper profile is the higher envelop one that is linked by the highest projecting spots on the x-z plane, the middle profile is the mean one of all profiles, and the below profile is the lower envelop one that is linked by the lowest projecting spot on the x-z plane. The area, which is surrounded by the higher envelop profile, the lower envelop profile and the boundary lines, is called envelop area. The envelop area is divided into two parts by the mean profile. The part on the mean profile that is easily worn away during shear tests has obvious influence on the shear strength of joint. The two halves of coupled joint have approximately equal envelop area and the same mean profiles by contrast of the two envelop areas which is shown in Fig. 10.

5)   Profile mean angle
Mean angle of profile and the weighted average mean angle of joint surface were developed to quantify the roughness of joint surface by Homand *et al.* (2001). The equation of θ_p in the k-direction of a decided profile of joint surface is

$$[\theta_p]_k = \arctan\left(\frac{1}{N_k - 1} \sum_{i=1}^{N_k-1} \left|\frac{z_{i+1} - z_i}{\varDelta k}\right|\right)$$

where k is the direction of the profile; i is the serial number of point in the k-direction, N_k is the total number of points, z_i is the height of profile on the i-th point, $\varDelta k$ is the span between two adjacent points, N_k-1 is the total number of $\varDelta k$.

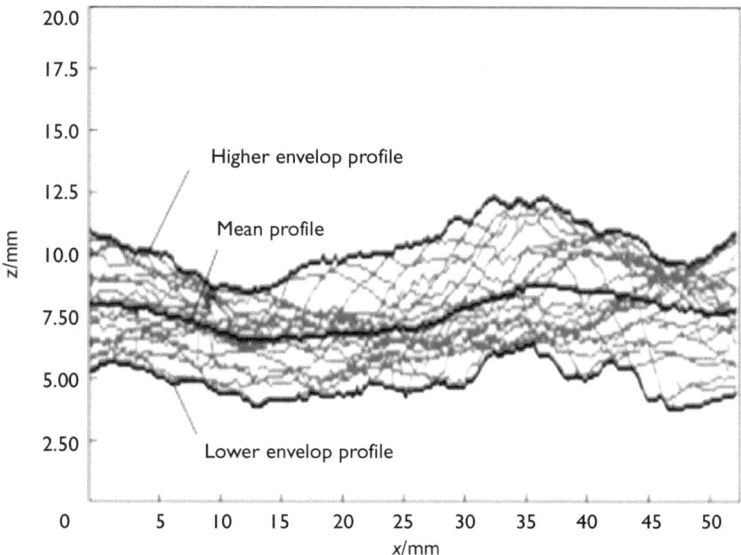

Figure 9 Profiles of join surface.

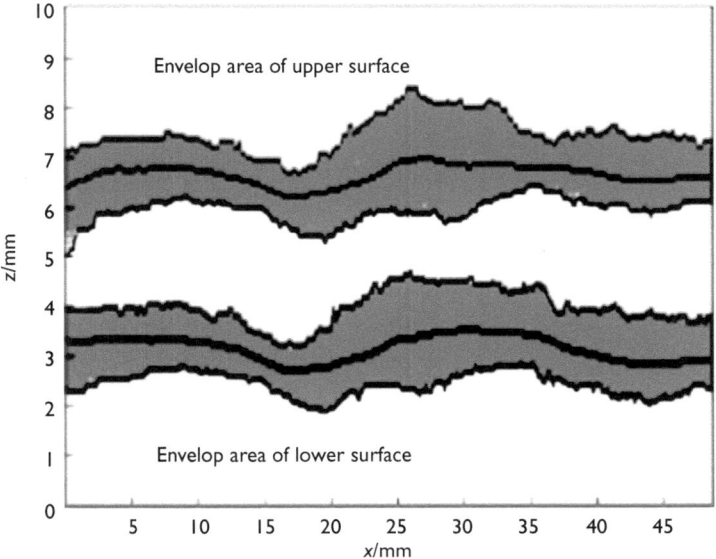

Figure 10 Envelop areas of two surfaces of joint.

Table 3 Weighted averages of mean profile angles ($\overline{\theta_p}$) of joint surface samples (°).

Sample	1	3	4	5	7
Upper	10.285	4.123	5.3752	6.889	4.710
Lower	10.007	40229	5.737	6.999	4.765
Sample	8	11	14	16	20
Upper	8.633	8.256	9.213	8.792	11.401
Lower	8.562	8.561	9.544	8.437	11.780

The equation used to compute the weighted average of the mean profile angles is expressed as

$$[\theta_p]_k = \arctan\left(\frac{\sum_{j=1}^{M_k} l_k^j \frac{1}{N_k^j - 1} \sum_{i=1}^{N_k^j - 1} \left| \frac{z_{i+1} - z_i}{\Delta k} \right|}{\sum_{j=1}^{M_k} l_k^j} \right)$$

where j is the serial number of profile, $j=1, 2, 3, ..., M_k$, l_k^j is the nominal length of the j-th in the k-direction, M_k is the total number of point in the k-direction.

Joint surface is composed of stochastic and irregular peaks. The height and volume of peaks decide the roughness of joint surface. The shear strength is related to undulation of joint surface. Through computing both surfaces of each coupled joint, it is found that $\overline{\theta_p}$ of the upper surface is approximately equal to that of the lower surface. That is to say, the one surface of well coupled joint has similar roughness with the other, which is shown from the data in Table 3. The better the joint matching is, the closer the value of $\overline{\theta_p}$ performs. The values of different coupled joints have big difference while the difference seems to be the result of different peak strengths and different curves of shear strength and displacement. Cao *et al.* (2011) investigated the changing law of $\overline{\theta_p}$ during shear tests and found that $\overline{\theta_p}$ is related to the shear strength, with verifying that $\overline{\theta_p}$ will decrease with the increase of times of shear tests. Due to that the $\overline{\theta_p}$ of two halves of well coupled joint are quite close, which can predict the $\overline{\theta_p}$ value of one surface of coupled joint based on the other one, one can also identify whether the surface is rough, flat, or smooth according to the $\overline{\theta_p}$ value.

3.2 Effects of water adsorption on the morphological parameters of the coupled joints by tensile tests

To investigate the effect of water adsorption on tensile strength and morphological characteristics of the coupled joints by Brazilian tests, in present section, the red sandstone specimens were soaked in water for different time, the water absorptions were obtained based on the weight variances before and after soaking in water. Subsequently, the Brazilian disc tests of the red sandstone with different water

Table 4 Water absorption ratio of red sandstone.

Soaking time/h	0	2	4	6	8	10
Quality before soaking in water/g	218	217	211	214	215	214
Quality after soaking in water/g	218	224	218	222	224	224
Water absorption/%	0	3.22	3.31	3.73	4.18	4.67
Number	a	b	c	d	e	f

absorptions were conducted, the maximum loads were recorded, and the tensile strengths of the red sandstone with different water absorptions were estimated. Furthermore, the joint surfaces of the Brazilian discs were scanned by Talysurf CLI 2000, with the aid of Talymap Gold software, some statistical parameters were calculated, moreover, the morphological analysis of fracture surfaces was conducted.

3.2.1 Test design and procedures

1) Water absorption measurement
   In the laboratory experiments, the red sandstone specimens of those the diameter and thickness are both 50mm have been adopted. Before soaking in water, quality of Brazilian discs was weighted by electronic balances, then the red sandstone specimens were soaked in water for 0, 2, 4, 6, 8, 10 hours, respectively. And then corresponding qualities of the Brazilian discs were recorded after soaking in water. Thereafter, the water absorption of red sandstone can be obtained.

$$w_a = \frac{m_a - m_d}{m_d} \times 100\%$$

   where, m_a is the quality of the Brazilian disc after soaking in water, m_d is the quality of Brazilian disc before soaking in water, w_a is the water absorption. The specimens' water absorptions were listed in Table 4.

2) Brazilian tests and morphology scanning
   The testing procedures of Brazilian tests and the scanning methods are similar to those in section 3.1.1. The obtained joint surfaces are shown in Fig. 11.

3.2.2 Results and discussions

1) Influence of water absorption on the tensile strength of red sandstone
   Based on the Brazilian disc tests, the tensile strengths of red sandstone with different water absorption were obtained. The tensile strength obtained by Brazilian disc tests can be expressed as follows:

$$\sigma_t = \frac{2p}{\pi D t}$$

   where σ_t is the tensile strength, p is the applied load, D is the diameter of Brazilian disc, t is the thickness of Brazilian disc.

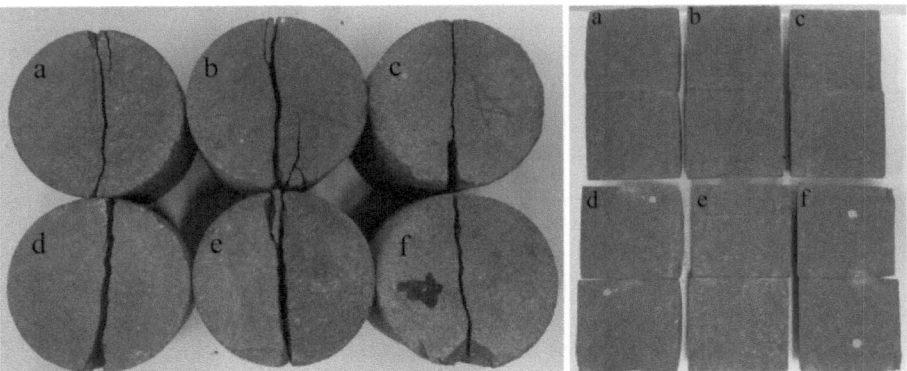

Figure 11 Failure of the Brazilian discs and their fracture surfaces (a is the failure specimen of red sandstone with water absorption 0%, b 3.22%, c 3.31%, d 3.73%, e 4.18%, f 4.67%).

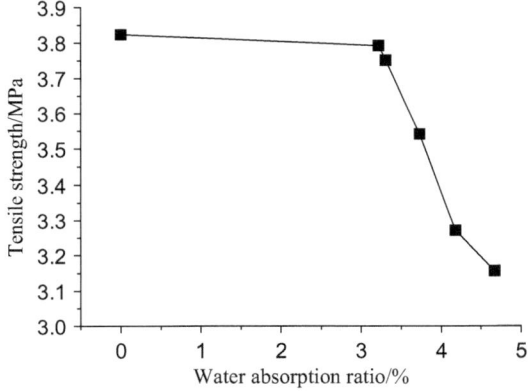

Figure 12 The tensile strength of red sandstone with increasing water absorption.

According to above equation, the tensile strength changing trend with variation of water absorption is shown in Fig. 12.

As illustrated in Fig. 12, the tensile strength of red sandstone exhibits a downward trend with increasing water absorption. Red sandstone is mainly composed of particles and pores, when the red sandstone was soaked in water, water seepage into the pores gradually, some cements of sandstone dissolved into water, hence the particles cementation of particles in sandstone was weaken, which given rise to the weakness of tensile strength.

2)   Morphological analysis on fracture surfaces of Brazilian discs with different water absorptions

   a)   Statistical parameters of the fracture surfaces

      The surfaces morphology were analyzed with Talymap Gold software, the calculation method can be found in users' manual of Talymap Gold software, and the definition for the parameters is based on ISO25178. Through the scanning of Talysurf CLI 2000 scanner, the 3D profiles of fracture surfaces were displayed in Fig. 13.

Based on the aforementioned morphology parameters, corresponding parameters for red sandstone joints were estimated, which are listed in Table 5.

Through analysis of Table 5, it is indicated that the parameters S_p, S_m and S_h of the fracture surfaces exhibited the same changing trend with increasing water absorptions, it has a regular decreased tendency, water absorption influenced the fracture surfaces roughness a lot, with water absorption increasing, the fracture surface would be more flattened. The same changing trend was mainly determined by their definition, the magnitude of S_p and S_m are mainly determined by the distance between the mean plane and the highest peak or the deepest valley, however, the magnitude of S_h is the sum of S_p and S_m. The parameters S_q, S_{sk} and S_{ku} are the parameters to describe the discreteness of height distribution, hence, the more rough of fracture surfaces are, the larger of absolute value of these parameters are, based on the data in Table 5, the absolute value of these parameters decrease with increasing water absorption. As regards to the parameter S_a, it is used to describing the roughness of fracture surfaces, it represents a downward trend with increasing water absorption.

b) Fractal dimensions of the fracture surfaces

The fractal geometry has been invented by the French mathematician Benoit Mandelbrot (1983; Giri *et al.*, 2012; Xie *et al.*, 2011) in order to describe the phenomena of scale invariance. From pure theory, fractals have become progressively important tools in a large number of scientific fields. The fractal dimension concept enables to describe the complexity of a surface under the form of a single number. Euclid's geometry teaches that a plane has a dimension 2 and that a volume has a dimension 3. However, the fractal dimension allows the use of fractal geometric dimensions, for instance for real surfaces of dimension between 2 and 3.

There are several ways to calculate the fractal dimension of a surface or a profile. Each method has advantages and drawbacks. The Talymap Gold software includes two calculation methods: including boxes method and morphological envelopes method. In this paper, including boxes method was used to calculate the fractal dimension of the fracture surfaces.

The including boxes method consists of enclosing each section of a profile by a box of width ε and calculating the area Aε of the boxes endorsing the whole profile, this procedure is iterated with boxes of different widths to build a graph $\ln(A\varepsilon)/\ln(\varepsilon)$, then the slope for line $\ln(A\varepsilon)/\ln(\varepsilon)$ would be the fractal dimension of the fracture surface. Fig. 14 gives the fractal dimension of the fracture surfaces of red sandstone with water absorption 3.31%.

Moreover, based on the including boxes method, the fractal dimensions of the fracture surfaces were calculated, the fractal dimensions of red sandstone fracture surfaces are listed in Fig. 15.

As shown in Fig. 15, the fractal dimensions of the red sandstone fracture surfaces are between 2 and 3. In the other words, these fracture surfaces are

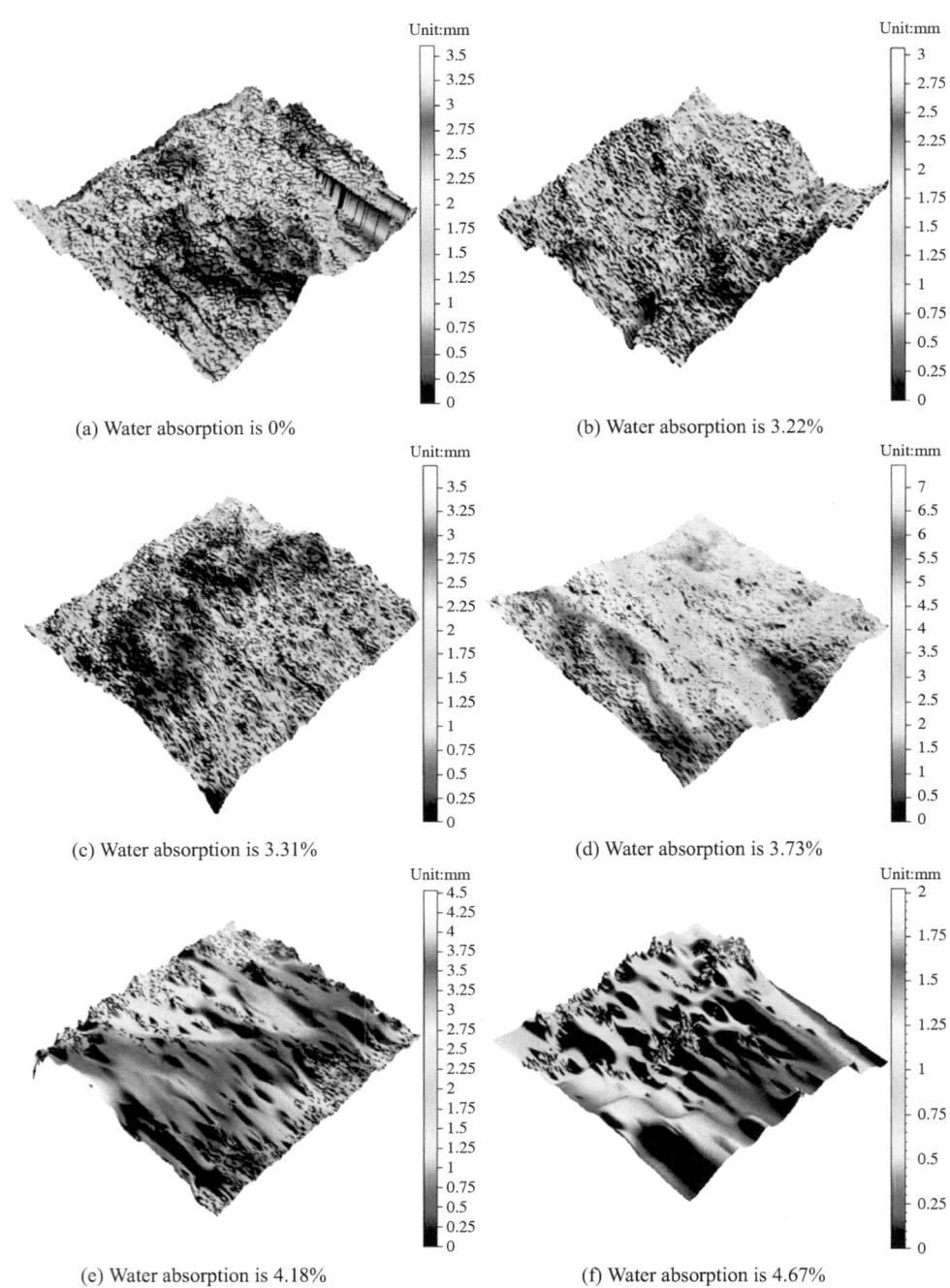

(a) Water absorption is 0%

(b) Water absorption is 3.22%

(c) Water absorption is 3.31%

(d) Water absorption is 3.73%

(e) Water absorption is 4.18%

(f) Water absorption is 4.67%

Figure 13 3D profiles for fracture surfaces of red sandstone with different water absorption.

Table 5 7 groups of parameters for the red sandstone fracture surfaces.

	Water absorption ratio/%					
	0	3.22	3.31	3.73	4.18	4.67
$S_{p/mm}$	1.710	1.490	1.610	3.890	1.390	0.903
$S_{m/mm}$	1.890	1.570	2.120	3.570	3.150	1.120
$S_{h/mm}$	3.600	3.060	3.730	7.460	4.540	2.030
$S_{q/mm}$	0.675	0.459	0.5654	1.680	0.434	0.405
S_{sk}	−0.280	0.129	−0.002	0.547	−0.528	−0.271
S_{ku}	2.740	2.910	2.580	2.160	4.180	2.580
$S_{a/mm}$	0.531	0.369	0.461	1.430	0.347	0.336

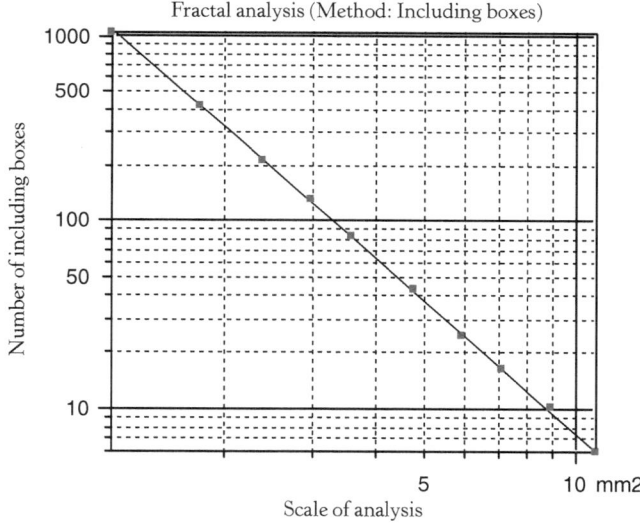

Figure 14 Fractal dimension calculation of the fracture surface of red sandstone with water absorption 3.31% and the fractal dimension is 2.33.

'near' to a plane (2D) than to a volume (3D). Specifically speaking, the water absorption of red sandstone is 0%, 3.22%, 3.31%, 3.73%, 4.18%, 4.67%, the fractal dimensions of the corresponding fractal surfaces are 2.45, 2.44, 2.33, 2.29, 2.26 and 2.25, respectively. By using the fractal dimension, the complexities of the fractal surfaces were reflected accurately. Furthermore, with increasing water absorption ratio, the fractal dimension decreased monotonically. It is concluded that, the fracture surfaces become simpler with increasing water absorption.

c)   Depths histogram and Abbott-Firestone curve of the fracture surfaces

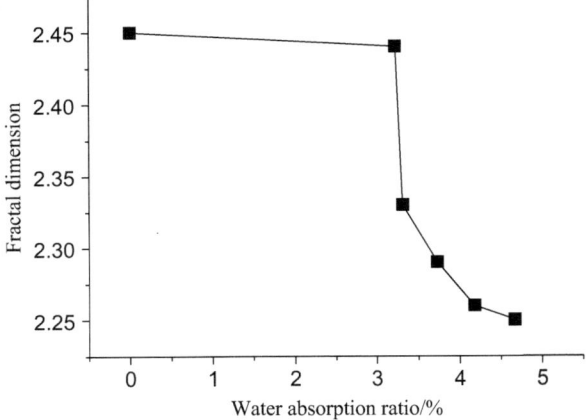

Figure 15 Fractal dimensions variation with increasing water absorption.

The depths histogram allows you to observe the density of the distribution of the data points in the profile being studied. The vertical axis is graduated in depths. The horizontal axis is graduated in % of the whole population.

The Abbott-Firestone curve presents the bearing ratio curve, *i.e.* for a given depth, the percentage of material traversed in relation to the area covered. This function is the cumulating function of the amplitude distribution function. The horizontal axis represents the bearing ratio (in %) and the vertical axis the depths (in the measurement unit).

With increasing water absorption, the depths distribution for fracture surfaces of red sandstone become more uniform, and the corresponding Abbott-Firestone curve increase slower, in the other words, the absolute value of slope for Abbott-Firestone curve decrease. For example, Fig. 16 gives the depths histogram and Abbott-Firestone curve of fracture surfaces of red sandstone with 0% and 3.73% water absorption.

As shown in Fig. 16, the depths distribution of fracture surface of red sandstone with 3.73% water absorption was more uniform than that of 0%, and its Abbott-Firestone curve changed more slowly than that of 0%. Generally speaking, the fracture surfaces of red sandstone become flatter with increasing water absorption.

d)   Frequency spectrum study on the fracture surfaces
Spectral analysis enables you to determine the periodicity and orientation of certain motifs that exist in the spectrum. This spectrum is obtained using the Fourier Transform. The Fourier Transform is used in many fields of science and engineering. The Fourier Transform is a mathematical operation enabling you to visualize the frequencies (or wavelengths) of a signal. It is used as a mathematical or physical tool to transform a problem difficult to solve (in the time or spatial domain) into one that can be easily solved (in the

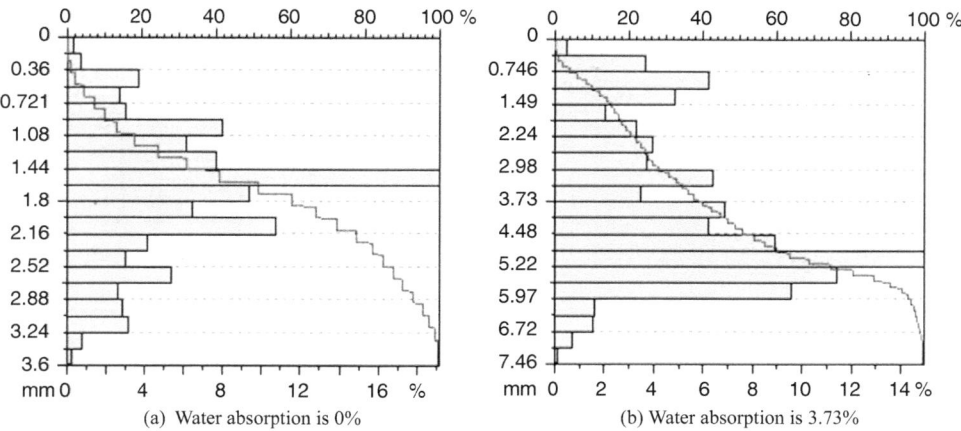

Figure 16 Depths histograms and Abbott-Firestone curves for fracture surfaces of red sandstone with 0% and 3.73% water absorption.

frequency domain). The FFT (Fast Fourier Transform) is a Fourier Transform algorithm optimized for a number of points equal to a power of 2.

For averaged power spectral density (PSD), the horizontal axis is graduated in wavelengths. The values above the peaks show the dominant wavelengths; the corresponding amplitude is displayed between brackets. The vertical axis displays the amplitude to a power of 2. The averaged power spectral density can be obtained by both the all directions method and the horizontal method. For all directions method, the PSD curve corresponds to the mean spectrum calculated from the individual spectrum curves for all directions. However, only motifs that are visible on the x-axis (example: vertical furrows) will be shown by the horizontal method. Horizontal furrows are not visible, as the PSD curve is calculated on each line (x-profile); the results are then added together. In this paper, the averaged power spectral density of fracture surface was obtained by the horizontal method.

In averaging and smoothing curve, the inverted axis creates a problem: the wavelengths found in the spectrum are not regularly spaced out anymore. For a profile of length L, the frequency $f(1)$ corresponds to the wavelength $\lambda(1)=L$, the frequency $f(2)$ to the wavelength $\lambda(2)=L/2$ etc., the frequency f(n) to the wavelength $\lambda(n)=L/n$. You can notice immediately that there are few high wavelengths to be found in the spectrum whereas there are lots of low ones.

In order to display the PSD in the form of a smooth curve that is as detailed in the high wavelengths as in the low ones, the spectrum is calculated several times using the Fourier Transform on portions of the profile having different lengths. The contributions of the various spectrums are added.

Sometimes it is necessary to increase the size of the representation of the low frequencies (large wavelengths) which provide less information. If you

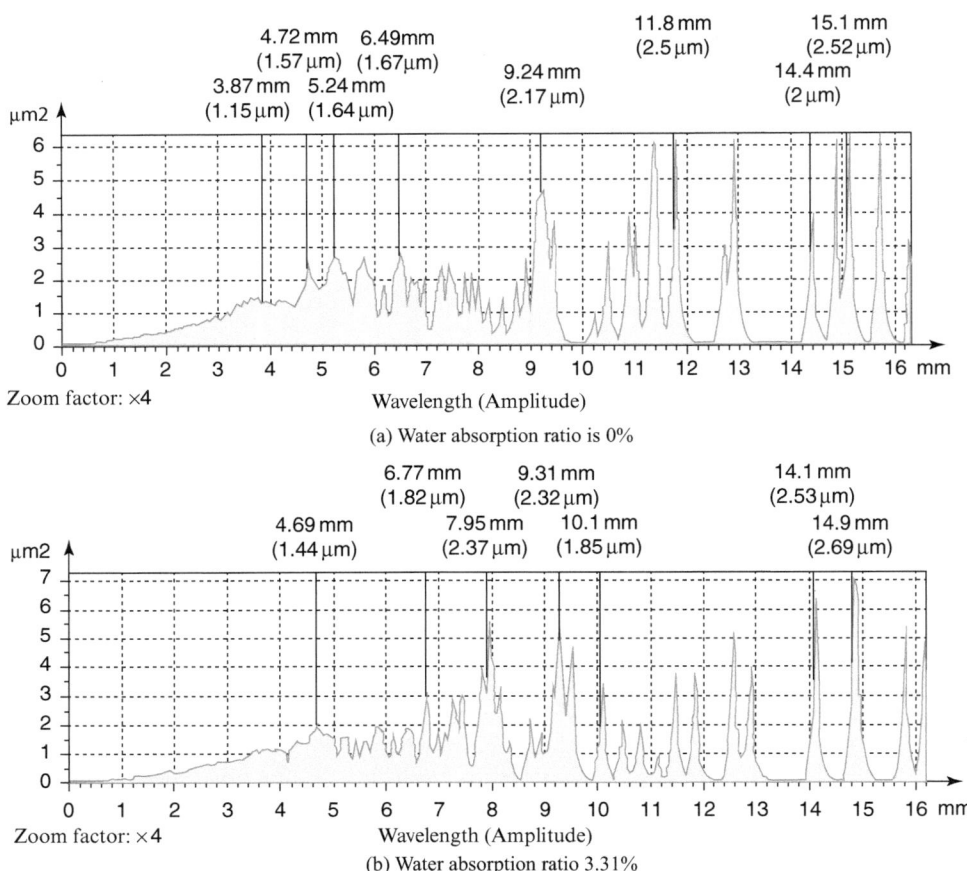

Figure 17 Power spectral diagram of fracture surface of sandstone with different water absorption ratio.

want to see high frequencies (small wavelengths), you will have to use a zoom factor as these frequencies may be invisible when other zoom factors are used. Fig. 17 gives power spectrum diagram of the fracture surfaces of red sandstone with water absorption 0% and 3.31%.

As shown in Fig. 17, with increasing water absorption ratio, the maximum dominant wavelength decreased, and thus the corresponding amplitude increased. Specifically speaking, when water absorption ratio is 0%, its maximum dominant wavelength of fracture surface is 15.1 mm, and its corresponding amplitude is 2.52 μm, while water absorption ratio is 3.31%, its maximum dominant wavelength is 14.9 mm, and amplitude is 2.69 μm. The maximum dominant wavelength is a parameter to reflect the magnitude of anisotropy for fracture surfaces, magnitude of anisotropy is negatively correlated with the maximum dominant wavelength, hence, with water absorption increasing, the magnitude of anisotropy of fracture surfaces decreases, that's to say, the fracture surfaces become more isotropy with increasing water absorption.

4 INVESTIGATIONS ON THE EFFECTS OF WATER-ROCK INTERACTION ON MORPHOLOGIES OF COUPLED JOINTS BY SHEAR TESTS

Direct compressive shear tests can also be used to generate the coupled joint surfaces successfully too. In present section, the results of the uniaxial compressive shear tests of lherzolite, peridotite, dolomite marble, migmatite and amphibolites specimens, found in China, with the contrast of water–rock interaction have been investigated. To elucidate the basic morphology characteristics and distribution, the micro-morphology of the fracture surfaces was measured by a 3D laser instrument (Talysurf CLI 20 0 0) with high-resolution (0.5μm). Twenty different parameters of height feature, texture feature, fractal geometry and frequency spectrum, including some functional parameters utilized from automotive and metal industries for the first time, were proposed innovatively for overall interpretation and analysis.

4.1 Test design and procedure

4.1.1 Rock samples and solution

Five kinds of rock samples with different petrographic, physical and mechanical properties were collected from Jinchuan no. 2 mine area. Jinchuan mine area is the largest production source of nickel and cobalt in China. Peridotite is a type of coarse-grained igneous rock and consists mostly of olivine and pyroxene. Lherzolite is a type of coarse-grained igneous rock consists mostly of olivine. Marble is a non-foliated metamorphic rock composed of calcite or dolomite. Amphibolite is a metamorphic rock consisting mainly of amphibole, especially the species hornblende and actinolite. Migmatite is a rock that is a mixture of metamorphic rock and igneous rock. The complex environment with water has become one of the most important problem s to be solved.

Before the experiment, rocks were cut and polished into cubes (sized 50 mm×50 mm×50 mm) with a cutting machine. The solution for the water–rock experiments was obtained from the depth portion of Jinchuan no. 2 mine area, where pH value is 7.1. The ion minerals in solution are Ni, Cu, Zn, Cr and SO_4^{2-}. The mechanical parameters of the rock samples are shown in Table 6. Before the uniaxial compressive shear tests, half of the samples were dried under 30° for 30 days, and the others were immersed in the same PVC containers filled with solution (pH ¼ 7.1) for 30 days. In order to obtain the actual engineering results, the sealing measurement was not carried out because of the interaction between the rock and the natural medium. As the solution for the experiments was obtained from the deep mine, it contains some minerals. Over 30 days of immersion, the residual solution of samples was fully absorbed with pieces of test paper. The immersion method used here is different from the ISRM suggested

Table 6 Parameters of rock specimens.

Set	Lithology	State	Quantity	Sample size (mm×mm×mm)
A	Lherzolite	Dry	3	50×50×50
B		Saturated	3	50×50×50
C	Peridotite	Dry	3	50×50×50
D		Saturated	3	50×50×50
E	Dolomite Marble	Dry	3	50×50×50
F		Saturated	3	50×50×50
G	Migmatite	Dry	3	50×50×50
H		Saturated	3	50×50×50
I	Amphibolite	Dry	3	50×50×50
J		Saturated	3	50×50×50

Figure 18 The spherical seats used for shearing which can be rotated at different angles.

method (*e.g.* vacuum immersion). In this study, we defined the rock specimens with open-type immersion as 'saturated specimens'. The saturated specimens were tested to contrast with dry specimens.

4.1.2 Uniaxial compressive shear tests and morphology scanning

A special shear testing device (SANS SHT4000 as shown in Fig. 18) for uniaxial compressive shear test was used in this experiment. The rock sample was placed inside the sample holders which were installed between two spherical seats. The spherical seats can be rotated to obtain different angles of shearing (θ). The device was installed

Figure 19 Different lithological specimens after uniaxial compressive shear tests: (a) lherzolite; (b) peridotite; (c) dolomite marble; (d) migmatite; (e) amphibolite.

in a servo-controlled testing machine. The load applied to the shearing device with a constant displacement rate of 0.005mm/s was measured by a load cell which was attached to the upper cross-head of the testing machine. Uniaxial compressive shear tests were conducted at three different angles of shearing: 30°, 45° and 60°. Such an arrangement allowed the samples to be investigated under different normal forces with water–rock interaction.

The displacements along normal and tangential directions of the joint were measured by the displacement transducers. A microcomputer was used for data acquisition, reduction and processing. The normal force F_n and tangential force F_t acting on the joint plane are calculated from the relations:

$$F_n = P\cos \theta$$
$$F_t = P\sin \theta$$

where P is the vertical load and θ is the shear orientation angle of the joint plane relative to the horizontal direction. By changing the shear angle, different ratios of tangential to normal force can be obtained. Different lithological specimens after uniaxial compressive shear tests are shown in Fig. 19.

This study only focuses on the morphological analysis of dry and saturated failure surfaces after shearing loading. For further consideration on shear strength, numerical method can be carried. The numerical model should include stress field, geometry, boundary and some other important elements.

When water pressure is present in a rock, the surfaces of the discontinuities are forced apart. Under steady state conditions, there is sufficient time for the water pressures in the rock to reach equilibrium. The advantage result of this experiment is that it makes both economical and practical sense to carry out a number of small scale laboratory shear tests, using equipment such as that illustrated in this paper, to determine the micro-morphology of sheared rock s under water–rock interaction. The morphology scanning process is similar to those in previous sections.

4.2 Results and analysis

4.2.1 Morphology characterization and analysis

Statistical parameters from asperities and joint roughness on the surfaces were governed by both the mechanical and hydraulic behaviors. To obtain the 3D model of each sample, each point data of surface is calculated with respect to the mean surface. Fig. 20 indicates the micro-morphology clearly with 3D view. The first properties to consider are S_p, S_m and S_b. By comparing with them, we can conclude that the water–rock interaction has played an important role. It can be seen from Table 7 that the maximum peak height (S_p) has a remarkable increase while the maximum depth of valleys (S_m) reduces markedly. More specifically, the Abbott–Firestone curve (Fig. 21) study can display the depth statistical distribution of the points on the surface. The blue depth histogram allows us to observe the distribution density of the data points on the rock surface being studied. The vertical axis is graduated in depths; the horizontal axis is graduated in % of the whole population. As shown in Fig. 21, the greatest proportion of depths of the sample (A-2) is 22.5%, and the depths lie between 9.68 6 mm and 10.57 mm. By contrast, almost 14.4% of the points of the sample (B-1) surface have a depth that lies between 5.823 mm and 6.47 mm. As for other categories, the Abbott – Firestone curve presents the bearing ratio curve, i.e. for a given depth. The upper horizontal axis represents the bearing ratio (in %) and the vertical axis the depths (in the measurement unit). According to the first red line (A-2), 80% of the points of the surface have a depth that lies between

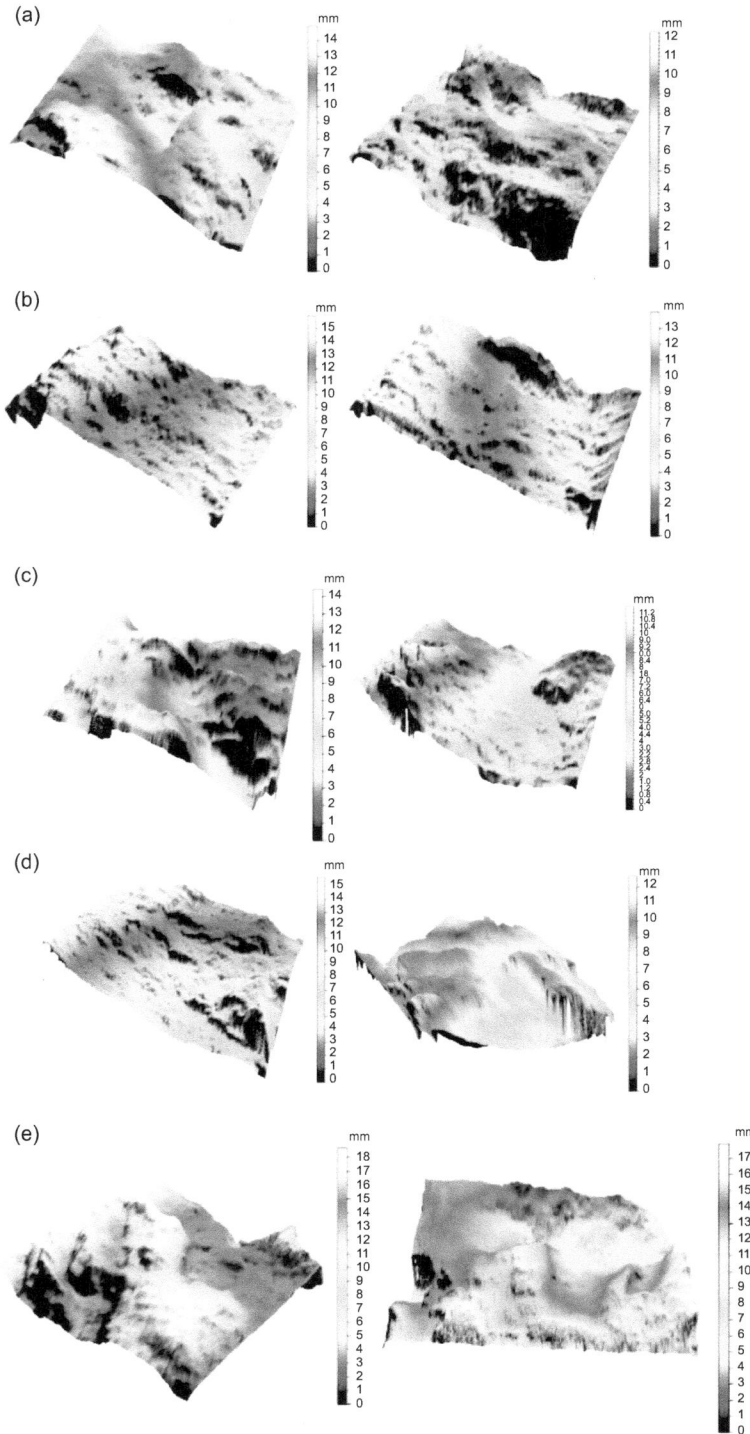

Figure 20 Micro-morphology diagram of sheared rock samples with and without water– rock interaction: (a) lherzolite; (b) peridotite; (c) dolomite marble; (d) migmatite; (e) amphibolite.

Table 7 Height feature parameters of surface morphology.

Lithology	Average S_p (μm)		Average S_m (μm)		Average S_h (μm)		Average S_a (μm)		Average S_q		Average S_{sk}		Average S_{ku}	
	Dry	Saturated	Dry	Saturated	Dry	Saturated	Dry	Saturated	Dry	Saturated	Dry	Saturated	Dry	Saturated
Lherzolite	3.521	4.728	10.49	5.632	14.01	10.36	1.121	1.572	1.399	1.978	-0.822	0.015	3.226	5.392
Peridotite	2.989	3.752	9.840	5.778	12.83	9.530	1.124	1.365	1.414	1.773	-0.898	0.439	4.138	5.750
Dolomite marble	3.341	3.394	8.130	7.293	11.47	10.69	1.173	2.571	1.492	1.627	-0.330	0.189	4.361	6.538
Migmatite	4.732	4.746	9.666	7.164	14.40	11.91	1.845	1.972	2.284	2.562	-0.548	0.324	2.971	3.458
Amphibolite	2.893	2.903	11.60	9.131	14.44	12.03	0.995	1.054	1.335	1.411	-1.102	1.028	7.173	8.468

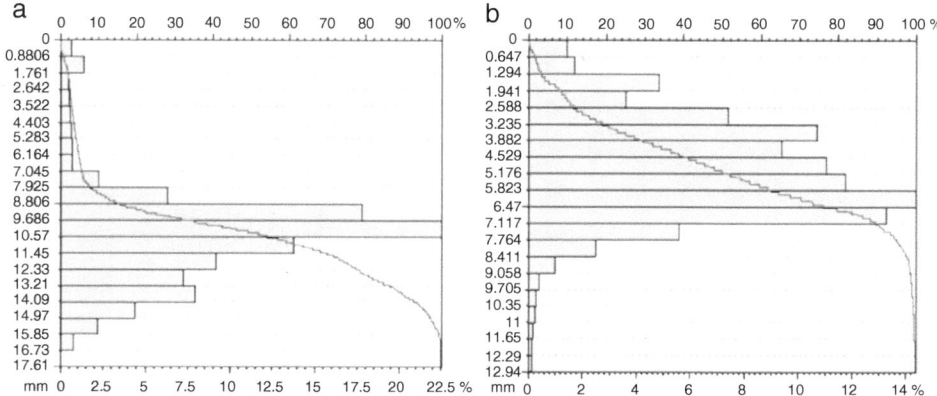

Figure 21 Abbott–Firestone curve of the lherzolite sample: (a) sample (A-2) without water – rock interaction and (b) sample (B-1) with water–rock interaction. (For interpretation of the references to color in this figure, the reader is referred to the web version of this article.)

0 and 12.50 mm approximately. However, 80% of the points of the surface (B-1) have a depth that lies between about 0 and 6.62 mm. It should be note that this value is the sum of the values of the histogram's bars between 0 and 6.62 mm. The height contour (S_h) reduces significantly in the range of 6.80 – 26.05%. The degree of decrease of the lherzolite is most obvious, whereas the degree of decrease in the dolomite marble is the smallest, indicating that the expansion effect of hydration is greater than the effect of hydrolysis and dissolution. The arithmetical mean height (S_a) reflects the average absolute value of random height distribution of surfaces. S_q stands for the discreteness and waviness of surface. It is sensitive to the higher and lower value of height particularly. Different S_q grows up and, more obviously, S_q value of migmatite rises from 2.284 to 2.562. The discreteness of rock surface height and the deviation of datum plane have increased with water–rock interaction. Skewness of the height distribution (S_{sk}) becomes the fundamental parameter in surface analysis since it reflects the probability of height distribution directly. Values of S_{sk}, without water–rock interaction, are negative, demonstrating that the maximum frequency of height appearing on the surface was negative. On the contrary, S_{sk} of each kind of saturated sample becomes positive illustrating that an opposite trend of maximum frequency of height appearing on the surface. Under water conditions, the distributed height becomes more symmetrical with decreasing absolute values of y. Lherzolite is so special that the value of S_{sk} ranges from 0.822 to 0.015, the change rate of which reaches 98.18% while the symmetry of amphibolite has not so much changed. In terms of S_{ku} value, S_{ku} of amphibolite reaches 8.468 from 7.173 with high kurtosis. Summarizing data for five rock samples, amphibolite has the most concentrated probability of height distribution. Furthermore, the increase of S_{ku} of a saturated sample relative to that of a dry sample provides a quantitative measure of the water effect in relation to shear test.

Table 8 Texture feature parameters of surface morphology.

Lithology	Average S_{dq}		Average S_{pd} $(/mm^2)$		Average S_{pc}		Average S_{dr} (%)		Average S_{bi}		Average S_{ci}		Average S_{vi}	
	Dry	Saturated	Dry	Saturated	Dry	Saturated	Dry	Saturated	Dry	Saturated	Dry	Saturated	Dry	Saturated
Lherzolite	4.274	3.385	0.0084	0.0132	0.4319	0.2958	58.89	49.23	2.354	1.973	1.286	0.8619	0.6578	0.5052
Peridotite	4.728	4.365	0.0050	0.0058	0.7961	0.1777	46.35	35.46	2.790	2.167	0.9765	0.6189	0.7451	0.5352
Dolomite marble	4.852	3.109	0.0045	0.0078	0.7426	0.6253	62.31	57.89	2.075	1.703	1.164	1.052	0.5745	0.5332
Migmatite	5.225	4.238	0.0012	0.0085	0.5074	0.4015	57.28	36.71	2.817	1.985	1.126	1.102	0.5557	0.5469
Amphibolite	5.260	3.112	0.0031	0.0036	0.9242	0.5881	69.32	54.27	2.110	1.907	0.8701	0.8098	0.6698	0.6422

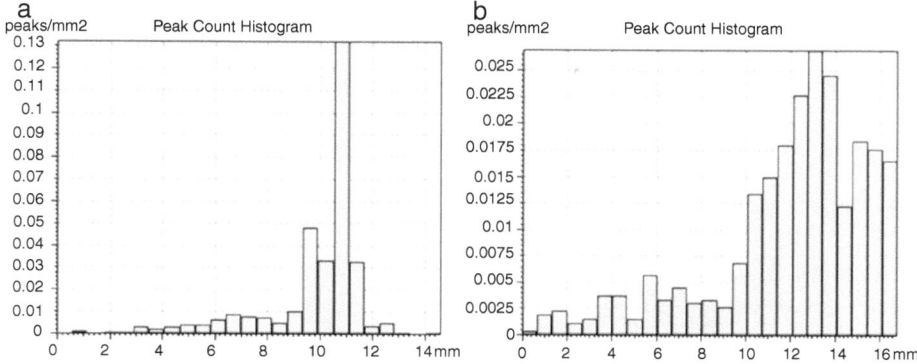

Figure 22 Peak count distribution histogram of the migmatite sample: (a) sample (G-1) without water – rock interaction and (b) sample (H-1) with water – rock interaction.

Table 8 indicates that the developed interfacial area ratio (S_{dr}) has decreasing tendency. It demonstrates that the degree of roughness and the case of ups and downs are more complex when the samples are dry. The root-mean-square slope of the surface (S_{dq}) is connected with the shape and tilted state of surface contour. This parameter is very sensitive to the large slope. S_{dq} values of different rock s have decreased generally. The feature parameters (S_{pd}, S_y) are a new family of parameters that is integrated in standard. Feature parameters are derived from the segmentation of a surface into motifs (hills and dales). Segmentation is carried out in accordance with the watersheds algorithm. The density of peak s (S_{pd}) increased regularly by comparing with all samples. It can be seen from the peak count distribution histogram (Fig. 22) that the horizontal axis is graduated in height from the lowest point to the highest point. The vertical axis can indicate the density of a rock sample. Especially, average S_{pd} value of migmatite ranges from 0.0012 to 0.0085, the change rate of which is 30 4.76%. The S_{pc} parameter which reflects the overall curvature of surface also has an obvious decreasing trend. Irrespective of whether the original density of peak s is high or low, the water–rock interaction makes the height of rock surfaces more coordinated.

The parameters $(S_{bi}, S_{ci}$ and $S_{vi})$ are a class of surface finish parameters characterizing the functional aspects of lubrication and grind. In this study, they were innovatively utilized to digitize the surface micro-morphology of a rock surface from automotive and metal industries for the first time. The obtained S_{bi} parameters showed that there were much less wearing shelves on the saturated rock surfaces. S_{bi} values of migmatite, however, reduced from 2.817 to 1.985. On the contrary, values of S_{ci} and S_{vi} indicated that dry sheared rock surface will hold a better degree for fluid retention at the core zone and valley zone, respectively. It should be noted that, due to the water lubrication, the degree of wear for rock surface reduced significantly (Zimmerman *et al.*, 1991). Hence, more and more liquid will stay on the rock surface by the impact of complex peaks and valleys.

(a)

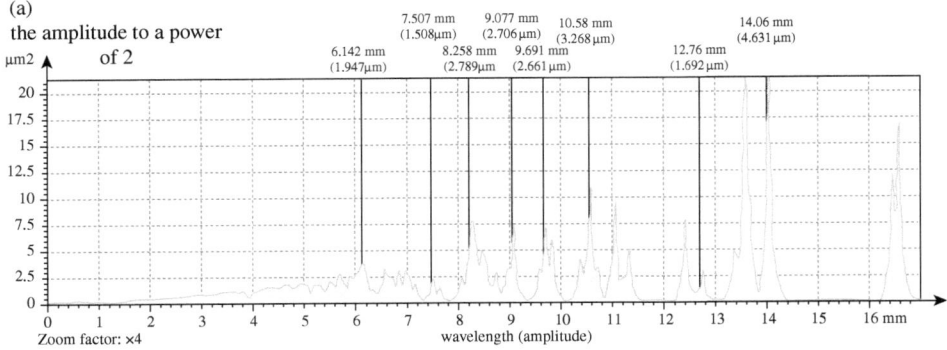

(b)

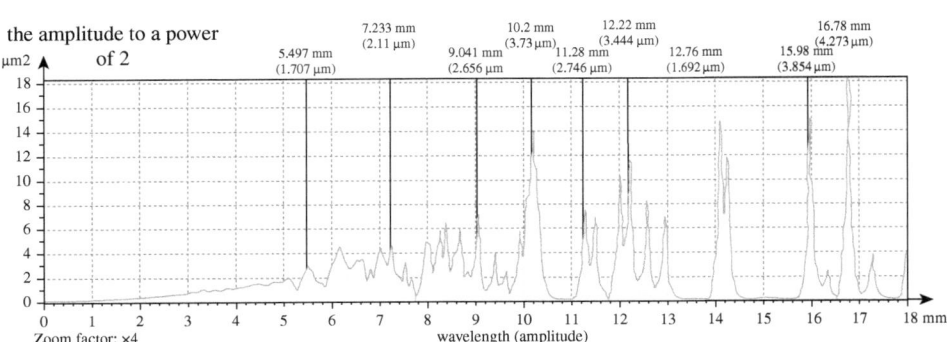

Figure 23 Power spectral density diagram of the migmatite sample: (a) sample (G-1) without water–rock interaction and (b) sample (H-1) with water – rock interaction.

4.2.2 Frequency spectrum analysis of micro-morphology

Each point in the spectrum corresponds to a frequency. The spectral representation is bi-directional and symmetrical in relation to the center of the image. The original surface has an equation of the type $z = f(x, y)$. This surface may contain periodicities in the directions of the X and Y axes or in any other directions. In order to be able to display the PSD in the form of a smooth curve that is as detailed in the high wavelengths as in the low ones, the spectrum is calculated several times by using the Fourier Transform on portions of the profile with different lengths. The contributions of the various spectra are added. The comparison of the PSD study is shown in Fig. 23. The horizontal axis ($\lambda = 1/f$) is graduated in wavelengths. The values above the peak s show the dominant wavelengths; the corresponding amplitude is displayed between brackets. The vertical axis displays the amplitude to a power of 2. When the different spectra obtained from iteration are added together, a comb effect is sometimes obtained.

The power spectral density gives a measure of the surface roughness at a certain scale. It is clear from Table 9 that the water–rock interaction has played an important role.

Table 9 Frequency spectrum parameters of samples.

Lithology	Average S_{tr}		Average dominant wavelength (mm)		Average corresponding amplitude (μm)	
	Dry	Saturated	Dry	Saturated	Dry	Saturated
Lherzolite	0.372	0.225	14.71	16.95	4.617	3.863
Peridotite	0.359	0.310	13.08	15.26	4.487	4.204
Dolomite marble	0.389	0.257	15.82	16.47	5.247	4.806
Migmatite	0.427	0.264	13.97	16.53	4.695	4.309
Amphibolite	0.442	0.378	12.89	14.22	3.714	3.592

More precisely, the average dominant wavelength increases. However, the average corresponding amplitude presents a declining tendency. With the range of S_{tr}, it is easier to point out anisotropy than isotropy because, in a natural setting, it is less obvious to find a real isotropic surface (Belem *et al.*, 2000). Consequently, these values of S_{tr} will be used as basis for a classification of anisotropic morphologies of rock surfaces. All values of S_{tr} decrease with water–rock interaction. To be more specific, S_{tr} of migmatite drops from 0.427 to 0.26 4, with a range of 38.17%. Based on the obtained results, we conclude that whether the water–rock interaction has happened or not, the sheared rock surface is considered as more or less anisotropic. It should be noticed that the sheared rock surface is much closer to the anisotropic state with water–rock interaction.

5 CONCLUSIONS

Based on the Brazilian split tests and the direct compressive shearing tests conducted on SANS SHT4000, the coupled joints are obtained. Combining morphology scanning on Talysurf CLI 2000 with the typical morphology parameters, the morphological characteristics of the coupled joints from tensile tests have been investigated. The effects of water adsorption on morphological characteristics of the coupled joints by tensile tests have also been investigated on red sandstones. And the influence of water-rock interaction on morphological characteristics of the coupled joints by shear tests has been investigated in further. Following results have been obtained:

Morphological investigation on coupled joints by tensile tests shows that the upper and lower surfaces of coupled joints have approximately equal values of S_h, S_a and S_q, but one surface's S_{sk} value is different from that of the other surface, and one surface's S_{sk} value of the coupled joints is positive while that of the other one is negative. Different joint surfaces have different values of S_h, S_a and S_q, which indicates the various height distribution of joint surfaces. Secondly, The *Sal* parameter values of both surfaces of each coupled joints are quite close, and the S_{tr} and S_{dr} values have the same situation to the S_{al} parameter, but the same parameters of different surfaces have big differences, which illustrates its own characteristics of each joint. Coupled joint surfaces have similar values of S_{dq}, whereas, all surfaces' values of S_{dq} are not significant. Thirdly, it also can be found by comparison that the upper and lower envelop areas are roughly equal and they have the same mean profile. The joints highest envelop profile of one surface is coupled to the lowest envelop profile of the other one within one joint and the coupled conditions determine the mechanical and hydraulic properties of joins in certain degrees. The last but not the least, the two surfaces of each coupled joints have similar values of $\overline{\theta_p}$. One-to-one corresponding analysis of $\overline{\theta_p}$ helps to identify that the values of $\overline{\theta_p}$ vary with different joint surface roughness, which can well reflect the roughness of joint surfaces.

The study on the effects of water adsorption on the morphological characteristics of coupled joints by tensile tests indicates that maximum peak height, maximum pit height and maximum height of the fracture surfaces exhibited the same changing trend with increasing water absorption of red sandstone, the changing trend is approximate decreasing, The height distribution parameters S_q and S_{ku} of the fracture surfaces show a downward trend with increasing water absorption. The fractal dimensions of fracture surfaces were calculated, the fractal dimensions decreased with increasing water absorption, *i.e.* the fracture surfaces become simpler with increasing water absorption. Trough analysis of depths histogram and Abbott-Firestone curve, the fracture surfaces become more uniform with increasing water absorption. The frequency spectrum of fracture surfaces gives insight into the anisotropy of the fracture surfaces, with increasing water absorption, the fracture surface become more isotropic.

The analysis of the influence of water-rock interaction on morphological characteristics of the coupled joints by shear tests presents that the height contour, height discreteness and deviation of datum plane increase with water–rock interaction. However, the height skewness, which becomes positive from negative under water condition, indicates that the height distribution is more concentrated and symmetrical. The degree of roughness and the case of ups and downs are more complex when the samples are dry. The overall curvature decreases obviously as the water–rock interaction makes the height of rock surfaces more coordinated. In addition, three functional parameters were innovatively utilized from automotive and metal industries for the first time. Dry sheared rock surface holds better degree for fluid retention at the core zone and valley zone. It demonstrates that the wear of rock surface reduces because of the water lubrication. Hence, the processes of hydration, hydrolysis and dissolution are simultaneous, whereas the expansion effect of hydration is greater than the effect of hydrolysis and dissolution. Secondly, the average dominant wavelength increases when the average corresponding amplitude decreases. Regardless of whether the water – rock interaction has happened or not, the sheared rock surface is considered as more or less anisotropic through the analysis of S_{tr} parameters. On the other hand, the sheared rock surface is much closer to the anisotropic state with water effect.

REFERENCES

Andrade PS, Saraiva AA. 2008. Estimating the joint roughness coefficient of discontinuities found in metamorphic rocks. Bulletin of Engineering Geology and the Environment, 67(8): 425–434.

Barton N, Choubey V. 1977. The shear strength of rock and rock joints in theory and practice. Rock Mechanics, 10(1): 1–54.

Beer AJ, Stead D, Coggan JS. 2002. Estimation of the joint roughness coefficient (JRC) by visual comparison. Rock Mechanics and Rock Engineering, 35(1): 65–74.

Belem T, Homand F, Souley M. 2000. Quantitative parameters for rock joint surface roughness. Rock Mechanics and Rock Engineering, 33(4): 217–242.

Cao Ping, Fan Xiang, Pu Cheng-zhi, Chen Rui, Zhou Han, Zheng Xing-ping, Zhang Chun-yang, Zhang Chun-yang. 2011. Shear test of joint and analysis of morphology characteristic evolution of joint surface. Chinese Journal of Rock Mechanics and Engineering, 30(3): 480–485. (in Chinese)

Carr JR, Warriner JB. 1989. Relationship between the fractal dimension and joint roughness coefficient. Bulletin of the Association of Engineering Geologist, 28(2): 253–263.

Chen Yu, Cao Ping, Chen Rui, Teng Yun. 2010. Effect of water-rock interaction on the morphology of a rock surface. International Journal of Rock Mechanics and Mining Sciences, 47: 816–822.

Colback PSB, Wild BI. 1965. Influence of moisture content on the compress strength of rock. In: Proceedings of third Canadian rock mechanics symposium. University of Toronto, 385–391.

Du Shi-Gui, Hu Yun-Jin, Hu Xiao-Fei. 2009. Measurement of joint roughness coefficient by using profilograph and roughness ruler. Journal of Earth Science, 20(5): 890–896.

Giri A, Tarafdar S, Gouze P, et al. 2012. Fractal geometry of sedimentary rocks: simulation in 3-D using a relaxed bidisperse ballistic deposition model. Geophysical Journal international, 192(3): 1059–1069. doi: 10.1093/gji/ggs084

Grasselli G. 2006. Shear strength of rock joints based on quantified surface description. Rock Mechanics and Rock Engineering, 39(4): 295–314.

Grasselli G, Egger P. 2003. Constitutive law for the shear strength of rock joints based on three-dimensional surface parameters. International Journal of Rock Mechanics and Mining Sciences, 40(1): 25–40.

Grasselli G, Wirth J, Egger P. 2002. Quantitative three-dimensional description of a rough surface and parameter evolution with shearing. International Journal of Rock Mechanics and Mining Sciences, 39(6): 789–800.

Hoek E, Diederichs MS. 2006. Empirical estimation of rock mass modulus. International Journal of Rock Mechanics and Mining Sciences, 43: 203–215.

Homand F, Belem T, Souley M. 2001. Friction and degradation of rock joint surfaces under shear loads. International Journal for Numerical and Analytical Methods in Geomechanics, 25(10): 973–999.

Indraratna B, Ranjith P, Gale W. 1999. Single phase water flow through rock fractures. Geotechnical and Geological Engineering, 17: 211–240.

Jae-Joon S. 2006. Estimation of a joint diameter distribution by an implicit scheme and interpolation technique. International Journal of Rock Mechanics and Mining Sciences, 43: 512–519.

Jang BA, Jang HS, Park HJ. 2006. A new method for determination of joint roughness coefficient [C]. Proceedings of the IAEG: Engineering Geology for Tomorrow's Cities, Nottingham, UK: Geological Society of London, 1–9.

Jiang Y, Xiao J, Tanabashi Y, Mizokami T. 2004. Development of an automated servo-controlled direct shear apparatus applying a constant normal stiffness condition. International Journal of Rock Mechanics and Mining Sciences, 41: 275–286.

Jiang Yu-Jing, Li Bo, Tanabashi Y. 2006. Estimating the relation between surface roughness and mechanical properties of rock joints. International Journal of Rock Mechanics and Mining Sciences, 43(6): 837–846.

Kulatilake PHSW, Shou G, Huang TH, Morgan RM. 1995. New peak shear strength criteria for anisotropic rock joints. International Journal of Rock Mechanics and Mining Sciences, 32(7): 673–679.

Lee HS, Park YJ, Cho TF, You KH. 2001. Influence of asperity degradation on the mechanical behavior of rough rock joints under cyclic shear loading. International Journal of Rock Mechanics and Mining Sciences, 38(7): 967–980.

Mandelbrot BB. 1983. The fractal geometry of nature [M]. San Francisco, CA: Freeman Company, 6–80.

Melo F, Vivanco F, Fuentes C, Apablaza V. 2007. On drawbody shapes: from Bergmark-Roos to kinematic models. International Journal of Rock Mechanics and Mining Sciences, 44: 77–86.

Melo F, Vivanco F, Fuentes C, Apablaza V. 2008. Kinematic model for quasi static granular displacements in block caving: Dilatancy effect on drawbody shapes. International Journal of Rock Mechanics and Mining Sciences, 45: 248–259.

Odling NE. 1994. Natural fractal profiles, fractal dimension and joint roughness coefficients. Rock Mechanics and Rock Engineering, 27(3): 135–153.

Pyrak-Nolte LJ, Morris JP. 2000. Single fractures under normal stress: the relation between fracture specific stiffness and fluid flow. International Journal of Rock Mechanics and Mining Sciences, 37: 245–262.

Rebinder PA, Schreiner LA, Zhigach KF. 1994. Hardness reducers in drilling: A physico-chemical method of facilitating mechanical destruction of rocks during drilling. Moscow: Akad Naunk. [Tansl by CSIRO, Melbourne, Australia].

Sari M, Karpuz C. 2006. Rock variability and establishing confining pressure levels for triaxial tests on rocks. International Journal of Rock Mechanics and Mining Sciences, 43: 328–335.

Tatone BSA, Grasselli G. 2010. A new 2D discontinuity roughness parameter and its correlation with JRC. International Journal of Rock Mechanics and Mining Sciences, 47(8): 1391–1400.

Tse R, Cruden DM. 1979. Estimating joint roughness coefficients. International Journal of Rock Mechanics and Mining Science, 16(5): 303–307.

Xia Cai-chu. 1996. A study on the surface morphological features of rock structure faces. Journal of Engineering Geology, 4(3): 71–78. (in Chinese).

Xie He-Ping, Wang Jin-An, Xie Wei-Hong. 1997. Fractal effects of surface roughness on the mechanical behavior of rock joints. Chaos, Solitons & Fractals, 8(2): 221–252.

Xie HP, Liu JF, Ju Y, et al. 2011. Fractal property of spatial distribution of acoustic emissions during the failure process of bedded rock salt. International Journal of Rock Mechanics and Mining Sciences, 48(8): 1344–1351. doi: 10.1016/j.ijrmms.2011.09.014

Yang ZY, Lo SC, Di CC. 2001. Reassessing the joint roughness coefficient (JRC) estimation using Z2. Rock Mechanics and Rock Engineering, 34(3): 243–251.

Zhang L, Einstein HH, Dershowitz WS. 2002. Stereological relationship between trace length distribution and size distribution of elliptical discontinuities. Geotechnique, 52: 419–433.

Zhao Jian. 1997. Joint surface matching and shear strength Part A: Joint matching coefficient. International Journal of Rock Mechanics and Mining Sciences, 34(2): 173–178.

Zhao XB, Zhao J, Cai JG. 2006. P-wave transmission across fractures with nonlinear deformational behaviour. International Journal for Numerical and Analytical Methods in Geomechanics, 30: 1097–1112.

Zimmerman RW, Al-Yaarubi A, Pain CC, Grattoni CA. 2004. Non-linear regimes of fluid flow in rock fractures. International Journal of Rock Mechanics and Mining Sciences, 41: 384.

Zimmerman RW, Kumar S, Bodvarsson GS. 1991. Lubrication theory analysis of the permeability of rough-walled fractures. International Journal of Rock Mechanics and Mining Sciences and Geomechanics Abstracts, 28: 325–331.

Chapter 7

Rock joints shearing testing system

Yujing Jiang
Department of Civil Engineering, School of Engineering, Nagasaki University, Nagasaki, Japan

Abstract: This chapter explains rock joints shearing testing system including the development of serve-controlled test apparatus, shear behavior of single rock joint under CNL and CNS conditions, laboratory direct shear tests and coupled shear-flow-tracer tests on rock joints with various surface characteristics.

I INTRODUCTION

The rock mass as a host ground of an underground excavation is generally not a continuum due to the presence of discontinuities, such as bedding, joints, faults and fractures, and the performance of the underground rock structure is principally ruled by the mechanical behaviors of the discontinuities in the vicinity of the excavation. Water flow, which could alter the mechanical and hydro-geological properties of rock mass, can be another important factor affecting the stability and safety of an underground excavation. The hydro-mechanical properties of a rock mass are to a large extent determined by the properties of rock discontinuities, as the discontinuities are usually weaker and more permeable than the intact rock (Hakami, 1995), and they are extremely important for long-term safety assessments of civil and environmental engineering works especially for underground radioactive waste repositories.

In general, the shear behavior of rock joints is usually investigated in laboratory tests using direct shear apparatus, where the normal load is kept constant during the shear process (*i.e.* CNL: Constant Normal Load condition) (*e.g.* Goodman, 1970; Kanji, 1974; Ladanyi & Archambault, 1977; Lama, 1978; Barla *et al.*, 1985; Pereira, 1990; Huang, X. *et al.*, 1993). As far as the rock structures in deep underground are concerned, however, shear tests under CNL condition may not be appropriate, and more representative behavior of rock fracture would correspond to a boundary condition of Constant Normal Stiffness (CNS) (Jiang *et al.*, 2004c).

In the previous CNS direct shear apparatuses, the springs were inserted between the normal load cell and the specimen of rock joint to reproduce the effect of the normal stiffness during the shear process (Brahim & Gerard, 1989; Ohnishi & Dharmaratne, 1990; Indraratna *et al.*, 1999). The normal stiffness of a set of springs, however, is very difficult to be changed according to the deformability of the surrounding rock mass. In addition, the joint surfaces are easily damaged when the springs are too strong.

In this chapter, a new direct shear apparatus and coupled shear-flow test apparatus for rock joints using virtual instrument (VI) software are developed to accommodate the change in normal stress with dilation under CNS boundary condition. A rational experimental procedure is described for the determination of the shear and shear-flow behavior of rock joints. The normal stiffness can be set automatically according to the deformational capacity of the surrounding rock masses. Shear tests of the artificial joint specimens are carried out with the developed apparatus in order to clarify the influence of the boundary conditions (*i.e.* normal load and normal stiffness) on the shear and shear-flow behavior of rock joints.

2 SHEAR BEHAVIOR OF SINGLE ROCK FRACTURE WITH NORMAL RESTRICTIONS

2.1 Development of digital controlled shear test apparatus

2.1.1 Boundary conditions applied by surrounding rock mass

The shear behavior of rock joints is usually investigated using a direct shear apparatus wherein the forces or stresses acting normal to the direction of shear displacement are maintained constant. The response of shear behavior, however, depends on boundary conditions that applied to joint surfaces by the surrounding rock mass. These boundary conditions can exist in a variety of forms as shown in Figure 1. A constant normal load (stress) boundary condition can be used, for instance, in rock slope stability problem to model the sliding of a block along a critical joint plane (see Fig. 1a. In the case of deep underground (see Fig. 1b), the forces or stresses acting normal to the direction of shear are not necessarily constant. The joints may slide on the asperities, causing dilation which acts against the normal stiffness of the surrounding rock mass. Therefore, the forces or stresses on the interface increase as the dilation increases.

On the other hand, if a system of rock bolts or cables is installed to stabilize the same block, the dilatancy of the joint is now constrained and controlled by the stiffness of the reinforcement. Similarly, a rock block constrained between dilatants joint that slides into an underground excavation, as shown in Figure 1 (b), does not move as freely as in Figure 1 (a). As the block moves, dilation of the joints is restricted by the surrounding rock masses, and the joint shear behavior is controlled by the rock masses stiffness, *i.e.*, the capacity of the rock masses to deform.

2.1.2 Normal stiffness acting on rock joint

The normal stiffness of the surrounding rock mass, k_n, in the CNS boundary condition is illustrated in Fig. 1(b) and can be obtained from expanding the infinite cylinder theory as

$$k_n = E/(1 + v)r \qquad (1)$$

where E and v are the modulus and the Poisson's ratio of rock mass, respectively, and r is the influenced radius (Johnston *et al.*, 1993; Jiang *et al.*, 2001). The value of k_n varies with the deformational properties of the surrounding rock masses. Since both E and v are reasonably constant for the stress range considered, the normal stiffness is constant.

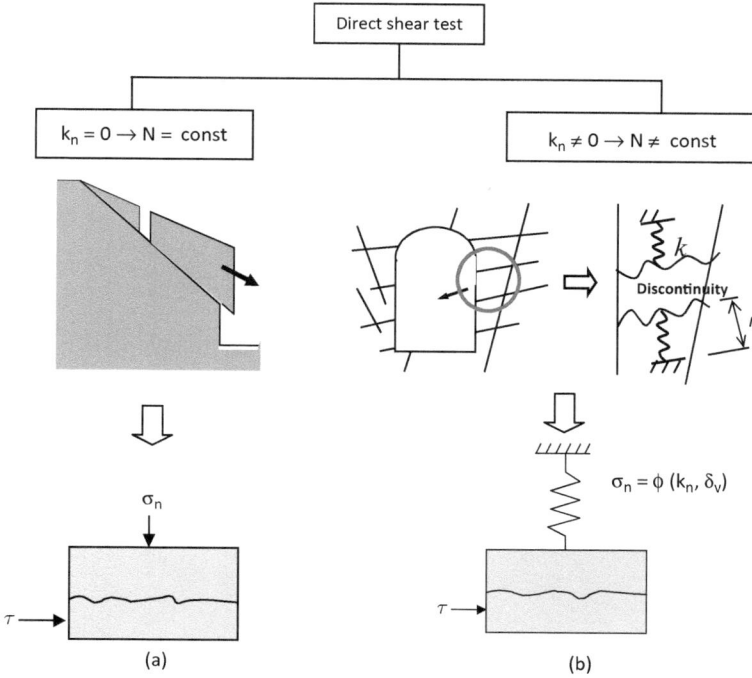

Figure 1 Different *in-situ* boundary conditions in direct shear test of rock joint: (a) Joint behavior of slope: (b) Joint behavior near an underground opening at depth and the evaluation of normal stiffness (k_n) acted on joint surface.

2.1.3 Hardware of apparatus

A novel servo-controlled direct shear apparatus, shown in Photo 1, is designed and fabricated for the purpose of testing both natural and artificial rock joints under various boundary conditions (Jiang *et al.*, 2001). The outline of the fundamental hardware configuration of this apparatus is described in Fig. 2 and consists of the following three units:

a) *A hydraulic-servo actuator unit*
    This device consists essentially of two load jacks that apply almost uniform normal stress on the shear plane. Both normal and shear forces are applied by hydraulic cylinders through a hydraulic pump which is servo-controlled. The loading capacity is 400 kN in both the normal and shear directions. The shear forces are supported through the reaction forces on two horizontal holding arms. The applied normal stress can range from 0 to 20MPa, which simulates field conditions from the ground surface to depths of about 800m.

b) *An instrument package unit*
    In this system, three digital load cells (tension-contraction type, capacity 200 kN (normal), 400 kN (shear)) for measuring shear and normal loads have been set with the rods connected at two sides of the shear box (Fig. 2a). Displacements are monitored and measured through LVDTs (linear variation displacement transducers), which are attached to the top of the upper shear box.

Photo 1 Direct shear test apparatus.

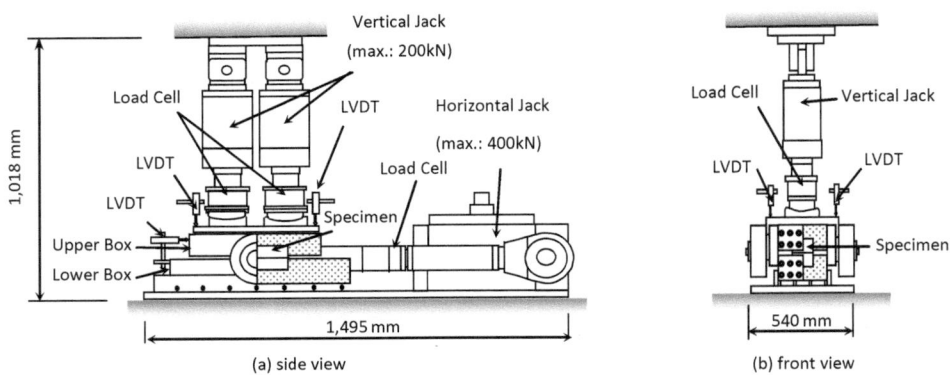

Figure 2 Sketch view of the digital-controlled shear testing apparatus.

c) *A mounting shear box unit*

As shown in Fig. 3, the shear box consists of lower and upper parts, the upper box is fixed and the joint is sheared by moving the lower box. The upper box is connected to a pair of tie rods to allow vertical movement and rotation (Photo 2). It is corresponding with a range of lengths of specimen between 100 and 500mm.

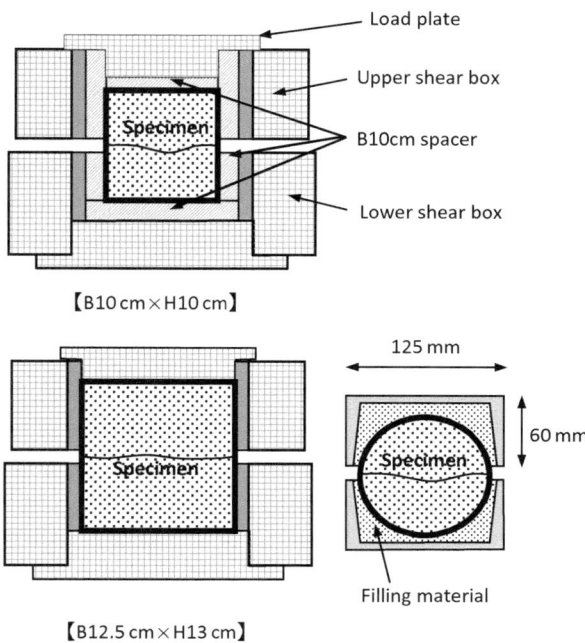

Load plate
Upper shear box
B10cm spacer
Lower shear box

【B10 cm×H10 cm】

125 mm

60 mm

Specimen

Filling material

【B12.5 cm×H13 cm】

Figure 3 Structure of shear box and preparation of artificial joint sample.

Photo 2 Shear test box.

2.1.4 *Digitally controlled data acquisition and storage system*

In the novel apparatus, the constant and variable normal stiffness control conditions are reproduced by digital closed loop control with electrical and hydraulic servos (Fig. 4). It is a nonlinear feedback and measurement is carried out on a PC through a multifunction analog-to-digital, digital-to-analog and digital input/output (A/D, D/A

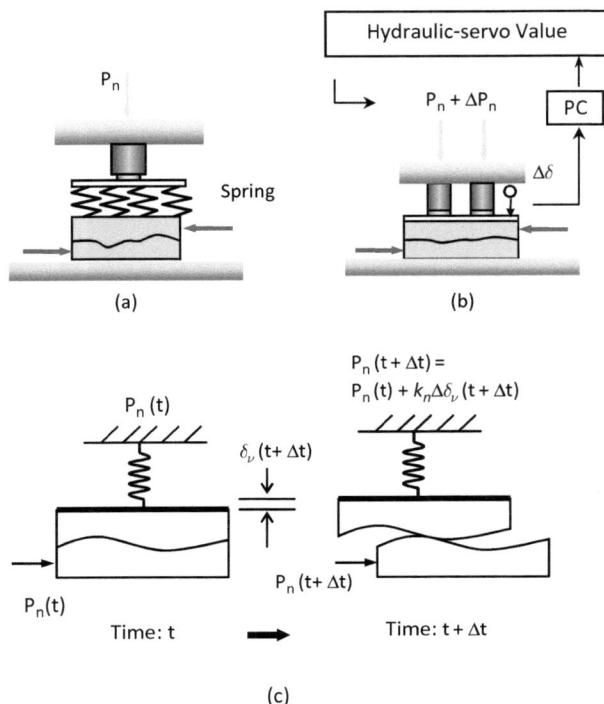

Figure 4 Principle of reproducing the CNS condition during the shear process in the novel direct shear apparatus. (a) CNS test in conventional apparatus, (b) CNS test in the novel apparatus, (c) method for the representation of CNS condition.

and DIO) board, using the graphical programming language LabVIEW, a custom built 'virtual instrument' (VI) and a PID control toolkit. Data acquisition and normal stiffness setting are carried out digitally by using the 16bit A/D and D/A boards (SPEC: AD: single-end/16CH; DA: single-end/2CH; DIO: single-end/8CH) which are inserted in a computer. The introduction to the AD/DA/DIO board, together with faster PC processors, has allowed the system to scan 16 input channels and 32 output channels. This performance challenges some of the dedicated instruments that are available for multichannel scanning. Collected test data includes the normal and shear forces, the corresponding displacements and the strokes of the vertical and horizontal loading cylinders. Digital control program was designed by using LabVIEW programming language for building data acquisition and instrumentation systems. With LabVIEW, the interactive control of the system can be created quickly. In this system, the time interval that is set in 100Hz is twice faster than the past ones.

Fig. 4 shows that the novel digital direct shear apparatus (Fig. 4(b), (c)) differs from the direct shear apparatus that involves a set of springs (Fig. 4 (a)), the normal load is corrected by the feedback hydraulic-servo controlled value. The normal loads are monitored by flat compression load cells and are to be changed by comparing the

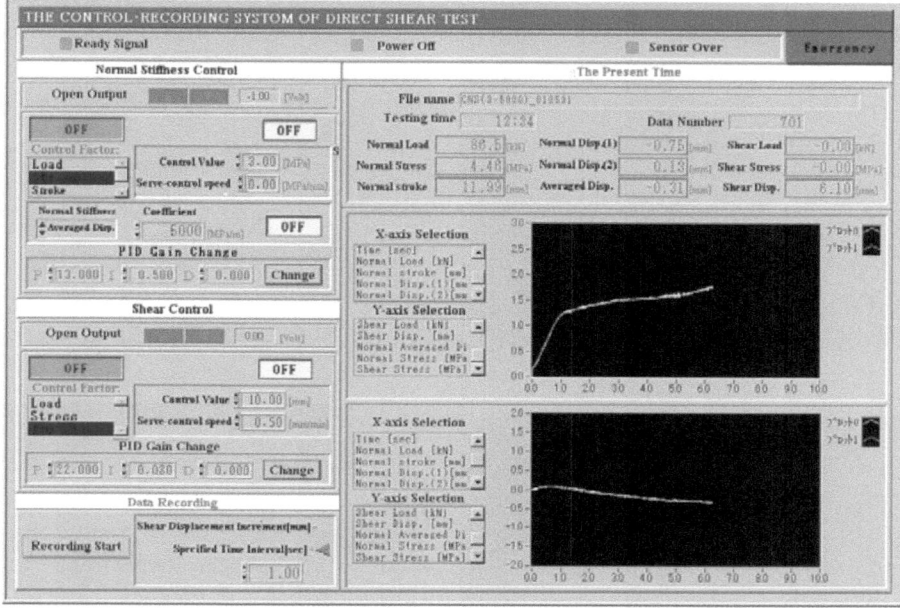

Figure 5 A LabVIEW screen showing the front panel of the 'shear main screen' VI. The user controls
and sets the CNS operation on this screen. Status information and the collected data curves
are presented on indicators.

vertical displacement of joints during the shear process together with the sign of the
feedback, which is calculated based on the normal stiffness value, k_n, as follows:

$$\Delta P_n = k_n \cdot \Delta \delta_n \tag{2}$$

$$P_n(t + \Delta t) = P_n(t) + \Delta P_n \tag{3}$$

where ΔP_n and $\Delta \delta_n$ are the changes in normal load and vertical displacement, respectively. Data acquisition and normal stiffness setting are carried out digitally using
16-bit A/D and D/A boards installed in a computer.

Fig. 5 is a LabVIEW screen showing the front panel of the 'shear main screen' VI. The
user controls and sets the CNS operation on this screen. Status information and the
collected data curves are presented on indicators.

2.2 Sample preparation

Two groups of samples were prepared for direct shear tests for different purposes. The
first group was tested for the evaluation of the different behaviors of rock fractures
under CNL and CNS boundary conditions. The latter group was tested to assess the
relations of surface roughness of rock fractures to their shear behaviors.

In both groups, specimens are 100 mm in width, 200 mm in length and 100 mm in
height with a synthetic joint. First, the top and bottom molds must be detached from the

(a)

(b)

(c)

(d)

Photo 3 Photographs of prepared samples. (a), (b) pair of molds before shearing; (c), (d) surface profiles of joints after shearing under low and high normal stresses.

shear apparatus before casting the upper and lower portions. Subsequently, the bottom mold was filled with the mixture and left for at least 60 minutes to ensure adequate hardening. After one surface of the joint profile was cast, the top mold was then placed over the bottom mold and filled with the plaster mixture. Thus, fully mated joint specimens were prepared for shearing. Photos 3 (a) and (b) show the prepared specimen, a pair of molds before shearing, and Photos 3 (c) and (d) show the surface profiles of joints after shearing under low and high normal stresses, respectively. Photo 4 shows three sample replicas of the surface roughness of rock joints made in three nearly parallel joint planes in granite, located at an underground pumped-storage power station.

For the samples in the first group, three types of rock-like materials are applied to describe soft rock, medium-hard rock and hard rock, respectively. The mix material (TR1), a mixture of plaster, water and retardant in a weight ratio of 1:0.2:0.005, was used to describe soft rock. The synthetic rock material (TR2) similar to medium-hard rock, used in this investigation, was a mixture of plaster, sand, water and retardant in a weight proportion of 1:1:0.28:0.005. Resin concrete (RC) was selected as test material as

Table 1 Physico-mechanical Properties of TRI, TR2 and RC.

Physico-mechanical properties	Index	Unit	TRI (Soft rock)	TR2 (Medium-hard rock)	RC (Hard rock)
Density	ρ	g/cm^3	2.066	2.069	2.247
Compressive strength	σ_c	MPa	47.4	89.5	1.07.7
Modulus of elasticity	E_s	MPa	28700	26200	27100
Poisson's ratio	v	–	0.23	0.29	0.24
Tensile strength	σ_t	MPa	2.5	4.5	10.3
Cohesion	c	MPa	5.3	9.9	15.9
Basic friction angle	φ	degree	63.3	64.4	57.0

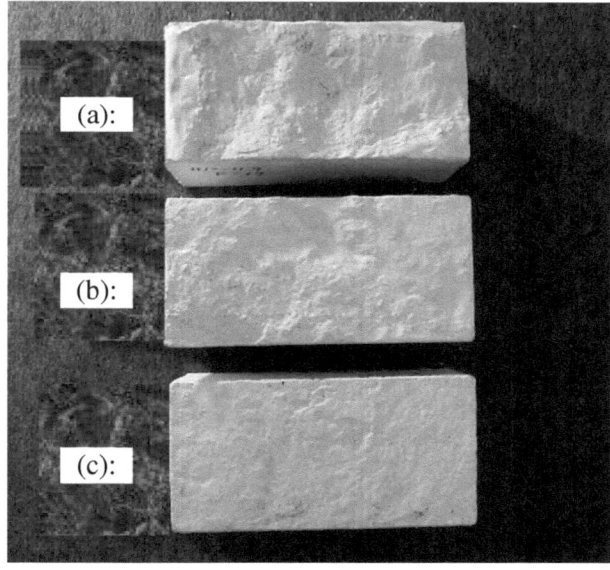

Photo 4 Three sample replicas of the surface roughness of rock joints are made in three nearly parallel joint planes in granite, at a depth of 158 m, at an underground pumped-storage power station.

it has properties similar to those of hard rock. Note that the plaster used here is specially designed and has been verified effective to simulate the mechanical behavior of rock. Table 1 shows the physical properties of three types of rock-like materials. The uniaxial compressive strengths of TR1, TR2 and RC are 47.4 MPa, 89.5 MPa and 107.7 MPa, respectively. Three standard JRC roughness profiles (Case 1: JRC value from 4 to 6, Case 2: JRC value from 8 to 10, Case 3: JRC value from 12 to 14) were selected for TR1 and TR2 of artificial rock joints. Totally 54 artificial rock-like joints were prepared and tested using the developed direct shear apparatus under three different normal stresses (2 MPa; 5 MPa and 10 MPa). Of these, 27 cases of soft-rock-like (TR1) specimens were subjected to shear tests, performed under the CNL condition ($k_n = 0$ (nine cases)) and under the CNS condition (k_n set to 3 GPa/m (nine cases) and 7 GPa/m (nine cases)). For the

Table 2 Different Rock-like Artificial Joints under CNS and CNL Conditions.

Sample	Boundary condition		Roughness	Initial normal stresses (MPa)
TR1(Soft rock)	CNS	CNL k_n=3MPa/m k_n=7MPa/m	JRC = 4~6 JRC = 8~10 JRC = 12~14	
TR2 (medium hard rock)	CNS	CNL k_n=7MPa/m k_n=14MPa/m		2; 5; 10
RC (Hard rock)	CNS	CNL k_n=5.4MPa/m	JRC = 1.5 JRC = 3.6 RC = 8.5	

medium-hard-rock-like joints, the tests were performed using the same three normal stress values. Table 2 shows different rock-like artificial joints under both CNS and CNL conditions and different initial normal stresses.

2.3 Confirmation of the accuracy of apparatus

In order to confirm the accuracy of the novel apparatus, three cases of tests are firstly conducted under CNL and CNS conditions. The normal stiffness, k_n, of the surrounding rock mass in the CNS test is set to 2GPa/m and 15GPa/m based on the Equation 1. An initial normal stress before test begins is set to 0.5MPa for all cases, and the servo-controlled shear speed is set to 0.5mm/min.

Figure 6 illustrates the stress-strain behavior of the three cases. In the CNL test, the normal stress σ_{no} is kept constant during shear process. Correspondingly, in the two cases (k_n=2GPa/m; k_n=15GPa/m) of the CNS tests, the normal stress and normal displacement are well proportioned in the process according to the input value of normal stiffness k_n. It is proved that both CNS and CNL conditions were realized at the good accuracy.

2.4 Comparison of shear behaviors of rock joints under CNL and CNS conditions

In order to compare the shear behaviors of rock joints with the natural roughness profiles under both CNS and CNL conditions, shear tests under initial normal stresses of 2, 5, 10 MPa were conducted on the RC specimens described in the previous section. Fig. 7(a) illustrates the shear stress-shear displacement relations under different roughness profiles and initial normal stresses. CNL tests always show the low peak shear strength at a small shear displacement. In CNS tests, the normal stiffness of the surrounding rock mass, k_n = 5.4 GPa/m, was selected for use in clarifying the effects of the normal stiffness on the shear behavior of joints. The peak shear stress is found to be greater than that obtained in the CNL tests as a result of the increased normal stress during the shear process (Fig. 7(b)). It is also clear that increasing the normal stiffness

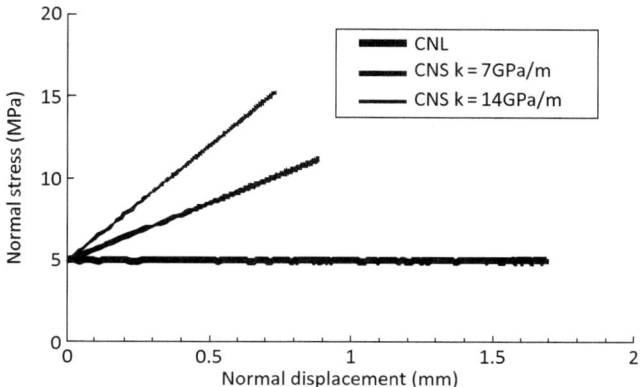

Figure 6 The relation between normal stress and normal displacement under CNL and CNS conditions (JRC: 12~14 and $\sigma_{n0} = 5$ MPa).

increases the normal stress, thus reducing the dilation of the rough joints. A more detailed investigation into the shear results shows some dependence between the evaluated parameters and the change in the initial normal stress and normal stiffness. The change in the normal displacement during shear process, *i.e.* the dilation angle, shows no change, whereas the peak shear stress is significantly affected by both the initial normal stress and the normal stiffness.

The relationships between normal displacement (volume change) and shear displacement are shown in Fig. 7(c). The mechanical aperture of the joint is considered to be altered significantly when a normal stiffness is applied to the sample and increased with shear displacement. The change in the mechanical aperture, which is smaller in the CNS tests than in the CNL tests, is also influenced by the value of JRC.

The normal stress versus shear stress paths with different profile samples under both CNL and CNS conditions are shown in Fig. 8. In each case, the shear stress initially rises. In CNL tests, the shear stress decreases when the peak shear strength is reached. The normal stress is constant so that the shear stress path is vertical. However, in the CNS tests, the shear stress in the residual shear process rises slowly when the peak shear stress is reached. The first break points on the stress paths under the CNS conditions are situated almost on the regression line of the peak shear stress under the CNL conditions.

2.5 Influence of surface roughness on shear behaviors of rock fracture

2.5.1 Sample preparation

The cases tested in group two are shown in Table 3. The physico-mechanical properties of the samples in group two are coincident with the sample TR 1 in group 1. Three standard JRC profiles (J1: JRC value from 4 to 6, J2: JRC value from 8 to 10, J3: JRC value from 12 to 14) were chosen as models to manufacture artificial rock fractures, identical to the cases 1, 2, 3 in group 1.

Table 3 Experiment Cases under CNS and CNL Conditions.

Specimen	Roughness (JRC range)	Initial normal stresses	Boundary condition
J1	4~6	2MPa	
J2	8~10	5MPa	CNL
J3	12~14	10MPa	CNS(k_n =3GPa/m)
J4	0~2	1MPa	CNS(k_n =3GPa/m)
		2MPa	
		5MPa	

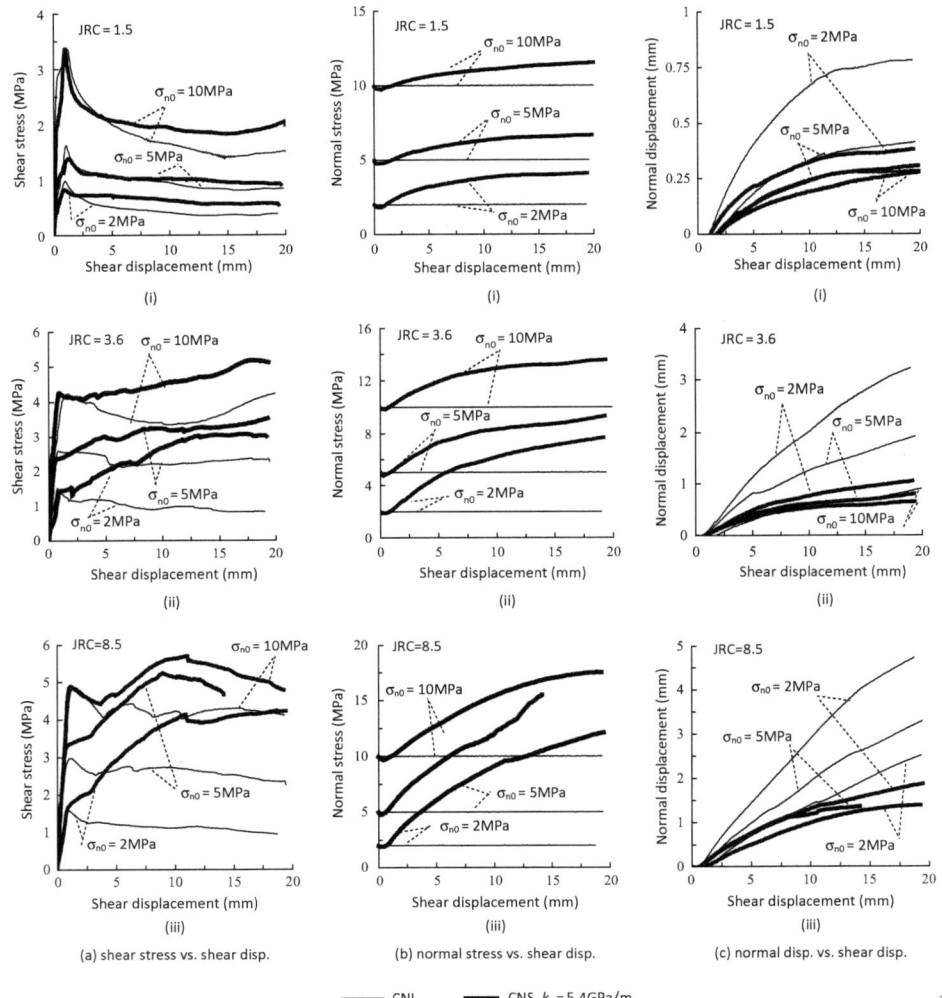

Figure 7 Shear behavior of natural rock fractures under CNL and CNS conditions.

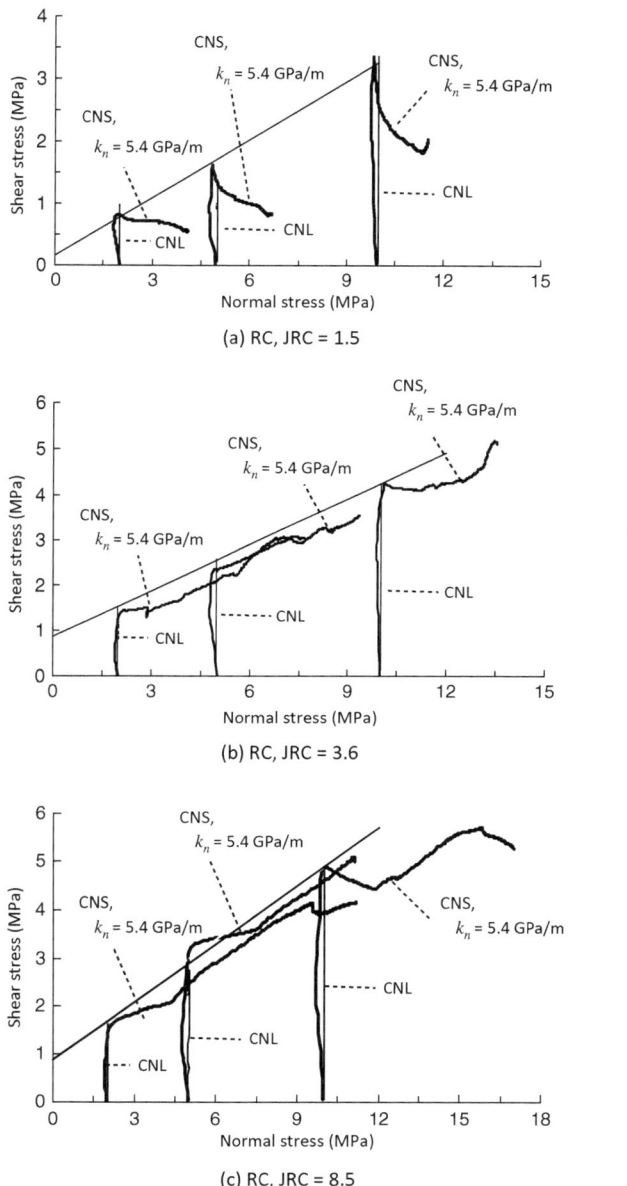

Figure 8 Shear stress versus normal stress paths for different JRC under CNL and CNS conditions.

The principal difference of the two test groups is that the surfaces of rock fractures before and after shearing in group 2 are measured and used in the estimations of the relations of surface roughness and shear strength, while tests in group 1 focus on the shear behavior of rock fractures under different boundary conditions.

2.5.2 Measurement of surface roughness

To obtain the topographical data of rock fracture surfaces, a three-dimensional laser scanning profilometer system, which has an accuracy of ±20 μm and a resolution of 10 μm, was employed. A X-Y positioning table can move automatically under the sample by pre-programmed paths to measure the desired portion of the sample surface. A PC computer performs data collecting and processing in real time.

Joint profiles of specimens J1, J2 and J3 were made based on the standard JRC profiles. Therefore, only one line on the surface assumed to represent the whole roughness was needed in theory. In this study, 5 lines with a constant distance along the shear direction on specimens were measured with a 0.3125mm interval and the mean values at every point are calculated to create 2D profiles that represent the mean surface shapes. The specimen J4 was measured in both shear direction and the direction perpendicular to the shear direction with the same sampling interval for obtaining sufficient data for 3-D fractal evaluation (Jiang *et al.*, 2004b, 2006).

2.5.3 Correlations of shear strength with surface roughness of rock joints

Fig. 9 shows two examples of the shear test results, one is the test with initial normal stress of 2MPa and normal stiffness of 0, 3, 7GPa/m on J2; another is the test result of J4 when initial normal stress is 1MPa and normal stiffness are 0, 3, 7GPa/m, respectively. For all tests, the shear stresses are greater under CNS condition because of the increased normal stress caused by dilatance during shearing. Also, the peak shear stress under CNL condition is smaller than that under CNS condition. Comparing all of the test results, it can be found that the shear stress increases with a larger initial normal stress and a higher value of JRC. Moreover, it is shown in Figs. 9 (b) and (e) that the normal displacement controls the mechanical aperture of joint, and changes significantly when a normal stiffness is applied to the sample. This is due to the normal displacement increases with shear displacement. The change in normal displacement in CNS tests is small in comparison with that in CNL tests.

Since the specimens used here are manufactured based on known JRC profiles, it is of interest to compare the shear results with the predicted value by using Equation 4 (Barton & Choubey, 1977) as follows.

$$\tau_p = \sigma_n \tan[JRC \log_{10}(JCS/\sigma_n) + \phi_r] \tag{4}$$

where σ_n is effective normal stress, JCS is joint wall compressive strength, which is equal to the unconfined compressive strength of the intact rock for unweathered fracture surfaces but which has a much lower value for weathered surfaces.

In this equation, the values of four variables should be investigated to predict shear strength. Herein, JRC and σ_n have certain values at initial conditions and the measured JCS by using Schmidt rebound hammer is 39MPa. ϕ_b can be obtained directly from shear tests. Fig. 10 shows the values of ϕ_b for three specimens.

As shown in Fig. 11, the experimental results agree well with the predicted values by Barton's empirical equation at low normal stresses. Predicted values tend to be higher than the experimental results as normal stress increases. These two kinds of values also differ considerably from J1 to J3, that is from low to higher JRC values. These differences show that Barton' equation may overestimate the shear strength at a relatively high normal stress, as 10MPa shown here or higher, and for very rough joints. However, shear strength would be influenced by a number of other factors such

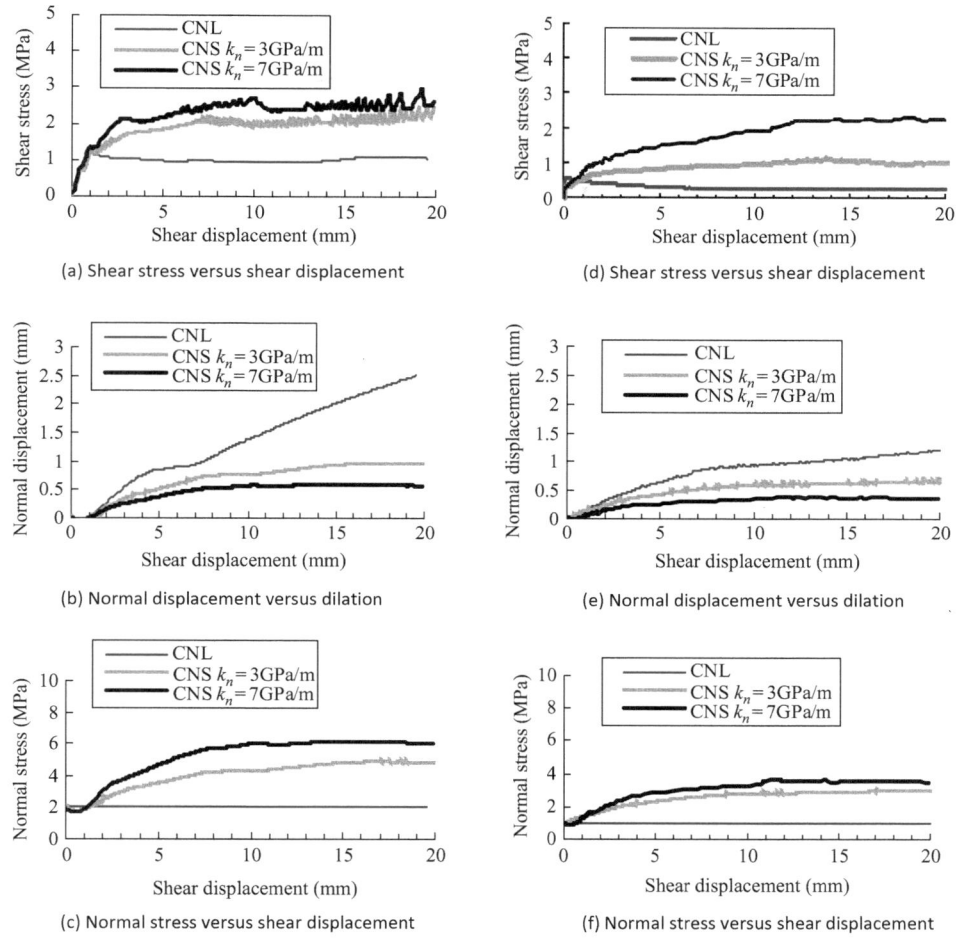

Figure 9 Shear and dilation behaviors of rock joints. (a), (b), (c) J2, $\sigma_{n0} = 2\,\text{MPa}$; (d), (e), (f) J4, $\sigma_{n0} = 1\,\text{MPa}$.

as shear direction and normal stiffness, a small number of experimental results may not give sufficient evidences to prove these points.

3 COUPLED SHEAR-FLOW-TRACER TEST FOR SINGLE ROCK FRACTURES

3.1 Development of a new coupled shear-flow-tracer test apparatus

3.1.1 Fundamental hardware configuration

Fluid flows through a rough fracture following connected channels by passing the contact areas with tortuosity. These phenomena cannot be directly observed in

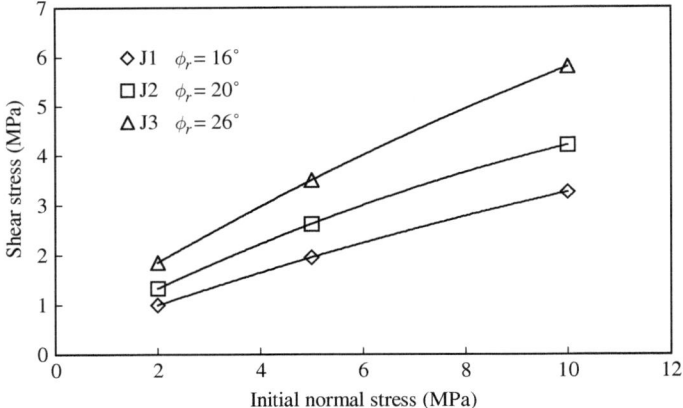

Figure 10 Evaluation of the basic friction angle of the three specimens.

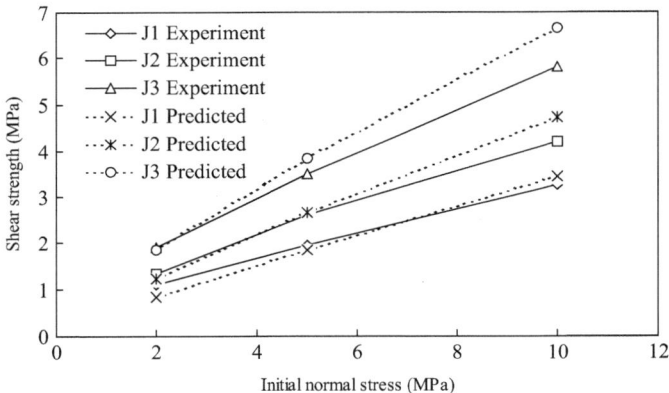

Figure 11 Comparison of shear strength from experiment tests (under CNL condition) and predicted with Barton's empirical equation.

ordinary laboratory tests without a visualization device. In this chapter, a laboratory visualization system of shear-flow tests under the CNS condition is developed. The outline of the fundamental hardware configuration of this apparatus is described in Fig. 12. It consists of the following five units:

(1)   A hydraulic-servo actuator unit. This device consists essentially of two load jacks and two sets of linear-guides for applying uniform stresses on the upper and lower blocks of rock fracture specimens. Both normal and shear forces are applied by hydraulic cylinders through a servo-controlled hydraulic pump. The loading capacity is 100kN in both normal and shear directions. The shear force is applied on the specimen holder through two horizontal holding arms.

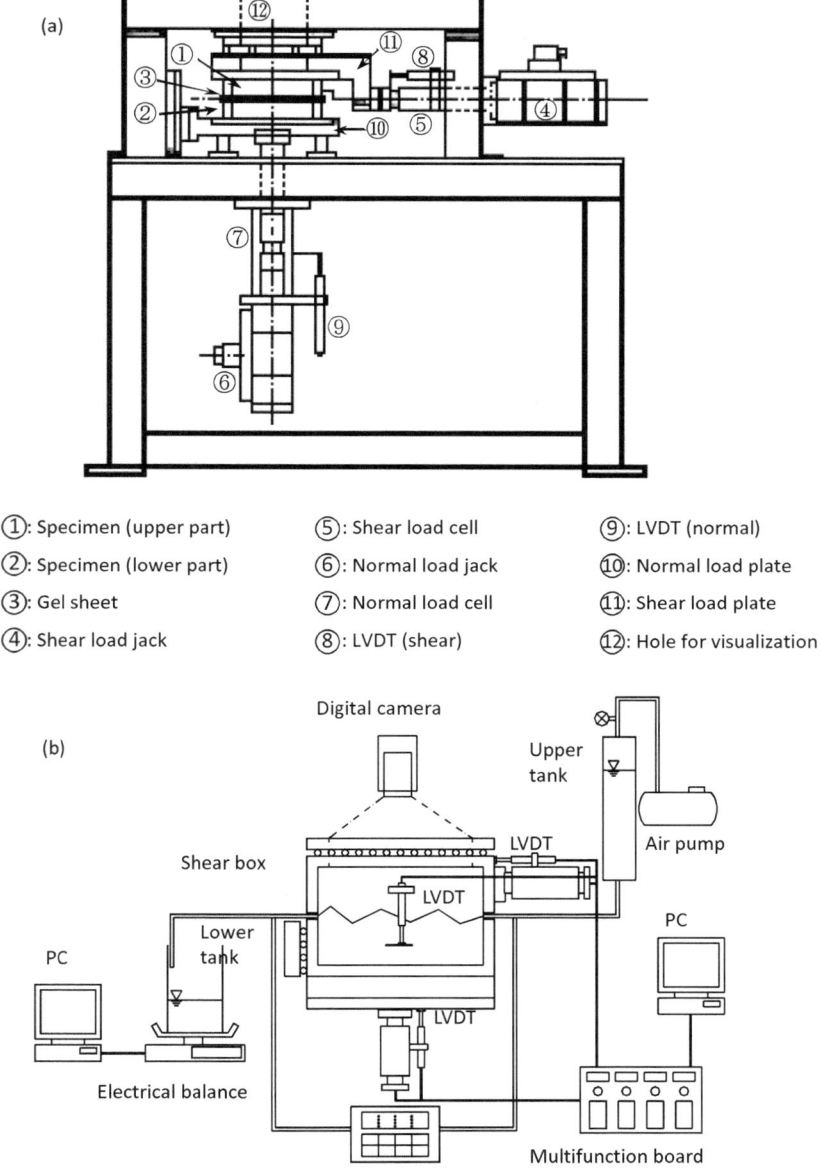

(a)

①: Specimen (upper part)   ⑤: Shear load cell   ⑨: LVDT (normal)

②: Specimen (lower part)   ⑥: Normal load jack   ⑩: Normal load plate

③: Gel sheet   ⑦: Normal load cell   ⑪: Shear load plate

④: Shear load jack   ⑧: LVDT (shear)   ⑫: Hole for visualization

(b)

Figure 12 Schematic view of the coupled shear-flow test apparatus. (a) Normal and shear load units, (b) hydraulic test mechanism.

(2) An instrument package unit. This system contains two digital load cells for measuring shear and normal loads. Displacements are measured through two LVDTs (linear variation displacement transducers), in which the one for measuring shear displacement is attached between load cells and load jacks and the other one for measuring normal displacement is set between the upper and lower blocks of the specimen.

(3) A mounting shear plate unit. This unit consists of a lower and an upper plate. The upper plate connects to a slide guide that can only move in the horizontal direction. It ensures the minimum friction and bending movement when the upper shear box moves during shearing. The lower plate connects to another slide guide who can only move in the vertical direction and the specimen holder is set between these two plates.

(4) A water supplying, sealing and measurement unit. Constant water pressure is obtained from an air compressor connecting to a water vessel. The water pressure is controlled with a regulator ranging from 0 to 1 MPa. The two sides of specimen parallel to the shear direction are sealed with gel sheets, which are very flexible with perfect sealing effect and minimum effect to the mechanical behavior of the shear testing. The weight of water flowing out of the fracture is measured by an electrical balance in real time.

(5) A visualization unit. When acrylic transparent replicas of rock fractures with natural surface features are used as the upper block of a fracture specimen in tests, the images of the fluid flow in the fractures are captured by a CCD camera through an observation hole on the upper shear plate. Colored water can be used to enhance the visibility of the flow paths.

The basic hardware to implement CNS boundary conditions and the digital serve and control systems are based on the direct shear test apparatus as described in section 2.1. The digital control program was also designed by using the LabVIEW programming language (Jiang *et al.*, 2004a, 2004c).

3.1.2 Application of visualization technique

A close view of the visualization unit is shown in Fig. 13. One significant change has been made on the loading unit to facilitate the visualization function. As can be seen in the figure, the normal loading is applied upwards from the bottom of lower shear box and one observation hole was opened on the plate of upper shear box so that the CCD camera above can directly capture the flow images in the rock fracture. In tracer test, utilizations of transparent acrylic specimen and dyed water provide high-quality flow images that can be used in image processing to digitize the flow characteristics.

In a flow image, the chroma of the dye color changes with the thickness of dyed water point by point. In order to find out the relationship between the chroma of flow images and the aperture (thickness of dyed water and also of void spaces), a few prior flow tests were conducted on two parallel plate specimens with inclined opening widths as demonstrated in Fig. 14. In these tests, the aperture between the upper acrylic specimen and the lower plaster specimen increases linearly from 0 to some specific values (*i.e.* 1mm, 2mm). Then, normal water and dyed water were injected into the fracture respectively until filling all the void spaces. The flow images of normal water and dyed water were captured by CCD camera and analyzed by image processing software. Herein, each image is

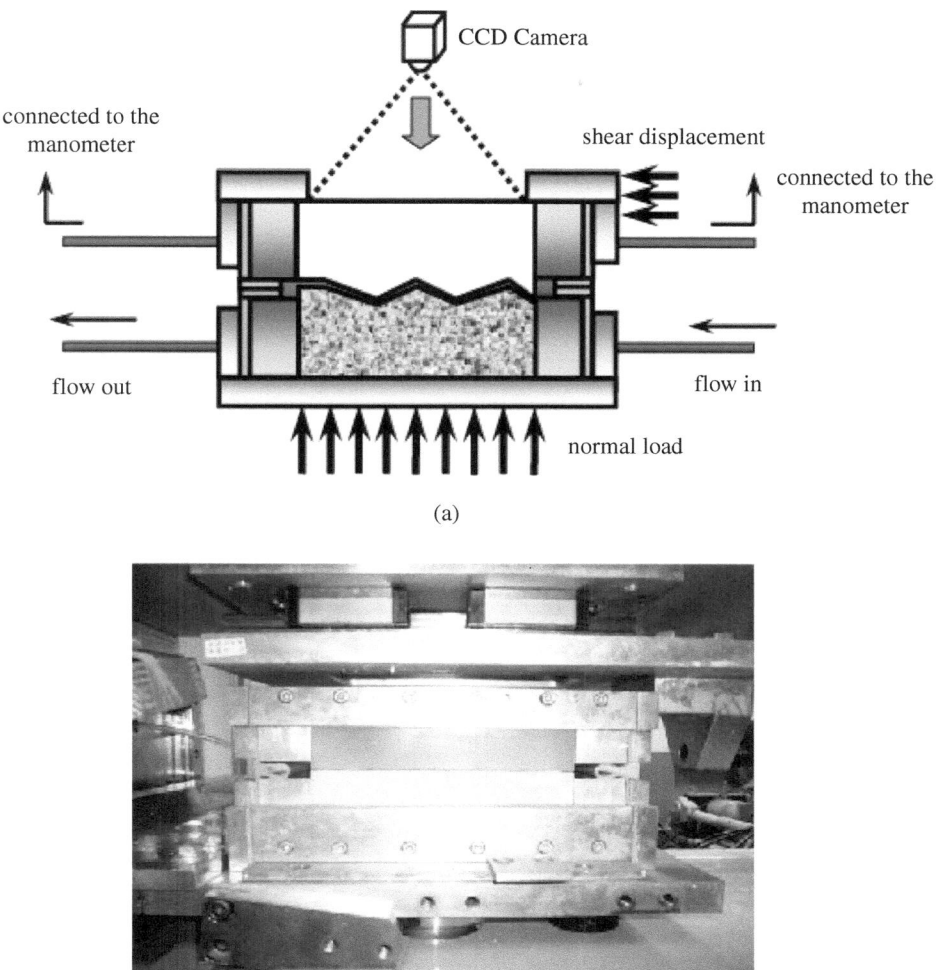

(a)

(b)

Figure 13 Modified shear box for coupled shear-flow-tracer test. (a) Schematic view of the visualization process during coupled shear-flow tests, (b) sample set of transparent acrylic and plaster replicas for upper and lower parts, respectively.

divided into 1024×1024 elements and the chroma value of each element was calculated. The difference of the chroma values obtained from normal water flow and dye flow is the increment of chroma at each element induced by the dye. Therefore, by taking the difference of normal water flow image and dye flow image, the background of fracture surface and the reflection on the surface of acrylic specimen can be eliminated (Li, B., Jiang *et al.*, 2008), remaining only the chroma increment introduced by dye at each element. One test result of the relationship of chroma and aperture evaluated from flow images can be represented by a mathematical equation as follows.

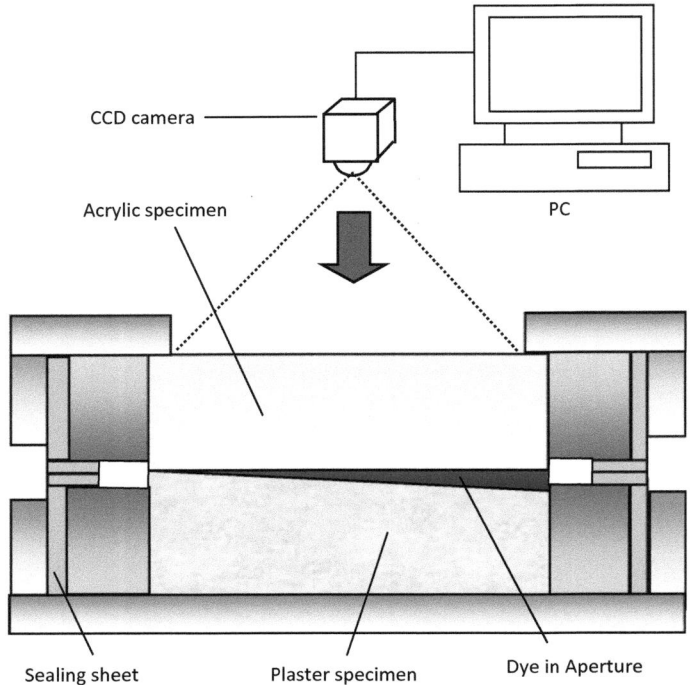

Figure 14 Demonstration of the cuneal aperture used for assessing the relation of mechanical aperture and chroma value.

$$e_v = 0.0318 C^{1.2062} \qquad\qquad (5)$$

where e_v is the aperture evaluated by flow images, C is chroma of dye.

3.2 Fracture sample preparation

A series of fracture samples with different surface characteristics were used in coupled shear-flow and/or shear-flow-tracer tests, respectively.

Three rock fracture specimens, labeled as J1, J2 and J3 (Fig. 15 a), b), c)), were taken from the construction site of Omaru power plant in Miyazaki prefecture in Japan (Jiang *et al.*, 2005; Li *et al.*, 2006), and were used as prototypes to producing artificial replicas of rock fractures which were used in the coupled shear-flow tests. Among these fracture specimens, J1 is flat with very few major asperities on its surface. The surface of J2 is smooth but a major asperity exists at the center, and a few other large asperities on other locations. J3 is very rough with no major asperities but plenty of small ones. The specimens (replicas) are 100mm in width, 200mm in length and 100mm in height, and were made of mixtures of plaster, water and retardant with weight ratios of 1: 0.2: 0.005. The surfaces of the natural rock fractures were firstly re-cast by using the resin material, then the two parts of a

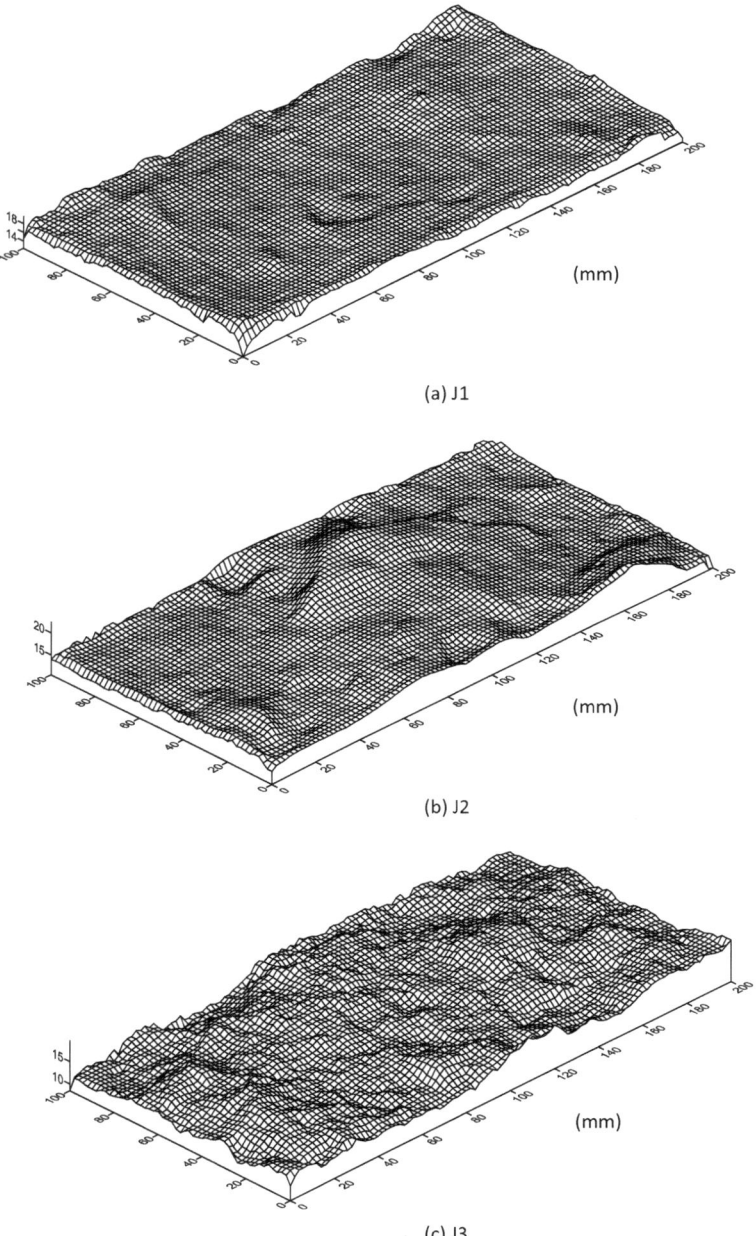

(a) J1

(b) J2

(c) J3

Figure 15 3-D models of surface topographies of specimens J1–J3 based on the measured topographical data. It should be noted that the size of mesh used in this figure is 2mm different from the one used in measurement 0.2mm.

specimen were manufactured based on the resin replica. The figure models constituted from the scanning data of the rock replicas re-cast from the same resin model are
well matched even to the small asperities in a scale of 0.2mm. Therefore, the two parts
of each specimen are almost perfectly mated.

Brazilian test was applied on sandstone and granite blocks to generate fresh artificial
fractures. During the test, normal load was firstly applied to some appropriate values
like 10kN (induce no destruction to the block), and then lateral loads were applied
through the wedges. The magnitude of the lateral loads needs several attempts for
different kinds of rocks to ensure that the fracture can be generated. Keeping the lateral
loads unchanged, the normal load then was gradually decreased until the generation of
tensile fracture was accomplished. The two opposing surfaces of the tensile rock
fracture were firstly re-cast by using resin and silicon rubber separately and then the
upper and lower parts of a fracture specimen were manufactured based on the silicon
rubber and resin replicas, respectively. The transparent upper part of the fracture
specimen was made of acrylic and the lower part was plaster sample. The curing time
of the upper acrylic specimen was carefully chosen to ensure its uniaxial compressive
strength is close to the lower plaster specimen. By doing so, the artificial fracture using
acrylic-plaster pair exhibits reasonably close mechanical behavior with the normal
plaster-plaster pair. The two parts of fracture specimen used in this chapter are not
perfectly mated as the initial condition due to the sample damage observed during
creating tensile fractures by Brazilian test and unavoidable relocation errors before
testing.

The surface topographical data of these samples were measured by using the laser
scanning profilometer system and were used to estimate the mechanical aperture of
rock fracture during coupled shear-flow process.

3.3 Experimental procedure of coupled shear-flow-tracer test

The boundary conditions (normal stress/normal stiffness) applied in the tests are
summarized in Table 4. In all of the tests, the flow direction is parallel to the shear
direction and the cubic law was used to evaluate the transmissivities based on the

Table 4 Experiment Cases under CNL and CNS Conditions.

Specimen	Case No.	Roughness (JRC range)	Boundary condition	
			Initial normal stresses σ_n (MPa)	Normal stiffness k_n (GPa/m)
JI	JI-I	0~2	1.0	0
	JI-2		1.0	0
	JI-3		1.5	0.5
J2	J2-I	12~14	1.0	0
	J2-2		2.0	0
	J2-3		1.0	0.5
J3	J3-I	16~18	1.0	0
	J3-2		1.0	0.2
	J3-3		1.0	0.5

measured flow rates. A water head of 0.1m was applied as a basic pattern during the tests. When the shear goes on, dilation of fracture takes place, remarkably increasing the flow rate and so as Reynolds number. To avoid the appearance of turbulent flow, the water head applied to fracture is better to be decreased gradually. Therefore, 3 patterns of water head were applied at each measurement with 0.1m unchanged and the other two patterns decreasing gradually with shear displacement. The water head was applied at an interval of 1mm during the shear process. When the water head was applied, the shear was temporarily stopped until the measurement of water volume flowing out of fracture was finished. The measurement of water volume was carried out by electrical balance and the weight of water was measured at each 1 second up to 50 seconds to get the mean value in the steady stage of flow. The total shear displacements for J1, J2 and J3 is 18mm.

As the shearing goes on, the effective shear length (the length of the upper and lower parts of specimen facing to each other) will decrease, thus increasing the hydraulic gradient when the water head keeps constant. This effect has been considered in calculation of the transmissivity by decreasing the hydraulic gradient corresponding to the shear displacement.

3.4 Experimental investigation of the hydro-mechanical behavior of rock joints

3.4.1 Coupled shear-flow tests

The shear behaviors of the tested fracture specimens are illustrated in Fig. 15. For a fracture, larger shear stresses could be obtained under either higher normal stress or higher normal stiffness at the same initial normal stress, depending on the surface roughness and stiffness of asperities. Normal displacement is the most important behavior in the coupled shear-flow tests for quantifying the change of transmissivity. Normal displacement is usually called dilation because it is primarily an increasing process during a shearing. As shown in the figure, for a fracture specimen, the larger normal stress or normal stiffness is, the larger magnitude of normal displacement could be inhibited. Normal behavior of a fracture depends also on the roughness of the fracture surface. Generally, the rougher the fracture surface is, the larger normal displacement could be obtained.

As shown in Fig. 17, the changes of transmissivities exhibit an obvious two-phase behavior. For all test cases, the transmissivities increase gradually in a relatively high gradient in the first several millimeters of shear displacement and then continue to increase but with a lower gradient gradually reaching to zero. The experimental results indicate that a rougher fracture may have higher gradient in the first phase and the second phase comes sooner. Under the same stress environment, a rougher fracture would produce larger normal displacement during shear so that it could obtain higher transmissivities in the second phase. The peak shear stress generally comes earlier than the turning point of transmissivity as shown in Fig. 15 and Fig. 17, which could be explained by the damage process of asperities on the fracture walls during shear. The Re numbers increase from almost 0 to as high as 1000 (depend on the hydraulic gradients) during the shear as the increase of flow rate. To avoid the occurrence of turbulent components in the fluid flow, the hydraulic gradients were carefully controlled to keep the Re numbers in an empirical range for laminar flow on the current test

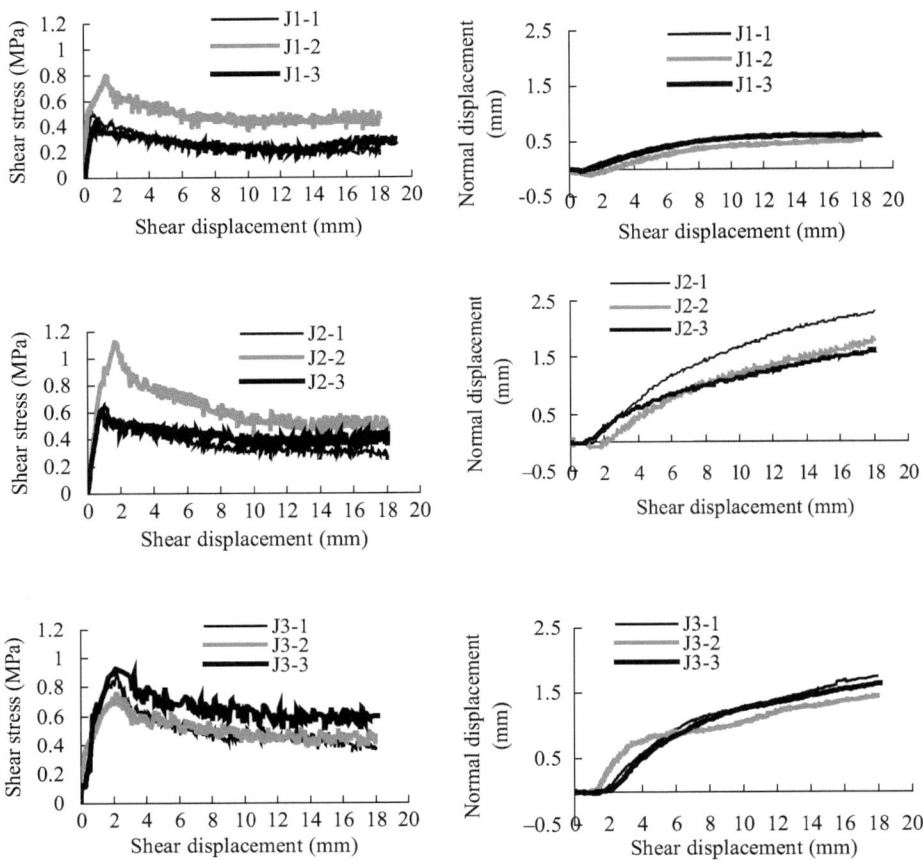

Figure 16 Direct shear test results on specimens J1, J2 and J3. The left three figures show the shear stress versus shear displacement and the right ones are normal displacement versus shear displacement.

apparatus. The hydraulic gradients used in the shear-flow tests are in a range of 0.25–10 and they were decreased during the shear to inhibit the fast increases of flow rate and Re number. The maximum Re numbers (obtained from the last few millimeters of shear displacement) of the hydraulic data used for calculating the transmissivities of J1, J2 and J3 are 229, 240.6 and 225, respectively. Further improvement of the test apparatus supporting lower hydraulic gradients to avoid the turbulent flow without decreasing the accuracy of measurement is under construction.

Fig. 18 shows the changes of aperture distributions of testing cases J1-1, J2-1 and J3-1 during shearing, respectively. The aperture fields change remarkably when a shear starts, *i.e.* in stage 1. After that, the change trends to become smaller and steady, due to the graduate reduction of dilation gradient as shown in Fig. 15 in stage 2. The contact ratio changes reversely to the transmissivity change in a shearing, which represents an

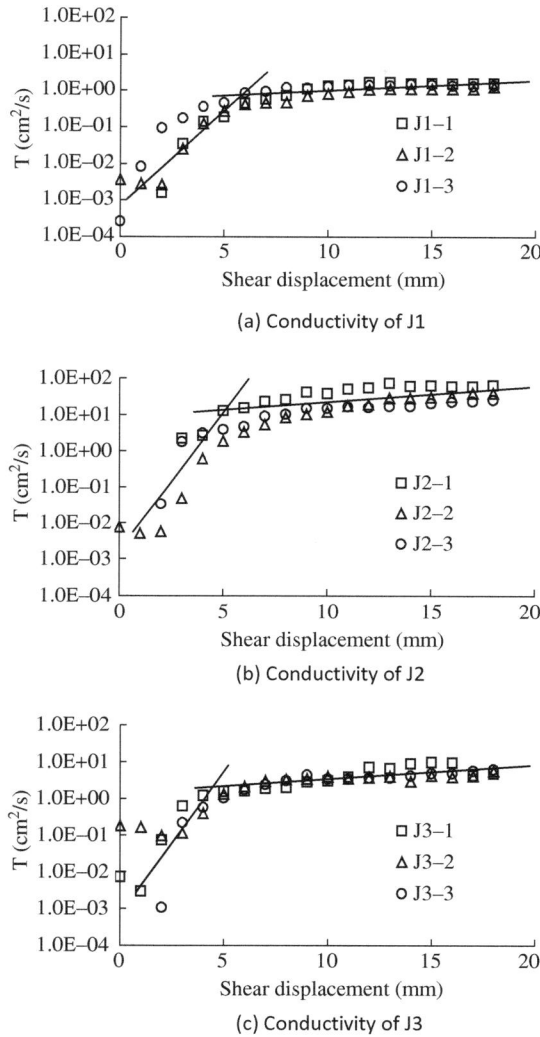

Figure 17 Two-phase behavior of the change of conductivity during shearing. A minus dilation of fracture occurs when a shear starts which causes the decrease of conductivity in phase I. After this, a rapid increase happens till the second phase in which the conductivity trends to keep constant.

opposite effect of contact area on the transmissivity. There is a rapid drop of contact ratio in stage 1 and then it keeps a small value in stage 2. For a rougher fracture, such reduction of contact ratio will be more significant. The peak shear stress occurs when the major asperities on the fracture surface lose their resistance to the shear and being destructed, while most asperities are undamaged and few gouge materials are generated. After that, the remaining asperities are crushed gradually, generating plenty of gouge materials and increasing the contact ratio. Therefore, the turning point of

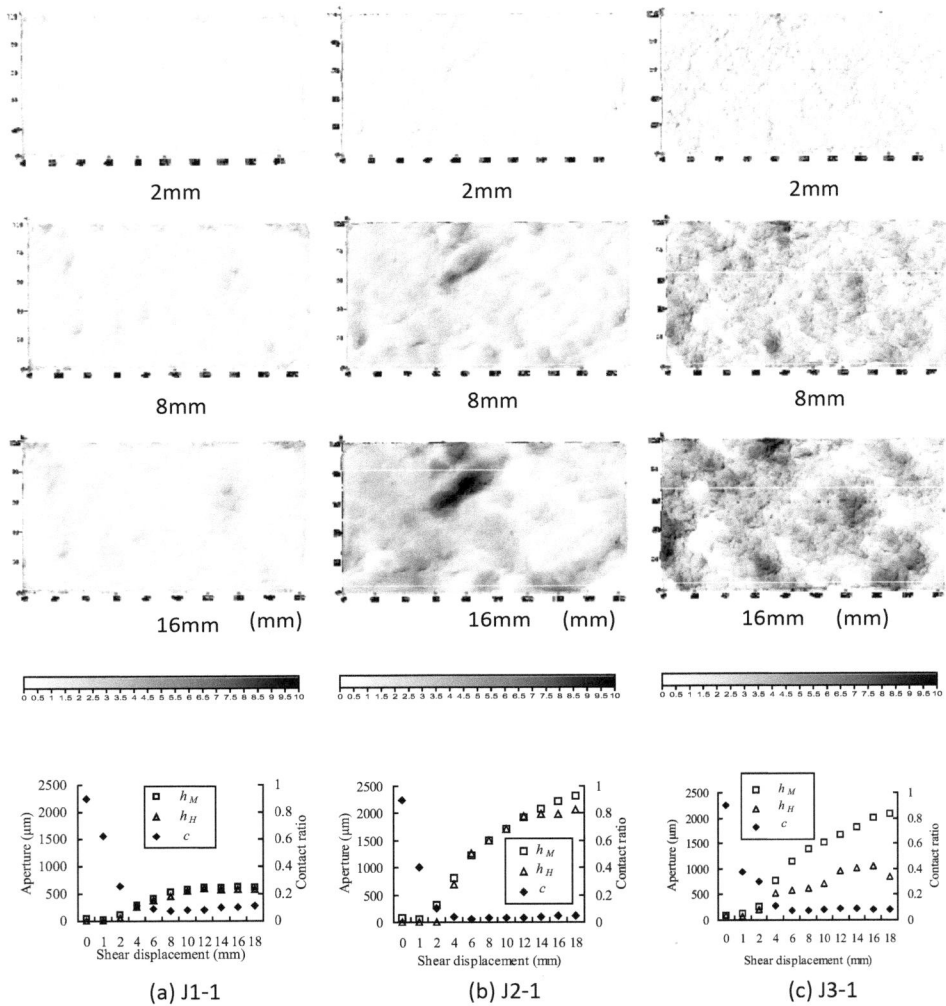

Figure 18 Comparison of the change of mechanical aperture h_M, hydraulic aperture h_H and contact ratio c of specimens J1-1, J2-1 and J3-1 during shearing. The upper three figures in figures (a), (b) and (c) show the distributions of mechanical apertures at shear offsets of 2mm, 8mm and 16mm, respectively. The white parts in these figures are the contact areas. The contact ratios at the initial state (0 shear displacement) for each case were assumed to be 0.9 since the fracture specimens were perfectly mated.

contact ratio occurs at almost the same time with that of the transmissivity and the effect of contact areas on the transmissivity of rock fracture is confirmed.

The influences of morphological behaviors of rock fractures on the evolution of aperture distributions are also reflected in Fig. 18. The surface of specimen J1 is smooth and flat. Therefore, its contact ratio is relatively high and its apertures distribute evenly over the fracture specimen. The few large asperities on the two parts of specimen J2 tended to climb over each other during shear, which decreased the contact ratio significantly and produced a large void space after a section of shear displacement (see the third figure in Fig. 18 (b)). For specimen J3, the widely distributed asperities developed a complicated void space geometry, which causes complex structure of transmissivity field (Fig. 18 (c)).

Zimmerman *et al.* (1996) revealed that, in general, reasonably accurate predictions of conductivity could be made by combining either the perturbation results, Equation 6, or the geometric mean, Equation 7, with the tortuosity factor given by Li *et al.* (2008), written as

$$h_H^3 \approx \langle h \rangle^3 [1 - 1.5\sigma_h^2/\langle h \rangle^2 + \cdots] \tag{6}$$

$$h_H^3 = k_{eff} \approx k_G = e^{\langle \ln k \rangle} = e^{\langle \ln(h^3) \rangle} = e^{3\langle \ln k \rangle} = (e^{\langle \ln h \rangle})^3 = h_G^3 \tag{7}$$

where h_H is the hydraulic aperture, $\langle h \rangle$ is the arithmetic mean value of h, σ_h is the standard deviation of h, k_{eff} is the overall effective transmissivity, k_G is the geometric mean of the transmissivity distribution, and c is the contact ratio, respectively.

A series of combinations of different forms of aperture predictors with the tortuosity factor $(1-2c)$ were evaluated as follows:

1) Predictor (1): $\langle h^3 \rangle (1 - 2c)$
2) Predictor (2): $\langle h \rangle^3 (1 - 2c)$
3) Predictor (3): $h_G^3 (1 - 2c)$
4) Predictor (4): $\langle h \rangle^3 [1 - 1.5\sigma_h^2/\langle h \rangle^2](1 - 2c)$

Results are shown in Fig. 19 for the tests on three kinds of fracture specimens under the CNL (σ_n=1MPa) boundary conditions, respectively. Herein, the "transmissivity" is not the ordinary T, but the cubic of mechanical aperture with a unit of 10^{-12}m^3. The results show that $\langle h \rangle^3$ is a more accurate predictor than $\langle h^3 \rangle$ for predicting hydraulic transmissivity. For the test case of J1-1, the mechanical aperture $\langle h \rangle$ itself agrees well with the hydraulic aperture h_H as back-calculated using the cubic law. Further modifications such as presented by predictor (3) or predictor (4) would underestimate the transmissivity. When the roughness of fracture increases, the predictors (3) and (4) give the closer predictions to the experiment data. For the test case of J3-1, the hydraulic aperture is much lower than the mechanical aperture, due to the influence of tortuosity produced by the complicated structure of void space and contact area. The tortuosity factor $(1-2c)$ plays a significant role when combined with predictor (2) in Fig. 13 (a) and (b), and with predictor (4) in Fig. 19 (c).

3.4.2 Coupled shear-flow-tracer tests

Accompanied with the development of the visualization system in the test apparatus, the coupled shear-flow-tracer tests applied on J4 and J5 used different photographing techniques. Tests on J4 were carried out earlier than J5 and the flow images were taken by normal digital camera with 10.1 megapixel. After that, the apparatus was improved when conducting tests on J5 by application of a high resolution CCD camera and its related image processing software so that flow images can be digitized to obtain the numerical data of the flow features like aperture distributions and flow rates.

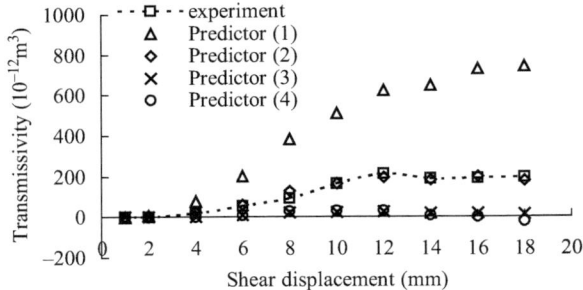

(a) Transmissivity versus shear displacement for J1-1

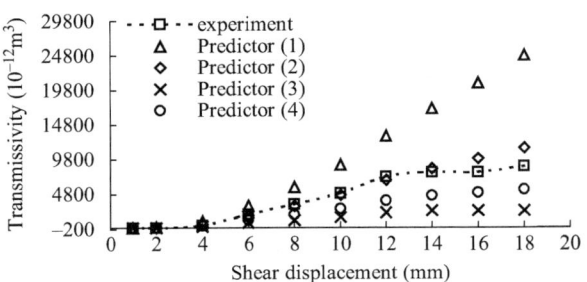

(b) Transmissivity versus shear displacement for J2-1

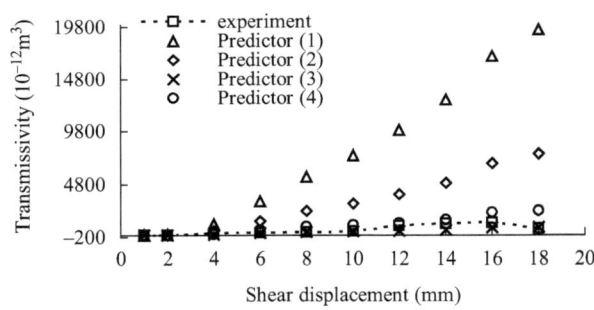

(c) Transmissivity versus shear displacement for J3-1

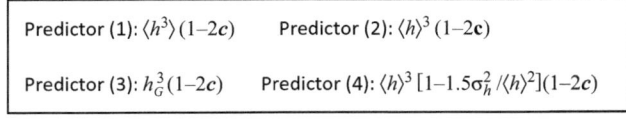

| Predictor (1): $\langle h^3 \rangle (1-2c)$ | Predictor (2): $\langle h \rangle^3 (1-2c)$ |
| Predictor (3): $h_G^3(1-2c)$ | Predictor (4): $\langle h \rangle^3 [1-1.5\sigma_h^2/\langle h \rangle^2](1-2c)$ |

Figure 19 Comparisons of various predictors to experiment results for test cases J1-1, J2-1 and J3-1. Predictor (3) and (4) are more accurate predictors when a fracture is rough enough to effectively produce complicated void geometry.

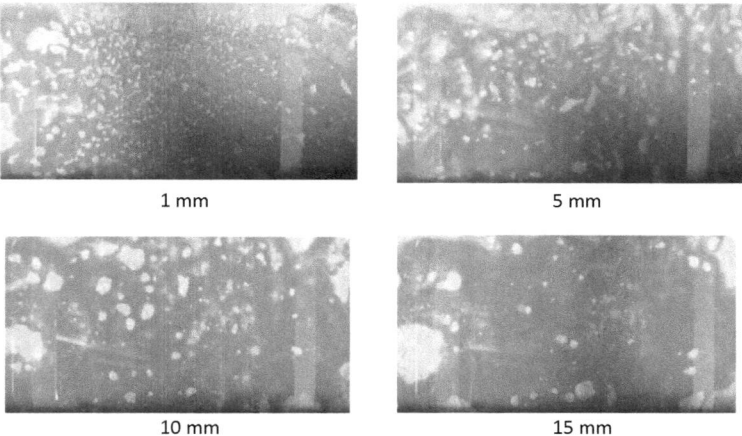

1 mm   5 mm

10 mm   15 mm

Figure 20 Flow images of J4 at different shear displacement of 1, 5, 10, 15mm.

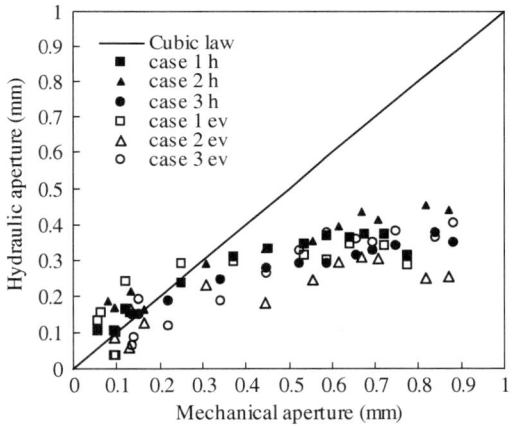

Figure 21 Relations of mechanical aperture E, hydraulic aperture h and aperture evaluated from flow images e_v.

Fig. 20 shows the images of flow fields captured by digital camera during the tests at different shear displacement of 1, 5, 10 and 15mm on J4. These pictures can clearly show the contact area distribution and its change during shear. The contact areas were localized with decreasing its ratio during shear and the dye flows bypassing the contact areas.

The relations of mechanical aperture E, hydraulic aperture h and aperture evaluated from flow images e_v are shown in Fig. 21. Comparisons of the flow images (photos), the aperture e_v distributions and mechanical aperture distributions are shown in Fig. 22. It should be noted that the dimension of image is 180mm×90mm, a little smaller than the

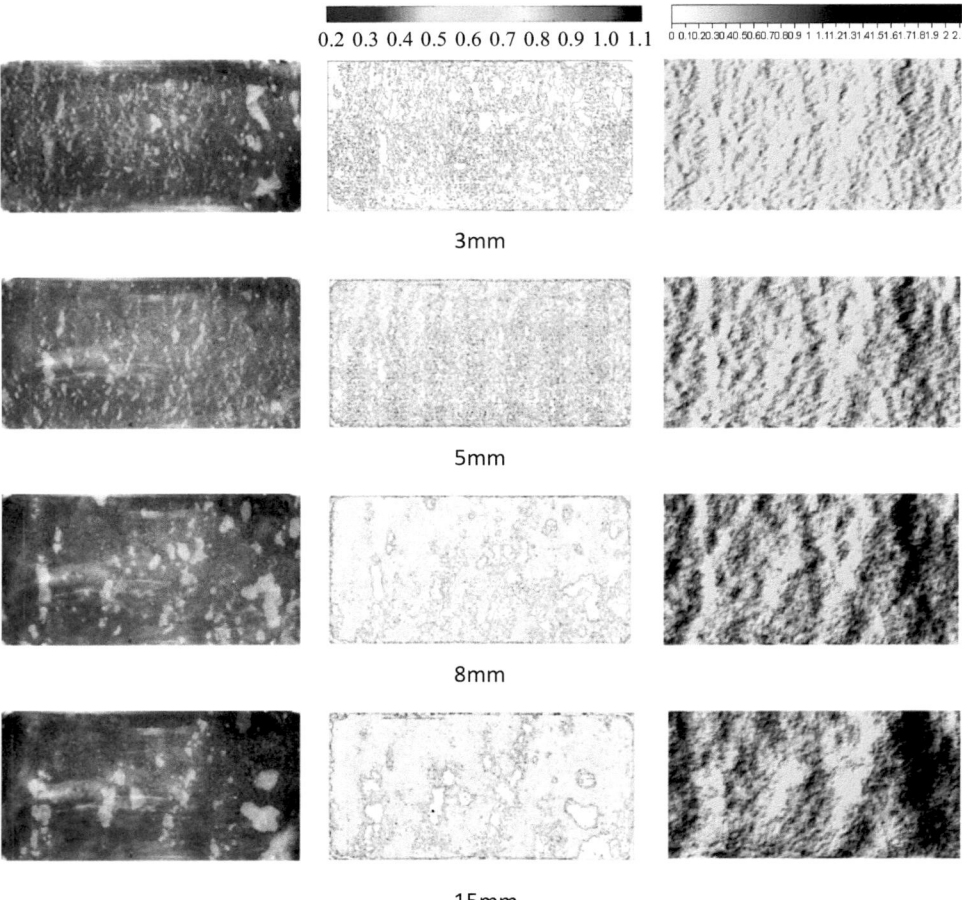

0.2 0.3 0.4 0.5 0.6 0.7 0.8 0.9 1.0 1.1

0 0.10.20.30.40.50.60.70.80.9 1 1.11.21.31.41.51.61.71.81.9 2 2.1

3mm

5mm

8mm

15mm

Figure 22 Flow field and aperture evolution during shear process of case 2. (a) Flow images, (b) aperture obtained by image processing on flow images, (c) mechanical aperture distribution by means of topographical data of fracture surface.

dimension of fracture surface (200mm×100mm). That means the flow images in the edge portions of fracture surface cannot be captured by the CCD camera on the current apparatus. In the edge portions, fluid flow channels generally exist, and if they are not taken into account, aperture e_v obtained from flow images may underestimate the mean aperture of a fracture. As demonstrated in Fig. 21, generally, e_v has close values with h in all cases. The evolution of apertures in a shear process generally exhibits a two-stage behavior (Li *et al.*, 2008). After minus dilation in the initial stage of shear, the mechanical apertures increase constantly with slight decrease of gradient. The hydraulic apertures increase in the first 5~7mm to the values 2~3 times of the initial ones, after that, the gradients trend to 0. In the first stage of shear when contact areas dominate the fracture surface, the mechanical apertures are small and water flows through a fracture from many connected small channels as shown in Fig. 22 (3mm). As shear advances, the dilation of fracture produces more void spaces in a fracture, providing abundant channels for water to pass through. At the same time, a few major channels begin to emerge (5mm). After entering stage 2 in a shear process, the fluid flows through a fracture bypassing a few major contact areas, concentrating in a few major channels (8-15mm). Deviation of mechanical apertures and hydraulic apertures begins to develop at shear displacements 3~5mm, increasing up to 2~3 times of differences at final shear displacements. If the water could fill all the void spaces in a fracture, the aperture e_v may provide close value to the mechanical aperture since it essentially measures the widths of void spaces occupied by water flow in a fracture. Close values of e_v and h reveal that in the void spaces that water flows have taken place mechanical aperture agrees well with hydraulic aperture.

4 SUMMARY

In this chapter, the shear behavior, coupled shear-flow behavior of single rock fractures, and the mechanical behaviors of fractured rock mass encompassing rock structures like underground opening have been studied through laboratory direct shear/coupled shear-flow-tracer tests. Some results of these studies are summarized as follows:

1) Direct shear tests
    - Peak shear stress can be observed for tests conducted under the CNL condition but not for all the tests conducted under the CNS condition, since the shear stress can continuously increase with the increase of shear displacement in some tests under the CNS condition.
    - The peak shear stress occurs when the major asperities on the fracture surface lose their resistance to the shear, while most asperities with lower importance are undamaged. After that, the remaining asperities are crushed gradually, increasing the contact ratio and generating plenty of gouge materials.
    - The shear strength is observed to increase with the increases in initial normal stress and normal stiffness. As the dilation occurs for rough fractures under CNS condition, the normal stress acting on the interface increases, thus contributing to an increased value of shear stress. This increase of normal stress also leads to smaller dilation under CNS condition than CNL condition at the same initial normal stress.
    - For all the shear tests on the rock fractures based on standard JRC profiles or with natural roughness surfaces, Barton's criterion of shear strength provides good predictions to the experimental results.
2) Coupled shear-flow-tracer tests
    - It has been observed and confirmed that fluid flows through a rough fracture following connected channels bypassing the contact areas with tortuosity, not only by theoretical predictions or numerical simulations but also from direct observation of flow images by the utilization of visualization system.
    - Since the CNS condition can inhibit dilation in a shear process, the transmissivity of a rock fracture under CNS condition in a shear can be much smaller

than that under CNL condition depending on the initial normal stress, normal stiffness and surface roughness of the tested fracture.

- The 'minus dilation' happening at the initial shear displacement (usually less than 2mm) could significantly decrease the transmissivity of a rock fracture. After that, the transmissivity increases quickly (stage 1) until a threshold, and the gradient of transmissivity trends to 0 subsequently (stage 2). Compared to a flat fracture, a rough fracture could obtain higher value of transmissivity in stage 2 and the threshold of stage 1 would come earlier. The change of contact ratio in a shear is just opposite to that of transmissivity.

- The cubic law performs reasonably well without need for any modifications for most tests on rough rock fractures. However, dispersedly distributed contact areas could remarkably decrease the threshold for the validity of cubic law. That means modifications may still be needed when estimating the hydromechanical behavior of the very rough rock fractures such us the last few profiles in JRC system with large JRC values.

REFERENCES

Brahim, B. and Gerard, B. (1989). Laboratory of shear behavior of rock joints under constant normal stiffness conditions. *Rock Mechanics as a Guide for Efficient Utilization of Natural Resources*, Khair (ed.), Balkema, Rotterdam, 899–906.

Barton, N. and Choubey, V. (1977). The shear strength of rock joints in theory and practice. *Rock Mech.*, 10, 1–54.

Barla, G., Forlati, F. and Zaninetti, A. (1985). Shear behaviour of filled discontinuities. *Proceedings of the International Symposium on Fundamentals of Rock Joints*, Bjorkliden, 163–172.

Esaki, T., Du, S., Mitani, Y., Ikusada K. and Jing L. (1988). Development of a shear-flow test apparatus and determination of coupled properties for a single rock joint. *Int. J. Rock Mech.* 36, 641–650, 1999. Feder, J. Fractals. Plenum Press, New York.

Goodman, R.E. (1970). Indetermination of the in-situ modulus of deformation of rocks. *The deformability of joints*. Special technical publication, 477, 174–196.

Hakami, E. (1995). Joint aperture measurements – An experimental technique. *Fractured and Jointed Rock Masses*, Myer, L.R., Cook, N.G.W., Goodman, R.E. and Tsang, C.F. (eds.), Balkema, Rotterdam, 453–456.

Hakami, E., Einstein, H.H., Genitier, S. and Iwano, M. (1995). Characterisation of fracture apertures – Methods and parameters. *Proceedings of the 8th International Congress on Rock Mechanics*, Tokyo, 751–754.

Huang, X., Haimson, B.C., Plesha, M.E. and Qiu, X. (1993). An investigation of the mechanics of rock joints-Part 1. Laboratory investigation. *Int. J. Rock Mech. Min. Sci. & Geomech. Abstr.*, 30(B3), 257–269.

Indraratna, B., Haque, A., and Aziz, N. (1999). Shear behavior of idealized infilled joints under constant normal stiffness. *Geotechnique*, 49(B3), 331–355.

Johns, R.A., Steude, J.S., Castanier, L.M. and Roberts, P.V. (1993). Nondestructive measurements of fracture aperture in crystalline rock cores using X-lay computed tomography. *J. Geophys. Res.*, 98(B2), 1889–1900.

Jiang, Y., Tanabashi, Y. and Mizokami, T. (2001). Shear behavior of joints under constant normal stiffness conditions. *Proceedings of the Second Asian Rock Mechanics Symposium* (ISRM2001-2nd ARMS), Wang, S., Fu, B. and Li Z. (eds), A.A. Balkema Publishers, Lisse, 247–250.

Jiang, Y., Tanabashi, Y., Xiao, J. and Nagaie, K. (2004a). An improved shear-flow test apparatus and its application to deep underground construction. *Int. J. Rock Mech. Min. Sci.* (SINOROCK2004 Paper 1A28), 41, 385–386.

Jiang, Y., Tanabashi, Y., Nagaie, K., Li, B. and Xiao, J. (2004b). Relationship between surface fractal characteristic and hydro-mechanical behaviour of rock joints. *Contribution of Rock Mechanics to the New Century, Proceedings of the 3rd Asian Rock* Mechanics Symposium, Ohnishi, Y. and Aoki, K. (eds.), Millpress, Rotterdam, 831–836.

Jiang, Y., Xiao, J., Tanabashi, Y., Mizokami, T. (2004c). Development of an automated servo-controlled direct shear apparatus applying a constant normal stiffness condition. *Int. J. Rock Mech. Min. Sci.*, 41(2), 275–286.

Jiang, Y., Tanabashi, Y., Nagaie, K., Yamashita, Y. and Andou, I. (2005). Stability analysis for underground pumped powerhouse cavern in jointed rock mass. *Proceedings of the 11th International Conference of IACMAG2005*, Torino, 671–677.

Jiang, Y., Li, B. and Tanabashi, Y. (2006a). Estimating the relation between surface roughness and mechanical properties of rock joints. *Int. J. Rock Mech. Min. Sci.*, 43(6), 837–846.

Jiang, Y., Saho, R., Tasaku, Y., Li, B. and Tanabashi, Y. (2007). Estimating the influence of surface characteristics of rock join on shear behavior and coupled shear-flow characteristics. *J. of the Society of Materials Science*, Japan, 56(9), 796–802 (In Japanese).

Kanji, M.A. (1974). Unconventional laboratory tests for the shear strength of soil-rock contacts, *Proceedings of the 3rd Congress of the International Society for Rock Mechanics*, Denver 2, 241–247.

Ladanyi, H.K. and Archambault, G. (1977). Shear strength and deformability of filled indented joints. *Proceedings of the International Symposium on Geotechnical Structural Complex Formations*, Capri, 317–326.

Lama, R.D. (1978). Influence of clay fillings on shear behaviour of joints. *Proceedings of the 3rd Congress of the International Association of Engineering Geology*, Madrid 2, 27–34.

Li, B., Jiang, Y., Saho, R., Tasaku, Y. and Tanabashi, Y. (2006). An investigation of hydro-mechanical behaviour and transportability of rock joints. *Rock Mechanics in Underground Construction, Proceedings of the 4th Asian Rock Mechanics Symposium*, Leung, C.F.Y and Zhou, Y.X. (eds.), World Scientific, 321.

Li, B., Jiang, Y., Koyama, T., Jing, L. and Tanabashi, Y. (2008). Experimental study on hydro-mechanical behaviour of rock joints by using parallel-plates model containing contact area and artificial fractures. *Int. J. Rock Mech. Min. Sci.*, 45(3), 362–375.

Ohnishi, Y. and Dharmaratne, P.G.R. (1990). Shear behaviour of physical model of rock joints under constant normal stiffness condition. *Proceeding of the International Conference on Rock Joints*, Barton, N. and Stephannsson, O. (eds.), Balkerma, Rotterdam, 267–273.

Olsson, R. and Barton, N. (2001). An improved model for hydromechanical coupling during shearing of rock joints. *Int. J. Rock Mech.* 38, 317–329.

Pereira, J.P. (1990). Shear strength of filled discontinuities. *Proceedings of the International Conference On Rock Joints*, Barton, N. and Stephansson, O. (eds.), Balkema, Rotterdam, 283–287.

Zimmerman, R.W. and Bodvarsson, G.S. (1996). Hydraulic conductivity of rock fractures. *Transp. Porous Media.*, 23, 1–30.

Dynamic and Creep Tests

Chapter 8

Coupled static and dynamic test

X.B. Li, M. Tao, L. Weng & Z.L. Zhou

School of Resources and Safety Engineering, Central South University, Chansha, Hunan, China

Abstract: Mechanical properties of rock under coupled static and dynamic stresses are apparently different from that under either static or dynamic load. Experimental techniques for testing rock under coupled static and dynamic loads are introduced in this chapter. A new testing technique relates the physical stress state of rock subjected to simultaneous coupled static and dynamic stresses is presented. The method involves modification of a split Hopkinson bar and INSTRON electro-hydraulic servo controlled testing system. The test specimen is subjected to coupled axial static pre-stress or triaxial pre-stress, and axial dynamic loading. The experimental results generally indicate that the coupled static-dynamic strength is higher than its static strength and its dynamic strength.

1 INTRODUCTION

The experimental study of rock mechanics has achieved great developments (Sun & Wang, 2000; Brown, 2002) and the progress of this subject has been widely applied to the practices in mining engineering, underground construction (caverns and tunnels) and resource exploitation. Generally, experimental investigations are conducted either in static or dynamic state. Static state experiments may be investigated using equipment such as INSTRON and MTS, and are mainly concerned with characteristics of rock under low strain rate loading, such as static constitutive relations, fracture and damage (Hudson & Harrison, 2000). Experimental study on dynamic properties of rock, however, is often related to fragmentation and dynamic strength, and specialized test equipments, such as the Hopkinson bar (Li & Gu, 1994; Zhao & Gary, 1996; Zhao *et al.*, 1999; Lok *et al.*, 2002; Li *et al.*, 2000; Li *et al.*, 2005) or drop hammer (Okubo *et al.*, 1997; Whittles *et al.*, 2006), is developed to this end.

On the other hand, rocks are naturally stressed and deformed by their mass and tectonic movements. When a dynamic loading conducted in rock with initial stress, it will induce stress and displacements incremental, the magnitude of incremental have both relationship with its initial stress and dynamic loading (Tao *et al.*, 2012; Tao *et al.*, 2013). The influence of the initial stress on the rock elastic coefficients and wave propagation has been explored by a number of scholars (Ogden & Sotiropoulos, 1997; Tolstoy, 1982; Dowaikh & Ogden, 1980). It is known that traditional theory cannot well describe the rock behaviors at depth (Grobbelaar, 1958–1959; Adams & Jager, 1980; Malan, 1999) with respect to many new engineering activities, such as deposition of nuclear wastes, oil exploitation, and mining in the deep underground. For example, the rock at depth shown in Figure 1(a) is

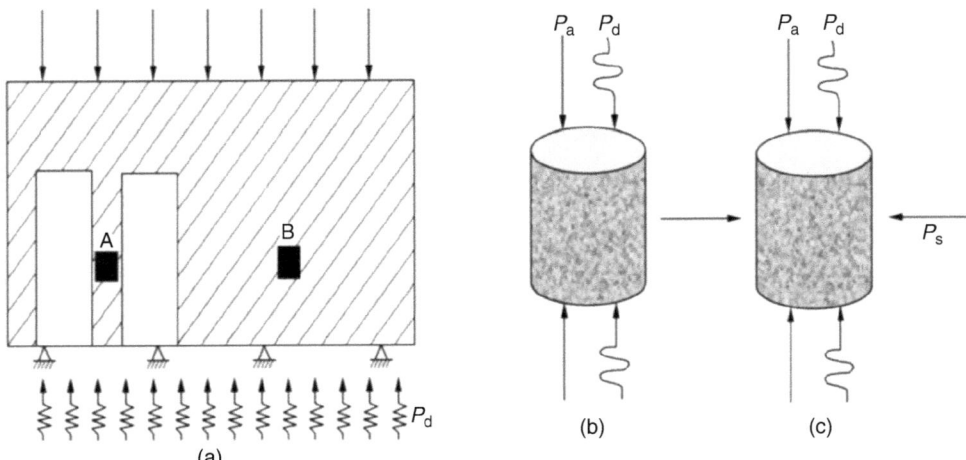

Figure 1 Schematic diagram of rock stress models at deep level: (a) schematic of underground cavern; (b) one-dimensional coupling load; and (c) multi-dimensional coupling load.

subjected to high confined vertical and horizontal static stresses coupled with dynamic loading arising from blasting and boring. Figure 1(b) and 1(c) outline the stress model at different locations subjected to different stress states, *i.e.* one-dimensional and three-dimensional static-dynamic coupling loads. Under coupling loads, the rock behaves differently compared to the material subjected separately to either static stress or impact loading. However, research on such stress models with simultaneous static and dynamic loads is scarce, even though these stress states commonly occur in deep underground. Thus, there is a need to investigate the behavior of rocks under such conditions with the increasing depths of underground mining and civil engineering tunneling projects.

The focus of this chapter is to introduce effective experimental techniques for testing rocks under coupled static and dynamic loads, including one-dimensional coupled loads and three-dimensional coupled loads. Experiments on rocks under different combinations of coupled loads are conducted by using self-developed apparatus, and illustrations of experimental results are presented, including the relation of mechanical characteristics with pre-axial stress and confining pressure, experimental validation of constitutive model of rock and induced rockburst mechanism.

2 TEST SYSTEM FOR COUPLED STATIC AND DYNAMIC LOADS BASED ON SHPB

2.1 Conventional split Hopkinson pressure bar apparatus

The split Hopkinson pressure bar (SHPB) was first invented by Bertram Hopkinson (1914) in 1914 to obtain the pressure-time curve with the dynamic load exerted by detonation. However, because of the limited measurement technique, he could not obtain or observe stress wave directly. Until 1948, Davies (1948) improved the measurement technique by utilizing an electrical method. Later on, Kolsky (1949) modified

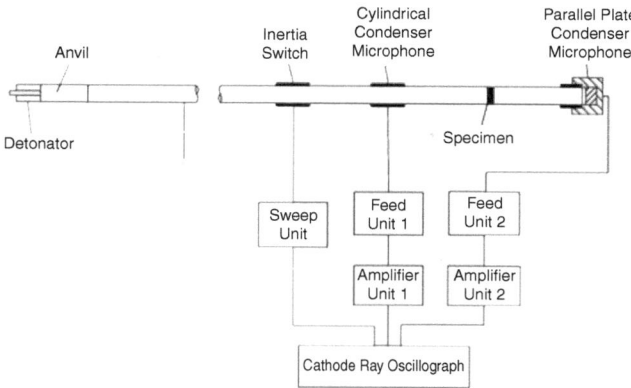

Figure 2 The original Kolsky bar.

Davies's split bar system by sandwiching a specimen between in two bars, and obtained the dynamic relationship of stress and strain for several materials with condenser microphones. The original Kolsky bar is shown in Figure 2, which can be used to characterize the mechanical response of materials deforming at high strain rates extent from 10^2 to 10^4s^{-1}. Then, Krafft *et al.* (1954) adopted strain gauge to measure the stress wave in the incident bar, and applied a striker bar to produce a repeatable impact stress wave in the incident bar.

Until 1968, Kumar (1968) first introduced SHPB system into dynamic strength experiment on short rock specimen. Shortly after that, Hakailehto (1969) finished his PhD thesis entitled "The behavior of rock under impulse loads—A study using the Hopkinson split bar method" by employing the split Hopkinson pressure bar apparatus. In 1972, Christensen *et al.* (1972) developed a triaxial SHPB system with confining pressure component, which was used to conduct dynamic impact tests on rock specimen under different confining stresses. Afterwards, Lindholm *et al.* (1972) determined the triaxial compressive strength of rock-like materials. Then, it was continuously modified and improved to precisely determine triaxial compressive strength of rock and rock-like materials (Malvern & Jenkins, 1990; Gary & Bailly, 1998; Li *et al.*, 2008).

2.2 Modified split Hopkinson pressure bar apparatus

Based on the standard SHPB apparatus, a new SHPB system for coupling load experiment was successfully constructed at Central South University (Li *et al.*, 2008), as shown in Figure 3. The system consists of elastic bars, a striker launcher, axial precompression inducer, confining pressure inducer and data processing unit. The diameter of the input and output bar is 50 mm; the lengths of the input bar and output bar are 2.00 m and 1.50 m, respectively. Strain gauges are glued on the surface of the middle of elastic bars to measure strain histories induced by the stress waves propagating along the elastic bars. Cylindrical rock specimen is sandwiched between the input and output bar, and can be surrounded by rubber sleeve of confining pressure component during the tests. Dynamic loading comes from the impact of the striker driven by high-pressure gas. The coupled static and dynamic loads testing system has two

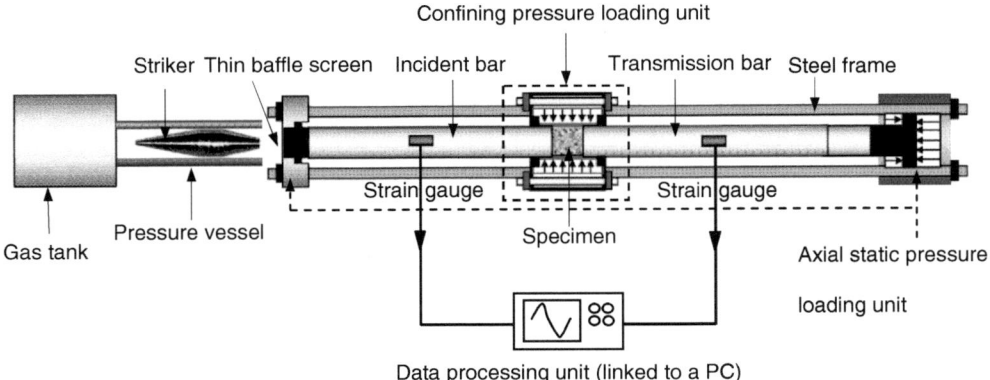

Figure 3 Configuration of the modified SHPB system.

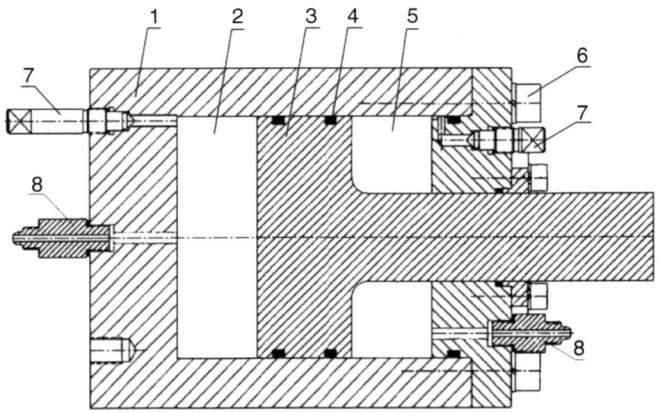

1- oil cylinder; 2- left chamber; 3- bidirectional piston; 4- seal ring;
5- right chamber; 6- bolt; 7- air outlet; 8- oil inlet/outlet

Figure 4 Axial static pre-compression stress component.

components, *i.e.* static axial initial stress component and initial confining pressure component, to realize loading static axial stress and static confining stress.

The static axial initial stress component includes a steel frame, baffled a one end with a pressure loading unit, as shown in Figure 4. The two ends of the oil cylinder are connected through hand pumps by oil inlet/outlet valves. When axial initial stress is needed, the left pump shown in Figure 4 starts to increase the oil pressure in the left chamber, and the piston moves to the right, then, stress onto the elastic bars and specimen.

The initial confining pressure component consists of a steel frame, oil chamber, rubber sleeve, as shown in Figure 5. Between the ends of the elastic bars and specimen,

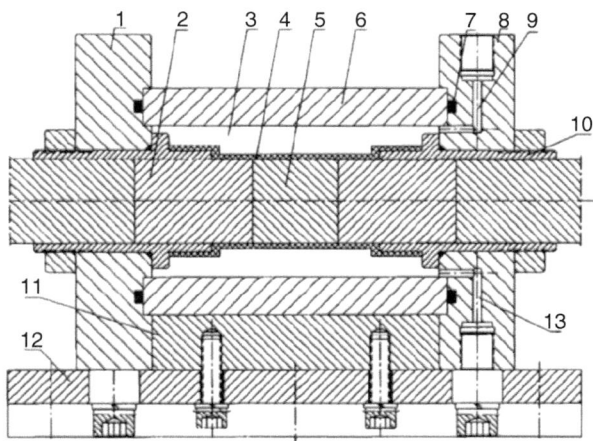

(1-steel box; 2-protective block; 3-oil chamber; 4-rubber sleeve; 5-specimen; 6-oil cylinder; 7- ring seal; 8-oil manometer joint; 9-connect channel; 10-steel sleeve; 11-fill block; 12-base frame; 13-oil inlet/outlet)

Figure 5 Lateral confining pressure component.

protective steel blocks are installed to prevent damage to the faces of the elastic bars. And the rubber sleeve plays a role in sealing and makes the specimen under uniform confining pressure. When confining pressure is needed, the hydraulic oil enters the oil cylinder through a manual pump, which drives the gas in the pump releasing from the gas vent. When the pressure reaches the required level, keep it to maintain a stable confining pressure.

2.3 Critical test technique of modified SHPB

2.3.1 Wave shaping by special shaped striker

With merits of easy operation, good repeatability and accurate results, SHPB has become a more and more popular experimental technique to test materials out of metals. Due to the advantages of SHPB in dynamic tests, it was gradually extended to the studies of brittle materials including rock, concrete and ceramic. Generally, traditional SHPB with rectangular incident wave can provide useful experimental results for metals, whose compressive flow stresses happen at strains larger than a few percent. By contrast, brittle materials such as rocks, ceramics and concrete normally fail at strains less than 0.5 percent. Because of the steep front of rectangular wave by the cylindrical strikers, a specimen of rock-like materials always fails before its stress equilibrium, which makes the test signals useless according to SHPB assumptions. Hence, in order to avoid the premature failure of specimen, slowly rising input waves are needed in SHPB tests for rock materials. Thus, a half-sine wave loading generated by a shaped striker were proposed to overcome the premature failure of specimen (Li *et al.*, 2008; Li & Zhou, 2009; Zhou *et al.*, 2013).

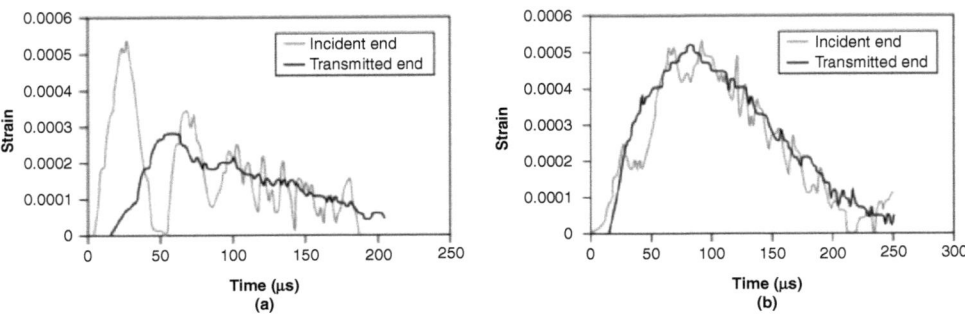

Figure 6 Stress at both ends of specimen with different incident wave: (a) rectangular incident wave; (b) half-sine incident wave.

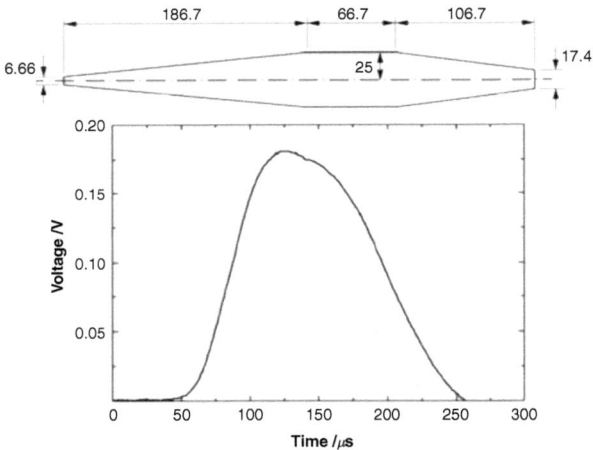

Figure 7 Typical special shaped striker.

For a half-sine wave with slow rising slope, the response of specimens differs greatly from the cases when the specimens are subjected to a steep rising incident wave like a rectangular wave. As for half-wave loading, upon arrival of the wave front, the incident end has slightly higher stress than the other end. With wave reflection in the specimen, the stress at the transmitted end increases gradually and accumulates gradually. After several reflections of wave in the specimen, the stress at both ends of the specimen reaches equilibrium with an average value still less than the failure stress of the specimen. Figure 6 gives typical stress histories at both ends of the specimen with rectangular and approximate half-sine wave from a shaped striker. It can be seen that the case with half-sine wave gives better stress equilibrium during specimen deformation, while the stress in specimen with a rectangular incident wave shows great deviation at the two ends during specimen deformation and failure, which violates the assumption of SHPB technique. A striker with the configuration shown in Figure 7 generating the approximate half-sine stress wave of Figure 7 is suggested.

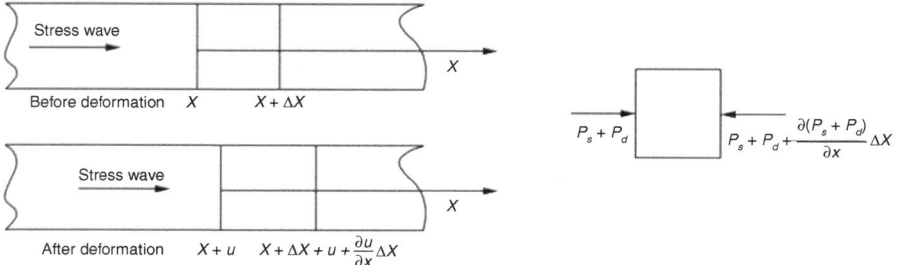

Figure 8 Effect of longitudinal wave on axially pre-stressed elastic bar.

2.3.2 Theoretical validation for coupled static and dynamic loads

The SHPB has been successfully used to investigate the dynamic characteristics of rock, and its principles are based on one-dimensional wave theory. However, this theory needs reexamining when axial pre-compression stress exists.

When compressive longitudinal wave propagates along the bar and the specimen in the presence of axial static stress at both ends, the deformation is described in Figure 8. It is assumed that there is no deflection in elastic bars and specimen during tests. Now, consider an infinitesimal segment with the force acting on the segment, the equation of motion is written as

$$-\frac{\partial(P_{\mathrm{s}} + P_{\mathrm{d}})}{\partial x}\Delta x = \rho A \Delta x \frac{\partial^2 u}{\partial t^2} \tag{1}$$

where A is the area of the cross-section, ρ is the density of the material, u is the axial translational displacement, P_{s} is the axial static load and P_{d} is the impact loading.

According to solid mechanics and wave theory,

$$\sigma = \frac{P_{\mathrm{s}} + P_{\mathrm{d}}}{A} \tag{2}$$

$$\varepsilon = -\frac{\partial u}{\partial x} \tag{3}$$

$$C = \sqrt{\frac{1}{\rho}\frac{\mathrm{d}\sigma}{\mathrm{d}\varepsilon}} \tag{4}$$

The axial static load is constant along at all times, so

$$\frac{\partial P_{\mathrm{s}}}{\partial x} = \frac{\partial P_{\mathrm{s}}}{\partial t} = 0 \tag{5}$$

Thus,

$$\frac{\partial^2 u}{\partial t^2} = C^2 \frac{\partial^2 u}{\partial x^2} \tag{6}$$

where C is the wave velocity in the pre-compression material.

Figure 9 1D coupled static and slight dynamic loads system.

Equation 6 is the wave equation governing the wave propagation characteristics of pre-compression bars and specimen. In the data processing of the new system, all the velocity values should use the wave velocity of material under specific pre-compression.

3 TEST SYSTEM ON CUBIC ROCK UNDER COUPLED STATIC-DYNAMIC LOADS

3.1 Test technique under coupled static and slight dynamic loads

A coupled static and slight dynamic loading system was developed based on INSTRON electro-hydraulic servo controlled testing system (Li *et al.*, 2004), as shown in Figure 9. Load-displacement and load-time curves are recorded using low cycle fatigue software (SAX). In order to obtain static-dynamic loading, an initial static load P_s is added first to the specimen along the direction of layer, and then a dynamic load $P_d(t)$ with medium strain rate is superimposed in the same direction, as shown in Figure 10. In this regard, the coupled static and dynamic loading $P_c(t)$ can be expressed as

$$P_c(t) = P_s + P_d(t) \tag{7}$$

where $P_d(t) = A \sin \omega t$, A is the amplitude of sine wave.

The axial and transversal strains are measured by two strain gauges (3 mm×5 mm) mounted to central surface of the rock specimen. Strain values from these gauges are recorded by DH-5932 data logging device and then analyzed by a computer. The sampling frequency is 500 Hz. The average strain rate can be calculated from the obtained strain history curves as follows,

$$\dot{\varepsilon} = \frac{\Delta\varepsilon}{\Delta t} = \frac{\varepsilon_f}{t_f} \tag{8}$$

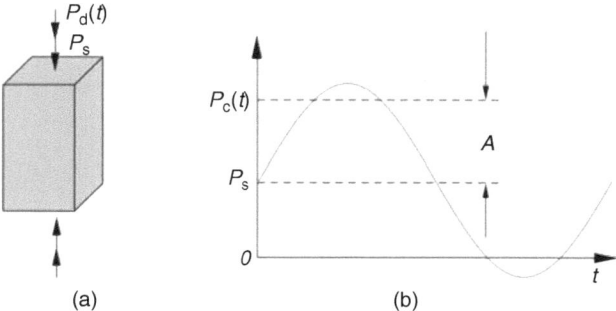

Figure 10 Schematic diagram for coupled static and slight dynamic loads.

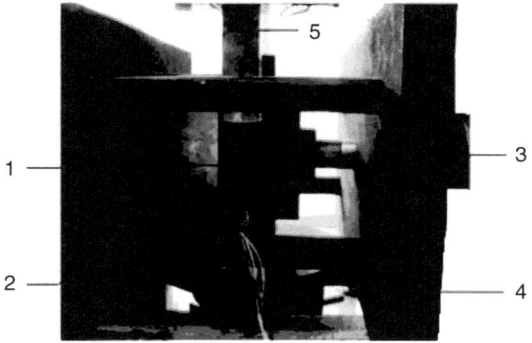

(1-specimen, 2-actuator of Instron; 3-servo valve; 4-lateral loading plate; 5-pressure head of Instron)

Figure 11 2D coupled static and slight dynamic loads apparatus.

where ε_f is the maximum strain measured at failure load and t_f is the time from beginning of test to the maximum load.

Particularly, underground mining or excavation is in fact an artificial unloading process of rock, in which the three-dimensional compressive stress state turns into a two-dimensional compressive stress state. This consideration leads to the evolution of the above test system from one-dimensional coupling loading to two-dimensional coupling loading. The 2D coupled loads apparatus was developed from the previous INSTRON electro-hydraulic servo controlled testing system, by assembling a pair of horizontal loading actuators, as shown in Figure 11.

3.2 Test technique under true triaxial coupled static-dynamic loads

Now that underground rock mass is generally subjected to true triaxial compressive stress $(\sigma_1 > \sigma_2 > \sigma_3)$, the consideration on mechanical behaviors of rock under true

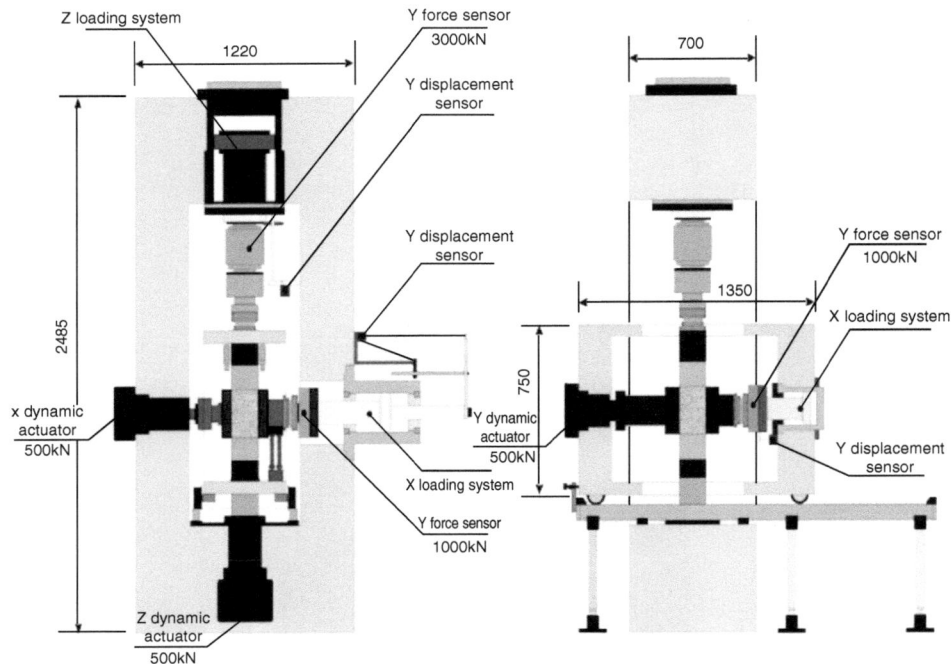

Figure 12 Schematic diagram of the true triaxial testing apparatus with coupled static and dynamic loads.

triaxial stress state is obviously more significant to practical operation during underground excavating. In this regard, a true triaxial dynamic disturbance test system was designed and produced for testing mechanical properties of cubic rock specimen under various loading and unloading conditions (Du *et al.*, 2015; Li *et al.*, 2015). The system is a true triaxial high-frequency coupled static and dynamic loads apparatus, which is capable of applying dynamic disturbance loading and quick unloading on specimens The TRW-3000 true triaxial electro-hydraulic servo dynamic disturbance test system is sketched in Figure 12. The test system consists of six independent loading units for X, Y and Z directions, including three static loading actuators and three dynamic loading actuators, as shown in Figure 13.

The available specimen sizes accommodated by the triaxial cell include 100×100×100, 200×200×200 and 300×300×300 mm. The loads are independently applied to the cubic rock specimen through solid pistons driven by oil pressure in the three orthogonal directions X, Y and Z. The maximum load capacity is 3000 kN in the vertical (Z) direction and 2000 kN in the horizontal (X and Y) direction. The loads on the specimen can be instantly removed in X or Y directions by a rotating dropping system to simulate the unloading process in underground excavation. In addition, the specimen can be applied dynamic loads in all the three directions to simulate dynamic disturbances caused by blasting or some other types of mechanical

Table 1 Main technical specifications of the true triaxial test system TRW-3000.

Loading mode	Items	Parameters		
		X direction	Y direction	Z direction
Static load	Maximum load	2000 kN	2000 kN	3000 kN
	Accuracy	±1%	±1%	±1%
	Resolution	15 N	15 N	20 N
	Loading rate	10N/s–10kN/s	10N/s–10kN/s	10N/s–10kN/s
Displacement control [a]	Range	0–200 mm	0–200 mm	0–200 mm
	Accuracy	<±1 mm	<±1 mm	<±1 mm
	Resolution	0.001 mm	0.001 mm	0.001 mm
Deformation control [b]	Range	0–10 mm	0–10 mm	0–10 mm
	Accuracy	<±0.05 mm	<±0.05 mm	<±0.05 mm
	Resolution	0.0005 mm	0.0005 mm	0.0005 mm
Dynamic load	Load range	0–500 kN	0–500 kN	0–500 kN
	Accuracy	±2.5 kN	±2.5 kN	±2.5 kN
	Frequency	0–70 Hz	0–70 Hz	0–70 Hz

Note: [a] Control by the displacement transducer installed in the triaxial test system;
   [b] Control by external displacement transducers

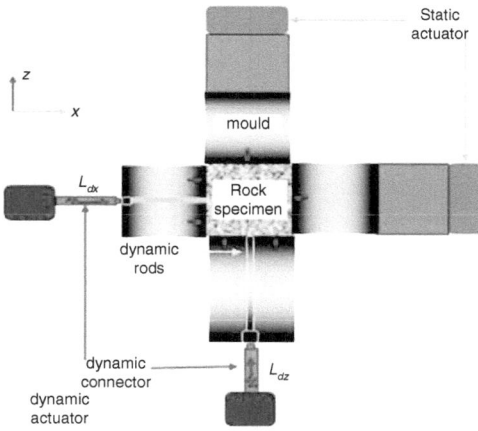

Figure 13 Schematic diagram of loading actuators.

vibrations. The test system consists of a computer, a controller, actuators, sensors, etc., which compose a closed loop digital system to automatically control the applied loads or displacements. The main technical specifications of the system are provided in Table 1. The test system TRW-3000 can be used to examine the behavior of the rock in either static or dynamic loading conditions. The mechanical response of the rock surrounding deep underground excavations can be simulated by the system.

4 FAILURE CHARACTERISTICS OF ROCK UNDER COUPLED STATIC AND DYNAMIC LOADS

4.1 One-dimensional static and dynamic coupling loads

4.1.1 Influence of pre-stress

An impact loading with peak stress of 200 MPa was kept constant but with the axial pre-compression stress varying at 18, 36, 45, 54, 72 and 81 MPa. These axial pre-stress are equivalent to 20%, 40%, 50%, 60%, 80% and 90% of the static strength, respectively. Five specimens were allocated for the test in each set. In order to provide contrasting results for different sets, one representative curve was chosen in every set. The representative curve should have the average characters of the other four curves in the set.

Figure 14 shows the stress-strain curves of rock subjected to the same dynamic loading but different axial pre-stresses. From Figure 14, it is evident that with coupling loads, the strength of siltstone is higher than its static strength and dynamic strength within the same strain rate, increasing 120% of its static strength and 30% of its dynamic strength. However, the strength under coupling loads declines when the static pre-stress reaches a level of about 54 MPa.

With constant impact loading and different axial pre-compression stress, the strength of rock varies as shown in Figure 15. When the rock is deforming in the elastic phase (axial static stress is higher than 15% but less than 70% of its static strength), the strength under coupling loads can be 40–60 MPa higher than its dynamic strength. Once the axial static stress of specimen is greater than 80% of its static strength, the rock is at impending yielding or near to yield point, and the rock strength with coupling loads decreases rapidly.

When axial static stress on the rock specimen is greater than 90% of its static strength, the rock is at near breakage and the system appears unstable. So, the derived stress-strain curve oscillates as shown by curve 6 in Figure 14. It can be seen that the

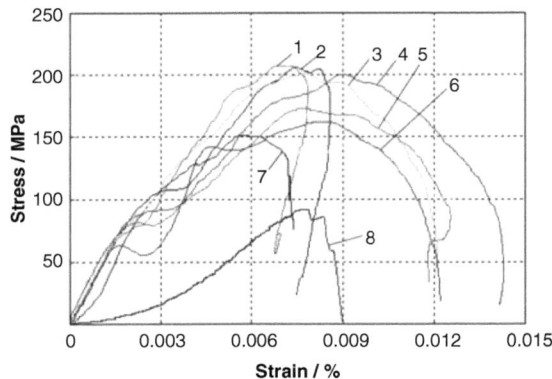

Figure 14 Stress–strain curves of siltstone subjected to same impact loading with peak stress of 200 MPa but different axial static pre-compression stress. (axial pre-compression stress: 1—18 MPa; 2—36 MPa; 3—45 MPa; 4—54 MPa; 5—72 MPa; and 6—81 MPa).

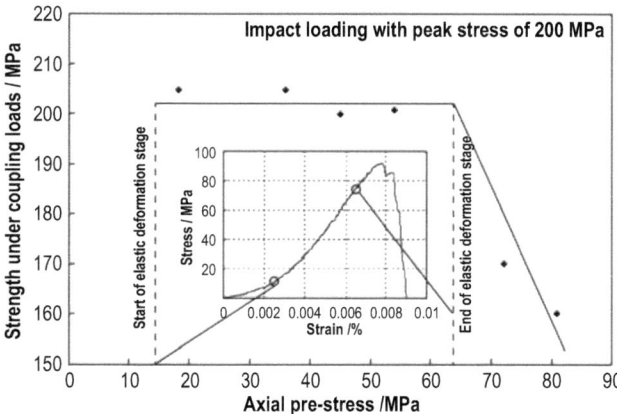

Figure 15 Strength of siltstone subjected to same impact loading but different axial pre-compression stress.

slope of the stress-strain curve of the specimen decreases abruptly when the strain of the specimen approaches 0.002. Then, the slope increases again. The phenomenon may be due to particle sliding within the specimen. The deterioration of the specimen at this instant is not sufficiently critical to cause the specimen to fracture entirely; the specimen is still able to sustain loads until a failure strain of 0.008.

4.1.2 Influence of dynamic loads

The axial static pre-compression stress was retained at 63 MPa (70% of the static compression strength of the siltstone) and the peak stress of the impact loading was varied at 150, 200, 250, 300 and 330 MPa. Then, Figure 16 shows the stress-strain curves of specimens with the same static pre-stress but with different impact loadings, and Figure 17 presents the strength change of the siltstone under the same pre-stress but various impact loads. It can be seen that with coupling loads, the strength of siltstone increases with increasing impact loading , or the increasing of strain rate of specimen. When the peak stress of impact loading is 150, 200, 250, 300 and 330 MPa, the strain rate of the specimen is about 50 s^{-1}, 80 s^{-1}, 120 s^{-1}, 150 s^{-1} and 180 s^{-1}, respectively. This phenomenon reinforces the original understanding that the material strength increases with increasing strain rate.

When the peak stress of impact loading is 150 MPa, the specimens fracture after 15 repeated impacts. In this case, the obtained strength is less than its dynamic strength. The reason may be due to the breakage of the rock in fatigue failure and the cumulative damage weakened the specimen before breakage.

4.2 Three-dimensional static and dynamic coupling loading

A series of tests are conducted on sandstone specimens by the improved SHPB testing system to investigate the mechanical behaviors of rock during the

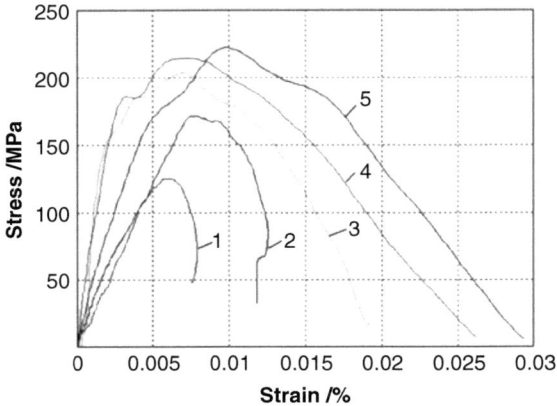

Figure 16 Stress–strain curves of siltstone subjected to constant axial pre-stress of 63 MPa but different impact loading. (peak stress of impact loading: 1—150 MPa; 2—200 MPa; 3—250 MPa; 4—300 MPa; and 5—330 MPa).

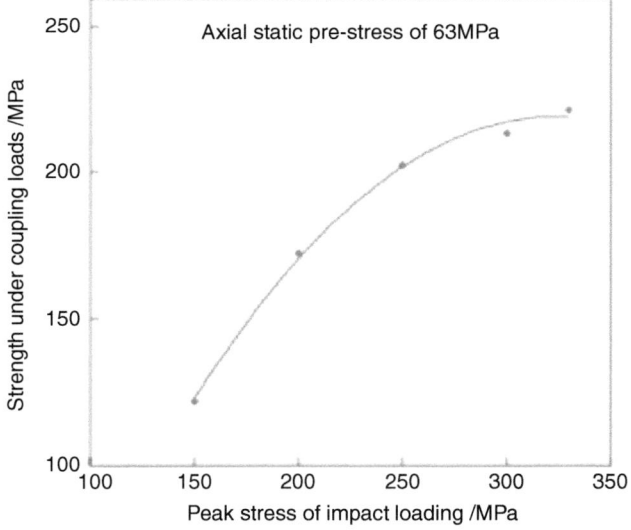

Figure 17 Strength of siltstone subjected to constant axial pre-compression stress of 63 MPa and different impact loading.

exploitation of mineral resources in deep. The loading conditions for the experiments are grouped into two sets, *i.e.* one is three-dimensional static-dynamic coupling loads with fixed confined pressure, and the other one is that with fixed axial static pressure.

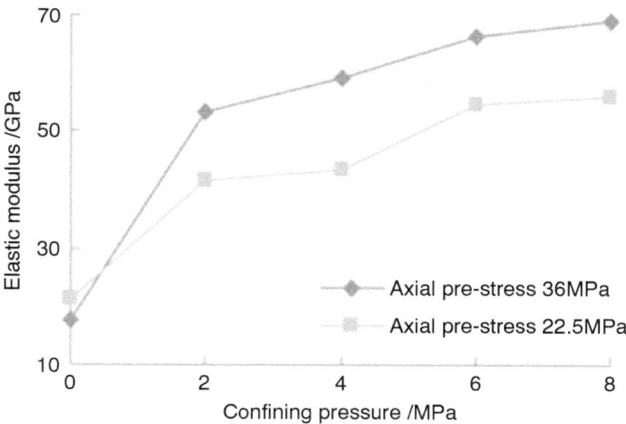

Figure 18 Relationship between elastic modulus and confining stress.

4.2.1 Relation between Young's modulus and triaxial pressure

When the confining pressure is constant, with the increase of axial static pressure, rock becomes denser and as a results, its Young's modulus increases. When the increasing axial static pressure exceeds the yielding stress, the closed microcracks inside the rock start to open. In this regard, the rock is damaged with the decrease of Young's modulus. As shown in Figure 18, when the rock is still in elastic deformation phase with axial static pressure of 22.5 MPa and 36 MPa, the rock's Young's modulus rises with increasing the confining pressure. Meanwhile, the Young's modulus of rock under high static axial pressure is generally greater than that of rock under low static axial pressure. Figure 19 presents the relation between Young's modulus and axial static pressure under confining pressure of 4 MPa and 8 MPa. It is evidential that the Young's modulus increases with increasing the axial static pressure when rock deforms in elastic phase, while the Young's modulus decreases in damage phase.

4.2.2 Relation between strain rate and triaxial pressure

Strain rate is an index that measures the rate of change in deformation of rock with respect to time, which also reflects the change rate of stress. Average strain rate can demonstrates the change rate of the whole deformation and fracture process of rock, while the maximum strain rate demonstrates the rock mechanical characteristics at the moment when largest deformation change rate occurs. Figures 20 and 21 show the relation of strain rate with confining pressure under fixed axial stress and axial stress under fixed confining pressure. The maximum strain rate is not obviously related with confining pressure under the fixed axial stress of 22.5 MPa and 36 MPa, while the corresponding average strain rate decreases with increasing the confining pressure. On the other hand, the maximum strain rate and average strain rate of rock both

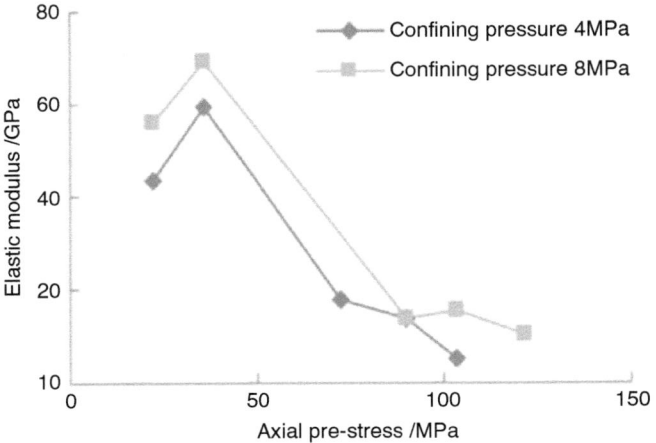

Figure 19 Relationship between elastic modulus and axis static stress.

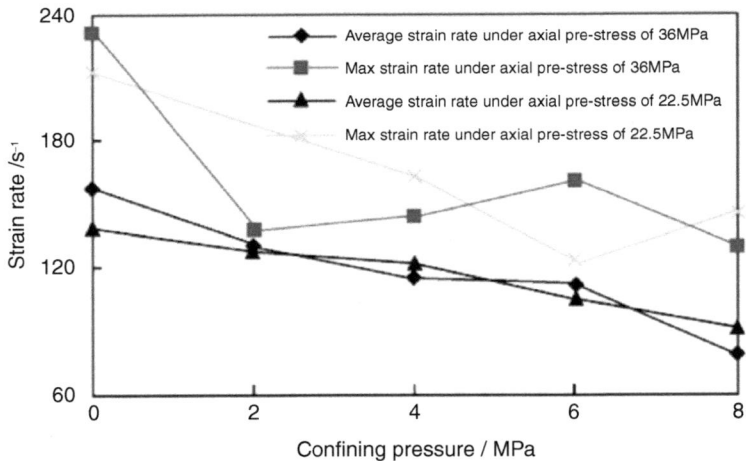

Figure 20 Relationship between strain rate and confining stress.

decrease at elastic phase and increase at damage phase, with the increase of axial stress.

4.2.3 Strength characteristics of rock under triaxial stress and impact loading

Figures 22 and 23 present the strength of sandstone under different coupled triaxial stresses. When rock is at elastic stage due to axial stress, the static-dynamic coupling

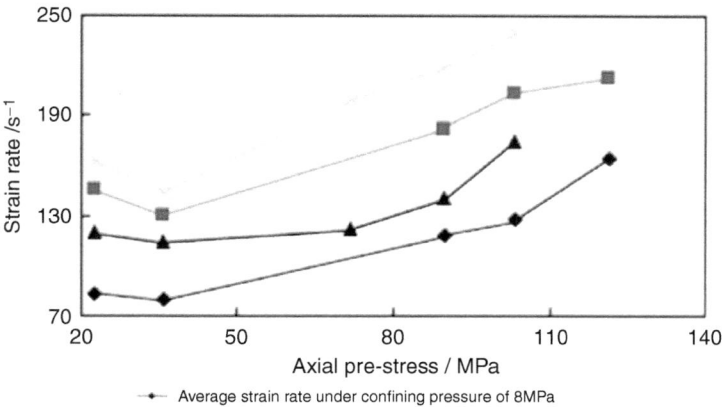

Figure 21 Relationship between strain rate and axis static stress.

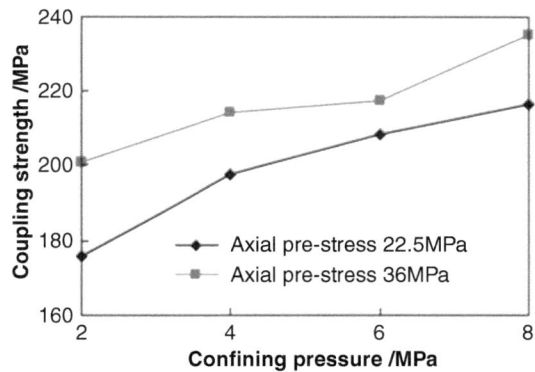

Figure 22 Relationship between coupled compressive strength and confining stress.

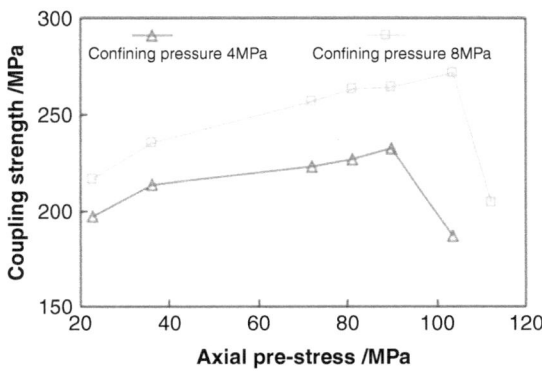

Figure 23 Relationship between coupled compressive strength and axis stress.

strength keeps increasing with the increase of confining pressure. However, the static-dynamic strength declines with increasing confining pressure when rock is at plastic deforming stage. What's more, under the constant confining pressure condition, it is found that the coupling strength increases first and decreases with increasing the axial stress. The turning point of axial stress for this change is about 80% of its corresponding triaxial compressive strength.

4.3 Constitutive model of rock under coupled static and dynamic loads

4.3.1 Rock unit model

Suppose that rock occupies the characteristics of both statistical damage and visco-liquid, and can be regarded as a combination of damage unit D_a and visco-piston unit η_b, as illustrated in Figure 24.

In the rock unit model, damage unit D_a characterizes isotropy of damage and linear elasticity before damage, whose strength obeys to the statistical distribution with parameters (m, a). The damage variable D can be expressed as the following two types according to the stress state of rock:

For 1D loading,

$$D = 1 - [(\varepsilon_a/\alpha)^m + 1]\exp[-(\varepsilon_a/\alpha)^m] \quad (\varepsilon_a \geq 0) \tag{9a}$$

For 2D and 3D loading,

$$D = 1 - \exp[-(\varepsilon_a/\alpha)^m] \quad (\varepsilon_a \geq 0) \tag{9b}$$

Constitutive relation $\sigma - \varepsilon$ may be expressed as

$$\sigma_a = E\varepsilon_a(1-D) \quad (\varepsilon_a \geq 0) \tag{10}$$

Additionally, visco-piston unit, with non-damage property, is supposed to satisfy to the following relation

$$\sigma_b = \eta d\varepsilon_b/dt \tag{11}$$

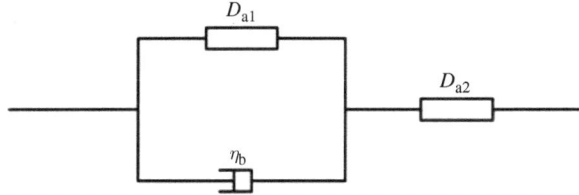

Figure 24 Rock unit model.

4.3.2 Constitutive model under one-dimensional coupled loads

The relation between stress and strain of combination mass is obedient to

$$\begin{cases} \sigma = \sigma_{a1} + \sigma_b = \sigma_{a2} \\ \varepsilon = \varepsilon_1 + \varepsilon_2 \\ \varepsilon_1 = \varepsilon_b \end{cases} \tag{12}$$

Substituting the constitutive relation of damage mass and visco-piston into Equation 12, the constitutive relation of combination is expressed

$$\eta\dot{\sigma} + [E_1(1-D) + E_2(1-D)]\sigma = E_2(1-D)[\eta\dot{\varepsilon} + E_1(1-D)\varepsilon] \tag{13}$$

In order to evaluate Equation 13, the damage variable is not considered at first, and then the viscoelastic constitutive equation of combination mass is written

$$\eta\dot{\sigma} + (E_1 + E_2)\sigma = E_2(\eta\dot{\varepsilon} + E_1\varepsilon) \tag{14}$$

Now that rock is subjected to static load before dynamic impact, the initial condition can be assumed $t = 0$, $\varepsilon(0) = \varepsilon_0$ and $\sigma(0) = S$. In this regard, the stress is derived by Laplace transformation

$$\sigma(t + t_0) = E_2\varepsilon(t + t_0) - \frac{E_2^2}{\eta} \int_0^t \varepsilon(\tau + t_0) \cdot e^{-\frac{E_1 + E_2}{\eta}(t + t_0 - \tau)} d\tau \tag{15}$$

Provided $\varepsilon(t + t_0) = \varepsilon_0 + \varepsilon_r(t) = \varepsilon_0 + ct$, Equation 15 can be written

$$\sigma(t + t_0) = [\varepsilon_0 + \varepsilon_r(t)]E_2\left(1 - \frac{E_2}{E_1 + E_2}e^{-\frac{E_1 + E_2}{\eta}t_0}\right)$$

$$+ \frac{E_2^2}{E_1 + E_2}\left(\varepsilon_0 - \frac{\eta c}{E_1 + E_2}\right)e^{-\frac{E_1 + E_2}{\eta}\left[\frac{\varepsilon_r(t)}{c} + t_0\right]} + \frac{E_2^2\eta c}{(E_1 + E_2)^2}e^{-\frac{E_1 + E_2}{\eta}t_0} \tag{16}$$

While Equation 9a is updated as

$$D(t + t_0) = D = 1 - \left[\left(\frac{\varepsilon_0 + \varepsilon_r(t)}{\alpha}\right)^m + 1\right]\exp\left[-\left(\frac{\varepsilon_0 + \varepsilon_r(t)}{\alpha}\right)^m\right] \tag{17}$$

By substituting E_1 and E_2 in Equation 15 with $E_1[1 - D(t + t_0)]$ and $E_2[1 - D(t + t_0)]$, the uniaxial constitutive equation of isotropy damage mass undergoing static-dynamic coupling load is obtained

$$\sigma(t + t_0) = (1 - D)[\varepsilon_0 + \varepsilon_r(t)]E_2\left(1 - \frac{E_2}{E_1 + E_2}e^{-\frac{(1-D)E_1 + E_2}{\eta}t_0}\right)$$

$$+ (1 - D)\frac{E_2^2}{E_1 + E_2}\left[\varepsilon_0 - \frac{\eta c}{(1 - D)(E_1 + E_2)}\right]e^{-\frac{(1-D)(E_1 + E_2)}{\eta}\left[\frac{\varepsilon_r(t)}{c} + t_0\right]}$$

$$+ \frac{E_2^2\eta c}{(E_1 + E_2)^2}e^{-\frac{(1-D)E_1 + E_2}{\eta}t_0} \tag{18}$$

In particular when $\varepsilon_0 = t$, $t_0 = 0$, Equation 18 can represent the constitutive relation of rock under uniaxial dynamic load, as written

$$\sigma(t) = (1 - D)\frac{E_1 E_2 \varepsilon_r(t)}{E_1 + E_2} + \frac{E_2^2 \eta c}{(E_1 + E_2)^2} \cdot \left\{1 - \exp\left[-(1 - D)\frac{\varepsilon_r(t)}{c}\frac{(E_1 + E_2)}{\eta}\right]\right\}$$

$$(19)$$

4.3.3 Constitutive model under three-dimensional coupled loads

1) Constitutive relation of unit mass before damage
    According to hypothesis, the stress-strain relation of a 3D rock mass can be expressed as

$$S_{ij} = 2Ge_{ij} \tag{20}$$

$$\sigma_m = 3K\varepsilon_m \tag{21}$$

where S_{ij} is the partial tensor of stress. The relation between S_{ij}, stress tensor σ_{ij} and spherical stress tensor σ_m is

$$\sigma_{ij} = S_{ij} + \delta_{ij}\sigma_m$$

where e_{ij} is the partial tenor of strain, the relation between e_{ij}, strain tensor ε_{ij} and spherical strain tensor ε_m is

$$\varepsilon_{ij} = e_{ij} + \delta_{ij}\varepsilon_m \tag{23}$$

where δ_{ij} is the sign of DIRAC, G is shear modulus and K is volume modulus.
According to the above principle, the 3D constitutive equation of viscoelastic mass under static-dynamic coupling loading is derived as

$$\sigma_z(t + t_0) = \frac{9KE_2}{(3K + E_2)\eta}\left\{\eta[\varepsilon_{z0} + \varepsilon, (t)] + \frac{E_1 - \beta\eta}{\beta}\cdot[\varepsilon_{z0} + \varepsilon_r(t) - c]e^{-\beta t_0}.\right.$$

$$\left. -\frac{E_1 - \beta\eta}{\beta}(\varepsilon_{z0} - c)e^{-\beta\left[\frac{\varepsilon_r(t)}{c} + t_0\right]}\right\}$$

$$+ \frac{S_{x0} + S_{y0}}{2(3K + E_2)\eta}\left\{\gamma + \delta e^{-\beta\left[\frac{\varepsilon_r(t)}{c} + t_0\right]}\right\} \tag{24}$$

in which

$$\beta = \frac{3K(E_1 + E_2) + E_1 E_2}{(3K + E_2)\eta} \tag{25}$$

$$\gamma = \frac{3K(E_1 + E_2) - 2E_1 E_2}{\beta} \tag{26}$$

$$\delta = (3K - 2E_2)\eta - \gamma \tag{27}$$

When $S_{x0} = 0$ or $S_{y0} = 0$, the 3D constitutive equation above can be expressed as constitutive relation of two dimension statically loaded rock subjected to dynamic load.

When $S_{x0} = S_{y0} = S_{z0}$, the 3D constitutive equation above can be expressed as the constitutive relation of triaxial dynamically loaded rock during confining pressure is loaded by triaxial experimental machine.

2)   Constitutive relation of unit mass after damage

The fragmentation format of rock behaves commonly shear yield under condition of triaxial stress. Suppose that the fragmentation of unit mass complies with Coulomb criterion, the damage variable of rock under 3D static-dynamic coupling loading is expressed from Equations 9 and 10,

$$D = 1 - \exp\left\{ -\left[\frac{\varepsilon_z E_2 - \left(\frac{1 + \sin\varphi}{1 - \sin\varphi} - 2\upsilon\right)\left(\frac{\sigma_{x0} + \sigma_{y0}}{2}\right)}{E_2 \alpha} \right]^m \right\} \tag{28}$$

where φ, υ is interior friction angle and Poisson's ratio respectively, while other signs are denoted as the same before.

$$\begin{cases} D = 1 - \exp\left\{ -\left[\frac{\varepsilon_z E_2 - \left(\frac{1 + \sin\varphi}{1 - \sin\varphi} - 2\upsilon\right)\left(\frac{\sigma_{x0} + \sigma_{y0}}{2}\right)}{E_2 \alpha} \right]^m \right\}, & \varepsilon_z > \left(\frac{1 + \sin\varphi}{1 - \sin\varphi} - 2\upsilon\right)\frac{\sigma_{x0} + \sigma_{y0}}{2E_2} \\ \\ D = 0, & \varepsilon_z \leq \left(\frac{1 + \sin\varphi}{1 - \sin\varphi} - 2\upsilon\right)\frac{\sigma_{x0} + \sigma_{y0}}{2E_2} \end{cases} \tag{29}$$

According to hypothesis, Lemaitre principle of strain equivalent is applicable. Applying Equation 29 to Equations 24–27 in which E_1, E_2 and K are substituted by $E_1[1 - D(t + t_0)]$, $E_2[1 - D(t + t_0)]$ and $K[1 - D(t + t_0)]$ respectively, isotropy 3D damage constitutive equation of viscoelastic material under static-dynamic coupling loading is written

$$D(t + t_0) = D = 1 - \exp\left\{ -\left[\frac{[\varepsilon_{z0} + \varepsilon_r(t)]/\alpha - \left(\frac{1 + \sin\varphi}{1 - \sin\varphi} - 2\upsilon\right)\left(\frac{\sigma_{x0} + \sigma_{y0}}{2}\right)}{E_2 \alpha} \right]^m \right\} \tag{30}$$

Equations 28–30 are also applicable to determine the constitutive relation of two dimensional statically loaded rock under dynamic load.

4.4 Experimental verification of the constitutive model of rock under coupled loads

4.4.1 One-dimensional coupled static and slight dynamic loads

The verification tests were carried out on red sandstone using INSTRON electro-hydraulic servo controlled testing system. Figure 25 presents the comparisons of the stress-strain curves from experimental and theoretical results. The theoretical curve is well fitting to the experimental one before rock failure when the static stress S equals to 2, 4 and 6 MPa. However, with increasing the static stress, the difference between the two curves becomes greater. The peak stress from theoretical calculation is a little smaller than that of experimental value when the static stress is small. With the increase of static stress, the theoretical peak stress is becoming closer to experimental result.

4.4.2 Two-dimensional coupled static and slight loads

The verification experiment of rock under 2D static-dynamic coupling load was done by low cycle fatigue loading. The constitutive relation of red sandstone

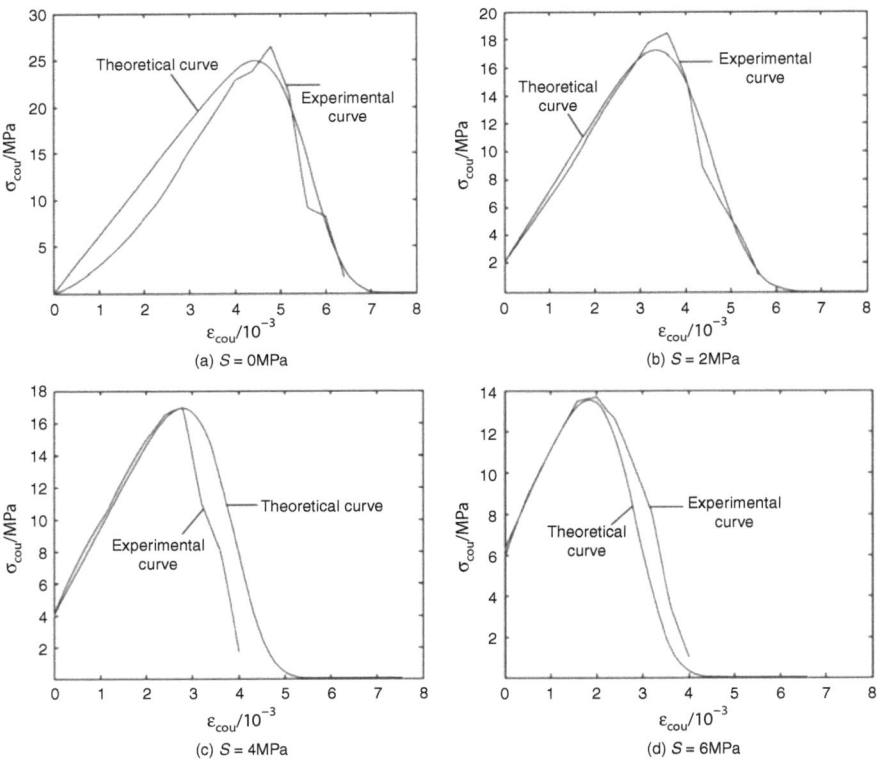

Figure 25 Experimental and 1D theoretical stress-strain curves of red sandstone experiencing different static loads.

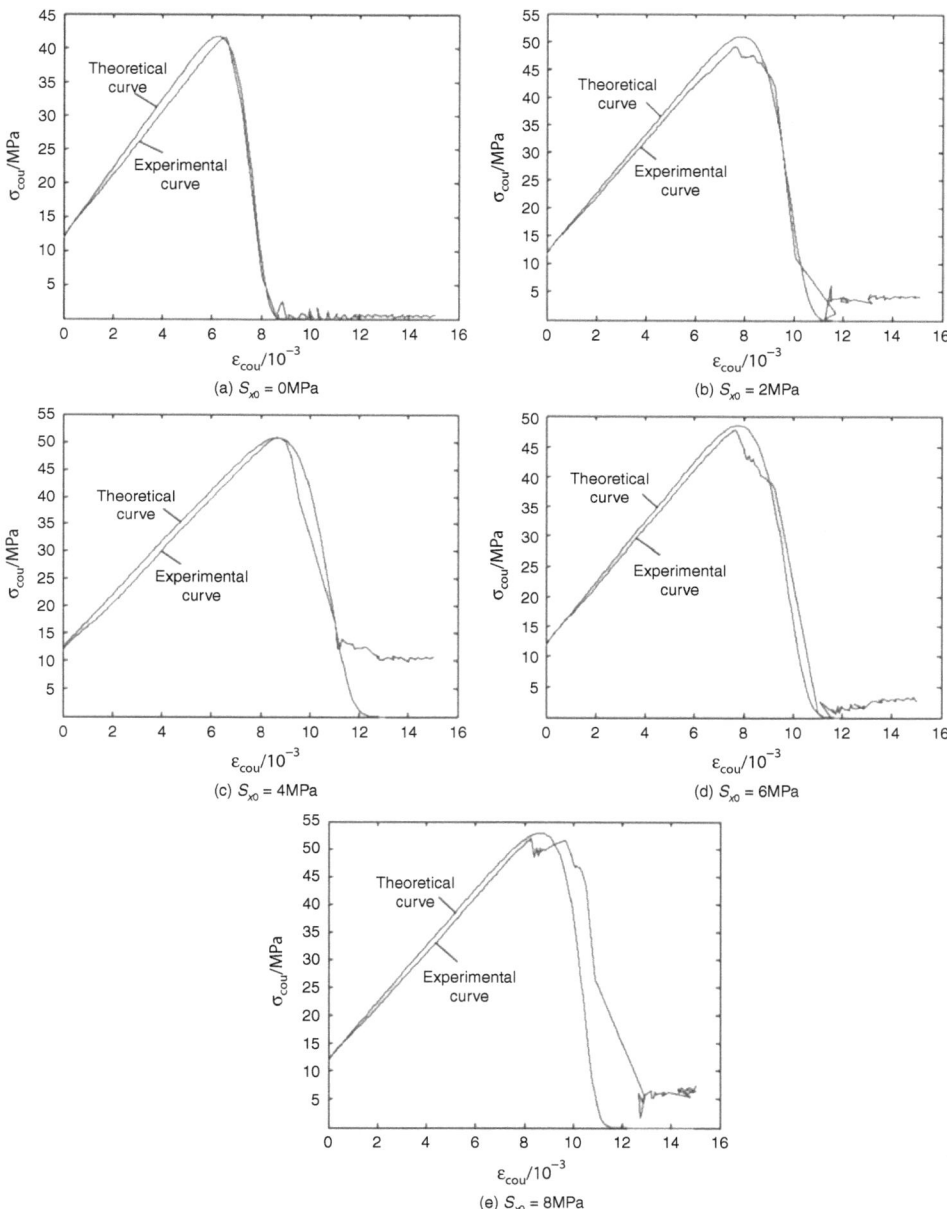

Figure 26 Experimental and theoretical 2D stress-strain curves of red sandstone experiencing different horizontal static stresses.

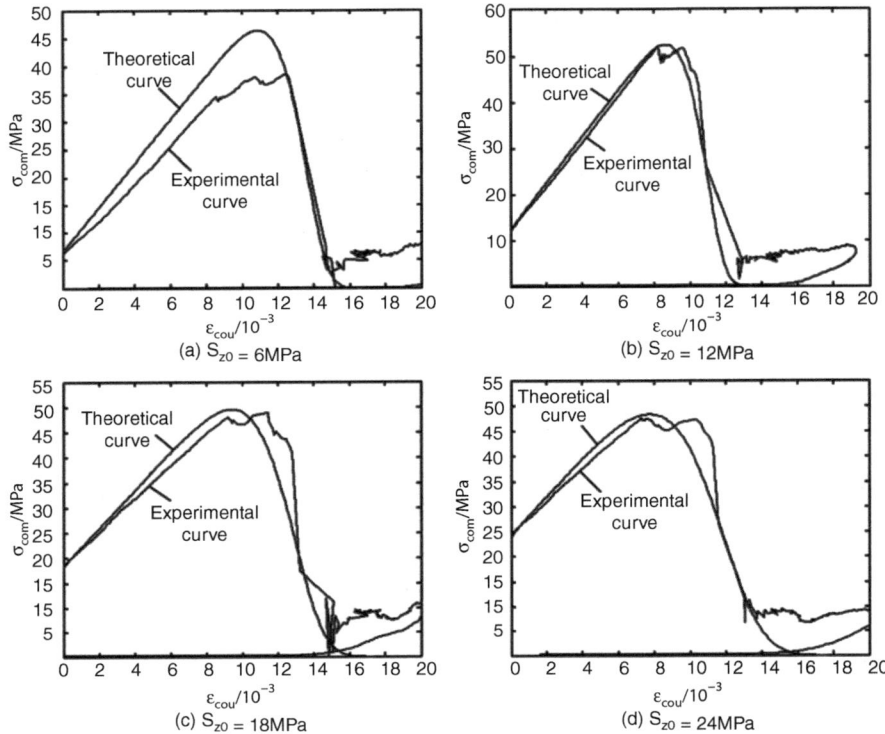

Figure 27 Experimental and theoretical 2D stress-strain curves of red sandstone experiencing different vertical static stresses.

experiencing 2D static-dynamic coupling loads was studied under different horizontal static pressures and different vertical static pressures. The results are shown in Figures 26 and 27.

Figure 26 shows the experimental and theoretical stress-strain curves of red sandstone under 2D static-dynamic coupling loading with vertical static stresses of 12 MPa and different horizontal static stresses of 0, 2, 4, 6 and 8MPa, respectively. There is disturbance near the peak value of stress of tested stress-strain curves, which is caused by shearing and sliding of rock and is not able to be expressed using theoretical stress-strain curves. The value of elastic module of theoretical stress-strain curves are commonly little more than that of experimental stress-strain curves. With the increase of static stress, the theoretical stress-strain curves after peak stress move commonly to left because of shearing and sliding. After peak stress, tested curves change largely, while the theoretical curves are relatively smoother.

Figure 27 shows the experimental and theoretical stress-strain curves of red sandstone under 2D static-dynamic coupling loading with horizontal static stress of 8 MPa and vertical static stresses of 6, 12, 18 and 24 MPa, respectively. It indicates that there is larger disturbance near the peak stress for the tested stress-strain curves, which is attributed to shearing and sliding of rock and is not able to be expressed from

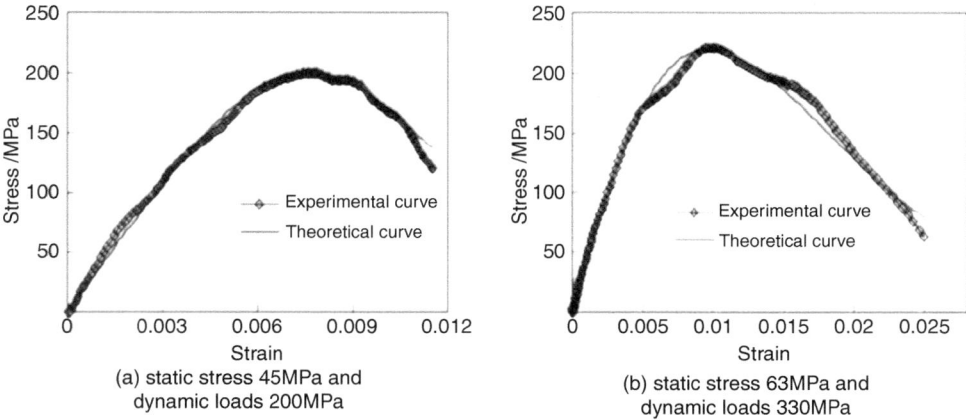

Figure 28 Comparisons of stress-strain curves from constitutive model and experiments.

theoretical stress-strain curves. Due to shearing and sliding, the theoretical stress-strain curves move commonly to the left. In order to narrow the error, the average strain corresponding to that of shearing and sliding should be used as the fitting parameter α of theoretical constitutive curves. In addition, the value of elastic modulus of theoretical stress-strain curves are commonly greater than that of experimental stress-strain curves, while the peak stress of the experimental and theoretical stress-strain curves are close. After peak stress, the tested curves change largely, while the theoretical curves are relatively smoother.

4.4.3 Coupled static and impact loading

From the results of coupled static and impact loading experiment in Section 1.4.2, the parameters for the rock constitutive model can be obtained. The stress-strain curve from constructive model is compared with the experimental stress-strain curve, as shown in Figure 28. It is obvious that the constitutive model with proper parameters can well fit to the experimental stress-strain relation.

4.5 Laboratory simulation of rockburst in highly stresses rock under dynamic disturbance

As one of the most destructive disasters in underground excavation, rockburst will bring potential danger to mineworkers at the face, damage to equipments and delays to the mine operation once it occurs (Tang, 2000). Of all the factors that induce a rockburst, dynamic disturbance is a non-ignorable external factor to induce rock burst, especially for metal mines where drill-and-blast method is used to tunnel. This section will discuss rockburst failure of highly stressed rock under dynamic disturbance by conducting laboratory experiments. The experimental apparatus has been described in Section 1.3.3. The loading process of rock specimen for the experiment consists of three stages, *i.e.* loading to in situ stress state stage, unloading the

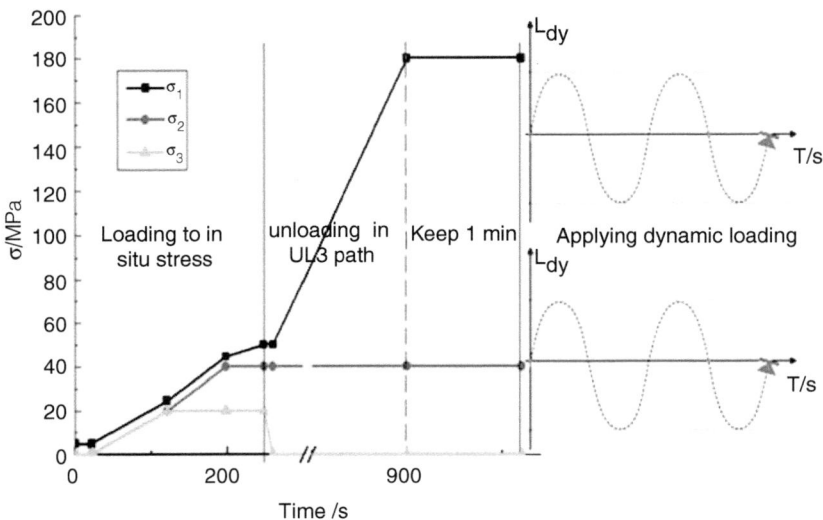

Figure 29 Loading path for rockburst experiment.

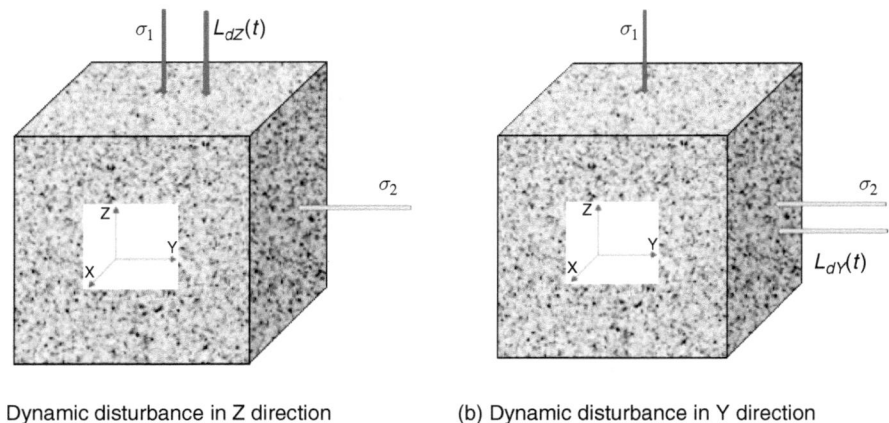

(a) Dynamic disturbance in Z direction   (b) Dynamic disturbance in Y direction

Figure 30 Schematic diagram of dynamic disturbance applied on rock.

minimum principal stress σ_3 stage and applying dynamic loading stage, as illustrated in Figure 29.

The in situ stress state is set as σ_1=50 MPa, σ_2=40 MPa and σ_3=20 MPa, and the loading rate is 2000 N/s. Then the minimum principal stress is unloaded instantaneously to zero, while the maximum principal stress keeps loading to a high level with a loading rate of 2000N/s. In the stage of dynamic disturbance, a stress wave $L_d(t)$ with frequency of 5 Hz is applied to rock specimen either in vertical direction (Z) or in horizontal direction (Y), as shown in Figure 30. The amplitude of the disturbance wave

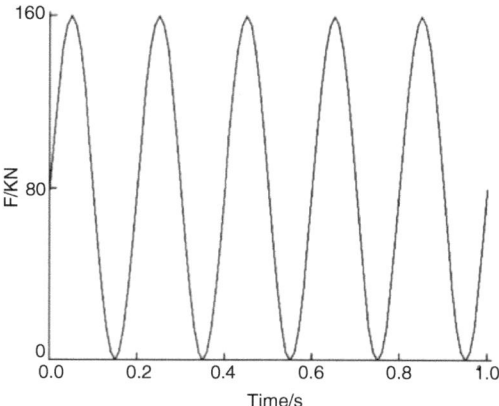

Figure 31 Typical dynamic disturbance wave.

varies from 300 kN, 400 kN, 500 kN and 600 kN. The typical dynamic disturbance wave is presented in Figure 31, and is expressed as

$$F = (P_A \sin(10\pi t) + P_A)/2 \qquad (31)$$

where F is the dynamic loading and P_A is the disturbance amplitude.

4.5.1 Influence of disturbance amplitude

Dynamic loading $L_{dZ}(t)$ is applied on the granite specimen in Z direction to observe rock failure behavior. The results show that under low dynamic amplitude, the specimen will not be fractured. With increasing the amplitude of dynamic disturbance, rockburst will be induced together with ejection of rock pieces. Also, fierce rockburst shall happen when the specimen is compressed in 80% of its corresponding spalling strength subjected to suitable disturbance amplitude, as shown in Figure 32.

4.5.2 Influence of disturbance direction

In underground excavation, especially in deep mining engineering, dynamic disturbance acted on rock mass could be in any directions. Therefore, it is necessary to examine the influence of disturbance direction on the occurrence of induced rockburst. Two different directions of Z and Y are chosen to apply on the specimen, while other experimental conditions are the same. The experimental results are listed in Table 2.

It can be seen from Table 2 that rockburst shall occur once the amplitude of dynamic disturbance reaches to a larger level (600 kN for this case in the study), no matter which direction the dynamic loading is from. When the amplitude is lower, the rock specimen

Table 2 Experimental results.

Specimen	Initial stress			Unloading parameters		Disturbance parameters				Rockburst
	σ_{1}/ MPa	σ_{2}/ MPa	σ_{3}/ MPa	f	$\sigma_{\theta max}$/ MPa	Direction	Frequency/ Hz	Amplitude/ kN	Acting time	
g-10-6-a	50	40	20	3.6	180	Z	5	300	I min	False
g-10-6-a1	50	40	20	3.6	180	Y	5	300	I min	False
g-10-6-b	50	40	20	3.6	180	Z	5	400	I min	False
g-10-6-b1	50	40	20	3.6	180	Y	5	400	I min	False
g-10-6-c	50	40	20	3.6	180	Z	5	500	I min	False
g-10-6-c1	50	40	20	3.6	180	Y	5	500	I min	False
g-10-6-d	50	40	20	3.6	180	Z	5	600	–	True
g-10-6-d1	50	40	20	3.6	180	Y	5	600	–	True

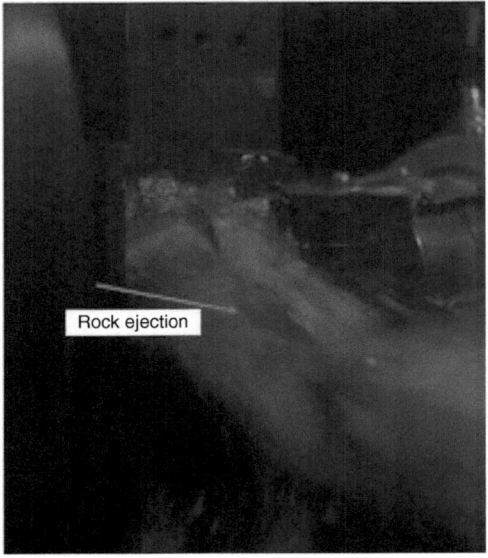

Figure 32 Ejection of rock pieces.

stays stable even if the acting time lasts for one minute. However, with increasing the amplitude, the rock fractures in local under dynamic disturbance with the ejection of pieces, and fails as rockburst finally. In addition, rockburst induced by dynamic disturbance is much fiercer than rock failure under compressive stress.

Meanwhile, acoustic emission characteristics were monitored during the loading and dynamic disturbance process, as shown in Figure 33. When the granite specimen was loaded close to its 80% of peak strength, the acoustic emission events climb dramatically, indicating that cracks are initiated inside the rock. In this occasion, when

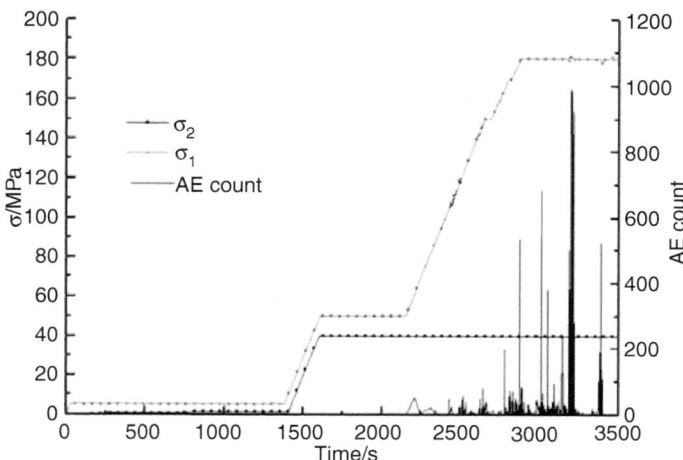

Figure 33 Acoustic emission characteristics for specimen g-10-6-d.

dynamic disturbance is applied on rock, the AE increases sharply once the rockburst occurs.

5 CONCLUSIONS

Testing apparatus of coupling static and dynamic loads were developed, and were employed to conduct a series of coupling static-dynamic loading experiments on rock. A half-sine form wave generated through a special shaped striker was used to load rock specimens. The experimental results generally indicate that the strength of siltstone with coupling loads is higher than its static strength and its dynamic strength. The coupled static-dynamic strength of siltstone increases with increasing impact loading, or with increasing loading strain rate of specimen.

From three-dimensional static and dynamic coupling loading tests, it is concluded that the change of Young's modulus and strain rate is dependent on the deformation stage of rock. With the increase of static axial stress, the Young's modulus of rock increases at elastic stage but decreases at damage stage, while the strain rate decreases at elastic stage but increase at damage stage.

Based on some hypothesis from damage mechanics, the constitutive models of rock under one-dimensional and three-dimensional coupling static and dynamic loads are theoretically demonstrated. The experimental results verify that the constitutive models can well demonstrate rock behaviors under coupled static and dynamic loads. At last, the influence of dynamic disturbance on failure characteristics of highly-stressed rock is deeply examined.

Acknowledgment: financial support from the National Natural Science Foundation of China (41272304, 11472311) and National Basic Research Program of China (2010CB732004) are greatly appreciated.

REFERENCES

Adams, G. R. & Jager, A. J. (1980) Petroscopic observations of rock fracturing ahead of stope faces in deep-level gold mines. *Journal of the South African Institute of Mining and Metallurgy*, 80 (6), 204–209.

Brown, E. T. (2002) Rock mechanics in Australia. *International Journal of Rock Mechanics and Mining Sciences*, 39 (5), 529–538.

Christensen, R. J., Swanson, S. R. & Brown, W. S. (1972) Split-Hopkinson-bar tests on rock under confining pressure. *Experimental Mechanics*, 12 (11), 508–513.

Davies, R. M. (1948) A critical study of the Hopkinson pressure bar. *Philosophical Transactions of the Royal Society of London. Series A, Mathematical and Physical Sciences*, 240 (821), 375–457.

Dowaikh, M. & Ogden, R. (1990) On surface waves and deformations in a pre-stressed incompressible elastic solid. *IMA Journal of Applied Mathematics*, 44, 261–284.

Du, K., Li, X. B., Li, D. Y. & Weng, L. (2015) Failure properties of rocks in true triaxial unloading compressive test. *Transactions of Nonferrous Metals Society of China*, 25 (2), 571–581.

Gary, G. & Bailly, P. (1998) Behaviour of quasi-brittle material at high strain rate. Experiment and modelling. *European Journal of Mechanics-A/Solids*, 17 (3), 403–420.

Grobbelaar, C. (1958–1959) Some properties of rock from a deep-level mine of the central Witwatersrand. *Association of Mine Managers South Africa-Paper Discuss*, 311–330.

Hakalehto, H. O. (1969) The behaviour of rock under impulse loads. A study using the Hopkinson split bar method, CTA Polytechnica Scandinavica, Chemistry including Metallurgy Series No. 81.

Hudson, J. A. & Harrison, J. P. (2000) *Engineering Rock Mechanics – An Introduction to the Principles*. Amsterdam, Elsevier Science.

Hopkinson, B. (1914) A method of measuring the pressure produced in the detonation of high explosives or by the impact of bullets. Philosophical Transactions of the Royal Society of London. Series A, Containing Papers of a Mathematical or Physical Character, 213, 437–456.

Kolsky, H. (1949) An investigation of the mechanical properties of materials at very high rates of loading. *Proceedings of the Physical Society Section B*, 62 (11), 676.

Krafft, J. M., Sullivan, A. M. & Tipper, C. F. (1954) The effect of static and dynamic loading and temperature on the yield stress of iron and mild steel in compression. *Proceedings of the Royal Society of London. Series A. Mathematical and Physical Sciences*, 221 (1144), 114–127.

Kumar, A. (1968) The effect of stress rate and temperature on the strength of basalt and granite. *Geophysics*, 33 (3), 501–510.

Li, X. B., Du, K., & Li, D. Y. (2015). True triaxial strength and failure modes of cubic rock specimens with unloading the minor principal stress. *Rock Mechanics and Rock Engineering*, 48 (6), 2185–2196.

Li, X. B. & Gu, D. S. (1994) *Rock Impact Dynamics*. Changsha, Central South University of Technology Press.

Li, X. B., Lok, T. S. & Zhao, J. (2005) Dynamic characteristics of granite subjected to intermediate loading rate. *Rock Mechanics and Rock Engineering*, 38 (1), 21–39.

Li, X. B., Lok, T. S., Zhao, J. & Zhao, P. J. (2000) Oscillation elimination in the Hopkinson bar apparatus and resultant complete dynamic stress–strain curves for rocks. *International Journal of Rock Mechanics and Mining Sciences*, 37 (7), 1055–1060.

Li, X. B., Ma, C. D., Chen, F. & Xu, J. C. (2004) Experimental study of dynamic response and failure behavior of rock under coupled static-dynamic loading. In: Ohnishi, Y. & Aoki, K. (eds.), *ARMS 2004: Contribution of the Rock Mechanics to the New Century: Proceedings of the ISRM International Symposium 3rd ARMS, ARMS 2004, 30 Nov–2 Dec 2004, Kyoto, Japan*. Rotterdam, Mill Press, 891–895.

Li, X. B., Yao, J. R. & Du, K. (2013) Preliminary study for induced fracture and non-explosive continuous mining in high-geostress hard rock mine – A case study of Kaiyang Phosphate mine. *Chinese Journal of Rock Mechanics and Engineering*, 32 (6), 1101–1111.

Li, X. B., Yao, J. R. & Gong, F. Q. (2011) Dynamic problems in deep exploitation of hard rock metal mines. *The Chinese Journal of Nonferrous Metals*, 21 (10), 2552–2562.

Li, X. B. & Zhou Z. L. (2009) Large diameter SHPB tests with special shaped striker. *ISRM News Journal*, 12, 76–79.

Li, X. B., Zhou, Z. L., Lok, T. S., Hong, L. & Yin, T. B. (2008) Innovative testing technique of rock subjected to coupled static and dynamic loads. *International Journal of Rock Mechanics and Mining Sciences*, 45 (5), 739–748.

Lindholm, U. S., Yeakley, L. M. & Nagy, A. *A Study of the Dynamic Strength and Fracture Properties of Rock*. San Antonio, TX, Southwest Research Institute, 1972.

Lok, T. S., Li, X. B., Liu, D. S. & Zhao, P. J. (2002) Testing and response of large diameter brittle materials subjected to high strain rate. *Journal of Materials in Civil Engineering*, 14 (3), 262–269.

Malan, D. F. (1999) Time-dependent behaviour of deep level tabular excavations in hard rock. *Rock Mechanics and Rock Engineering*, 32(2), 123–155.

Malvern, L. E. & Jenkins, D. A. *Dynamic Testing of Laterally Confined Concrete*. Department of Information Sciences, California Institute of Technology, Pasadena, CA , 1990.

Ogden, R. & Sotiropoulos, D. (1997) The effect of pre-stress on the propagation and reflection of plane waves in incompressible elastic solids. *IMA Journal of Applied Mathematics*, 59, 95–121.

Okubo, S., Fukui, K. & Kawakami, J. (1997) Large-scale penetration test using a drop hammer. *Rock Mechanics and Rock Engineering*, 30 (2), 13–118.

Sun, J. & Wang, S. (2000) Rock mechanics and rock engineering in China: Developments and current state-of-the-art. *International Journal of Rock Mechanics and Mining Sciences*, 37 (3), 447–465.

Tang, B.Y. *Rockburst Control using Destress Blasting*. Ph.D. thesis, Department of Mining and Metallurgical Engineering, McGill University, 2000.

Tao, M., Li, X. B. & Li, D. Y. (2013) Rock failure induced by dynamic unloading under 3D stress state. *Theoretical and Applied Fracture Mechanics*, 65, 47–54.

Tao, M., Li, X. B. & Wu, C. Q. (2012) Characteristics of the unloading process of rocks under high initial stress. *Computers and Geotechnics*, 45, 83–92.

Tao, M., Li, X. B. & Wu, C. Q. (2013) 3D numerical model for dynamic loading-induced multiple fracture zones around underground cavity faces. *Computers and Geotechnics*, 54, 33–45.

Tolstoy, I. (1982) On elastic waves in prestressed solids. *Journal of Geophysical Research: Solid Earth (1978–2012)*, 87, 6823–6827.

Whittles, D. N., Kingman, S., Lowndes, I. & Jackson, K. (2006) Laboratory and numerical investigation into the characteristics of rock fragmentation. *Minerals Engineering*, 19 (14), 1418–1429.

Zhao, H. & Gary, G. (1996) On the use of SHPB techniques to determine the dynamic behavior of materials in the range of small strains. *International Journal of Solids and structures*, 33 (23), 3363–3375.

Zhao, J. (2000) Applicability of Mohr–Coulomb and Hoek-Brown strength criteria to the dynamic strength of brittle rock. *International Journal of Rock Mechanics and Mining Sciences*, 37 (7), 1115–1121.

Zhao, J., Li, H. B., Wu, M. B. & Li, T. J. (1999) Dynamic uniaxial compression tests on a granite. *International Journal of Rock Mechanics and Mining Sciences*, 36 (2), 273–277.

Zhou, Z. L., Li, X. B., Zou, Y., Jiang, Y. H. & Li, G. N. (2014) Dynamic Brazilian tests of granite under coupled static and dynamic loads. *Rock Mechanics and Rock Engineering*, 47 (2), 495–505.

Zou, Y., Li, X. B., Zhou, Z. L., Yin, T. B. & Yin, Z. Q. (2012) Energy evolution and stress redistribution of high-stress rock mass under excavation distribution. *Chinese Journal of Geotechnical Engineering*, 34 (9), 1677–1684.

Chapter 9

Dynamic behavior

C. Menna[1], D. Asprone[1] & E. Cadoni[2]
[1]Department of Structures for Engineering and Architecture, University of Naples Federico II, Napoli, Italy
[2]SUPSI, University of Applied Sciences of Southern Switzerland, Lugano, Switzerland

Abstract: This chapter deals with the experimental techniques commonly employed to characterize the dynamic behavior of rock materials. The dynamic testing machines are illustrated, focusing on the strain rate level achievable and the loading scheme. The main dynamic properties of common rock materials are discussed, with reference to the corresponding dynamic increase factors, which affect compression and tensile behavior.

1 INTRODUCTION

The study of the dynamic behavior of rocks has several sources of complexity but, at the same time, it is essential to analyze and develop predictive models of many engineering problems, such as: earthquakes, blast events, drilling, underground excavations (tunneling), quarrying, rock cutting penetration etc.

A major source of complexity lies in the fact that rocks are inhomogeneous materials characterized by possible interacting damage mechanisms during failure. In addition, due to their physico-chemical composition and microstructure, structural properties of rocks are rate-dependent. Indeed, mechanical properties of rocks under dynamic loadings are different to the corresponding static values. The fracture process under dynamic regime is much more complicated than the static one due to the inertia effects and the stress wave propagation. As general trend, rocks show an improvement of the mechanical properties (tensile and compression strength, failure strain and fracture energy) with increasing strain-rates (Cadoni, 2013); fracture process is responsible for that different behavior. In particular, at low strain-rate regime the fracture process begin from existing micro-cracks and material discontinuities, and chooses to develop along the path of least resistance. At high strain rate a large amount of energy in a very short time induces existing cracks to develop along the shortest paths to higher resistance (Cadoni, 2013).

Dynamic behavior of rocks is influenced not only by intrinsic rocks properties but also by numerous external factors such as confining pressure, temperature and ground water (Zhang & Zhao, 2013). As an example, Figure 1 shows the influence of the mentioned effects involved in a dynamic problem for an underground cavern. Moreover the figure shows that the wave propagation is influenced first by the microstructure and then by the macro properties (jointed rock mass), there are also problems related to the interface between two layers of rock.

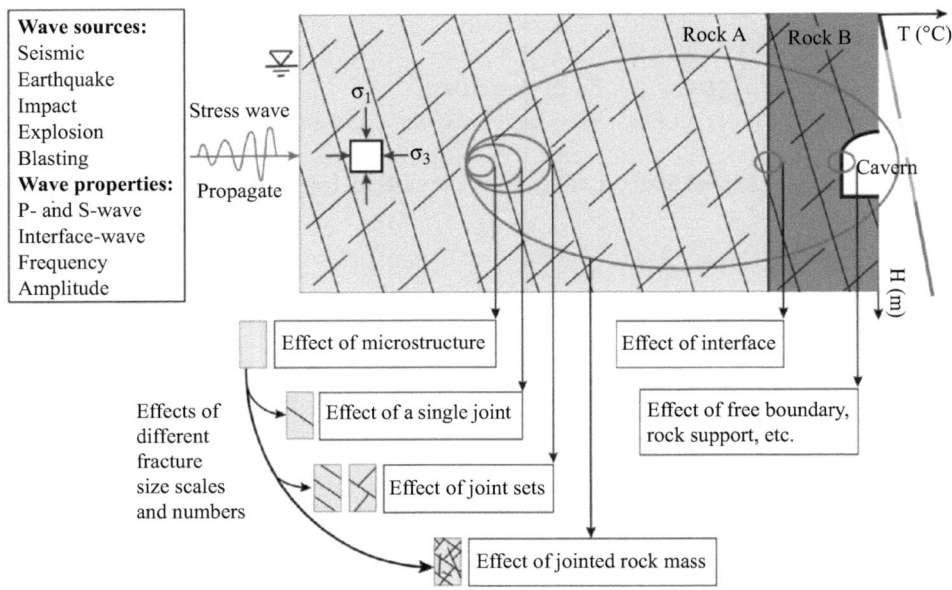

Figure 1 Main issues about dynamic behavior of rocks in underground engineering design (Zhao *et al.*, 1999).

In the last decades, many theoretical and experimental studies have been conducted to characterize the dynamic behavior of the rocks and, at the same time, to developed adequate setup system to test the resistance of different types of rocks in different load conditions. Indeed, as mentioned above, rocks are inhomogeneous materials, and for this reason the dynamic behavior assessment usually requires very extensive experimentations that include different types of rocks. The high variability of rock types (and consequently the need for several dynamic characterization experimental studies) is related to the fact that rocks are not industrial products but they have a chemical composition that can vary depending on geographical location. This has forced the research community in this field to conduct research activities focused on experimentation and numerical modeling of local rocks. Some examples of dynamic studies on local rocks are provided as following. Zhao *et al.* (1999) conducted dynamic uniaxial compression tests on Bukit Timah granite in Singapore. Cho *et al.* (2003) studied strain-rate effects on Inada granite and Tage tuff. Cai *et al.* (2007) conducted a study on the dynamic behavior of the Meuse/ Haute-Marne argillite. Kubota *et al.* (2008) presented experimental tests on Kimachi sandstone for measuring dynamic tensile strength. Xia *et al.* (2008) investigate the effect of microcrack-induced anisotropy on the dynamic response of Barre granite rock. Cadoni (2010) used for testing an Orthogneiss rock from the Onsernone Valley (Swiss Alps) of the Cantone Ticino (Switzerland). Zhao *et al.* (2014) studied the mechanisms of strain rate dependency of dynamic tensile strength in Gosford sandstone.

The general target of the experimental campaigns conducted on rocks under dynamic regime is to evaluate the variation of mechanical parameters as a function of strain-rate. One of the most useful parameter investigated is the tensile strength as a function of strain-rate due to the importance of this parameter in many application such as blast resistant materials and structures. Dynamic mechanical properties can be obtained directly or indirectly through different experimental methods such as drop weight machines, split Hopkinson pressure bar (SPHB), gas gun, etc. This chapter is made of three sections: after this section of Introduction, section two describes experimental techniques used to study the dynamic behavior of rocks. Section three compares dynamic properties of rocks with static one under different stress conditions.

2 EXPERIMENTAL TECHNIQUES

Available experimental techniques allow the reproduction of dynamic phenomena that take place in real cases through particular (ad hoc) setup and machines. However, laboratory experiments are performed on small scale specimens that usually limit the loading conditions during the test. Indeed, the size of the specimen to be tested under dynamic conditions is related to the microstructure of the materials which, in turns, is linked to the characteristic length/wave propagation in the material medium. A specimen should contain at least 1000 grains or crystals to be mechanically representative of the material (in general, not only for rocks) (Armstrong, 1961, 2001). In addition, while a coarse-grained material implies a larger specimen, for fine-grained materials, experimentation is easier because setup is smaller. This material characteristic has to be related to the rate of load application in order to utilize the proper testing technique. For example, with very fine-grained metals it is possible to deform under 1 D stress state at strain-rates close to 10^5 s^{-1} using scaled down Hopkinson bars (3 mm diameter) and 1 mm sized specimens (Gorham, 1980; Gorham et al., 1992). Figure 2 summarizes

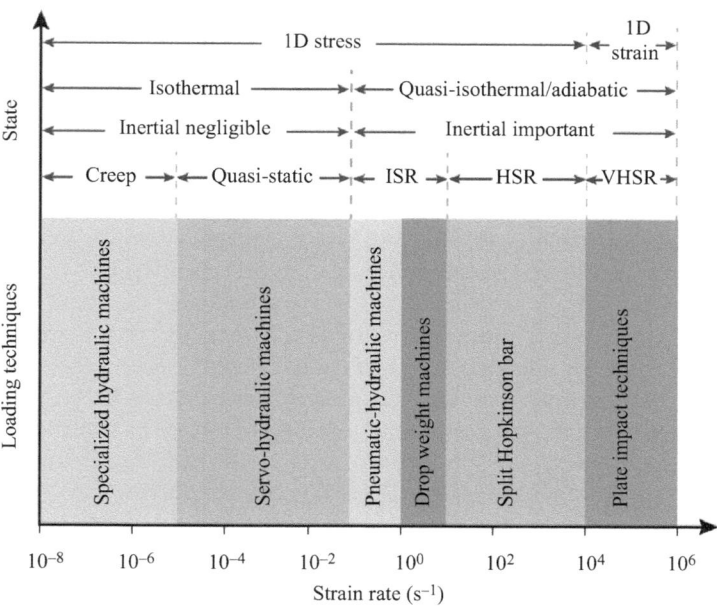

Figure 2 Strain rate range and related experimental technique (Zhang & Zhao, 2014).

strain rate ranges (in reciprocal seconds) that are associated with the experimental technique along with the loading regime.

The strain rate range of the whole spectrum is 14 orders of magnitude. Lower strain rates refer to the creep and stress relaxation regimes. The creep behavior experimental characterization requires very slow loading rates that are typically achieved with specialized hydraulic machines. From 10^{-5} to $10^{-1} s^{-1}$ of strain rate, the load application is comparable to a static process which is referred to as a quasi-static regime; servo-hydraulic machines are used to investigate the quasi-static strain range. At higher strain rates (10^{-1} to $10^{0} s^{-1}$) fast pumps and valves are employed to increase the flow rate of hydraulic oil (pneumatic-hydraulic machines). Drop weight machine is used for a strain rate range from 10^{0} to $10^{1} s^{-1}$. The strain rate range between 10^{-1} and $10^{1} s^{-1}$ is identified with "intermediate strain rate" (ISR) or "medium strain rate" (Green & Perkins, 1968; Logan & Handin, 1970) or "quasi dynamic strain rate" (Logan & Handin, 1970). For high strain rate (HSR) range ($10^{1} - 10^{4} s^{-1}$), the most widely used experimental setup is Split Hopkinson Pressure Bar (SHPB). With such a strain rate, factors as inertia and stress wave propagation can affect the experimental results. For very high strain rate (VHSR) range $10^{4} - 10^{6} s^{-1}$, plate impact techniques have been successfully employed. This strain rate regime affects the experimental results due to inertia effects and wave propagation characteristics. As shown in Figure 2, experimental techniques related to strain rates up to $10^{4} s^{-1}$ allow uniaxial stress state in the material (1 D); conversely, shock wave techniques are able to reproduce 1D strain states.

For the purposes of this chapter, the following sections focus on ISR and HSR techniques typically employed to characterize the dynamic behavior of rocks and introduce VHSR techniques. Techniques are discussed in the following order:

- Hydro-pneumatic machine
- Drop weight machine
- Split Hopkinson pressure bar
- Other VHSR techniques

2.1 Hydro-pneumatic machine

The hydro-pneumatic machine (HPM) functioning scheme is showed in Figure 3. At the beginning of the test, a sealed piston divides a cylindrical tank into two chambers, one being filled with gas at high pressure (*e.g.* 150 bars) and the other with water. At first, equal pressure is established in the water and gas chambers so that forces acting on the two piston faces are in equilibrium. The test starts when the second chamber discharges the water through a calibrated orifice, activated by a fast electro-valve. Then the piston starts moving, expelling the gas through a sealed opening; the end of the piston shaft is connected to the specimen. The specimen is linked to the piston shaft and to one end of an elastic bar, whose other end is rigidly fixed to a supporting structure. When the piston shaft moves, the specimen is pulled at a fixed strain-rate level, depending on the velocity of the gas expelled from the chamber. The elastic bar is instrumented with a strain-gauge that provides, through the elastic properties of the bar, the force acting on the specimen during the test. Two targets are attached on both ends of the specimen and their movements are measured by two contact-less displacement transducers.

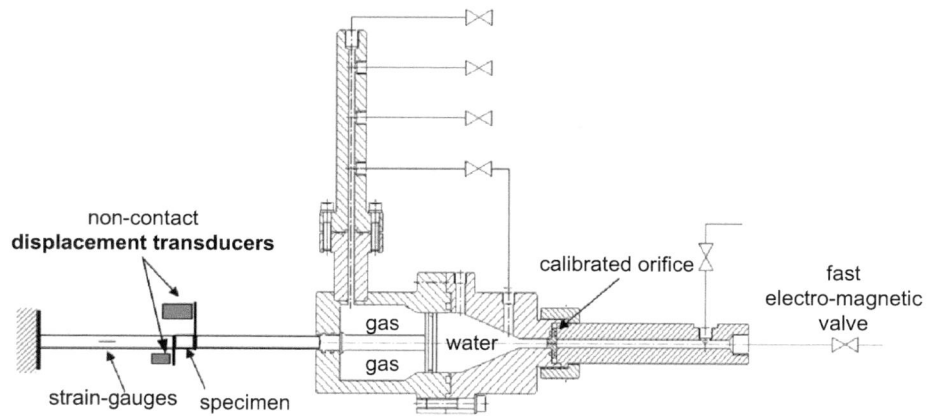

Figure 3 Hydro-pneumatic machine scheme (Asprone *et al.*, 2009).

It can be evidenced that a constant speed movement of the piston guarantees the constancy of the strain rate during the test; this depends mainly on the constancy of the force exerted by the gas pressure on the piston face. A good result in this sense was obtained with a small change of gas volume during the test in order to have small gas pressure decrease and consequently small piston force decrease. Furthermore, the load P resisted by the specimen is measured by the dynamometric elastic bar, whereas the specimen elongation ΔL (strain) is measured by the displacement transducers (strain gauge), sensing the displacement of the plates target fixed to both specimen ends. Such acquisitions allow the stress versus strain relationship to be obtained at a given strain rate level, achieved during the test. Generally hydro-pneumatic machine tests rock materials in direct tension (Asprone *et al.*, 2009; Cadoni & Albertini, 2011; Li *et al.*, 2013) uniaxial compression (Friedman *et al.*, 1970; Zhao *et al.*, 1999), and triaxial compression (Logan & Handin, 1970; Perkins *et al.*, 1970; Gran *et al.*, 1989; Li *et al.*, 1999).

2.2 Drop weight machine

Drop weight machine working principle is based on a hammer that, due to the gravitational force, impacts on specimen (Figure 4). Height and weight of specimen are known, and energy calculations depend on the momentum impulse of the falling weight and the resultant temporal impact velocity. By placing a shock- absorbing material on the specimen (such as a rubber layer), it is possible to vary the strain rate range from 10^0 to 10^1 s^{-1} (Zhang & Zhao, 2014).

Using a gas gun is possible to accelerate the drop weight and increase the strain rate up to 10^2 s^{-1}, but such deformation speed can be obtained directly from split Hopkinson pressure bar which considers also waves propagation effects. This experimental technique has simpler setup compared with the others, but many parameters cannot be controlled. During the impact, machine deformation could be affect the deformation recorded on the specimen. Moreover, overlapping between wave propagation and machine vibration requires a careful data analysis. With drop weight

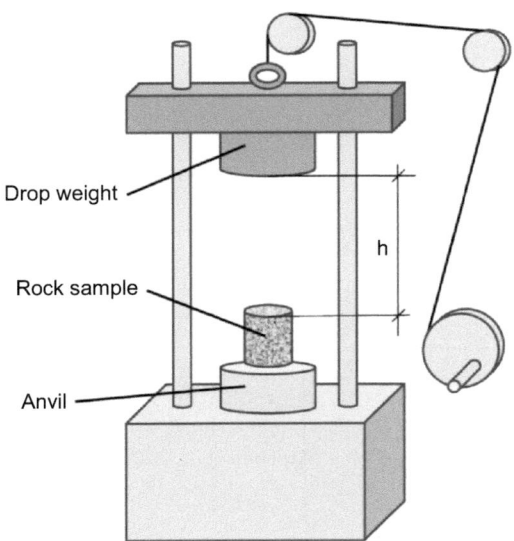

Figure 4 Drop weight machine scheme (Mwanga *et al.*, 2015).

machine multiaxial tests are unreliable because loading rate cannot be well controlled (Zhang & Zhao, 2014). Because of these problems, only few studies have been conducted on rocks combined with this technique (Whittles *et al.*, 2006; Hogan *et al.*, 2012) and fracture toughness (Yang *et al.*, 2009; Islam & Bindiganavile, 2012).

2.3 Split Hopkinson pressure bar

The traditional SHPB is used mainly in uniaxial compression dynamic tests and consist in three parts (see Figure 5): an impactor system like gas gun or a simple striker, an input bar which transmits the impact from the striker to the specimen, and an output bar, with a specimen sandwiched between the input and output bars. When the incident pulse reaches the specimen, part of it is reflected by the specimen whereas another part passes through the specimen propagating into the output bar.

Striker bar (or projectile) impacting on input bar generates compressive waves that propagates in the specimen. On both bars the incident pulse (ε_I), reflected pulse (ε_R) and

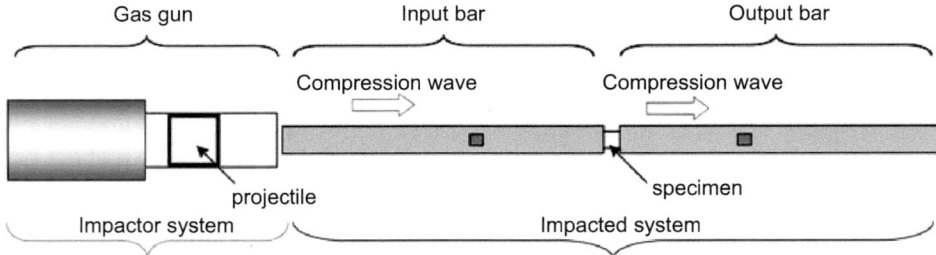

Figure 5 Traditional split Hopkinson pressure bar for compression test (Cadoni, 2010).

transmitted pulse (ε_T) are propagated. Strain-gauges glued on the input and output bars of the SHPB are used for the measurement of the elastic deformation (as a function of time). The application of the elastic uniaxial stress wave propagation theory of the Hopkinson bar system (Davies, 1948; Kolsky, 1949) allows the calculation of the forces F_1 and F_2 and the displacements δ_1 and δ_2 acting on the two faces of the specimen in contact with the input and output bars. If the specimen is short, propagation time of the wave through the specimen is small compared to the duration of the test, the specimen can be considered in equilibrium at its ends, and in homogeneous stress state. These experimental conditions allow the calculation of the average stress–strain characteristics of the specimen material at different strain rates.

2.3.1 JRC-Split Hopkinson Tensile Bar (JRC-SHTB)

Innovative modification of Hopkinson bar has been developed by the researcher of the Joint Researcher Centre of the European Commission (Albertini & Montagnani, 1974; Cadoni *et al.*, 1997). The traditional Hopkinson bar generally works only in compression and consist principally in a projectile that impacts on an input bar which transmits loads to a specimen inserted between input and output bars, as shows previously in Figure 5. The innovation of the Hopkinson bar introduced at Joint Researcher Centre of the European Commission consist in the substitution of the projectile, normally used to generate the impact loading pulse, with a statically elastic pre-stressed bar which is the physical continuation of the input bar. This modification is capable to perform impact precision tests in tension, compression and shear using the same loading device and the same measuring equipment and instrumentation. The scheme of JRC-Hopkinson bar in compression and tension respectively, is proposed in the two figures below.

In this version of the Hopkinson bar, the elastic energy is stored in the pre-stressed loading bar by statically tensioning the length of this bar comprised between a blocking ring placed at the extremity of the pre-stressed bar continuous to the input bar and a brittle intermediate piece placed at the other extremity connected to the hydraulic actuator. A rectangular stress wave pulse is generated by suddenly breaking the brittle intermediate piece and propagates through the input bar, the specimen and the output bar, provoking a state of compression stress in the specimen because the particles move from the left to the right. Records are taken by the strain-gauge stations glued on the input and output bars of the elastic deformation.

The main advantage of the JRC-SHTB with respect to the traditional one, consists mainly in the fact that the generation of the loading stress pulse is performed by means of a statically pre-stressed bar which is the physical continuation of the input bar,

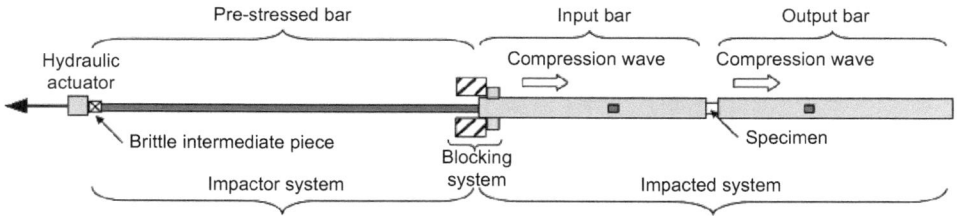

Figure 6 Modified SHPB for rocks specimen in compression (Cadoni & Albertini, 2011).

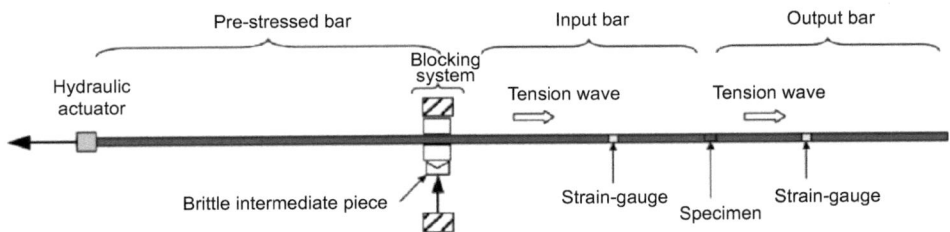

Figure 7 Modified SHPB for rocks specimen in tension (Albertini & Montagnani, 1974, 1977).

therefore avoiding the difficulties connected to the launching and impacting of projec-tiles. This characteristic has allowed the generation of very long loading pulses. A pulse of 40 ms duration obtained by 100 m (Albertini *et al.*, 1999; Cadoni *et al.*, 1997, 2001a) length of pre-stressed bar is a feature that cannot be achieved with the projectile impactor technique. Such long duration loading pulses are required for testing very ductile materials and structural components. The JRC-SHTB permits to perform dynamic tension tests at high strain rates that are very difficult to conduct for rocks.

2.3.2 Issues of SHPB

As described before, specimen is placed between input bar and output bar. Friction effect between surfaces can generate a complex multiaxial stress state, and results may not represent the effective behavior of the material. End friction effects can be mini-mized with lubrication: many authors compared test result in different conditions *i.e.* lubricated, dry and bonded using high-strength adhesive (Dai *et al.*, 2010). An example of the results is shown in Figures 8 and 9:

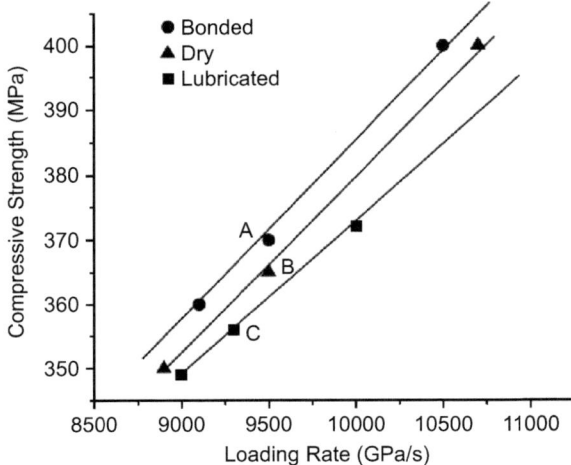

Figure 8 Dynamic compressive strengths with loading rates measured on bonded, dry and lubricated specimen (Dai *et al.*, 2010).

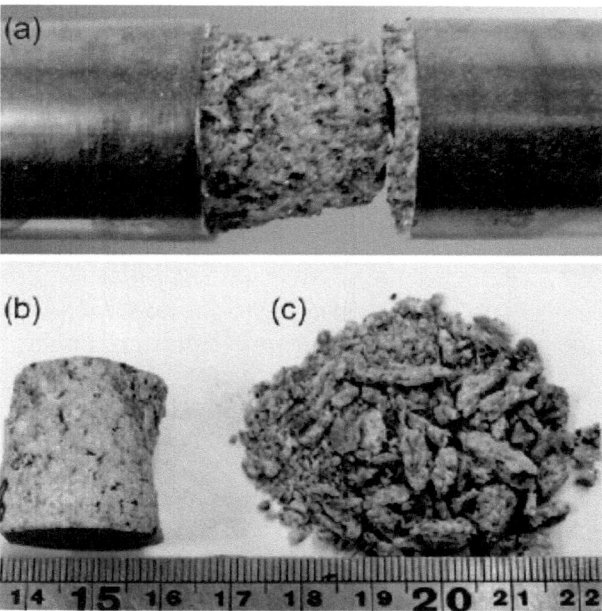

Figure 9 Photograph of recovered samples with a) bonded, b) dry, c) lubricated bar/sample interfaces showing (Dai *et al.*, 2010).

Dai *et al.* (2010) tests were performed on a constant L/D ratio (L/D=1). Thanks to these studies Gray (2000) suggested that friction and inertia effects can be reduced by minimizing the area mismatch between the specimen and the bars ($D_{specim} \approx 0.8 D_{bar}$) and choosing for the specimen the ratio L/D between 0.50–1.0. In SHTB another problem is related to the Poisson's ratio that causes inertia influence on measurement data (Davies & Hunter, 1963). The flow stress of the specimen overlaps with flow stress associated to axial inertia and radial inertia. To minimize this effect there is an optimal L/D ratio suggested by Davies & Hunter (1963) (L/D=$\sqrt{3}v/2$). A further problem of the SHPB is related to the dispersion of the waves. This problem is more relevant when bar diameter increases. Analytical, numerical and experimental corrections for wave dispersion were proposed for testing of ductile materials by Follansbee & Frantz (1983) and Gorham (1983), for concrete by Gong *et al.* (1990) and for rock by Li *et al.* (2000b).

2.4 Other techniques: Taylor impact and shockloading by plate impact

In the 1930s, G.I. Taylor and his co-worker developed a method to calculate the dynamic behavior of ductile materials in compression. Set up is made up of two parts: a flyer plate of material to investigate and a target specimen. The planar impact of flyer plate onto a target produces shock waves in both elements. In this method, a cylindrical specimen is shooting against a rigid surface. Measuring specimen deformation, through an equation, is possible to obtain the dynamic behavior of material. This method lacks precision so is rarely used for rocks. Shockloading by plate impact technique permits to investigate very high strain rate range 10^4–10^8 (VHSR). Strain rate is given by $\tau\, u_p/U_s$, where u_p represent the particle velocity, U_s is the shock velocity and τ is the rise time of shock.

3 DYNAMIC PROPERTIES OF ROCKS

3.1 Compression behavior

3.1.1 Uniaxial loading

By means of experimental tests described in the previous sections, the stress–strain curves of a given rock material under dynamic compression can be extracted directly from load transducers and on-specimen strain gauges; starting from these curves, the dynamic mechanical parameters required for a proper characterization of dynamic behavior of rocks can be determined. There are four main theoretical methods available to evaluate dynamic parameters from experiments: one-wave analysis, three-wave analysis, direct estimate and foot-shifting methods), providing different levels of accuracy of the results. When dealing with rock materials the one-wave analysis method has been commonly used for determining uniaxial stress strain behavior; in particular, the method is based on the one-dimensional stress theory and allows the determination of the histories of strain rate $\dot{\varepsilon}(t)$, strain $\varepsilon(t)$, and stress $\sigma(t)$ within the samples.

As general feature of the uniaxial dynamic, it has been widely shown that Young's modulus of rock-like materials increases with an increase in strain rate, even though some studies revealed that initial tangent modulus can be unaffected by strain or dependent on measurements; indeed, its determination relies on the stress equilibrium condition which, actually, cannot be satisfied in the initial range of small strain (Gray, 2000). The effects of strain rate on the stress–strain curve of rock materials in uniaxial compression also regard possible increase or decrease in critical strain with increasing strain rate, corresponding to a more ductile or brittle behavior at higher strain rate, respectively.

In the study by Xia et al. (2008) barre granite block were tested by means of split Hopkinson pressure bar (SHPB). The authors investigated the effect of microcrack-induced anisotropy on the dynamic response of such a rock material under dynamic compression loads. It was shown that the maximum dynamic stress achieved was not sensitive to the sample microstructures, resulting in rate dependence for all directions in the range 1–150 s^{-1} (Figure 10).

Li et al. (2000a) applied a sliding crack model (based growth and nucleation phases) to study the mechanical properties of the Bukit Timah granite under dynamic uniaxial compressive loads at strain rates from 10^{-4} to $10^{\wedge 0}s^{-1}$. By comparing with experimental results, it was observed that the strength generally increases with increasing strain at different initial crack lengths and crack spacing.

Cai et al. (2007) conducted an experimental campaign on Meuse/Haute-Marne argillite by means of compression split Hopkinson pressure bar (SHPB). They found that the average dynamic increase factors compressive strength was about 2.4 with the specimens which deformed and fail uniformly around the circumference of the specimen, by a spalling process.

Several studies have been also conducted to investigate the hydrostatic-stress influence on the compressive dynamic behavior of rock materials. For instance, Li & Meng

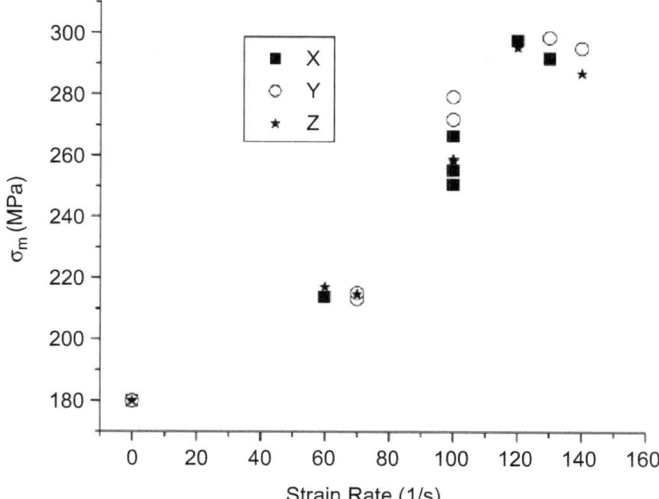

Figure 10 Strain rate effects for X, Y and Z samples (Xia *et al.*, 2008).

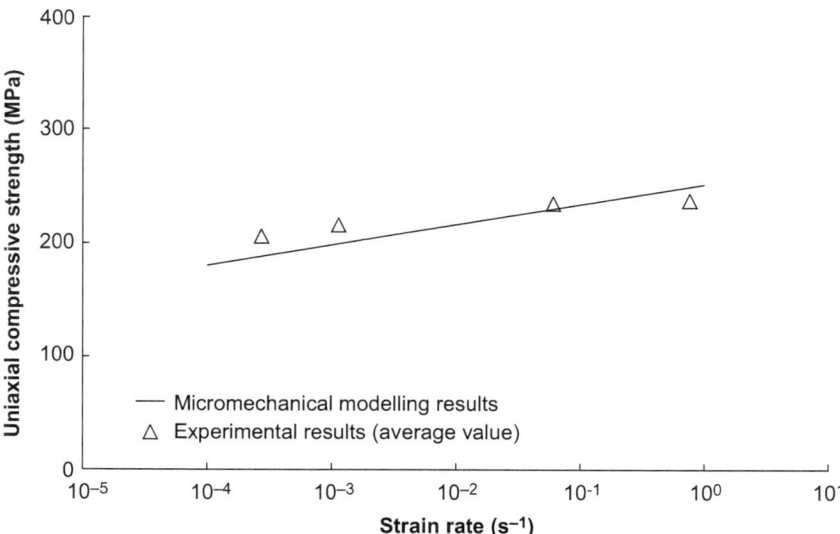

Figure 11 Relation of uniaxial compressive strength with strain rate of Bukit Timah granite, micromechanical modelling and experimental results (Li *et al.*, 2000a).

(2003) utilized SHPB to determine the dynamic strength of concrete-like materials along with the hydrostatic-stress-dependency of compressive strength. It was shown and demonstrated with numerical analyses that the apparent dynamic strength enhancement beyond the strain-rate of 10^2 s^{-1} was strongly influenced by the hydrostatic stress effect due to the lateral inertia confinement in a SHPB test (Figure 12).

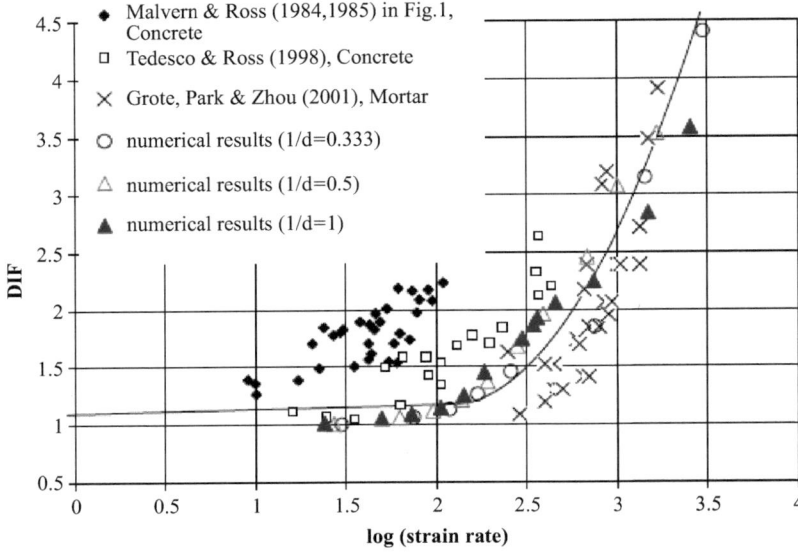

Figure 12 Comparison between the predicted DIF and the measured DIF from SHPB tests (Li & Meng, 2003).

3.1.2 Triaxial loading

The fully understanding and the development of proper constitutive models for rock materials require experimental data under various stress and/ or strain paths, including confinement conditions and impact loadings. Indeed, in common applications/physical conditions, the mechanical loads are applied to rock materials under loading conditions which are not uniaxial but mostly in terms of high pressures and high strain rates related to impacts. Generally, rock-like materials are very sensitive to the confining pressure both under quasi-static loads and dynamic ones. As general trend, the dynamic triaxial compressive strength increases with increasing strain rate at constant confining pressure. However, the complexities related to testing equipment and measurement often result in different results depending on the rock material tested. Moreover, it has been shown that the multiaxial loading condition also affects the criterion to be applied for the dynamic triaxial strength determination. In particular, the dynamic triaxial strength can be represented by the Hoek– Brown and Mohr–Coulomb criterion at both low-high and low confining pressures, respectively (Bailly *et al.*, 2011, Hao & Hao, 2013).

3.2 Tension test

Relationships between tensile stress and strain curves at different strain rates can be obtained by means of direct and indirect experimental tension tests (Asprone *et al.*, 2009) and using the equipment described in Section 2.3.1. As general features of the tensile dynamic behavior of rock materials, it has been observed that the tensile strength increases with increasing strain rate, whereas the corresponding strain to failure decreases. This behavior indicates that rock materials typically exhibit a more brittle feature in higher strain rate tests. However, non-linear relationships between strength and strain rates are typically achieved for rocks and, consequently, suitable tests and formulations are necessary to describe the DIF evolution in relation to the strain rate magnitude.

Recently, Li *et al.* (2013) carried out dynamic tensile tests, under and without side confinement, on Gypsum and Granite rocks by means of an air and oil hydraulically driven dynamic loading machine at strain rates ranging from 10^{-5} to 10^{-2} s^{-1}. It was found that the direct tensile strength of Gypsum samples without side confinement increased by about 80% with increasing strain rate from 10^{-5} to 10^{-2} s^{-1}, similar to the increase obtained from the indirect tensile tests. However, in the presence of side confinement, the increase rate of the tensile strength with the strain rate reduced with the augmentation of the side confinement. Granite rock samples without confinement exhibited an increment of the tensile strength of about 60% when the strain rate varies from 10^{-5} to 10^{-2} s^{-1} (also in this case, similar to indirect tensile test results). In contrast to Gypsum samples, the tensile strength of the Granite rock samples Granite under side confinement showed significant increases with the increase of the strain rate. Kubota, *et al.* (2008) investigated the tensile dynamic behavior of Kimachi sandstone using underwater shock waves. The experimental machine was made of an emulsion explosive used as the source of dynamic loading, and a pipe filled with water arranged between the explosive and a cylindrical specimen (Figure 13). The study indicated that the dynamic tensile strength of Kimachi sandstone varied as the 1/3 power within strain rates of 10–40 s^{-1}.

The influence of the loading direction with respect to the schistosity of rock materials was investigated by Cadoni (2010) and Cadoni *et al.* (2011). In details, dynamic characterization tests were performed on Onsernone Orthogneiss for loading directions 0°, 45° and 90° (with respect to the schistosity) at three different strain rate levels: 0.1, 10, 100 s^{-1}. Two special apparatus were utilized, *i.e.* the split Hopkinson tension bar and the hydro-pneumatic machine which entailed the determination of the entire stress–strain curves of such a brittle material, including the softening branch. The experimental results showed a non-linear relationship between tensile strength and strain rates which was influenced by the loading direction referred to the schistosity of the rock (Figure 14).

Figure 13 Main part of the setup for fracture test (Kubota *et al.*, 2008).

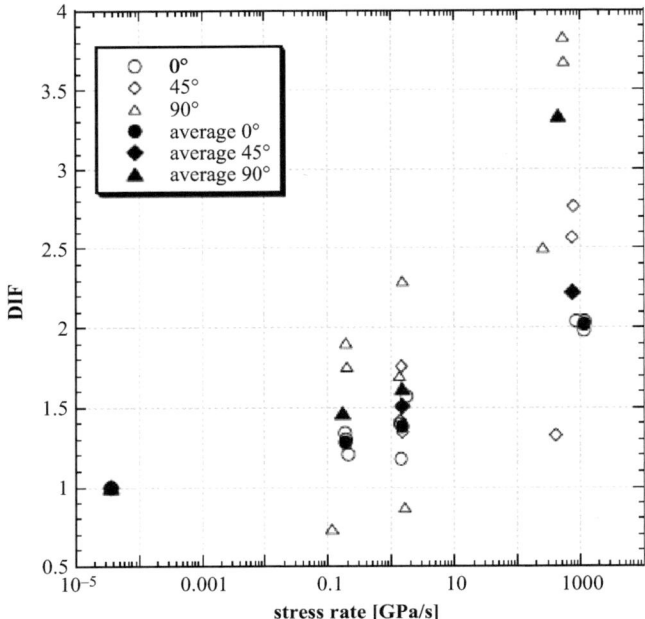

Figure 14 Dif versus stress rate for the Orthogneiss at θ= 0°, 45° and 90° (Cadoni, 2010).

Besides direct tensile tests, different indirect methods have been adopted to charac-terize the tensile dynamic behavior of rock materials. Zhao & Li (2000) investigated the dynamic behavior of Bukit Timah granite rocks up to 10^4 MPa/s by means of both Brazil and 3-point flexural methods. The results highlighted that the tensile strength of the granite increased with increasing loading rate as well as the slope of the tensile strength vs loading rate curve. However, the two different methods provided slightly different results in terms of magnitude of strain rate effect on tensile strength and Young's modulus (Figure 15 and Figure 16).

In the study by Cai *et al.* (2007), indirect tensile tests were carried out on Meuse/ Haute-Marne argillite material by means of SHPB. A high-speed video camera was used to visualize the initiation of failure and subsequent deformation of the specimens. They observed that failure occurred in tension along the line of load application while radial fractures were also revealed. Dynamic splitting experiments were conducted by Gomez *et al.* (2001) using a Split Hopkinson Pressure Bar in order to characterize the tensile dynamic behavior of Granite and concrete material in relation to the effects of induced levels. They found that, in the static loading conditions, the splitting strength was highly dependent on the damage orientation with respect to the loading line. On the contrary, the dynamic splitting strengths were not affected by the random crack orientation and decreased with increasing damage for both concrete and granite (Figure 17 and Figure 18).

Wang *et al.* (2009) utilized a different experimental configuration to perform dynamic split tensile tests on brittle marble material. Flattened Brazilian Disc (FBD) specimens were employed in their experimental campaign and impacted diametrically

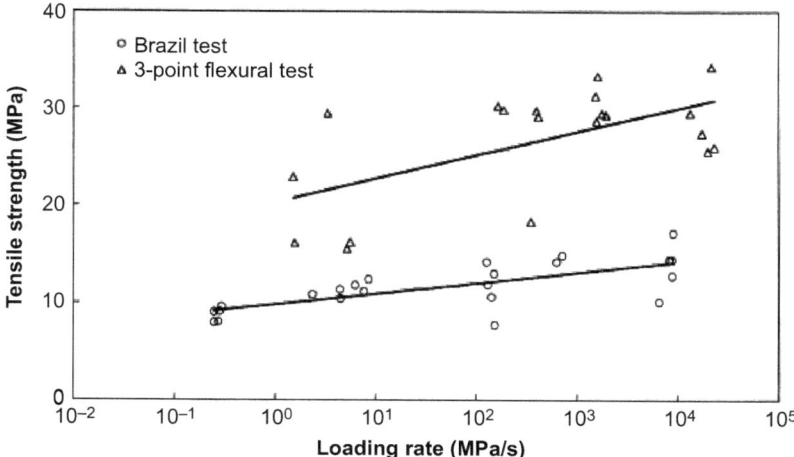

Figure 15 The Brazil and the 3-point flexural tensile strength at the different loading rates (Zhao & Li, 2000).

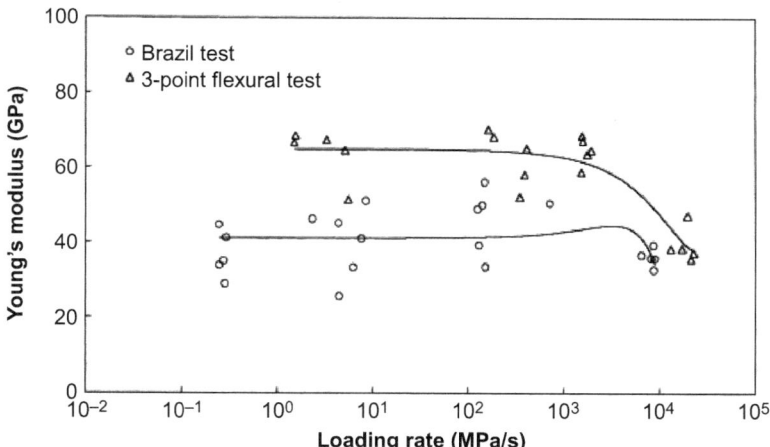

Figure 16 The Brazil and the 3-point flexural tensile Young's moduli at the different loading rates (Zhao & Li, 2000).

by a pulse shaping split Hopkinson pressure bar to measure dynamic tensile strength of the brittle rock. By using this technique, they found that the dynamic tensile strength of marble at the strain rate of about $22 \, s^{-1}$ to $25 \, s^{-1}$ increased from 5 MPa to approximately 22–27 MPa, leading to a DIF of about 5.

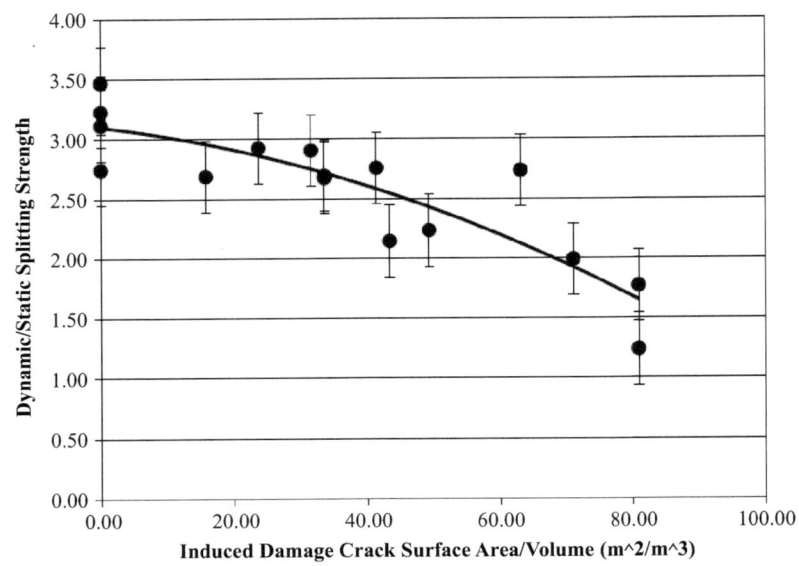

Figure 17 Dynamic splitting strength of G-mix concrete as a function of induced damage (Gomez *et al.*, 2001).

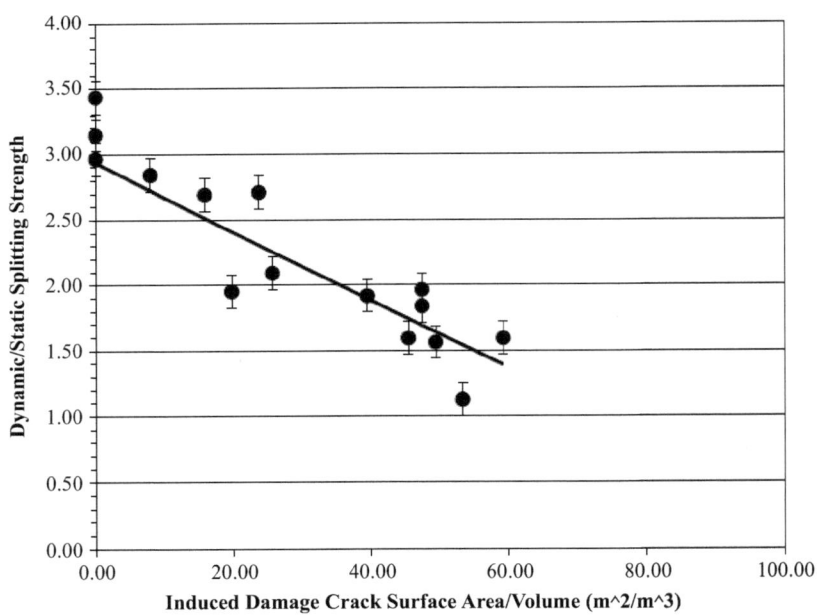

Figure 18 Dynamic splitting strength of Barre granite as a function of induced damage (Gomez *et al.*, 2001).

REFERENCES

Albertini, C., Cadoni, E. & Labibes, K., 1999. Study of the mechanical properties of plain concrete under dynamic loading. *Experimental Mechanics*, 39(2), pp. 137–141.

Albertini, C. & Montagnani, M., 1974. Testing techniques based on the split Hopkinson bar. In: J. Harding (ed.), *Mechanical Properties at High Rates of Strain*. London: Institute of Physics, pp. 22–32.

Albertini, C. & Montagnani, M., 1977. Dynamic material properties of several steels for fast breeder reactor safety analysis. *Commission of the European Communities*. Brussels/Luxemburg: Joint Research Centre Ispra Establishment – Italy Applied Mechanics Division.

Armstrong, R., 1961. On size effect in polycrystal plasticity. *Journal of the Mechanics and Physics of Solids*, 9(3), pp. 196–199.

Armstrong, R., 2001. Plasticity: Grain size effects. In: K. Buschow, R. Cahn & M. Flemings (eds.), a cura di *Encyclopedia of Materials: Science and Technology*. Amsterdam: Elsevier.

Asprone, D., Cadoni, E., Prota, A. & Manfredi, G., 2009. Dynamic behavior of a Mediterranean natural stone under tensile loading. *International Journal of Rock Mechanics and Mining Sciences*, 46(3), pp. 514–520.

Bailly, P. *et al.*, 2011. Dynamic behavior of an aggregate material at simultaneous high pressure and strain rate: SHPB triaxial tests. *International Journal of Impact Engineering*, 38(2–3), pp. 73–84.

Cadoni, E., 2010. Dynamic characterization of orthogneiss rock subjected to intermediate and high strain rates in tension. *Rock Mechanics and Rock Engineering*, 43(6), pp. 667–676.

Cadoni, E., 2013. Mechanical characterization of rock materials at high strain-rate. In: Zhao & Li (eds.), *Rock Dynamics and Applications – State of the Art*. Leiden: CRC Press/Balkema, pp. 137–148.

Cadoni, E. & Albertini, C., 2011. Modified Hopkinson bar technologies applied to the high strain rate rock tests. *Advances in Rock Dynamics and Applications*. Leiden: CRC Press/Balkema, pp. 79–104.

Cadoni, E., Albertini, C., Labibes, K. & Solomos, G., 2001a. Behavior of plain concrete subjected to tensile loading at high strain-rate. In de Borst *et al.* (eds), *Proceeding of the Fourth International Conference on Fracture Mechanics of Concrete and Concrete Structures*. Lisse: Swets & Zeitlinger, pp. 341–348.

Cadoni, E., Antonietti, S., Dotta, M. & Forni, D., 2011. Strain rate behaviour of three rocks in tension. *Engineering Transactions*, 59(3), pp. 197–210.

Cadoni, E. *et al.*, 2001b. Strain-rate effect on the tensile behaviour of concrete at different relative humidity levels. *Materials and Structures*, 34(1), pp. 21–26.

Cadoni, E., Labibes, K., Solomos, G. & Albertini, C., 1997. Mechanical response in tension of plain concrete in a large range of strain-rates. *Technical Note*.

Cai, M., Kaiser, P. K., Suorineni, F. & Su, K., 2007. A study on the dynamic behavior of the Meuse/Haute-Marne argillite. *Physics and Chemistry of the Earth*, 32(8–14), pp. 907–916.

Cho, S. H., Ogata, Y. & Kaneko, K., 2003. Strain-rate dependency of the dynamic tensile strength of rock. *International Journal of Rock Mechanics and Mining Sciences*, 40(5), pp. 763–777.

Dai, F., Huang, S., Xia, K. & Tan, Z., 2010. Some fundamental issues in dynamic compression and tension tests of rocks using split Hopkinson pressure bar. *Rock Mechanics and Rock Engineering*, 43(6), pp. 657–666.

Davies, E. D. H. & Hunter, S. C., 1963. The dynamic compression testing of solids by the method of the split Hopkinson pressure bar. *Journal of the Mechanics and Physics of Solids*, 11(3), pp. 155–179.

Davies, R. M., 1948. A critical study of the Hopkinson pressure bar. *Philosophical Transactions of the Royal Society of London A: Mathematical, Physical and Engineering Sciences*, 240(821), pp. 375–457.

Delvare, F., Hanus, J. L. & Bailly, P., 2010. A non-equilibrium approach to processing Hopkinson Bar bending test data: Application to quasi-brittle materials. *International Journal of Impact Engineering*, 37(12), pp. 1170–1179.

Follansbee, P. S. & Frantz, C., 1983. Wave propagation in the split Hopkinson pressure bar. *Journal of Engineering Materials and Technology*, 105(1), pp. 61–66.

Friedman, M., Perkins, R. D. & Green, S. J., 1970. Observation of brittle-deformation features at the maximum stress of Westerly granite and Solenhofen limestone. *International Journal of Rock Mechanics and Mining Sciences & Geomechanics Abstracts*, 7(3), pp. 297–302.

Gomez, J. T., Shukla, A. & Sharma, A., 2001. Static and dynamic behavior of concrete and granite in tension with damage. *Theoretical and Applied Fracture Mechanics*, 36(1), pp. 37–49.

Gong, J. C., Malvern, L. E. & Jenkins, D. A., 1990. Dispersion investigation in the split Hopkinson pressure bar. *Journal of Engineering Materials and Technology*, 112(3), pp. 309–314.

Gorham, D., 1980. Measurement of stress-strain properties of strong metals at very high strain rates. *Journal of Physics: Conference Series*, 47, pp. 16–24.

Gorham, D. A., 1983. A numerical method for the correction of dispersion in pressure bar signals. *Journal of Physics E: Scientific Instruments*, 16(6), p. 477.

Gorham, D. A., Pope, P. H. & Field, J. E., 1992. An improved method for compressive stress-strain measurements at very high strain rates. *Proceedings of the Royal Society of London A: Mathematical, Physical and Engineering Sciences*, pp. 153–170.

Gran, J. K., Florence, A. L. & Colton, J. D., 1989. Dynamic triaxial tests of high strength concrete. *Journal of Engineering Mechanics*, 115(5), pp. 891–904.

Gray, G. I., 2000. Classic split Hopkinson pressure bar testing. ASM Handbook Volume 8, Mechanical Testing and Evaluation (ASM International), Materials Park: IMS, pp. 462–476.

Green, S. J. & Perkins, R. D., 1968. In: Ray, K.E. (ed.), *Uniaxial Compression Tests at Varying Strain Rates on Three Geologic Materials*. The 10th U.S. Symposium on Rock Mechanics (USRMS). New York: AIME, pp. 35–54.

Hao, Y. & Hao, H., 2013. Numerical investigation of the dynamic compressive behaviour of rock materials at high strain rate. *Rock Mechanics and Rock Engineering*, 46(2), pp. 373–388.

Hogan, J. D., Rogers, R. J., Spray, J. G. & Boonsue, S., 2012. Dynamic fragmentation of granite for impact energies of 6–28J. *Engineering Fracture Mechanics*, 79, pp. 103–125.

Islam, M. T. & Bindiganavile, V., 2012. Stress rate sensitivity of Paskapoo sandstone under flexure. *Canadian Journal of Civil Engineering*, 39(11), pp. 1184–1192.

Kolsky, H., 1949. An investigation of the mechanical properties of materials at very high rates of loading. *Proceedings of the Physical Society. Section B*, 62(11), pp. 676–700.

Kubota, S. *et al.*, 2008. Estimation of dynamic tensile strength of sandstone. *International Journal of Rock Mechanics & Mining Sciences*, 45(3), pp. 397–406.

Lajtai, E. Z., Duncan, E. S. & Carter, B. J., 1991. The effect of strain rate on rock strength. *Rock Mechanics and Rock Engineering*, 24(2), pp. 99–109.

Li, H. B., Zhao, J. & Li, T. J., 1999. Triaxial compression tests on a granite at different strain rates and confining pressures. *International Journal of Rock Mechanics and Mining Sciences*, 36(8), pp. 1057–1063.

Li, H. B., Zhao, J. & Li, T. J., 2000a. Micromechanical modelling of the mechanical properties of a granite under dynamic uniaxial compressive loads. *International Journal of Rock Mechanics and Mining Sciences*, 37(6), pp. 923–935.

Li, H. *et al.*, 2013. Direct tension test for rock material under different strain rates at quasi-static loads. *Rock Mechanics and Rock Engineering*, 46(5), pp. 1247–1254.

Li, Q. M. & Meng, H., 2003. About the dynamic strength enhancement of concrete-like materials in a split Hopkinson pressure bar test. *International Journal of Solids and Structures*, 40(2), pp. 343–360.

Li, X. B., Lok, T. S., Zhao, J. & Zhao, P. J., 2000b. Oscillation elimination in the Hopkinson bar apparatus and resultant complete dynamic stress–strain curves for rocks. *International Journal of Rock Mechanics and Mining Sciences*, 37(7), pp. 1055–1060.

Logan, J. M. & Handin, J., 1970. In: Clark GB (ed.), *Triaxial Compression Testing at Intermediate Strain Rates*. Dynamic Rock Mechanics. Proceeding of the 12th Symposium on Rock Mechanics. New York: A.I.M.E.

Mwanga, A., Rosenkranz, J. & Lamberg, P., 2015. Testing of ore comminution behavior in the geometallurgical context – A review. *Minerals*, 5, pp. 276–297.

Perkins, R. D., Green, S. J. & Friedman, M., 1970. Uniaxial stress behavior of porphyritic tonalite at strain rates to 103/Second. *International Journal of Rock Mechanics and Mining Sciences & Geomechanics Abstracts*, 7(5), pp. 527–535.

Serdengecti, S. & Boozer, G. D., 1961. The effects of strain rate and temperature on the behavior of rocks subjected to triaxial compression. *Proceedings of the Fourth Symposium on Rock Mechanics*, Bulleton No. 76. University Park: Mineral Industries Experiment Station, pp. 83–97.

Wang, Q. Z., Li, W. & Xie, H. P., 2009. Dynamic split tensile test of flattened Brazilian disc of rock with SHPB setup. *Mechanics of Materials*, 41(3), pp. 252–260.

Whittles, D. N., Kingman, S., Lowndes, I. & Jackson, K., 2006. Laboratory and numerical investigation into the characteristics of rock fragmentation. *Minerals Engineering*, 19(14), pp. 1418–1429.

Xia, K. *et al.*, 2008. Effects of microstructures on dynamic compression of Barre granite. *International Journal of Rock Mechanics and Mining Sciences*, 45(6), pp. 879–887.

Yang, R. S. *et al.*, 2009. Dynamic fracture behavior of rock under impact load using the caustics method.. *Mining Science and Technology (China)*, 19, pp. 79–83.

Zhang, Q. B. & Zhao, J., 2013. A review of dynamic experimental techniques and mechanical behaviour of rock materials. *Rock Mechanics and Rock Engineering*, 47(4), pp. 1411–1478.

Zhang, Q. B. & Zhao, J., 2014. Quasi-static and dynamic fracture behaviour of rock materials: phenomena and mechanisms. *International Journal of Fracture*, 189(1), pp. 1–32.

Zhao, G.-F., Russell, A. R., Zhao, X. & Khalili, N., 2014. Strain rate dependency of uniaxial tensile strength in Gosford sandstone by the Distinct Lattice Spring Model with X-ray micro CT. *International Journal of Solids and Structures*, 51(7–8), pp. 1587–1600.

Zhao, J. & Li, H. B., 2000. Experimental determination of dynamic tensile properties of a granite. *International Journal of Rock Mechanics and Mining Sciences*, 37(5), pp. 861–866.

Zhao, J., Li, H., Wu, M. & Li, T., 1999. Dynamic uniaxial compression tests on a granite. *International Journal of Rock Mechanics and Mining Sciences*, 36(2), pp. 273–277.

Zhao, J. & Li, J., 2013. *Rock Dynamics and Applications – State of the Art*. Leiden: CRC Press/Balkema.

Zhao, J. *et al.*, 1999. Rock dynamics research related to cavern development for ammunition storage. *Tunnelling and Underground Space Technology*, 14(4), pp. 513–526.

Chapter 10

Dynamic rock failure and its containment

T.R. Stacey

School of Mining Engineering, University of the Witwatersrand, Johannesburg, South Africa

Abstract: Rockbursts are dynamic failure events that continue to be a scourge in the mining industry, and, more recently, in the tunneling industry. They can be the cause of, or result of, seismic events, and are usually associated with high stress conditions. These dynamic events occur essentially in two categories: those in which the rockburst damage location and the location of seismicity are coincident; and those in which the seismic source and the resulting damage may be separated by a substantial distance. In either of these categories, rock can be ejected from excavation walls, often at high velocity. Conventionally, the solution to such problems lies in appropriate engineering design. The logical approach is: firstly, design of layouts and geometries of excavations, with the aim of minimizing the occurrence of seismicity and rockbursts; secondly, implementation of destressing or preconditioning, with the aim of minimizing the occurrence of rockbursts; finally, if rockbursts cannot be prevented, the damage they cause must be contained by appropriately designed rock support. Conventional design of support requires knowledge of the demand on the support system imposed in a rockburst event and the capacity of the support system. Regrettably, demand and capacity parameters are usually not known with confidence. Conventional design of rock support for these conditions is therefore problematic. Physical testing of rock support for over 40 years, and observed behavior underground, have shown that available rock support components can absorb large amounts of energy. This represents an alternative approach to design of support – empirical design, based on test results and observations of performance of support in rockburst events.

1 INTRODUCTION

In many countries in the world excavation and construction in rock is venturing to greater depths. Mining is taking place in the region of 4km below surface, and being planned to 5km, and civil tunnels are being driven at depths exceeding 2.5km. When high horizontal stress fields are taken into account, such as have been commonly experienced in mining in several countries, depth is not the only factor of influence. In deep and high stress environments, dynamic failure of the rock, in the form of rockbursts, is being experienced more frequently. Rockbursting has, for many years, been a common occurrence in the mines of South Africa (Ortlepp, 1997), Chile (Rojas *et al.*, 2000; Araneda & Sougarret, 2007), Canada (Simser *et al.*, 2002) and Western Australia (Potvin, 2009), and in tunnels in Norway (Broch & Sørheim, 1984). It has

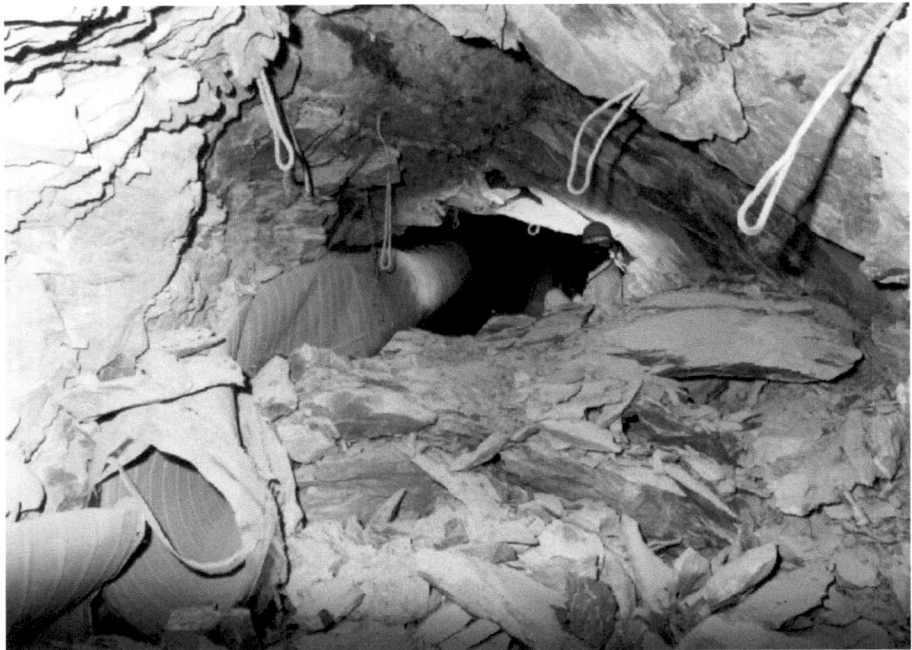

Figure 1 Rockburst damage and fragmentation (Stacey & Rojas, 2013).

been reported in excavations in other countries – USA, China (Gong *et al.*, 2012; Zhang *et al.*, 2012a), Sweden, Switzerland (Kaiser & Kim, 2008) and Russia. It could be concluded that rockbursting is now a universal problem.

Rockbursts are very violent events that commonly result in considerable damage to excavations (Ortlepp, 1997). Rock is usually ejected, and when this is the case, the ejected rock is commonly observed to be fragmented into relatively small blocks and slabs, as illustrated in Figure 1. Rockbolts, and surface support elements such as wire mesh and shotcrete, often fail. In such events, conventional rockbolts and cables often exhibit brittle failures. Another common observation is that, when the surface support fails, the ejection of rock often leaves the reinforcement elements exposed, protruding out of the rock mass as shown in Figures 1 and 2.

Gravity does not play a significant role in rockburst events, and ejection can be in any direction. Sidewall ejections (Figures 1 and 2) and floor-heave (Figure 3) are common. It can be seen from these illustrations that rockbursts are very violent events, with unpredictable damage.

2 TYPES OF ROCKBURSTS

A rockburst may be understood to be a seismic event which causes violent and significant damage to tunnels and other excavations in the mine. There are no constraints on the magnitude of the seismic event. Thus, the event can range from a

Figure 2 Exposed, protruding cables after a rockburst (Stacey & Rojas, 2013).

Figure 3 Floor heave of nearly 2m caused by a rockburst (Stacey & Rojas, 2013).

strainburst, in which superficial surface spalling with violent ejection of fragments occurs, to a mining-induced "earthquake" involving slip along a fault plane.

Dynamic failures of rock in excavations, in the form of rockbursts, have been studied for many years by many researchers. Ortlepp (1997) studied rockbursts over many years, and summarized descriptions and interpretations of numerous events. Ortlepp (1992a) identified five rockburst source mechanisms. However, for the purposes of this Chapter, only two categories need to be considered: those in which the rockburst damage location and location of seismicity are coincident, commonly known as strainbursts; and those in which the seismic source and resulting rockburst damage may be separated by a substantial distance, known as rupture type rockbursts. These two categories, and the conditions that favor their occurrence, are dealt with below.

2.1 Strainbursts

It is considered that three parameters must be present simultaneously for strainbursts to occur:

-   The rock material in which the excavation is created must have the *capacity* to store strain energy, and to release it violently on failure of the rock.
-   The rock mass surrounding the excavation must have the *capability* of storing sufficient strain energy to result in a rockburst. This capability involves the rock mass characteristics: its competence, massiveness or degree of jointing, stiffness and brittleness, as well as the excavation geometry, including the extent of exposed surface area. The rock mass competence in the immediate vicinity (walls) of the opening will be influenced by excavation method.
-   The *conditions* must be conducive to the occurrence of dynamic failure. For example, the boundary conditions in the form of in situ and induced stresses must be sufficiently high for dynamic failure to occur. This does not imply that high stress levels are necessary condition, since rockbursts have been experienced in very low to moderate stress level environments (Stacey *et al.*, 2007).

The consequence of the dynamic failure is the rockburst, and this dynamic failure of the rock is the source of the coincident seismicity. These types of rockbursts are commonly called strainbursts.

The capacity of a rock or rock mass to store strain energy depends on its strength, its stiffness and its brittleness. Since the strain energy is effectively given by the product of the stress and the strain, the lower the strength, the less the energy that the rock can store. Doubling the strength quadruples the energy available, and this corresponds with observations in a tunnel, in which strainbursting activity increased with increasing rock strength (Broch & Sørheim, 1984). The complete stress-strain behavior of the rock is useful in indicating the strain energy capacity of the rock (ISRM, 1999). Rocks with a negative post-peak slope (Class I) require the addition of energy to cause failure, whereas Class II rocks demonstrate uncontrolled failure beyond their peak strength.

The brittleness of the rock dictates its ability to release the stored strain energy violently. A good measure of this is the brittleness index k defined by Tarasov & Potvin (2013), using the pre- and post-peak characteristics of the rock determined in a servo-controlled testing machine:

$$\text{Brittleness index } k = (M - E)/M$$

where   M is the post-peak modulus in the servo-controlled test
E is the unloading elastic modulus

The post-peak modulus of a Class II rock is positive. As the value of M approaches that of E, the value of the brittleness index becomes very small. When M = E, the value of the index is zero, representing absolute brittleness. The more brittle the rock, the more likely it is that it will exhibit strainbursting when it is stressed.

For strainbursting behavior to be likely, the rock mass must also have the capability of storing energy. The capability depends on several factors:

- The rock mass quality – a greater number of planes of weakness (joints) will result in a less brittle rock mass with a lower stiffness and hence less potential for strainbursting.
- Excavation by blasting is likely to "soften" the rock in the surface of the resulting opening due to the creation of blasting fractures, therefore resulting in a lower potential for bursting. Conversely, mechanical excavation minimizes rock damage, and therefore the boundary of the opening will have greater bursting potential.
- The smaller the size of the excavation, the less the potential for storing a large amount of energy, and hence the less the risk of an event. Big openings on the other hand, with large exposed surfaces, present greater potential for energy storage and violent release, with greater volumes of failure due to the excavation size. Bardet (1990), Dyskin et al. (1993) and McGarr (1997) have considered buckling as a rockbursting mechanism. The critical buckling loads for plates and beams depend directly on the rock stiffness (modulus), directly on the cube of the thickness, and indirectly on the square of the span. Larger exposed areas, with greater "beam" or "plate" lengths, will therefore facilitate buckling failure. Therefore, size of excavation does matter.

Strainbursting cannot occur unless the levels of in situ and induced stresses are sufficient to cause such bursting. The orientations of the principal stresses and their relative magnitudes will have a significant influence on strainbursting behavior. For example, in a tunnel, a significant axial stress has an inhibiting effect on the size of extension zones ahead of the tunnel face. Further, because of the orientations and relative magnitudes of the principal stresses, a shaft and tunnel at an identical point in rock mass, for example, may have completely different responses to the stress conditions.

Absolute stress levels do not need to be high for strainbursting to occur. Stress-induced failure of rock can occur when the in situ stress level is as low as 4% (but more commonly about 10%) of the rock uniaxial compressive strength (UCS) (Stacey & Yathavan, 2003). There are many examples of rock failure at "very low" stress levels (Grimstad, 1986; Stacey et al., 2007). Examples of strainburst damage that occurred at low stress levels are given in Figures 4 and 5. The first shows the result of a very violent strainburst event in a coal mine at a depth of only 22m below surface. Figure 5 shows a fracture that occurred violently in a dimension stone quarrying operation: reportedly, a drill was thrown several meters into the air.

Figure 4 Illustration of violent failure in competent rock at very low stress levels (Stacey et al., 2007).

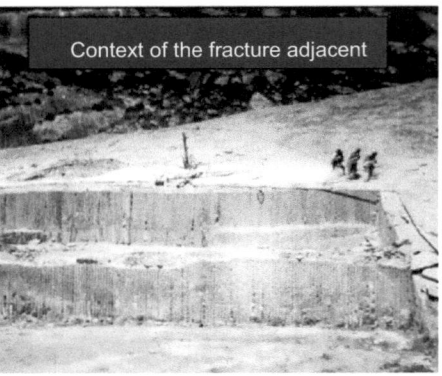

Figure 5 Illustration of violent fracture development in a quarry (Stacey et al., 2007).

Strainbursting is more likely to occur in strong, massive rocks, and therefore is more likely in machine-bored excavations than in blasted excavations, and is more likely to occur in bigger excavations than smaller.

2.2 Rupture type rockbursts

In the second category of dynamic failure, rupture type rockbursts, the in situ or induced stress conditions must be sufficiently high to cause fault/contact slip, or a new shear rupture through the intact rock mass. The fault slip or shear rupture represent the seismic source. Rockburst damage may occur at multiple locations as a result of a single seismic rupture source.

For rupture type rockbursts to occur, high in situ or induced stress levels are necessary, sufficient to re-activate movement on existing fault surfaces, or to activate movement on geological contacts, or to generate ruptures (new "faults") within the rock mass. The energy associated with rupture type rockbursts will usually be much greater than that associated with strainbursts. It is possible, but not usually likely, that a major strainburst, the first type of dynamic event described above, could be a trigger for the second type.

Seismic waves travel outwards through the rock mass from the source. It is not the passage of the waves that damages the rock, but the interaction of the waves with excavations in the rock mass. It is this interaction that may lead to rockburst damage, hence the common occurrence of damage at multiple locations. Rockburst damage resulting from rupture type events is usually far more extensive than in strainburst events. In a mining environment it is not uncommon for many tens, or even hundreds, of meters of tunnels to be damaged in a single event. Damage may be severe, and excavations may be completely closed by the damage material. The rockburst damage mechanism is violent ejection of a volume or mass of rock from the excavation surface.

3 DESIGN CONSIDERATIONS TO COMBAT STRAINBURSTS AND ROCKBURSTS

As an introduction the consideration of design approaches to combat burst occurrence and damage, it is appropriate to describe an appropriate design process.

3.1 Design process

Engineering design usually involves the development of a "solution" (the design) to a known "problem". The key input to design is required in the early stages of planning a project, when key thinking, and most important decision-making, take place. These provide the groundwork in the definition of the requirements for the design.

Satisfactory engineering design involves a design process. Bieniawski (1991, 1992) defined six design principles that encompass a design methodology: Clarity of design objectives and functional requirements; Minimum uncertainty of geological conditions; Simplicity of design components; State of the art practice; Optimization; Constructability. He described a 10 step design methodology or process corresponding with these design principles, and this has been expressed in a circular format called the wheel of design (Stacey, 2009), as shown in Figure 6. This circular format draws from

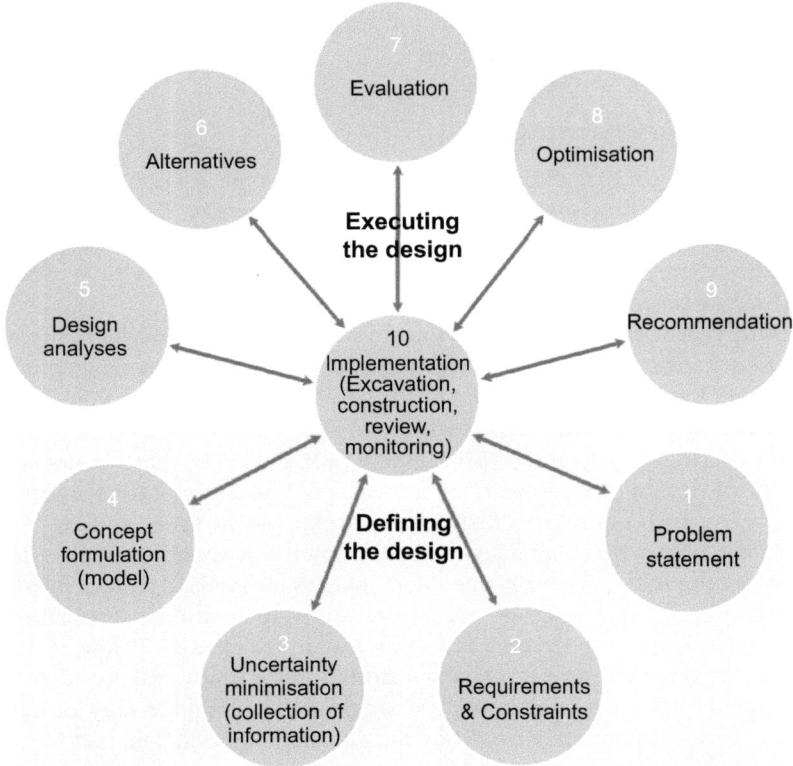

Figure 6 Engineering wheel of design (Stacey, 2009).

the similarity with the strategic planning process introduced by Ilbury & Sunter (2005) and demonstrates the very close correlation between the processes of logical strategic planning and thorough engineering design. Logically, review and monitoring are positioned at the center of the wheel. Review should take place at every step, as indicated by the spokes of the wheel, to ensure that the stated objectives of the design are being satisfied at each step.

Diligent application of this defined process or methodology is a form of quality control, or checklist, which will ensure that all aspects that *should be* taken into account in the design, *are* taken into account.

3.2 Design considerations in rockbursting situations

Based on the design process described above, the design objectives in situations prone to strainbursting and rockbursting must be:

- To prevent strainbursts and rockbursts if possible.
- If this is not possible, to reduce the potential for rockbursts by introducing ameliorating measures.

–   If neither of the above options are satisfactorily possible, to design rock support measures that will contain the rockburst damage.

Each of these objectives will be dealt with below.

3.2.1 Prevention of strainbursts and rockbursts

The design objective of "prevention of strainbursts and rockbursts" implies that their prediction is possible. This objective is equivalent to the possibility of prediction of earthquakes. Despite many years of research, this has not been possible, and the same conclusion applies to rockbursts and strainbursts. There has been some success in predicting where an earthquake can be expected at some time in the future (for example Ruegg et al., 2009), and predicting likely locations of seismic sources in mining environments (Ryder & Jager, 2002; Vieira & Durrheim, 2001). However, the when, where, and with what magnitude a seismic event will occur, unfortunately cannot be predicted with adequate certainty for satisfactory design purposes. If the actual locations, magnitudes, and times of occurrence of rockbursts and strainbursts cannot be predicted, except in a general sense, there is therefore no possibility of their prevention. The only way in which excavation-induced seismicity can be prevented is by stopping excavation, and this will not normally be a feasible option. It may therefore be concluded that it is not possible to prevent the occurrence of bursts.

3.2.2 Design of ameliorating measures to reduce the potential for occurrence of strainbursts and rockbursts

Minimizing the potential for the occurrence of strainbursts involves influencing the capacity, the capability and the conditions associated with the rock mass (see Section 2.1 above). Excavation in rock has no alternative but to deal with the rock types present. Consequently, there is often little that can be done to alter the capacity of the rock and rock mass to store strain energy. Methods that have shown benefits in affecting capacity are destressing and preconditioning (Toper et al., 2000; Zhang et al., 2012b). Destress blasting will "soften" the rock by the introduction of blast-induced fractures, decreasing the stiffness of the rock mass. This decreases the ability of the rock mass to store strain energy and hence decreases its strainbursting potential. With regard to rupture type rockbursts, preconditioning by hydraulic fracturing has shown significant benefits in a mining environment by reducing levels of seismicity and its effects (Araneda & Sougarret, 2007).

3.2.3 Design of the geometries of excavations, mining layout geometry and mining sequence

Optimizing the geometries of excavations, and design of the excavation and mining sequence and resulting layout, will help to reduce the occurrence of seismicity and hence strainbursts and rockbursts. The geometry of an excavation will have an influence on the potential for strainbursting to occur from its surfaces. Long, straight surfaces will be much more susceptible to strainburst occurrence. Circular excavations are known to be more stable than square or rectangular excavations: stress concentrations in the sharp corners of the latter excavations can be the locations of initiation of failure. Doming of excavation roofs and sidewalls can provide benefit in reducing strainbursting potential. In practice, excavation geometries will be dictated by usage requirements and, in a mining environment, by the geometry of the orebody and the mining method. It may therefore often not be possible to adopt a particular excavation shape.

The overall mining geometry can have a large influence on the stress concentrations due to interactions between adjacent excavations. Therefore, the proximity of excavations must be taken into account in the design. The overall mining geometry can also have a large influence on the occurrence of seismicity that could cause rupture type rockbursts. For example, an extensive longwall face, parallel to the strike of a fault, and advancing towards the fault, has the potential to cause significant releases of seismicity. In contrast, if the face is much shorter, or if the face was to be orientated at an angle of >35° to the fault strike, the potential for significant seismicity will be substantially reduced. Sequencing and layout designs are routinely carried out using three-dimensional numerical modelling programs, with the application of appropriate design criteria (Vieira & Budavari, 2003; Vieira & Durrheim, 2001). These will often be empirical, based on mine experience, and also on the results of quantitative monitoring of behavior of the rock mass, for example, seismic monitoring, displacement monitoring, and stress monitoring. The aim of the design will be to avoid adverse geometries (ie reduce the capability) and hence minimize the occurrence of high induced stresses (ie avoid adverse conditions) that might lead to rock failure.

When seismicity is expected in a mining or civil engineering project, a programme of microseismic monitoring, involving the installation of appropriately-spaced sensors with appropriate sensitivities will assist in the characterization of the rock mass, the identification of zones that are more, or less, responsive with regard to seismicity, and the identification of geological planes that are more responsive to seismicity. Early, detailed monitoring of seismicity will be of value since it will provide a picture of the background seismic behavior in the project location. Any subsequent changes in behavior will therefore be able to be evaluated at the earliest opportunity. A *designed* monitoring programme is an important step in a systematic design process and is essential in ensuring that the design is implemented correctly. It will usually involve seismic monitoring, geotechnical monitoring and quality control monitoring. A monitoring programme such as this could possibly provide empirical design criteria for application in layout and sequence design, and ultimately will provide data to confirm that the design has been correctly implemented.

Although appropriate design of excavations, layout geometry, and mining sequence, is probably the most important step in the design process with regard to *reducing the occurrence of seismicity*, it cannot guarantee that bursts will not occur. Its execution is likely to identify potential problem areas in advance of excavation, which can therefore be addressed in the design. Additional design measures will be necessary to contain or minimize burst damage, and these measures involve the design of suitable rock support for bursting conditions.

4 DESIGN OF DYNAMICALLY-CAPABLE ROCK SUPPORT

It is instructive to examine the behavior of rock support in a burst event. Figure 2 shows damage that occurred in a rupture type rockburst. The support consisted of 40T cable anchors (fully grouted, non-yielding), high capacity chain link mesh, shotcrete and rockbolts. The following behavior can be seen: a "curtain" of mesh-reinforced shotcrete has remained relatively coherent; in numerous cases, cable faceplates have pulled through the mesh and shotcrete; the lengths of cable anchors exposed indicates the extent of the ejected mass; although not visible in the photograph, failures of cables also occurred; the ejected mass of rock can be observed to be rather "pulverized" (this is not representative of the in situ fragmentation, and indicates that this pulverization occurred in the rockburst event).

These observations show that even support with very high *strength* capacity can fail in a rockburst event. Other common mechanisms of support failure that can be observed in dynamic events are: "brittle" failure of conventional, fully grouted rockbolts and cables; failure of rockbolts at threads; bending of faceplates around the nut, facilitating pulling through mesh and pulling over nuts; faceplates with sharp edges inducing early failure of mesh; faceplates too small and not spanning a sufficient number of mesh segments; weldmesh failing at welds; ejection of fragments/discs of shotcrete when not contained by mesh; etc. The weakest link principle will usually apply: if a single component fails, it is then likely that the support *system* will be incapable of containing the damage.

In contrast with the above, dynamically-capable rock support may contain burst damage to such an extent that damage is barely visible, if at all. In such a case, a rockburst would then not be recorded. The aim therefore, is to design such dynamically-capable rock support that is effective in reducing the effects of strainburst and rockburst events so as to maximize safety and to minimize damage, should an event occur. To achieve a satisfactory rock support design, there must be sufficient input data available to carry out the design, and the following must be understood and known: the mechanisms of loading; the mechanisms of rock and rock mass failure that the support is being designed to combat; and the mechanisms of action of the support elements and their interactions in the support system. If this information is known, it is usually possible to determine the demands imposed on the support elements. The capacities of support elements and support systems can usually be calculated, and these data can then be used to design the required support. In concept therefore, the design of support is a straightforward engineering process. However, this is not the case for rockbursting conditions.

Attention has been given to rock support and its design for rockbursting conditions over several decades. As early as the 1960's Ortlepp (1968, 1969) recognized that yielding support was necessary in dynamic, rockbursting conditions. He developed a yielding rockbolt, and demonstrated its effectiveness under simulated rockbursting conditions (Ortlepp, 1969; Ortlepp & Reed, 1969). Further attention has been given to rock support and its design for rockbursting conditions over many years (Ortlepp *et al.*, 1975; Wagner, 1982; Ortlepp, 1983; Roberts & Brummer, 1988; Jager, 1992; Ortlepp, 1992a; Kaiser, 1993; Ortlepp, 1994; Kaiser *et al.*, 1996; Ortlepp & Stacey, 1997; Stacey & Ortlepp, 2002b; Stacey, 2011; Cai & Kaiser, 2011; Kaiser & Cai, 2013; Potvin & Wesseloo, 2013; Stacey, 2013). This list of references on the subject is by no means exhaustive.

A straightforward engineering design process conventionally makes use of the concepts of stress and strength. However, in dynamic loading situations it has been found that these concepts are inappropriate for rock support design, and an energy-based design method is more satisfactory. Ortlepp (1992a) presented an energy-based design rationale, concluding, "It is neither practicable or economically possible to contain severe rockburst damage by increasing the *strength* of the tunnel support." And, "Designed yieldability or compliance is essential to prevent support components being broken by rockbursts." The ejection velocity of the rock walls associated with the seismicity is the key factor with regard to support performance, and hence design requirements. For design purposes using this energy approach, it is necessary to determine or predict ejection velocities. These dictate the demand to which the rock support will be subjected, and hence the capacity of the support system necessary to contain the rockburst damage. However, in practice, the determination of both the demand and the capacity under rockbursting conditions is problematic. These two aspects are dealt with in more detail below.

4.1 Support demand

The determination of "demand" requires knowledge not only of the ejection velocity, but also of the direction of ejection and of the mass of rock that is involved in the ejection.

4.1.1 Volume and mass of rock involved in a burst

The "thickness" or mass of rock ejected, and the ejection velocity, will determine the kinetic energy involved in the event that must be contained if safe and stable conditions are to be maintained. With regard to safety, it is important to note that even small fragments of rock (small mass), ejected at high velocities, can be hazardous. It is commonly observed that a thickness of rock of approximately 1m is ejected in a burst. The volume, or thickness, of rock that is involved in the burst can be estimated from numerical analyses, and from empirical evidence. Kaiser and Cai (2013) give a simple equation, based on a semi-empirically predicted depth of fracturing and the stress level, to calculate the depth of failure, and they give bulking factors relevant for strainbursting. Although the volume (mass) can be determined definitively from examination after the event, if rock support was installed, it may have had a significant influence on the extent of rock failure; the energy absorbed by the support will not be quantifiable. In brittle rock progressive failure may also occur, making the depth of failure difficult to determine. Ortlepp and Stacey (2000) described an example of such progressive failure for a self-mined tunnel.

In rupture type rockburst events, the ejected rock will involve rock that has been fractured due to high stress levels. However, such stress fracturing is not a determinant of the depth of failure. Owing to seismic wave interaction at the surface of the excavation, new rock failure will occur and be involved in the ejected material. In this regard, the composition of the walls, and the extent and geometry of fracturing within them, will also have an effect on the depth of failure.

The direction of ejection will also have an influence on the depth of rock ejected. Ejections from the floor of a tunnel are considered to be less likely than from the roof or

Figure 7 Rockbolt inclinations indicating upwards direction of ejection (Stacey & Rojas, 2013).

walls with regard to strainbursting. In contrast, in rupture type events, ejection of the floor (floor heave) and sidewalls is commonly observed (Figures 1, 2 and 3). In Figure 7, the rockbolts in the sidewall of the tunnel are inclined upwards after the upwards ejection of material in a rockburst. Observations indicated that rock from the sidewall was ejected upwards and across the tunnel, some of that rock impacting the roof of the tunnel.

From the above, it can be concluded that, although the thickness of rock prone to ejection may be "calculated", there can be little confidence in such a prediction. This is not a satisfactory situation as far as rock support design is concerned.

4.1.2 Velocity of ejection

Back-analyses of observations of rockburst damage provide definitive data on ejection velocities. McGarr (1997) referred to "... numerous observations, in nearby damaged tunnels ... imply wall-rock velocities of the order of 10m/s and greater". Ortlepp (1993) back-calculated a velocity exceeding 50m/s for the ejection of a small rock fragment; and 8m/s in a case involving a lump of concrete with a size of 0.2 to 0.3m. The floor concrete that can be observed in Figure 3 is estimated to have been "ejected" upwards at a velocity perhaps as high as 9m/s (Stacey & Rojas, 2013). The interesting case illustrated in Figure 7, shows upwards ejection of rock from the sidewall of a 4m high tunnel (Stacey & Rojas, 2013). The ejected rock impacted on the roof of the tunnel, and a simple back-analysis of the ejection velocity of this rock gives a value of 8m/s. The "bent" geometry of the rockbolts indicates that they probably decelerated the rock, and hence the actual ejection velocity could have been about 10m/s.

The use of PPV, determined from seismic data, has been suggested as a basis for the prediction of ejection velocities. For this process, Kaiser *et al.* (1996) recommended that seismicity records be examined to establish the spatial and temporal distributions of seismic events; that a location of a design event and design magnitude be chosen; and, using appropriate scaling law parameters, that PPV and ejection velocity can be predicted. Essentially the same process is suggested in a recent publication (Kaiser & Cai, 2012). This recommended design approach is dependent on the relationship between PPV and ejection velocity.

However, back-calculated ejection velocities tend to be much higher than would be predicted from a PPV approach. Milev *et al.* (1999) considered that a PPV amplification factor of between 4 and 10 could be applicable. Durrheim (2012) suggested the same range of amplification factors. From measurements of ground movements in the rock mass and on the surfaces of excavations, Cichowicz *et al.* (2000) indicated that the PPV on the surface can be up to five times that in the rock mass. It is probable that this conclusion is an underestimate of the amplification since, in "extreme" events such as rockbursts, monitoring equipment is likely to be destroyed. Therefore, such "extreme" PPV data will probably never be measured in the field with monitoring instruments installed on the surface of an excavation. Maximum amplifications are therefore likely to be greater than those determined from non-extreme conditions. This opinion is perhaps confirmed by the in situ measurements in mines reported by Milev *et al.* (2002), which indicated amplification factors of between 1 and 25. From back-analyses of about 60 events in mining operations, which occurred over a 12 year period, ejection velocities were determined (Bacco, 2010), and, for six selected events, amplification factors ranged between 8 and 52, with an average value of 32 (Stacey & Rojas, 2013).

Alternative explanations for the amplifications have been given by several researchers. Linkov & Durrheim (1998) considered wave amplification as a phenomenon of energy release due to softening. The implication of this mechanism is that it applies to stressed rock conditions. Thus it might be an explanation in certain cases of stressed rock surrounding excavations, but it cannot explain the ejection behavior observed in Figure 3, in which the unstressed concrete is ejected. Linkov & Durrheim (1998) also referred to the proposition suggested by McGarr (1997) in which buckling of stressed rock slabs on the boundary of an excavation is considered as an amplification mechanism. Again, this mechanism refers to stressed rock conditions, and therefore would not be an explanation for the behavior observed in Figure 3, or in Figure 7, in which the direction of ejection does not correspond with what could be expected in buckling behavior. These alternative explanations for increased amplification factors also do not apply in other cases (Stacey & Rojas, 2013).

The implication from the above information is that at this stage there can be no confidence in the prediction of ejection velocities from seismic data. Therefore, since neither the volume (mass) of rock involved in an ejection, nor the ejection velocity, can be predicted satisfactorily, *for design purposes the demand is effectively unknown.* Kaiser & Cai (2013) have suggested that it is time to rethink the design principles for rock support in burst-prone ground.

4.2 Capacity of rock support and support systems

As stated above, a formal design of support for underground excavations requires knowledge of the demand to which support will be subjected and the capacity of that support. The behaviors of various types of support in rockburst events have been graphically shown above (Figures 1, 2 and 3). The capacities of individual elements of support such as rockbolts, wire mesh and shotcrete can be calculated from their mechanical properties and the loading conditions to which they will be subjected. Support performance also depends on installation quality, such as the effectiveness of anchoring, the extent of grouting, the strength of the grout, shotcrete quality, strength, thickness variation, mesh overlaps, face-plate and mesh interaction, etc. The individual performance of support elements is rarely of much relevance, however, since it is the performance of the rock support *system* that is of importance. A rock support system is a combination of individual support components that work together to retain and contain the rock. In doing this, the components are subjected to loading by the rock and to interactive loading between one component and another. Therefore, a rockbolt could be subjected to a combination of tensile, shear, bending and torsional loading by the rock under static and, particularly, dynamic conditions. Similarly, other components of support – wire mesh, shotcrete, fiber-reinforced shotcrete, face plates, straps, lacing, etc. – could be subjected to combinations of loading mechanisms. Connection between the rockbolts and the surface support also implies that the surface support will impose loadings on the rockbolts, and vice versa. Owing to these complex situations, whilst calculation of the capacity of individual support components is easy, theoretical determination of the capacity of a *support system* is very unlikely to be successful.

4.2.1 Physical testing of rock support

An alternative approach to determining the capacity of rock support systems is to carry out physical testing of support components and support systems. Testing in relation to support for rockburst conditions has been carried out for many years.

A summary of dynamic testing of rock support carried out in various countries has been presented by Hadjigeorgiou & Potvin (2007), and an interpretation of all the results obtained has been presented by Potvin et al. (2010). These two papers deal with testing using blasting and using drop weight impacts to represent rockburst loading. Ortlepp (1968; 1969) carried out two blast loading tests on rockbolt and mesh support systems installed in a tunnel, one with conventional rockbolts, and the other with yielding rockbolts that he had developed. This blast-load testing demonstrated the effectiveness of the yielding support system, and the ineffectiveness of the conventional support system. Ortlepp (1992b) repeated this type of blast loading test in a different mining environment, and the result was similar. Measurements of the ejection velocity of the wall supported with conventional, non-yielding support showed a value of 10m/s. A short while after completion of this test, a nearby tunnel was damaged in an actual rockburst, and the damage observed was indistinguishable from that in the blasting test.

More recent blasting "rockburst" tests carried out by several researchers are described by Hadjigeorgiou & Potvin (2007). These include the test carried out by the CSIR in South Africa, summarized by Hagan *et al.* (2001). Tests carried out in Canada are described by Espley *et al.* (2002), Archibald *et al.* (2003), and Tannant *et al.* (1993), and those in Australia by Heal and Potvin (2007). The results are summarized by Potvin *et al.* (2010). This summary of results indicates that gas pressure "was a problem" in some of the tests. The most recent blast testing of rock support is that reported by Shirzadegan *et al.* (2011).

The testing described by Hagan *et al.* (2001) minimized gas loading and indicated ejection velocities were in the range of 0.7 to 2.5m/s. Ground velocities of 3.3m/s were recorded by an accelerometer. Rock support involved in the test consisted of fully cement grouted rockbolts only. "Rockburst" damage occurred on the tunnel wall where the PPV exceeded 0.7m/s. High intensity damage occurred where the ground velocity of 3.3m/s was recorded.

A similar blasting geometry was used by Potvin & Heal (2010) to ensure that the dynamic testing of the rock support was not influenced by gas pressure. In their first test they measured PPVs in the range of 0.3 to 2.4m/s. Two support systems were used: yielding rockbolts (cone bolts) with High Energy Absorption (HEA) mesh; and cone bolts with mesh and fiber-reinforced shotcrete. Minor damage of the support was observed. The same location was used for the second test, with the implication that the rock mass was possibly "damaged" (fractured) by the first blast. In this second test, PPVs of 0.6 to 3.0m/s were recorded, and significant damage occurred. A mass of rock of about 100 tons was ejected, with both support systems sustaining damage.

Owing to the difficulty of carrying out blasting tests underground, Ortlepp (1992a) proposed the use of a "synthetic concrete sidewall" for ejection. This concept was implemented in a quarry, with vertical ejection of the concrete mass (Ortlepp, 1994) by means of explosives. Ejection velocities of the order of 12m/s were measured in the tests. It was demonstrated that 16mm diameter yielding cone bolts performed satisfactorily at these velocities without breaking, displacements being of the order of 0.5m. In contrast, much stronger, fully grouted rebar bolts failed in the tests with low energy absorption. These tests were of groups of rockbolts only, not of support systems involving a combination of support elements. In addition, they involved tensile loading only, and bolts were not subjected to shear, or combinations of stresses.

One may question whether blast loading is a satisfactory representation of rockburst loading, since shock waves and subsequently, and substantially, gas pressure provide the loads. In contrast, in a rockburst, a mass of rock is suddenly accelerated, with no gas pressure involved. Alternative methods have been developed for evaluation of rock support, usually involving some form of drop weight system. Such "laboratory rockburst" testing of rock support components and systems has been carried out in several countries using somewhat different testing methods (Ortlepp & Stacey, 1994; Yi & Kaiser, 1994; Kaiser *et al.*, 1996; Ortlepp & Stacey, 1997, 1998; Ortlepp *et al.*, 1999; Stacey & Ortlepp, 1999, 2001, 2002a,b; Gaudreau *et al.*, 2004; Li, 2011; Li & Charette, 2010; Player *et al.*, 2004, 2008a,b; Plouffe *et al.*, 2008. Further, the references and bibliography provided by Kaiser *et al.* (1996) indicate numerous unpublished reports of testing authored mainly by Tannant. Most of these test methods and results have been described by Hadjigeorgiou & Potvin (2007) and the results are summarized in Figure 8 (Potvin *et al.*, 2010).

It can be seen that very significant levels of energy can be absorbed by appropriate support, provided that yield, or displacement, can take place. The value of wire rope

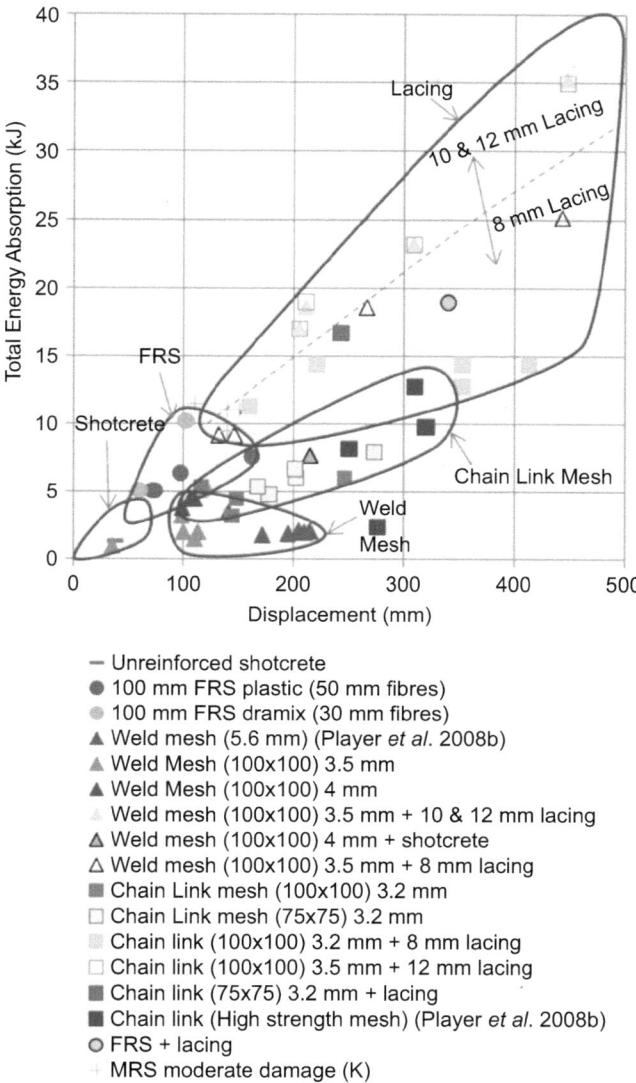

Figure 8 Performance of surface support systems under dynamic loading (Potvin *et al.*, 2010).

lacing in absorbing energy is apparent from Figure 8, a contribution that was specifically identified by Stacey & Ortlepp (2002a). The tests demonstrated that it could possibly enhance the capacity of mesh and shotcrete support by a factor of 7. It is likely that similar gains in performance will be supplied by tendon straps or mesh straps. Straps also provide a buffer between face plate and mesh, protecting the mesh, and due to their width, they also reduce the exposed spans of rock, shotcrete and mesh. The link between a rockbolt and containment support such as mesh usually involves a steel face plate on the bolt. Such plates often fail because of irregular rock bearing surfaces or non-axial loading on the bolt. Tests in which loading simulated "real" conditions rather than idealized flat-surface bearing conditions, demonstrated that plate capacities were much less than their specified values (Van Sint Jan & Palape, 2007) because the plates failed in a folding mode. Nut failures were also observed in these tests.

The blasting approach pioneered by Ortlepp (1969) more than 40 years ago, probably still provides the greatest validity as a severe test of rockburst support capabilities even though it does not simulate a rockburst.

The results in Figure 8 give the capacities of containment support (mesh, shotcrete, liners, straps, lacing), but do not provide data on capacities of *support systems*, which are combinations of retainment and containment support elements, as well as the connecting components (nuts, faceplates, etc.). The performance of a *support system* will depend on the performance of all of these components. Thus, in summary, this review of alternative testing methods has shown that, whilst data are available on individual *support components*, knowledge of the capacities of *rock support systems*, from theoretical calculations or in the form of data from practical testing programmes, is absent.

4.3 Conclusions regarding rock support design

It will be clear from the information above that the occurrence of a rockburst cannot be predicted confidently with regard to time, location, and magnitude. Further, neither the demand that will be imposed on rock support in the rockburst, nor the capacities of rock support systems are known. Therefore, with regard to a conventional design process for rock support, neither of the two essential parameters required to enable a robust rock support design to be carried out, namely capacity and demand, are known. This is a clear case of design indeterminacy, and it is therefore impossible to determine the required dynamically-capable rock support using the classical engineering design approach.

In contrast with this negative conclusion, there is positive information that results from the physical testing of the capacity of support elements. This is that yielding rockbolts have been proven to have the capability of yielding at high velocities; and that containment support has proven capacity of absorbing large amounts of energy (Potvin *et al.*, 2010; Villaescusa *et al.*, 2014). This information is of use in a proposed way forward regarding the containment of dynamic rock failure. Added to this is the information from observations of yielding support performance in rockburst events (Simser *et al.*, 2002; Heal *et al.*, 2004; Stacey & Rojas, 2013) confirming the effectiveness of such support. There is now significant experience with the use of yielding rockbolts, and other components, in rockbursting situations.

4.4 Specification of rock support

Since the use of the conventional rock support design approach is not possible, a suggested philosophy in determining support requirements for situations in which rockbursting occurs or may occur is to *specify the support* rather than attempt to design it. This approach may be considered to be a simple form of the observational approach (Peck, 1969) which, formally, requires measurement and quantified observation of behavior, from which design and construction decisions can be made. This will ensure that "appropriate" rock support will be installed. Since the design parameters are unknown, and safety is paramount, the resulting conservatism is justified.

In the following, rock support for strainbursts and rupture type rockbursts will not be differentiated. The result in both cases is ejection of rock at significant velocities, and therefore an energy/specification approach is considered to be valid for both.

If the design measures outlined above regarding geometry, layout and sequence of excavation, and amelioration measures, are diligently applied, then the occurrence of strainbursts and rockbursts will be minimized. However, their occurrence will not be prevented completely, and rockburst resistant support must therefore be installed to limit the extent of damage if an event occurs. Rock support systems for rockbursts usually involve components such as rockbolts with plates and nuts, cables, wire mesh, shotcrete, fiber shotcrete, straps, lacing, etc. A range of progressive levels of rockburst support can be considered regarding the latter part of this pragmatic approach, and suggested levels are described below. A primary and essential requirement of the support system is that no single component of the system can be allowed to fail. Failure of a single component will almost certainly lead to failure of the support system and lack of containment of the burst. For example, if a face plate fails, mesh and shotcrete will not be restrained, and the burst will not be contained. Similarly, if a bolt fails, the surface support will not be retained, and the rock will be free to eject. The same applies to all other components.

4.4.1 Yielding rockbolts

The energy involved in a rockburst event may be of the order of 150kJ/m^2. The capacity of conventional rebar type rockbolts is about 5 to 10kJ, which indicates why yielding bolts with much greater energy absorbing capacities are necessary. Yielding rockbolts are the primary rockburst support component, and contribute by far the most as far as energy absorption is concerned. There are now many alternative yielding bolts commercially available, and these bolts are capable of absorbing very large amounts of energy without failing. Reference to yielding rockbolts in this Chapter may be considered to include yielding cables.

Rockbolts fulfill two important functions: they limit the span of any potential rock plate or beam to that of the bolt spacing, and thus buckling behavior will be inhibited. In this way the incidence and magnitude of strainburst events will be lessened. The second function is to yield in the bulking of the rock and, particularly in the case of ejections resulting from rupture type rockbursts, to absorb the energy involved in the ejection. Testing of yielding rockbolts has demonstrated that most types will be capable of yielding in tension at high velocities (exceeding 10m/s and as high as 20m/s (Ortlepp & Stacey, 1997). Back analyses of rockburst damage have shown that ejection velocities of the order of 10m/s are not uncommon. From a pragmatic point of view, therefore, it is suggested that the spacing between bolts should be determined assuming a 10m/s ejection velocity. Since bolts do not yield as well under shear loading as in tension, it is suggested that the spacing should be conservatively small. To cater for severe cases of shear loading, orientations of adjacent bolts could be varied to promote tensile loading of at least some of the bolts and minimize shear failures. Rockbolt plates extend the effect of the bolts themselves.

4.4.2 Sprayed liners and wire mesh

Surface support provides some confinement to the rock, inhibiting loosening and therefore reducing the capability of the excavation with regard to strainbursting. Shotcrete, both unreinforced and reinforced, is very commonly used for surface support. It is a brittle material which tends to crack in rockbursting conditions. Nevertheless, shotcrete and thin spray-on liners (TSLs) may provide some benefit when applied in addition to yielding rockbolts (bolting through the liner). The liner serves to contain the rock between the bolts and to enhance the area of the plate. The level of benefit provided by unreinforced shotcrete is small, as can be seen in Figure 8. Also in Figure 8 it can be seen that fiber reinforced shotcrete has greater energy absorption capacity than unreinforced shotcrete. Application of a TSL has proved to be beneficial in rockbursting conditions (Carstens, 2005). It has also been found that application of a TSL over shotcrete enhances the performance of shotcrete (Simser, 2014).

Wire mesh, retained by yielding rockbolts, is a more effective rockburst support than a sprayed liner. Weld mesh does not perform very well, mainly because failure commonly occurs at the welds. Chain link mesh performs better owing to its greater flexibility. Test results are shown in Figure 8. Cases in which mesh is cut by sharp-edged face plates have been observed, and such plates should be avoided. In an environment in which mechanical equipment is used, protection of the mesh and bolts against mechanical damage is essential, and this can be achieved by applying shotcrete over the mesh. This is satisfactory in static conditions, but in rockbursting environments the shotcrete tends to spall off (eject) easily, and this may be hazardous. Mesh over shotcrete is an effective rockburst support. The shotcrete inhibits movement of the rock and may be considered to act as a large, continuous face plate. Again, however, additional shotcrete may be required to protect the mesh against mechanical damage.

For optimum performance of a rockbolt and mesh support system, with or without shotcrete, the bolts, mesh, plates and nuts must all have matched capacities to ensure that the mesh/plate/nut components do not fail prematurely. They must have sufficient capacity to cause the bolts to yield under rockburst loading. There must be no weak link.

4.4.3 Mesh straps, tendon straps, wire rope lacing

Mesh straps, tendon straps and wire rope lacing, retained by yielding rockbolts and faceplates, can be applied directly over the rock, or over mesh or a sprayed liner. They have the effect of reducing the unsupported spans between rockbolts. They will enhance the performance of mesh and shotcrete since they effectively represent larger face plates. In themselves, they also have large energy absorption capacities. It has been estimated that wire rope lacing can increase the energy absorption capacity of wire mesh by up to seven times (Stacey & Ortlepp, 2004). This support will also require protection against mechanical damage by underground equipment. Simser *et al.* (2002) showed that this type of support system (yielding rockbolts, heavy chain link mesh, and mesh straps) was successful in containing rockburst damage in a severe event.

4.4.4 Sacrificial support

In numerous rockburst cases it has been observed that floor damage has not occurred when water and mud have been present. A particular case is described by Stacey & Rojas (2013). An area of clean concrete was observed to have been heaved upwards, and the steel arch founded in the concrete showed significant damage. A distance of a meter on either side were two other steel arches, completely undamaged. The concrete floor at the bases of these two arches was covered with mud, and it appears that the mud provided protection against the rockburst.

In other rockburst events, cases were observed in which concrete support panels, retained against the sidewall by grouted cables, were ejected, leaving the rock behind them essentially undamaged. This behavior suggested the concept of sacrificial support (Stacey & Rojas, 2013), in which some form of non-hazardous sacrificial layer, which would be expected to be ejected in a rockburst event, would be applied over other support. Although there are observations that prove that the concept has worked, further research is necessary before the concept can be converted to a practical engineering solution.

4.4.5 Recommended rock support for rockbursting conditions

For situations in which rockbursting is expected, or has occurred, the following are pragmatic rock support specifications:

- Fiber-reinforced shotcrete (50mm or greater); yielding rockbolts (typically 20mm diameter) at typically 1m spacings; plus heavy chain link mesh. Apply fiber-reinforced shotcrete over mesh if required for mechanical protection. Avoid too great a thickness of such shotcrete, to minimize the risk of ejection of shotcrete fragments.
- If any failures of bolts are experienced, change bolt type to bolts with greater yield capacity; if failure of mesh is experienced, add zero gauge mesh straps, or equivalent, to the support system. Such straps should be retained under the rockbolt faceplates. If wire rope lacing is used, it should pass under the rockbolt faceplates, and, to ensure yield capability, should not turn through an angle of greater than 90° around the bolt.
- If any rockbolt failures are experienced in the above system, replace rockbolts with yielding cables; if mesh failure is experienced, install additional mesh straps at different orientations, to reduce exposed mesh spans even further.

It must be emphasized that failure of the connecting components (faceplates and nuts, barrel and wedge, and failure of bolts on the threads) *must not occur*. It is expected that the above rock support systems will contain even severe rockbursts. In the event that this specified support is still not sufficient, other measures that could be implemented are reduced bolt or cable spacings, and varied orientations in the drilling of rockbolt holes, to cater for variations in directions of ejection and possible shear loading on rockbolts.

Continued observations of rock support behavior are essential so that continued improvement and optimization of the specified support can be achieved using the observational "design" approach.

4.5 Risk considerations

Risk can be quantified as the product of the probability of occurrence of an event and the consequences of occurrence of the event. If rockbursts are expected, or known to occur, on a civil engineering or mining project, then it can be assumed that the probability of occurrence is high. Therefore it is essential that the consequences are addressed, using all the measures described above, to minimize the safety risk. The installation of rockburst support will reduce this risk further, but is likely to increase the overall support costs and hence may increase the financial risk. However, the costs of individual support components are a small proportion of total support costs, and the difference in cost between rockburst and conventional support materials will be small (Ortlepp & Stacey, 1995). Support component costs are only one of many costs associated with the consequences of rockbursts. Other consequential costs include accidents and associated costs; work stoppages; clean-up costs; rehabilitation costs; costs of loss of production; costs due to reassignment of crews; costs associated with dilution of ore; costs due to loss of ore reserves; and costs that are difficult to quantify, such as public perception, reduction of operating company share price, reduced worker morale, and labor unrest. Indirect costs will usually be much greater than direct costs, with loss of production likely to be the major indirect cost (Rwodzi, 2010). The introduction of rockburst support, and containment of damage, will therefore reduce total costs, and will probably even create value for the project. The decision on whether rockburst support is justified or not should therefore not be taken on the basis of direct costs only, but rather on a risk basis.

5 CONCLUSION

Rockbursts are very hazardous and usually damaging occurrences, which have been experienced in mining excavations for many years, and which are being increasingly experienced in highly stressed civil engineering excavations. The occurrence of rockbursting can be reduced by appropriate design of excavation geometries and extraction sequences, as well as amelioration measures such as destressing and preconditioning. However, it is unlikely that rockbursts can be prevented. Containment of rockburst damage is therefore essential for safety, and for reduction of economic risk. Conventional design of containment measures is difficult, if not impossible, since rockburst magnitude, location and time of occurrence cannot be predicted. In addition, the *demand* on rock support in such an event, and the *capacity* of the support system, *are both unknown*. Observations of support performance during physical testing, and in rockburst events, represent the basis for an "observational design" approach (Peck, 1969). This has been used to specify rock support systems that are likely to be able to contain rockburst damage. This may appear to be a conservative and costly approach. However, safety is of paramount importance, and in most cases, if one serious accident or one serious burst or collapse can be prevented, then the support costs will have been justified. Further, cost savings due to the non-occurrence of rockburst damage are likely to be greater than the direct support costs, and therefore it is probable that rockburst support will actually create value for the project rather than being a negative cost. Ongoing observations will be essential to ensure that rock support designs for rockbursting conditions can be further improved in the future.

ACKNOWLEDGMENT

The research by the author of this chapter is supported in part by the National Research Foundation of South Africa (Grant specific unique reference number (UID) 85971). The Grantholder acknowledges that opinions, findings and conclusions or recommendations expressed in any publication generated by the NRF supported research are that of the author, and that the NRF accepts no liability whatsoever in this regard.

REFERENCES

Araneda, O. & Sougarret, A. (2007) Lessons learned in cave mining: El Teniente 1997–2007, *Proc. 1st Int. Symp. on Block and Sub-level Caving, Cape Town, 2007*, S. Afr. Inst. Min. Metall., pp. 1–13.

Archibald, J.F., Baidoe, J.P. & Katsabanis, P.T. (2003) Rockburst damage mitigation benefits deriving from use of spray-on rock linings, *Proc. 3rd Int. Seminar on Surface Support Liners: Thin Spray-on Liners, Shotcrete and Mesh, Quebec City, 2003*, Universite Laval, Section 19.

Bardet, J.P. (1990) Numerical modeling of a rockburst as surface buckling, In: Fairhurst (ed.) *Rockbursts and Seismicity in Mines*, Balkema, Rotterdam, pp. 81–85.

Bieniawski, Z.T. (1988) Towards a creative design process in mining, *Min. Eng.*, 40, 1040–1044.

Bieniawski, Z.T (1991) In search of a design methodology for rock mechanics, In Roegiers (ed.) *Rock Mechanics as a Multidisciplinary Science, Proc. 32nd U S Symp. on Rock Mech.*, Balkema, pp. 1027–1036.

Bieniawski, Z.T. (1992) Invited Paper: Principles of engineering design for rock mechanics, In Tillerson & Wawersik (eds.) *Rock Mechanics, Proc. 33rd U S Symp. on Rock Mech.*, Balkema, Rotterdam, pp. 1031–1040.

Broch, E. & Sørheim, S. (1984) Experiences from the planning, construction and supporting of a road tunnel subjected to heavy rockbursting, *Rock Mech. Rock Eng.*, 17, 15–35.

Cai, M. & Kaiser, P.K. (2011) Rock support for deep tunnels in highly stressed rocks, *Proc 12th Int. Cong. Int. Soc. Rock Mech., Beijing, China*, CRC Press/Balkema, pp. 1467–1472.

Carstens, R. (2005) Thin sprayed liners – Adding or destroying value?, *Proc. 1st Int. Seminar on Strategic vs Tactical Approaches in Mining*, S. Afr. Inst. Min. Metall., pp. 327–335.

Cichowicz, A., Milev, A.M. & Durrheim, R.J. (2000) Rock mass behaviour under seismic loading in a deep mine environment: Implications for stope support. *Jl S. Afr. Inst. Min. Metall.*, 100, 121–128.

Durrheim, R. (2012) Functional specifications for in-stope support based on seismic and rockburst observations in South African mines, In *Deep Mining 2012, Proc. 7th Int. Seminar on Deep and High Stress Mining, Perth, Australian Centre for Geomechanics*, pp. 41–56.

Dyskin, A.V., Ustinov, K.B. & Germanovich, L.N. (1993) A 2-D model of skin rock burst and its application to rock burst monitoring, In Szwedzicki (ed.) *Geotechnical Instrumentation and Monitoring in Open Pit and Underground Mining*, Balkema, Rotterdam, pp. 133–141.

Espley, S.J., Heilig, J. & Moreau, L.H. (2002) Assessment of the dynamic capacity of liners for application in highly-stressed mining environments at INCO Limited, In Potvin, Stacey & Hadjigeorgiou (eds.), *Surface Support in Mining*, Australian Centre for Geomechanics, pp. 187–192.

Gaudreau, D., Aubertin, M. & Simon, R. (2004) Performance of tendon support systems submitted to dynamic loading, In Villaescusa & Potvin (eds.), *Ground Support in Mining and Underground Construction*, Balkema, pp. 299–312.

Gong, Q.M., Yin, L.J., Wu, S.Y., Zhao, J. & Ting, Y. (2012) Rock burst and slabbing failure and its influence on TBM excavation at headrace tunnels in Jinping II hydropower station, *Eng. Geol.*, 124, 98–108.

Grimstad, E. (1986) Rock-burst problems in road tunnels, in Norwegian Road Tunnelling, *Pub. No. 4, Norwegian Soil and Rock Engineering Association*, Tapir Publishers, University of Trondheim.

Hadjigeorgiou, J. & Potvin, Y. (2007) Overview of dynamic testing of ground support, In Potvin (ed.) *Deep Mining 07*, Australian Centre for Geomechanics, Perth, pp. 349–371.

Hagan, T.O., Milev, A.M., Spottiswoode, S.M., Hildyard, M.W., Grodner, M., Rorke, A.J., Finnie, G.J., Reddy, N., Haile, A.T., Le Bron, K.B. & Grave, D.M. (2001) Simulated rockburst experiment – an overview, *Jl S. Afr. Inst. Min. Metall.*, 101 (5), 217–222.

Heal, D., Hudyma, M. & Potvin, Y. 2004. Assessing the in-situ performance of ground support systems subjected to dynamic loading, in Villaescusa & Potvin (eds), *Ground Support in Mining and Underground Construction, Perth*, Balkema, pp. 319–326.

Heal, D. & Potvin, Y. (2007) In-situ dynamic testing of ground support using simulated rockbursts, In Potvin (ed.), *Deep Mining 07*, Australian Centre for Geomechanics, Perth, pp. 373–394.

Jager, A.J. (1992) Two new support units for the control of rockburst damage, *Proc. Int. Symp. on Rock Support, Sudbury*, pp. 621–631.

Kaiser, P.K. (1993) Keynote Address: Support of tunnels in burst-prone ground – toward a rational design methodology, in Young (ed.) *Rockbursts and Seismicity in Mines*, Balkema, Rotterdam, pp. 13–27.

Kaiser, P.K. & Cai, M. (2012) Design of rock support system under rockburst condition, *J Rock Mech. Geotech. Eng.*, 4 (3), 215–227.

Kaiser, P.K. & Cai, M. (2013) Critical review of design principles for rock support in burst-prone ground – time to rethink!, In Potvin, Y. & Brady, B. (eds.) *Proc. 7th International Symposium on Ground Support in Mining and Underground Construction, Ground Support 2013*, Australian Centre for Geomechanics, pp. 3–38.

Kaiser, P.K. & Kim, B-H. (2008) Rock mechanics advances for underground construction in civil engineering and mining, Keynote lecture, *Korea Rock Mechanics Symposium, Seoul*, pp. 1–16.

Kaiser, P.K., McCreath, D.R. & Tannant, D.D. (1996) *Canadian Rockburst Support Handbook: Volume 2: Rockburst Support*, Canadian Rockburst Research Program 1990–1995, CAMIRO.

Li, C.C. (2011) Dynamic test of a high energy-absorbing bolt, *Proc. 12th Int. Cong. Int. Soc. Rock Mech., Beijing*, CRC Press/Balkema, pp. 1227–1230.

Li, C.C. & Charette, F. (2010) Dynamic performance of the D-Bolt, In Van Sint Jan & Potvin (eds.), *Proc. 5th Int. Seminar on Deep and High Stress Mining, Santiago, Chile*, Australian Centre for Geomechanics, pp. 321–328.

McGarr, A. (1997) A mechanism for high wall-rock velocities in rockbursts, *Pure Appl. Geophys.*, 150, 381–391.

Milev, A.M., Spottiswoode, S.M., Noble, B.R., Linzer, L.M., van Zyl, M., Daehnke, A. & Archeampong, E. (2002) *The meaningful use of peak particle velocities at excavation surfaces for the optimisation of the rockburst criteria for tunnels and stopes*, Safety in Mines Research Advisory Committee, SIMRAC, GAP Project 709, Final Report.

Milev, A.M., Spottiswoode, S.M. & Stewart, R.D. (1999) Dynamic response of the rock surrounding deep level mining excavations, *Proc. 9th Int. Cong. Int. Soc. Rock Mech., Paris, France*, pp. 1109–1114.

Ortlepp, W.D. (1968) A yielding rockbolt, *Research Organisation Bulletin of the Chamber of Mines of South Africa*, No 14, pp. 6–8.

Ortlepp, W.D. (1969) An empirical determination of the effectiveness of rockbolt support under impulse loading, In Brekke & Jorstad (eds.), *Proc. Int. Symp. on Large Permanent Underground Openings, Oslo, September 1969*, Universitats-forlaget, pp. 197–205.

Ortlepp, W.D. (1983) Considerations in the design of support for deep hard-rock tunnels, *Proc. 5th Int. Cong. Int. Soc. Rock Mech., Melbourne*, pp. D179–D187.

Ortlepp, W.D. (1992a) Invited lecture: The design of support for the containment of rockburst damage in tunnels, In Kaiser & McCreath (eds.) *Rock Support in Mining and Underground Construction*, Balkema, Rotterdam, pp. 593–609.

Ortlepp, W.D. (1992b) Implosive-load testing of tunnel support, In Kaiser & McCreath (eds.), *Rock Support in Mining and Underground Construction*, Balkema, Rotterdam, pp. 675–682.

Ortlepp, W.D. (1993) High ground displacement velocities associated with rockburst damage, In Young (ed.), *Rockburst and Seismicity in Mines*, Balkema, Rotterdam, pp. 101–106.

Ortlepp, W.D. (1994) Grouted rock-studs as rockburst support: A simple design approach and an effective test procedure, *Jl S. Afr. Inst. Min. Metall.*, 94, 47–63.

Ortlepp, W.D. (1997) *Rock Fracture and Rockbursts – an illustrative study*, S. Afr. Inst. Min. Metall., 255p.

Ortlepp, W.D., More O'Ferrall, R.C. & Wilson, J.W. (1975) Support methods in tunnels, *Symp. on Strata Control & Rockburst Problems of the S A Goldfields*, Association of Mine Managers of S. Afr. 1972–73, Johannesburg, pp. 167–195.

Ortlepp, W.D. & Reed, J.J. (1969) Yieldable rock bolts for shock loading and grouted rockbolts for faster stabilization, *Proc. AIME Intermountain Minerals Conference, Vail, Colorado*, July, 21p.

Ortlepp, W.D. & Stacey, T.R. (1994) The need for yielding support in rockburst conditions, and realistic testing of rockbolts, *Proc. Int. Workshop on Applied Rockburst Research, Santiago, Chile*, SOCHIGE, pp. 265–275.

Ortlepp, W.D. & Stacey, T.R. (1995) The spacing of support - safety and cost implications, *Jl S. Afr. Inst. Min. Metall.*, 95, 141–146.

Ortlepp, W.D. & Stacey, T.R. (1997) *Testing of tunnel support: dynamic load testing of rock support containment systems*, Safety in Mines Research Advisory Committee, SIMRAC GAP Project 221.

Ortlepp, W.D. & Stacey, T.R. (1998) Performance of tunnel support under large deformation static and dynamic loading, *Tunnelling and Underground Space Technol.*, 13 (1), 15–21.

Ortlepp, W.D. & Stacey, T.R. (2000) A self-mined tunnel at a depth of nearly 3 km, *Tribune, ITA Newsletter*, No 14, April, pp. 6–7.

Ortlepp, W.D., Stacey, T.R. & Kirsten, H.A.D. (1999) Containment support for large static and dynamic deformations in mines, *Proc. Int. Symp. Rock Support and Reinforcement Practice in Mining, Kalgoorlie, Australia*, March, Balkema, pp. 359–364.

Peck, R.B. (1969) Advantages and limitations of the observational method in applied soil mechanics, Geotechnique, 19, pp. 171–187.

Player, J.R., Villaescusa, E. & Thompson, A.G. (2004) Dynamic testing of rock reinforcement using the momentum transfer concept, In Villaescusa & Potvin (eds.), *Ground Support in Mining and Underground Construction*, Balkema, pp. 3597–3622.

Player, J.R., Thompson, A.G. & Villaescusa, E. (2008a) Dynamic testing of reinforcement systems, in Stacey & Malan (eds.), *Proc. 6th Int. Symp. on Ground Support in Mining and Civil Engineering Construction, Cape Town*, S. Afr. Inst. Min. Metall., Symposium Series S51, pp. 581–595.

Player, J.R., Morton, E.C., Thompson, A.G. & Villaescusa, E. (2008b) Static and dynamic testing of steel wire mesh for mining applications of rock surface support, In Stacey & Malan (eds.), *Proc. 6th Int. Symp. on Ground Support in Mining and Civil Engineering Construction, Cape Town*, S. Afr. Inst. Min. Metall. Symposium Series S51, pp. 693–706.

Plouffe, M., Anderson, T. & Judge, K. (2008) Rock bolts testing under dynamic conditions at CANMET-MMSL, In Stacey & Malan (eds.), *Proc. 6th Int. Symp. on Ground Support in Mining and Civil Engineering Construction*, Cape Town, S. Afr. Inst. Min. Metall. Symposium Series S51, pp. 581–595.

Potvin, Y. (2009) Surface support in extreme ground conditions – HEA Mesh™, In Dight, P. (ed.), *Proc. 1st Int. Seminar on Safe and Rapid Development Mining, SRDM 2009*, Australian Centre for Geomechanics, pp. 111–119.

Potvin, Y. & Heal, D. (2010) Dynamic testing of High Energy Absorption (HEA) mesh, In Van Sint Jan & Potvin (eds.) *Proc 5th Int. Seminar on Deep and High Stress Mining, Santiago*, Australian Centre for Geomechanics, pp. 283–300.

Potvin, Y. & Wesseloo, J. (2013) Towards an understanding of dynamic demand on ground support, *Jl S. Afr. Inst. Min. Metall.*, 113 (12), 913–922.

Potvin, Y., Wesseloo, J. & Heal, D. (2010) An interpretation of ground support capacity submitted to dynamic loading, In Van Sint Jan & Potvin .(eds.) *Proc. 5th Int. Seminar on Deep and High Stress Mining, Santiago, Chile*, Australian Centre for Geomechanics, pp. 251–272.

Roberts, M.K.C. & Brummer, R.K. (1988) Support requirements in rockburst conditions, *Jl S. Afr. Inst. Metall.*, 88 (3), 97–104.

Rojas, E., Cavieres, P., Dunlop, R. & Gaete, S. (2000) Control of induced seismicity at El Teniente Mine, Codelco-Chile, *Proc. Mass. Min 2000, Brisbane*, pp.775–781.

Ryder, J A & Jager, A J (2002) *A textbook on rock mechanics for hard rock mines*, SIMRAC, Safety in Mines Research Advisory Committee, South Africa.

Ruegg, J.C., Rudloff, A., Vigny, C., Madariaga, R., De Chabalier, J.B., Campos, J., Kausel, E., Barrientos, S. & Dimitrov, D. (2009) Interseismic strain accumulation measured by GPS in the seismic gap between Constitución and Concepción in Chile, *Physics of the Earth and Planetary Interiors*, 175, 78–85.

Rwodzi, L. (2010) *Rockfall Risk: Quantification of the consequences of rockfalls*, MSc Eng. Dissertation, University of the Witwatersrand.

Shirzadegan, S., Nordlund, E., Nyberg, U., Zhang, P. & Malmgren, L. (2011) Rock support subjected to dynamic loading: Field testing of ground support using simulated rockburst, *Proc. 12th Int. Cong. Int. Soc. Rock Mech., Beijing, China*, CRC Press/Balkema, pp. 1269–1273.

Simser, B. (2014) Empirical experience with shotcrete in deep underground mines, WSN 2014 pdf.

Simser, B., Joughin, W.C. & Ortlepp, W.D. (2002) The performance of Brunswick Mine's rockburst support system during a severe seismic episode, *Jl S. Afr. Inst. Min. Metall.*, 102, 217–224.

Stacey, T.R. (2009) Design – a Strategic Issue, *Jl S. Afr. Inst. Min. Metall.*, 109 (3), 157–162.

Stacey, T.R. (2011) Support of excavations subjected to dynamic (rockburst) loading, *Keynote Paper, Proc. 12th Int. Cong. Int. Soc. Rock Mech., Beijing, China*, CRC Press/Balkema, pp. 137–145.

Stacey, T.R. (2012) A philosophical view on the testing of rock support for rockburst conditions, *Jl S. Afr. Inst. Min. Metall.*, 113 (7), 227–245.

Stacey, T.R. (2013) Dynamic rock failure and its containment – a Gordian Knot design problem, *Rock Dynamics and Applications State of the Art, Proc. Int Conf. RocDyn 1, Lausanne*, CRC Press/Balkema, Taylor & Francis Group, pp. 57–70.

Stacey, T.R., Ortlepp, W.D. & Ndlovu, X. (2007) Dynamic rock failures due to "high" stress at shallow depth, *Proc 4th Int. Seminar on Deep and High Stress Mining, Perth*, Australian Centre for Geomechanics, pp 193–204.

Stacey, T.R. & Ortlepp, W.D. (1999) Retainment support for dynamic events in mines, *Proc. Int. Symp. Rock Support and Reinforcement Practice in Mining, Kalgoorlie, Australia, March 1999*, Balkema, pp. 329–333.

Stacey, T.R. & Ortlepp, W.D. (2001) Tunnel surface support – capacities of various types of wire mesh and shotcrete under dynamic loading, *Jl S. Afr. Inst. Min. Metall.*, 101 (7), 337–342.

Stacey, T.R. & Ortlepp, W.D. (2002a) The contribution of wire rope lacing in surface support, *Proc. 2nd Int. Seminar on Surface Support Liners: Thin Sprayed Liners, Shotcrete, Mesh, Johannesburg*, S. Afr. Inst. Min. Metall., pp. 1–8.

Stacey, T.R. & Ortlepp, W.D. (2002b) Yielding rock support – the capacities of different types of support, and matching of support type to seismic demand, *Proc. Int. Seminar Deep and High Stress Mining, Perth, Australia, November 2002*, Australian Centre for Geomechanics, Section 38, 10p.

Stacey, T.R. & Rojas, E (2013) A potential method of containing rockburst damage and enhancing safety using a sacrificial layer, *Jl S. Afr. Inst. Min. Metall.*, 113, 565–573.

Stacey, T.R. & Yathavan, K. (2003) Examples of fracturing of rock at very low stress levels, *Proc. 10th Int. Cong. Int. Soc. Rock Mech, ISRM 2003 – Technology Roadmap for Rock Mechanics*, S. Afr. Inst, Min, Metall., pp 1155–1159.

Tannant, D.D., McDowell, G.M., Brummer, R.K. & Kaiser, P.K. (1993) Ejection velocities measured during a rockburst simulation experiment, *Proc. 3rd Int. Symp. on Rockbursts and Seismicity in Mines, Kingston*, A A Balkema, Rotterdam, pp. 129–133.

Tarasov, B. & Potvin, Y. (2013) Universal criteria for rock brittleness estimation under triaxial compression, *Int. J. Rock Mech. Min. Sci*, 59, 57–69.

Toper, A.Z., Kabongo, K.K., Stewart, R.D. & Daehnke, A. (2000) The mechanism, optimization and effects of preconditioning, *Jl S. Afr. Inst. Min. Metall.*, 100, 7–15.

Van Sint Jan, M. & Palape, M. (2007) Behaviour of steel plates during rockbursts, In Potvin (ed.), *Deep Mining 07*, Australian Centre for Geomechanics, Perth, pp. 405–412.

Vieira, F.M.C.C. & Budavari, S. (2003) Evaluation of four layouts for gold mining at a depth-range of 3500 and 5000m, *Proc. 10th Int. Cong. Int. Soc. Rock Mech. ISRM 2003 – Technology Roadmap for Rock Mechanics*, S. Afr. Inst. Min. Metall., 2, 1299–1310.

Vieira, F.M.C.C. & Durrheim, R.J. (2001) Probabilistic mine design methods to reduce rock-burst risk, *Rocbursts and Seismicity in Mines – RaSIM5*, S. Afr. Inst. Min. Metall., Symposium Series 27, 251–262.

Villaescusa, E.V., Player, J.R. & Thompson, A.G. (2014) A reinforcement design methodology for highly stressed rock masses, *ARMS8, Proc. 8th Asian Rock Mechanics Symposium, Sapporo, Japan, October*, 8p.

Wagner, H. (1982) Support requirements for rockburst conditions, In Gay & Wainwright (eds.), *Proc. 1st Int Symp. Rockbursts and Sesimicity in Mines*, S. Afr. Inst. Min. Metall. Symp. Series No 6 (1984), pp. 209–218.

Yi, X. & Kaiser, P.K. (1994) Impact testing for rockbolt design in rockburst conditions, *Int. J. Rock Mech. Min. Sci. and Geomech. Abstr.*, 31 (6), 67–68.

Zhang, C., Feng, X-T., Zhou, H., Qiu, S. & Wu, W. (2012a) Case histories of four extremely intense rockbursts in deep tunnels, *Rock Mech. Rock Eng.*, 45, 275–288.

Zhang, C., Feng, X-T., Zhou, H., Qiu, S. & Wu, W. (2012b) A top pilot tunnel preconditioning method for the prevention of extremely intense rockbursts in deep tunnels excavated by TBMs, *Rock Mech. Rock Eng.*, 45, 289–309.

Chapter 11

Tests on creep characteristics of rocks

Ö. Aydan[1], T. Ito[2] & F. Rassouli[3]

[1]Department of Civil Engineering and Architecture, University of the Ryukyus, Nishihara, Okinawa, Japan
[2]Department of Civil Engineering, National College of Technology, Toyota, Aichi, Japan
[3]Department of Earth Sciences, Stanford University, Stanford, USA

Abstract: Creep characteristics of rocks are very important for assessing the long-term stability of rock engineering structures. Creep tests have been carried out on soft rocks, medium hard rocks and hard rocks. These experiments are mostly carried out under compressive loading conditions. There are few studies on rocks for creep tests under a tensile loading regime and direct shear loading regime. Furthermore, the in-situ creep tests are rarely carried out. This chapter describes the state of art on creep tests under laboratory and in-situ conditions. In addition, the impression creep test is described as an index creep tests and its possible use for the evaluation of creep properties of rocks is discussed. Constitutive models are briefly described and their applications to actual experimental results are given.

1 INTRODUCTION

It is important to note that creep is one aspect of time-dependent behavior of rocks (*i.e.* Aydan *et al.*, 2013). In Figure 1, three cases are illustrated with respect to the complete stress-strain curve: creep, which is increasing strain when the stress is held constant; stress relaxation, which is decreasing stress when the strain is held constant; and a combination of both when the rock unloads along a chosen unloading path (*i.e.* Hagros *et al.*, 2008).

Creep tests have been carried out on soft rocks such as tuff, shale, lignite, and sandstone, medium hard rocks such as marble, limestone and rock salt and hard rocks such as granite and andesite (*i.e.* Akagi, 1976; Akagi *et al.*, 1984; Akai *et al.*, 1979; Ito & Akagi, 2001; Berest *et al.*, 2005; Doktan, 1983; Passaris, 1979; Serata *et al.*, 1968; Wawersik, 1983; Okubo *et al.*, 1991, 1993; Masuda *et al.*, 1987, 1988; Ishizuka *et al.*, 1993; Lockner & Byerlee, 1977; Bukharov *et al.*, 1995; Fabre & Pellet, 2006). These experiments were mostly carried out under compressive loading conditions. There are few studies on rocks for creep tests under a tensile loading regime (Ito & Sasajima, 1980, 1987; Ito *et al.*, 2008; Aydan *et al.*, 2011) and discontinuities and interfaces under direct shear loading regime (Amadei & Curran, 1982; Aydan *et al.*, 1994; Voegel *et al.*, 1998). Particularly, shallow underground openings may be subjected to sustained tensile stress regime, which require the creep behavior of rocks under such condition.

This chapter deals only with creep behavior of rocks, which is particularly relevant for cases where the applied load or stress is actually kept constant. First, fundamental

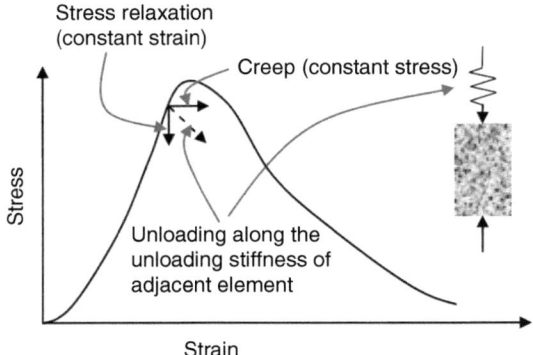

Figure 1 Possible stress-strain paths during testing for the time-dependent characteristics of rocks (from Hagros *et al.*, 2008).

definitions on the creep behavior of rocks are given. Then the utilization of impression creep testing technique as index test for creep responses of rocks is presented and its potential use in evaluating creep properties of rocks are discussed. And then experimental procedures under laboratory and in-situ conditions are described. Finally, constitutive models for modeling creep behavior of rocks are briefly summarized and their applications to actual tests are given.

2 CREEP RESPONSE OF GEOMATERIALS

The creep experiments are often used to determine the time-dependent strength and/or time-dependent deformation modulus of rocks. It has often been stated that the creep of rocks does not occur unless the load/stress level exceeds a certain threshold value, which is sometimes defined as the long-term strength of rocks (Ladanyi, 1974; Bieniawski, 1970). However, experiments carried on igneous rock (*i.e.* granite, gabbro etc.) beams by Ito (1991) for three decades show that a creep response definitely occurs even under very low stress levels. The threshold value suggested by Ladanyi (1993) may be associated with the initiation of dilatancy of volumetric strain as illustrated in Figure 2. The initiation of dilatancy generally corresponds to 40–60% of the stress level and the fracture propagation tends to become unstable when the applied stress level exceeds 70–80 % level of the ultimate deviatoric strength for given stress state (Aydan *et al.*, 1994; Hallbauer *et al.*, 1973) Therefore, the behavior below the threshold generally corresponds to visco-elastic behavior. Creep threshold according to Ladanyi (1974) corresponds to an elasto-visco-plastic response and it should not be possible to obtain visco-elastic properties directly from measured responses.

The creep responses terminating in failure are generally divided into three stages as shown in Figure 3. These stages are defined as primary, secondary and tertiary creep stages. The secondary stage appears to be a linear response in time (but in a real sense, it is not a linear response). On the other hand, the tertiary stage is the stage in which the strain response increases exponentially resulting in the failure of Creep behavior. The

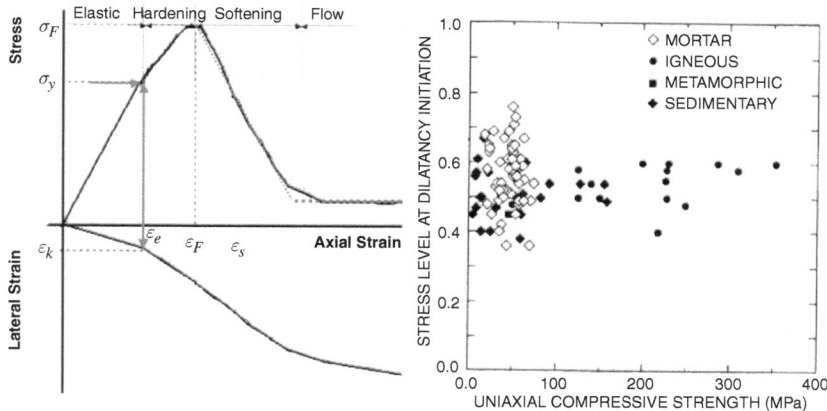

Figure 2 Illustration of threshold value for dilation and experimental results for different rocks (arranged from Aydan *et al.*, 1993, 1994).

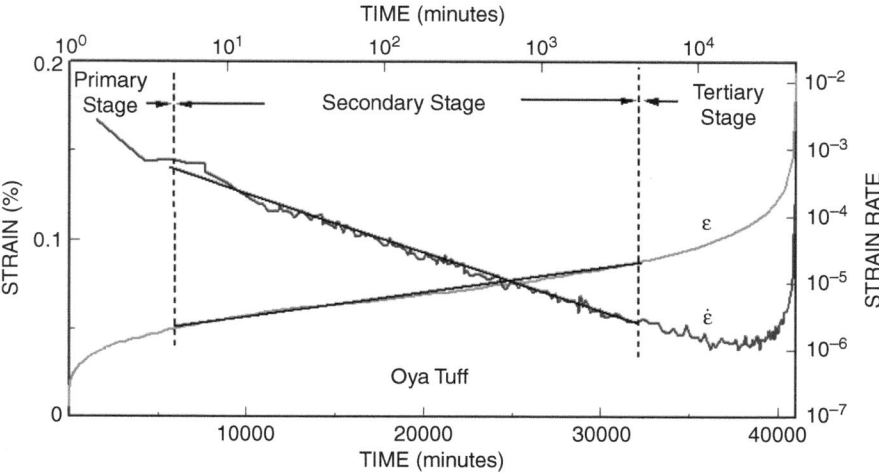

Figure 3 Strain and strain rate response of a creep experiment on Oya tuff (Japan).

modeling of this stage in the constitutive laws is an extremely difficult aspect as it also depends upon the boundary conditions.

3 LABORATORY CREEP TESTING DEVICES

Apparatuses for creep tests can be of the cantilever type or the load/displacement-controlled type. Although the details of each testing machine may differ, the features of apparatuses for creep tests are described herein.

3.1 Cantilever type testing device

The cantilever-type apparatus has been used in creep tests since early times (*i.e.* Serata *et al.*, 1968; Akagi, 1976; Farmer, 1983; Ito & Akagi, 2001) (Figure 4). It is in practice the most stable apparatus for creep tests because the load level can easily be kept constant with time. The greatest restrictions of this type apparatus are the level of applicable load, which depends upon the length of the cantilever arm and its oscillations during the application of the load. The cantilever-type apparatus utilizing a multi-arm lever overcomes the load limit restrictions (Okada, 2005, 2006). The oscillation is another technical problem to be dealt by the producers of the creep devices. If the load increase is manually done through putting deadweights in some creep testing devices, an utmost care must be undertaken during loading procedure in order to prevent undesirable oscillations.

The load is applied onto samples by attaching deadweights to the lever, which may be done manually for low-stress creep tests or mechanically for high-stress creep tests. In triaxial experiments, special load cells are required and the confining pressure is generally provided through oil pressure. The maximum care should be taken in keeping the confining pressure constant in terms of the continuous power supply for the compressor of the confining pressure system.

Deformation and strain measurements can be taken in several ways. The simple approach is to utilize a couple of LVDTs. When a triaxial creep experiment is carried out, the LVDTs may be fixed onto the sample and inserted into the triaxial chamber. In

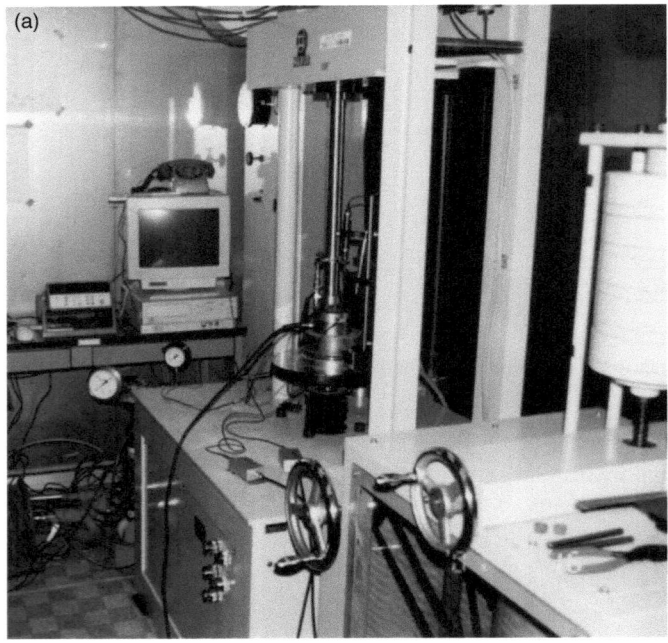

Figure 4 Examples of cantilever type creep apparatuses: (a) Single arm cantilever type creep apparatus, (b) multi-lever arm cantilever type creep apparatus (pictures by Ö.A.)

such a case, special precautions is necessary for the accurate measurement of displacements. Strain gauges may be used; however, the strain gauges glued onto samples are required to be capable of measuring strain over a long period of time without any debonding. For lateral deformation or strain measurements, diametric or circumferential sensors are used.

3.2 Load/displacement controlled apparatus

Loading testing system is a servo-controlled testing machine that is capable of applying high constant loads onto samples (Figure 5). The most critical aspect of this experiment is to keep very high axial stresses acting on a sample constant, which will require continuous monitoring of the load and its automatic adjustment (*i.e.* Peng, 1973). The load applied onto samples is maintained to within ±1% of the specified load. The deformation or strain measurements are measured in the same way as in the cantilever type creep experiments. Vibration associated with the constant high-speed closed-loop operation is a matter of concern.

There are also true triaxial testing apparatuses (loading is performed independently in three directions on cubic or prismatic samples) to perform creep tests under true triaxial stress conditions (Serata *et al.*, 1968; Adachi *et al.*, 1969). Three principal stresses can be controlled independently in such triaxial testing apparatuses. New technologies make such tests to be performed much easier.

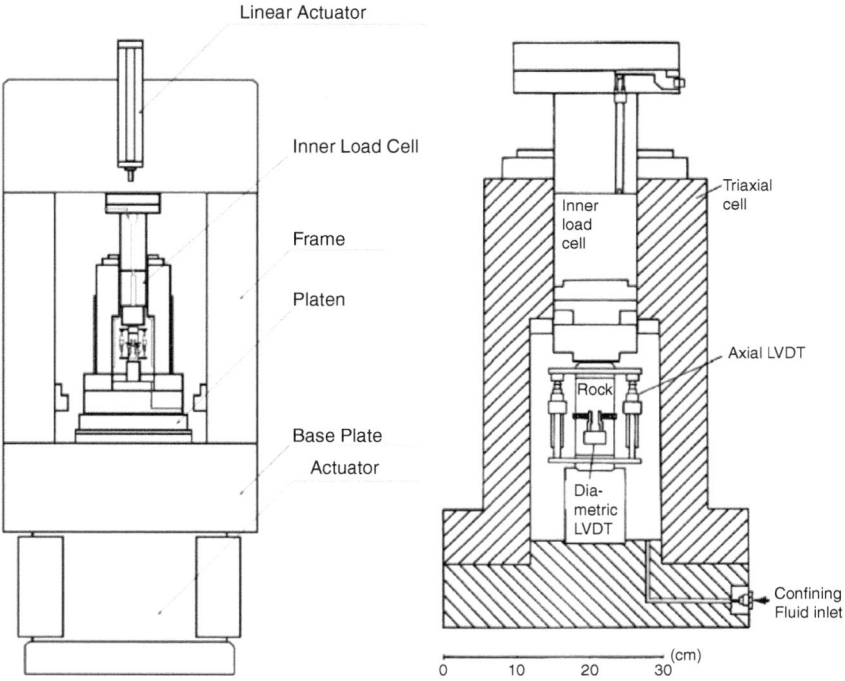

Figure 5 Load/displacement controlled apparatus (from Ishizuka *et al.*, 1993).

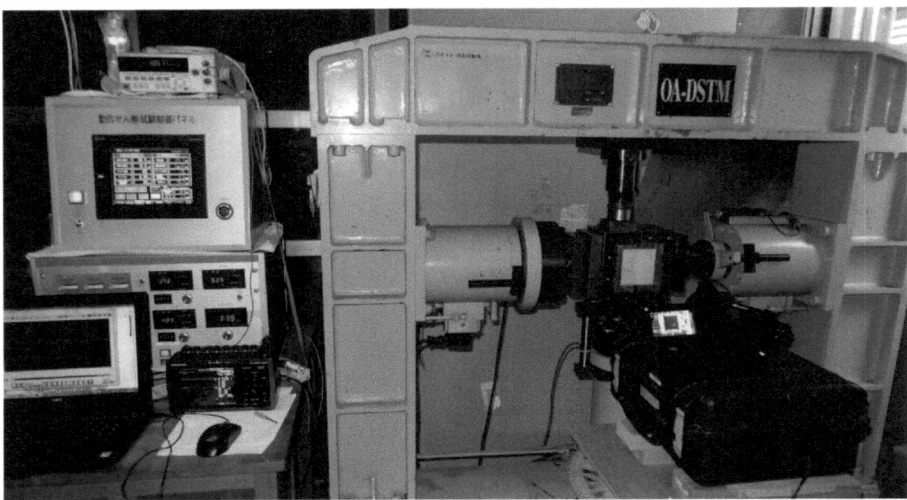

Figure 6 Multi-purpose dynamic shear-testing machine with an ability to perform creep tests on rocks, discontinuities and interfaces at the University of the Ryukyus.

Creep tests under direct shear stress condition on rocks, discontinuities and interfaces are also carried out using a servo-control loading system (*i.e.* Amadei & Curran, 1982; Aydan *et al.*, 1994, 2016; Voegler *et al.*, 1998; Larson & Wade, 2000). Figure 6 shows the multi-purpose dynamic shear-testing machine with an ability to perform creep tests on rocks, discontinuities and interfaces at the University of the Ryukyus. The device was originally developed for conventional direct shear creep test and cyclic shear tests and has been recently upgraded to perform dynamic shear testing (Aydan *et al.*, 1994, 2016).

4 LABORATORY CREEP TESTS

In this section, some examples of creep tests performed under different environmental conditions are described. Although there are numerous experimental studies on the creep response of rocks (*i.e.* Berest *et al.*, 2005; Doktan, 1983; Passaris, 1979; Serata *et al.*, 1968; Wawersik, 1983 etc.), the creep experiments on tuff samples from Oya region of Japan and Cappadocia region of Turkey would be referred. Nevertheless, the experimental results and conclusions are relevant for creep tests on other rocks.

4.1 Uniaxial compression tests

The procedure described in the method suggested by the ISRM (ISRM, 2007) to test for uniaxial compression strength should be followed unless the size of the samples differs from the conventional size. The displacement is measured continuously or periodically (seconds, minutes, hours or days depending upon the stress level applied on samples) as suggested in the ISRM Suggested Methods (ISRM, 2007; Aydan *et al.*, 2013). The load application rate might be higher than that used in the ISRM Suggested Methods when a

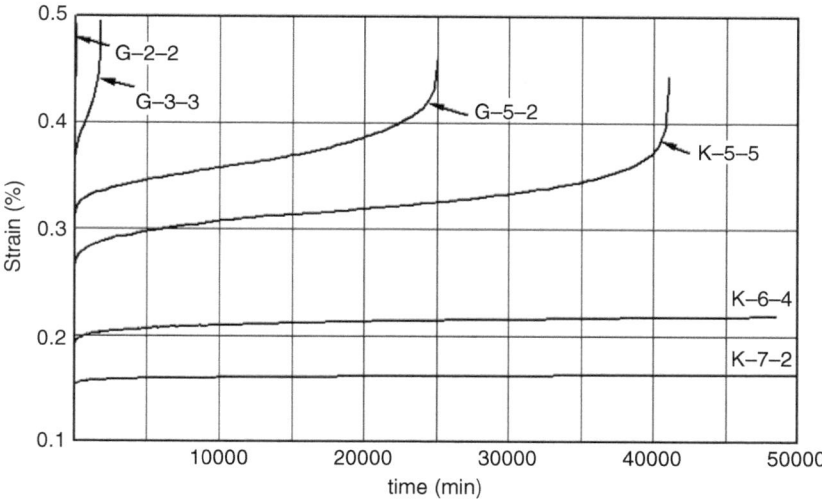

Figure 7 Uniaxial compression creep response of Oya tuff under dry condition (modified from Ito & Akagi, 2001).

cantilever type apparatus is used. Once the load reaches the designated load level, it should be kept constant. If the experiments are to be carried out under saturated conditions, the sample should be put in a special water-filled cell. Results of creep experiments are generally presented in the space of time and strain for different combinations of experimental conditions. Rock specimens used in experiments presented here are obtained from tuff blocks from Oya region in Japan and Cappadocia region in Turkey.

4.1.1 Creep tests under dry conditions

Creep tests on Oya tuff carried out by Ito & Akagi (2001) under dry conditions are plotted in Figure 7. As noted from Figure 7, some of responses terminate with failure while the others become asymptotic to certain strain levels depending the applied stress ratio (SR), which is defined as the ratio of applied stress to the short-term strength. The responses terminating in failure are generally divided into three stages as shown in Figure 3. The transitions from the primary stage to the secondary stage and from the secondary stage to the tertiary stage are generally determined from the deviation of a linearly decreasing or increasing strain rate plotted in a logarithmic time space. Generally, it should, however, be noted that strain data must be smoothed before its interpretation. Direct derivation of strain data containing actual responses as well as electronic noise may produce entirely different results.

4.1.2 Effect of saturation on uniaxial compression creep tests

When rocks have water absorption ability, their strength tends to decrease compared with that under dry condition (*i.e.* Aydan, 1993; Aydan & Ulusay, 2002, 2013). Particularly, the strength of soft rocks like tuffs decreases drastically and the strength

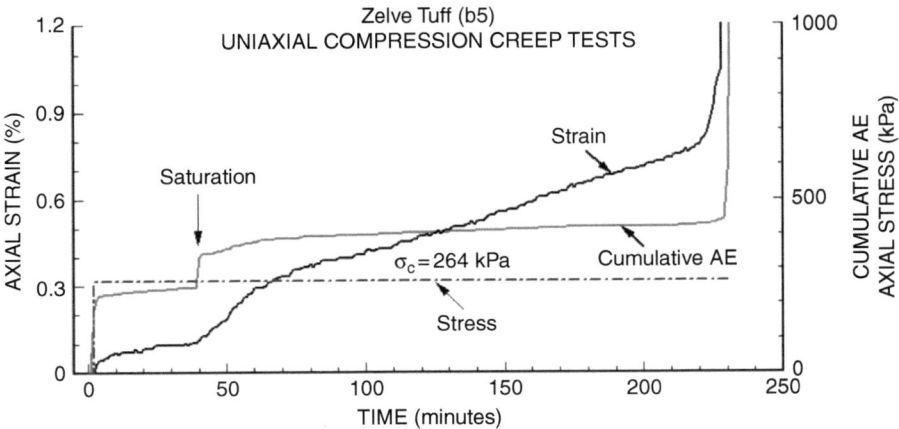

Figure 8 Responses of initially dry later saturated tuff samples from Zelve (b5) during uniaxial compression creep tests (arranged from Ito *et al.*, 2008).

reduction is generally greater than 60%. In some cases, soft rocks may disintegrate upon water absorption and the resulting strength reduction may be up to 100%. Several researchers investigated the effect of saturation on uniaxial compression creep tests (*i.e.* Ito *et al.*, 2008; Okubo & Chu, 1994; Okubo *et al.*, 2005; Aydan *et al.*, 2013). Figure 8 shows an example of creep response of a tuff sample from Zelve, Cappadocia (Turkey). The sample was initially subjected to a creep loading at a level of about 16% of its uniaxial compressive strength under dry conditions. The sample was fully saturated 40 minutes after the start of the creep test. The stress ratio becomes about 95% of the uniaxial compressive strength under saturated condition. As the stress ratio increased, the sample failed about 190 minutes after the saturation. Creep experiments carried out on tuff samples from Derinkuyu, Avanos and Ürgüp yielded similar results (*i.e.* Aydan and Ulusay, 2013; Ulusay *et al.*, 2013).

4.1.3 Effect of temperature on uniaxial compression creep tests

Effect of temperature on creep response of various rocks is investigated by various researchers (*i.e.* Shibata *et al.*, 2007; Okada, 2005, 2006; Cristescu & Hunsche, 1998; Hunsche & Hampell, 1999). It is well known that the strength of rocks decreases with temperature (*i.e.* Handin, 1966; Shimada, 1993; Hirth & Tullis, 1994; Brace & Kohlstedt, 1980). Figure 9 shows plots of responses during uniaxial compression creep tests on Oya tuff and its failure time determined at different temperatures. As noted from the figure, the creep response is accelerated and the long-term uniaxial compression strength of Oya tuff decreases.

4.2 Triaxial compression tests

Triaxial compression creep experiments are quite limited as compared with uniaxial compression creep experiments due to sophistication of equipments and costs. Nevertheless, there were several attempts to conduct such tests (*i.e.* Serata *et al.*,

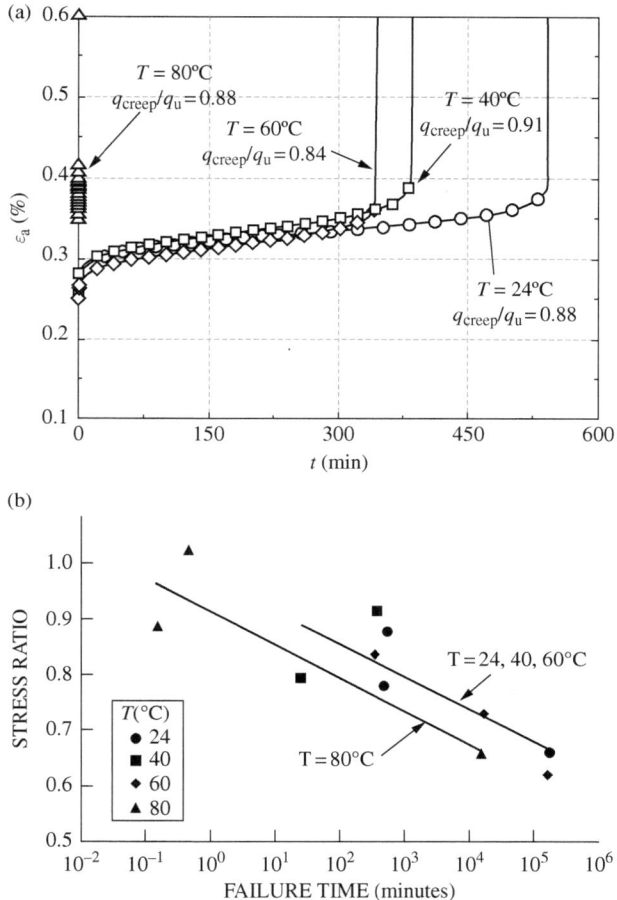

Figure 9 (a) Creep response of Oya tuff and (b) Relationship between stress ratio and failure time at various temperatures (arranged from Shibata *et al.*, 2007).

1968; Lockner & Byerlee, 1977; Waversik, 1983; Masuda *et al.*, 1987; Okada, 2005; Ito *et al.*, 1999). Provided that friction angle is not rate-dependent, the stress ratio under triaxial compression creep test are defined in an analogy to that in uniaxial compression creep tests as:

$$SR = \frac{\sigma_1 - \sigma_3}{2c\cos\phi + (\sigma_1 + \sigma_3)\sin\phi} \tag{1}$$

where c, ϕ, σ_1 and σ_3 are cohesion, friction angle and maximum applied and confining stresses, respectively. If friction angle is rate-dependent, the ratio of the applied deviatoric stress to the deviatoric strength is used as stress ratio. However, the experimental results confirm that the rate-dependency of friction angle is negligible according to Aydan & Nawrocki (1998).

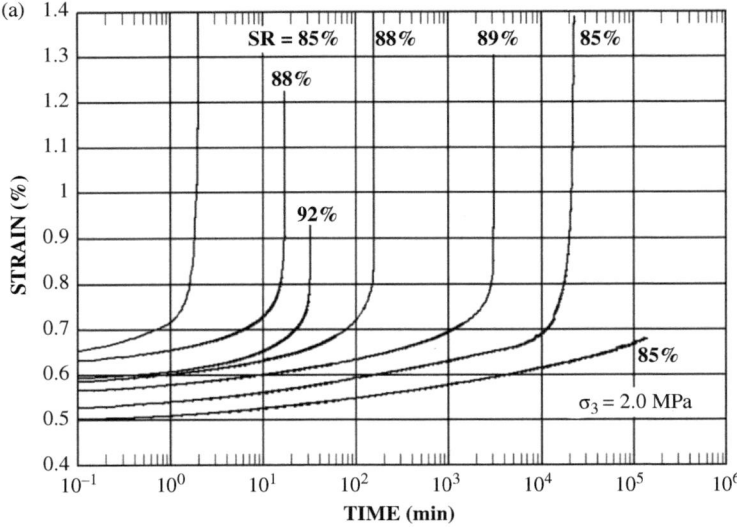

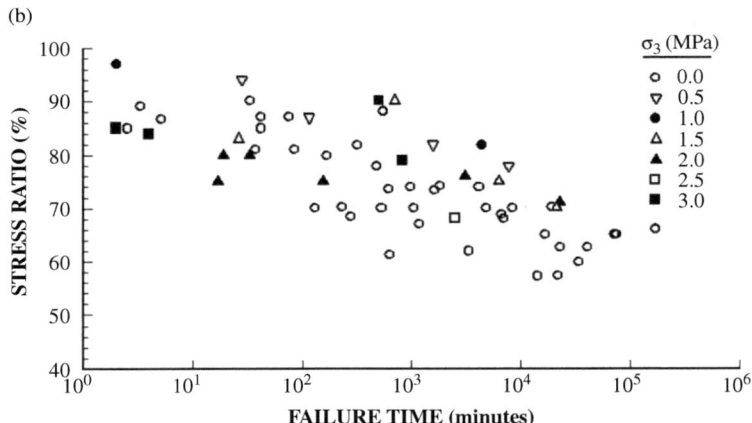

Figure 10 (a) Creep response at confining pressure of 2 MPa, (b) Creep failure time of Oya tuff uniaxial and triaxial compression creep tests (arranged from Ito *et al.*, 1999; Shibata *et al.*, 2007; Akai *et al.*, 1979).

Figure 10 shows the creep response under a confining stress of 2 MPa and the failure time of compression creep tests under both uniaxial and triaxial compression environment. It is interesting to note that the overall tendency obtained in triaxial creep tests are basically similar to those of uniaxial compression creep tests irrespective of confining pressure.

4.3 Brazilian creep tests

There are not many studies on tensile creep behavior of rocks using Brazilian creep tests. However, rock may be subjected to tensile stresses in nature such as cliffs with toe erosion and roof layers above underground openings excavated in sedimentary rocks. Aydan *et al.* (2011, 2013), Agan *et al.* (2013) and Ulusay *et al.* (2013) have recently reported some Brazilian creep tests on tuff samples. The tensile strength of the specimen is calculated using the well-known following formula:

$$\sigma_t = \frac{2}{\pi} \frac{P}{Dt} \tag{2}$$

where P is the load at failure, D is the diameter of the test specimen (mm), t is the thickness of the test specimen measured at its center (mm). The nominal strain of the Brazilian tensile test sample can be given as (for details see Hondros, 1959; Jaeger & Cook, 1979)

$$\varepsilon_t = 2\left[1 - \frac{\pi}{4}(1 - v)\right]\frac{\sigma_t}{E} \quad \text{with} \quad \varepsilon_t = \frac{\delta}{D} \tag{3}$$

where δ is diametrical displacement in loading direction.

If Poisson's ratio of rock is unknown, it is reasonable to choose Poisson's ratio as 0.25. Thus, the formula given above can be simplified to the following form (*i.e.* Aydan *et al.*, 2011)

$$\varepsilon_t = 0.82\frac{\sigma_t}{E} \tag{4}$$

Here we quote some experimental results from Ito *et al.* (2008) and Aydan *et al.* (2011). The diameter of samples was 46mm and their thickness ranged between 14 and 25 mm. All samples were subjected to creep loading level at a chosen period of time under dry conditions. After reaching the ultimate loading level, the samples were saturated. Figure 11 shows some of the measured response of a sample in Brazilian creep experiments on Oya tuff. Oya tuff sample numbered SN1-W3 was tested under fully saturated conditions at a stress ratio of 87%. As noted from the figures, acoustic emission occurs at each load increase, simultaneously.

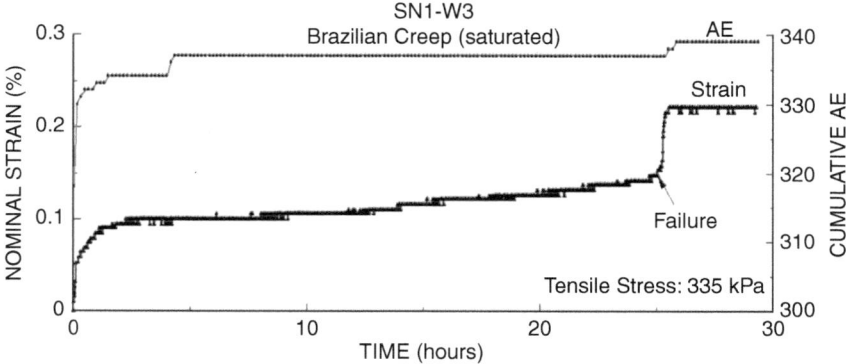

Figure 11 Brazilian creep response of SN1-W3 sample.

4.4 Direct shear creep tests

Direct shear creep tests on rocks, discontinuities and interfaces are also quite rare. Amadei & Curran (1982) performed direct shear creep tests on rock discontinuities. The direct shear tests by Aydan *et al.* (1994, 2016), Voegler *et al.* (1998), Larson & Wade (2000) may be counted in addition the initial tests performed by Amadei & Curran (1982). We present the experimental results by Aydan *et al.* (1994) performed on the interfaces and grouting material in rock anchor systems. Figure 12 shows the direct shear creep experiment on tendon-grout interface under a normal stress of 2 MPa. The stress ratio was about 95%. The overall response is similar to those of uniaxial and triaxial compression and Brazilian creep tests.

Figure 13 shows the creep responses of grouting material of rock anchor systems tested under direct shear condition. The initial instantaneous displacements are

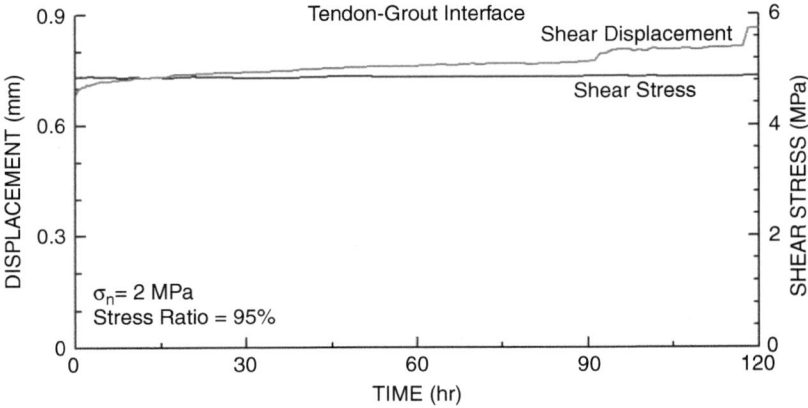

Figure 12 Direct shear creep test on tendon-grout interface.

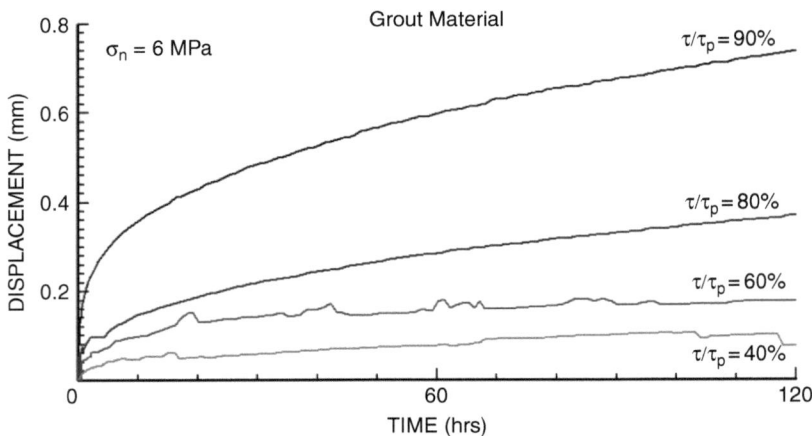

Figure 13 Responses of grouting material measured during direct shear creep test at various stress ratios.

subtracted from displacement response for each stress ratio. Similarly, the creep displacement increases as the stress ratio becomes higher.

5 IMPRESSION CREEP TEST AS AN INDEX CREEP METHOD

Impression creep experiments are relatively easy to perform and the capacity of loading equipments is relatively small compared to conventional creep experiments. The critical issue with this technique is the definition of strain and stress, which can be associated with conventional creep experiments. There are several proposals on how to correlate impression creep experiments to conventional creep experiments, which are summarized in Table 1 (*i.e.* Hyde *et al.*, 1996; Timoshenko & Goodier, 1970; Sastry, 2005; Rassouli *et al.*, 2010; Aydan *et al.*, 2011). If applied load is assumed to be the same, all equations in Table 1 imply that corresponding strains would be smaller so that plastic behavior would occur at higher loading levels.

The loading in impression creep tests is achieved through dead weights and/or hydraulic jacks. Figure 14 shows two examples of impression creep testing device. Indenters may have different forms. Mousavi *et al.* (2008), Rassouli *et al.* (2010) and Aydan *et al.* (2011, 2012, 2013) are probably first pioneers to utilize this index technique in rock mechanics and rock engineering. Mousavi *et al.* (2008) and Rassouli *et al.* (2010) utilized flat-ended cylindrical indenters. The preferable diameter was 3 mm. Aydan *et al.* (2008, 2011) also used an indenter having a diameter ranging from 1 to 3mm. They concluded that the indenter with a diameter of 3mm was preferable, which are in accordance with the conclusion of Mousavi *et al.* (2008) and Rassouli *et al.* (2010). Aydan *et al.* (2012, 2013) also utilized the indenter of the needle penetration index test device (Aydan, 2012, 2013; Ulusay *et al.*, 2013).

The experimental results are presented in this subsection using the device shown in Figure 14(a) with a flat-ended indenter having a diameter of 3 mm. The device is capable of inducing loads, which is 10 times the applied load at the end of the arm. The device was equipped with a displacement transducer and an acoustic emission sensor. However, electric potential measurement system could be included in the monitoring system under dry condition. Figure 15 shows the results of an impression creep test on Oya tuff sample denoted WEZ-4 under saturated condition. The saturated strength of Oya tuff is about 40–50% of that under dry condition and yielding level is expected to be more than 14 MPa. The response becomes stable following the applied nominal pressure of 12.2 MPa. However, the sample fails when the applied pressure is 21 MPa. The stress ratio is about 61% in view of the short-term indentation tests.

An impression creep experiment carried out on rocksalt sample from Tuzköy in Cappadocia region of Turkey. The short-term and long-term properties of this rocksalt

Table 1 Proposed correlations between impression creep experiments and conventional uniaxial compression creep experiments.

Reference	Stress	Strain
Hyde *et al.*, 1980	$\sigma = \eta p$	$\varepsilon = \frac{1}{\beta} \cdot \frac{\delta}{D}$
Timoshenko & Goodier, 1970	p	$\varepsilon = \frac{\delta}{D} = \frac{\pi(1-\nu^2)}{4E} p$
Aydan *et al.*, 2008	p	$\varepsilon = \frac{\delta}{D} = \frac{1+\nu}{2E} p$

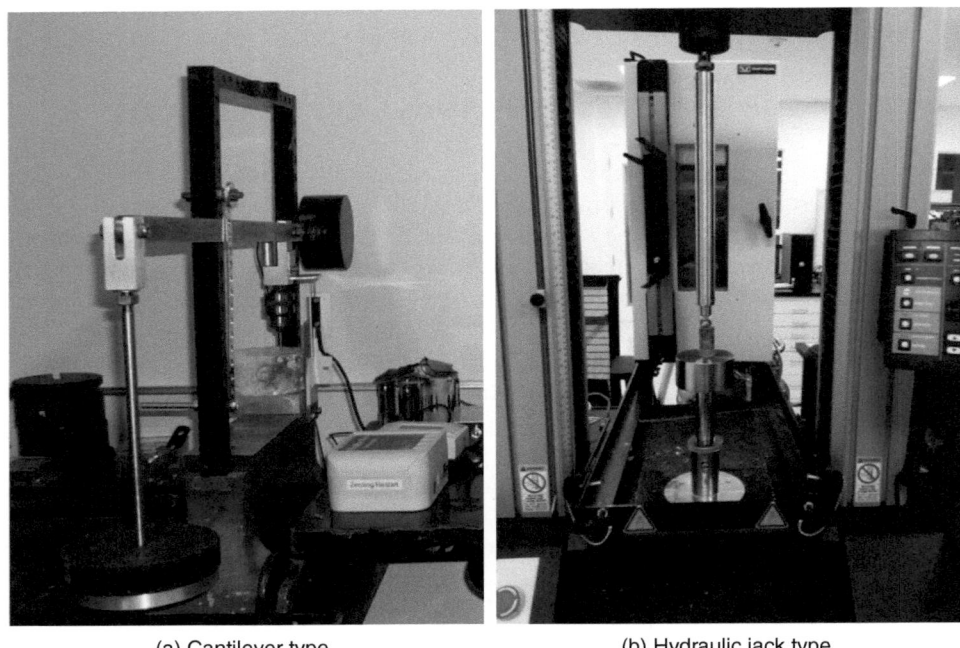

(a) Cantilever type   (b) Hydraulic jack type

Figure 14 Two examples of impression creep devices.

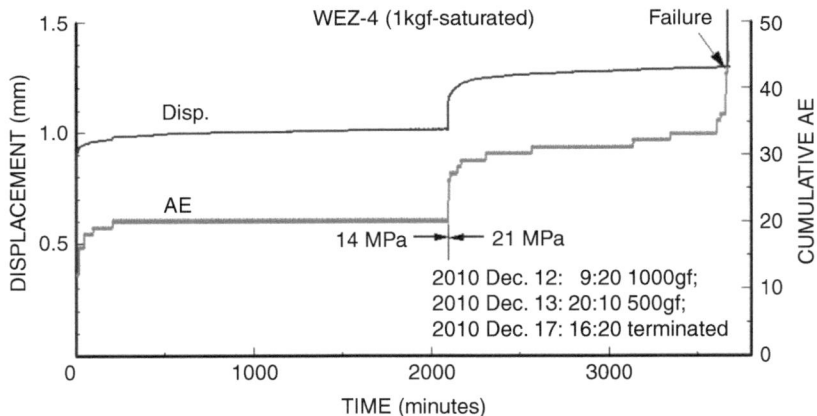

Figure 15 Impression creep response of saturated Oya tuff sample denoted WEZ-4.

was investigated by Özkan *et al.* (2009) and Özşen *et al.* (2014) under uniaxial compression creep test. The short-term average uniaxial compressive strength of Tuzköy rock salt is about 26.5 MPa. Figure 16 shows the response obtained from the impression creep test. The load level was gradually increased in steps up to 85 MPa. In

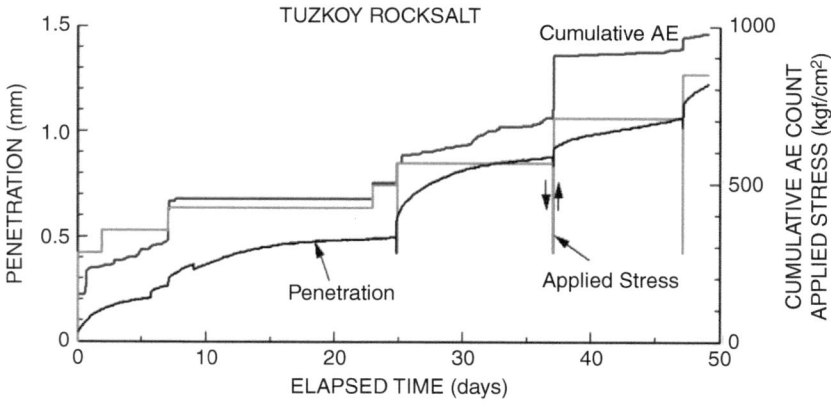

Figure 16 Response of Tuzköy rocksalt during the impression creep test.

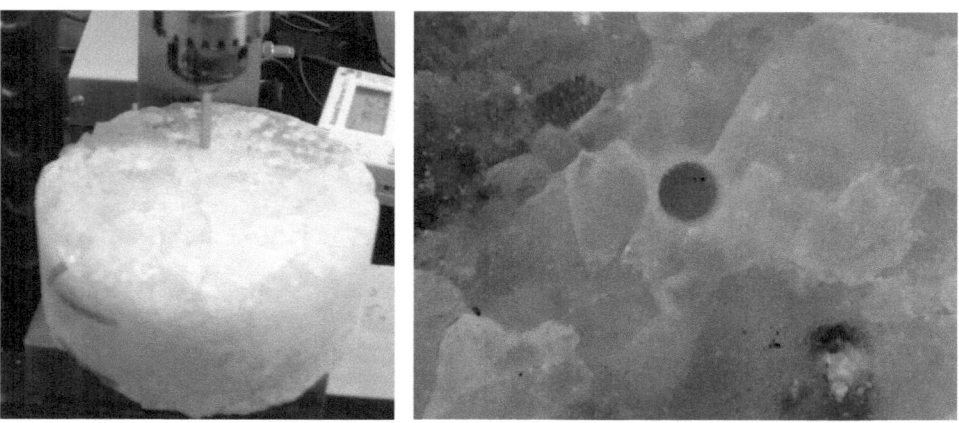

Figure 17 Views of Tuzköy rocksalt sample during and after impression test.

the last three steps, the amplitude of load was decreased to 28 MPa first and increased to designated level greater than the previous state. It was noted that the elastic recovery was very small and the behavior of rocksalt was almost visco-plastic. Upon unloading at the end of the test, a circular hole was observed as a result of permanent deformation. Furthermore, some radial fractures around the hole were formed (Figure 17).

6 LONG-TERM STRENGTH AND CORRELATIONS AMONG VARIOUS CREEP TEST RESULTS

The strength of rocks is generally assumed to be hardening type. However, it is well known that the long-term strength $(\sigma_a(t))$ of rocks decreases with time and it is expressed in the following forms

Aydan *et al.* (1996)

$$\frac{\sigma_a(t)}{\sigma_{co}} = \alpha + (1 - \alpha)e^{-b(t^* - 1)} \tag{5}$$

Aydan & Nawrocki (1998)

$$\frac{\sigma_a(t)}{\sigma_{co}} = 1 - b\ln(t^*) \tag{6}$$

Aydan *et al.* (2011) proposed the following function, which combines both functions above

$$\frac{\sigma_a(t)}{\sigma_{co}} = \alpha + (1 - \alpha)\frac{t^*}{1 + \beta(t^* - 1)} \tag{7}$$

where

α: The ultimate normalized strength of rock,

τ: The duration of short-term strength (σ_{co}) test

b: empirical constant and $t^* = \frac{t}{\tau}$.

Figure 18 compares the failure time of samples tested in Brazilian, impression and uniaxial compression creep experiments under dry and saturated conditions. From experimental results, it is very interesting to note that if the stress ratio remains same, the failure time of dry and saturated samples are very close to each other. Furthermore, the failure times of samples tested under uniaxial compression and Brazilian creep experiments are also similar to those of impression creep experiments.

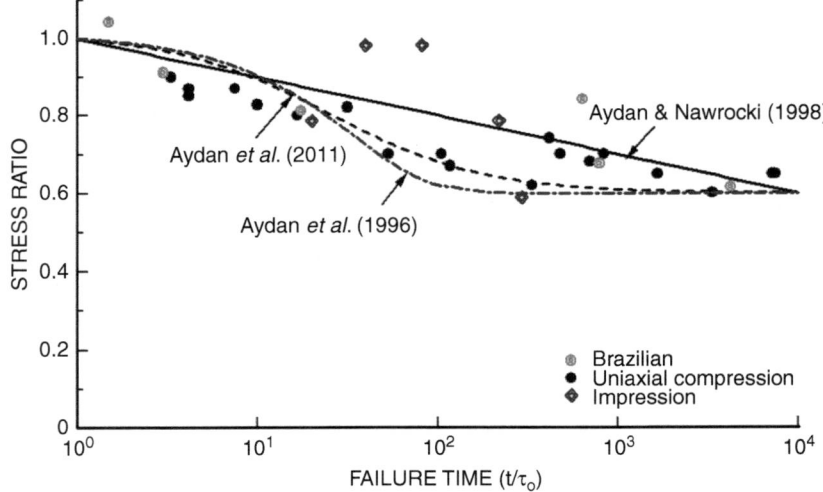

Figure 18 Comparison of failure time of various creep experiments and empirical relations by Aydan *et al.* (1996, 2011) and Aydan & Nawrocki (1998).

7 IN-SITU CREEP TESTS

Results of in-situ creep test method are used to predict time-dependent deformation characteristics of rock mass resulting from loading. This test method may be useful in structural design analysis where loading is applied over an extensive period. This test method is normally performed at ambient temperature, but equipment can be modified or substituted for operations at other temperatures. There are applications of this test technique in pillars of rock salt mines. In-situ creep tests are generally plate-bearing tests and direct shear creep tests (Figure 19). In this section, some examples from construction sites in Japan are introduced.

7.1 Plate-bearing creep tests

The diameter of platens used in plate-bearing tests is 300mm and the maximum load is about 500 kN. The maximum nominal pressure is about 7.2 MPa. The deformation modulus is obtained from the following relation based on Boussinesq's solution:

$$E_0 = \frac{1 - v^2}{D} \cdot \frac{F}{\delta_0} \tag{8}$$

where E, v, δ_0, D and F are deformation modulus, Poisson's ratio, instantaneous settlement, diameter of platen and applied load. Poisson's ratio is generally assumed to be 0.2 or 0.25. The total displacement is the sum of initial displacement and delayed creep displacement given as:

$$\delta_t = \delta_0 + \delta_c \tag{9}$$

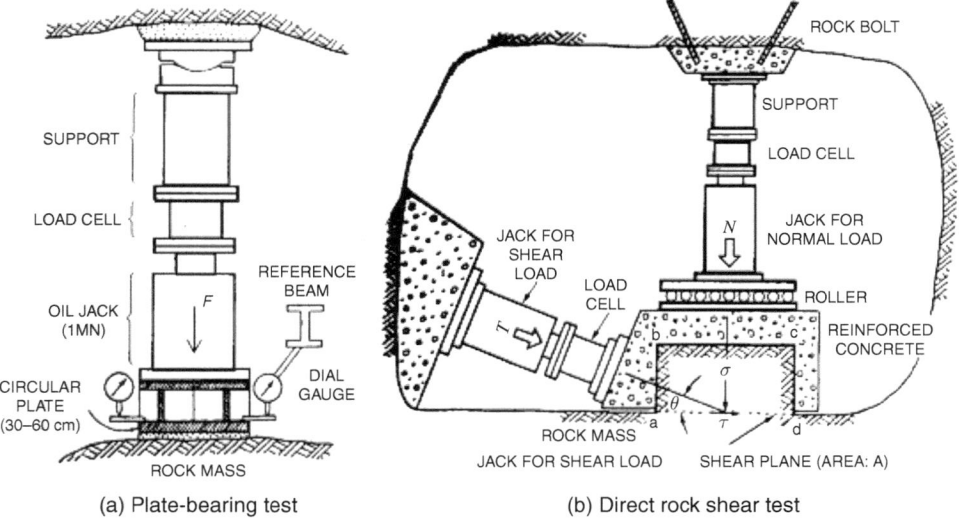

(a) Plate-bearing test   (b) Direct rock shear test

Figure 19 Illustration of in-situ tests.

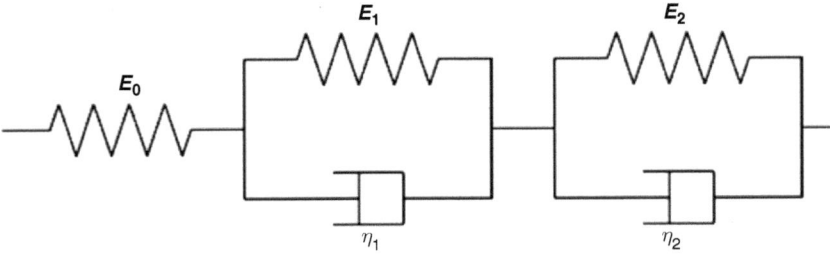

Figure 20 Five element generalized Voigt-Kelvin model.

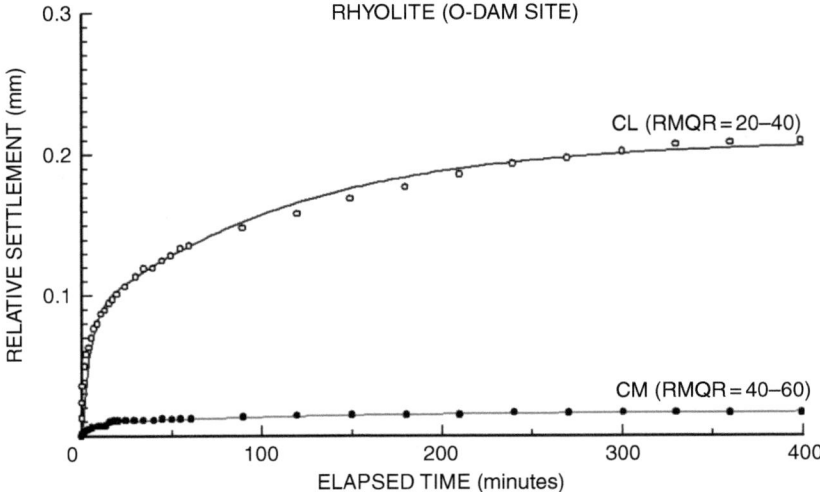

Figure 21 Creep displacement of Rhyolite rock mass at a dam site in Central Japan.

The creep displacement is given as a fraction of the total displacement using a 5 element generalized Voigt-Kelvin model as (Figure 20):

$$\delta_t = \delta_0 \left(1 + \frac{E_1}{E_0} \left(1 - \exp(-\frac{E_1}{\eta_1}t) \right) + \frac{E_2}{E_0} \left(1 - \exp(-\frac{E_2}{\eta_2}t) \right) \right) \tag{10}$$

Thus, the creep displacement would be given as

$$\delta_c = \frac{1 - v^2}{D} \cdot \frac{F}{E_0} \left[\frac{E_1}{E_0} \left(1 - \exp(-\frac{E_1}{\eta_1}t) \right) + \frac{E_2}{E_0} \left(1 - \exp(-\frac{E_2}{\eta_2}t) \right) \right] \tag{11}$$

Figure 21 shows examples of plate-bearing creep tests on rhyolite foundation of a dam-site in Central Japan. Rock-mass classified as CL and CM in rock mass classification system (DENKEN) of the Central Research Institute of Electric Power Companies of Japan. The values of parameters for creep responses shown in Figure 21 are listed in Table 2.

Table 2 The values of parameters for creep responses measured in plate bearing test.

Rock Class		δ_c/δ_e	E_0 (MPa)	E_1/E_0	E_2/E_0	E_1/η_1 (1/min)	E_2/η_2 (1/min)
DENKEN	RMQR						
CL	20–40	0.08	534	0.4	0.6	0.238	0.009
CM	40–60	0.03	3332	0.5173	0.4828	0.300	0.008

7.2 Direct rock shear creep tests

In-situ direct shear rock test set-up shown in Figure 19(b) is utilized for creep tests. The shearing area is 600 × 600 mm and the height of the sample is 300mm. Direct shear creep experiments are done in two stages. The first stage is called "Primary creep" stage and the specimen is loaded at a level of 1/3 of the ultimate peak shear strength at a given normal load. The duration of the primary creep stage is generally more than 90 minutes. The second stage is called the secondary creep stage and the specimen is loaded at a level of 2/3 of the ultimate peak shear strength at a given normal load. The duration of the primary creep stage is generally more than 120 minutes. However, the duration of the creep tests may be several days to months depending upon the importance of the structure. A generalized Voigt-Kelvin model having three elements are generally used as models for the primary and secondary creep stages

$$w_t = w_e + w_c = w_e\left(1 + a\left(1 - \exp(-\beta t)\right)\right); a = \frac{K_e}{K_v}; \beta = \frac{K_v}{\eta_v}$$

where τ, K_e, K_v and η_v are applied shear stress, Hookean stiffness; Kelvinean stiffness and Kelvinean viscosity.

Figure 22 shows examples of responses during direct shear creep tests on rhyolite foundation of a dam-site in Central Japan. Rock-mass classified as CM in rock mass classification system (DENKEN) of the Central Research Institute of Electric Power Companies of Japan. The values of parameters for creep responses shown in Figure 20 are listed in Table 3.

8 CONSTITUTIVE MODELLING OF RESULTS OF CREEP TESTS

8.1 Uni-dimensional constitutive models

Constitutive models essentially are based on responses obtained from experiments and it is fundamentally a fitting procedure of some functions to experimental results. Therefore, they cannot be purely derived from a certain theory. Nevertheless, they must satisfy certain rules established in constitutive modeling of material science. Uni-dimensional constitutive models can also be broadly divided into two categories; intuitive models and rheological models. Table 4 summarizes some of the well-known intuitive models, while Table 5 summarizes linear rheological models (*i.e.* Mirza, 1978; Doktan, 1994; Farmer, 1983). Figure 23 compares the experimental responses with those from intuitive and rheological models. As noted from these figures, each model has its own merits and demerits; the user decides, which one to adopt for his/her purpose in view of experimental results.

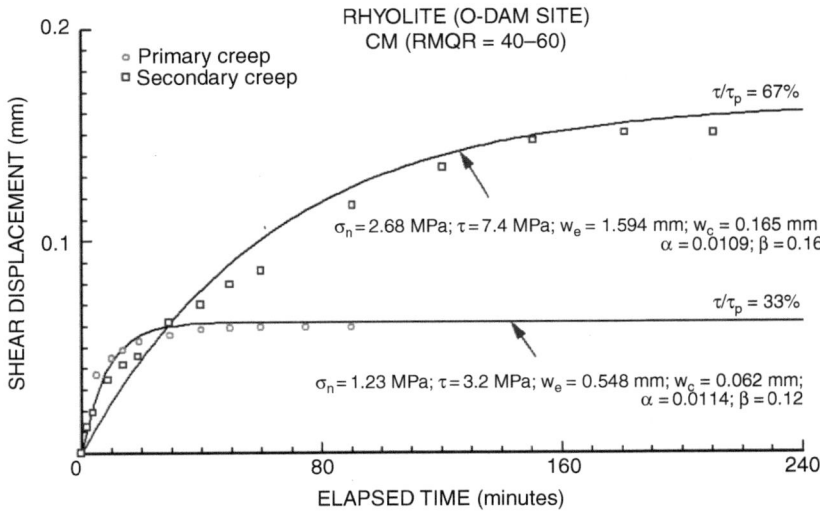

Figure 22 Direct rock shear creep displacement of Rhyolite rock mass at a dam site in Central Japan.

Table 3 The values of parameters for creep responses measured in direct shear test.

Stages	τ (MPa)	αw_e (mm)	K_e MPa/mm	w_e (mm)	$\alpha = \frac{K_e}{K_v}$	$\beta = \frac{K_v}{\eta_v}$ (1/min)
Primary creep	3.2	0.062	5.839	0.548	0.114	0.12
Secondary Creep	7.4	0.165	4.888	1.594	0.109	0.16

Table 4 Intuitive uni-dimensional creep models (except Aydan *et al.* (2003), the references can be found in Farmer (1983)).

Proposed by	Formula	Comments
Andrade (1910, 1914)	$\varepsilon_c = Bt^{1/\beta}$	Applicable to primary stage; $\beta = 3$;
Lomnitz (1956, 1957)	$\varepsilon_c = A\ln(1 + \alpha t)$	Applicable to primary stage
Modified Lomnitz	$\varepsilon_c = A + B \log(t) + t$	Primary and secondary stages
Norton's law	$\varepsilon_c = A\sigma_a^n t$ or $\dot{\varepsilon}_c = A\sigma_a^n$	Applicable to secondary stage and $n = 4$–5
Modified Norton	$\varepsilon_c = B\left\langle \dfrac{\sigma_a}{\sigma_{ct}} - 1 \right\rangle^n t$ or $\dot{\varepsilon}_c = B\left\langle \dfrac{\sigma_a}{\sigma_{ct}} - 1 \right\rangle^n$	Applicable to secondary stage and σ_{ct} is the stress threshold to induce steady state creep response.
Griggs & Coles (1958)	$\varepsilon_c = A + Bt^2$	Applicable to tertiary stage
Aydan *et al.* (2003)	$\varepsilon_c = A\left(1 - e^{-t/\tau_1}\right) + B\left(e^{t/\tau_2} - 1\right)$	Applicable to all stages creep leading to failure

$A, B, C, \alpha, \tau_1, \tau_2$ and n are constants to be determined from experimental results. $\sigma_a, \varepsilon_c, \dot{\varepsilon}_c$ and t are applied stress, creep strain, strain rate and time, respectively, hereafter.

Table 5 Rheological models uni-dimensional constitute modeling.

Model	Formula	Geometrical Illustration
Hooke	$\sigma_a = E\varepsilon$	
Newton	$\sigma_a = \eta\dot{\varepsilon}$	
Voigt-Kelvin	$\varepsilon = \dfrac{\sigma_a}{E}\left(1 - e^{-t/t_r}\right); t_r = \dfrac{\eta}{E}$	
Maxwell	$\varepsilon = \dfrac{\sigma_a}{E} + \dfrac{\sigma_a}{\eta}t$	
Generalized Voigt-Kelvin	$\varepsilon = \dfrac{\sigma_a}{E_h} + \dfrac{\sigma_a}{E_k}\left(1 - e^{-t/t_r}\right); t_r = \dfrac{\eta}{E_k}$	
Hill-Maxwell model	$\varepsilon = \dfrac{\sigma_a}{E_h}\left[1 - \dfrac{E_m}{E_h + E_m}e^{-t/t_r}\right]$	
Burgers	$\varepsilon = \dfrac{\sigma_a}{E_m} + \dfrac{\sigma_a}{E_k}\left(1 - e^{-t/t_k}\right) + \dfrac{\sigma_a}{\eta_m}t;$ $t_k = \dfrac{\eta_k}{E_k}$	

E and η are elastic and viscosity moduli, respectively. Suffixes h, k and m corresponds to moduli of Hooke, Kelvin and Maxwell units. ε is the total strain.

When the behavior of rock includes irrecoverable (permanent) strain, non-linear rheological models are also developed and some of them are listed in Table 6 and responses are compared in Figure 24. Expressions for elasto-visco-plastic models can be developed in a similar manner. However, they tend to be rather complicated. It should be, however, noted that such models require the determination of irrecoverable response from experiments. This would definitely require the implementation of loading and unloading procedures.

8.2 Multi-dimensional constitutive models

Some of linear models are listed in Table 7. Particularly, the rheological models presented in the previous section can be extended to the multi-dimensional situation. But the algebra involved in developing relations between total stress and total strain may be quite cumbersome. For an isotropic homogenous rock material, if the

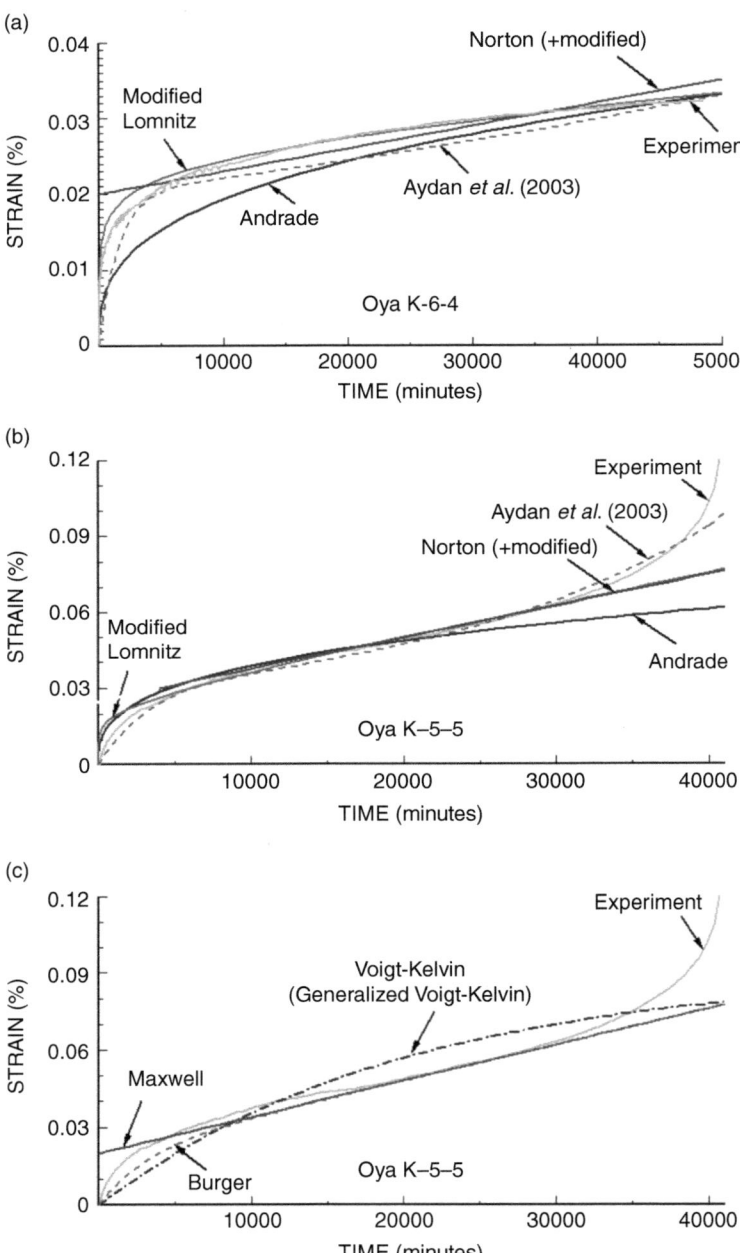

Figure 23 Comparison of intuitive and rheological models with experimental responses: (a) Asymptotic response – Intuitive models, (b) Response terminating with failure – Intuitive models, (c) Response terminating with failure – Rheological models.

Table 6 Rheological uni-dimensional non-linear creep models.

Model	Formula	Geometrical Illustration
Bingham model: Elastic-Perfectly Visco-plastic Model	$\varepsilon = \frac{\sigma_a}{E}$ if $\sigma_a \leq \sigma_Y$ $\varepsilon = \frac{\sigma_a - \sigma_Y}{\eta} t + \frac{\sigma_a}{E}$ if $\sigma_a > \sigma_Y$ σ_Y: yield threshold $\gamma = \frac{1}{\eta}$: fluidity coefficient	$\sigma_a \leq \sigma_Y$ $\sigma_a > \sigma_Y$
Elastic-Visco-Plastic Model of Hardening Type (Owen-Hinton, 1980)	$\sigma_a = E\varepsilon$ if $\sigma_a \leq Y$ $Y = \sigma_Y + H\varepsilon_{vp};$ $\varepsilon = \varepsilon_e + \varepsilon_{vp}$ $\sigma_a = \sigma_Y + H\varepsilon_{vp} + C_p \frac{d\varepsilon_{vp}}{dt}$ $\varepsilon = \frac{\sigma_a}{E} + \frac{(\sigma_a - \sigma_Y)}{H}\left(1 - e^{-\frac{H}{C_p}t}\right)$	$\sigma_a \leq \sigma_Y + H\varepsilon_{vp}$ $\sigma_a > \sigma_Y + H\varepsilon_{vp}$

ε_e and ε_{vp} are elastic and visco-plastic component of strain. H and C_p are plastic hardening modulus and visco-plastic modulus, respectively.

Table 7 Linear models.

Model	Formula	Comments
Newton type	$\sigma_{ij} = C_{ijkl}\dot{\varepsilon}_{kl}$ or $\sigma_{ij} = 2\mu * \dot{\varepsilon}_{ij} + \lambda^* \delta_{ij}\dot{\varepsilon}_{kk}$	C_{ijkl}: viscosity tensor λ^* and μ^* are viscous Lame-like coefficients
Voigt-Kelvin type	$\sigma_{ij} = D_{ijkl}\varepsilon_{kl} + C_{ijkl}\dot{\varepsilon}_{kl}$ or $\sigma_{ij} = 2\mu\varepsilon_{ij} + \lambda\delta_{ij}\varepsilon_{kk} + 2\mu^*\dot{\varepsilon}_{ij} + \lambda * \delta_{ij}\dot{\varepsilon}_{kk}$	D_{ijkl}: elasticity tensor λ and μ are elastic Lame coefficient
Maxwell type	$\dot{\varepsilon}_{ij} = E_{ijkl}\sigma_{kl} + F_{ijkl}\dot{\sigma}_{kl}$	E_{ijkl} and F_{ijkl} are elasticity and viscosity compliance tensors

σ_{ij}, ε_{kl} and $\dot{\varepsilon}_{kl}$ are stress, strain and strain rate tensors, respectively.

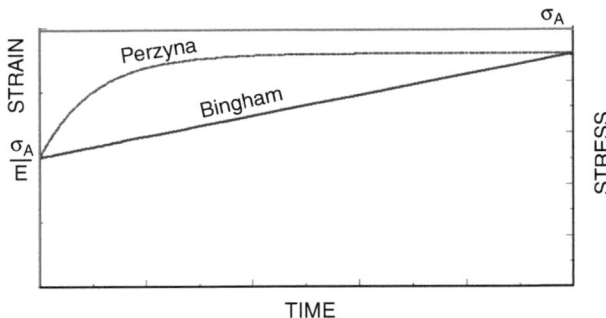

Figure 24 Comparison of Bingham and Perzyna type visco-plastic responses.

coefficients to determine lateral components for a given direction are time-independent, it may be possible to develop constitutive relations between total stress and total strain in analogy to those presented in the previous sections. However, the general situation would require some numerical integration and complex algebra.

Non-linear behavior involving irrecoverable (permanent) responses is more difficult to model by constitutive models. Particularly, it is tedious to determine parameters of constitutive models from experimental data. Therefore, it is quite common to introduce the effective stress (σ_e) and effective strain (ε_e) concepts (these are different concepts from that used for the effect of pore water pressure on the stress tensor) if the irrecoverable part of the strain tensor is independent of the volumetric component. They are defined as follow:

$$\sigma_e = \sqrt{\frac{3}{2}\mathbf{s}\cdot\mathbf{s}} \quad \text{and} \quad \varepsilon_e = \sqrt{\mathbf{e}_p \cdot \mathbf{e}_p} \tag{12}$$

where $\mathbf{s}$ and $\mathbf{e}$ are deviatoric stress and deviatoric strain tensors, respectively, as given below

$$\mathbf{s} = \boldsymbol{\sigma} - \frac{tr(\boldsymbol{\sigma})}{3}\mathbf{I} \quad \text{and} \quad \mathbf{e}_p = \boldsymbol{\varepsilon}_p - \frac{tr(\boldsymbol{\varepsilon}_p)}{3}\mathbf{I} \quad \text{with} \quad tr(\boldsymbol{\varepsilon}_p) = 0 \tag{13}$$

It is interesting to note that the effective stress and strain correspond to those at a uniaxial state, that is,

$$\sigma_e = \sigma_1 \text{ and } \varepsilon_e = \varepsilon_1 \tag{14}$$

This is a very convenient conclusion that the non-linear response can be evaluated under a uniaxial state and it can be easily extended to the multi-dimensional state without any triaxial testing. However, it should be noted that this is only valid when the volumetric components are negligible in the overall mechanical behavior.

Some of non-linear models are listed in Table 8. The visco elasto-plastic model by Aydan & Nawrocki (1998) is illustrated in Figure 25 for a one-dimensional situation.

8.3 Yield functions

There is no yield (failure) criterion directly incorporating the effect of creep experiments (Aydan & Nawrocki, 1998; Aydan *et al.*, 2012), although the basic concept has been presented previously (*i.e.* Ladanyi, 1974) for the time-dependent response of tunnels. Based on the yielding concept shown in Figure 2, some of the yield criteria are listed in Table 9. The general form of the plastic potential functions is also assumed to be similar to yield criteria. If a plastic potential function is assumed to be the same as a yield criterion, it corresponds to the associated flow rule.

Aydan & Nawrocki (1998) discussed how to incorporate the results of creep experiments in yield functions on the basis of results of rare creep triaxial experiments. On the basis of experimental results on various rocks by several researchers (Ishizuka *et al.*, 1993; Kawakita *et al.*, 1981; Masuda *et al.*, 1987; Aydan *et al.*, 1995), Aydan & Nawrocki (1998) concluded that time-dependency of friction angle is quite negligible and the time-dependency of the cohesive component of yield criteria should be sufficient for incorporation of results of creep experiments. The creep experiments would generally yield the decrease of deviatoric strength in time in view of experimental results, it would correspond to the shrinkage of the yield surface in time as shown in Figure 26.

Based on the concept given above, the time dependent uniaxial compressive strength ($\sigma_c(t)$) of Oya tuff (Japan) and Cappadocia tuff (Turkey) is represented in terms of their uniaxial compressive strength (σ_{cs}) and the duration (τ) of a short-term experiment by the following function through the utilization of Equation 6 (*i.e.* Aydan & Nawrocki, 1998; Aydan & Ulusay, 2013)

$$\frac{\sigma_c(t)}{\sigma_{cs}} = 1.0 - b\ln\left(\frac{t}{\tau}\right) \tag{15}$$

The value of b in Equation 15 for Oya tuff and Cappadocia tuffs is 0.08 and 0.05, respectively. However, functions different from that given by Equation 15 can be used provided that they fit to the experimental results (*i.e.* Aydan *et al.*, 2011).

9 SUMMARY

Creep behavior of rocks, which is particularly relevant for cases where the applied load or stress is actually kept constant, is presented. Fundamental definitions on the creep behavior of rocks are first given and experimental procedures under laboratory and in-situ conditions are described. Experimental procedures involve the creep tests under

Table 8 Non-linear models.

Model		Formula	Comment
Elastic-visco-plastic	Power Law	$\dfrac{d\varepsilon_{vp}}{dt} = \left(\dfrac{\sigma_{eq}}{\sigma_o}\right)^n \dfrac{\partial \sigma_{eq}}{\partial \boldsymbol{\sigma}}$ with $e_{vp} = \varepsilon_{vp}$	λ is interpreted called as fluidity coefficient. Flow rule implies that any plastic straining is time-dependent
	Perzyna model	$\dfrac{d\varepsilon_{vp}}{dt} = \lambda s$ and λ is determined from experimental response using the following relation $\lambda = \dfrac{\dot{\varepsilon}_c}{\sigma}$ Flow rule $d\dot{\boldsymbol{\varepsilon}}^p = \lambda\dfrac{\partial G}{\partial \boldsymbol{\sigma}}$	
Elasto-visco-plastic	Aydan – Nawrocki model	$d\boldsymbol{\sigma} = \mathbf{D}^{rp}\,d + \mathbf{C}^{rp}\,d\dot{\boldsymbol{\varepsilon}}$ $\mathbf{D}^{rp} = \mathbf{D}^r - \dfrac{\mathbf{D}^r\frac{\partial G}{\partial \boldsymbol{\sigma}}\otimes\frac{\partial F}{\partial \boldsymbol{\sigma}}\mathbf{D}^r}{h_{rp} + \frac{\partial F}{\partial \boldsymbol{\sigma}}\cdot\left(\mathbf{D}^r\frac{\partial G}{\partial \boldsymbol{\sigma}}\right) + \frac{\partial F}{\partial \boldsymbol{\sigma}}\cdot\left(\mathbf{C}^r\frac{\partial G}{\partial \boldsymbol{\sigma}}\right)}$ $\mathbf{C}^{rp} = \mathbf{C}^r - \dfrac{\mathbf{C}^r\frac{\partial G}{\partial \boldsymbol{\sigma}}\otimes\frac{\partial F}{\partial \boldsymbol{\sigma}}\mathbf{C}^r}{h_{rp} + \frac{\partial F}{\partial \boldsymbol{\sigma}}\cdot\left(\mathbf{D}^r\frac{\partial G}{\partial \boldsymbol{\sigma}}\right) + \frac{\partial F}{\partial \boldsymbol{\sigma}}\cdot\left(\mathbf{C}^r\frac{\partial G}{\partial \boldsymbol{\sigma}}\right)}$ Flow rule $d\dot{\boldsymbol{\varepsilon}}^p = \lambda\dfrac{\partial G}{\partial \boldsymbol{\sigma}}$	Flow rule implies that the plastic potential function expands or shrinks in time domain and strain increments consist of time-dependent and time independent parts

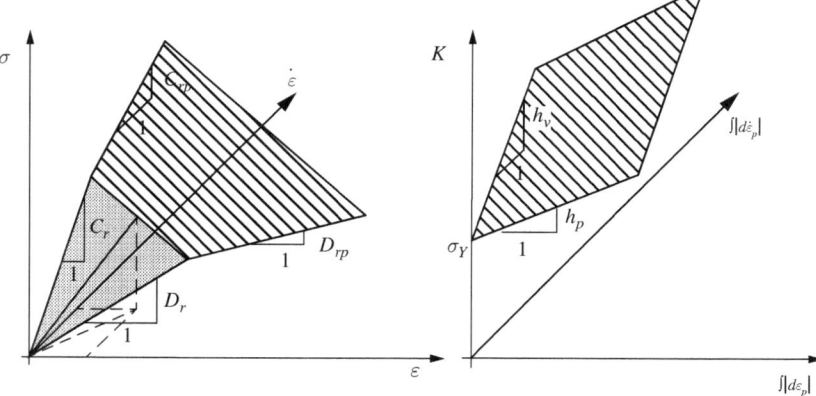

Figure 25 Illustration of visco-elasto-plastic model by Aydan & Nawrocki (1998) for a one dimensional-situation.

Table 9 Yield criteria (see Aydan *et al.*, 2012 for details).

Model	Formula	Comments
Mohr-Coulomb	$\tau = c + \sigma_n \tan\phi$ or $\sigma_1 = \sigma_c + q\sigma_3$ $\sigma_c = \dfrac{2c\cos\phi}{1-\sin\phi}; \sigma_t = \dfrac{2c\cos\phi}{1+\sin\phi}; q = \dfrac{1+\sin\phi}{1-\sin\phi}$	c: cohesion ϕ: friction angle σ_t: tensile strength σ_c: uniaxial compressive strength
Drucker-Prager (Drucker & Prager, 1952)	$\alpha I_1 + \sqrt{J_2} = k$ $I_1 = \sigma_1 + \sigma_2 + \sigma_3$ $J_2 = \frac{1}{6}((\sigma_1-\sigma_2)^2 + (\sigma_2-\sigma_3)^2 + (\sigma_3-\sigma_1)^2$	$\alpha = \dfrac{2\sin\phi}{\sqrt{3}(3 \pm \sin\phi)}$ $k = \dfrac{6c\cos\phi}{\sqrt{3}(3 \pm \sin\phi)}$ $-$: outer apexes $+$: inner apexes
Hoek-Brown (1980)	$\sigma_1 = \sigma_3 + \sqrt{m\sigma_c\sigma_3 + \sigma_c^2}$	$m = \dfrac{\sigma_c^2 - \sigma_t^2}{\sigma_c\sigma_t}$
Aydan (1995)	$\sigma_1 = \sigma_3 + [S_\infty - (S_\infty - \sigma_c)e^{-b_1\sigma_3}]e^{-b_2 T}$	σ_∞: the ultimate deviatoric strength T: temperature b_1, b_2: constants.

uniaxial and triaxial compression, Brazilian and direct shear loading regimes and the basic components of the procedures are explained. Then the utilization of impression creep testing technique as index test for creep responses of rocks is presented and its potential use in evaluating creep properties of rocks are discussed. The experimental

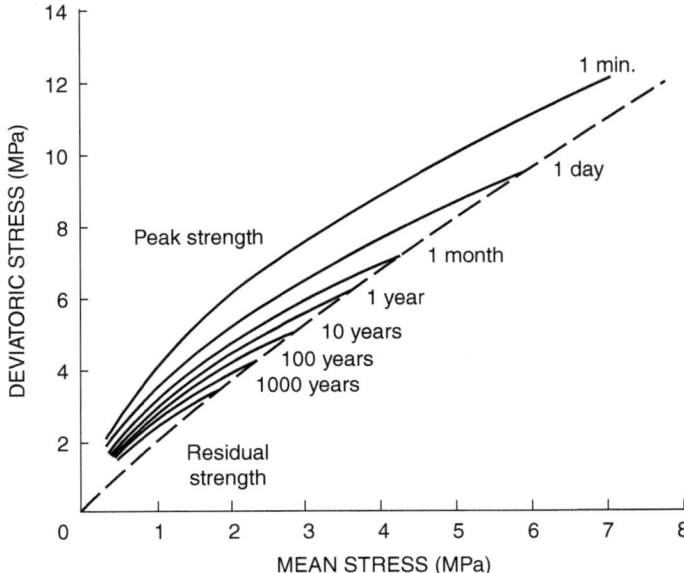

Figure 26 An illustration of the failure criterion for fully saturated Oya tuff at various times estimated by Adachi & Takase (1981).

results for each creep testing procedures are compared and it is concluded that the creep response of rocks would be fundamentally quite similar to each other. In addition, the utilization of in-situ plate bearing and direct shear tests for determining creep tests are described and some actual examples are given. Finally, the constitutive models for modeling creep behavior of rocks are briefly presented and their applications to actual tests are given.

REFERENCES

Adachi T, Serata S, Sakurai S (1969) Determination of underground stress field based on inelastic properties of rocks. *Proc. 11th Symposium on Rock Mechanics*, University of California, Berkeley, 293–328.

Adachi T, Takase A (1981) Prediction of long term strength of soft rocks. *Proc. of International Symposium on Weak Rock*, Tokyo, Balkema, 99–104.

Ağan C, Yeşilnacar MI, Geniş M, Kulaksız S, Ulusay R, Aydan Ö (2013) A geoengineering assessment of Bazda Antique Underground Quarries in Şanlıurfa, Turkey. *The 2013 ISRM EUROCK International Symposium*, Wroclaw, CRC, 93–98

Akai K, Adachi T, Nishi K (1979) Time dependent characteristics and constitutive equations of soft sedimentary rocks (porous tuff). *Proc. JSCE*, 282-2, 75–87 (in Japanese).

Akagi T (1976) An analytical research on the visco-elastic behaviour of civil engineering structures. *Doctorate Thesis*, Nagoya University (in Japanese).

Akagi T, Ichikawa Y, Kuroda T, Kawamoto T (1984) A non-linear rheological analysis of deeply located tunnels. *International Journal for Numerical and Analytical Methods in Geomechanics*, 8, 107–120.

Amadei B, Curran JH, (1982) Creep behaviour of rock joints. In: Underground rock engineering: 13th Canadian Rock Mechanics Symposium. *Transactions of the Canadian Institute of Mining and Metallurgy*, 22, 146–150.

Aydan Ö (1995) The stress state of the earth and the earth's crust due to the gravitational pull. *The 35th US Rock Mechanics Symposium*, Lake Tahoe, 237–243.

Aydan Ö, Akagi T, Ito T, Ito J, Sato, J (1995) Prediction of deformation behaviour of a tunnel in squeezing rock with time-dependent characteristics. *Proc. of Numerical Models in Geomechanics* NUMOG V, 463–469.

Aydan Ö, Akagi T, Kawamoto T (1993) Squeezing potential of rocks around tunnels; theory and prediction. *Rock Mechanics and Rock Engineering*, 26(2), 137–163

Aydan Ö, Akagi T, Okuda H, Kawamoto T (1994) The cyclic shear behaviour of interfaces of rock anchors and its effect on the long term behaviour of rock anchors. *Int. Symp. on New Developments in Rock Mechanics and Rock Engineering*, Shenyang, 15–22.

Aydan Ö, Ito I, Özbay U, Kwasniewski M, Shahriar K, Okuno T, Özgenoğlu A, Malan DF, Okada T (2014) ISRM suggested methods for determining the creep characteristics of rock. *Rock Mechanics and Rock Engineering*, 47(1), 275–290.

Aydan Ö, Nawrocki P (1998) Rate-dependent deformability and strength characteristics of rocks. *Int. Symp. on the Geotechnics of Hard Soils-Soft Rocks, Napoli*, 1, 403–411.

Aydan Ö, Rassouli F, Ito T (2011) Multi-parameter responses of Oya tuff during experiments on its time-dependent characteristics. *Proc. of the 45th US Rock Mechanics / Geomechanics Symposium*, San Francisco, ARMA 11–294.

Aydan Ö, Tokashiki N, Geniş M (2012) Some considerations on yield (failure) criteria in rock mechanics ARMA 12-640. *Proc. of 46th US Rock Mechanics / Geomechanics Symposium*, Chicago, 10p (on CD).

Aydan Ö, Tokashiki N, Iwata N and Adachi K. (2016) The development of a servo-control testing machine for dynamic shear testing of rock discontinuities and soft rocks. Eurock Ürgüp, CRC-Balkema, 791–796.

Aydan Ö, Ulusay R (2013) Studies on Derinkuyu antique underground city and its implications in geo-engineering. *Rock Mechanics and Rock Engineering*, DOI 10.1007/s00603-012-0301-7.

Berest P, Blum P, Charpentier J, Gharbi H, Vales F (2005) Very slow creep tests on rock samples. *International Journal of Rock Mechanics and Mining Sciences*, 42, 569–576.

Bieniawski ZT (1970) Time-dependent behaviour of fractured rock. *Rock Mechanics*, 2, 123–137.

Boukharov GN, Chandi MW, Boukharov NG (1995) The three processes of brittle crystalline rock creep. *International Journal of Rock Mechanics and Mining Sciences & Geomechanics Abstracts*, 32 (4), 325–335.

Chan KS (1997) A damage mechanics treatment of creep failure in rock salt. *International Journal of Damage Mechanics*, 6, 122–152.

Cristescu ND, Hunsche U (1998) *Time Effects in Rock Mechanics*. John Wiley and Sons, New York.

Doktan M (1983) The longterm stability of room and pillar workings in a Gypsum mine. *Ph.D. Thesis*, Newcastle University, Newcastle upon Tyne.

Drucker DC, Prager W (1952). Soil mechanics and plastic analysis for limit design. *Quarterly of Applied Mathematics*, 10 (2), 157–165.

Farmer I (1983). *Engineering Behaviour of Rocks*. 2nd ed. Chapman and Hall, London.

Fabre G, Pellet F (2006) Creep and time-dependent damage in argillaceous rocks. *International Journal of Rock Mechanics and Mining Sciences*, 43 (6), 950–960.

Hagros A, Johanson E, Hudson JA (2008) *Time Dependency in the Mechanical Properties of Crystalline Rocks: A Literature Survey*. Possiva OY, Finland.

Hallbauer DK, Wagner H, Cook NGW (1973) Some observations concerning the microscopic and mechanical behaviour of quartzite specimens in stiff, triaxial compression tests. *International Journal of Rock Mechanics and Mining Sciences & Geomechanics Abstracts*, 10, 713–726.

Handin, J. (1966) Strength and ductility. *Geological Society of America Memoirs*, 97, 223–289.

Hirth G, Tullis J (1994) The brittle-plastic transition in experimentally deformed quartz aggregates. *Journal of Geophysical Research*, 99 (11), 731–747.

Hoek E, Brown ET (1980) Empirical strength criterion for rock masses. *Journal of the Geotechnical Engineering Division*, ASCE, 106 (GT9), 1013–1035.

Hondros G (1959) The evaluation of Poisson's ratio and the modulus of materials of low tensile resistance by the Brazilian (indirect tensile) tests with particular reference to concrete. *Australian Journal of Applied Sciences*, 10, 243–268.

Hunsche U (1992). True triaxial failure tests on cubic rock salt samples – experimental methods and results. *Proc. IUTAM Symp. on Finite Inelastic Deformations – Theory and Applications*, Hannover, Springer Verlag, 525–538.

Hunsche U, Hampel A (1999) Rock salt – the mechanical properties of the host rock material for a radioactive waste repository. *Engineering Geology*, 52, 271–291.

Hyde TH, Sun W, Becker AA (1996) Analysis of the impression creep test method using a rectangular indenter for determining the creep properties in welds. *International Journal of Mechanical Sciences*, 38, 1089–1102.

Ishizuka Y, Koyama H, Komura S (1993) Effect of strain rate on strength and frequency dependence of fatigue failure of rocks. *Proc. of Assessment and Prevention of Failure Phenomena in Rock Engineering*. Istanbul, Balkema, 321–327.

ISRM (2007) The complete ISRM suggested methods for rock characterization, testing and monitoring: 1974–2006. In: R. Ulusay and J.A. Hudson (eds.), *Suggested Methods Prepared by the Commission on Testing Methods, International Society for Rock Mechanics*. Compilation Arranged by the ISRM Turkish National Group, Ankara, Turkey.

Ito H (1991) On rheological behaviour of in situ rock based on long-term creep experiments. *Proc. of 7th ISRM Congress*, Aachen, Germany, 1, 265–268.

Ito H, Sasajima, S (1980) Long-term creep experiment on some rocks observed over three years. *Tectonophysics*, 62 (3–4), 219–232.

Ito H, Sasajima S (1987) A ten year creep experiment on small rock specimens. *International Journal of Rock Mechanics and Mining Sciences & Geomechanics Abstracts*, 24, 113–121.

Ito T, Akagi T (2001) Methods to predict the time of creep failure. *Proc. of the 31st Symposium on Rock Mechanics of Japan*, 77–81 (in Japanese).

Ito T, Aydan Ö, Ulusay R, Kaşmer Ö (2008) Creep characteristics of tuff in the vicinity of Zelve antique settlement in Cappadocia region of Turkey. *Proc. of 5th Asian Rock Mechanics Symposium (ARMS5)*, Tehran, 337–344.

Ito T, Fujiwara T, Akagi T (1999) Triaxial creep characteristics of soft rocks. *Proc. of the 29th Symposium on Rock Mechanics of Japan*, 126–130 (in Japanese).

Jaeger JC, Cook NGW (1979) *Fundamentals of Rock Mechanics*. 3rd ed. Chapman & Hall, London, 79 and 311.

Kawakita M, Sato K, Kinoshita, S (1981) The dynamic fracture properties of rocks under confining pressure. *Memoirs of the Faculty of Engineering*, Hokkaido University, 15 (4), 467–478.

Ladanyi B (1974) Use of the long-term strength concept in the determination of ground pressure on tunnel linings. *Proc. of 3rd Congress International Society on Rock Mechanics*, Denver, Vol. 2B, 1150–1165.

Ladanyi B (1993) Time-dependent response of rock around tunnels. In: C. Fairhurst (ed.), *Comprehensive Rock Engineering: Principles, Practice & Projects*. Vol. 2, *Analysis and Design Methods*. Pergamon, Oxford, 77–112.

Larson MK, Wade RG (2000) Creep along weak planes in roof and how it affects stability. *Society for Mining Engineers Annual Meeting and Exhibit*, Salt Lake City, UT, Feb 29–Mar 1, 2000, Littleton, CO, 8pp.

Lockner DA, Byerlee JD (1977) Acoustic emission and creep in rock at high confining pressure and differential stress. *Bulletin of Seismological Society of America*, 67, 247–258.

Masuda K, Mizutani H, Yamada I (1987). Experimental study of strain-rate dependence and pressure dependence of failure properties of granite. *Journal of Physics of the Earth*, 35, 37–66.

Masuda K, Mizutani H, Yamada I, Imanishi Y (1988) Effects of water on time-dependent behavior of granite. *Journal of Physics of the Earth*, 36, 291–313.

Mirza UA (1978) Investigation into the design criteria for underground openings in rocks which exhibit rheological behaviour. *PhD thesis*, Newcastle University, Newcastle upon Tyne.

Moosavi M, Jafari M, Rassouli FS (2008) The impression creep test as new method for creep measuring the soft rocks. In A. Majdi, A. Ghazvinian (eds.), *ARMS 5th International Symposium, Tehran, 24–26 November 2008*. Iranian Society for Rock Mechanics, Iran, 407–413.

Mottahed P, Szeki A (1982) The collapse of room and pillar workings in a Shaley Gypsum mine due to dynamic loading. *Symp. on Strata Mechanics*, Newcastle, 260–264.

Okada T (2005) Mechanical properties of sedimentary soft rock at high temperatures (part 1) evaluation of temperature dependency based on triaxial compression test. *Central Research Institute of Electric Power Industry* 04026, Chiba, Japan, 1–26 (in Japanese).

Okada T (2006) Mechanical properties of sedimentary soft rock at high temperatures (part 2) – evaluation of temperature dependency of creep behavior based on unconfined compression test. *Central Research Institute of Electric Power Industry* 05057, Chiba, Japan, 1–26 (in Japanese).

Okubo S, Chu SY (1994) Uniaxial compression creep of Tage and Oya Tuff in air-dried and water-saturated conditions (in Japanese). *Journal of the Society of Materials Science (Japan)*, 43 (490), 819–825.

Okubo S, Fukui K, Hashiba, K (2010) Long-term creep of water-saturated tuff under uniaxial compression. *International Journal of Rock Mechanics and Mining Sciences & Geomechanics Abstracts*, 47, 839–844.

Okubo S, Fukui K, Nishimatsu Y (1993) Control performance of servocontrolled testing machines in compression and creep tests. *International Journal of Rock Mechanics and Mining Sciences & Geomechanics Abstracts*, 30, 247–255.

Okubo S, Nishimatsu Y, Fukui K (1991) Complete creep curves under uniaxial compression. *International Journal of Rock Mechanics and Mining Sciences & Geomechanics Abstracts*, 28, 77–82.

Owen DRJ, Hinton E (1980) *Finite Element in Plasticity: Theory and Practice.*, Pineridge Press Ltd, Swansea.

Özşen H, Özkan İ, Şensöğüt C (2014) Measurement and mathematical modelling of the creep behaviour of Tuzköy rock salt. *International Journal of Rock Mechanics & Mining Sciences*, 66, 128–135.

Özkan İ, Özarslan A, Geniş M, Özşen H (2009) Assessment of scale effects on uniaxial compressive strength in rock salt. *Environmental & Engineering Geoscience*, 15 (2), 91–100.

Passaris EKS (1979) The rheological behaviour of rocksalt as determined in an *in situ* pressurized test cavity. *Fourth International Congress on Rock Mechanics*, Balkema, Rotterdam, 257–264.

Peng S (1973) Time-dependent aspects of rock behavior as measured by a servo-controlled hydraulic testing machine. *International Journal of Rock Mechanics and Mining Sciences & Geomechanics Abstracts*, 10, 235–246.

Perzyna P (1966) Fundamental problems in viscoplasticity. *Advances in Applied Mechanics*, 9 (2), 244–368.

Rassouli FS, Moosavi M, Mehranpour MH (2010) The effects of different boundary conditions on creep behavior of soft rocks. *The 44th U.S. Rock Mechanics Symposium & 5th U.S. Canada Symposium*, Salt Lake City, Utah.

Sastry DH (2005) Impression creep technique – An overview. *Journal of Materials Science and Engineering*, 409, 67–75.

Serata S, Sakurai S, Adachi, T (1968) Theory of aggregate rock behavior based on absolute three-dimensional testing (ATT) of rock salt. *Proc. 10th Symposium on Rock Mechanics*, University of Texas at Austin, 431–473.

Shibata K, Tani K, Okada T (2007) Creep behaviour of tuffaceous rock under high temperature observed in uniaxial compression test. *Soil and Foundations*, 47 (1), 1–10.

Shimada M (1993) Two types of brittle fracture of silicate rocks and scale effect on rock strength: their implications in the earth crust. *Scale Effects in Rock Masses*, Lisbon, 55–62.

Slizowski J, Lankof L (2003) Salt-mudstones and rock-salt suitabilities for radioactive-waste storage systems: Rheological properties. *Applied Energy*, 75 (1/2), 137–144.

Timoshenko SP, Goodier JN (1970) *Theory of Elasticity*. 3rd ed. Singapore, McGraw-Hill Int. Book Company, 506.

Ulusay R, Ito T, Akagi T, Seiki T, Yüzer E, Aydan Ö (1999) Long term mechanical characteristics of Cappadocia tuff. *The 9th Int. Rock Mechanics Congress*, Paris, 687–690.

Vogler UW, Malan DF, Drescher K (1998) Development of shear testing equipment to investigate the creep of discontinuities in hard rock. In H.P. Rossmanith (ed.), *Proc. Mechanics of Faulted and Jointed Rock, MJFR-3*. Balkema, Rotterdam, 229–234.

Wawersik WR (1983) Determination of steady state creep rates and activation parameters for rock salt. In: *High Pressure Testing of Rock, Special Technical Publication of ASTM*, STP869, 72–91.

Yang CH, Daemen JJK, Yin JH (1999) Experimental investigation of creep behavior of salt rock. *International Journal of Rock Mechanics and Mining Sciences*, 36 (2), 233–242.

Physical Modeling Tests

Chapter 12

Physical, empirical and numerical modeling of jointed rock mass strength

P.H.S.W. Kulatilake

Professor and Director, Rock Mass Modeling and Computational Rock Mechanics Laboratories, University of Arizona, USA

Abstract: The presence of complicated discontinuity patterns, the inherent statistical nature of their geometrical parameters, and the uncertainties involved in the estimation of their geometrical and geo-mechanical properties and in-situ stress make accurate prediction of rock mass strength a very difficult task. It has been a great challenge for the rock mechanics and rock engineering profession to develop a rock mass strength criterion in three dimensions (3-D) which captures the scale effects and anisotropic properties. Rock mechanics and rock engineering researchers have dealt with this topic for more than 50 years. The aim of this paper is to provide the state-of-the-art on estimation of jointed rock mass strength using physical, empirical and numerical modeling methodologies.

I INTRODUCTION

Most naturally occurring discontinuous rock masses comprise of intact rock interspaced with different types of discontinuities. Such discontinuities include fissures, fractures, joints, faults, bedding planes, shear zones and dikes. Discontinuities may be divided into major and minor depending on the feature size. Large features can be considered as major discontinuities and the rest of the small features can be considered as minor discontinuities. For most of the civil and mining engineering projects, at the upper most level, rock masses contain only a few major discontinuities. For such projects, major discontinuities can be considered as single features and their geometry may be represented deterministically. On the other hand, due to its large number and inherent statistical nature, the geometry of minor discontinuities should be characterized statistically. Henceforth, the minor discontinuities are referred to as either "joints" or "fractures" in this paper. In civil and mining engineering, the engineers face design and construction tasks associated with geotechnical systems that are in or on discontinuous rock masses. Some examples for such geotechnical systems are tunnels for hydropower and transport, dams, foundations, natural and man-made slopes, surface and underground excavations made for mineral extraction, underground caverns for oil and gas storage and hazardous waste isolation caverns. In these rock engineering systems, one comes across stability concerns of the rock structures. Rock mass strength plays a vital role on stability of these structures. Rock mass strength depends on the (a) lithology, (b) discontinuity network, (c) geo-mechanical properties of the discontinuities, (d) geo-mechanical properties of the intact rock, (e) in situ stress system, (f) size of the rock block, (g) shape of the rock block (h) loading/unloading stress path, (i) loading rate, (j) pore pressure in the rock mass and (j) environmental conditions (such as temperature, humidity etc.) of the rock mass. A good understanding of rock mass strength is vital to arrive at safe and economical designs for structures built in and on rock masses. Due to the presence of discontinuities, discontinuous rock masses show scale (size) dependent and anisotropic properties. The presence of complicated discontinuity patterns, the inherent statistical nature of their geometrical parameters, and the uncertainties involved in the estimation of their geometrical and geo-mechanical properties and in-situ stress make accurate prediction of rock mass strength a very difficult task. It has been a great challenge for the rock mechanics and rock engineering profession to develop a rock mass strength criterion in three dimensions (3-D) which captures the scale effects and anisotropic properties. The rock mechanics and rock engineering researchers have dealt with this topic for more than 50 years. The aim of this paper is to provide the progress the rock mechanics profession has made so far in achieving the said task.

To obtain intact rock mechanical properties, usually 2-inch diameter samples are tested. At this size, the distribution of micro-cracks with the location of the sample has influence on the scatter for the considered rock block property. Further increase of sample size will include the influence of one or more fractures. When the sample size contains only a few fractures, the sample property varies considerably from one sample to another, reflecting the statistical inhomogeneity with respect to the influence of fractures. This variability decreases as the sample size is further increased to contain more fractures and to make the sample more statistically homogeneous with respect to the fractures. Beyond a certain minimum volume, the mass property of the rock may not change significantly (in practical point of view) with respect to the effect of fractures. This minimum volume may be used as the element size to represent the equivalent mass property of a statistically homogeneous rock mass volume that contains a significant number of fractures. This volume may be termed as a "Representative Elementary Volume" (REV) and may be of great importance for engineering purposes. Each rock mass property may have a different REV value. For some rock masses, the REV sizes may be small compared to the size of the problem domain of interest. For such cases, REV sizes and associated properties will be very useful for engineering applications. For some other rock masses, REV sizes may be large compared to the size of the problem domain. In such cases, if we know how the rock mass property of interest varies between the intact rock value and the REV value with respect to joint geometry parameters and block size that will be quite useful for dealing with the latter category of rock masses in engineering applications. It is important to keep in mind that the variability of the rock mass property at the REV size or greater than the REV size will be smaller compared to the rock mass property at sizes less than the REV size. If the sample size is increased further to allow major discontinuities to enter into the sample, then the rock mass property will vary reflecting the influence of major discontinuities. In summary, the REV sizes and the corresponding REV mechanical properties can be used to represent the combined equivalent continuum behavior of minor discontinuities (joints) and intact rock. To this system of the rock mass, the major discontinuities can be added as single features to complete the representation of the whole rock mass.

2 PHYSICAL MODELING OF ROCK MASS STRENGTH

2.1 Large scale in-situ tests

Currently, the ways to estimate strength properties of jointed rock masses can be categorized into direct and indirect methods. Direct methods include laboratory and in situ tests. Laboratory results obtained from small-sized specimens that include only micro-joints are very different from the results obtained from large-scale blocks because the laboratory samples cannot accommodate the whole spectrum of different size discontinuity network which is present in the field. A significant number of in-situ tests have been performed on various rock types to study the effect of size on rock mass compressive strength using various sample sizes having different width to height (w/h) ratios (Greenwald et al., 1939, 1941; Nose, 1964; Jahns, 1966; Gimm et al., 1966; De Reeper, 1966; Bieniawski, 1968; Georgi et al., 1970; Chaoui et al., 1970; Cook et al., 1971; Lama, 1971; Pratt et al., 1972; Wagner, 1974; Bieniawski & Van Heerden, 1975). Some of these investigators also have performed corresponding laboratory tests on the same rock types. Fig. 1 shows the large scale in situ test conducted on a 2 m cube sample and reported by Bieniawski (1968). Bieniawski & Van Heerden (1975), and Heuze (1980) have reviewed the work done prior to their publications on scale effects on rock mass strength. The reported results of these investigations clearly show the reduction of rock mass strength with increasing size up to a certain size, beyond which change becomes insignificant in practical point of view. Bieniawski (1968) observed this size to be about 1.5 m for coal. Jahns (1966) and Pratt et al. (1972) noted this size to be about 1.0 m for diorite and iron ore, respectively. Bieniawski (1968) has observed strength of the largest sample size (2 m) to be 0.15 of the smallest laboratory sample size. Pratt et al. (1972) have noted the aforementioned ratio to be 0.1. Bieniawski & Van Heerden (1975) provide valuable guidelines for experimental procedures to perform large scale in-situ tests. They also suggested various empirical equations to relate rock mass uniaxial strength to w/h ratio. It is important to note that the relations developed from in-situ tests in the above stated studies primarily depend on the discontinuity network of the tested rock masses. However, unfortunately, in these early investigations, no attempt had been made to map the discontinuity network before subjecting the rock mass to mechanical behavior testing. Therefore, the reported relations are highly site dependent and have qualitative value only.

2.2 Jointed block testing with a significant number of fractures in the laboratory

To obtain realistic results for jointed rock mass mechanical properties, many large volumes of rock of different sizes having a number of different known joint configurations should be tested at significant stress levels under different stress paths. Such an experiment program is almost impossible to carry out in the laboratory. With in situ tests, such an experimental program would be very difficult, time-consuming and expensive. Results of laboratory model studies on rock-like materials (Ladanyi & Archambault, 1969; Brown & Trollope, 1970; Einstein & Hirschfeld, 1973; Chappell, 1974; Heuze, 1980; Kulatilake et al., 1997; Yang et al., 1998; Kulatilake et al., 2001a, 2006) have shown that many different failure modes are possible with

Figure 1   A large size coal specimen with loading system and deformation measuring equipment in place (from Bieniawski, 1968).

jointed rock and that the internal distribution of stresses and strains within a jointed rock mass can be highly complex. Yang *et al.* (1998), Kulatilake *et al.* (1997, 2001a, 2006) found that the failure modes can be divided into three types, namely splitting failure through the rock blocks, sliding along joint planes and mixed splitting-sliding mode. They also have demonstrated that the failure strength exhibits pronounced anisotropy mainly due to the effect of joint orientation. Even though these jointed blocks included a significant number of joints, all the included joints were persistent joints. Only a few experimental studies have been done using a significant number of non-persistent joints (Mughieda, 1997; Prudencio & Van Sint Jan, 2007; Chen *et al.*, 2011, 2012). Prudencio & Van Sint Jan's (2007) paper presents the results of biaxial tests performed on physical models of rock with non-persistent joints. The failure modes and maximum strengths developed were found to depend on, among other variables, the geometry of the joint systems, the orientation of the principal stresses, and the ratio between the intermediate principal stress and intact material compressive strength· These test results showed three basic failure modes: failure through a planar surface, stepped failure, and failure by rotation of new blocks. Planar failure and stepped failure were associated with high strength behavior and small failure strains, whereas rotational failure was associated with a very low strength, ductile behavior, and large deformation. Chen *et al.* (2012) investigated the combined influence of joint inclination angle and joint continuity factor (lengths of the joints divided by the total length of the block) on strength and deformation behavior of jointed rock mass for gypsum specimens with a set of non-persistent open flaws in uniaxial compression. Complete axial stress-strain curves were classified into four types, *i.e.*, single peak, softening after multi-peak yield platform, hardening after multi-peak yield platform and multi-peak during softening. To investigate the brittleness of the specimens, the ratio of the residual strength to the maximum peak strength as well as the first and last peak strains were studied. At the same joint inclination angle, the ratios between the residual strength and the maximum peak strength and the last peak strains were observed to increase while the first peak strain decreased with the increase of joint continuity factor. At the same joint continuity factor, the curves of the three brittleness parameters versus joint inclination angle were found to be either concave or convex single-peak or wave-shaped. In summary, the mechanical behavior of jointed blocks having a significant number of non-persistent joints was found to be more complicated and significantly different to that having persistent joints. The mechanical behavior depends on the number of joints, joint orientation, size, density, spacing, arrangement and whether the joints are open or closed. It is important to note that as the joint density increases, the behavior around each joint is affected by the presence of the rest of the joints in the jointed block.

He *et al.* (2014) performed CT scanning tests to detect the pre-existing fracture systems of 36 cubical coal blocks. Systematic procedures were developed to construct the pre-existing fracture geometry of the cubical coal blocks from the CT images. Based on the constructed fracture systems, the fracture tensor components were calculated to capture the directional effects of the fracture system of the cubical coal blocks. The effect of the pre-existing fracture system and different combinations of confining stresses on the strength of coal masses was investigated at a preliminary level using some true triaxial test results and the computed fracture tensor components (see Kulatilake *et al.* (1993) for details on the fracture tensor components). The strength of the cubical coal blocks in the loading direction was found to be closely related to the summation of the fracture tensor components in the two perpendicular directions to the loading direction and the level of the confining stress system. Further research on this aspect is currently in progress.

2.3 Jointed block testing with a few fractures in the laboratory to study rock fracture mechanics behavior

Many investigators who perform research in the rock fracture mechanics field have performed experiments on different materials such as glass, PMMA (Poly Methyl Meth Acrylate), resins, Plaster of Paris, molded gypsum and different rock types to study fracture initiation, propagation and coalescence resulting from one to three two-dimensional pre-existing flaws under uniaxial loading (Hoek & Bieniawski, 1965; Bombolakis, 1968; Lajtai, 1971; Bobet & Einstein, 1998; Wong & Einstein, 2009; Lee & Jeon, 2011; Wong *et al.*, 2001). Almost all studies have observed initiation of primary tensile wing cracks at flaw tips or at other locations and propagation of that toward parallel to the loading direction. In addition, some studies have observed initiation of tensile or shear secondary cracks. One of the studies (Lee & Jeon, 2011) has reported not seeing wing cracks when the flaw is parallel to the loading direction. Many different coalescence possibilities have been reported in these studies. It seems that even in the same material, initiation, propagation and especially the coalescence are very much dependent on the flaw geometry and heterogeneity. In addition, different materials have produced different initiation, propagation and coalescence patterns. Coalescence produced through the linkage of shear cracks only have occurred at a higher stress level than the coalescence produced by a combination of shear and wing cracks. The smallest coalescence stress has resulted only by wing cracks (Park & Bobet, 2009). Experimental results from open and closed flaws have shown that the initiation stresses and coalescence are higher for closed flaws than for open (Park & Bobet, 2009). This is explained by the existence of friction along the closed flaws, which needs to be overcome before a crack can initiate, and also by the capability of closed flaws to transmit normal stresses.

Uniaxial laboratory testing of a single 3-D flaw in a transparent Material have shown that unlike 2-D cracking, there are intrinsic limits on 3-D crack growth and the maximum possible size of wings is about the size of the initial crack sprouting them (Dyskin *et al.*, 1999). In other words a different type of behavior has been observed with 3-D flaws compared to 2-D flaws under the same type of loading. Biaxial loading on a few 2-D flaws have shown that the crack initiation, propagation and coalescence behavior is significantly different to that under uniaxial conditions (Bobet & Einstein, 1998). Based on the experiments performed, Sahouryeh *et al.* (2002) have stated that the experimental results on 2-D crack growth in uniaxial and biaxial compression cannot be used to model fracture in biaxial compression in 3-D for the following reasons: firstly, in the case of a 2-D shear crack (through crack) under biaxial compression, the loads act on the surface and the tips of the crack. In the 3-D case, the loads act only on the contour of the crack, *i.e.* the surface is free from loading. Secondly, splitting routinely observed in uniaxial compression of plates or blocks with through cracks (the 2-D problem) is not observed in uniaxial compression tests on blocks with internal 3-D cracks. In summary, even though much research has been conducted for more than 45 years, still it is not possible to predict fracture initiation, propagation and coalescence accurately for a given material with a given flaw geometry.

3 EMPIRICAL STRENGTH CRITERIA

3.1 Mohr-Coulomb and Hoek-Brown strength criteria

An indirect approach to obtain estimates of the strength of a jointed rock mass is by empirical correlation. In this approach, the rock mass properties determined through laboratory or field tests are linked to a representative rock mass classification index or rock mass joint geometry properties which reflect rock mass quality. Several empirical rock mass strength criteria have been suggested in the literature. Out of these, the oldest and most widely used strength criteria are Mohr-Coulomb and Hoek-Brown. The Mohr-Coulomb criterion is given in Equation 1 as follows:

$$\sigma'_1 = \sigma_c + k\sigma'_3 \tag{1a}$$

$$\sigma_c = (2c'\cos\phi')/(1 - \sin\phi') \tag{1b}$$

and

$$k = (1 + \sin\phi')/(1 - \sin\phi') \tag{1c}$$

where σ'_1, σ'_3, σ_c, c' and ϕ' are respectively the effective major principal stress, effective minor principal stress, uniaxial compressive strength, effective cohesion and effective angle of internal friction for the rock mass. Equation 1 shows that it is a linear criterion which does not include the intermediate principal stress (σ_2). On the other hand the experimental test results indicate that the rock mass strength exhibits non-linear properties and dependence on intermediate principal stress. Also the intercept given by the Mohr-Coulomb criterion on the negative σ_3-axis does not provide a reasonable value for the tensile strength. Therefore, usually a tension cut off is used in using the Mohr-Coulomb criterion. In addition to the above shortcomings no guidelines are available to relate the two strength parameters given in Equation 1 to discontinuity geometry and mechanical parameters. It has been used in practice mainly because of simplicity.

Hoek and Brown rock mass strength criterion (Hoek & Brown, 1980) was introduced in 1980 as an attempt to provide input data for the analyses required for the design of underground excavations in hard rock. It is an empirical criterion developed through trial and error curve fitting of different parabolic functions to triaxial test data. Choose of a parabolic function seems to have originated from one part of the equation given for Griffith's crack theory (Griffith, 1920, 1924). It also involved Hoek's experience on research results with respect to the brittle failure of intact rock (Hoek, 1968) and Brown's experience on research results of model studies of jointed rock mass behavior (Brown, 1970). The empirical parameters of the criterion were first developed for intact rock and then those parameter values were reduced by linking them to Bieniawski's (1976) Rock Mass Rating (RMR) classification system. The criterion developed in 1980 has been updated several times (Hoek *et al.*, 1992; Hoek & Brown 1997; Hoek *et al.*, 2002). The most current one seems to be the Generalized Hoek-Brown criterion (Hoek *et al.*, 2002). Note that the empirical parameters of the criterion were lately directly linked to the Geological Strength Index (GSI) introduced by Hoek *et al.* (1992, 1995). GSI is determined based on the structure of the rock blocks and the surface conditions of the joints. Number of joint sets and fracturing level are considered in evaluating the structure of the rock blocks. Roughness/smoothness, degree of weathering, degree of alteration and presence and type of filling are considered in evaluating the surface conditions of the joints. Cai *et al.* (2004) modified the descriptive term "the structure of the rock blocks" to quantitative block volume and the descriptive term "the joint surface condition" to quantitative joint condition factor in estimating the GSI value for a rock mass. Cai *et al.* (2007) have proposed a methodology to extend the GSI system for the estimation of a rock mass's residual strength. It was proposed to adjust the peak GSI to the residual GSI$_r$ value based on the two major controlling factors in the GSI system—the residual block volume V^r_b and the residual joint condition factor J^r_c. Methods to estimate the residual block volume and joint condition factor are presented. The Generalized Hoek-Brown criterion is expressed by the following equation:

$$\sigma_1' = \sigma_3' + \sigma_{ci}\left(m_b\frac{\sigma_3'}{\sigma_{ci}} + s\right)^a \tag{2a}$$

where σ_{ci} is the uniaxial compressive strength of the intact rock; m_b is a reduced value of the material constant m_i given in the Hoek-Brown equation for intact rock, and the relation between the two is given by

$$m_b = m_i\exp\left(\frac{GSI - 100}{28 - 14D}\right) \tag{2b}$$

s and a are constants for the rock mass given by the following relations:

$$s = \exp\left(\frac{GSI - 100}{9 - 3D}\right) \tag{2c}$$

$$a = \frac{1}{2} + \frac{1}{6}\left(e^{-GSI/15} - e^{-20/3}\right) \tag{2d}$$

D is a factor which depends upon the degree of disturbance to which the rock mass has been subjected by blast damage and stress relaxation. It varies from 0 for undisturbed in situ rock masses to 1 for very disturbed rock masses. Guidelines for the selection of D are given in Hoek *et al.* (2002).

Relations are also available between GSI and RMR as well as GSI and Q system (Barton *et al.*, 1974) as given below:

(a)   For better quality rock masses (GSI > 25): for $RMR_{76} > 18$, GSI = RMR_{76} and for $RMR_{89} > 23$, GSI = $RMR_{89} - 5$

(b)   For rock masses having GSI values less than 25, Hoek *et al.* (1995) suggested using the Q system (Barton *et al.*, 1974) to obtain the GSI value

$$GSI = 9\ln Q + 44 \tag{3}$$

Because it is a non-linear criterion, conceptually, Hoek-Brown criterion is better than the Mohr-Coulomb criterion. In addition, a lot of information is available to estimate the empirical parameters given in the Hoek-Brown criterion. Both the Mohr-Coulomb and Hoek-Brown are easy to use simple empirical equations. Therefore, in practice, Hoek-Brown criterion has been used more often than the Mohr-Coulomb criterion. One of the shortcomings of the Hoek-Brown criterion is the absence of the intermediate principal stress. A few investigators (Pan & Hudson, 1988; Priest, 2005; Zhang & Zhu, 2007; Zhang *et al.*, 2013; Melkoumian *et al.*, 2009) have extended the Hoek-Brown criterion to three dimensions by including the intermediate principal stress. All these criteria are limited to rock masses having isotropic strength. It is important to note that the Generalized Hoek-Brown criterion as well as the modified 3-D Hoek-Brown strength criteria do not include the orientation of discontinuity sets as well as discontinuity size/rock block size explicitly. Because these discontinuity geometry parameters are not included in GSI, RMR and Q systems explicitly, these rock mass criteria do not have the capability of predicting the anisotropic strength and the strength reduction that takes place as a function of increasing block size that are observed in majority of the rock masses.

These criteria may have applicability for highly fractured rock masses having discontinuity orientations in many directions. Such rock masses have the potential to show isotropic strength characteristics. Also, it is important to point out that both RMR and Q systems include RQD. For the same rock mass, infinite many values can be obtained for RQD by considering different directions. This creates significant uncertainty and variability in estimating a single value for RMR or Q. Note that both RQD and fracture frequency are included in the RMR system. These two parameters are dependent. That means double counting for fracture frequency exists in estimating a RMR value.

3.2 Yudhbir et al. strength criterion

Yudhbir *et al.* (1983) proposed the following equation based on some test results obtained from gypsum-celite-water mixtures. It is a slightly modified version of Bieniawski's empirical strength criterion. The constant a has been linked to the degree of jointedness and varies between 0 and 1. The constant b depends on the rock type, ranges between 2 and 5 and is independent of jointedness. The exponent α is a constant having a value less than 1 and is independent of the rock type and jointedness. It is important to note that this equation does not exist in the tensile quadrant. This criterion does not include the intermediate principal stress. In addition, it is important to note that all the shortcomings mentioned for Generalized Hoek-Brown criterion are applicable for this criterion too.

$$\sigma_1/\sigma_c = a + b(\sigma_3/\sigma_c)^\alpha \tag{4}$$

3.3 Sheorey et al. strength criterion

Sheorey *et al.* (1989) proposed the following equation to represent the intact rock strength. It was fitted successfully to 23 sets of triaxial test data covering coal, sandstone, shale, shaley sandstone, marble, granite and slate rocks.

$$\sigma_1 = \sigma_c \left(1 + \frac{\sigma_3}{\sigma_t}\right)^b \tag{5}$$

In Equation 5, σ_c and σ_t (tensile strength) are related to the cohesion and coefficient of friction. The parameter b is related to σ_c, σ_t and the coefficient of friction. Equation 5 was extended to Equation 6 to represent the rock mass strength.

$$\sigma_1 = \sigma_{cn}\sigma_c \left(1 + \frac{\sigma_3}{\sigma_{tn}\sigma_c}\right)^b \tag{6}$$

where σ_c = unconfined intact rock strength

$$\sigma_{cn} = \sigma_{cj}/\sigma_c$$
$$\sigma_{tn} = \sigma_{tj}/\sigma_c$$

Estimations of σ_{cn}, σ_{tn}, b are linked to the Q system. σ_{cj} and σ_{tj} are the uniaxial compressive and tensile strengths for the rock mass, respectively. It is important to note that all the shortcomings mentioned for Generalized Hoek-Brown criterion are applicable for this criterion too. In addition this criterion does not include the intermediate principal stress.

3.4 Ramamurthy *et al.* strength criterion

Ramamurthy (2001) proposed the following rock mass strength criterion:

$$\frac{\sigma'_1 - \sigma'_3}{\sigma'_3} = B_j \left(\frac{\sigma_{cj}}{\sigma'_3}\right)^{\alpha_j} \tag{7}$$

where σ'_1 and σ'_3 are the effective major and minor principal stresses, respectively, and σ_{cj} is the uniaxial compressive strength of the jointed specimen obtained from Equation 8 given below (Ramamurthy & Arora, 1994). B_j and α_j are the empirical strength parameters of the jointed rock.

$$\sigma_{cj}/\sigma_{ci} = \exp[-0.008J_f] \tag{8}$$

In Equation 8, σ_{ci} is the uniaxial compressive strength of the intact rock and the joint factor J_f is expressed as:

$$J_f = J_n/rn \tag{9}$$

where, J_n is the joint frequency, *i.e.* the number of joints per meter depth of rock in the direction of loading (major principal stress), n is a coefficient for the effect of joint inclination and r is the coefficient of friction on the sliding joint or joint set. Table 8 of Ramamurthy's (2001) paper provides the n value corresponding to the joint inclination angle. The values of α_j and B_j are estimated according Equations 10 and 11.

$$\alpha_j/\alpha_i = \left(\sigma_{cj}/\sigma_{ci}\right)^{0.5} \tag{10}$$

$$B_i/B_j = 0.13\exp[2.04\alpha_j/\alpha_i] \tag{11}$$

In Equations 10 and 11, α_i and B_i are the values of the strength parameters obtained from the triaxial tests on intact rock specimens. Compared to the previously mentioned strength criteria, one positive aspect of this criterion is the inclusion of the effect of the orientation of a sliding joint or joint set. In most of the rock masses, usually more than one joint set exist. Orientation and intensities of all these joint sets make a contribution to the rock mass strength. This aspect is not included in Ramamurthy's rock mass strength criterion. Also note that the joint orientation is not included explicitly in the strength criterion. Therefore, it cannot predict the rock mass strength in different directions to ultimately find the direction that provides the minimum strength. The strength criterion also does not include the joint size/block size. Therefore, the strength criterion does not have the capability of predicting rock mass strength variation with block size. In addition this criterion does not include the intermediate principal stress.

3.5 Kulatilake *et al.* strength criterion

Kulatilake *et al.* (2006) performed uniaxial and biaxial compression tests on model material intact prismatic samples of size $35.6 \times 17.8 \times 2.5$ cm to determine the strength of the intact model material. The failure process was captured by a video camera to investigate the failure mode of each sample. For intact prismatic samples, the main mechanism of failure was the tensile failure through the intact model material. This failure mode is also known as the splitting mode. Tensile fractures almost perpendicular to σ_2 (intermediate principal stress) direction were observed for samples subjected to low levels of $\sigma_2/\sigma_u \leq 0.03$. Note that σ_u is the uniaxial compressive strength of the intact model material. Fracturing parallel, or slightly inclined, to the σ_1 (major principal stress)-σ_2 plane (*i.e.* normal to σ_3 (minor principal stress) = 0) was observed for most of the intact samples that were subjected to high levels of σ_2 ($\sigma_2/\sigma_u > 0.03$). A new intact rock failure criterion in 3-D given by Equation 12 was proposed by Kulatilake *et al.* (2006). In the equation, J_1 and J_{2D} represent the mean stress (one third of the first invariant of stress tensor) and the second invariant of deviatoric stress tensor, respectively.

$$J_{2D} = A + BJ_1 + CJ_I^2 \tag{12a}$$

where

$$J_1 = \frac{1}{3}\left(\sigma_{1,I} + \sigma_2 + \sigma_3\right) \tag{12b}$$

$$J_{2D} = \frac{1}{6}\left(\left(\sigma_{1,I} - \sigma_2\right)^2 + \left(\sigma_2 - \sigma_3\right)^2 + \left(\sigma_3 - \sigma_1\right)^2\right) \tag{12c}$$

and A, B and C are empirical material constants.

In Equation 12b and 12c, $\sigma_{1,I}$, σ_2, and σ_3 represent respectively, the major principal stress applied on intact rock sample at failure, intermediate principal stress, and minor principal stress. In the case of biaxial state of stress, J1 and J2D reduce to $1/2 \times \left(\sigma_{1,I} + \sigma_2\right)$ and $1/3 \times \left(\sigma_{1,I}^2 - \sigma_{1,I}\sigma_2 + \sigma_2^2\right)$, respectively. Therefore, the new intact rock failure criterion in the case of biaxial state of stress can be expressed as:

$$\left(\sigma_{1,I}^2 - \sigma_{1,I}\sigma_2 + \sigma_2^2\right) = 3A + \frac{3B}{2}\left(\sigma_{1,I} + \sigma_2\right) + \frac{3C}{4}\left(\sigma_{1,I} + \sigma_2\right)^2 \tag{13}$$

A very good fit (regression coefficient of determination of 0.97) of Equation 13 was obtained for the intact material strength data of glastone. Figure 2 shows the results of the fit obtained for the new proposed strength criterion along with the fits obtained for the existing failure criteria such as Drucker & Prager (1952), Mogi (1971) and modified Wiebols & Cook (Zhou, 1994). It is clear from Figure 2 that the intact material strength data of glastone fit the new criterion the best followed up with an almost equally strong fit for modified Wiebols & Cook criterion. However, Drucker & Prager and Mogi criteria show poor fits for the considered data.

Uniaxial and biaxial compression tests on prismatic jointed blocks of size $35.6 \times 17.8 \times 2.5$ cm were performed using the same biaxial load frame used to test intact prismatic samples. Both persistent and impersistent joint configurations were included in producing jointed model material blocks. A novel method was used in producing blocks with impersistent joints. Note that all the joints of the produced jointed model material blocks were smooth. In other words, effect of roughness of joints on the mechanical behavior of

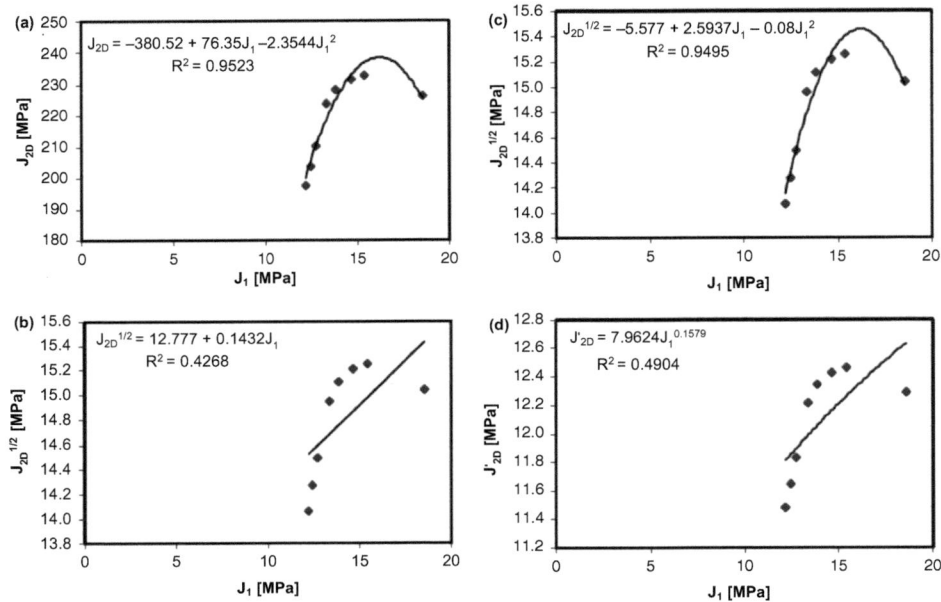

Figure 2 Fitting of different intact rock failure criteria: (a) proposed new criterion; (b) Drucker & Prager (1952); (c) modified Wiebols & Cook (Zhou, 1994) and (d) Mogi (1971) at the two dimensional level to data obtained from biaxial compression tests performed on intact blocks of glastone model material (from Kulatilake *et al.*, 2006).

jointed model material blocks was not investigated. The failure process was captured by a video camera to investigate the failure mode of samples. Results indicated three failure modes for jointed blocks: (a) tensile failure through intact material; (b) combined shear and tensile failure or only shear failure on joints; and (c) mixed failure of the above two modes depending on the joint geometry. The fracture tensor component (Kulatilake *et al.*, 1993) was used to quantify the directional effect of the joint geometry, including the number of fracture sets, fracture density, and probability distributions for size and orientation of the fracture sets. The strength results obtained for the jointed glastone blocks were used along with the strength results obtained for the intact glastone blocks in developing the following new rock mass failure criterion (see Equation 14) for biaxial loading conditions. In Equation 14a, $\sigma_{1,b}$ and F_{22} are respectively, the jointed rock block strength under biaxial loading and the fracture tensor component in σ_2 direction; $\sigma_{1,I}$ is the intact rock strength under biaxial loading condition and is estimated using Equation 12 or 13. In Equation 14b, ω_0, a and b are parameters of the rock mass strength criterion.

$$\sigma_{1,b}/\sigma_{1,I} = \exp(-\omega F_{22}) \tag{14a}$$

$$\text{where } \omega = \omega_0 / \left(a(\sigma_2/\sigma_{u,I})^b + 1 \right) \tag{14b}$$

The parameter ω_0 is obtained from the relation given in Equation 15 between the uniaxial compressive strength of the jointed block, $\sigma_{u,b}$, F_{22} and $\sigma_{u,I}$.

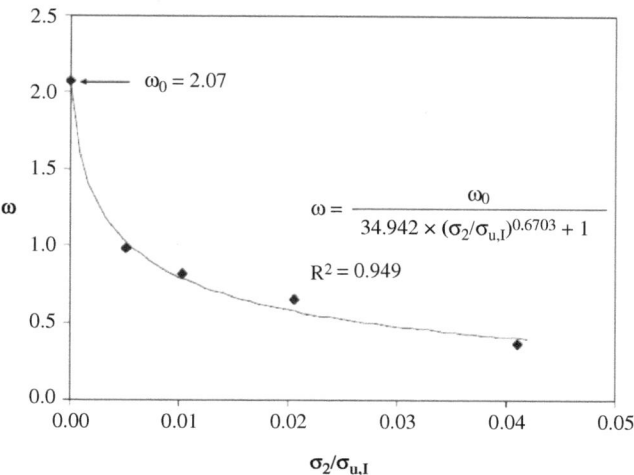

Figure 3 Variation of the decay parameter, ω, with $\sigma_2/\sigma_{u,I}$ (from Kulatilake *et al.*, 2006).

$$\sigma_{u,b}/\sigma_{u,I} = \exp(-\omega_0 F_{22}) \tag{15}$$

Figure 3 shows the comparison obtained between the predictions and observed values for the relation between ω and $\sigma_2/\sigma_{u,I}$ for glastone material. Figure 4 shows the rock mass failure criterion on $\sigma_1/\sigma_{u,I}$ versus $\sigma_2/\sigma_{u,I}$ space for different F_{22} values. This figure clearly shows the importance of σ_2 on jointed rock mass strength.

Extension of this research to development of a new 3-D rock mass strength criterion based on true triaxial data is currently in progress.

4 ANALYTICAL DECOMPOSITION TECHNIQUE

The second indirect approach to study rock mass strength behavior is through the analytical decomposition technique. Jennings (1970) expressed the combined strength of joint and rock bridges through the following equation by simple linear weighing of the strength contributed by each fraction of material:

$$\tau = k\left(c_j + \sigma \tan\phi_j\right) + (1 - k)(c_r + \sigma \tan\phi_r) \tag{16a}$$

In Equation 16a, (c_j, ϕ_j) and (c_r, ϕ_r) represent the cohesion and friction angle of the joint and intact rock, respectively, and k is the joint continuity factor given by

$$k = L_j/(L_j + L_r) \tag{16b}$$

where L_j and L_r are the length of the joint and rock bridge, respectively. Note that Equation 16 ignores the interaction between the joints and intact rock and assumes simultaneous failure of the intact rock and joints. That means it disregards the possibility of progressive failure. Amadei (1988) has also used analytical superposition theorem to

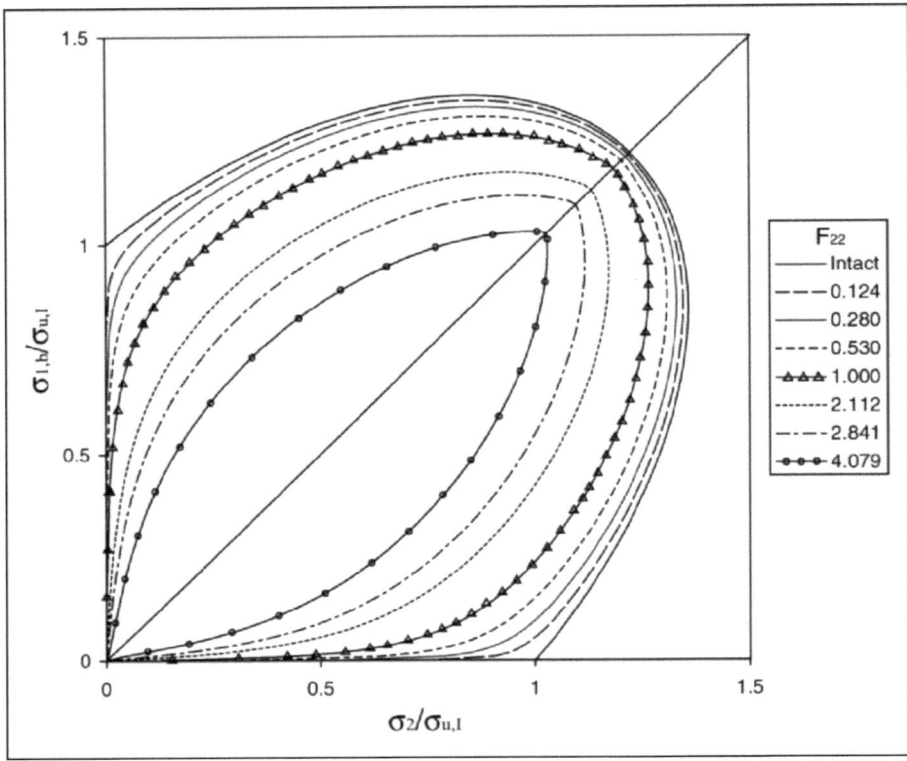

Figure 4 Obtained biaxial failure criterion for glastone jointed blocks on $\sigma_{1,b}/\sigma_{u,I}$ versus $\sigma_2/\sigma_{u,I}$ space (from Kulatilake *et al.*, 2006).

describe the strength of a regularly jointed rock mass under the full range of states of stress that are likely to be encountered in situ with one or several stress components being tensile. The intact rock was modeled using the Hoek and Brown criterion. The joints were modeled with no tensile strength and the shear strength described by the Coulomb criterion. Again the interactions between the intact rock and joints were ignored in coming up with the derivation. In these derivations, simplified fracture systems have been used with constant deterministic spacing and deterministic size and orientations for joint sets. In reality, joints are of finite size and all the joint geometry parameters are inherently statistical. Therefore, even though these early research has given some analytical thinking with respect to developing rock mass strength criteria, the assumptions made in these models are quite difficult to satisfy for most in situ rock masses.

5 NUMERICAL MODELING

The third indirect approach available is the numerical decomposition technique. This technique calculates strength of rock masses using a numerical method, incorporating the strength and deformability properties of the intact rock and joints. This technique allows incorporation of any joint network to the jointed rock mass and also includes interaction between joints and intact rock. Mainly the following three types of numerical modeling have been used to model rock mass strength: (a) Finite Element Modeling (FEM); (b) Particle Flow Modeling through PFC2D or PFC3D (Itasca, 2003; Ivars *et al.*, 2011) and (c) Distinct Element Modeling through UDEC and 3DEC (Itasca, 2008). In addition to that, even though the rock mass strength is not directly addressed, some fracture mechanics based analytical and numerical modeling have been used to study the crack initiation, propagation and coalescence in jointed blocks having a few fractures with extremely low fracture density. Because this aspect is important in modeling rock mass strength, it is briefly covered in this paper. The following sections provide a summary of the literature available on the aforementioned numerical modeling categories.

5.1 FEM based numerical modeling

A pioneering scale effect research performed using this approach at the two dimensional level using the finite element method (Kulatilake, 1985) with Goodman's joint element Goodman *et al.* (1968) has shown anisotropic, scale dependent mechanical behavior for jointed rock masses. The REV and the equivalent continuum behavior for mechanical properties have been investigated at the 2-D level using the finite element method (Kulatilake *et al.*, 1985; Pouya & Ghoreychi, 2001; Chalhoub & Pouya, 2008; Yang *et al.*, 2015). However, since the finite element method is based on continuum mechanics, simulation of large displacements and large rotations is difficult with the method, even though they may occur in jointed rock masses.

5.2 Fracture mechanics based analytical and numerical modeling

Several analytical and numerical procedures have been suggested to simulate crack initiation, propagation and coalescence in blocks having very few flaws with extremely low flaw density (Dyskin *et al.*, 1999; Chan *et al.*, 1990; Tang *et al.*, 2001). These numerical methods include the finite element method (FEM), boundary element method (BEM), and displacement discontinuity method (DDM). During the last few decades, various criteria have been proposed for crack initiation and propagation at the flaw tips. Three main criteria suggested are: the maximum tangential stress theory (Erdogan & Sih, 1990), maximum energy release rate theory (Hussain *et al.*, 1974) and minimum energy density theory (Sih, 1974). The FEM based numerical simulation code, RFPA2D (Rock Failure Process Analysis), has been used to simulate crack propagation and coalescence in a rock bridge area (Tang *et al.*, 2001). All of the aforementioned analytical and numerical procedures have made strong material property assumptions with respect to homogeneity/heterogeneity and isotropy/anisotropy of the medium around the flaws. One of the main questions is the applicability of these assumptions especially when the crack density increases in the considered medium.

5.3 PFC based numerical modeling

To avoid the assumptions used in fracture mechanics approaches, some investigators have resorted to particle flow codes (PFC^{2D} and PFC^{3D}) (Itasca, 2003; Potyondy, 2007;

Ivars *et al.*, 2011) to model jointed rock behavior under uniaxial loading (Kulatilake *et al.*, 2001b; Koyama & Jing, 2007; Lee & Jeon, 2011; Zhang & Wong, 2012; Bahaaddini *et al.*, 2013; Gao *et al.*, 2014; Zhang *et al.*, 2015; Fan *et al.*, 2015). PFC^{3D} allows one to study the mechanical interaction behavior between intact rock and joints incorporating a significant number of joints without making unrealistic assumptions about the surrounding medium around each joint. In addition, it allows failure through both the intact rock and joints under both tensile and shear modes leading to progressive failure which usually occur in jointed blocks having non-persistent joints. Kulatilake *et al.* (2001b) performed pioneering research in providing a realistic calibration procedure for micro-mechanical parameters of PFC^{3D} for a contact bonded particle flow model. Using this model they have studied jointed rock behavior of blocks having persistent joints under uniaxial loading. They included spherical particles to model both intact material and joints. In other words they considered closed flaws. Their focus was on global mechanical behavior of jointed blocks and possible failure modes. Lee & Jeon (2011) and Zhang & Wong (2012) have used PFC^{2D} to study crack initiation, propagation and coalescence using one or two flaws. In their models they have removed particles to simulate open flaws. They have reported about the successes, failures and difficulties they encountered in comparing their PFC results with experimental results.

By selecting appropriate micro-mechanical parameter values through a trial and error procedure, Fan *et al.* (2015) used the computer code PFC^{3D} to study the macro-mechanical behavior of jointed blocks having multi-non-persistent joints with high joint density under uniaxial loading. The focus was to study the effect of joint orientation, size and joint mechanical properties on jointed block strength, deformability, stress-strain relation and failure modes at the jointed block level. The uniaxial compressive strength of the block, UCS$_B$, was found to depend heavily on the joint dip angle, β, and joint continuity factor, k (see Fig. 5). The jointed blocks produced three types of stress-strain curves labeled as Type I through Type III (see Fig. 6). A relation was found to exist between the types of curves and β and k values (see Fig. 6 and Section 4 of Fan *et al.* (2015)). The dominance of tensile failures over the shear failures was observed for all three types of curves based on the micro-mechanical parameter values used in the paper. The UCS$_B$, rate of bond failures and the number of bond breakages were found to decrease as the curve type moves from Type I to Type III through Type II (see Fig. 7). The jointed blocks resulted in 4 failure modes as follows: (1) splitting failure; (2) plane failure; (3) stepped path and (4) intact material failure. These failure modes were compared with the corresponding experimental results obtained from Chen *et al.* (2011, 2012) (see Fig. 8). The main features of each failure mode and possible relations between the failure modes, UCS$_B$ and β and k values are given in Section 5 of Fan *et al.* (2015).

5.4 Distinct element based numerical modeling

The distinct element method introduced by Cundall (1971) and further developed by Lemos *et al.* (1985), Cundall (1988) and Hart *et al.* (1988) is a powerful technique to perform stress analyses in blocky rock masses formed by persistent discontinuities. In this method, the rock mass is modeled as an assemblage of rigid or deformable blocks. Discontinuities are considered as distinct boundary interactions

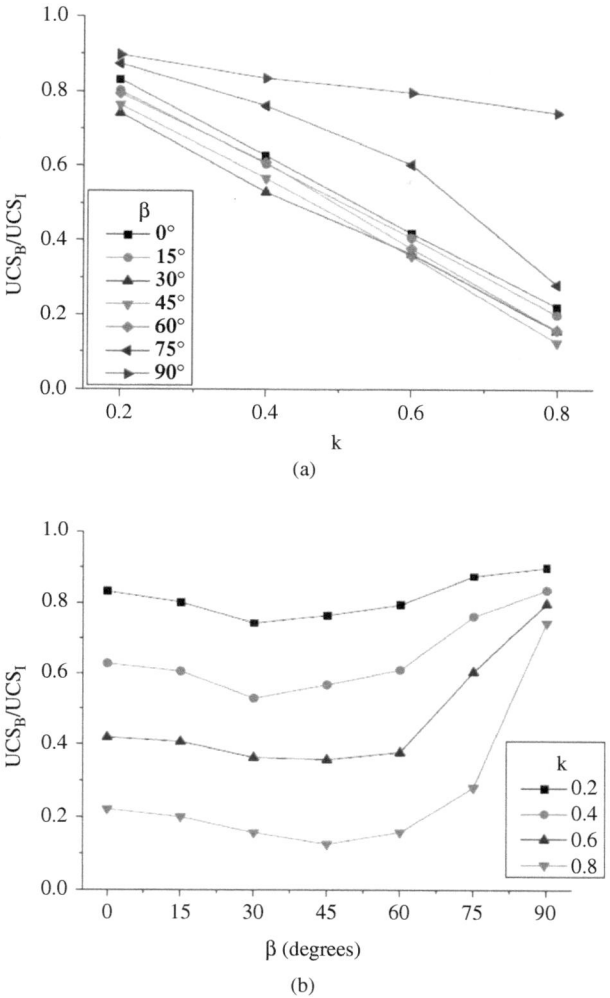

Figure 5 Effect of β, and k on UCS$_B$ (from Fan *et al.*, 2015).

between these blocks; joint behavior is prescribed for these interactions. The distinct element algorithm includes not only continuum theory representation for the blocks, but also force displacement laws which specify forces between blocks and a motion law which specifies the motion of each block due to unbalanced forces acting on the block. By taking into account the interaction of intact blocks and joints, the distinct element method can effectively calculate the mechanical behavior of block systems under different stress and displacement boundary conditions. The method employs an explicit solution procedure. An advantage of the explicit method is that, because matrices are never formed, large displacements, rotations and complex constitutive behavior for both intact material and joints are possible with no additional computing effort.

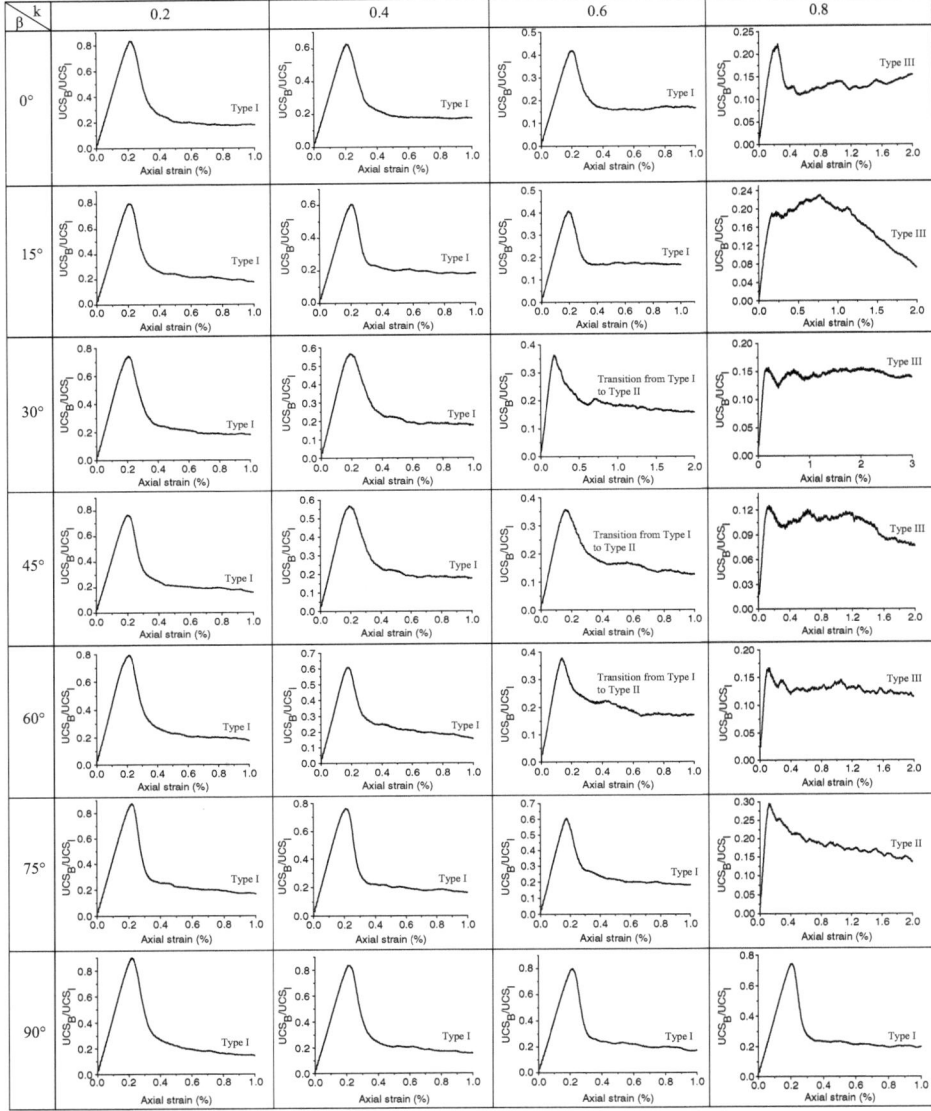

Figure 6 Stress-strain curves obtained for all jointed blocks studied through numerical modeling (from Fan *et al.*, 2015).

To use the distinct element method for stress analysis, first the problem domain should be discretized into polygons in 2-D and polyhedra in 3-D. To achieve that for rock masses having finite size actual joint configurations, Kulatilake *et al.* (1992) and Wang & Kulatilake (1993) introduced some fictitious joints that behave as intact rock to the domain to interact with actual joints. After performing a detailed study

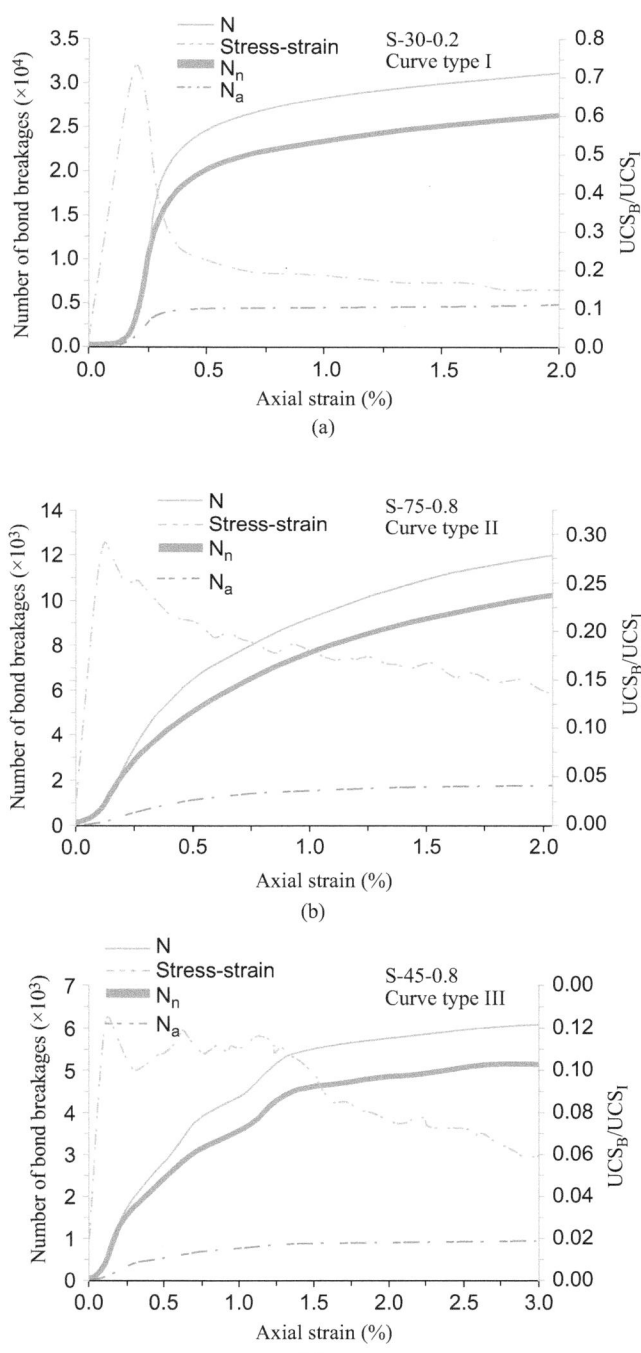

Figure 7 Relation between axial strain level of the stress-strain curve and the bond breakages for the three typical stress-strain curve types (from Fan *et al.*, 2015).

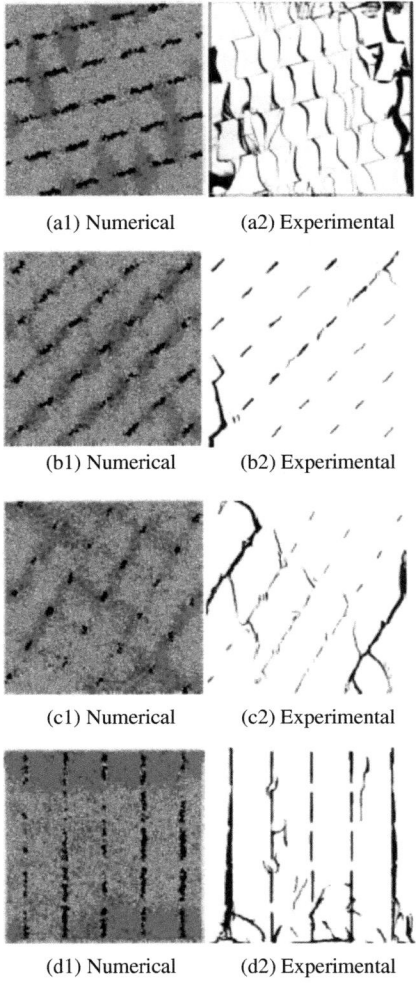

(a1) Numerical   (a2) Experimental

(b1) Numerical   (b2) Experimental

(c1) Numerical   (c2) Experimental

(d1) Numerical   (d2) Experimental

Figure 8 Failure mode comparisons between experimental and numerical results: (a) Sample S-15-0.6 for failure mode I (splitting failure mode); (b) Sample S-45-0.4 for failure mode II (plane failure mode); c) Sample S-60-0.2 for failure mode III (step path failure mode); (d) Sample S-90-0.8 for failure mode IV (intact material failure mode) (from Fan *et al.*, 2015).

under different stress paths, Kulatilake *et al.* (1992) have provided recommendations to select proper mechanical property values for these fictitious joints to reflect the intact rock behavior. Using these techniques, pioneering detailed investigations have been performed to study the effect of finite size joint geometry networks on the deformability and strength of jointed rock blocks, REV size and equivalent conti-nuum behavior at the 2-D (Kulatilake *et al.*, 1994) and 3-D levels (Kulatilake *et al.*, 1993, 2004). Results of these studies have shown anisotropic, scale dependent mechanical behavior for jointed rock masses. Also, an incrementally linear elastic,

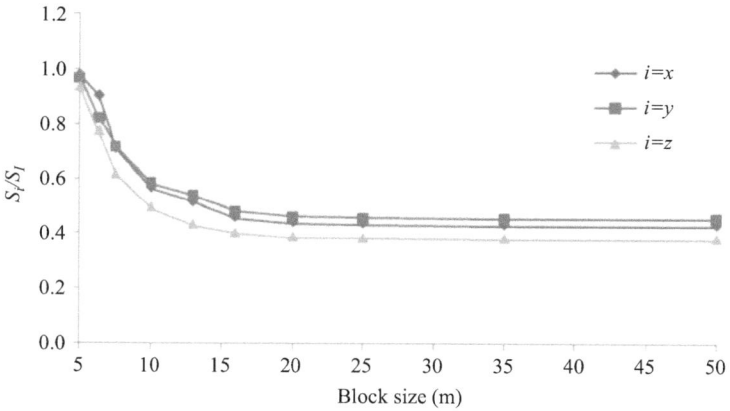

Figure 9 Relation between rock block strength and block size (Wu & Kulatilake, 2012).

orthotropic constitutive model has been suggested at the 3-D level to represent the pre-failure mechanical behavior of jointed rock blocks (Kulatilake *et al.*, 1993). This constitutive model has the capability to capture the anisotropic, scale-dependent behavior of jointed rock blocks. In that model, the effect of the joint geometry network in the rock mass is incorporated in terms of fracture tensor components which captures the effect of all the joint geometry parameters- the number of joint sets, joint density, orientation and size.

In a recent study, Wu & Kulatilake (2012) and Kulatilake & Wu (2013) applied similar procedures as stated above to investigate the effect of a fracture system that exist in a limestone rock mass located at the Yujian River Dam site, China on rock block strength and deformability, REV sizes for mechanical properties and equivalent continuum behavior by performing numerical stress analyses in three perpendicular directions (x, y and z) on different block sizes ranging between 5 m and 50 m. Fig. 9 shows the relation obtained between jointed rock block strength in the i direction/intact rock block strength (S_i/S_I) and block size for the investigated limestone rock mass. This plot shows the rock mass strength scale effect, anisotropy, REV behavior and equivalent continuum behavior. Fig. 10 shows the unique relation obtained between the jointed rock block strength in the i direction/intact rock block strength (S_i/S_I) and the addition of the fracture tensor components in the j and k directions, which are perpendicular to the i direction. The fracture tensor component in a certain direction combines the effect of orientation and size probability distributions and mean 3-D intensity of fractures coming from all the fracture sets in the rock mass in the selected direction (Kulatilake *et al.*, 1993). It is important to note that in Fig. 10, the calculated values obtained from three perpendicular directions are given. Also, an incrementally linear elastic, orthotropic constitutive model has been suggested to represent the pre-failure mechanical behavior of the jointed rock mass at a refined level. In summary, these numerical studies showed the possibility of developing relations between the rock mass strength and rock mass joint geometry properties.

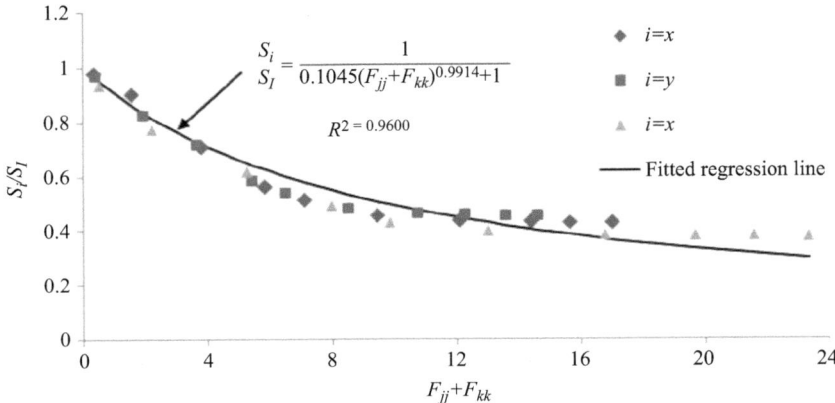

Figure 10 Relation between rock block strength and relevant fracture tensor parameters (Wu & Kulatilake, 2012).

Currently, research is underway to develop a new 3-D rock mass strength criterion by extending the rock mass strength models discussed in Kulatilake *et al.* (2006), Wu & Kulatilake (2012) and Kulatilake & Wu (2013) by using both experimental and numerical modeling data in 3-D.

6 SUMMARY

The intermediate principal stress does not exist in most of the available strength criteria. However, the intermediate principal stress plays a major role with respect to rock mass strength. Majority of the available strength criteria can be applied only to isotropic rock masses. These can be applied only to rock masses that are heavily fractured and do not have preferred discontinuity orientation directions. On the other hand most of the rock masses contain a few fracture sets which provide distinct discontinuity orientation directions. Such rock masses exhibit anisotropic rock mass strength. Majority of the available strength criteria do not include joint size/block size as a parameter in the strength criterion equation. These strength criteria do not have the capability to predict change of rock mass strength with the block size. This author feels that the rock mass strength criteria should be developed at the 3-D level incorporating the intermediate principal stress. In addition, each of the strength criteria should include the intensity of all the fracture sets and probability distributions of orientations and size explicitly in the strength criterion. Significant progress has been made in this direction (see Kulatilake *et al.*, 1993, 2004, 2006; Wu & Kulatilake, 2012; Kulatilake & Wu, 2013; He *et al.*, 2014). Research is in progress at present to extend these developed strength criteria.

ACKNOWLEDGMENT

The preparation of this paper was partially funded by the Centers for Disease Control and Prevention (Contract No. 200-2011-39886).

REFERENCES

Amadei, B. (1988) Strength of a regularly jointed rock mass under biaxial and axisymmetric loading conditions. *International Journal of Rock Mechanics and Mining Sciences & Geomechanics Abstracts*, 25 (1), 3–13.

Bahaaddini, M., Sharrock, G. & Hebblewhite, B.K. (2013) Numerical investigation of the effect of joint geometrical parameters on the mechanical properties of a non-persistent jointed rock mass under uniaxial compression. *Computers and Geotechnics*, 49, 206–225.

Barton, N.R., Lien, R. & Lunde, J. (1974) Engineering classification of rock masses for the design of tunnel support. *Rock Mechanics*, 6, 189–239.

Bidgoli, M.N., Zhao, Z.H. & Jing, L.R. (2013) Numerical evaluation of strength and deformability of fractured rocks. *Journal of Rock Mechanics and Geotechnical Engineering*, 5 (6), 419–430.

Bieniawski, Z.T. (1968) The effect of specimen size on compressive strength of coal. *International Journal of Rock Mechanics and Mining Sciences & Geomechanics Abstracts*, 5, 325–335.

Bieniawski, Z.T. (1976) Engineering classification in rock engineering. In: *Proceedings of the Symposium on Exploration for Rock Engineering, Johannesburgh*. pp. 97–106.

Bieniawski, Z.T. (1989) *Engineering Rock Mass Classification: A Manual*. New York, Wiley.

Bieniawski, Z.T. & Van Heerden, W.L. (1975) The significance of in situ tests on large rock specimens. *International Journal of Rock Mechanics and Mining Sciences & Geomechanics Abstracts*, 12, 101–103.

Bobet, A. & Einstein, H.H. (1998) Fracture coalescence in rock-type materials under uniaxial and biaxial compression. *International Journal of Rock Mechanics and Mining Sciences*, 35 (7), 863–888.

Bombolakis, E.G. (1968) Photoelastic study of initial stages of brittle fracture in compression. *Tectonophysics*, 6 (6), 461–473.

Brown, E.T. (1970) Strength of models of rock with intermittent joints. *Journal of Soil Mechanics and Foundation Division ASCE*, 96, 1935–1949.

Brown, E.T. (1974) Fracture of rock under uniform biaxial compression. In: *Advances in Rock Mechanics: Proceedings of the 3rd Congress, International Society for Rock Mechanics, 1974, Denver*. pp. 2A, 111–117.

Chen, X., Liao, Z.H. & Li, D.J. (2011) Experimental study on the effect of joint orientation and persistence on the strength and deformation properties of rock masses under uniaxial compression. *Chinese Journal of Mechanical Engineering*, 30 (4), 781–789.

Chen, X., Liao, Z.H. & Peng, X. (2012) Deformability characteristics of jointed rock masses under uniaxial compression. *International Journal of Mining Science and Technology*, 22 (2), 213–221.

Cai, M., Kaiser, P.K., Tasaka, Y. & Minami, M. (2007) Determination of residual strength parameters of jointed rock masses using the GSI system. *International Journal of Rock Mechanics and Mining Sciences*, 44 (2), 247–265.

Cai, M., Kaiser, P.K., Uno, H., Tasaka, Y. & Minami, M. (2004) Estimation of rock mass deformation modulus and strength of jointed hard rock masses using the GSI system. *International Journal of Rock Mechanics and Mining Sciences*, 41 (1), 3–19.

Chalhoub, M. & Pouya, A. (2008) Numerical homogenization of a fractured rock mass: a geometrical approach to determine the mechanical representative elementary volume. *Electronic Journal of Geotechnical Engineering*, 13, 1–12.

Chan, H.C.M., Li, V. & Einstein, H.H. (1990) A hybridized displacement discontinuity and indirect boundary element method to model fracture propagation. *International Journal of Fracture*, 45 (4), 263–282.

Chaoui, A., Mariotti, M. & Orliac, M. (1970) In situ calcareous marls strain and shear strength: comparison between different test characteristics. In: *Proceedings of the 2nd ISRM Congress Vol. 2, Belgrade*. pp. 3–50.

Chappel, B.A. (1974) Load distribution and deformational response in discontinua. *Geotechnique*, 24, 641–654.

Cook, N.G.W., Hodgson, K. & Hojem, J.P.M.A. (1971) 100-MN jacking system for testing coal pillars underground. *Journal of the South African Institute of Mining and Metallurgy*, 71, 215–224.

Cundall, P.A. (1971) A computer model for simulating progressive large-scale movements in blocky rock system. In: *Proceedings of the Symposium of the International Society for Rock Mechanics*. 2 (8), 129.

Cundall, P.A. (1988) Formulation of a three-dimensional distinct element model-Part 1. A scheme to detect and represent contacts in a system composed of many polyhedral blocks. *International Journal of Rock Mechanics and Mining Sciences & Geomechanics Abstracts*, 25, 107–116.

De Reeper, F. (1966) Design and execution of field pressure tests up to pressures of 200 kg/cm^2. In: *Proceedings of the 1st ISRM Congress, Lisbon*. pp. 613–619.

Drucker, D. & Prager, W. (1952) Soil mechanics and plastic analysis or limit design. *Quarterly of Applied Mathematics*, 10, 157–165.

Dyskin, A.V., Germanovich, L.N. & Ustinov, K.B. (1999) A 3-D model of wing crack growth and interaction. *Engineering Fracture Mechanics*, 63 (1), 81–110.

Einstein, H.H. & Hirschfeld, R.C. (1973) Model studies on mechanics of jointed rock. *Journal of Soil Mechanics and Foundation Division ASCE*, 99, 229–248.

Erdogan, F. & Sih, C.C. (1963) On the crack extension path in plates under plane loading and transverse shear. *Journal of Basic Engineering*, 85 (4), 519–527.

Gao, F.Q., Stead, D. & Kang, H.P. (2014) Numerical investigation of the scale effect and anisotropy in the strength and deformability of coal. *International Journal of Coal Geology*, 136, 25–37.

Georgi, F., Hofer, K.H., Knoll, P., Menzel, W. & Thomas, K. (1970) Investigation about the fracture and deformation behavior of rock masses. In: *Proceedings of the 2nd ISRM Congress, Belgrade*. pp. 2, 3–43.

Gimm, W.A.R., Richter, E. & Rosetz, G.P.A. (1966) A study of the deformation and strength properties of rocks by block tests in situ in iron-ore mines. In: *Proceedings of the 1st ISRM Congress, Lisbon*. pp. 1, 457–463.

Goodman, R.E., Taylor, R.L. & Brekke, T.L. (1968) A model for the mechanics of jointed rock. *J. Soil Mech. Found. Div.*, 94 (3), 637–659.

Greenwald, H.P., Howarth, H.C. & Hartman, I. (1939) *Experiments on strength of small pillars of coal in the Pittsburgh bed*. U.S. Bureau of Mines. Technical Paper number: 605.

Greenwald, H.P., Howarth, H.C. & Hartman, I. (1941) *Progress report: Experiments on strength of small pillars of coal in the Pittsburgh bed*. U.S. Bureau of Mines. Report number: 3575.

Griffith, A.A. (1920) The phenomena of rupture and flow in solids. *Philosophical Transactions of the Royal Society of London, Series A: Mathematical, Physical and Engineering Sciences*, 221 (587), 163–198.

Griffith, A.A. (1924) The theory of rupture. In: Biezeno, C.-B. & Burgers, J.-M. (eds.): *Proceedings of the First International Congress for Applied Mechanics, Delft*. Delft, J. Waltman Jr. pp. 55–63.

Hart, R., Cundall, P.A. & Lemos, J. (1988) Foundation of a three dimensional distinct element model-Part II. Mechanical calculations for motion and interaction of a system composed of many polyhedral blocks. *International Journal of Rock Mechanics and Mining Sciences & Geomechanics Abstracts*, 25, 117–126.

He, P., Kulatilake, P.H.S.W., He, M. & Liu, D. (2014) Experimental observations and interpretations on fracture networks, strength and deformability of coal masses under three-dimensional loading conditions. In: *Proceedings of the 33rd International Conference on Ground Control in Mining, Morgantown, West Virginia*. pp. 13–21.

Heuze, F.E. (1980) Scale effects in the determination of rock mass strength and deformability. *Rock Mechanics and Rock Engineering*, 12, 167–192.

Hoek, E. (1968) Brittle failure of rock. In: Stagg, K.G. & Zienkiewicz, O.C. (eds.) *Rock mechanics in engineering practice*. New York, Wiley. pp. 99–124.

Hoek, E. & Brown, E.T. (1980) Empirical strength criterion for rock masses. *Journal of Geotechnical Engineering Division ASCE*, 106, 1013–1035.

Hoek, E. & Brown, E.T. (1997) Practical estimates of rock mass strength. *International Journal of Rock Mechanics and Mining Sciences & Geomechanics Abstracts*, 34, 1165–1186.

Hoek, E., Kaiser, P.K. & Bawden, W.F. (1995) *Support of underground excavations in hard rock*. Rotterdam, A.A. Balkema.

Hoek, E., Wood, D. & Shah, S. (1992) A modified Hoek–Brown criterion for jointed rock masses. In: Hudson, J.-A. (ed.) *Eurock 1992: Rock characterization: ISRM Symposium, EUROCK 1992, Chester, UK*. London, Thomas Telford. pp. 209–213.

Hussain, M.A., Pu, S.L. & Underwood, J. (1974) Strain energy release rate for a crack under combined Mode I and Mode II. *Fracture Analysis*, 560 (1), 2–28.

Itasca. (2003) *PFC3D user's manual, version 4.0*. Minneapolis, Minnesota, Itasca Consulting Group Inc.

Itasca. (2008) *3DEC user's guide*. Minneapolis, Minnesota, Itasca Consulting Group Inc.

Ivars, D.M., Pierce, M.E., Darcel, C., Montes, J.R., Potyondy, D.O., Young, R.P. & Cundall, P.A. (2011) The synthetic rock mass approach for jointed rock mass modelling. *International Journal of Rock Mechanics and Mining Sciences*, 48 (2), 219–244.

Jahns, H. (1966) Measuring the strength of rock in situ at an increasing scale. In: *Proceedings of the 1st ISRM Congress, Lisbon*. pp. 1, 477–482.

John, K.W. (1970) Civil engineering approach to evaluate strength and deformability of closely jointed rock. In: *Rock mechanics-Theory and Practice: Proceedings of the 11th Symposium on Rock Mechanics AIME, New York*. pp. 69–80.

Koyama, T. & Jing, L. (2007) Effects of model scale and particle size on micro-mechanical properties and failure processes of rocks – A particle mechanics approach. *Engineering Analysis with Boundary Elements*, 31 (5), 458–472.

Kulatilake, P.H.S.W. (1985) Estimating Elastic Constants and Strength of Discontinuous Rock. *Journal of Geotechnical Engineering ASCE*, 111 (7), 847–864.

Kulatilake, P.H.S.W., Liang, J. & Gao, H. (2001) Experimental and numerical simulations of jointed rock block strength under uniaxial loading. *Journal of Engineering Mechanics ASCE*, 127 (12), 1240–1247.

Kulatilake, P.H.S.W., Malama, B. & Wang, J. (2001) Physical and particle flow modeling of jointed rock block behavior. *International Journal of Rock Mechanics*, 38 (5), 641–657.

Kulatilake, P.H.S.W., Park, J. & Malama, B. (2006) A new rock mass strength criterion for biaxial loading conditions. *International Journal of Geotechnical and Geological Engineering*, 24 (4), 871–888.

Kulatilake, P.H.S.W., Park, J. & Um, J. (2004) Estimation of rock mass strength and deformability in 3-D for a 30m cube at a depth of 485m at Äspö Hard Rock Laboratory, Sweden. *Internaltional Journal of Geotechnical and Geological Engineering*, 22 (3), 313–330.

Kulatilake, P.H.S.W., Ucpirti, H., Wang, S., Radberg, G. & Stephansson, O. (1992) Use of the Distinct Element Method to perform stress analysis in rock with non-persistent joints and to study the effect of joint geometry parameters on the strength and deformability of rock masses. *Rock Mechanics and Rock Engineering*, 25 (4), 253–274.

Kulatilake, P.H.S.W., Ucpirti, H. & Stephansson, O. (1994) Effect of Finite Size Joints on the Deformability of Jointed Rock at the Two Dimensional Level. *Canadian Geotechnical Journal*, 31, 364–374.

Kulatilake, P.H.S.W., Wang, S. & Stephansson, O. (1993) Effect of Finite Size Joints on Deformability of Jointed Rock at the Three Dimensional Level. *International Journal of Rock Mechanics and Mining Sciences*, 30 (5), 479–501.

Kulatilake, P.H.S.W. & Wu, Q. (2013) REV and equivalent continuum/discontinuum 3-D stability analyses of a tunnel. In: *Proceedings of the 3rd International FLAC-DEM Symposium, October 2013, China*. Paper selected for a Peter Cundall Award.

Ladanyi, B. & Archambault, G. (1980) *Direct and indirect determination of shear strength of a jointed rock mass*. Society of Mining Engineers of AIME for presentation at the AIME Annual Meeting, Las Vegas, Nevada. Report number: 80–25. pp. 24–28.

Lajtai, E.Z. (1971) A theoretical and experimental evaluation of the Griffith theory of brittle fracture. *Tectonophysics*, 11 (2), 129–156.

Lama, R.D. (1971) In situ and Laboratory strength of coal. In: *Proc. 12th Symp. Rock Mech., University of Missori, AIME, New York*. pp. 265–300.

Lee, H. & Jeon, S. (2011) An experimental and numerical study of fracture coalescence in pre-cracked specimens under uniaxial compression. *International Journal of Solids and Structures*, 48 (6), 979–999.

Lemos, J.V., Hart, R.D. & Cundall, P.A. (1985) A generalized distinct element program for modelling jointed rock mass. In: *Proc. Symp. On Foundamentals of Rock Joints, Bjorkliden, Sweden*. pp. 335–343.

Melkoumian, N., Priest, S.D. & Hunt, S.P. (2009) Further development of the three-dimensional Hoek–Brown yield criterion. *Rock Mechanics and Rock Engineering*, 42 (6), 835–847.

Mogi, K. (1971) Fracture and flow of rocks under high triaxial compression. *Journal of Geophysical Research*, 76, 1255–1269.

Mughieda, O. & Omar, M.T. (2008) Stress analysis for rock mass failure with offset joints. *Geotechnical and Geological Engineering* 26 (5), 543–552.

Nose, M. (1964) Rock Tests in situ. In: *Transactions 6th Congress Large Dams, ICOLD*.

Pan, X.D. & Hudson, J.A. (1988) A simplified three-dimensional Hoek–Brown yield criterion. In: Romana, M. (eds.): *Rock mechanics and power plants*. Rotterdam, A.A. Balkema. pp. 95–103.

Park, C.H. & Bobet, A. (2009) Crack coalescence in specimens with open and closed flaws: a comparison. *International Journal of Rock Mechanics and Mining Sciences*, 46 (5), 819–829.

Potyondy, D.O. (2007) Simulating stress corrosion with a bonded-particle model for rock. *International Journal of Rock Mechanics and Mining Sciences*, 44 (5), 677–691.

Pouya, A. & Ghoreychi, M. (2001) Determination of rock mass strength properties by homogenization. *Int. J. Numer. Anal Methods*, 25, 1285–1303.

Pratt, H.R., Black, A.D., Brown, W.S. & Brace, W.F. (1972) The effect of specimen size on the mechanical properties of unjointed diorite. *International Journal of Rock Mechanics and Mining Sciences & Geomechanics Abstracts*, 9, 513–529.

Priest, S.D. (2005) Determination of shear strength and three-dimensional yield strength for the Hoek–Brown criterion. *Rock Mechanics and Rock Engineering.*, 38 (4), 299–327.

Prudencio, M. & Van Sint Jan, M. (2007) Strength and failure modes of rock mass models with non-persistent joints. *International Journal of Rock Mechanics and Mining Sciences*, 44 (6), 890–902.

Ramamurthy, T. (2001) Shear strength response of some geological materials in triaxial compression. *International Journal of Rock Mechanics and Mining Sciences*, 38, 683–697.

Ramamurthy, T. & Arora, V.K. (1994) Strength predictions for jointed rocks in confined and unconfined states. *International Journal of Rock Mechanics and Mining Sciences & Geomechanics Abstracts*, 31 (1), 9–22.

Sahouryeh, E., Dyskin, A.V. & Germanovich, L. (2002) Crack growth under biaxial compression. *Engineering Fracture Mechanics*, 69 (18), 2187–2198.

Sheorey, P.R., Biswas, A.K. & Choubey, V.D. (1989) An empirical failure criterion for rocks and jointed rock masses. *Engineering Geology.*, 26, 141–159.

Sih, G.C. (1974) Strain-energy-density factor applied to mixed mode crack problems. *International Journal of fracture*, 10 (3), 305–321.

Tang, C.A., Lin, P., Wong, R.H.C. & Chau, K.T. (2001) Analysis of crack coalescence in rock-like materials containing three flaws – Part II: numerical approach. *International Journal of Rock Mechanics and Mining Sciences*, 38 (7), 925–939.

Wagner, H. (1974) Determination of the complete load-deformation characteristics of coal pillars. *Proceedings of the Congress of the International Society for Rock Mechanics*. 3, Vol. 2, Part B, Advances in rock mechanics; reports of current research. pp. 1076–1081.

Wang, S. & Kulatilake, P.H.S.W. (1993) Linking between joint geometry models and a distinct element method in three dimensions to perform stress analyses in rock masses containing finite size joints. *Japanese Society of Soil Mechanics and Foundation Engineering*, 33 (4), 88–98.

Wiebols, G. & Cook, N. (1968) An energy criterion for the strength of rock in polyaxial compression. *International Journal of Rock Mechanics and Mining Sciences & Geomechanics Abstracts*, 5, 529–549.

Wong, L.N.Y. & Einstein, H.H. (2009) Crack coalescence in molded gypsum and Carrara marble – Part 1. Macroscopic observations and interpretation. *Rock Mechanics and Rock Engineering*, 42 (3), 475–511.

Wong, R.H.C., Chau, K.T., Tang, C.A. & Lin, P. (2001) Analysis of crack coalescence in rock-like materials containing three flaws – Part I: experimental approach. *International Journal of Rock Mechanics and Mining Sciences*, 38 (7), 909–924.

Wu, Q. & Kulatilake, P.H.S.W. (2012) REV and its properties on fracture system and mechanical properties, and an orthotropic constitutive model for a jointed rock mass in a dam site in China. *Int. Jour. of Computers and Geotechnics*, 43, 124–142.

Yang, Z.Y., Chen, J.M. & Huang, T.H. (1998) Effect of joint sets on the strength and deformation of rock mass models. *International Journal of Rock Mechanics and Mining Sciences*, 35 (1), 75–84.

Yang, J.P., Chen, W.Z., Yang, D.S. & Yang, J.Q. (2015) Numerical determination of strength and deformability of fractured rock mass by FEM modeling. *Computers and Geotechnics*, 64, 20–31.

Yudhbir, L. W. & Prinzl, F. (1983) An empirical failure criterion for rock masses. In: *Proceedings of the 5th International Congress on Rock Mechanics, Melbourne, Balkema, Rotterdam*. vol. 1, pp. B1–B8.

Zhou, S. (1994) A program to model the initial shape and extent of borehole breakout. *Computational Geosciences*, 20, 1143–1160.

Zhang, L. & Zhu, H. (2007) Three-dimensional Hoek–Brown strength criterion for rocks. *J Geotechnical and Geoenvironmental Engineering ASCE*, 133 (9), 1128–1135.

Zhang, X.P. & Wong, L.N.Y. (2012) Cracking processes in rock-like material containing a single flaw under uniaxial compression: a numerical study based on parallel bonded-particle model approach. *Rock Mechanics and Rock Engineering*, 45 (5), 711–737.

Zhang, Q., Zhu, H. & Zhang, L. (2013) Modification of a generalized three-dimensional Hoek–Brown strength criterion. *International Journal of Rock Mechanics and Mining Sciences*, 59, 80–96.

Zhang, Y., Stead, D. & Elmo, D. (2015) Characterization of strength and damage of hard rock pillars using a synthetic rock mass method. *Computers and Geotechnics*, 65, 56–72.

Chapter 13

Some recent progress in physical modeling of rock stability of tunnels or underground caverns in China

W.S. Zhu, Q.Y. Zhang, L.P. Li & Y. Li

Geotechnical and Structural Engineering Research Center, Shandong University, Jinan Shandong, P. R. China

Abstract: In this chapter, first the similarity principal of the geological mechanical model test is introduced, and then the research and production method for a new advanced materials model are introduced. After that, as an example of its application and as a case study, three practical projects are simulated by model testing. The first simulates the phenomenon of zonal disintegration in rock surrounding a deep mine tunnel; the second is a simulation study of the stability of rock surrounding an underground power house cavern group under the condition of high initial stress; the third is a study on a seepage flow from rock surrounding a subsea tunnel. In most of the tests, a new type of true three-dimensional stress state load system and many advanced measurement test technologies are used. The results of the experiments were verified by field monitoring or by numerical simulation.

1 INTRODUCTION

A large variety of numerical methods have been used to study the stability of rock masses surrounding large underground cavern complexes. These include the finite element method (FEM), the finite difference method (FDM), and the discrete element method (DEM). These methods have contributed to improvements in design, especially where scale effects and heterogeneities are present and the response of the rock mass is non-linear. However, as the physical settings of underground projects become increasingly more complex, it is becoming increasingly difficult to represent responses using numerical methods with any degree of confidence. This is because the complexities of the physics involved in complex processes such as rock bursting or splitting or spalling failures are difficult to accommodate with confidence. So although some discontinuum-based numerical methods (3DEC, DDA and PFC3D) can be applied to simulate and analyze failure modes and processes, they remain innately limited by computational constraints and inadequate understanding of correct physical response of complex media under complex loading conditions. Although physical model testing is not a panacea, well designed experiments may yield important insights into behavior that are not available from numerical models.

A variety of testing methods have developed to allow high fidelity physical modeling of structures in brittle rocks, including the monitoring of various signals of stress, deformation, the evolution of fluid pressures, and identification of separations and differential movements and rotations. These methods have evolved since the 1960s and include experiments on a variety of structures on, and in, rock. Heuer & Hendron (1971) conducted geomechanical model tests on the excavation of underground caverns under static loading. However, to represent the true behavior of underground structures in rock, the effects of reinforcement must also be considered, including loading to failure.

This has been the focus of some experimental work in China where significant extensions to previous work have resulted. Li *et al.* (2003, 2004) developed several new techniques in physical experimentation. These methods have been successfully applied in three-dimensional physical models of the underground complex of the Xiluodu hydropower station. Chen *et al.* (2004) and Chen (1994) developed a new, three-dimensional steel frame, which can be used to apply tri-axial loading with either a uniform or an inverted trapezoid distribution. Zhang (2007) developed another type of steel structural frame for three-dimensional geomechanical experiments. This frame combines high stiffness and rigidity with the flexibility of assembly and easy adjustment of cell dimensions. Wang (2002), and Wang (2006) developed a new type of modeling material (called IBSCM) which can simulate a broad variety of rock types and rock masses. Zeng (2001) conducted a physical model experiment on underground caverns in a hydropower station and obtained the stress distributions, displacement distributions, and failure modes and mechanisms of the surrounding rock masses. All of these physical model tests have not confronted the difficult problems of large overburden depths, high *in situ* stresses, and fractured and weak rock masses common in contemporary design and construction, and most of them did not consider the seepage effect of underground water. The authors explore the application of physical model experimentation in the following: in relation to a deeply-embedded gallery in the Huainan Coal Mining project in Anhui Province, China; a cavern complex's stability in the Shuangjiangkou Hydropower station in Sichuan Province, China; and a subsea tunnel in Qingdao, Shandong Province, China.

2 THE SIMILARITY PRINCIPLE AND MODEL MATERIAL SELECTION

The similarity requirements of a geomechanics model test can be expressed as if the model and the prototype are two similarity systems, and their geometric features – as well as various physical quantities – need to keep a certain proportional relationship. These similarity proportion relations are the similarity requirements of the ratios of a geomechanics model test.

2.1 Similarity scale

We put the ratios of the same physical quantities between the prototype (P) and the model (M), called the similarity scale. The following symbols represent the various physical or geometric parameters in various scales of Ci on behalf of the above, that are: L (length), γ (density), δ (displacement), σ (stress), ε (strain), σ^t (tensile strength), σ_c (compressive strength), c (cohesion), φ friction angle, μ (Poisson's ratio), f (friction coefficient), $\overline{X}$ (boundary surface force), X (volume force) and T (time). Corresponding geometric; stress; strain; the elastic modulus; Poisson's ratio; density; the coefficient of friction; the internal friction angle and displacement; the boundary surface; volume forces; and time of the similarity scale definition have the following form, viz. $C_i = \frac{C_p}{C_m}$ in which C_p, C_m are respectively the model of the physical or geometric parameters. See Zhang *et al.* (2005)

2.2 Establishment of the similarity conditions

According to the equilibrium equation; geometric equation; physical equation; and the stress boundary condition and displacement boundary conditions of the prototype and the model, the similarity condition of the geological mechanical model test can be established (Zhang *et al.*, 2005).

1) Establishing the similarity conditions
   According to the equilibrium equation for the model and prototype (Zhang *et al.*, 2005) the relationship between the stress similarity ratio C_σ, the bulk similarity ratio C_r and the geometric similarity ratio C_L can be derived as:

$$\frac{C_r C_L}{C_\sigma} = 1 \tag{1}$$

2) Setting up the geometrical similarity conditions of the prototype and the model
   According to the geometric equations in Zhang *et al.* (2005), the relationship between the displacement similarity ratio C_δ, the geometrical similarity ratio C_L and the strain similarity ratio C_ε can be derived as:

$$\frac{C_L C_\varepsilon}{C_\delta} = 1 \tag{2}$$

3) Setting up the stress similarity conditions of the prototype and the model
   According to the physical equations of the prototype and the model, the similarity relationship between the ratios of stress, strain and the elastic modulus can be obtained as:

$$\frac{C_\sigma}{C_\varepsilon C_E} = 1 \tag{3}$$

At the same time, the Poisson's similarity ratio, which is non-dimensional, can be obtained as $C_\mu = 1$.

4) Time similarity conditions
   In the model experiments, to consider the influence of the time factor, a formula on the basis of Newton's second law can be derived by a dimensional analysis method of the time similarity ratio C_t from the relationship between the geometric similarity ratio C_L, and the time similarity ratio under the effect of gravity and inertia force, from which the similarity condition of the prototype and the model of time can be obtained (Zhang *et al.*, 2005)

$$C_t = \frac{T_P}{T_M} = \sqrt{\frac{L_P}{L_M}} = \sqrt{C_L} \tag{4}$$

2.3 Similarity conditions of geomechanics model test

1) The relationship between the stress similarity ratio C_σ, the bulk density similarity ratio C_r and the geometric similarity ratio is C_L

$$C_\sigma = C_r C_L \tag{5}$$

2) The relationship between the displacement similarity ratio C_δ, the geometrical similarity ratio C_L and the strain similarity ratio is C_ε

$$C_\delta = C_\varepsilon C_L \tag{6}$$

3) The relationship between the stress similarity ratio C_σ, the elastic modulus similarity ratio C_E and the strain similarity ratio is C_ε

$$C_\sigma = C_\varepsilon C_E \tag{7}$$

4) A geomechanics model test requires that the similarity ratio of all dimensionless parameters (such as strain, internal friction angle, friction coefficient, Poisson ratio, etc.) should be equal to 1, namely:

$$C_\varepsilon = 1, \ C_f = 1, \ C_\varphi = 1, \ C_\mu = 1 \tag{8}$$

5) The relationship between time similarity ratio C_t and geometric similarity scale C_L should be:

$$C_t = \sqrt{C_L} \tag{9}$$

2.4 Basic principles of model material development

The precondition of a geomechanical model test is to choose a material similar to the similarity requirements conditions. For the model test it is impossible to satisfy all the similarity criteria; therefore, satisfy only the main parameters of similarity. The secondary parameters should be matched as closely as possible, dependent upon what is required. For an underground cavern project, the prototypes of the main material of rock mass must be considered. Rock mass is often a very complex anisotropic geological material, and its mechanical characteristics may change over a wide range. In order to simulate the wide range of stiffness and strength characteristics of an actual rock mass, mechanical properties need to be developed in the model to reflect changes over a big enough range, and it needs to have similar enough stable mechanical properties.

A geomechanics model test requires that the constitution of the prototype and the model material are as similar as possible, and that the model's gravity should simulate the prototype material density as closely as possible. For a geomechanics model test, the key is the choice of model material.

The relationship of the similarity ratio for the model for stress, the elastic modulus, bulk density and geometry parameters should be:

$$\frac{C_E}{C_r \cdot C_L} = \frac{C_\sigma}{C_r \cdot C_L} = 1 \tag{10}$$

It can be seen that, in order to make the model simulate more closely the qualities of the prototype rock mass, in terms of geometrical scale, it is necessary to increase the value C_L. To meet the requirements of formula (1, 2), a corresponding increase in stress similarity coefficient C_σ and reduction in the bulk density similarity coefficient, C_r, are needed. Therefore, in order to satisfy the similarity criteria, the model material developed should meet the high density, low strength, low modulus of deformation, but, unfortunately, this kind of the material does not exist in nature; the only way is fabricate of a new one by artificial means.

The basic principle in developing model materials is to use particle cementing materials which have a reasonable aggregate gradation. Model materials have the characteristics of low strength, low modulus of deformation, speedy drying, low cost, convenient processing, stable mechanical property, flexible material preparation, and nontoxicity as well as harmlessness. At home and abroad, the range of common modeling materials with the mechanical parameters of with low bulk density, slow drying time, poor stability of mechanical properties etc. is very narrow. So new materials with a superior performance need to be developed. After testing more than 300 combinations of materials, giving nearly 1000 test specimens with different material mechanical parameters, the author himself developed a new material using pure iron powder, barite powder, quartz sand, gypsum powder, and a proportion of rosin and alcohol solution evenly mixed for compaction, which became a kind of composite material, called iron-sand-cemented rock similarity material (Zhang *et al.*, 2008). The material has pure iron powder, barite powder, and quartz sand as aggregate, gypsum powder as a regulator, and the rosin–alcohol solution as an adhesive. By changing the component content and the concentration of the binder the material mechanical parameters can be greatly adjusted, which has the advantages of mechanical parameters with a wide change range of properties, stable behavior, a low cost, fast drying, and simple, non-toxic and harmless, etc. to make.

2.5 Materials selection and specimens making

In the experiment, we chose the barite powder, pure iron powder, quartz sand, rosin, a concentration of 95% above the level of medical alcohol, and gypsum for raw materials. As pure iron powder and barite powder are too thin on their own, to optimize grading and adjust the mechanical properties of the materials, the quartz sand is added as aggregate (Figure 1).

In order to test the physical and mechanical parameters of the materials, two sets of molds were designed and processed: one, a 50 × 100 mm round double open steel mound, used to compress a 50 × 100 mm standard specimen; the other a 62.8 × 200 mm round double open steel mound used to compress a 62.8 × 200 mm specimen for direct shear test of the specimen.

The detailed material production process of the specimens in this paper is shown as below: firstly, the pure iron powder, barite powder, quartz sand, and a small amount of gypsum powder are mixed together. Secondly, according to the design ratio of specimen made of weighing, we pour them into the mound, and place demolding specimens in a ventilated place to dry in 2–3 days at room temperature. Then the specimens can then be used to laboratory tests for physico-mechanical parameters. Results show that, at room temperature, the specimens can be completely dried in 60 hours (shown in Figure 2).

Figure 1 Model of raw material.

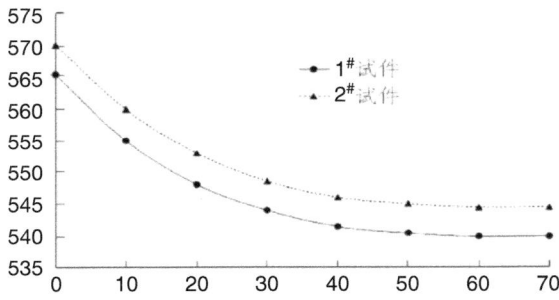

Figure 2 Specimens under room temperature of 20 °C drying time curve.

For more than 300 sets of materials prepared in different proportions for the tests, nearly 1000 specimens were created for testing mechanical parameters.

2.6 Mechanical properties of iron-crystal-sand cemented rock similar material

Figures 3 and 4 show the material stress–strain curve and Moore strength envelope, respectively.

By testing many different material proportions for their mechanical parameters, the effect of material content on the mechanical behavior of similarity material can be summarized as follows.

(1)   Through adjusting the molar concentration of the cementing agent – *i.e.* the rosin and alcohol solution – to adjust the compressive strength and elastic modulus of materials and their cohesion, the higher the molar concentration of the cementing agent, the higher the compressive strength of similarity materials, elastic modulus and the cohesion.

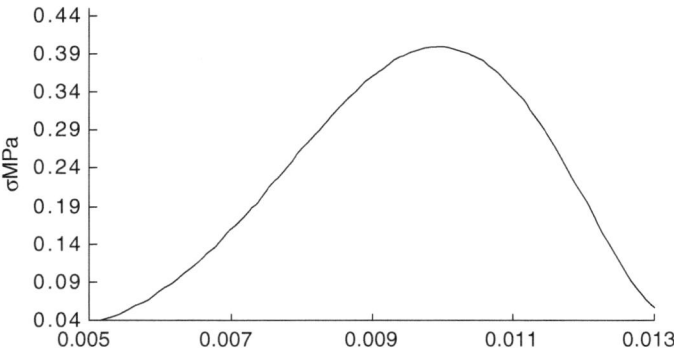

Figure 3 Typical stress–strain relationship curve of materials.

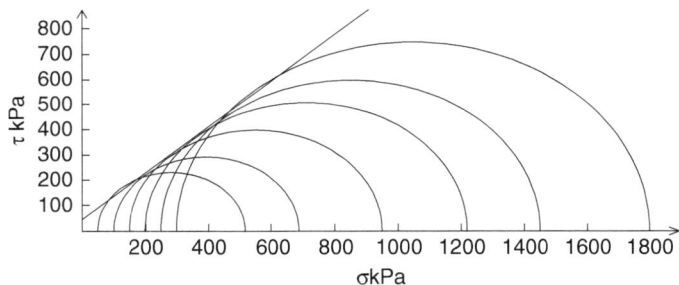

Figure 4 Typical Mohr-Coulomb envelope.

(2) Pure iron powder content has the greatest influence on the material density: with an increase of the content of fine iron powder the material density also increases. However, this is not a linear relationship; by adjusting the proportion, the material density will reach up to 29 kN/m³. In addition, when adjusting the proportions of iron powder and barite powder the material's elastic modulus may reach maximum.

(3) Quartz sand mesh will affect the material cohesion and internal friction angle of the material; the bigger the quartz sand mesh, the greater the cohesion, but the smaller the angle of internal friction.

The test results of the material physical and mechanical parameters of the similarity material show us that the crystal–iron–sand similarity material has the widest range of physical and mechanical parameters: the material density range is 23–30 kN/m³, the modulus of elasticity change range is 30–1400 MPa, the compressive strength change range is 0.3–5.0 MPa, the cohesive change range is 20–350 kPa, the internal friction angle range is 27–50°. Therefore, it can be used to simulate most of the middle-hard to hard rock materials with hard-brittle behavior. The material has advantages as follows.

(1) A wide range of material physico-mechanical parameters, greatly simplifying the difficulty in finding the appropriate material for the model test.
(2) The material's mechanics characteristics are stable, so are not affected by temperature and humidity.
(3) The main raw materials are cheap.
(4) Alcohol is easy to volatilize, so the material has a short drying time.
(5) Materials are non-toxic and harmless.
(6) The material synthesized formed blocks that are easy to cut, and easy to manufacture in any form.
(7) If the block surface is wetted with alcohol, it has an adhesive ability, so seams of blocks can be self-adhesive.
(8) Materials with the same contents can be used repeatedly, greatly improving the utilization rate of material.

3.1 Introduction

The condition of surrounding rock with an increase in mining depth will be one of significant nonlinear deformation and damage intensity, and zonal fracture phenomenon is one of the typical characters of nonlinear deformation and fractures of deep rock mass excavation. When a cavity or roadway is excavated in deep rock, it can produce an alternating fracture zone and zonal fracture on both sides and the front of the surrounding rock, a phenomenon known as partition burst phenomenon. This phenomenon is confirmed in many deep cavern excavations at home and abroad, through a variety of physical detection methods. Authors such as Shemyakin in the 1980s (Shemyakin, 1986) researched in deep mines in the Taimyrskii mining field and found the zonal fractures phenomenon using resistivity meter cracking for rock (Figure 5).

Adams & Jager (1980) at Witwatersrand gold mine in South Africa observed the zonal rupture of a deep stope from 2000 m to 3000 m between using a borehole periscope.

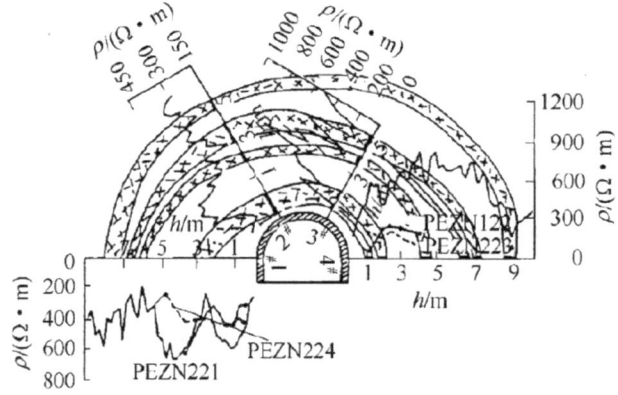

Figure 5 Mine zonal fractures phenomenon at Taimyrskii (Qian *et al.*, 2008).

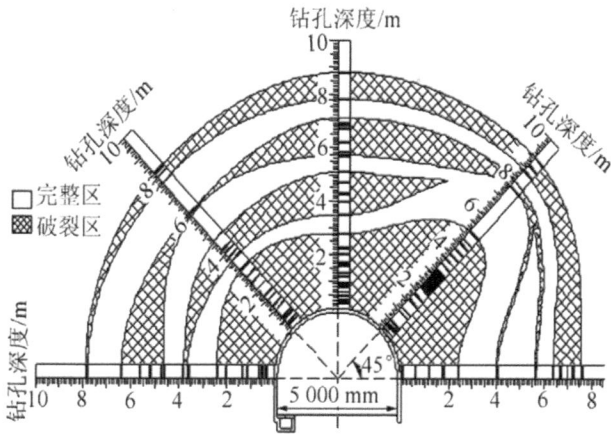

Figure 6 Zonal fractures phenomenon at Huainan coal mine roadway (Qian *et al.*, 2008).

Li Shucai *et al.* (2008) observed at a deep roadway surrounding rock in a mine in the Huainan mining area, using a kind of borehole TV instrument, the breakdown phenomenon of zonal fractures. Therefore, the existence of the zonal fracture of a deep roadway surrounding rock has been verified again (Figure 6).

In biggest buried for now as deep as 2375 m of Jinpin II hydropower station at phase of the auxiliary tunnel for loose circle, the observed data show that the surrounding rock loose circle includes several fracture zone and the non-fracture section appeared alternatively (Qian, 2008) (Figure 3).

The mechanism that forms zonal fractures in rock mass is unclear, so it is necessary to study through experiment their formation, *i.e.* the mechanism and process of their formation, and their development by the geomechanics model test. Based on the known engineering background of the deep mine at Huainan coal mine roadway, the newly developed similarity material is applied. A high-stress, true

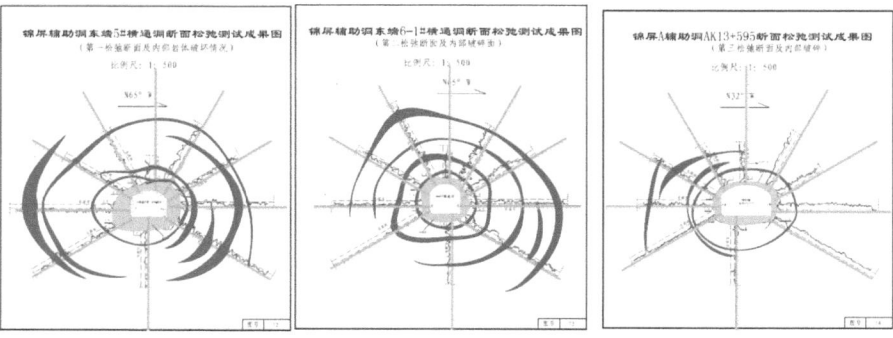

Figure 7 Zonal fracture phenomenon of Jinpin II hydropower project auxiliary diversion tunnel at east end cross sectional (Qian *et al.*, 2008).

three-dimensional loading model test system and a displacement monitoring system (Zhang Qiangyong, 2010) are also adopted here. Therefore, a geomechanics model test is conducted successfully, which reflects a true three-dimensional roadway excavation under high in-situ stress conditions. The result shows the partition of similar zonal fracture phenomenon appearing in the surrounding rock of Huainan mine project during excavations. It is the first time to obtain the radial strain and the interval distribution of peaks and troughs in the surrounding rock mass during excavations. (Zhang Qiangyong, 2009).

3.2 Model construction and modeling conduction

(1) Model discussion

In the model test for the Huainan coal mine roadway project, the roadway cross section shape is of semicircle arch, where the cross section size is 5000 x 3880 mm and it is 910 m in depth. The surrounding rocks are mainly sandstone and a silty sand rock stratum, the mining ground stress field is mainly composed of horizontal tectonic stress.

The model geometry is similar, taking $C_L = 50$, consider density scale similar $C_r = 1$, depending on the similarity principle and the original physical and mechanical parameters of the rock roadway model available material physical and mechanical parameters, see Table 1.

The axis of the tunnel, *i.e.* y-direction, to height (along the height direction, *i.e.* the z-direction) = 30 x 30 x 30 m. At a 1:50 geometric scale the model sizes are, accordingly, for length x height x width, 0.6 x 0.6 x 0.6 m. The tunnel cross-section size in the model is 100.0 x 77.6 mm (Figure 9).

(2) Construction process of the model

The test model was formed using a layer-by-layer compaction method. In the test bed frame each layer of material was laid down, compacted, and dried. Different measuring gauges were embedded within the model, and the process repeated until the whole model was finished.

(3) Measuring gauges arrangement in the model

In order to effective observing the surrounding rock of zonal disintegration phenomenon, along the far distance from the roofs or side wall surfaces of the tunnel. There are setting with certain spacing planted produced a variety of measurement gauges, including: strain brick by the strain gage adhesive, high-precision micro extensometers, fiber grating strain sensors.

Table 1 The physical and mechanical parameters of prototype rock and model materials.

Material type	Bulk density (kN/m³)	Young's modulus (MPa)	Cohesion (MPa)	Friction angle (°)	Compressive strength (MPa)	Poisson's ratio
The original rock	26.2	12970	10–15	40–43	88.55	0.268
Prototype	26.2	259.4	0.2–0.3	40–43	1.771	0.268

Prototype simulation range: long (along the axis of the tunnel, *i.e.* x-direction) to width (perpendicular, *i.e.* y-direction).

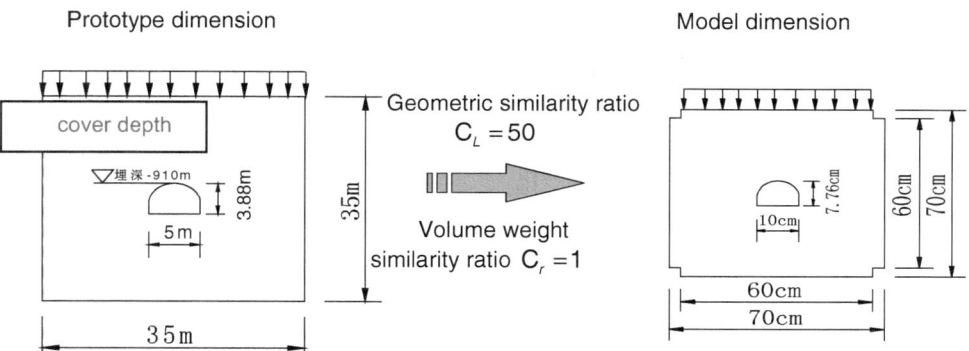

Figure 8 Prototype, and model, dimensions.

Figure 9 Site of tunnel excavation test photographs.

(4)   Modeling test of tunnel excavation

In order to ensure tunnel excavation in true three-dimensional stress conditions, the tunnel model should have all the loading in three directions applied first, then the excavation conducted. With regard to the true three-dimensional loading of the model, the maximum principal stress σy is applied at the tunnel axis direction, and kept to σy > σx > σz always.

When the loading is completed and maintaining an unchanged load, then begin to excavate using artificial excavation drilling to form the roadway section. Whenever a footage (40 mm) of the excavation is reached, the relevant measuring instruments would automatically record the changes on the stress and deformation of the surrounding rock mass. A footage of excavation would be continued and the monitoring data would be recorded again until the tunnel excavation is completed. With the help of the micro endoscope high-speed camera system and high-precision digital photogrammetry system making dynamic real-time observations, the zonal fracture processes are recorded. At the end of the experiment, the model of the tunnel is cut into sections along the lateral and longitudinal axes. Through horizontal and vertical profiles the observations of the full range of the splitting fracture phenomenon and the laws of deformation and failure are observed. Fractures are recorded very clearly; Figure 9 is a photograph showing the testing field.

3.3 Analysis of model test results of zonal disintegration

(1)   Variation of radial strain around tunnel
      Figure 10 shows the radial strain changes around the tunnel; as can be seen from the figure, around the tunnel the radial strain appeared in regularly spaced peaks and troughs attenuated regularly, indicating the peripheral existence of ruptured and unruptured alternating zonal fractures after excavation in the near area of tunnel.

(2)   Fracture morphology around the tunnel
      Figure 10 shows some characteristics along the lateral axis of the tunnel at different distances away from the surfaces of tunnel as follows.

(3)   With the depth increase, the zonal fracture phenomenon is more and more obvious; it is appeared of obvious rupture and non-rupture in spacing distribution of the zonal fracture phenomenon.

(4)   The four fracture zones and the four non-fracture zone are distributed at intervals, see Figure 11. The outermost rupture zone from the opening surface conversion into a prototype of about 7 m, which compares to the field value measured value of fracture layers and rupture range being consistent, indicating generally that the model test verifies effectively the zonal disintegration phenomenon for the deep roadway of surrounding rock.

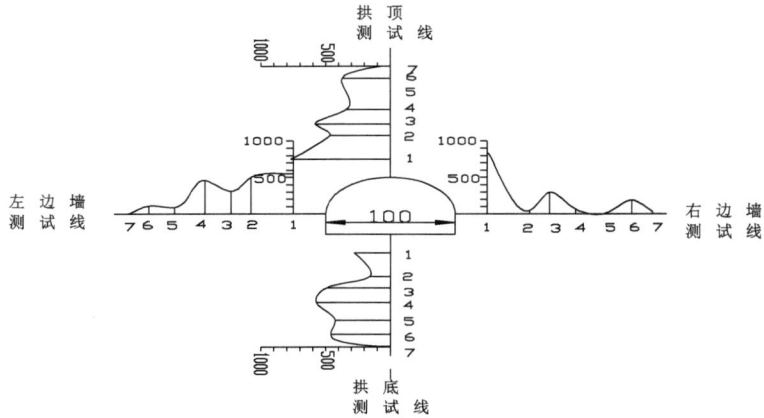

Figure 10 Radial strain around tunnel.

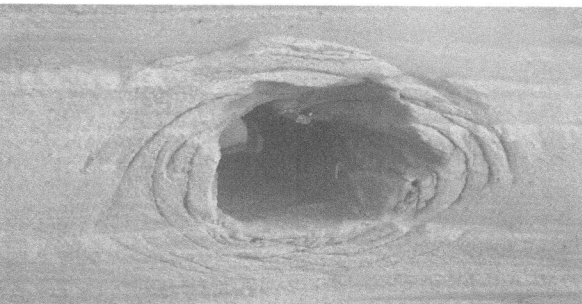

Figure 11 At 40 cm from the mouth of the tunnel.

3.4 Discussion of the model test

Using similarity material and three-dimensional geomechanical model testing to represent the Huainan deep mine roadway, we can reach the following conclusions from the study

(1) Using the iron-crystal-cemented sand model of rock similarity materials manufactured in the laboratory, and an accurate high loading force three-dimensional model testing system, and by conducting a three dimensional geomechanical model test, the zonal fracture of rock around the deep roadway is successfully reproduced.

(2) Under high stress conditions the radial strain around a deep tunnel showed a characteristic distribution of peaks and troughs, with intervals of shock attenuation variation, which verifies the existence of the zonal fracture phenomenon which is reproduced by zonal disintegration in the laboratory.

(3) The key conditions that cause the zonal disintegration phenomenon are the initial maximum principal stress, parallel to the tunnel axis, and the amount by which this value exceeds the rock compressive strength.

(4) The amount of zonal fracture layers and the rupture zone distribution range obtained from the model testing, compared with the field measured ones, are very close. This indicates that the geomechanical model test can effectively simulate the deep roadway in true three-dimensional high stress conditions with regard to nonlinear deformation characteristics and the mechanical failure of the surrounding rock.

4. MODELING FOR CAVERN COMPLEX CASE STUDY II

4.1 Quasi-three-dimensional physical model tests on a cavern complex under high *in situ* stresses

4.1.1 Project descriptions

The Shuangjiangkou Hydropower Station is located on the Dadu River in Sichuan Province, China. The stream is in a valley with slope heights to 1000 m and inclinations of between 35° and 60°. *In situ* stress fields near the underground cavern complex are strongly influenced by the incised terrain, and active tectonics, with *in situ* stresses reaching 38 MPa at a depth of only 600 m. The rock mass comprises medium- to fine-grained granites.

The underground power house contains four turbines with a total capacity of 2 GW. The underground cavern complex consists mainly of the main power house, the transformer house, and the surge chamber. The spacing is 45 m between the main power house and the transformer house, and 40 m between the transformer house and the surge chamber. For the main power house, the transformer house and the surge chamber, the heights are 67.05 m, 26.5 m and 80.2 m, respectively. Their spans are 28.3 m, 18 m and 20 m, respectively.

4.1.1.1 Quasi-three-dimensional model tests on the cavern complex of the Shuangjiangkou project

Physical model tests were conducted to investigate the stability of this cavern complex, which contains a power house, a transformer house, a tail water surge chamber and other openings under high *in situ* stresses. Figure 12 shows the layout of the center part of cavern complex and the dimensions of the whole model.

The dimensions of the physical model are 2.5 m (length) × 0.5 m (width) × 2.0 m (height).

4.1.1.2 Geomechanical analogous material

After considering the above factors in an integrated way, the similarity constant for geometry $C_L = 200$ was determined. Table 2 shows the physico-mechanical parameters of both the prototype and the model.

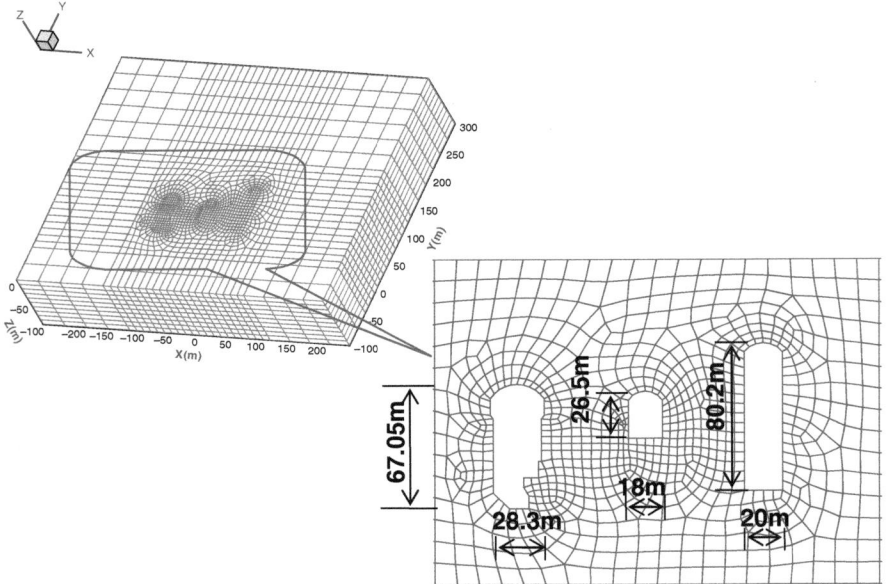

Figure 12 The layout of the whole model with the cavern complex.

Table 2 The physico-mechanical parameters of the prototype and the model.

Type of Material	Density (kN/m³)	Young's Modulus (MPa)	Cohesion (MPa)	Friction Angle (°)	Compressive Strength (MPa)	Poisson's Ratio
Rock Mass	26.5	3000	2	40.36	80	0.2
Analog Material	26.5	15	0.01	40.36	0.4	0.2

4.1.1.3 Measurement technologies and instrumentation

The instruments used in this study include strain bricks, internal caliper gauges, high-accuracy mini extensometers, and a digital vision displacement measuring system. Only a few of them are introduced below.

4.1.1.4 The high-accuracy mini extensometer system

In order to perform real-time displacement measurements of the surrounding rock masses in the underground complex in the excavation procedure of the modeling, two monitoring sections are set up along the axis direction of the caverns (see Figure 13). The measurement is focused on the principal displacements on the pivotal points around the main power house. In total there are 15 measuring points in each section, three are above the arch crown and 12 near the side walls. The precision of this measuring technology is $1\mu m$.

The new mini extensometer system is made up of the displacement sensors, the displacement transfer device, the data receiving device, and the data processing software. The high-accuracy grating fiber sensor is selected as the displacement sensor. The processing software can automatically receive and save the real-time data and generate all types of displacement curves.

4.1.1.5 The digital video convergence measuring system

A digital video convergence measuring system provides non-contact displacement measurements. In the process of excavation, the variations in the co-ordinate values

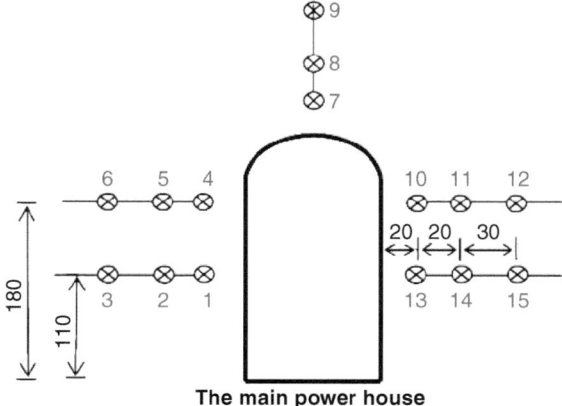

Figure 13 Layout of mini extensometers at Section I of the main power house (Unit: mm).

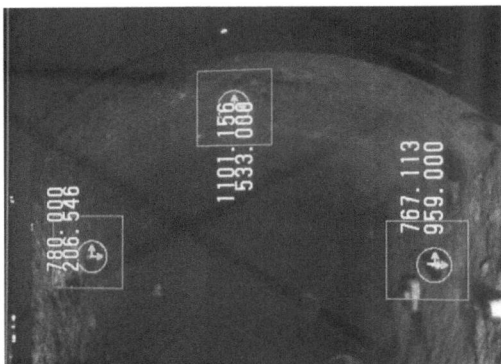

Figure 14 Co-ordinates of three monitoring points of the main power house in the digital video image.

of the monitoring points are automatically saved. Figure 14 shows the co-ordinates of the three monitoring points in the digital video image.

4.1.2 Development of rock bolts and pre-stressed cables

(1)   The rock bolts
To simplify the construction process, one bolt in the model is used to simulate six bolts in field through the method of equal rigidity. After evaluating the physico-mechanical parameters of a number of materials, bamboo is identified being an appropriate material for the bolts in the model. The spacing of the horizontal and vertical bolts is 5 cm. For simplification of the difficult drilling in modeling to form the boreholes for the bolts, some columns are used, which are made of an analogous material and pre-embedded at the position of the bolts. When the excavation reaches the columns, the columns are pulled out and the bamboo bolts are placed and grouted in the hole as the model bolts.

(2)   The pre-stressed cables
The specifications of the pre-stressed cables used in the field are as follows: the pre-stressed force T is 2000 kN; with the length and the spacing of 25 m and 4.5 m, respectively. For the feasibility of the tests, one cable in the model is equivalent to four cables in the field, based on the mechanical equivalence principle. To make the mini cable, a small spring is welded and connected to a screw at the end, which can be tightened afterwards. A strain gauge is installed on the cable to measure the stress. The cables are pre-embedded in the model. When excavated to the location of screw, the screw is tightened to apply a pre-stress to the cable. The figure below shows the cables used in the model tests.

4.1.3 The construction of the model and the design of loading

The model is constructed within the steel structural frame. The analogous material is placed in the frame and compacted by a hammer. In the process of construction forming the model, the measuring sensors are embedded at the designated locations. The whole

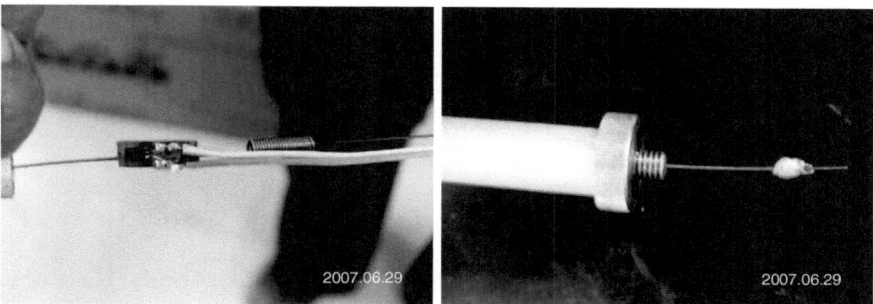

Figure 15 Fabrication of the pre-stressed cables used in the model test.

model is a rectangular solid with a dimension of 2.0x2.5x0.5m³. The model is 0.5 m thick and has two generating sets. The caverns include the main power house, the transformer house, the surge chamber, the busbar chamber, and partial tailrace tunnels (see Figure 16). The whole model can be considered as a quasi-three-dimensional model. The vertical depth of the caverns to be modeled is approximately 600 m. Two loading schemes are utilized in the model test. One scheme is to simulate the actual excavation process, and the vertical load on the model is equal to the dead weight of 600 m depth rock masses. According to the *in situ* stress field, the coefficient of horizontal pressure is determined to be $K_o = 1.5$. The loads applied by five lateral hydraulic loading systems are trapezoidally distributed, simulating the real *in situ* stresses.

The other scheme is to evaluate the behavior of the model under higher *in situ* stresses by overloading. The higher *in situ* stresses are equal to the dead weights of rock masses at 1000, 1300, 1600, 1900, and 2200 m under the ground surface. When the vertical

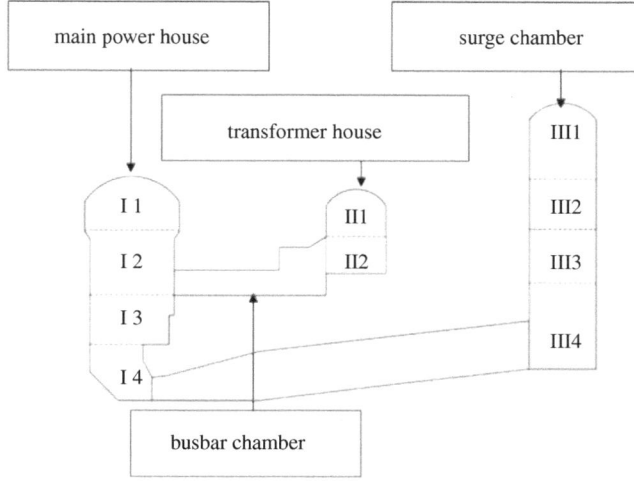

Figure 16 Layout and excavating sub-sequences of the cavern complex.

load reached 2200 m overburden stresses, the loading is terminated. At this stage, a few cracks appear around the caverns even though they do not completely fail. Then the model is unloaded gradually.

4.1.4 Cavern excavation and measurement of the model

Due to the space restriction of the model, twenty excavation sub-sequences in the field were simplified to ten sub-sequences in the model test (see Figure 16). They are I1, II1, III1, I2, II2, III2, I3, III3, I4 and III4.

The actual caverns were excavated by drilling tools with sharp heads. The amount of excavation for every sub-step is 5 cm. The whole excavation sub-sequences can be divided into 10 circles, and the last two sub-steps are to excavate the busbar chambers and the other openings, so there are 102 steps in total. During the excavation, the internal digital video system monitored the whole excavation and the pre-installed sensors took monitoring data automatically. When a circle was completed, the column bolts pre-embedded for bolt holes were pulled out, the rock bolts were placed, and grouting was applied (the column bolts are the same size as the rock bolts, made of slim iron rods). All the pre-installed sensors recorded the data during the excavation. The next circle continued until the whole excavation was completed. Figure 17 shows the layout of the convergence measurement points, the head of pre-stressed cables, and the head of rock bolts inside the main power house after excavation.

4.1.5 Test results and numerical analyses

Numerical simulation is carried out and used to compare with the model test results. In the numerical analysis, a three-dimensional finite difference analytical model was set up using FLAC3D for the underground cavern complex. The whole numerical model has 20754 nodes and 96662 elements.

It is worth noting that the first scheme (the actual load test) can be greatly simulated by FLAC3D, but the second scheme (the overload test) cannot be absolutely simulated by FLAC3D and other continuum-based numerical methods. As during the overload

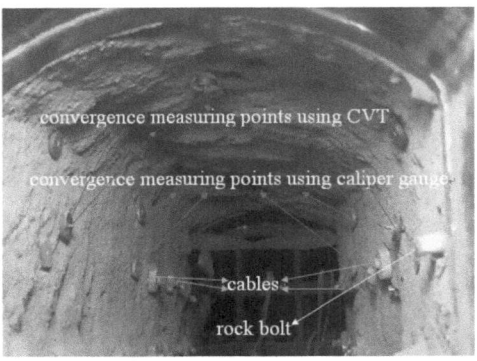

Figure 17 Layout of the convergence measurement points, the pre-stressed cables, and the rock bolts inside the main power house after excavation.

test, the splitting or spalling failure phenomena appear. The deformation of the key points in the surrounding rock mass in the process of the overload test is not only the elasto-plastic displacement, but also the opening displacement induced by the splitting or spalling failures. However, it is difficult to simulate the failures using the present continuum-based numerical methods. That is the reason why the physical model tests should be necessarily performed in some complicated circumstances.

The test results analyses of the mini extensometers

The displacements of the typical key points, No. 7 and No. 4, are shown in Figure 18. No. 7 is at the arch crown of the main power house, and No. 4 is at the side walls. Figure 18 shows the displacement variation of the arch crown of the main power house. It shows that the variation trends of the test and numerical results are consistent, and their ultimate total displacements are almost the same.

Based on the maximum displacements presented in Figure 19, the maximum displacement in the field can be predicted to be 2 cm. The figures show that the deformation of the surrounding rock mass is regular, indicating that the reinforced surrounding rock mass is stable.

(1) The test results analyses of the convergence measuring system

Figure 20 shows the layout of the convergence measurement points in the main power house. From the experiences during the test, the convergence measuring system is greatly influenced by human factors, for instance, the visual errors and the installed locations of the five marks. After finishing half of the excavations of the main power house, the five marks are installed at the designated locations. Therefore, the monitoring data starts from excavation step 10. In numerical analyses, the displacement of excavation step 10 is assumed to be zero and then the displacements induced by the subsequent excavation steps are compared with the measured values. From the comparison analyses, the curves of monitoring No. 2 point display reasonable closeness, as

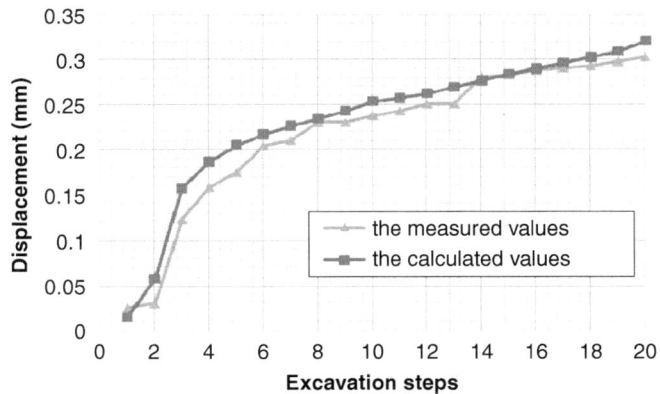

Figure 18 Comparison between the measured by extensometer and calculated displacements at the No. 7 measurement point at the arch crown of the main power house.

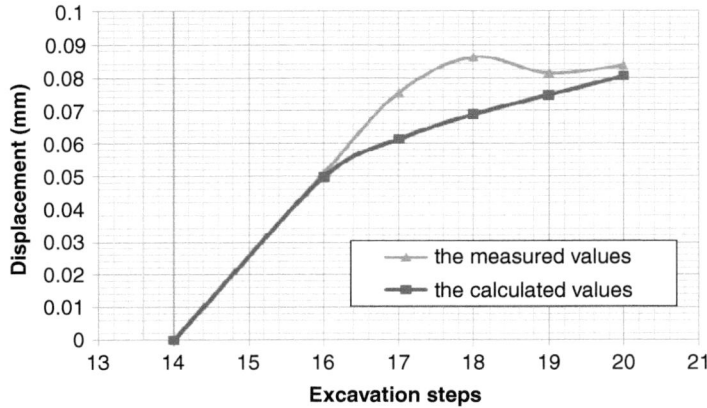

Figure 19 Comparison between the measured and calculated displacements at the No. 4 measurement point on the upstream side wall of the main power house during the key period of the excavation.

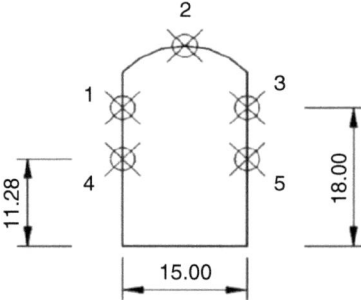

Figure 20 Layout of the convergence measurement points in the main power house (unit: cm).

shown in Figure 21. It is concluded that the measured values are greater than the calculated values generally, which can demonstrate that some splitting failures have appeared in the surface layer of the surrounding rock mass and the additional displacement cannot be simulated by the present continuous deformation numerical methods.

4.1.6 Summary and discussions

(1)   In this study, a steel structural frame for physical model tests and hydraulic loading systems for simulating high *in situ* stresses and great overburden depth (2200 m) are successfully developed. At the front and back sides of the steel frame, two transparent tempered glass windows are installed so that the cracks in the surrounding rock masses can be monitored.

(2)   For the newly-developed mini extensometer, high-precision fiber grating sensors, linear transfer system and relevant tension technology, and a fast data processing

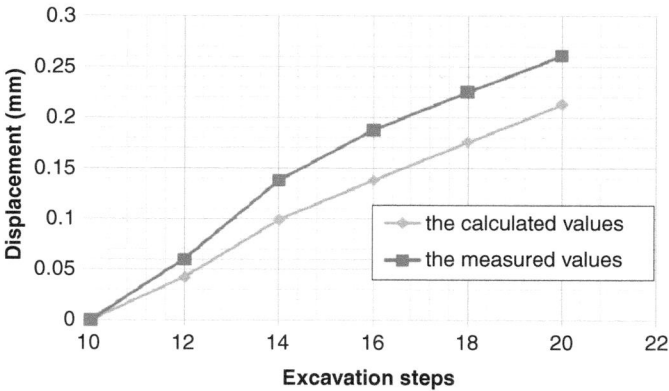

Figure 21 Comparison between the measured and calculated displacements at No. 2 point in Figure 20.

technique is adopted. The digital video measuring system is successfully employed for the first time to measure the displacement convergence around the cavern, and excellent results are achieved.

(3)   For the simulation of anchoring, pre-stressed mini cables and a unique grouting technique are developed and successfully used in the model test. It is also concluded that the physical model test has more advantages than numerical methods in studying the reinforcement effect of the surrounding rock mass.

(4)   The model test results and the numerical modeling results are found to be consistent with each other under the condition of the actual load test.

(5)   The test results show that if the overloading increases by three times or higher, the displacement in the surrounding rock will be relatively large; however, no apparent failure occurs. Therefore, the anchoring reinforcement is considered to be effective.

(6)   The interim results for model tests under high *in situ* stresses are presented in the paper. Some techniques and methods need to be further improved. However, the study has achieved satisfactory results and can provide guidance for the design and construction of the Shuangjiangkou project and future research.

4.2   Large-scale geomechanical model testing of an underground cavern group in a true three-dimensional (3-D) stress state

In the previous section, the physical model test is conducted in a quasi-three-dimensional state. Here a true three-dimensional physical model test would be introduced based on the latter physical model test.

4.2.1   Development of the steel structural frame

A new self-balanced true three-dimensional loading is developed. The steel structural frame for the three-dimensional model tests is shown in Figure 22. The frame

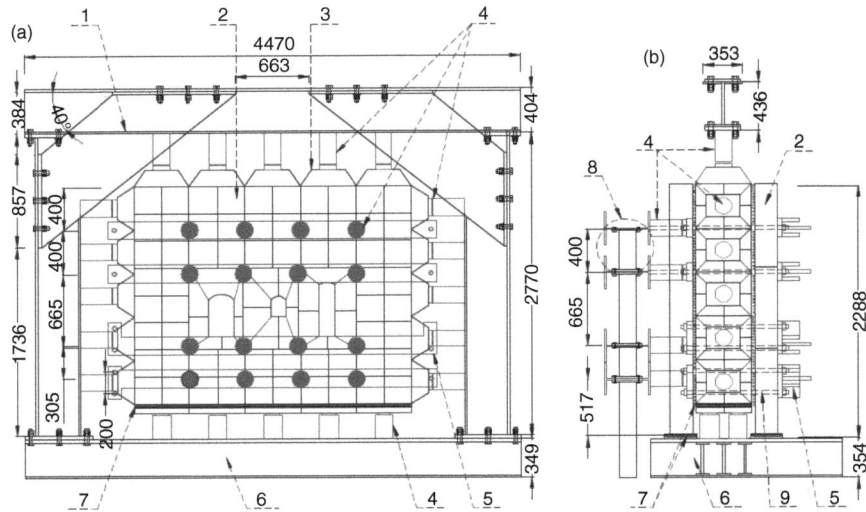

Figure 22 Design drawing of the steel frame for three-dimensional geomechanical model tests (unit: mm): (a) front view; (b) side view. 1, door-shaped reaction frame; 2, structural wall; 3, transmission kit; 4, hydraulic jack; 5, reaction beam; 6, base; 7, combinational sliding wall; 8, layered reaction frame; 9, high-strength steel rod.

accommodated the model and served as a reaction device for loading. The frame consisted of a base, a door-shaped reaction frame, a layered reaction frame, structural walls, loading jacks, and a combination of sliding walls. Many combinational ball sliding blocks are installed between the model surfaces and the structural walls, as well as at the model bottom, to reduce the friction. The friction coefficient of the new sliding wall is measured to be less than 0.005 compared with the conventional method using PTFE films, therefore, the sliding blocks can reduce the friction by 30 times.

4.2.2 Advanced deformation measurement methods and techniques of surrounding rock masses

(1) Digital photogrammetric system

The close-range photogrammetric technique used in this study for determining displacement and strain has recently been applied to a number of geotechnical engineering problems.

In these model tests, the digital photogrammetric system consisted of four 10.2 megapixel Nikon D200 digital single-lens reflex cameras, twelve 200 W ordinary incandescent lamps, a number of color marks, and the digital image-processing software PhotoInfo.

Considering that the model material was brittle and that the location for displacement measurement was to be excavated, color marks were affixed to the model surface in a step by step manner.

(2)   FBG sensing bars for displacement monitoring

FBG sensing bars were adopted to measure the displacement of the surrounding rock masses. Fiber Bragg gratings are intrinsic sensing elements that are written into optical fibers through exposure to ultraviolet light. A grating acts as a selective spectral reflector at a characteristic wavelength, which is strain and temperature dependent. In this study, one piece of FBG that was not under the influence of any external force was added to compensate for temperature. FBG sensing bars were embedded into the model in order to measure the deformation of the surrounding rock masses. According to the mechanical properties of the model material, plastic bars with good elasticity and stiffness were selected as the parent material for the fiber Bragg grating. The designed bar diameter was 10 mm with a length of 1000 mm. Two fibers were affixed to the bar surface. Each fiber was connected by using five FBGs of different initial central wavelength (the spacing is 200 mm) forming a quasi-distributed strain-sensing array. Deflection was imposed simultaneously on different points of the sensing bar, and a number of linear variable displacement transducers (LVDT) were used to record the actual deflection. The results show that the sensing bar can measure the deflection distribution along the bar with a precision of 0.1 μm. During the production of the model, three holes were prepared for the installation of FBG sensing bars near the main power house and the surge chamber. The holes were 2000 mm in length and 18 mm in diameter. The center of the hole was 20 mm away from the side wall of the cavern, as shown in Figure 23. When the sensing bars were vertically embedded in the model, in

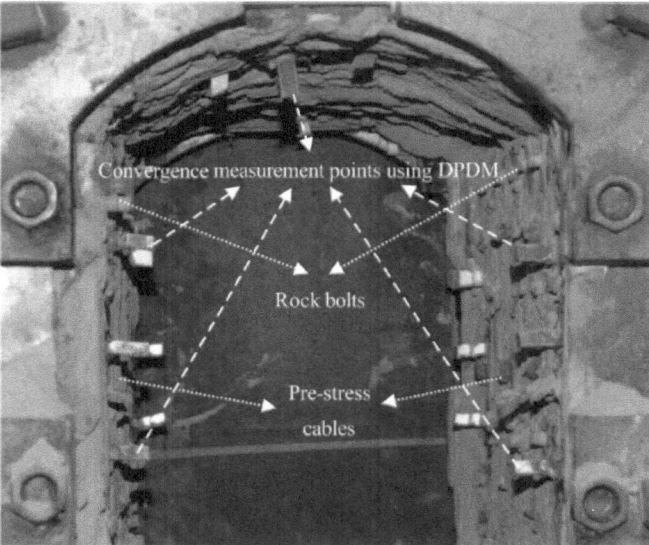

Figure 23 Layout of the convergence measurement points, pre-stressed cables, and rock bolts inside the power house. DPDM, digital photogrammetric deformation method.

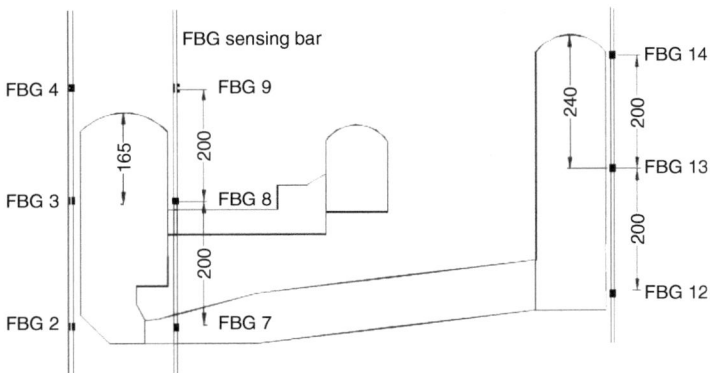

Figure 24 Layout of the fiber Bragg grating (FBG) sensing bars.

order to satisfy the deformation compatibility between the bar and the model, the bonding material had to be selected. Silica gel was selected as the bonding material following a large number of trials. After the 24-hour curing of the silica gel, the bonding between the sensors and the model was determined to be sufficiently strong.

4.2.3 *Construction and excavation of the model*

Many precast blocks made from the analogy material were used to construct the model, and monitoring holes were prepared beforehand in the relevant blocks. Then the individual blocks are adhered to each other with glue. In this way, the consistency of the model's mechanical properties can be guaranteed. The model dimensions (width × height × thickness) are 2.5×2 ×0.5 m. The caverns were excavated by manual drilling and boring. The entire excavation process was divided into twenty-one 10 cm steps (for example, the first step included excavation of I1, followed by the second excavation II1, and so forth, at 10 cm increments).

4.2.4 *Overloading tests*

During the construction of the modeling, different measures are recorded step by step. After finishing the excavations, supporting and monitoring work in the design loads, and the overloading tests are performed. Vertical and horizontal loads (with same coefficients of K_1 and K_2) were applied gradually in the overloading test. The load was equivalent to the overburden at depths of 800, 1000, 1200, 1400, 1600 and 1800 m. The deformation trend and failure process were monitored to investigate the stability of the caverns. The model was then unloaded in steps.

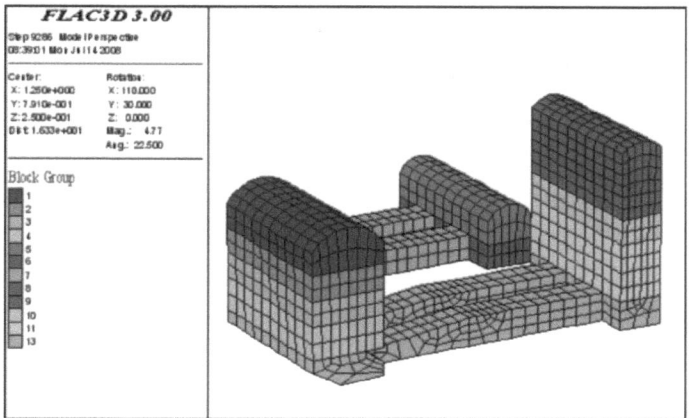

Figure 25 Model for numerical simulation.

4.3 Numerical simulations

In the numerical simulations, a three-dimensional finite element model, which had the same dimensions as the physical model, was built. The shotcrete layer and the surrounding rock masses were represented by elements with only axial stiffness taken into account. The model was divided into 39798 nodes and 35440 elements, as shown in Figure 25. The surrounding rock masses of the underground cavern group were of good integrity, joints did not develop, and seepage of water was not observed in the initial exploration adits. These factors were therefore not considered in the numerical simulations. The Drucker–Prager criterion was adopted and the material properties were obtained according to *in situ* physico-mechanical parameters provided by Hydro-China Chengdu Engineering Corporation and the scaling ratio was 150:1, as shown in Table 2. Stability analysis was performed by using the finite-difference method and the FLAC-3D code. The corresponding numerical simulations were applied to the simulated overloading tests.

4.3.1 Results and discussion

(1)   The scheme of the design engineering project
       The monitoring results were compared with the displacement values obtained from the numerical simulations. During excavation of the model, the surrounding rock masses were stable. No obvious cracks or catastrophic deformations were observed at the peripheries of the caverns. The displacement value at the actual project site can be estimated by multiplying the monitoring results by the similarity ratio. The maximum displacement of the cavern side wall at the project site is approximately 42 mm. The displacement measured in the model test was generally larger than that obtained by the numerical simulation. The displacement profile in the surrounding rock masses and at the key points on

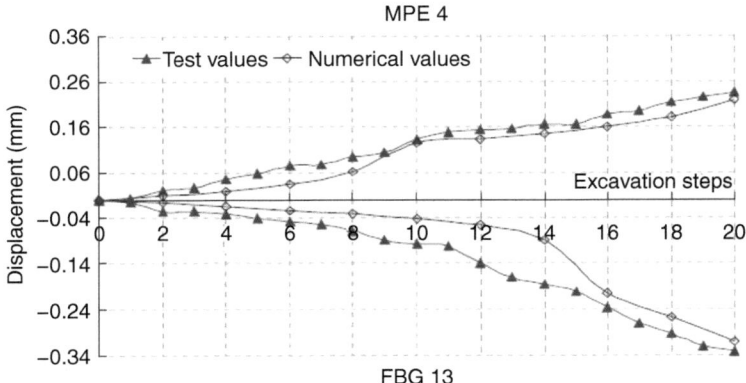

Figure 26 Comparison of calculated and measured deformation near the upstream side wall of the main power house using a multipoint extensometer and near the downstream side wall of the surge chamber using a fiber Bragg grating (FBG) bar.

the side walls were regular. The deformation characters shown in this model test provide valuable information for the design of this project.

(2)   The overloading tests

The deformation in the surrounding rock masses measured during the overloading test is shown in Figure 26.

As shown by the above results, the numerical values increased nearly linearly with an increase in overburden depth. The monitoring results, however, showed that the displacement rate increased dramatically when the overburden depth increased from 800 to 1000 m. According to the similarity ratio, the maximum relative displacement in the surrounding rock masses in the field should reach about 36 cm for the depths of 600 to 1800 m, compared with about 20 cm as indicated by the numerical simulations.

The software adopted for numerical simulations is based on the continuum method. The model material, however, was brittle and the number of micro-cracks increased gradually as the external load increased to extreme values. The micro-cracks open and spread with a gradually increased load, and eventually large splitting cracks occur. The opening displacements in the surrounding rock masses account for a large portion of the total deformation. A similar phenomenon was reported in other large-scale cavern groups. Opening displacement accounted for a large portion of the total deformation in the surrounding rock masses, and an even larger portion in igneous rock. The study showed that discontinuous deformation induced by cracks accounted for a large portion of the total deformation in the side wall.

Visible cracks during the overloading test are shown in Figure 27. When the loading surpassed the equivalent overburden depth of 800 m, split rock masses dropped from the crown and side walls of the power house and local cracks were observed in the surrounding rock masses of the surge chamber. When the overburden depth reached 1000 to 1400 m, small flakes fell from the side walls

Figure 27 Failure pattern of the main power house under an overburden depth of 1800 m.

of the power house. With a gradual increase in overburden pressure, the number of cracks in the surrounding rock masses increased, and cracks occurred in the crown and larger blocks collapse. Wide-ranging collapse occurred in the crown and side walls of the surge chamber. When the equivalent overburden depth was between 1600 and 1800 m, flake spalling in the side walls of the power house increased remarkably, but no wide-ranging collapse occurred. Cracking and collapsing were more severe in the crown. In the crown and side walls of the surge chamber, block collapses were prevalent. (No analogy rock bolts and cables were installed for the surge chamber in the model test.) This showed that the rock bolts and cables played an effective role in the reinforcement and stability of the caverns.

(3)   Discussion
In this study, a quasi three-dimensional test and a true three-dimensional geomechanical model test and numerical simulations of a large underground power house were conducted in order to investigate phenomena related to excavation, such as deformation and failure mechanisms. Based on experimental and simulation results, the following conclusions were obtained.

(a)   In the quasi three-dimensional physical model test study, a steel structural frame for physical model tests and hydraulic loading systems for simulating high *in situ* stresses and great overburden depth (2200 m) are successfully developed. In the front and back sides of the steel frame, two transparent tempered glass windows are installed so that the cracks in the surrounding rock masses can be monitored. In the true three-dimensional physical model test, a strong steel frame including a new type of sliding wall was developed and successfully applied in geomechanical model tests. The hydraulic loading system successfully simulated high *in situ* stresses and great overburden depth in true axial stress condition. High-precision mini multipoint extensometers were developed and applied for measurement of deformation in the surrounding rock masses. Digital photogrammetric technology and fibre Bragg-grating sensing bars were used to measure small deformations on the surface and inside of the surrounding rock masses. Reliable and accurate results have been obtained for engineering design applications.

(b)   For the newly-developed mini extensometer, high-precision fibre grating sensors, linear transfer system and relevant tension technology, and a fast data processing technique is adopted. The digital video measuring system is successfully employed for the first time to measure the displacement convergence around the cavern, and excellent results are achieved.

(c)   For the simulation of anchoring, pre-stressed mini cables and a unique grouting technique are developed and successfully used in the model test. It is also concluded that the physical model test has more advantages than numerical methods in studying the reinforcement effect of the surrounding rock mass.

(d)   The model test results and the numerical modeling results are found to be consistent with each other under the condition of the actual load test.

(e)   The test results show that if the overloading increase by three times or higher, the displacement in the surrounding rock will be relatively large, however, no apparent failure occurs. Therefore, the anchoring reinforcement is considered to be effective.

(f)   The interim results for model tests under high *in situ* stresses are presented in the paper. Some techniques and methods need to be further improved. However, the study has achieved satisfactory results and can provide guidance for the design and construction of the Shuangjiangkou project and future research.

5 MODELING OF A SUBSEA TUNNEL CONSIDERING FLUID–SOLID COUPLED EFFECT

5.1 Introduction

In recent years, underwater tunnel technology has been developing rapidly. Based on the Qingdao Jiaochow Bay tunnel, the fluid–solid interaction model test was carried out to capture the response of surrounding rock during subsea tunnel excavation.

During the excavation of the subsea tunnel, the original equilibrium state of the surrounding rock will be disturbed, causing change to the stress field and seepage field of the surrounding rock. That is, the seepage body force will be developed due to water seepage, and the permeability of surrounding rock will be changed because of the variation of stress state. Furthermore, the presence of water can also cause a decrease of rock mass strength and elastic modulus (Li, 2013).

Due to the complexity of the subsea tunnel, a fluid–solid coupling model experiment can be used to simulate the engineering actual situation. The current test commonly used is an airtight space composed of a single material, but the seepage and deformation characteristics of surrounding rock in the process of the test have poor visibility. In order to solve the shortcomings of the existing model test system, a fluid–solid coupling similarity material with controllable hydraulic characteristics was developed, and a model experimental study on the subsea tunnel construction process was carried out using the large-scale fluid–solid coupling model test system that was developed in the laboratory. The space–time change law of the surrounding rock displacement field, the stress field and the seepage field in the experimental process are analyzed in the present paper.

5.2 New similarity material for a fluid–solid coupling test

5.2.1 The similarity theory

According to the elastic mechanics equation and the dimensional analysis method combined with the fluid–solid coupling theory, the following similarity relationships were derived (Li, 2012; Yu, 2010)

$$C_\varepsilon = C_\varphi = C_\mu = 1 \tag{11}$$

$$C_\sigma = C_{\sigma_t} = C_{\sigma_c} = C_E = C_C \tag{12}$$

$$C_\varphi = C_\varepsilon \cdot C_L, \ C_\sigma = C_\gamma \cdot C_L, \ C_\sigma = C_\varepsilon \cdot C_E \tag{13}$$

$$C_K = \sqrt{C_L} \Big/ C_\gamma \tag{14}$$

where γ is bulk density; L is length; δ is displacement; ε is strain; E is elastic modulus; σ is stress; σ_t is tensile strength; σ_c is uniaxial compressive strength; c is cohesion; φ is friction angle; μ is Poisson ratio; and K is permeability coefficient.

5.2.2 New similarity material (Li, 2012)

Traditional geomechanics model tests use materials which most closely resemble natural ingredients, such as iron powder, gypsum, barite powder, quartz sand etc. as major core materials. These are combined with cementitious materials, or directly experiment on a certain proportion of mortar materials. Hence, a reasonable and reliable similarity material is the key to the success of a model test. The traditional simulation materials are all single solids which are liable to demonstrate softening or plastic deformation and as a result it is impossible for them to fulfill the demand of the fluid–solid coupling model test (Huang, 2010; Li, 2010).

Considering the requirements of permeability, new similarity material without hydrophilic organic gel materials as a cementing agent tend to be chosen. Based on the research achievements of predecessors, it combined with seawater and material coupling effect. We achieved effective control over material strength, deformation characteristics and hydraulic properties, and proposed a similarity material preparation over which we have overall control and retroregulation. The test material has cement as the main adhesion agent and uses standard sand as major core material. In this test, barite powder, talc, silicon oil and Vaseline are used as modifying elements. We strived to achieve a main index of materials that can be adjusted by two

Table 3 Tensile and compressive strength values of the similarity materials.

	Ratio of constituent substances in the similarity material (S:B:T:V:S:C)	Compressive strength σc /MPa	Strength of extension σt /MPa	σt /σc
1	1:0.12:0.08:0.06:0.06:0.065	0.61	0.058	1/10.5
2	1:0.12:0.08:0.08:0.04:0.065	0.48	0.053	1/9.1
3	1:0.12:0.08:0.08:0.04:0.10	1.16	0.114	1/9.4
4	1:0.18:0.12:0.08:0.04:0.065	0.51	0.052	1/9.8

Note: S:B:T:V:S:C show the quality ratio of sand, barite powder, talcum powder, Vaseline, silicone oil, cement.

components. In addition to cement, adjusting other parts of the composition have little impact on other material qualities, thus improving the similarity material for various analogue simulation tests.

According to the geological data and field sampling and experimental results of rock mass parameters, the similitude ratio of the similarity material test of physical and mechanical parameters was 1:35. Several group tests of similarity materials' physical and mechanical parameters and fluid–solid coupling effects were carried out. The tensile compressive strength value of the reference blocks are shown in Table 3.

Table 3 shows that the tension and compression strength ratio σ_t/σ_c of the similarity material is between 1/8 and 1/11. The ratio is very close to the average of the rock tension and compression strength ratio 1/10. This indicates that the present similarity materials can be used to simulate the tensile properties of the rock. Mechanical parameters are shown in Table 4, proportions of the model materials are shown in Table 5.

Table 4 Parameters of prototype and model material.

Medium	Bulk density (g/cm³)	Pressure-proof MPa	Elastic modulus (MPa)	Permeability coefficient (cm⁻¹)	Poisson ratio
Original rock	2.0–2.3	15–21	1300–1950	5.79×10^{-5}–1.18×10^{-4}	0.30–0.35
Model material	2.144	0.43	47	6.39×10^{-6}	0.31

Table 5 Proportions within the similarity model material.

Aggregate : colloid ratio	Sand : earth ratio	Talcum : barite powder ratio	Vaseline : silicon oil ratio	Sand : ash ratio
10:1	8:1	1:1.5	2:1	15:1

Figure 28 Test system illustration.

In detail, the colloid in the aggregate-to-colloid ratio mainly means the quality combination of silicon oil and Vaseline, while the earth in the sand-to-earth ratio refers to fine aggregates which consist of barite powder and talcum powder.

5.3 The development of the fluid–solid coupling model test system (Song, 2010)

The system for the fluid–solid coupling model test was composed of a steel structural frame, hydraulic loading system, water seepage flow collection device and multivariate information monitoring system, as shown in Figure 28.

5.3.1 Steel structural frame

The steel structural frame was built following the similarity scale of 1:35 and composed of a rack body with a steel structure and a test chamber with toughened glass, designed as 3400 mm in length, 3000 mm in height and 800 mm in width. The internal sketch of steel structure rack body was toughened glass test chamber. The size of the glass was equal to that of the corresponding alloy steel members, and explosion-proof film was used in the whole internal structure to ensure that the test chamber was leak-proof.

5.3.2 Hydraulic loading system

The hydraulic loading system was composed of the pressure tank, the centrifugal pump and pressure indicator, which together can provide enough water for model test.

5.3.3 Water seepage flow collection device

In order to measure the tunnel seepage quantity and seepage flow velocity changes, a water seepage flow collection device was built.

5.3.4 Multivariate information monitoring system

The multivariate information monitoring system is mainly composed of a fibre Bragg grating sensor, a micro pressure sensor, a demodulation instrument and an automatic data acquisition system. This system can realize parallel experiment information acquisition, and can achieve real-time monitoring of surrounding rock deformation, stress, osmotic pressure, temperature and pressure of rock mass.

Furthermore, a DZ-I miniature pressure sensor was adopted to obtain the pressure of the surrounding rock. The size of a DZ-I miniature pressure sensor is 16×7 mm.

5.4 Model test of Qingdao Subsea Tunnel

A fluid–solid coupling model test was used to assess the longitudinal deformation profile and seepage characteristics of rock surrounding the Qingdao Subsea Tunnel. The section located between ZK7+053.2 and ZK7+080 in the left line of Qingdao Subsea Tunnel was taken as a prototype. Rock in this section was mainly composed of slightly weathered diabase. The depth of water was 23 m, the height and width of the tunnel was 11.2 m to 12.0 m and 15.23 m to 16.03 m, respectively. The dimensions (length × width × height) of

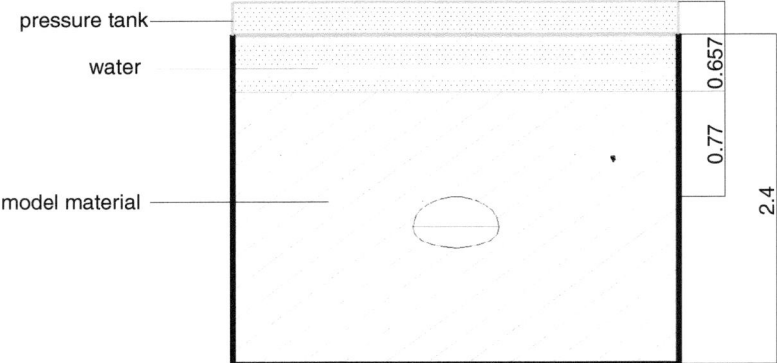

Figure 29 Layout of the model body.

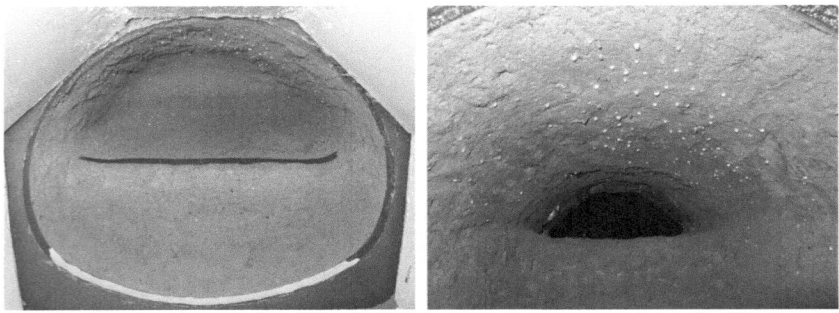

Figure 30 Photograph of test process; (a) model excavation (b) top bench link.

the model were 3.4×1.43×0.8 m. The direction of length was vertical with the axes of the tunnel and the direction of width parallel with the axes of the tunnel.

The model was filled by consolidated method. Sensors were embedded in the model around the tunnel during the process of filling the model with the similarity material.

The tunnel section was separated into two benches. Longitudinal excavation was divided into 18 steps with each step being 6 cm long. The tunnel was excavated in sequence from the top bench down to the bottom bench. The length of the top bench was 30 cm. After the top bench was excavated for five steps, the bottom bench was then excavated together with

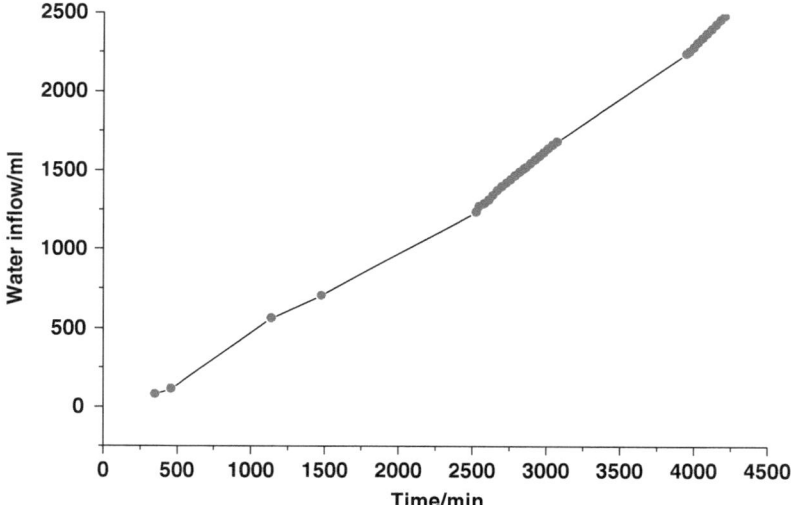

Figure 31 Water inflow with time growth.

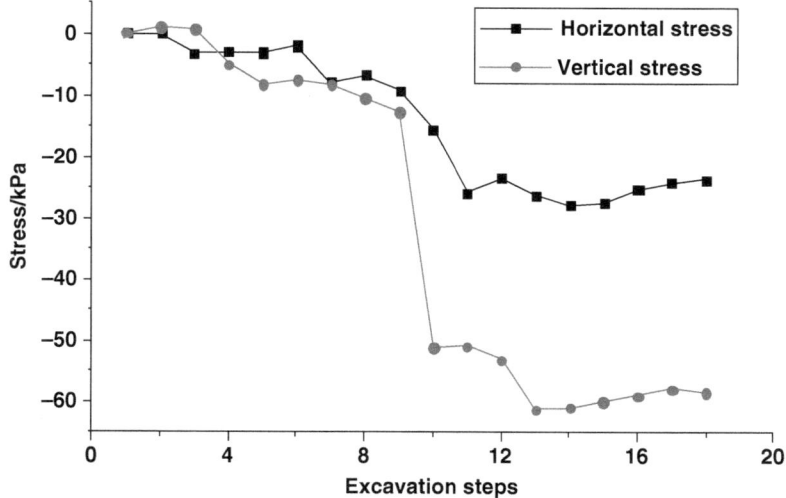

Figure 32 Stress curves with excavation steps.

the top bench when the cross section was 80 cm. The seepage collection device was arranged to collect the seepage. The following procedure is summarized as below.

5.5 Data analysis

5.5.1 Analysis of water inflow

The relationships between water inflow and time is shown in Figure 31. Figure 32 shows that water inflow increases with increasing time, while the growth rate slowed slightly. Water inflow increases obviously after 3071 min, which is because the hydraulic characteristics of the similarity material were changed.

5.5.2 Analysis of stress

Stress curves of crown and side wall with excavation steps are shown in Figure 32. Figure 5.5 shows that the stress reduces when the distance from the model tunnel face to the monitoring section is about 0.5 times the model tunnel diameter. When the excavation of the top bench reached the monitoring section, vertical stress entered into a phase of rapid change. When excavation of the bottom bench reached the monitoring section, vertical stress reached its minimum. When the model tunnel face across the monitoring section, the crown settlement tends to be stable.

5.5.3 Analysis of osmotic pressure

The relationships between osmotic pressure at the crown position and excavation steps are shown in Figure 33.

Figure 33 shows that osmotic pressure reduces as the excavation steps proceed. When excavation of the top bench reaches the monitoring section, vertical stress

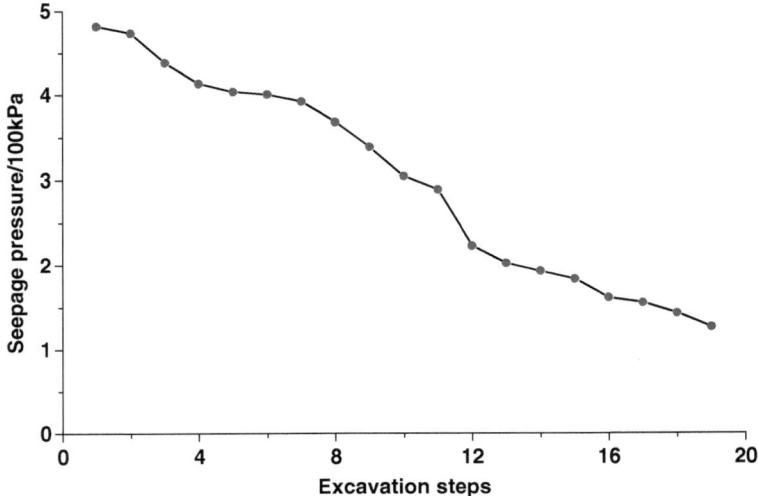

Figure 33 Osmotic pressure curves with excavation steps.

entered into a phase of rapid change, and the value of osmotic pressure changed from 0.43 MPa to 0.1 MPa; excavation of the bottom bench is the main factor affecting the reduction in osmotic pressure of the side wall position.

5.6 Discussion

(1)   A new, bigger fluid–solid coupling model test system was designed and produced in the laboratory, which can be assembled conveniently. This system was devised to capture the response of surrounding rock during subsea tunnel excavation.

(2)   Water inflow increases with increasing time, while the growth rate slowed slightly. Water inflow increases obviously since the height of sea water reached 76 cm. After that, water inflow increases sharply and presents a non-linear curve.

(3)   Vertical stress of the monitoring section experienced three stages with the excavation steps: slowly changing, sharply reducing and stable. When the excavation of the bottom bench reached the monitoring section, the value of vertical stress reached the minimum.

(4)   Osmotic pressure at the crown is mainly affected by the excavation of the top bench, while osmotic pressure at the side wall is mainly affected by the excavation of the bottom bench. The influence of excavation on the seepage field is about once the model tunnel diameter.

REFERENCES

Adams GD, Jager AJ (1980). Etroscopic observations of rock fracturing ahead of the stope faces in deep-level gold mines. Journal of the South Africa Institute of Mining and Metallurgy (2):115–127.

Chen AM, Gu JC, Shen J, Ming ZQ, Gu LY, Lu ZY (2004a). Application study on the geomechanical model experiment techniques. Chinese Journal of Rock Mechanics and Engineering 22(23): 3785–3789.

Chen AM, Gu JC, Shen J (2004b). Development and application of multifunctional apparatus for geomechanical engineering model test. Chinese Journal of Rock Mechanics and Engineering 23(3): 372–378.

Chen XL, Han BL, Liang KD (1994). Test research on stability for surrounding rock mass of underground opening. Chinese Journal of Wuhan University of Hydraulic and Electric Engineering 27(1): 17–23.

Heuer RE, Hendron AJ. Geo-mechanical model study of the behavior of underground openings in rock subjected to static loads (report 2)-tests on unlined openings in intact rock, 1971.

Huang Qingxiang, Zhang Wenzhong, Hou Zhicheng (2010). Study of simulation materials of aquifuge for solid-liquid coupling. Chinese Journal of Rock Mechanics and Engineering 29 (Supp.1): 2 813–2 818.

Li ZK, Lu DR, Hong L, Liu J (2004). Design and research on concealed excavation system for 3D geo-mechanical model test of large underground houses. Chinese Journal of Rock Mechanics and Engineering 23(2): 181–186.

Li ZK, Lu DR, Zhong SY, Xi JH, Sun JS (2003). Development and application of new technology for 3D geomechanics model test of large underground houses. Chinese Journal of Rock Mechanics and Engineering 22(9): 1430–1436.

Li Shucai, Han Peng Wang, Qian Qihu (2008). Deep roadway surrounding rock zonal disintegration phenomenon field monitoring studies. Chinese Journal of rock mechanics and engineering 27(8): 1545–1553.

Li Shu-cai, Li Li-ping, Li Shu-chen, Feng Xian-da, Li Guo-ying, Liu Bin, Wang Jing, XU Zhen-hao. (2010). Development and application of similar physical model test system for water inrush of underground engineering. Journal of Mining and Safety Engineering 27(3): 299–304.

Li Shucai, Song Shuguang, Li Liping, Zhang Qianqing, Wang Kai, Zhou Yi, Zhang Qian, Wang Qinghan. (2013). Development on subsea tunnel model test system for solid–fluid coupling and its application. Chinese Journal of Rock Mechanics and Engineering 32(5): 883–890.

Li Shucai, Zhou Yi, Li Liping, Zhang Qian, Song Shuguang, Li Jinglong, Wang Kai, Wang Qinghan. (2012). Research of a new similar material for underground engineering fluid–solid coupling model test and its application. Chinese Journal of Rock Mechanics and Engineering 31(6): 1128–1137.

Qian QH, Feng XT (2008). 21 issue of the new ideas, new theory Academic Salon. Zonal disintegration phenomenon and related theoretical problem report. *Beijing: China Association for Science and Technology. Chinese Journal of Rock Mechanics and Engineering Society.*

Song Shuguang. (2010). Study on seepage and deformation of surrounding rock and influence of overburden thickness in construction process of Qingdao subsea tunnel. Jinan: Shandong University.

E. I. Shemyakin, G. L. Fisenko, M. V. Kurlenya, V. N. Oparin, V. N. Reva, F. P. Glushikhin, M. A. Rozenbaum, É. A. Tropp, Yu. S. Kuznetsov (1986). Zonal disintegration of rocks around underground workings, part I: data of in-situ observations. Journal of Mining Science 22(3): 157–168.

Wang AM, Tao JK, Li ZK (2002). Design of the mini-size high-precision multipoint displacement indicator and its application in the 3D model experiment. Experimental Technology and Management 19(5): 21–26.

Wang Hanpeng, Li Shucai, Zhang Qiangyong, Li Yong, Guo Xiaohong (2006). Development of a new geomechanical similar material. Chinese Journal of Rock Mechanics and Engineering 25(9): 1842–1847.

Yu Liyuan (2010). Study on stability of surrounding rocks and selection of overburden thickness for underwater tunnels [PhD Thesis]. Jinan: Shandong University.

Zeng YW, Zhao ZY (2001). Model testing studies of underground openings. Chinese Journal of Rock Mechanics and Engineering 20(supp.): 1745–1749.

Zhang Qiangyong, Chen Xuguang, Lin Bo (2010). High power true three-dimensional loading model test system development and application. Chinese Journal of Geotechnical Engineering 32(10): 1588–1593.

Zhang Qiangyong, Chen Xuguang, Lin Bo, Liu Dejun, Zhang Ning (2009). Deep roadway surrounding rock zonal disintegration of 3D geomechanical model test. Chinese Journal of Rock Mechanics and Engineering 28(9): 1757–1766.

Zhang Qiang-yong, Li Shu-cai, Guo Xiao-hong, Li Yong, Wang Han-peng (2008). Iron crystal cemented sand new geotechnical similar material development and application. Rock and Soil Mechanics 29(8): 2126–2130.

Zhang Qiangyong, Li Shucai, Jiao Yuyong (2005). Rock mass numerical analysis method and geomechanical model test principle and engineering application. Beijing: China Water Conservancy and Hydropower Press.

Zhang QY, Li SC, You CA, Guo XH (2007). Development and application of new type combination 3D geomechanical model test rack apparatus. Chinese Journal of Rock Mechanics and Engineering 26(1): 143–148.

Zhu Weishen, Li Yong, Li Shucai, Wang Shugang, Zhang Qianbing (2011). Quasi-three-dimensional physical model tests on a cavern complex under high in-situ stresses. International Journal of Rock Mechanics and Mining Sciences 48(2): 199–209.

W. S. Zhu, Q. B. Zhang, H. H. Zhu, Y. Li, J.-H. Yin, S. C. Li, L. F. Sun, L. Zhang (2010). Large-scale geomechanical model test study of an underground cavern group in a true three-dimensional (3-D) stress state. Canadian Geotechnical Journal.

Field Testing and URLs

Chapter 14

Underground research laboratories

Joseph S.Y. Wang[1], Xia-Ting Feng[2] & John A. Hudson[3]
[1]*Lawrence Berkeley National Laboratory, Berkeley, CA, U.S.A.*
[2]*Institute of Rock and Soil Mechanics, Chinese Academy of Sciences, Wuhan, China*
[3]*Imperial College, London, U.K.*

Abstract: This overview of Underground Research Laboratories (URLs) is based on studies by the International Society for Rock Mechanics (ISRM) Commission on URL Networking, which was formed after an URL workshop associated with the ISRM Congress in 2011. We also held another URL workshop in 2015, and several meetings and sessions in between. Recent progress in heater tests in radioactive waste URLs, designs of large excavations in deep physics facilities, and other underground studies are reviewed in this article. Rock mechanics findings and multi-disciplinary studies are among topics of interest to the ISRM Commission. Heater tests for better understanding of the coupled thermal-hydro-mechanical-chemical processes are of interest to radioactive repository assessments and for other thermal storage and geothermal energy production projects. Large excavations in physics laboratories are driven by the needs associated with designing and housing next generation of experiments to detect rare events. Some existing physics laboratories are interested to use available spaces for geo-sciences studies, including microbiological search for deep life. Examples of energy/environmental and inter-disciplinary studies and networking activities are also discussed. We review the progress in these topics and welcome inputs on case histories and planned developments in URLs. Each URL could provide lessons learned and offer as an analog for other sites. The inputs from the geo-engineering, rock mechanics, geoscience, and physics communities are essential for our continuing efforts of the ISRM URL Networking Commission.

1 INTRODUCTION

From 2011 to 2015, we have attended and organized the following URL-related meetings. We started with a 3rd URL workshop associated with the 12th ISRM Congress at Beijing, China. In 2012, we attended the DEvelopment of COupled models and their VALidation against EXperiments (DECOVALEX) workshop at Berkeley, U.S.A.; the 3rd inter-Disciplinary Underground Science and Technology (iDUST) conference at Apt, France; the 2012 EUropean ROCK mechanics symposium (EUROCK) at Stockholm, Sweden; the 46th American Rock Mechanics Association (ARMA) symposium in Chicago, U.S.A.; the 2012 American Geophysical Union (AGU) meeting at San Francisco, U.S.A.; and the AStroparticle Physics European Research Area network (ASPERA) workshop at Durham, U.K. In 2013, we attended the Deep Underground Research Association (DURA) meeting at Palo Alto, U.S.A.; the 3rd Chinese ROCK mechanics symposium (SINOROCK) at Shanghai, China; 47th ARMA symposium at San Francisco, U.S.A; the 13th Topics in Astroparticle and Underground Physics (TAUP) conference at Asilimor, U.S.A.; the 2013 EUROCK symposium at Wroclaw, Poland; and the 2013 AGU meeting at San Francisco, U.S.A. In 2014, we attended the 4th iDUST conference at Apt, France and the 8th Asian Rock Mechanics Symposium (ARMS) at Sapporo, Japan. Note that the 3rd URL workshop in Beijing, China, follows the 1st URL workshop 2003 at Johannesburg, South Africa (Ellsworth *et al.*, 2003) and the 2nd workshop 2007 at Lisbon, Portugal. We had the 4th URL workshop 2015 at Montreal, Canada, associated with the 13th ISRM Congress. The abbreviations of meetings in this paragraph are used are used in following Sections and in the Reference List.

Wang *et al.* (2010) presented an early literature review at the 6th ARMS in New Delhi, India, for URL activities before the 3rd URL workshop. Wang *et al.* (2015) presented an overview of the 3rd URL workshop lectures and other subsequent meeting sessions in following years in the 13th ISRM Congress. Underground studies have been conducted primarily either to evaluate the capacities of different formations to isolate wastes, to explore resources, or to house detectors at depths. (For the ISRM URL Networking Commission, we use the term URL for any facility and sites dedicated to all these research activities.) We focus on recent advances in understanding various processes conducted in URLs, and on known plans and designs for expansions.

In the following three sections, we summarize Asian, European, North and South American URLs, alphabetically for countries in each continent. We discuss recent presentations and publications on rock mechanics and geoscience investigations, physics and multi-disciplinary interactions, and examples of energy/environmental and inter-disciplinary studies and networking activities in communities interested in underground studies, followed by a summary section.

2 RADIOACTIVE WASTE URLS

Hudson (2010) presented an overview of radioactive waste URLs. Hudson (2015b) discussed the contributions of rock mechanics to radioactive waste disposal in the 4th URL workshop, and Hudson *et al.* (2013) discussed the value of URLs in reducing rock engineering risk in the 3rd SINOROCK. In addition to laboratory and field studies, the reproducibility of numerical modeling, associate technical auditing, and validation of the models against experiments, are important for radioactive waste assessments. Existing and future URLs are required for in situ experiments to support design premises which need thorough and relevant suitable checking to ensure all aspects of the design are included and appropriately accommodated. The DECOVALEX project (Hudson *et al.*, 2011) addresses verification through benchmarking and validation against field experiments for nuclear waste assessments since 1992. Hudson (2015a) summarized the 2012–2015 phase of DECOVALEX. The DECOVALEX project contributes to networking among URLs by organizing validation efforts from different URLs.

Figure 1 shows that three of the test cases in DECOVALEX-2015 are using experiments conducted along horizontal tunnels and one of the test cases is in an URL accessible through vertical shafts, as illustrated in Figure 1. Currently the test cases for DECOVALEX-2015 project include heater tests from Japan's Horonobe URL and

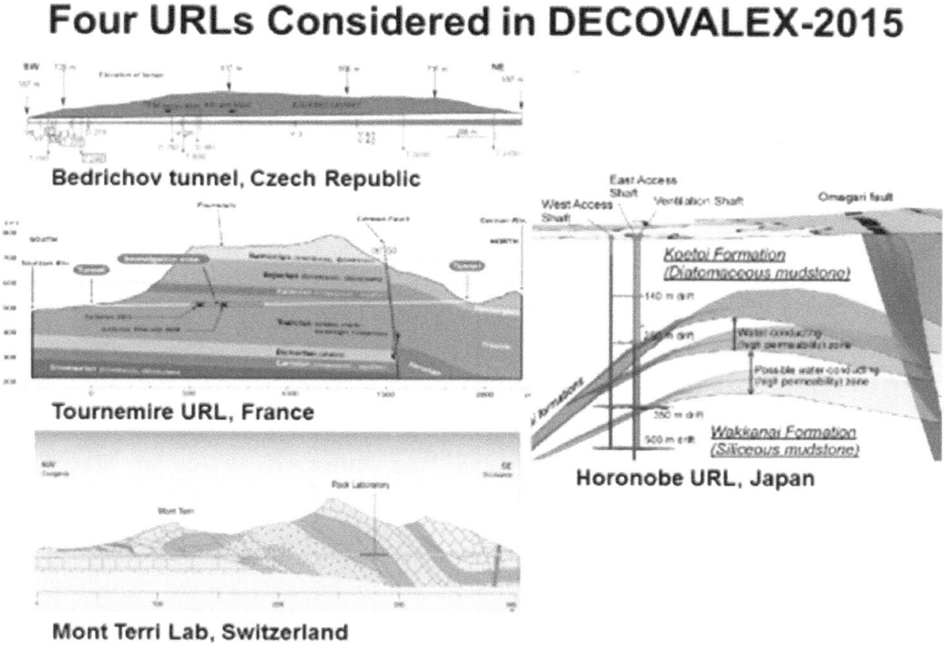

Four URLs Considered in DECOVALEX-2015

Bedrichov tunnel, Czech Republic

Tournemire URL, France

Horonobe URL, Japan

Mont Terri Lab, Switzerland

Figure 1 Four field test cases considered in ECOVALEX-2015 project, with URLs accessible through horizontal and roadways and vertical shafts. Modified from Wang *et al.* (2013) from inputs of Sugita & Nakama (2012), Gaus (2012), Barnichon & Millard (2012), and Hokr (2012).

Switzerland Mont Terri Laboratory, hydro-mechanical tests at France's Tournemire URL, water inflow distributions at the Czech Republic's Bedrichov tunnel, and a thermos-hydro-mechanical-mechanical coupled processes in fracture experiment.

2.1 Asian URLs

In Asia, we have URL activities in China, Japan, and Korea. Wang Ju (2010, 2015), Wang Ju & Chen (2015) regarded the Beishan granite site in China, for a planned URL operational in 2020-2030, as an area-specific URL. During 1999-2013, 11 deep boreholes and 8 shallow boreholes have been drilled and a series of pumping, injection tests, televiewer, radar surveys, water-sampling, and geo-stress measured were conducted. Wang Ju (2004) showed that the Beishan granite has cores over 2 m long. The Beishan granite massif was shown to have enough volume with favorable conditions.

To simulate heat from radioactive decays, electrical heater tests have been planned, in progress, or conducted in different URLs to understand thermally induced coupled processes, dependence on features/properties in response to temperature and pressure increases in different media, and designs of expansions in different waste emplacement configurations. Tsusaka *et al.* (2011) and Sugita & Nakama (2012) described an Engineered Barrier System (EBS) test at the Japan's Horonobe URL in sedimentary formations with levels to depths of 140, 250, 350, and to 500 m. Aoyagi *et al.* (2014)

and Asahina *et al.* (2014) described the testing and modeling of fracturing and excavation damage processes for the Horonobe URL. The heater is surrounded by an EBS of sorbing backfilled materials emplaced between the heater and the host rocks.

The Korea Atomic Energy Research Institute (KAERI) Underground Research Tunnel (KURL) with originally maximum depth of 90 m has two research tunnels total 75 m long, with the access tunnel 180 m long in granite. Lee and Jeon (2015) described an ongoing extension phase II for KURT which started the test and experiments in 2015. Ongoing tests include application of hydrochemical properties to streaming potential, geochemical investigation of redox disturbance, hydro-mechanical coupled behavior of bentonite buffer, and planned long-term sorption and diffusion experiment and monitoring of the THM evolution of the EBS.

2.2 European URLs

Figure 2 shows that the URLs in Europe are in hard crystalline rocks, in soft argillaceous rocks, and in rock salts. Belgium's HADES, France's Meuse/Haute-Marne, and German's Gorleben URLs are accessed through vertical shafts for access, Finland's ONKOLO and Sweden's HRL are accessed through a combination sloped *ramps*, and Switzerland's Monte Terri and GTS are through horizontal roadways.

In Europe, recent advances include the PRACLAY heater experiment in Belgium's HADES URL in Boom clay. The HADES URL is at a depth of 225 m. Li (2011b), Li *et al.* (2012b) and Sneyers and Li (2012) described the HADES URL site and the design

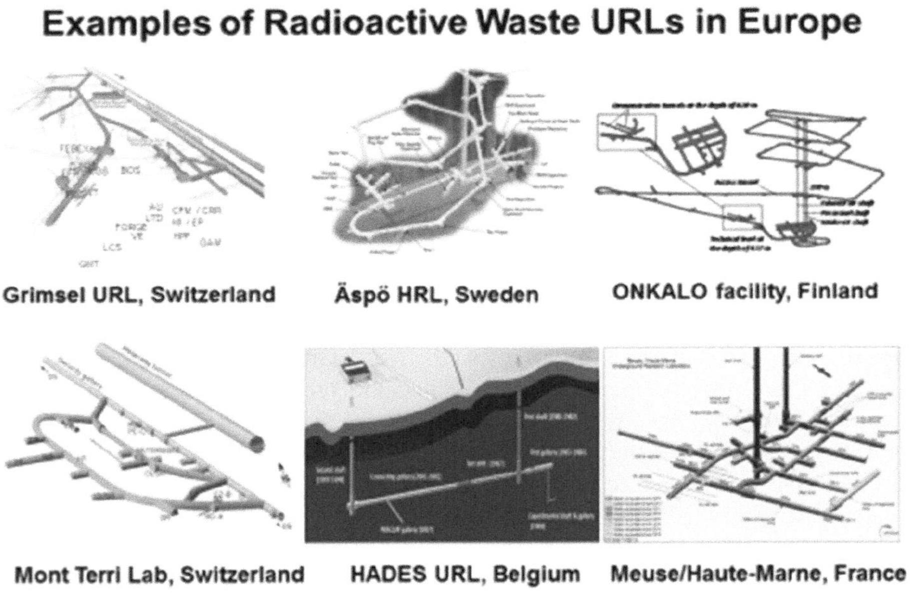

Examples of Radioactive Waste URLs in Europe

Grimsel URL, Switzerland   Äspö HRL, Sweden   ONKALO facility, Finland

Mont Terri Lab, Switzerland   HADES URL, Belgium   Meuse/Haute-Marne, France

Figure 2 European radioactive waste URLs with access via vertical shafts, via horizontal roadways, and via a combination of vertical shafts and sloped ramps. Modified from Wang *et al.* (2015), with inputs from Li *et al.* (2015a), Armand (2011), Fahland *et al.* (2015a), Svemar *et al.* (2011), Vomvoris (2011) and URL web sites.

of the PRACLAY heater experiment. The temperature at the lining-rock interface is designed to be held at 80°C during a 10-year heating phase (Sneyers & Li, 2012). Li *et al.* (2015a,b) described the first results of the heater test with heating started in November 2014.

Siren and Johansson (2011) discussed the ONKALO URL in crystalline gneiss and pegmatite in Finland. ONKALO facility has reached the depth of 455 m. In addition to the 1:10 sloped access ramp, there are also one personnel and two ventilation shafts excavated. ONKALO is regarded as an underground site characterization facility for in-situ testing for a future radioactive waste repository. Rock spalling is one of the processes being evaluated. Johansson *et al.* (2015a, b) described in-situ test of thermal diffusivity, buffer material, EDZ studies, POsiva's spalling experiment, hydrological interference tests, water matrix diffusion, through diffusion, gas matrix diffusion, sulfate reduction, prediction-outcome studies, rock suitability criteria, and a demonstration area for final disposal concept. Tests are conducted at ~140, 345, 360, 400, and 437 m levels.

Armand (2011) and Lebon (2012) presented the status of and experiments at the Meuse/Haute-Marne URL in French in Callovo Oxfordien claystone with galleries and experiment sites continuously excavated starting at 1999 until at least 2016. Meuse/Haute-Marne is 420-550 m in depth. Armand *et al.* (2015a,b) described the hydro-mechanical behavior of parallel drifts excavated by different construction methods, and pore pressure changes during excavation of drifts, micro tunnels, and a borehole. In-situ stress states and rate of excavation have impacts on pore pressures.

Heusermann *et al.* (2013) and Fahland & Heusermann (2013) described modeling of the Gorleben salt dome URL in Germany at depths of 860 to 1,170 m, and the analysis of the q were at depths deeper than many other radioactive waste URLs. Fahland *et al.* (2015a,b) and Vogel *et al.* (2015) described the geoscientific program, the three-dimensional geological and geomechanical modeling, and the temperature measurements in the Groleben site.

The first heater test in crystalline granite was first conducted in late 1970s at the Stripa mine in Sweden (Witherspoon *et al.*, 1980) (see also Wang & Hudson, 2012 for a review). Svemar *et al.* (2011) and Hakimi & Christansson (2012) discussed additional experiments at Sweden in the Äspo Hard Rock Laboratory (HRL) at depth from 220 to 460 m. Among many experiments, the rock spalling in canister borehole and pillar stability under thermal stress were evaluated. Tests were conducted on groundwater flow and radionuclide retention, comparison between blasted and bored drifts and distributions of disturbances, spalling induced by heating, counter force applied to prevent spalling, many completed and present projects for repository technology, vertical vs. horizontal emplacement configurations, gas injection test, canister retrieval test, and bentonite studies. There are international projects and NOVA R&D platform with universities and companies.

Sweden's Aspo Hard Rock Lab and Finland' ONKALO Facility share common interest in radioactive waste repository performance in granite. Both URLs have borehole spalling and pillar stability studies conducted and compared.

Vomvoris (2011), Gaus (2012), Nussbaum & Bossart (2013) described research program and key results of the Mont Terri laboratory, Switzerland, in Opalinus clay, with many experiments including the fifth phase of heater experiment HE-E. The experiments at Mont Terri include dry excavation with road header, pore pressure changes due to stress redistributions, monitoring of excavation damage zone, self-sealing of EDZ fracture network, engineered barrier experiment, full scale emplacement experiment, hydrogen transfer experiment, microbial activity experiment, and diffusion and retention of radionuclide experiment. Switzerland also has the Grimsel Test Site (GTS) in crystalline hard rocks with heater tests conducted. Blechschmidt & Martin (2012) presented in-situ tracer tests using radioactive elements and presented models developed to understand flow paths in a shear zone at the GTS.

2.3 North American URLs

In North America, Atomic Energy of Canada Limited (AECL) Whiteshell URL at Manitoba is closed and floored. The ongoing effort in Canada is in Ontario for low and intermediate waste site characterization. AECL Whiteshell URL was where 1992 observations of progressive failure in brittle rocks around a circular tunnel, brought about a change to a stable configuration with V-shaped notches developed after the rock falls in the minimum stress orientation (Martin et al., 1997, 2003), and 1987–2004 observations of induced seismicity from acoustic emissions, microseismic/ultrasonic detections of crack creations, above, below, and around the tunnel network at the 420 m level into initially unfractured granite (Young et al., 2001, 2004). Both these studies were major contributions to rock mechanics (Fairhurst, 2004). The characterization of the excavation-disturbed-zone was also carried out at the Mount Terri Rock Laboratory, in Switzerland, in a clay gallery parallel to a freeway tunnel under up to 400 m of overburden (Martin et al., 2003; Blümling et al., 2008).

Tsang et al. (2009) reviewed the single heater test and the drift scale test, with four-year-heating and four-year-cooling periods, in the U.S. Yucca Mountain Exploratory Study Facility (ESF) in unsaturated welded tuff at a depth of 300 m. ESF is currently closed. Alley & Alley (2013a,b) described the challenges in the science of nuclear waste disposal. The Waste Isolation Pilot Plant (WIPP) in New Mexico is open for transuranic waste disposal, and research in salt disposal systems.

Flow and transport through the rocks and coupled processes induced by the waste heat are sensitive to characterization of fractures in excavation disturbed and damaged zones surrounding the waste emplacement tunnels and in faults intersecting the tunnels connecting to the far field. These characterization studies and heater experiments are among issues to be evaluated and resolved associated with URLs. These issues are the focus of radioactive waste URLs worldwide, before radioactive waste problems can be addressed and nuclear waste repositories for high-level wastes can be licensed for geological time scale isolation.

2.4 International networking

For the radioactive waste URLs, the International Atomic Energy Agency (IAEA) has conducted for decades training sessions, technical exchanges, and established centers of excellence in countries with radioactive waste repository programs and for countries interested in regional and international repositories. Clearly there are challenges to solve the radioactive waste disposal problems in all counties as the future of nuclear energy depends on the solutions to disposing the wastes. The 2012 IAEA underground research facility network meeting at Albuquerque, U.S.A. in 3–7 December had presentations from Belgium, Switzerland, and U.S.A. and a field trip to the WIPP with participants from 19 member states including Philippines, Germany, South Africa, Canada, South Korea, Kazakhstan, the Czech Republic, U.K. and the Ukraine.

There are clay and salt clubs established among countries conducting and interested in housing radioactive waste repositories in different geological media. The clay club has the latest gathering at Belgium in March 23–26, 2015, with presentations from Belgium, France, Netherlands, Sweden, Switzerland, and U.S.A. on either large scale heater tests, or on mass transfer, alternation, hydro-mechanical studies in smectite, bentonite, and clay barriers. Belgium's HADES and Switzerland's Mont Terri Lab have collaborative microbiology studies for water collected in URLs. The salt club of Nuclear Energy Agency (NEA) has working group members from Germany, Netherlands, Poland, and U.S.A. Annual workshops and meetings on actinide and brine has occurred since the salt club kick-off meeting in Paris on 20 April, 2012.

3 PHYSICS LABS WITH EXPANSION PLANS AND WITH MULTI-DISCIPLINARY STUDIES

Bettini (2007, 2011, 2013) reviewed existing and new physics underground laboratories and facilities worldwide. With detectors for rare event detections growing in sizes, and demands for shielding increasing, there is a need to have large halls excavated to house experiments. Solid crystals, noble liquids, scintillators, and water are used for detector materials. Water is also used for shielding against neutron and radiations. Depth is a premier as the damping of cosmic ray derived muon increases with depth with the rock in the crust provides the shielding and decreases the background. Bettini (2014) summarized the expansion of deep laboratories worldwide, with some examples illustrated in Figure 3. The plans for LNGS in Italy, LSM in France, CUPP in Finland, LSC in Spain, SNOLab in Canada, SURF in U.S.A., INO in India, CJPL in China, CUNPA in Korea, and ANDES between Chiles and Argentina respond to the needs for spaces for housing underground detectors.

3.1 Asian DULs

In Asia, China currently has the deepest physics laboratory with horizontal access, along the JinPing tunnels excavated for a hydropower project at depths 1900–2525 m in marble. Feng (2011), Feng et al. (2012), Zhang et al. (2012), Li et al. (2012a, 2013c) measured excavation disturbed and damaged zones and rock bursts at the China JinPing tunnel complexes associated with a hydropower project at depths 1900–2525 m in marble. The Ya Long River in Sichuan province makes a big U-turn around the JinPing mountain. The JinPing tunnels were excavated to connect opposing sides of the mountain before and after the U-turn, in order to exploit the difference in river water levels over the length of the tunnels, to generate electric power in a hydropower station. The project includes four headrace tunnels, two traffic tunnels, and one water drainage tunnel, excavated initially by tunnel boring machines (TBM) and later by drill and blast methods after some damages to the TBMs. Most rock bursts were observed associated with tension failures immediate during excavation unloading, and some shear-failure rock bursts were observed with time-delayed over days and tens of days and located up to hundreds of meter away from newly excavated faces. A rock sliding micrometer and

Examples of DULs with Spaces Planned and Existed

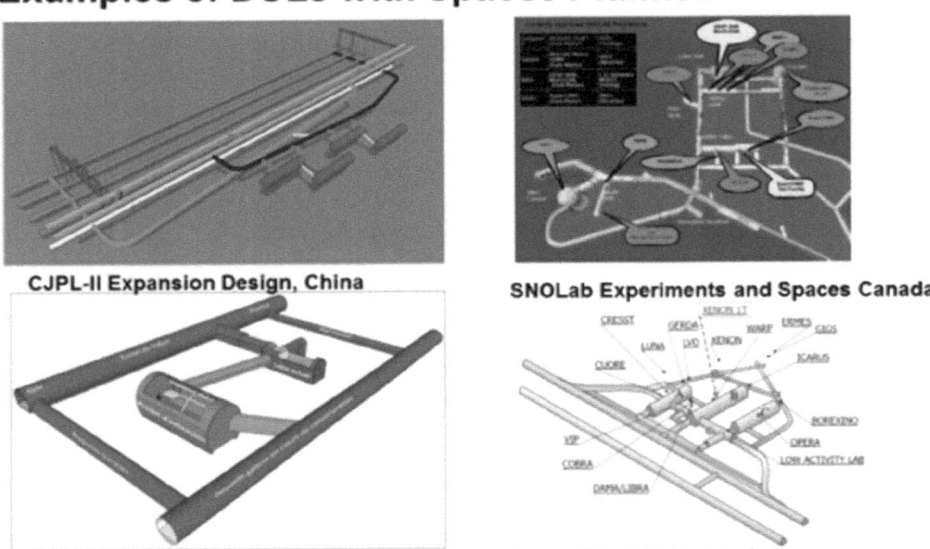

CJPL-II Expansion Design, China

SNOLab Experiments and Spaces Canada

LSM Expansion Plan, France

LNGS Experiments, Italy

Figure 3 shows that DULs with plans for recent excavations or spaces for new experiments.

a borehole televiewer were used in and developed into two ISRM suggested methods to characterize the excavation disturbed and damaged zones (Li *et al.*, 2013b,d). The challenges and progress in understanding rock burst, water inrush, and squeezing processes in tunneling and underground construction at the JinPing II hydropower station were discussed by Feng (2014, 2015).

One of the experiment hall excavated for rock mechanics studies for tunneling, 6.5 × 6.5 × 40 m, ~4,000 m^3, was converted to the China JinPing Laboratory (CJPL) for dark matter search experiments (Li, 2011a). There is an extension of CJPL planned and executed by additional excavations staring in 2015 along a bypass tunnel in the vicinity of CJPL. Li (2013, 2015a) and Li *et al.* (2013a) discussed the CJPL-II, which plans to have eight additional halls, each up to 14 m wide, 14 m high, and over 60 m long, being excavated in 2015 (Li, 2015b) to accommodate next generation dark matter search experiments, and have spaces available for additional dark matter, neutrino-less double beta decay, solar neutrino, astrophysics accelerator, and other geoscience experiments accommodated, as discussed in the 1st CJPL-II Workshop associated with the 2013 TAUP Conference (Li *et al.*, 2013a).

Dighe (2013) described that the India-based Neutrino Observatory (INO) is currently being excavated by tunneling 2 km into the Bodi West Hills in southern India. The vertical cover is 1,289 m, with all around cover of ~ 1,000 m. Cavern 1 with dimension of 26 m × 32.5 m × 132 m will host a 50 kt (space for 100 kt) magnetized iron calorimeter (ICAL) detector. There are three other caverns of dimensions 55 × 12.5 × 8.6 m, 40 × 10 × 10 m, and 10 × 10 × 10 m for multiple dark matter and neutrono-less double beta decay experiments.

The existing Super-Kamiokande (Super-K) cavern, in Japan's Kamioka Observatory at depth of 1,000 m, is 40 m in diameter and 55 m in height. Yamatomi (2013) described the Hyper-Kamiokande (Hyper-K), a proposed megaton water Cherenkov detector for long-baseline neutrino physics, atmospheric neutrino detection, and proton decay detection, has currently a design in the configuration of twin cylindrical caverns aligned horizontally and side-by-side, each 48 m wide, 54 m high, and 247.5 m long. Such a wide-span design is very challenging, even at the relatively moderate depths ~ 650 m planned for Hyper-K. The Hyper-K detector will be the world largest underground physics detector. Jung (2015) described the physics and the experiments ongoing at Super-K and planned at Hyper-K as detectors in the Tokai to Kamioka (T2K) long baseline experiment. In Kamioka Observatory, there are two 3 km long tunnels being excavated for large scale cryogenic gravitational wave telescope, using laser interferometers. There are also a laser strain meter and a superconducting gravity meter for geophysics studies.

The Centre for Underground Nuclear and Particle Astrophysics (CUNPA) in Korea has plans to expand the existing space at the YangYang Laboratory (Y2L) at 700 m depth in three alternative locations: 1,000 m^2 in area and 7 m in height at the Y2L site, at a site 1,050 m deep accessible by a 1.6 km accessible tunnel, or at an operating mine. The dark matter search, double beta decay, nuclear astrophysics, and low temperature detector research and development are programs evaluated for the expansion (Bettini, 2013).

3.2 European DULs

In Europe, there are underground laboratories in Finland, France, Italy, Spain, and U.K. with physics experiments. The Centre for Underground Physics in Pyhasalmi (CUPP) in Finland, 1,444 m in depth and with spaces in other mine levels, has cosmic-ray and other geology and biology experiments, and is a candidate site for the European LAGANA Long Baseline Neutrino Observatory (LBNO). Nuijten (2013) described the designs for LBNO that include an ellipsoidal cavern to house the LBNO liquid argon detector with dimensions 64 m wide, 51 m high, and 103 m long, and another cavern for liquid scintillator 44 m wide, 120 m high, and 71 m long at 1,450 m depth. These designed caverns are wider in spans than most excavations at shallower depths.

Piquermal (2013, 2015) described the Laboratoire Souterrain de Modane (LSM) in France which is Europe's deepest lab at a depth of 1,700 m, with existing spaces for dark matter search, neutrino-less double beta decay, and low-radioactivity counting experiments. LSM plans to excavate an additional multi-disciplinary underground sciences hall, 19 m wide, 16 m high, and 40 m long, 12,000 m^3 in volume, in an area adjacent to the existing laboratory and between the Modane highway tunnel and a new safety gallery that will be bored parallel to the highway tunnel between France and Italy. The existing non-physics experiments at LSM include the use of gamma spectroscopy for biological studies and ^{137}Cs dating to determine if a win bottle is before or after Chernobyl nuclear power plant accident.

Piquemal (2015), with inputs from Ragazzi (2015), Ianni (2015) and Paling (2015), described all activities in four European DULs. The Deep Underground Laboratory Integrated Activity (DULIA) aims to form toward a deep underground European distributed platform. Agralioti (2015) described multi-disciplinary studies in European deep underground laboratories organized by the Astroparticle Physics European Consortium (APPEC), with examples from all four European DULs., and the low noise inter-disciplinary LSBB URL at France. The ASPERA-2 workshop in 2012 in a great source for multi-disciplinary studies at Europe.

Ragazzi (2015) described the Laboratori Nazionali del Gran Sasso (LNGS) in Italy. LNGS has three halls 20 m wide, 18 m high, 100 m long, along a highway tunnel. LNGS has a rich program and is the world largest underground physics laboratory with the total volume of 180,000 m^3, for many dark matter, neutrino-less double beta decay, long-baseline neutrino beam, solar and geoneutrino experiments in the main halls, and geoscience experiments in surrounding connecting tunnels. Some of these existing large spaces are currently being made available for many next generation experiments considered by an international scientific committee (Bettini, 2014). In 2013, the geosciences experiments: including ^{222}Rn – ^{14}C – ^{3}He monitoring of environmental radioactivity, crustal deformation, the maintenance of cryo-preserved biological materials in reduced-radiation environments, studies of aseismic creep strain episodes that might be associated with earthquakes, seismographic monitoring, and the effects of reduced-radioactivity environments on the biochemical behavior of living organisms. One example of geoscience study is the radioactivity measurements of water samples in Wolfgango (2012).

Bettini (2013), Innai (2015) described the Laboratorio Subterráneo de Canfranc (LSC) in Spain which has expanded to house dark matter search, neutrino-less double beta decay, and low background assay experiments. LSC is at depth of 850 m along a tunnel between Spain and France, with area 1,560 m^2 and volume 10,500 m^3. Hall A has dimension of $14.5 \times 10 \times 40$ m^3 and can accommodate experiments considering by a scientific program, including dark matter search, neutrino-less double beta decay, and detector research and development. The GEODYN earth sciences observatory around LSC uses seismographs and laser strain meters mounted on ground surfaces and within the tunnel to monitor the entire geodynamical spectrum, measuring strain, velocity and acceleration. GEODYN is part of the European Plate Observation System (Diaz, 2012; Bettini, 2013). LSC has nearby tunnels for microbiology studies (Ianni, 2015).

Paling (2015) described the Boulby Underground Laboratory (BUL) in U.K., at depth of 1,100 m, 1,500 m^2 in area, within a potash and rock-salt mine. BUL hosts dark matter search experiments, a cosmoclimatology project, a muon tomography survey, a radio-ecology, and other geology and engineering projects. The Palmer Laboratory facility, 120 m long, where several dark matter and other physics experiments were located, has been converted into the International Subsurface Astrobiology Laboratory, where studies of the microbiology of deep subsurface salts are conducted (Cockell et al., 2012).

3.3 North and South American DULs

In North America, Smith (2013), Smith and Larsson (2015) described the Sudbury Neutrino Observatory (SNO) Laboratory (SNOLab) in Canada. SNOLab is the deepest lab in North America at depth of 2,070 m, with a rich program of dark matter searches with both cryogenic and room-temperature bubble chamber approaches.

SNOLAb has newly excavated spaces including the barrel-shaped cryo-pit with 18.3 m in diameter at the waist, 15.2 m in diameter at the base and top, and 19.8 m in height. The original SNO cavern with similar shaped is 22 m in diameter at its waist and 34 m in height, and is currently the site for SNO$^+$, a double beta decay and geo-neutrino detector currently under construction that will use Te-loaded scintillator. Similar barrel designs could be considered at CJPL-II or in other laboratories.

The U.S. has four facilities with physics experiments, one has recent expansions. Lesko (2013) described the Sanford Underground Science Facility (SURF) which is in the former Homestake gold mine, at the depth of 1,480m (4850 Level). Near the Yates shaft, the historic Davis Cavity was enlarged to 11 m wide, 13 m high, and 18 m long space for a dark matter search experiment, and a 16 m wide, 5 m high, and 43 m long space excavated for a neutrino-less double beta decay experiment. Vardiman (2012) and Roggenthen (2015) described the geotechnical and geological study of SURF. Seismic surveys in multiple levels at Homestake continue for a gravity wave study. Elsworth (2012) summarized other earlier bio-geo-engineering investigations in multiple levels at Homestake.

Nuitjen (2014) described the cost and other geoengineering comparisons between European LAGUNA-LBNO detector at Pyhasalmi mine and U.S.A. LBNE detector at Homestake. Jung (2015) described the physics of and the plan for a long baseline neutrino experiment from a westward neutrino beam from Fermi Laboratory near Chicago to SURF in South Dakota. The new collaboration of Deep Underground Neutrino Experiment (DUNE) is being considered to be located near the Ross Shaft, 1 km away from the Yates Shaft. Bettini (2011) and Laughton (2011, 2015) compared the costs for underground excavations between European and North American DULs. Laughton (2015) noted that the U.S.A underground costs were over twice as much as in Scandinavian countries due to likely union cost differences but near-surface excavations were comparable.

Marshak (2013) described the Sudan Underground Laboratory (SUL) that has a rectangular room 16 m wide, 14 m high, 89 m long, oriented along a northward neutrino beam from Fermi Laboratory, to house iron plate neutrino detectors. The other hall houses a low background counting facility. SUL conducted proton decay and dark matter search experiments, and also geology, geochemical and microbiological studies of salty water.

Link & Vogelar (2013) described the Kimballton Underground Facility (KURT) with spaces accessible in a drive-in limestone mine. The current experiments at KURT include neutron spectroscopy and neutrino-less double beta decay in a deep level, and a spallation isotope production in water experiments in a shallow level.

Gratta (2013) described the neutrino-less double beta decay experiments located away from the waste storage area in the Waste Isolation Pilot Plant (WIPP). WIPP has also experimental spaces available for other scientific investigations.

In South America, Dib (2013) described the future ANDES laboratory along the Argentina-Chiles Agua Negtashi Koshibara tunnel was. ANDES will provide an overburden of 1,750 m, has a large pit planned, 30 m in diameter and 42 m in height. There are other halls, caverns, and pit planned for neutrino detector, dark matter search, double beta decay, and nuclear astrophysics experiments. Geophysics experiments for this geo-active region are planned by linking seismograph networks in the region.

3.4 History of using mines and UULs in the neutrino sector

Frederick Reines was co-winner of the 1995 Nobel Prize for physics for the detection of neutrino. Atmospheric neutrinos were detected in 1995 in the East Rand gold mine in South Africa and in the Kolar Gold Field mine in India. Raymond Davis and Masatoshi Koshiba shared the 2002 Nobel Prize for the detection of solar neutrinos at the Homestake gold mine and the detection of supernovae neutrinos at the Kamioka mine. These discoveries led to the onset of the field of neutrono astrophysics. Neutrinos associated with electron, muon, and tau particle are now known to have masses, with the mass differences among neutrinos known, but the absolute masses unknown. Only one-third of electron neutrinos from fusions in the interior of sun reach the earth with the remaining two third converted to other two types of neutrinos.

Neutron studies are currently joined by the dark matter searches, neutrino-less double beta decay, proton decay, gravitational wave detection, and other rare event detections. The dark energy idea to explain the acceleration of the universe presents an even more challenging concept in fundamental understandings of the big band, evolution of the universe, and disappearance of anti-matters and creation of the matters. This is an exciting time for physics research in astrophysics and using deep mines and DULs.

4 EXAMPLES OF ENERGY/ENVIRONMENTAL, INTER-DISCLIPLINARY, AND NETWORKING STUDIES

In this section, we summarize several underground studies beyond radioactive waste URLs and physics deep underground facilities. These examples address additional rock mechanics, geoscience, and inter-disciplinary studies discussed in some recent meetings associated with AGU, iDUST, ISRM, and information in the literature.

4.1 Energy and environmental assessments

Brinkman (2012), Hitzman *et al.* (2012), Hitzman (2012), and Tester (2012), discussed various alternative energy evaluations to reduction of CO_2 by energy productions, including critical elements needed for renewable solar and wind energy and for car batteries, geothermal resources for electric generation and for district heating, and shale gas production through hydraulic fracturing and the associated environmental concerns. Increases in renewables, in liquid biofuels, and in natural gas are expected, while nuclear energy may not change, and decreases in coal and in petroleum and other liquid products are expected. McNutt (2012) discussed challenges in energy and resource recovery associated with undiscovered oil and gas reserves. Shephard (2013) discussed the development of shale reserves and water use strategies for a sustainable future. Wakimoto (2013) presented impacts on energy production from storms, hurricanes, earthquakes, tsunamis, and melting of the Arctic ice. Uhle & Anooshepoor (2012) described the aftermath of the Fukushima Daiichi Nuclear Power Plant in Japan and the associated regulatory concerns.

Elsworth (2011, 2013) discussed the uncertainties in sustaining energy recovery and the roles of underground experiments and URLs in resolving the issues. Elsworth *et al.* (2015) discussed the use of super-saturated and under-saturated gases in addition to water in hydraulic fracturing and in enhanced geothermal system and in CO_2 sequestration. Sosna *et al.* (2014) described a heating, cooling, and heating experiment with water heater and heat exchanger for temperature, stress and deformation changes in the Josef URL in Czech Republic for thermal storage evaluation.

The need for development of earth resource engineering was discussed by Fairhurst (2015). The development of shale gas and oil depends on hydraulic fracturing (or fracking) along mainly horizontal wells. Tutuncu & Gutierrez (2011) presented site characterization and injection/withdrawal assessment of a proposed in situ laboratory in shale gas and shale oil. Dusseault (2013) presented the theory, reality and uncertainty of hydraulic fracturing operations. Hydraulic fracturing may be used in underground mining, and in deep slurried solid injections. Dusseault (2011) explained that saline water and sold slurry injection practices into deep fractured and porous layers had been used in oilfields for over more than 20 years, could be practiced in China in identified areas, occurred at Alberta Canada in 1996, had a zero discharge oil field solid injection site in Indonesia, and applied to municipal biosolids in Los Angeles. Doe (2011) discussed flow quantifications in fractured rocks. Rutqvist (2011) presented the geo-mechanical aspects of CO_2 sequestration and modeling of the in Salah CO_2 storage project in Algebra, which is 1.8-1.9 km deep. The ground surface uplift and microseismic events were shown to be caused by tensile opening of faults intersecting the injection zone. Examples for injection and withdraw fluids through boreholes are illustrated in Figure 4.

Examples of Injection/Withdrawal in Boreholes

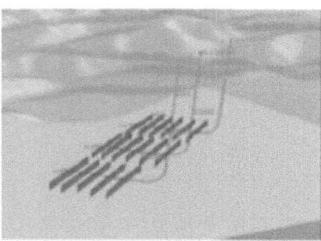

Hydraulic Fracture Growth

Double-Loop Uplift from CO_2 Sequestration

Slurry Injection for Zero Discharge

Proposed In Situ Lab for Unconventional Gas/Oil

Figure 4 Examples for energy and environmental studies, modified from Wang *et al.* (2013), with inputs from Doe (2011), Dusseault (2011), Rutqvist (2011), Tutuncu & Gutierrez (2011).

4.2 Inter-disciplinary studies

Gaffet (2012, 2013), Guglielmi (2011), Guglielmi *et al.* (2012), Waysand (2006), and Waysand *et al.* (2010), and Fourie *et al.* (2013) described various studies at the Laboratoire Souterrain à Bas Bruit (LSBB) in France with a depth of 518 m, with laboratory spaces along ~ 4 km horizontal tunnel spaces in a low anthropogenic environment away from through-going traffic. The infrastructure includes at the center of LSBB an electromagnetic shield capsule 8 m in diameter and 28 m long. The original experiments at LSBB include the use of [SQUID]2 detector for global and regional seismic-magnetic observations (Waysand, 2006), and the SIMPLE detector for dark matter search (Girard *et al.*, 2013).

LSBB is unique in practicing and promoting inter-disciplinary underground science and technology under low noise conditions. There are seismic sensors on the ground surface and along underground horizontal tunnel segments, with the planned MIGA tunnels for matter wave laser interferometer (Farah *et al.*, 2012), water and petroleum resource evaluation in a carbonic aquifer, HPPPCO2 project for fault zone characterization and CO2 sequestration evaluation (Jeanne *et al.*, 2012; Guglielmi *et al.*, 2012), and many other projects.

A step-rated injection method for fracture in-situ properties (SIMFIP) was accepted in 2013 as one of the ISRM Suggested Methods, as discussed in Guglielmi *et al.* (2013) (see also Wang *et al.*, 2013). Guglielmi (2015) described that the SIMFIP has been applied and planned to be deployed in studies of fault zones intersected in tunnels at LSBB and Tournemire URL in France and at Mont Terri Lab in Switzerland.

A France-South Africa network of SQUID toward global magnetic network for earthquake evaluation has been proposed (Febvre *et al.*, 2012; Fourie *et al.*, 2013). SQUID detectors are used to form toward global magnetic earthquake evaluation. Very long wavelength magnetic signals can detect earthquake signals from locations thousands of kilometers away. Underground installations of similar sensors in different URLs can form network for global evaluations. The i-DUST-2014 Proceedings presented many presentations on inter-disciplinary studies.

4.3 Examples of networking in mines, URLs, and DULs

Kaiser *et al.* (2011) and Duff & Kaiser (2015) discussed that a networked deep mine observatory could be formed from deep mines and from DULs, similar to what achieved among physics DULs. The Smart Underground Monitoring and Integrated Technologies (SUMIT) was planned for up to ten deep mines in Canada, Australia, and U.S.A., including in Canada the Creighton mine where SNOLab was excavated, NiRim South, Kidd mine, etc. SUMIT is part of the Center for Excellence in Mining Innovations (CEMI).

Sherwood Lollar (2013) described the microbiology studies in between Sweden's Aspo HRL, Finland's ONKALO, and Canada's SNOLab and many other URLs, DULs, and facilities worldwide contribute to understanding deep life in subsurface. Sherwood Lollar *et al.* (2014) described the global H$_2$ production and distribution in Precambrian continental lithosphere. Hydrogen, methane, and other fluids can be energy and food sources in the dark for underground microbes without photosynthesis processes. Holland *et al.* (2013) showed that oldest fracture fluid samples a billion years of age were collected in the Timmiins mine in Ontario, Canada, based on noble element isotope analyses. Deep levels in mines such as Creghton mine over 2. 5 km deep are of interests to the microbiology communities for deep life searches.

High pressure pockets were observed at CJPL in China at depth over 2 km. This may indicate that at such great depths, the hydrostatic-lithostatic transition occur and water pressure is no longer controlled by continuously-connected water paths from the surface to depths but by the high rock stressed states of rock formations. Perhaps deep facilities can be studied for this transition. The collection of uncontaminated water can contribute for understanding of deep life distributions.

4.4 Networking with underground nuclear power plant studies

Duffaut (2007, 2011, 2013) and Duffaut & Vasku (2014) pointed out that many hydropower plants were excavated underground worldwide; extensive studies occurred in 1960–1978 on underground nuclear power plants in Norway, Switzerland, France, Sweden, and studies in Canada, Three Mile Island, Chernobyl, and Fukushima were the accidents since these events that slow down the development of nuclear energy. There were 439 nuclear power reactors operated in 31 countries as at January 2013. Since some of reactors are over the typical 40 year design operation period, the future of nuclear energy may depend on siting the nuclear power plants underground, by excavated into hills near existing and licensed locations or in shallow depths to be covered by dirt to protect from natural hazards and terrorism. There also has to be a co-located source of cooling water. Perhaps underground sittings with dry cask waste storage near existing costal nuclear power plants will be one solution to nuclear energy and waste problems. Large caverns needed for underground nuclear power plants may be located near urban centers with high demands for electricity. Melnikov *et al.* (2014) pointed out the use of underground space could increase the safety of nuclear power plants. Varun & Fairhurst (2014) described seismic advantage of underground nuclear power plant. Sakurai (2014) presented case histories of seismic monitoring for underground tunnels in Japan after an earthquake.

5 SUMMARY

The ISRM URL Networking Commission, since its formation in 2011 after the URL Workshop in Beijing, China associated with the 12th ISRM Congress, has activities in 2012–2015 for meeting sessions, including the 4th URL Workshop in Montreal, Canada associated with the 13th ISRM Congress. We focus up to now on interacting with the communities and to collect presentations and publications related to underground studies in different URLs. These activities are summarized in the last three sections for an overview of available information for each URL's case history and planned development. Heater tests and quantifications of excavation damaged zones were discussed for radioactive waste URLs. Many deep underground laboratories for physics experiments plan to expansions in space. Some deep physics laboratories and facilities have additional earth and multi-disciplinary programs. We also present examples of energy/environmental, inter-disciplinary, and networking among URLs.

The URL Networking Commission currently relies on workshop and meetings organized in association with different rock mechanics, geosciences, underground physics, and multi-disciplinary conferences. In the foreseeable future, we plan to learn more about lessons learned from different URLs and uses of existing URLs as analogs for other related studies and/or at other sites. This effort will tailor to supplement many related presentations in the past Congresses and various meetings. We hope that we can in the long run contribute to better understanding of underground settings. Underground studies are critical for contributing to technologic advances. We hope that our periodic compilation of key references and presentations in key conferences contribute to better interactions and networking among different underground activities. We plan to continue gathering materials collected during studies conducted by the ISRM URL Commission to document for case histories and planned developments. Inputs from the Commissioners and Authors on lessons learned are crucial for the success for future activity for the continuance of the ISRM URL Commission.

ACKNOWLEDGMENTS

This manuscript has been authored by an author, J.S.Y. Wang, affiliated with the Lawrence Berkeley National Laboratory, which is under Contract No. DE-AC02-05CH11231 with the U.S. Department of Energy. The U.S. Government retains, and the publisher, by accepting the article for publication, acknowledges that the U.S. Government retains a non-exclusive, paid-up, irrevocable, worldwide license to publish or reproduce the published form of this manuscript, or allow others to do so, for U.S. Government purposes.

REFERENCES

Alley, W. M. & Alley, R. (2013a) Too hot to touch, 370 pages, Cambridge University Press.

Alley, W. M. & Alley, R. (2013b) Challenges in uncertainty and the science of nuclear waste disposal, Abstract U21A-03 presented at *2013 Fall Meeting, AGU*, San Francisco, CA, 9–13 December.

Aoyagi, K., Tsusaka, K., Nohara, S., Kubotsa, K., Kondo, K. & Inagaki, D. (2014) Hydrogeomechanical investigation of an excavation damaged zone in the Horonobe underground research laboratory, 8 pages, Paper 0127, *8th ARMS*, Sapporo, Japan, 14–16 October.

Armand, G. (2011) Status of the Meuse/Haute-Marne underground research laboratory, Presentation III-5, 20 slides, URL Workshop, *12th ISRM Congress*, Beijing, China, 17 October.

Armand, G., Derwonck, S., Bosgiraud, J.-M. & Richard-Panot, L. (2015a) Development and new research program in the Meuse Haute-Marne underground research laboratory (France), Paper 425, *13th ISRM Congress*, Montreal, Canada, 11–13 May.

Armand, G., Noiret, A. & Morel, J. (2015b) Pore pressure change during the excavation of deep tunn in the Callovo-Oxfordian claystone, Paper 820, *13th ISRM Congress*, Montreal, Canada, 11–13 May.

Asahina, D., Aoyagi, K., Tsusaka, K., Houseworth, J. E. & Birkholzer, J. T. (2014) Modeling damaged processes in laboratory tests at the Horonobe underground research laboratory, 9 pages, Paper 0372, *8th ARMS*, Sapporo, Japan, 14–16 October.

Barnichon, J. D. & Millard, A. (2012) Task A: The SEALEX in-situ experiments, 33 slides, *DECOVALEX-2015 Workshop*, Berkeley, CA, 17–19 April.

Bettini, A. (2007) The world underground scientific facilities, a compendium, 33 pages, *TAUP*.

Bettini, A. (2011) The world underground physics laboratories, Paper I–3, 20 pages, URL Workshop, *12th ISRM Congress*, Beijing, China, 17 October.

Bettini, A. (2013) New underground laboratories, 34 slides, *13th TAUP*, Asilomar, CA, 13 September.

Blechschmidt, I. & Martin, A. J. (2012) In-situ tracer tests and models developed to understand flow paths in a shear zone at the Grimsel test site, Switzerland, Abstract H33J-1472 presented at *2012 Fall Meeting, AGU*, San Francisco, CA, 3–7 December.

Blümling, P., Aranyossy, J.-F., Jing, L., Li, X. L., Marschall, P., Rothfuchs, T. & Vieter, T. (2008) Disturbed and damaged zones around underground openings – effects induced by construction and thermal loading, 20 Pages, *EURAWASTE*.

Brinkman, W. F. (2012) Energy Independence with sustainability, Abstract U24A-01 presented at *2012 Fall Meeting, AGU*, San Francisco, CA, 2 December.

Chen, C. S. & Jeon, S. W. (2015) KAERI Underground Research Tunnel, 32 slides, 4th URL Workshop, *13th ISRM Congress*, Montreal, Canada, 10 May.

Cockell, C., Paling, S., Pyrus, D., McLuckie, D., *et al.* (2012) Boulby international subsurface astrobiology laboratory (BISAL), 22 slides, *ASPERA-2*, Duhram, 17–19 December.

Diaz, J. (2012) GEODYN: A geodynamic facility at Canfranc laboratory, 52 slides, *ASPERA-2*, Duhram, UK, 17–19 December.

Dib, C. (2013) ANDES Agua Negra deep experiment site, 18 slides, *13th TAUP*, Asilomar, CA, 13 September.

Dighe, A. (2013) Physics goal and status of INO, 41 slides, *DAE-BRNS HEP Symposium*, Santiniketan, India, January.

Doe, T. (2011) Thirty years of contributions to fracture flow from underground radioactive waste research laboratories, 14 slides, 3rd URL workshop, *12th ISRM Congress*, 17 October.

Duff, D. & Kaiser, P. K. (2015). Mined network observatories – challenges 34 slides, 4th URL workshop, *13th ISRM Congress*, 10 May.

Duffaut, P. (2007) Safe nuclear power plants shall be built underground, 207–212, *11th ACUUS Conference*, Athens, 10–13 September.

Duffaut, P. (2011) Large caverns, design and construction, Paper I-3, 4 pages, URL Workshop, *12th ISRM Congress*, Beijing, China, 17 October.

Duffaut, P. (2013) State of art of underground nuclear power plant, Underground Nuclear Power Plant/URL meeting, EUROROCK, Wroclaw, Poland, 22 September.

Duffaut, P. & Vaskou, P. (2014) Geological and geographic criteria for underground siting of nuclear reactors, Paper 148, *8th ARMS*, Sapporo, Japan, 14–16 October.

Dusseault, M. (2011) Deep sedimentary basin solids placement, Presentation IV-4, 34 slides, URL Workshop, *12th ISRM Congress*, Beijing, China, 17 October.

Dusseault, M. (2013) Hydraulic fracturing: Theory – reality – uncertainty, *Distinguished Scientist Lecture*, Berkeley, 15 March.

Elsworth, D. (2011) Roles of URLs in exploring complex process feedbacks related to energy recovery and sequestration of energy wastes. Presentation V-1, 29 slides, URL Workshop, *12th ISRM Congress*, Beijing, China, 17 October.

Elsworth, D. (2012) DUSEL interdisciplinary science studies: Geoscience and geoengineering @ DUSEL, Presentation V-1, 27 slides, *ASPERA-2*, Duhram, UK, 17–19 December.

Elsworth, D. (2013) Resolving key uncertainties in energy recovery: On role of in-situ experimentation and URLs, Abstract H31G-09 presented at *2013 Fall Meeting, AGU*, San Francisco, 9–13 December.

Elsworth, D., Morone, C., Taran, J., Izadi, G., Gan, Q., Zhong, Z., Fang, Y. & Candella, T. (2015), Role of URLs in probing controls on induced seismicity and related permeability evolution, 36 slides, 4th URL Workshop, *13th ISRM Congress*, Montreal, Canada, 10 May.

Elsworth, D., Smeallie, P. & Heuze, F. (2003) Workshop on Deep Underground Science and Engineering Laboratories (DUSELs), 66p., in conjunction with the *10th ISRM Congress*, Johannesburg, South Africa, 11 September.

Fahland, S. & Heusermann, S. (2013) Geomechanical analysis of the integrity of waste disposal areas in the Morsleben repository, in rock characterization, modeling and engineering design methods, edited by Feng, X.-T., Hudson, J. A. Hudson, and F. Tan, 345–350, *the 3rd ISRM SINOROCK Symposium*, Shanghai, China, 18–20 June.

Fahland, S., Heusermann, S., Schafers, A., Behlau, J. & Hammer, J. (2015a) Three-dimensional geological and geomechanical modeling of a repository for waste disposal in deep rock salt formations, Paper 121, *13th ISRM Congress*, 11–13 May.

Fahland, S., Vogel, P. & Heusermann, S. (2015b) Geoscientific exploration of a candidate waste disposal site in rock salt, 31 slides, 4th URL Workshop, *13th ISRM Congress*, Montreal, Canada, 10 May.

Fairhurst, C. (2004) Nuclear waste disposal and rock mechanics: Contributions of the underground research laboratory (URL), Pinawa, Manitoba, Canada, Int. J. Rock Mech. Min. Sci., 37(1–2), 1221–1227.

Fairhurst, C. (2015) Mechanics and rock, *Winter 2015 ARMA e-Newsletter*, pp. 7–19.

Farah, T., Guerlin, C., Merlet, S., Landragin, A., Dos Santos, F. P., Bouyer, P., Auguste, M., Cavaillou, A., Boyer, D.,Poupeney, J., Sudre, C. & Gaffet, S. (2012) Absolute atomic gravimeter at LSBB, a first step to MIGA, 21 slides, *i-DUST Conference*, Apt, France, 9–11 May.

Febvre, P., Pozzo de Borgpo, E., Chambodut, A., Fourie, C., Saunderson, E., Gouws, D., Bernard, S., Cavillou, A., Poupeney, J., Boyer, D. & Sudre, C. (2012) Towards an international network of superconducting magnetometers for geomagnetic and earth-ionosphere studies, 1 page, *i-DUST Conference*, Apt, France, 9–11 May.

Feng, X.-T. (2011) From Sweden to China's competent, sparsely fractured sites and deep tunnels, 39 slides, Presentation I-2, URL Workshop, *12th ISRM Congress*, Beijing, China, 17 October.

Feng, X.-Y. (2014) Tunnelling and underground construction in China: Challenges and progress, 50 slides, *ASRM-8*, Sapporo, Japan, 14–16 October.

Feng, X.-T. (2015) Rock mechanics associated with deep and large excavations in China, 26 slides, 4th URL Workshop, *13th ISRM Congress*, Montreal, Canada, 10 May.

Feng, X.-T., Li, S., Zhang, C., Chen, B. & Zhou, H. (2012) Evolution and control of excavation damaged zone of large tunnels at overburden of 1900-2525m, ARMA-11-553, 7 pages, *45th ARMA*, San Francisco, 26–29 June.

Fourie, C., Van Vuuren, J., Lockner, E. T., Kwisanga, C., Febvre, P., Pozzo de Borgo, E., Waysand, G., Gouws, D. Sauderson, E., Matladi, T. & Henry, S. (2013) Field demonstration of transcontinental SQUID magnetometry, Abstract NH31B-1606, presented by S. Henry at *2013 Fall Meeting*, *AGU*, San Francisco, 9–13 December.

Gaffet, S. (2012) LSBB underground research laboratory a new platform for fundamental & applied low background noise inter-disciplinary underground science & technology, 20 slides, *ASPERA-2*, Duhram, UK, 17–19 December,

Gaffet, S. (2013) A new platform for fundamental & applied low background noise interdisciplinary underground science & technology, 12 slides, URL Meeting, *EUROROCK*, Wroclaw, Poland, 22 September.

Gaus, I. (2012) Introduction to Task B of DECOVALEX 2015: HE-E experiment: In-situ heater test, 38 slides, *DECOVALEX-2015 Workshop*, Berkeley, CA, 17–19 April.

Girard, T. A., Ramos, A. R., Roche, I. L. & Fernandes, A. C. (2014) Phase III (and may IV) of the SIMPLE dark matter search experiment at LSBB, 6 pages, *5th iDUST Conferences*, E3S Web of Conferences, 4, 01001, Apt, France, 5–7 May.

Gratta, G. (2013) Double beta decay: A very important experiment, 33 slides, *DURA meeting*, Palo Alto, 5 March.

Guglielmi, Y. (2011) Hydro-mechanical coupled tests and induced seismicity, Presentation V-4, 15 slides, URL Workshop, *12th ISRM Congress*, Beijing, China, 17 October.

Guglielmi, Y. (2015) Hydro-mechanical experiments in fault zones in shales and carbonate rocks, *13th ISRM Congress*, Montreal, Canada, 10 May.

Guglielmi, Y., Cappa, F., Derode, B., Jeanne, P. & Rutqvist, J. (2012) Underground testing of permeability and earthquake nucleation in fault zones, Abstract H32G-05 presented at *2012 Fall Meeting, AGU*, San Francisco, CA, 3 December.

Guglielmi, Y., Cappa, F., Lancon, H., Janowczyk, J. B., Rutqvist, J., Tsang, C. F. & Wang, J. S. Y. (2013) ISRM suggested method for step-rate injection method for fracture in-situ properties (SIMFIP): Using a 3-components borehole deformation sensor, Rock Mech. Rock Eng., 47(1), 301–311.

Hakami, E. & Christensson, R. (2012) Using induced borehole breakouts as a method for stress orientation determination in hard crystalline rocks, *ISRM EUROCK*, Stockholm, 28–30 May.

Heusermann, S., Eickemeier, R., Nipp, N. K. & Fahland, S. (2013) Three-dimensional Thermomechanical modeling of high-level emplacement in a salt dome, 647–652, in Rock characterization, modeling and engineering design methods, edited by X.-T. Feng, J. A. Hudson, and F. Tan, *3rd ISRM SINOROCK Symposium*, Shanghai, China, 18–20 June.

Hitzman, M. (2012) Necessity for industry-academic economic geology collaborations for energy critical minerals research and development, Abstract H32G-02 presented at *2012 Fall Meeting, AGU*, San Francisco, CA, 3 December.

Hitzman, M., Tester, J. W. & Zoback, M. D. (2012) Opportunities for fundamental university-based Research in Energy and Resource Recovery, Abstract H32G-06 presented at *2012 Fall Meeting, AGU*, San Francisco, CA, 3 December.

Hokr, M. (2012) Bedrichov tunnel test case proposal for DECOLALEX 2015, 52 slides, *DECOVALEX-2015* Workshop, Berkeley, CA, 17–19 April.

Holland, G., Sherwood Lollar, B., Li, L., Lacrampe-Couloumme, Slater, G. F. & Ballentine, C. J. (2013) Deep fracture fluids isolated in the crust since Precambrian era, Nature, 497, 357–360.

Hudson, J. A. (2010) Underground radioactive waste disposal: The rock mechanics contributions, ISRM International Symposium, *6th ARMS*, New Delhi, 23–27 October.

Hudson, J. A. (2015a) Validating computer programs: A description of the international DECOVALEX Project, *Winter 2015 ARMA e-Newsletter*, pp. 3–6.

Hudson, J. A. (2015b) Contributions of rock mechanics to nuclear waste disposal, 27 slides, 4th URL Workshop, *13th ISRM Congress*, 10 May.

Hudson, J. A., Andersson, J. C., Jing, L. & Tsang, C.-F. (2011) Studies of coupled THMC problems in the DECOVALEX project with support from URL field experiments, 18 pages, Paper V-2, URL Workshop, *12th ISRM Congress*, Beijing, China, 17 October.

Hudson, J. A., Wang, J. S. Y. & Jing, L. (2013) Value of underground research laboratories in reducing rock engineering risk, 33 slides, URL Meeting, *3rd ISRM SINROCK Symposium*, Shanghai, China, 17 June.

Ianni, A. (2015) Laboratorie Souterrain de Canfrac, 7slides, 4th URL Workshop, *13th ISRM Congress*, Montreal, Canada, 10 May.

Jeanne, P., Guglielmi, Y. & Cappa, F. (2012) Dissimilar properties of a small fault zone in a carbonate reservoir and their impact on the pressurization and leakage associated with CO_2 injection, 26 slides, *i-DUST*, Apt, France, 9–11 May.

Jung, C. K. (2015) T2K experiment and long baseline neutrino studies worldwide, 4th URL Workshop, *13th ISRM Congress*, Montreal, Canada, May 10.

Jung, C. K. (2015) Very large underground detectors for neutrino physics and nucleon decay searches: Recent discovery of eletron neutrino appearance from a muon beam in T2K and future outlook for discovery of CP violation in the lepton sector, 28 slides, 4th URL Workshop, *13th ISRM Congress*, Montreal, Canada, 10 May.

Johansson, E., Siren, T. & Kemppainen, K. (2015) ONKALO – underground rock characterization facility for in-situ testing for nuclear waste repository, 16 slides, 4th URL Workshop, *13th ISRM Congress*, 10 May.

Johansson, E., Siren, T. & Kemppainen, K. (2015) ONKALO – underground rock characterization facility for in-situ testing for nuclear waste repository, Paper 506, *13th ISRM Congress*, 11–13 May.

Kaiser, P. K., Duff, D. & Valley, B. (2011). Smart underground monitoring and integrated technologies in networked deep mining observatories, Paper IV-3, 6 pages, Presentation 14 slides, URL Workshop, *12th ISRM Congress*, Beijing, China, 17 October.

Laughton, C. (2011) Underground research laboratories: Siting criteria and engineering design considerations a rock engineer's top ten list, Presentation I-4, 16 slides, 3rd URL workshop, *e*, Beijin, China, 17 October.

Laughton, C. (2015) Another engineering top ten.. cost-effective development of URL facilities., 16 slides, 4th URL workshop, *13th ISRM Congress*, Montreal, Canada, 10 May.

Lee, C. S. & Jeon, S. K. (2015) Current status of KURT and its long-term research programme, 4th URL workshop, *13th ISRM Congress*, Montreal, Canada, 10 May.

Lebon, P. (2012) Bure underground research laboratory: Role in the radwaste management and 2012–2015 programme, 13 slides, *i-DUST*, Apt, France, 9–11 May.

Lesko, K. T. (2013) Update on progress in establishing research efforts at the Sanford underground research facility, 41 slides, *DURA meeting*, Palo Alto, March 5.

Li, J. M. (2011a) The status and plan of China JinPing underground lab (CJPL), Presentation II-2, 25 slides, 14 slides, URL Workshop, *12th ISRM Congress*, Beijing, China, 17 October.

Li, J. M. (2013) China JinPing Underground Lab. Radiation Background Measure, 26 slides, URL Meeting, *3rd ISRM SINOROCK Symposium*, Shanghai, China, 17 June.

Li, J. M. (2015a) New site for physics China Jinping underground laboratory, 38 slides, 4th URL Workshop, *13th ISRM Congress*, Montreal, Canada, 10 May.

Li, J. M., Ji, X., Haxton, W. & Wang, J. S. Y. (2013a). The second phase development of the China JinPing underground laboratory for physics rare-event detectors and multidisciplinary sensors, 33 slides, 13th TAUP, Asilomar, CA, 12 September.

Li, S. J. (2015b) Comprehensive monitoring of rock behavior for the deep underground laboratory of CJPL-II, 4th YRL workshop, *13th ISRM Congress*, Montreal, Canada, 10 May.

Li, S. J., Feng, X.-T. & Hudson, J. A. (2013b). ISRM suggested method for measuring rock displacement using a rock sliding micrometer, *Rock Mech. Rock Eng.*, 46(3), 645–653.

Li, S. J., Feng, X.-T., Li, Z., Chen, B., Zhang, C. & Zhou, H. (2012a) In-situ monitoring of rock burst nucleation and evolution in the deeply buried tunnel of JinPing II hydropower station, *Eng. Geol.*, 137–138, 85–96.

Li, S. J., Feng, X.-T., Li, Z., Zhang, C. & Chen, B. (2013c) Evolution of fractures in the excavation damaged zone of a deeply buried tunnel during TBM excavation, *Int. J. Rock Mech. Min. Sci.*, 25, 215–238.

Li, S. J., Feng, X.-T., Wang, C. Y. & Hudson, J. A. (2013d) ISRM suggested method for rock fractures observations using a borehole digital optical televiewer, *Rock Mech. Rock Eng.*, 46(3), 635–644.

Li, X. L. (2011b) The roles of the URL HADES in the Belgian RD & D program on radwaste geological disposal, Presentation III-6, 26 slides, URL Workshop, *12th ISRM Congress*, Beijing, China, 17 October.

Li, X. L., Chen, G. J., Dizier, A., Leysens, J., Troullinos, I. Verstricht, J., Bastiaens, W., Levasseur, S., Van Marcke, P. & Sillen, X. (2015a) The PARCLAY experiment at URL HADES in Mol, Belgium, 28 slides, 4th URL Workshop, *13th ISRM Congress*, Montreal, Canada.

Li, X. L., Chen, G. J., Verstricht, J., Troullinos, I., Dizier, A., Leysens, J. & Van Marcke, P. (2015b) The PARCLAY experiment at URL HADES, Mol, Belgium, Paper 213, *13th ISRM Congress*, Montreal, Canada.

Li, X. L., Volckaert, G. & Bastiaen, W. (2012b) R & D at the underground research laboratory, HADES, Belgium and the State of Art on THMC Behavior of Boom Clay, Abstract H33J-1471 presented at *2012 Fall Meeting*, *AGU*, San Francisco, CA, 3–7 December.

Link, J. for Vogelaar, B.(2013) Kimballton underground research facility (KURT), 35 slides, *Deep Underground Research Association (DURA) Meeting*, Palo Alto, 5 March.

Marshak, M. (2013) The Soudan underground laboratory, 37 slides, *DURA Meeting*, Palo Alto, 5 March.

Martin, C. D., Lanyon, G. W., Blümling, P. & Mayer, J.-C. (2003) The excavation disturbed zone around a test tunnel in the Opalinus clay. 121–125, Proc. *European Commission CLUSTER Conference*, Luxemburg, 3–5 November.

Martin, C. D., Read, R. S. & Martino, J. B. (1997) Observations of brittle failure around a circular test tunnel, *Int. J. Rock Mech. Min. Sci*, 34(7), 1065–1073, PII: S0148–9062(97) 00296–9

McNutt, M. K. (2012) The frontiers of resource-related scientific research, Abstract U24A-02 presented at *2012 Fall Meeting*, *AGU*, San Francisco, CA, 3 December.

Melnikov, N. N., Konukhin, V. P., Naumov, V. A. & Gusak, S. A. (2014) Use of underground space to increase safety of nuclear power engineering, Paper 231, *8th ARMS*, Sapporo, Japan, 14–16 October.

Nuijten, G. (2013) LUGANA, 30 slides, URL Meeting, *EUROROCK*, Wroclaw, Poland, 22 September.

Nuijten, G. (2014) Large underground experiments: Engineering point of view Pyhasalmi + Homestake, 58 slides, *15th International Workshop on Next Generation Nucleon Decay and Neutrino Detector (NNN14)*, Paris, France, 4–6 November.

Nussbaum, C. O. & Bossart, P. J. (2012) Mont Terri underground rock laboratory, Switzerland – Research program and key results, Abstract H32G-06 presented at *2012 Fall Meeting*, *AGU*, San Francisco, CA, 3 December.

Paling, S. M. (2015) Boulby underground laboratory, 7 slides, 4th URL Workshop, *13th ISRM Congress*, Montreal, Canada, 10 May.

Piquemal, F. (2013) Activities at Modane underground laboratory, *13th TAUP*, Asilomar, CA, 12 September.

Piquemal, F. (2015) Deep underground laboratories in Europe, 63 slides, 4th URL Workshop, 13th ISRM Congress, Montreal, Canada, 10 May.

Ragazzi, S. (2015) Laboratori Nationali del Gran Sasso (LNGS), 7 slides, 4th URL Workshop, 13th ISRM Congress, Montreal, Canada, 10 May.

Rutqvist, J. (2011) Geomechanical aspects of CO_2 sequestration and modeling, Presentation IV-3, 21 slides, URL Workshop, *12th ISRM Congress*, Beijing, China, 17 October.

Sakurai, S. (2014) Case studies on the dynamic behavior of tunnels caused by hyogoken Nanbu earthquake whose epicenter was very close to the tunnels, Paper 189, *8th ARMS*, Sapporo, Japan, 14–16 October.

Shephard, L. (2013) Developing America's shale reserve – Water strategies for a sustainable future, Abstract U21A-02 presented at *2013 Fall Meeting*, *AGU*, San Francisco, CA, 9–13 December.

Sherwood Lollar, B., Onstott, T. C., Lannampe-Couloume, G. & Ballentine, C. J. (2014) The contribution of the Precambrian continental lithosphere to global H_2 production, *Nature*, 516, 379–382.

Sherwood Lollar, B., Onstott, T. C., Van Heerden, E., Kieft, T. L. & Ballentine, C. J. (2013) Life in inner space: Subsurface microbiology investigations in underground research laboratories and deep mines, *AGU meeting*, San Francisco, CA, 9–13 December.

Siren, T. & Johansson, E. (2011) ONKALO – On-site, second generation URL: Its role in POSIVA's repository development program, Paper III-2, 6 pages, URL Workshop, *12th ISRM Congress*, Beijing, China, 17 October.

Smith, N. (2013) Status update of non-U.S. deep underground facilities, 42 slides, *DURA meeting*, Palo Alto, 5 March.

Smith, N. presented by Lawson, I. (2015) Status of non-U.S. underground facilities, 32 slides, 4th URL Workshop, *13th ISRM Congress*, Montreal, Canada, 10 May.

Sneyers, A. & Li, X. (2012) Update on the Belgian programme, *IAEA URL Network Partners and Annual Meeting*, Albuquerque, NM, 3–7 December.

Sosna, K., Franek, J., Vondroveic, L. & Zaruba, J. (2014) Stresses and strains induced by in-situ thermal loading in granitic massif, Paper 458, 9 pages, *8th ARMS*, Sapporo, Japan, 14–16 October.

Sugita, Y. & Nakama, S. (2012) Task B EBS Horonobe experiment, 34 slides, *DECOVALEX-2015 Workshop*, Berkeley, CA, 17–19 April.

Svemar, C., Christansson, R., Ohlsson, M., Eng, T. & Laaksoharju, M. (2011) SKB's hard rock laboratory – Current and possible future activities, Paper III-3, 7 pages, URL Workshop, *12th ISRM Congress*, Beijing, China, 17 October.

Tester, J. W. (2012) The potential of geothermal as a major supplier of U.S. primary energy Using EGS Technology, Abstract H32G-01 presented at *2012 Fall Meeting*, *AGU*, San Francisco, CA, 3 December.

Tsang, Y. W., Birkholzer, J. T. & Mukhopadhyay, S. (2009) Modeling of thermally-driven hydrological processes in partially saturated fractured rocks, 30 pages, *Rev. Geophys.*, 47(3), RG3004.

Tsusaka, K., Sugita, Y., Inagai, D., Nakayama, M., Yabuuchi, S., Yokata, H. & Tokiwa, T. (2011) Japan investigation of rock mechanics at Horonobe underground research laboratory project – Present status and future plan, 8 pages, Paper III-4, URL Workshop, *12th ISRM Congress*, Beijing, China, 17 October.

Tutuncu, A. N., presented by Gutierrez, M. (2011) Shale gas and shale oil reservoir site characterization and fluid injection/withdrawal assessment, 9 slides, Presentation IV-2, URL Workshop, *12th ISRM Congress*, Beijing, China, 17 October.

Uhle, J. L., presented by Anooshepoor, R. (2012) External hazard research at the U.S. Nuclear Regulatory Commission, Abstract U24A-03 presented at *2012 Fall Meeting*, *AGU*, San Francisco, CA, 3 December.

Vardiman, D. (2012) Sanford lab geotechnical state investigations, 18 slides, A*GU meeting*, San Francisco, CA, 7 December.

Varun, M. P., & Fairhurst, C. (2014) Underground nuclear power plants: The seismic advantage, Paper 363, 7 pages, *8th ARMS*, Sapporo, Japan, 14–16 October.

Vogel, P., Fahland, S. & Furche, M. (2015) Temperature field in the exploration area of the Gorleben site – geothermal borehole logging and numerical modeling model calculations, paper 296, *13th ISRM Congress*, Montreal, Canada, 11–13 May.

Vomvoris, S. (2011) Overview and highlights of the activities in the two international URLs in Switzerland, 13 p., Paper III-1, URL Workshop, *12th ISRM Congress*, Beijing, China, 17 October.

Wakimoto, R. M. (2013) Energy infrastructure and extreme events, Abstract U21A-01 presented at *2013 Fall Meeting*, *AGU*, San Francisco, CA, 9–13 December.

Wang, J. S. Y., Feng, X.-T. & Hudson, J. A. (2015) Evolution and progress in underground research laboratories, development facilities, and multi-disciplinary studies, 10 pages, 32 slides, *13th ISRM Congress*, Montreal, Canada, 10 May.

Wang, J. S. Y., Guglielmi, Y., Hudson, J. A. & Feng, X.-T. (2013) Progress in interactions among underground research laboratories, ARMA 13-718, 8p., *47th ARMA Symposium*, San Francisco, CA, 23–26 June.

Wang, J. S. Y. & Hudson, J. A. (2012) Laboratory and field studies of fracture flow and its extension in underground settings, Abstract H14A-03 presented at *2012 Fall Meeting*, *AGU*, San Francisco, CA, 3 December.

Wang, J. S. Y., Smeallie, P. H., Feng, X.-T. & Hudson, J. A. (2010). Evaluation of underground research laboratory for formulation of interdisciplinary global networks, 18 pages, *6th ARMS*, New Delhi, 23–27 October.

Wang, Ju (2004) Geological disposal program of high-level radioactive waste in China, 60 slides, *TCM*, Vienna, 6–10 December.

Wang, Ju (2014) On area-specific underground research laboratory for geological disposal of high-level radioactive waste in China, *J Rock Mech. Geotech. Eng.*, 6, 99–104.

Wang, Ju (2015) A quick review of the international workshop on rock mechanics in nuclear waste disposal, 40 slides, 4th URL Workshop, *13th ISRM Congress*, 10 May.

Wang, Ju & Chen, L. (2015) Area-specific URL for geological disposal in China, 48 slides, 4th URL Workshop, *13th ISRM Congress*, 10 May.

Waysand, G. (2006) The low noise underground laboratory at Rustrel-Pays d'Apt, *J. Phys. Conf. Ser.*, 39, 157–159.

Waysand, G., Pozzo di Borgo, E., Soule, S., Pyee, M., Marfaing, J., Yedlin, M., Blancon, R., Barroy, P. & Cavaillou, A. (2010) Azimuthal analysis in [SQUID]^2 for mesopause and sprites excitations. 2 pages, *i-DUST Conference*, Apt, France.

Witherspoon, P. A., Cook, N. G. W., & Gale, J. E. (1980) Geologic storage of radioactive waste: Field studies in Sweden, *Science*, 211, 894–900.

Wolfgango, P. (2012) Environmental radioactivity measurement for earth sciences @ INFN Gran Sasso National Laboratory, Italy, *ASPERA-2 Meeting*, Durham, UK, 18 December.

Yamatomi, J. (2013) The Hyper-KAMIOKANDE project in the Kimioka mine, Japan, 15 slides, URL Meeting, *3rd ISRM SINOROCK Symposium*, Shanghai, China, 17 June 2013.

Young, R. P. & Collins, D. S. (2001) Seismic studies of rock fracture at the underground research laboratory, Canada, *Int. J. Rock Mech. Min. Sci.*, 36(1–2), 787–799.

Young, R. P., Collins, D. S. & Reyes-Montes, J. M. (2004) Induced seismicity at the Canadian underground research laboratory 1987–2004. Abstract S41B-0963, *AGU*.

Zhang, C., Feng, X.-T., Zhou, H. Qui, S. & Wu, W. (2012) Case histories of four extremely intense rock bursts in deep tunnels, *Rock Mech. Rock Eng.*, 45, 275–288.

Chapter 15

The Mont Terri rock laboratory

Paul Bossart[1], Fabrice Burrus[2], David Jaeggi[1] & Christophe Nussbaum[1]
[1]*Swiss Geological Survey at swisstopo, Wabern, Switzerland*
[2]*GGT, Delémont, Switzerland*

Abstract: The Mont Terri Project is an international research project to investigate the hydrogeological, geochemical and geotechnical characterization of the Opalinus Clay formation. Clay formations are being considered in various countries for deep geological disposal of radioactive waste. Our research is carried out in the generic Mont Terri rock laboratory, an underground facility near the security gallery of the Mont Terri highway tunnel in northwestern Switzerland. Sixteen partners from eight countries participate in this project. The experiments are designed to better understand the processes and mechanisms in undisturbed argillaceous formations. These include experiments related to excavation- and repository-induced perturbations and to repository performance during operational and post-closure phases. Here we outline a timescale for key processes affecting repository performance and relate the ongoing experiments to this timescale. The main focus is placed on the applied in-situ testing methods and the corresponding equipment and materials. We start with excavation and drilling techniques, then proceed to hydrogeological, geochemical and geophysical experiments. Finally, a recent mine-by experiment is presented where acoustic emission signals were used to identify the temporal and spatial evolution of the excavation damaged zone.

I INTRODUCTION

The Mont Terri rock laboratory is located in the security gallery of the Mont Terri motorway tunnel in Northwest Switzerland (Figure 1a). This tunnel belongs to the A16 highway connecting Switzerland with France. The Mont Terri rock laboratory is located 300 m below the surface in the Opalinus Clay, a claystone formation of lower Jurassic age. In total 700 m of tunnels and niches have been excavated from 1996 to 2012 (Figure 1).

The aim of the Mont Terri research program is to analyze hydrogeological, geochemical, and rock-mechanical properties of the Opalinus Clay. In particular, we investigate the changes induced by excavation of galleries and by heating of the formation, including testing of sealing and canister emplacement techniques, and evaluating and improving appropriate investigation techniques and methodologies.

The research in the Mont Terri rock laboratory is being conducted and financed by sixteen organizations from five European countries, Canada, Japan, and the United States. These comprise ANDRA (Agence Nationale pour la Gestion des Déchets

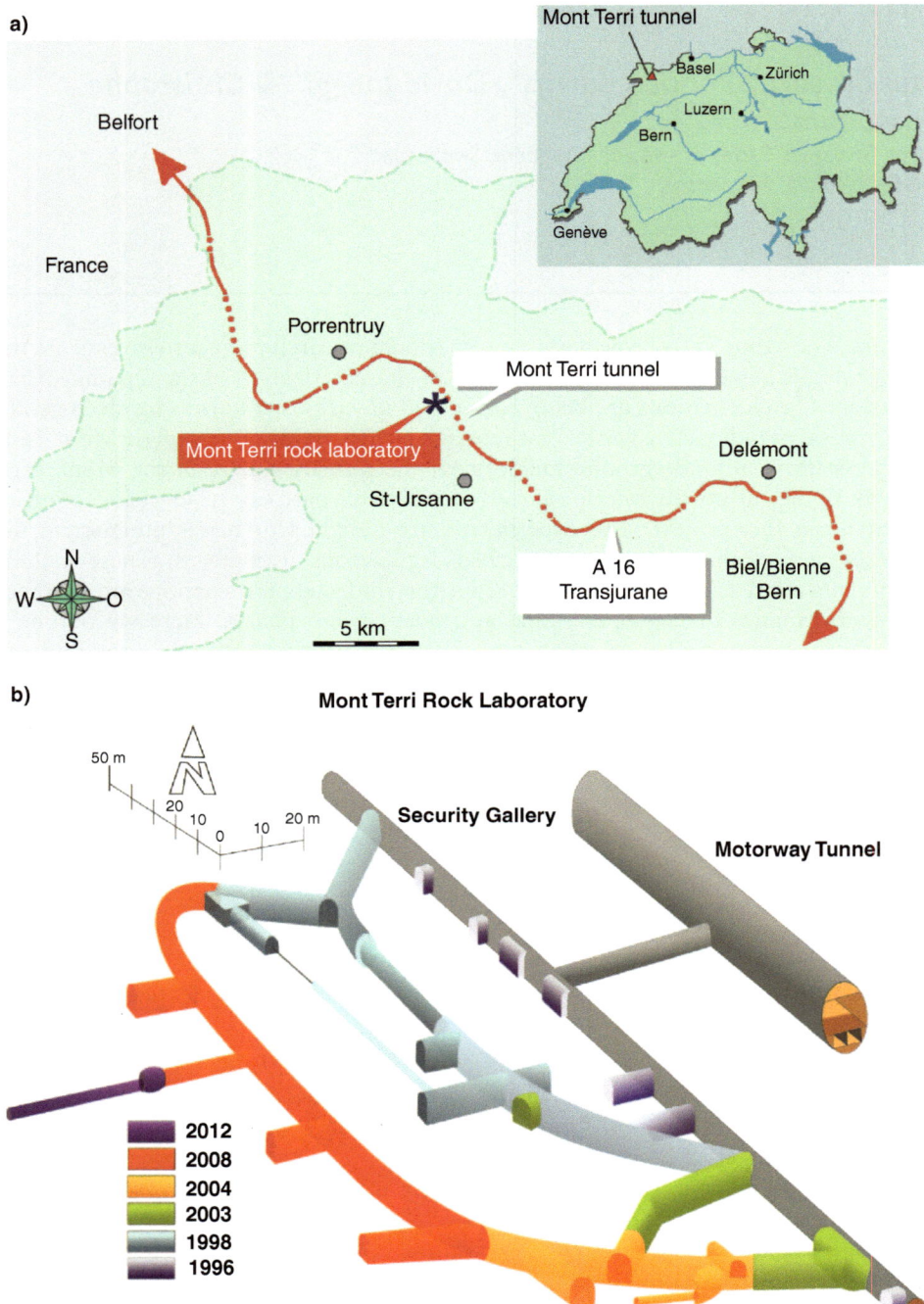

Figure 1 The Mont Terri rock laboratory a) geographic location and b) perspective view.

Radioactifs, France), BGR (Bundesanstalt für Geowissenschaften und Rohstoffe, Germany), CHEVRON (Chevron Energy Technology Company, Houston, USA), CRIEPI (Central Research Institute of Electric Power Industry, Japan), DOE (Department of Energy, Washington D.C., USA), ENRESA (Empresa Nacional de Residuos Radiactivos, Spain), ENSI (Swiss Federal Nuclear Safety Inspectorate), FANC (Federaal Agentschap Voor Nucleaire Controle, Belgium), GRS (Gesellschaft für Anlagen- und Reaktorsicherheit mbH, Germany), IRSN (Institut de Radioprotection et de Sûreté Nucléaire, France), JAEA (Japan Atomic Energy Agency), NAGRA (National Cooperative for the Disposal of Radioactive Waste, Switzerland), NWMO (Nuclear Waste Management Organisation, Toronto, Canada) OBAYASHI (Obayashi Corporation, Japan), SCK·CEN (Studiecentrum voor Kernenergie, Belgium) and the Swiss Geological Survey at SWISSTOPO (Federal Office of Topography), the latter is the rock laboratory operator and is also responsible for the underground safety. The owner of the underground facilities is the Republic and Canton of Jura (RCJU).

In the first chapter, we present the geological setting in relation to the possible performance of a repository for deep geological disposal in relation with the ongoing experiments. In the second chapter, we outline the in-situ testing methods, show compilations in tables and present some selected examples in more detail. In the third chapter, we describe the excavation damaged zone and discuss a recent mine-by test in more detail.

1.1 Geological setting

A geological profile along the tunnels of Mont Russelin and Mont Terri is shown in Figure 2a. The Mont Terri anticline was formed between 10 and 5 million years ago. This anticline was sheared off and thrust over the Tabular Jura. The Opalinus Clay is thus accessible by an existing horizontal security tunnel; a vertical shaft was not necessary, thus reducing access costs considerably. As shown in Figure 2b, the Opalinus Clay can be grouped into three main facies, a shaly facies, a sandy facies, and a carbonate-rich sandy facies (Nussbaum et al., 2011). These three facies are the result of different sedimentary environments in a shallow marine basin during the time of deposition. A larger fault zone, called the main fault, was encountered in the center of the Opalinus Clay.

The Opalinus Clay consists mainly of sheet silicates, framework silicates, and carbonates. The qualitative mineralogical composition is identical in all facies that occur at Mont Terri and includes calcite, dolomite, ankerite, siderite, quartz, albite, K-feldspar, muscovite, illite, illite/smectite mixed-layer phases, chlorite, kaolinite, pyrite, and organic carbon. Typical values for the shaly facies are: 66% clay minerals, 16% carbonates, 14% quartz, 2% feldspars, 1.1% pyrite and 0.8% organic carbon. Compared to the shaly facies, the sandy facies contains more quartz (25%) and less clay minerals (50%). Locally, there may be a considerable variability of these mineralogy values. Key parameters of the shaly and sandy facies are shown in Table 1.

We found two types of discontinuities: 1) tectonic fractures and 2) EDZ (excavation damaged zone) fractures. The first type of discontinuities, tectonic fractures, are shown in Figure 2b, consisting of a main fault and numerous single faults, the latter dispersed over the whole formation. The main fault has a thickness between 0.5–3 m, a lateral extent of several tens to a few 100s of m and is characterized by a large number of fault

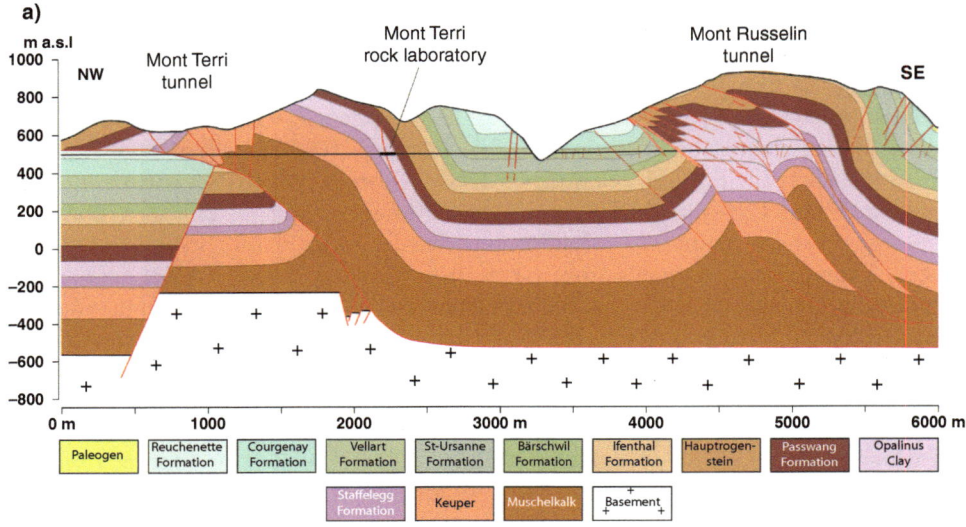

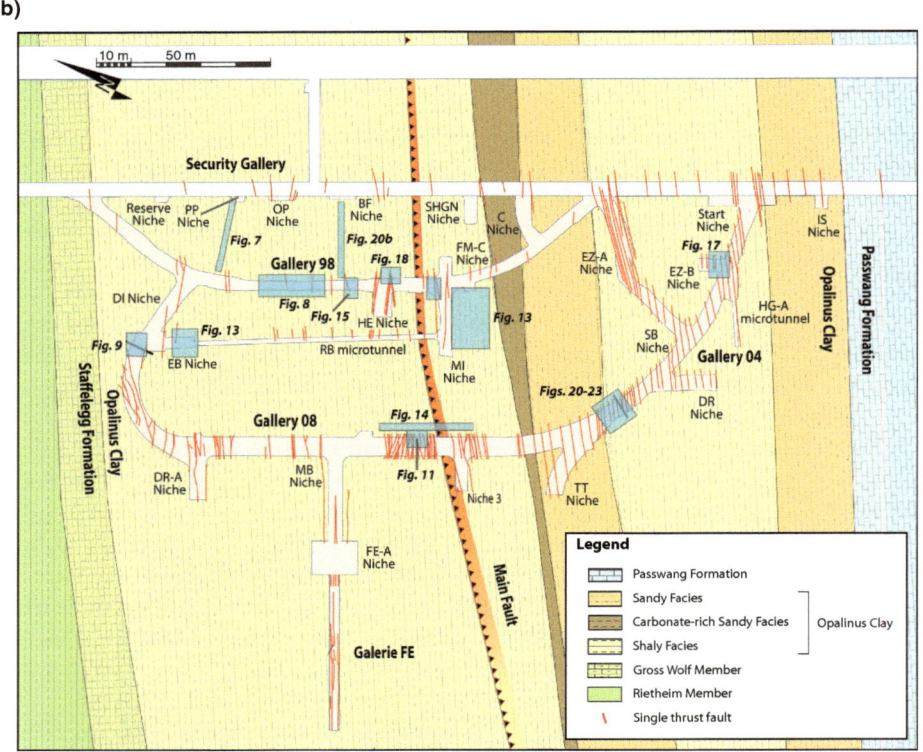

Figure 2 Geology of the Mont Terri a) profile along the Motorway tunnel (Caer *et al.*, 2015) and b) horizontal map of the rock laboratory. The figures of the different experiments presented on the following pages are indicated as rectangles in this figure together with the figure number.

Table 1 Key parameters of Opalinus Clay, derived in the Mont Terri rock laboratory (Jaeggi et al., 2014). The ranges and best estimates are presented for the shaly and sandy facies, respectively. Anisotropy is mentioned as P and N (parallel and normal to bedding, respectively). The number of analyses are shown in parenthesis.

Parameters	Shaly facies		Sandy facies	
	Range	Best estimate	Range	Best estimate
Water content [Weight%]	5.0–8.9 (22)	6.6	2–6 (112)	4
Density (wet) [g/cm³]	2.40–2.53 (239)	2.45	2.42–2.63 (65)	2.52
Total physical porosity [Vol%]	14–25 (17)	18	5.3–17.7 (17)	11.1
Water loss porosity [Vol%]	13–21	16	4.9–17.5 (19)	10.5
Seismic P-wave velocity Vp (N) [m/s]	2220–3020 (48)	2620	1470–4610 (61)	3280
Seismic P-wave velocity Vp (P) [m/s]	3170–3650 (111)	3410	2870–5940 (112)	3860
Hydraulic conductivity (N and P) [m/s]	$2 \cdot 10^{-14} - 1 \cdot 10^{-12}$ (57)	$2 \cdot 10^{-13}$	$1 \cdot 10^{-13} - 5 \cdot 10^{-12}$ (10)	$1 \cdot 10^{-12}$
Specific storage coefficient [m^{-1}]	$1 \cdot 10^{-7} - 5 \cdot 10^{-4}$ (6)	$2 \cdot 10^{-13}$	–	–
Uniaxial compressive strength, UCS (N) [MPa]	4–17 (22)	10.5	4–37 (60)	18.0
Uniaxial compressive strength, UCS (P) [MPa]	5–10 (19)	7	6–37 (51)	16
Elastic modulus, E-Modulus (N) [GPa]	2.1–3.5 (34)	2.8	0.4–19.0 (51)	6.0
Elastic modulus, E-Modulus (P) [GPa]	6.3–8.1 (39)	7.2	2.0–36.7 (60)	13.8
Poisson ratio (N) [–]	0.28–0.38 (73)	0.33	0.06–0.42 (51)	0.22
Poisson ratio (P) [–]	0.16–0.32 (73)	0.24	0.13–1.23 (59)	0.44
Thermal conductivity (N) [Wm^{-1}k^{-1}]	1.0–3.1 (9)	1.2	–	–
Thermal conductivity (P) [Wm^{-1}K^{-1}]	–	2.1	–	–
Heat capacity [JKg^{-1}k^{-1}]	–	860	–	–
Total cation exchange capacity CEC (Co-Hexamine, Ni-en, bold) [meq/100 g of rock]	9.4–13.4 (24)	16 (24)	7.3–21.9 (13)	14.4
	–	16 (24)		
Gas entry pressure [MPa]	1.2–3.2 (11)	2.2	1.8–2.5	2.2

planes running subparallel to the bedding. Most of these fault planes show slickensides and shear fibers on polished surfaces with sense of movement consistently indicating overthrusting. In parts with higher tectonic shear strain, we observe loss of cohesion, which results in crumbly fault gouges and scaly clays.

The second type of discontinuities, the EDZ fractures, were formed during excavation of the tunnels. The redistribution of stress during the excavation indicates that the tangential stresses in the tunnel walls are higher than the uniaxial compressive strength, leading to extensile fracturing. Lower compressive and tensile strengths of bedding planes result in bedding-parallel slip, which is often observed at the tunnel ceiling and floor. The EDZ fractures form an interconnected network within the outermost 1 m of the tunnel wall. The geometry of this network is characterized by high fracture frequency in the outermost 70 cm of the tunnel wall. Local observations, such as gypsum spots on fracture surfaces and resin injections into the network, indicate that the EDZ fractures are interconnected, directly related to the tunnel itself, and appear not to be water-saturated, but rather filled with air and water vapor.

The Opalinus Clay can be considered to be an aquiclude. Although its pore space is water-saturated, water circulation is almost absent. The Opalinus Clay is overlain by Middle Jurassic karstic aquifers (Passwang Formation and Hauptrogenstein, see Figure 2a) and underlain by Liassic marls and limestone (Staffelegg Formation). The latter are karstic aquifers with advective water flow. The average water content of an Opalinus Clay sample from Mont Terri is 6.6 wt% in the shaly facies and 4 wt% in the sandy facies (see Table 1). Around 55% of these water molecules are attached to the surface of clay aggregates or are incorporated into the clay minerals. The remaining 45% fill the available pore space, but cannot move freely in this space because the extremely fine pore diameters (generally in the range of several to some tens of nanometers) prevent water circulation. The physical or total porosity is 18% for the shaly and 11.1% for the sandy facies, water loss porosity is 16% and 10.5%, respectively (Table 1). Hydrotests carried out on rock samples, but also in different boreholes, result in hydraulic conductivities ranging between $2 \cdot 10^{-14}$ and $5 \cdot 10^{-12}$ m/s, with a mean of $2 \cdot 10^{-13}$ m/s for the shaly and $1 \cdot 10^{-12}$ m/s for the sandy facies. The main conclusion is that water barely flows – it stagnates in the pores.

1.2 Repository performance and on-going experiments

Experimental results provide input for assessing different phases of repository evolution and performance (Bossart, 2007). A generic repository evolution with causes and effects is shown in Figure 3, where we distinguish a disequilibrium and an equilibrium phase. During the construction of a repository (time period of several years), stress redistribution leads to the formation of an excavation damaged zone (EDZ) around the access and emplacement galleries. During the operational phase (waste emplacement, backfilling, monitoring and sealing, time period of up to 100 years), an unsaturated zone will evolve in the nearfield due to ventilation, and redox conditions will become oxidizing. In the first few hundred years after closure, significant changes in the repository evolution will occur, including heat transport from the canister across the bentonite toward the rock with canister surface temperatures up to 150 °C, and temperatures still reaching about 90 °C at the bentonite-rock interface (Nagra, 2002). Heat flow into the rock will cause excess pore water pressures and reduction

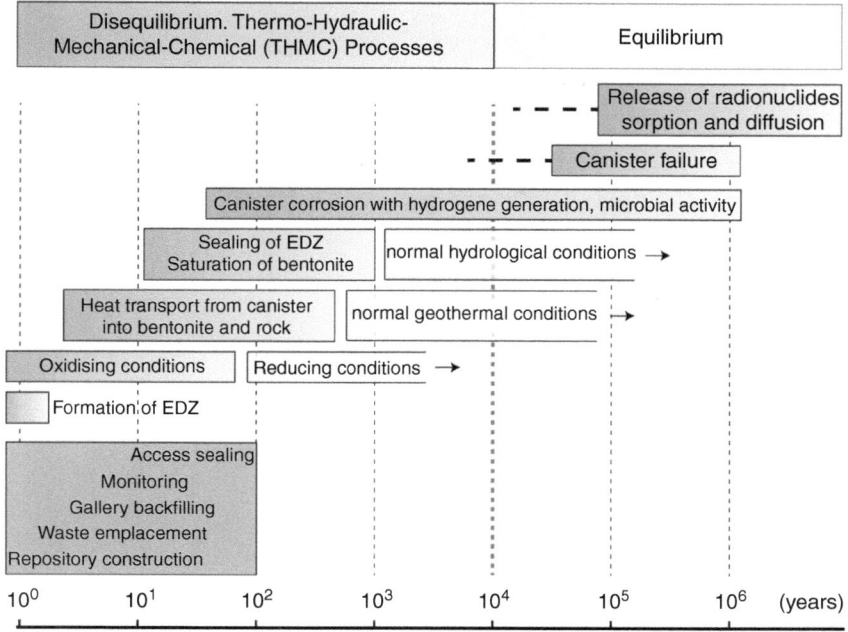

Figure 3 The generic performance of a repository.

of effective stresses in the nearfield, due to the fact that expansion of pore water is higher than expansion of the rock fabric. Later on, heat flow will decrease with time, leading to enhanced saturation in the nearfield and swelling of the bentonite backfill. These processes take place concurrently with the self-sealing of EDZ fractures. Both swelling of clay minerals in the EDZ fractures and mechanical fracture closure (swelling bentonite) contribute to self-sealing. Redox conditions in the nearfield are clearly reducing and anaerobic corrosion of the steel canisters prevails together with hydrogen production. During this period, microbial activity becomes important: bacteria are involved in redox reactions, on one hand, degrading hydrogen (pore water sulfates are reduced to sulfides such as H_2S) and, on the other hand, enhancing steel canister corrosion. Finally, the first steel canister failure is expected to occur after some 1000 to 10,000 years. This is the earliest time when radionuclides will be released into the backfill and host rock. This is also the time when equilibrium conditions in the claystone are reached and hydraulic and thermal conditions are comparable to those before the repository was constructed. Radionuclide sorption and diffusion into the bentonite and the natural clay barrier is then the final process in the repository evolution.

Most of the Mont Terri experiments are related to repository evolution and performance, as outlined in Figure 3. In Figure 4, the ongoing experiments in the Mont Terri rock laboratory are related to the different repository phases, starting from evaluation of safety parameters until the radionuclide release and transport at the end of a repository lifetime. These ongoing experiments can be further divided into observation and monitoring experiments over real timescales, such as

Repository evolution

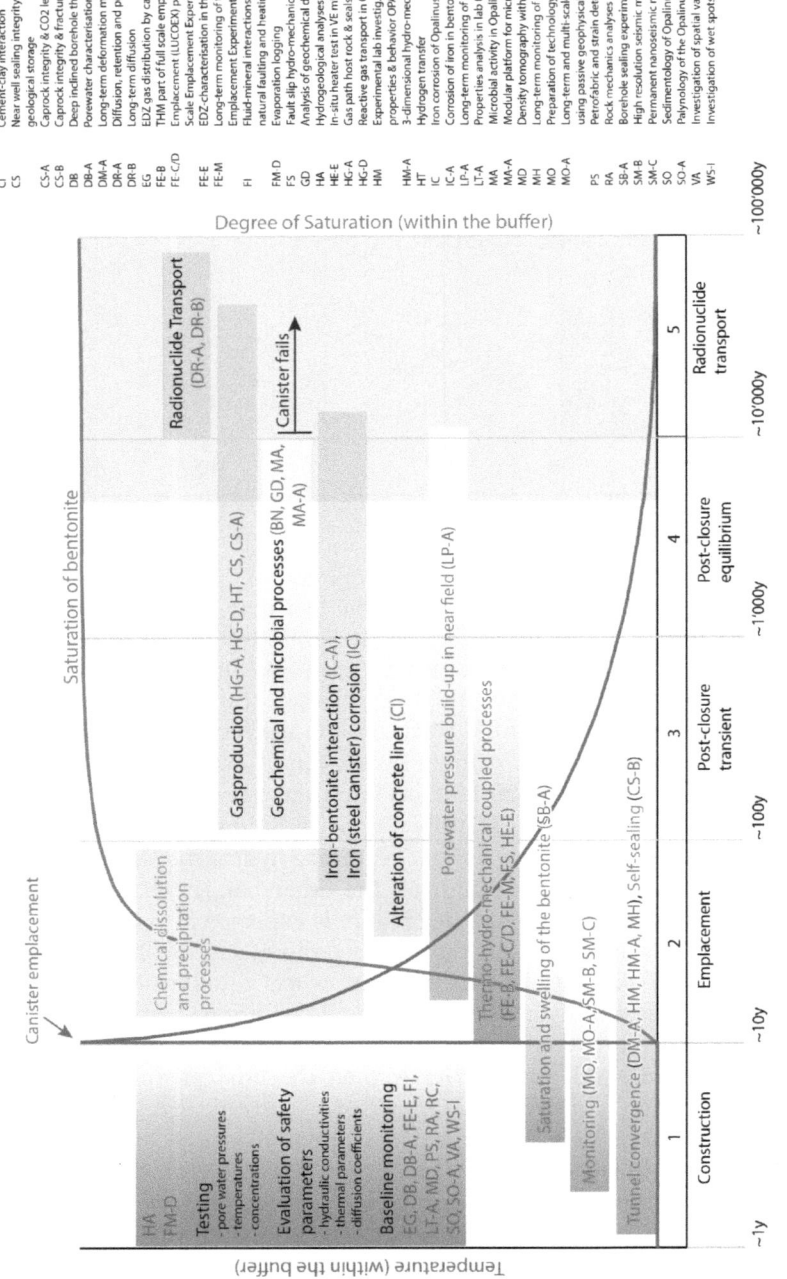

Figure 4 Ongoing experiments (2015) in the context of the repository evolution.

experiments dealing with repository construction, mine-by testing, backfilling and sealing. On the other hand, many experiments are designed to investigate long-term processes on timescales that are not directly accessible through conventional laboratory work. These are *e.g.* self-sealing of the excavation damaged zone, anaerobic corrosion, or sorption and diffusion. These experiments aim to identify and understand the processes and thus provide robust parameters. The results form the basis for long-term safety analyses.

Different countries have different repository construction and safety concepts with different conditions. Experiments can be planned and carried out by taking into account such concepts and conditions. Especially important are demonstration experiments where the interaction of differently designed and engineered barrier systems is assessed in a natural clay barrier environment, such as the Opalinus Clay in the Mont Terri rock laboratory.

2 APPLIED IN-SITU TESTING METHODS

In this chapter we present the applied in-situ testing methods. We start with excavation and drilling techniques. Then we describe the applied methods and equipment of the hydrogeological, geochemical, and rockmechanical experiments. Emphasis is placed on the material used for the different surface and downhole equipment. The tables in the subsequent subchapters compile differently applied testing methods and list the corresponding equipment.

2.1 Excavation and drilling techniques

2.1.1 Excavation methods

Several excavation methods were used for the construction of the Mont Terri underground rock laboratory. We adapted the techniques to the experimental requirements, namely stability of galleries, dimensions of the profile, length of the galleries, and costs and time constraints. The methods and their applied equipment are outlined in Table 2.

We applied mainly drilling and blasting excavation techniques for Gallery 98, a road header excavation for Gallery 04, and a pneumatic hammer technique for the niches (Figure 5a, for location see Figure 2b). A horizontal, raised boring technique was used to excavate the RB microtunnel in Gallery 98 and a steel auger was used to excavate the HG-A microtunnel in Gallery 04. Progress rates are in the order of 1–2 m per day for drilling and blasting and road heading, but slower for pneumatic hammering. In the Mont Terri laboratory, the lining consists of a shotcrete hull with a mean thickness of 15 cm. The strength of the shotcrete lining is increased with metal or plastic fibers. In sections of reduced stability, steel meshes are fixed on rock surfaces; anchors are installed around the sections and then covered with shotcrete. Particular sections require the installation of steel arches or a concrete pillar in order to allow transfer of the load to the invert. The gallery invert usually consists of a plane horizontal concrete bed with a principal slope toward the security gallery. No drainage pipes are installed in the invert because no water inflow occurs in the Opalinus Clay sections. The evacuation

Table 2 Applied methods and corresponding equipment used for excavation and drilling.

Method	Equipment
Drill- and blast technique	Drilling rig allows dryly drilled blasting holes, nitro-glycerine explosives electrical & electronic detonators.
Hydraulic or pneumatic hammer techniques	Pneumatic hammer installed on excavator or hydraulic shovel, different sizes, joystick technique for vertical excavations. No deduster necessary.
Road header technique	Longitudinal and transverse cutting heads, small head can be installed on hydraulic shovel. Deduster, air ventilation system. No use of water. See Figure 5.
Tunnel boring machine (TBM)	Not applied in Mont Terri rock laboratory (too expensive for short tunnels). Method has been applied elsewhere. See e.g. Li et al., 2012.
Horizontal raise boring	Drilling rig for central borehole, reamer, Robbins machine to pull reamer, aspiration system to evacuate cuttings.
Steel auger technique	Auger head reinforced with tungsten teeth and industrial diamond grains.
Core drilling, overcoring	Drilling rig with single, double, and triple-core-barrels of different diameters (from 35–600 mm), deduster at borehole mouth. Air drilling, no water used. See Figure 6.

of very fine clay dust requires a very powerful ventilator, which aspirates the dust at the tunnel face and removes it to a dust extraction system (Figure 5b). Thus the dust concentrations at the tunnel face can be kept to a minimum and the environmental requirements can be fulfilled. A safety concept was developed by the rock laboratory operator for monitoring tunnel stability after excavation. It consists of qualitative observation of deformation in the galleries, quantitative monitoring of convergence and regular maintenance of the galleries and niches. Shotcrete renovation on selected locations have to be carried out in the average once every decade.

2.1.2 Drilling techniques

Most of the experiments are performed in short boreholes of 5 to 20 m length. Special emphasis was thus placed on adapting drilling methods to a claystone environment. Single, double and triple core barrels with diameters between 10 and 600 mm were used. It became obvious that the triple core barrel delivered the best core quality, particularly when cores from tectonically deformed zones are required (Figure 6a). The overcoring technique, particularly the recovery of overcores in the Opalinus Clay, has been developed and improved. Now, overcoring to a depth of 12 m with diameters up to 600 mm is standardized (Figure 6b). Boreholes in claystones are usually drilled with air as the drilling medium because the use of water must be avoided due to the swelling of the claystone; use of drilling water would directly result in borehole collapse. Alternatively, other fluids such as oil-based fluids or nitrogen were used, particularly for experiments requiring saturated samples or samples which are not to be contaminated by bacteria from outside. The use of nitrogen required the development of a new ventilation and evacuation technique and special security measures. Drillcores are documented (mapped, photographed and sealed in plastic and aluminum foil) and, if not used for laboratory investigations, stored in the core library.

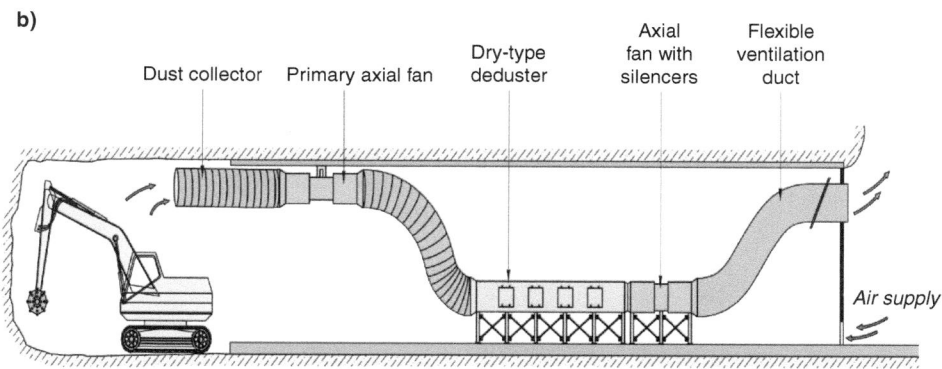

Figure 5 Excavation techniques applied in the Opalinus Clay a) road header at the gallery front, without utilization of any water and b) evacuation of dust.

2.2 Hydrogeological, gas and diffusion experiments

The major emphasis of hydrogeological, gas, and diffusion experiments was placed on the methodology, mainly test equipment, reliable measurements, parameter estimation, and test interpretation. Applied methods and correspondent equipment are compiled in Table 3. A good example is the PP (pore pressure)-piezometer, which was developed to measure pore water pressures in formations with very low hydraulic conductivity and very low free-water content (Volckaert & Fierz, 1999). The design of this PP-piezometer is shown in Figure 7a with the following requirements 1) prevention of bypass flow along borehole walls with the help of a resin seal, 2) minimization of dead volumes (the smaller

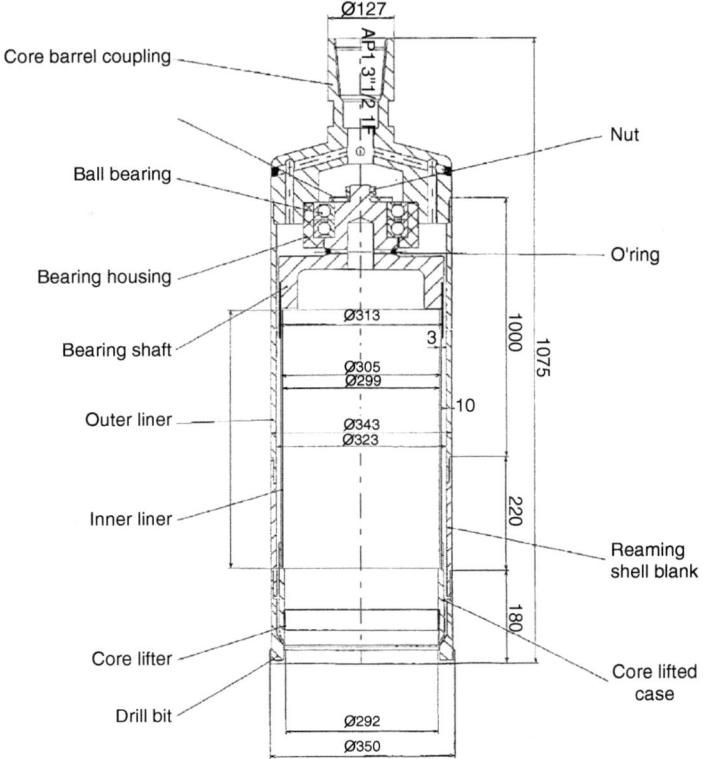

a)

Double core barrel
fore drilling and overcoring

Ø127

AP1 3"1/2 1F

Core barrel coupling

Nut

Ball bearing

Bearing housing

O'ring

Bearing shaft

Ø313

3

1000

1075

Ø305
Ø299

Outer liner

10

Ø343
Ø323

Inner liner

220

Reaming
shell blank

Core lifter

180

Drill bit

Core lifted
case

Ø292

Ø350

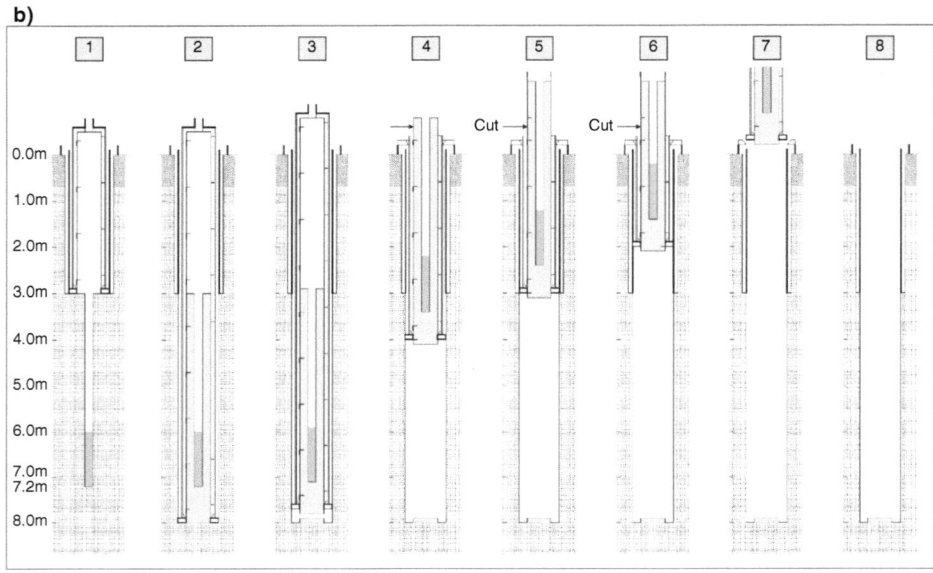

Figure 6 Drilling techniques a) double core barrel with inner and outer liner, and b) overcoring methodology: heaving and cutting sequence 1–8 of overcore until test interval section is obtained on surface.

Table 3 Methods and applied equipment for hydrogeological, gas and diffusion experiments.

Method	Equipment
Borehole logging for evaluation of bulk density, porosity and water content	Nuclear radiation logging tools (natural gamma logger, gamma-gamma density log, neutron-water content log) Time-domain-reflectometry (water content).
Vapor pressure measurements for evaluation of matrix potential	Thermo-couple psychrometer in small diameter boreholes.
In-situ pore pressure evaluation Hydraulic testing for parameter estimation – Constant head and rate tests, slug and pulse tests – Crosshole tests – Osmotic pressure tests – Evapometer tests	Inflatable single and multi-packer systems, PP-piezometer (Figure 7) and mini-packers, installed in boreholes. Resin-sealed systems are not retrievable. Flow meters, pressure sensors and data monitoring at surface. Concentration of borehole fluids should not differ from in-situ pore waters to avoid chemico-osmotic effects. Borehole evapometer to measure absolute humidity- and temperature-changes.
Gas testing – Pneumatic tests – Gas entry tests	Pneumatic test equipment shown in Figure 8, applicable in excavation damaged zone. Gas entry test with retrievable double-packer systems, which are moved along borehole.
Fluid logging, detection of advective inflows	Electric conductivity and temperature probe moved along fluid-filled vertical borehole.
Hydraulic fracturing, estimation of in-situ stress	Straddle packer moved along borehole. Fluid injections and pressure controlled at surface. Borehole TV and impression packer to identify orientation of new fracks.
Diffusion testing, estimation of in-situ diffusion parameters	Diffusion equipment shown in Figure 9a.

the volume, the faster hydraulic and chemical equilibrium is obtained), and 3) minimization of system-compressibility. The application of a two-interval PP-piezometer is shown in Figure 7b, which was installed in 1997 and which still works perfectly almost 20 years later. This PP-piezometer not only functions in different hydraulic gradient situations (pressure increase from 22 to 28 bar due to stress redistributions during the construction of the Gallery 97 in 1997, Figure 7b), but it was also used for in-situ water sampling and hydrotesting. For the proper estimation of hydraulic parameters, particular attention has to be paid to wellbore storage times, which can be quite long (days rather than hours). Experimental evidence was provided for a permeability increase at low effective stresses, known as pressure-dependent permeability. This process is reversible, *i.e.* reconsolidation of the rock leads to a return to the initial low hydraulic conductivities (mechanical self-sealing). The impact of chemico-osmotic phenomena on pore pressure measurements was also demonstrated with PP-piezometers. Nevertheless, the osmotic efficiency of the Opalinus Clay at Mont Terri is quite moderate (mean of 6%). The hydraulic conductivity of the Opalinus Clay ranges between $2 \cdot 10^{-14}$ and $5 \cdot 10^{-12}$ m/s, with a mean of $2 \cdot 10^{-13}$ m/s (Table 1). A significant change in hydraulic conductivity between the different facies was not observed. The tectonic fractures were also shown to be tight (sealed during their tectonic formation), *i.e.* they do not increase rock permeability. Due to the very low hydraulic conductivity of Opalinus Clay, molecular diffusion is the dominant transport process. Advection-dispersion can be ignored, provided that hydraulic gradients are not abnormally high.

a)

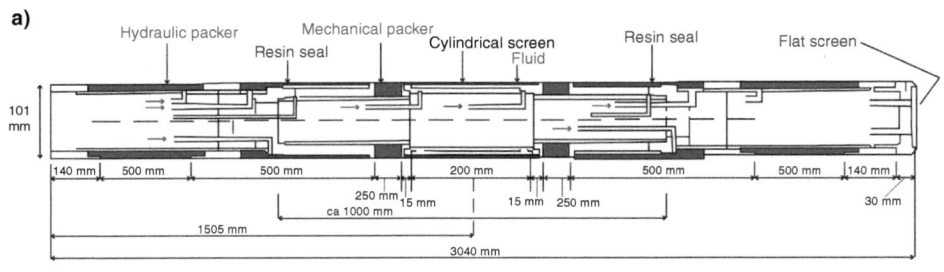

b)

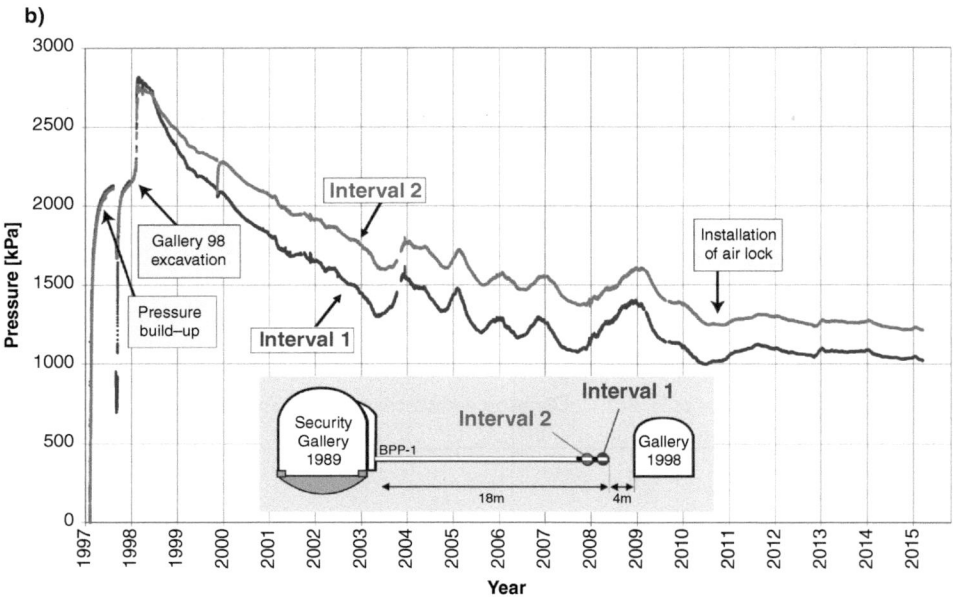

Figure 7 Hydraulic testing a) design of the PP-piezometer (not to scale), and b) example of long-term pore pressure evolution in borehole BPP-1. For location see Figure 2b.

In-situ methods for the determination of gas-related parameters, such as gas threshold pressure tests and multi-step gas injection tests, were elaborated and successfully refined (Marschall *et al.*, 2008). Also here, PP-piezometers were used, together with other removable multipacker systems. The favorable conditions in the Mont Terri rock laboratory allowed for long-term gas injection periods of many months. Optimization of test procedures made it possible to measure the gas entry pressure with a higher degree of confidence and to obtain a more complete picture of the two-phase flow characteristics of the Opalinus Clay, such as capillary pressures and relative permeability. Multiple evidence on different experimental scales was found for visco-capillary flow of gas in the existing pore space of the Opalinus Clay, suggesting that classical flow concepts of immiscible displacement in porous media can be applied to such problems when the gas entry pressure is less than the minimum principal effective stress acting within the rock. In contrast to classical porous media, gas transport through the Opalinus Clay does not cause a major desaturation of the rock. The shape of the capillary pressure curve suggests that high gas pressures are needed to displace pore water and only a small fraction of the total porosity of the Opalinus Clay is accessible by a gas phase, typically in the order of a few percent of the total porosity. Microscopic pathway dilation with an increase in gas permeability was observed at pressures around 2 MPa, which is considerably lower than the fracking pressure of around 9 MPa and the refrac pressure of 4 to 5 MPa.

We also carried out special role-play pneumatic gas injections, or withdrawal tests, in the excavation damaged zone (EDZ) in which air or nitrogen is injected in a borehole and the crosshole response observed in neighboring boreholes. The aim of such pneumatic tests is to obtain permeability profiles across the EDZ and to evaluate the heterogeneity and connectivity of the EDZ fracture network. The principle of such pneumatic testing is shown in Figure 8a. Injection tests were performed in very high permeability zones, but extraction tests are more appropriate for less permeable parts of the EDZ (Trick, 1999). Figure 8b shows an example of an EDZ permeability profile. Zones of very high permeability greater than $1 \cdot 10^{-14}$ m^2 are located in all boreholes within the first 10–20 cm after the shotcrete-rock interface. This is consistent with results from resin-filled EDZ fractures, which were sampled by overcoring.

We carried out several in-situ diffusion experiments, each using a single borehole, usually drilled downwards and sealed off with a packer system to isolate the test interval. A general design of such a test is shown in Figure 9a. Once hydraulic and chemical conditions in the test interval were stable and considered to be close to those of the in-situ pore water, we injected tracers into the circulation system. Evolution of tracer concentration in the circulation system was followed by sampling and analysis or by online analysis. After one to several years, the test interval was overcored to obtain the rock section around the interval. The tracer concentrations in the overcore were analyzed in the laboratory of the Paul Scherrer Institute PSI (Gimmi et al., 2014), followed by modelling and interpretation. Examples of tritiated water tracer profiles are shown in Figure 9b. The decline at near the injection borehole is due to pore water evaporation, when the retrieved overcore was exposed to air. We obtained diffusion parameters by inverse modelling of these tracer profiles (Palut et al., 2003). For tritiated water, the effective diffusion coefficient parallel to bedding is $5.4 \cdot 10^{-11}$ m^2/s and normal to bedding $1.5 \cdot 10^{-11}$ m^2/s respectively; effective porosity is 16 vol% (same as water loss porosity, see Table 1); sorption is negligible.

Diffusion is the governing transport process in the Opalinus Clay at Mont Terri, even in the highly fractured main fault zone. Advection-dispersion can be neglected. This is entirely in line with profile data along the Opalinus Clay formation, such as natural chloride or helium profiles (Mazurek et al., 2011). In general, data obtained from tracer evolution in boreholes and rock profiles showed good consistency between different diffusion experiments and also with diffusion data from laboratory-obtained results for small-scale samples. The derived diffusion data show that diffusive flux of anions (I$^-$, Br$^-$) is lower than that of water tracers (HTO, HDO), while that of cations (Na$^+$, Sr^{2+}, Cs$^+$) tends to be higher (Wersin et al., 2006). The reason for the lower anionic flux is anion exclusion. The higher effective diffusion coefficients of cations are explained by preferential entry into diffusive double layers of clay minerals. Sorption values for Na$^+$ and Sr^{2+} derived from in-situ experiments show good agreement with

(a)

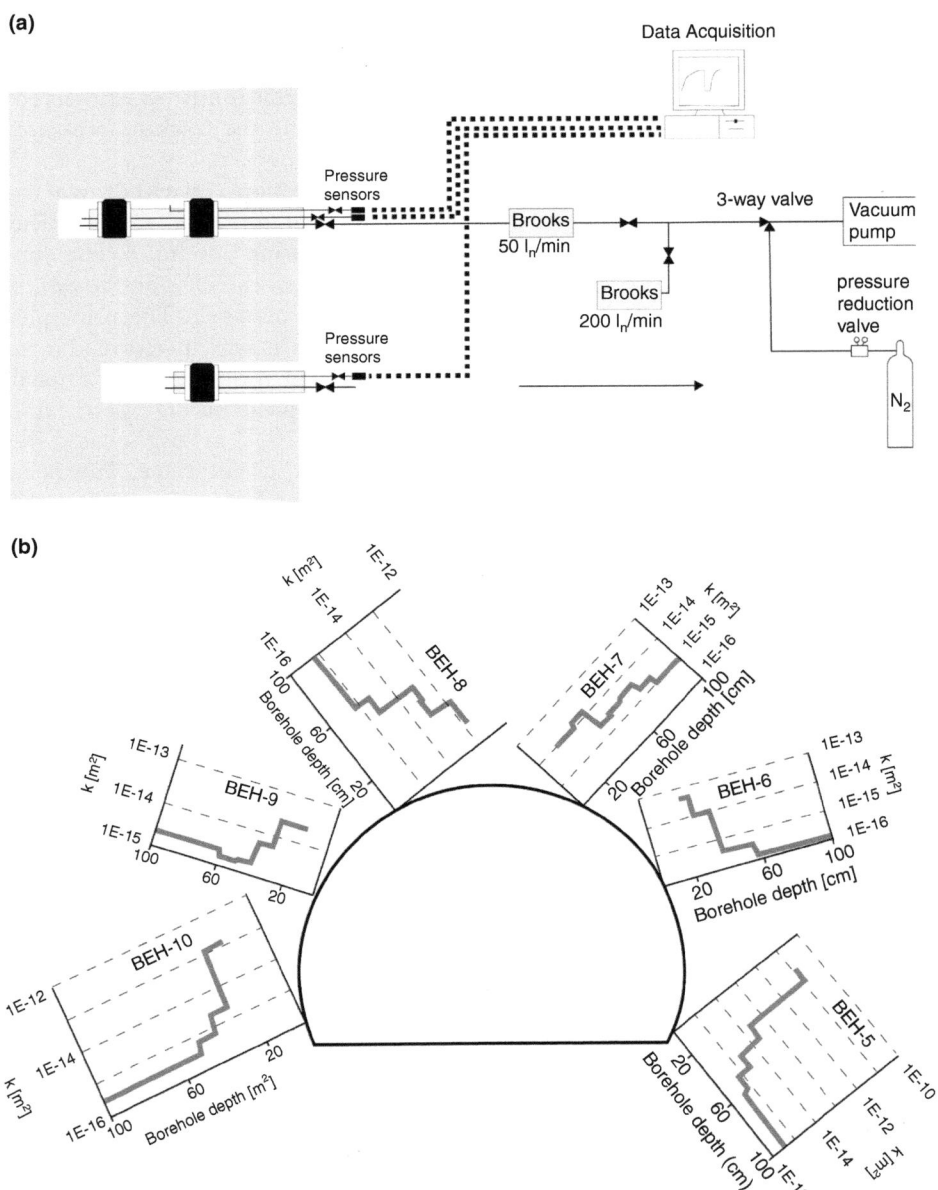

(b)

Figure 8 Pneumatic testing a) principle: injection borehole with double packer system and observation borehole with single packer, and b) permeability distribution. For location see Figure 2b.

those from laboratory studies. Diffusion data for the more strongly sorbing Cs$^+$ indicate a lower sorption capacity compared to batch sorption data, suggesting that fewer sorption sites in the claystones are available in-situ compared to powdered laboratory material. Profile data obtained from high-resolution spectroscopic methods

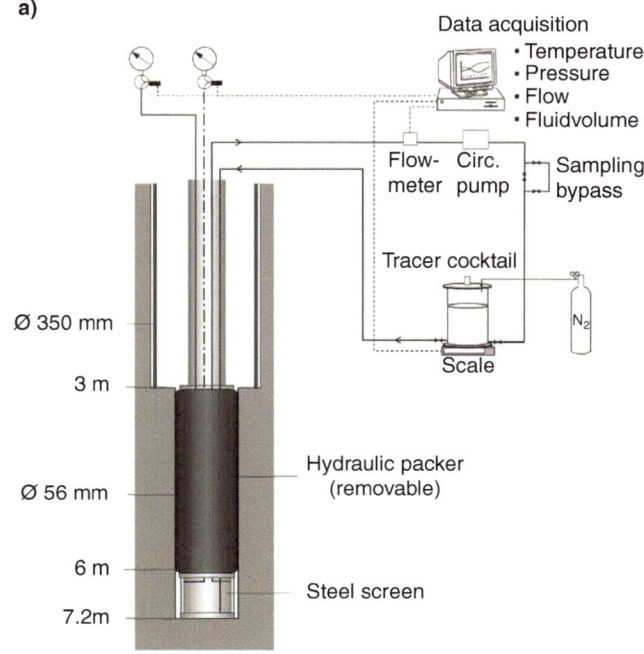

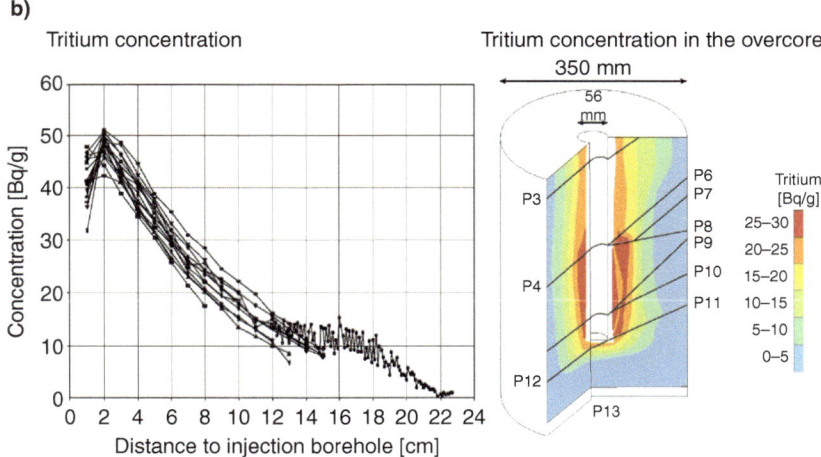

Figure 9 Diffusion experiment a) surface and downhole equipment and b) overcore with distribution of diffused HTO diffused during a 1-year period. For location see Figure 2b.

for Cs^+ and Co^{2+} indicate inhomogeneous spreading into the rock. As a general result, the concept for in-situ diffusion experiments developed and improved in the Mont Terri rock laboratory has proved to be successful for investigating diffusion of non-reactive and reactive tracers.

2.3 Geochemistry and microbiology experiments

The Mont Terri rock laboratory has provided a unique opportunity to measure complete hydro-chemical profiles across the entire low permeability Opalinus Clay formation that is normally accessible only by means of deep boreholes drilled from the surface. The accessibility of the formation to geochemists and microbiologists and the development of in-situ sampling and measurement techniques such as squeezing, aqueous leaching and diffusive equilibration allowed the collection of pore water samples and chemical data on fluids and solids under in-situ conditions. We focused our main efforts on geochemical characterization of water in the Opalinus Clay formation under undisturbed conditions by gradually improving sampling and measurement techniques (Pearson *et al.,* 2003). However, it is virtually impossible to have access to an undisturbed, pristine system despite taking considerable precautions. Oxidation by atmospheric oxygen was minimized, exchange of dissolved gases with those in the atmosphere was prevented, and sampled water was protected from contamination with undesirable borehole equipment and materials. Carefully controlled drilling conditions using inert nitrogen, and different techniques and equipment have made it possible to identify bacterially mediated perturbations. The resulting geochemical data can be used to perform geochemical modelling of local water-rock equilibria and to study the spatial distribution of chemical data along the security gallery, leading to an understanding of the long-term evolution of pore water.

The methods and applied equipment for in-situ geochemistry and microbiology experiments are compiled in Table 4. Many methods for pore water characterization, such as water squeezing and aqueous leaching, cannot be carried out in-situ and are reserved to laboratory investigation (Pearson *et al.,* 2003). Transporting unperturbed and equilibrated pore water and rock samples to the laboratory without chemical change is challenging. More promising are direct in-situ analyzing techniques to monitor the pore water at different sites and different. Such experiments are called in-situ diffusive equilibration experiments.

Solute transport in the Opalinus Clay occurs mainly through diffusion. Hydraulic conductivity with a mean of $2 \cdot 10^{-13}$ m/s is simply too small for detectable advective flows. The basic idea of an in-situ diffusive equilibration experiment is to circulate artificial water similar to that of the formation water into a packed-off borehole. This artificial water then equilibrates with the in-situ pore water by diffusion. Measuring pore water parameters in the circuit gives directly true in-situ values. Figure 10a shows the downhole and surface equipment of such an equilibration experiment. A 10 m long vertical borehole is filled with argon immediately after drilling. The downhole equipment consists of a 4.5 m long porous polyethylene screen with a hydraulic mechanical packer out of polyurethane on the top. The upper empty part of the borehole was isolated with epoxy resin. The surface equipment consists of a membrane pump for circulation, flow-through cells made of inert PEEK material and equipped with Xerolyt-pH and Eh electrodes, a graphite-type electrical conductivity electrode, pressure transducers, a data acquisition system, and three teflon coated sampling cylinders for more sophisticated laboratory analyses (De Cannière *et al.,* 2011). The whole surface equipment is contained in an argon-filled cabinet in order to avoid oxidation.

Synthetic pore water was circulated for 5 years and during this time we monitored the pH, Eh, electrical conductivity and temperature (Figure 10b). These measurements

Table 4 Methods and applied equipment of geochemistry and microbiology experiments.

Method	Equipment
Pore water sampling	Upwards-directed borehole, isolated (packed off) test interval. For design and selection of appropriate materials see Figure 16. Passive water inflow into borehole may need time (order of weeks to months).
Pore water monitoring	Vertical downward borehole drilled with argon, packed off test interval consisting of inert materials such as PVC, PEEK, polyamide, or polyethylene porous screens. Surface equipment: membrane pump for fluid circulation, flow through cells (PEEK) equipped with pH, Eh, and electrical conductivity electrodes (protected from atmosphere by argon cabinet), pressure transducer for monitoring test interval pressure. Sampling by Swagelok stainless-steel sampling cylinders. Design see Figure 10a.
Microbial sampling	Downhole equipment: borehole-drilling and design of test interval are same as those of pore water monitoring. Surface equipment: closed fluid circulation unit connected to test interval. Fluid circulation driven by peristaltic pump with low flow rates controlled by flowmeter. Monitoring of pH, Eh, and sulfides in circuit. Sterile sampling devices such as Plexiglas cylinders and needle valve sampling ports for bacterial probes. Balance for fluid samples weighing in blood bags. Optional: hydrogen injection device with dosage into fluid circulation by semi-permeable membrane. Surface equipment is protected from atmosphere by argon filled cabinet. For details see Figure 11a.
Natural gas monitoring Hydrogen injection and retrieval	Ascending borehole, drilled with argon, packed-off for gas circulation, and pore water sampling. Test interval consisting of PFA-coated stainless steel screen. Gas circulation module at surface consisting of gas circulation pump, Swagelok stainless-steel sampling cylinders (for lab analyses), and on-site gas analytics with Raman spectrometer. Monitoring of interval pressure, gas flow rate and pressure, permanent supervision of hydrogen gas concentrations. For appropriate materials and design see Figures 17a and Figures 17b, respectively.

show that diffusive equilibration with the surrounding pore water of the clay formation was reached after about 2 years. The interpretation of these measurements is outlined in Wersin *et al.* (2011), who came to the conclusion that the measured pH was considerably lower than indicated by complementary laboratory investigations and was also different from conditions in near-by boreholes. They explained these differences by microbiologically induced redox reactions occurring in the borehole that were caused by one or several degrading organic sources. These findings motivated experiments focusing on microbial activity.

The general aim of microbial activity experiments in the Mont Terri rock laboratory is to characterize the phylogenetic diversity of the microbial community in the Opalinus Clay as well as its metabolic potential (Bagnoud, 2015). The concept of such an experiment consists of installing an in-situ bioreactor, where artificial water is circulating into a borehole containing electron donors such as acetate, hydrogen or lactate. After the bioreactor equilibrates with the pore water and the clay, fluid samples are taken for geochemical analyses, such as hydrogen, sulfides, methane and total organic carbon, and for microbial analyses to assess the microbial communities in the Opalinus Clay when electron acceptors are depleted.

A design of downhole and surface equipment that we applied in such a microbial experiment is shown in Figure 11a. Lettry *et al.* (2012) describe the closed circuit for

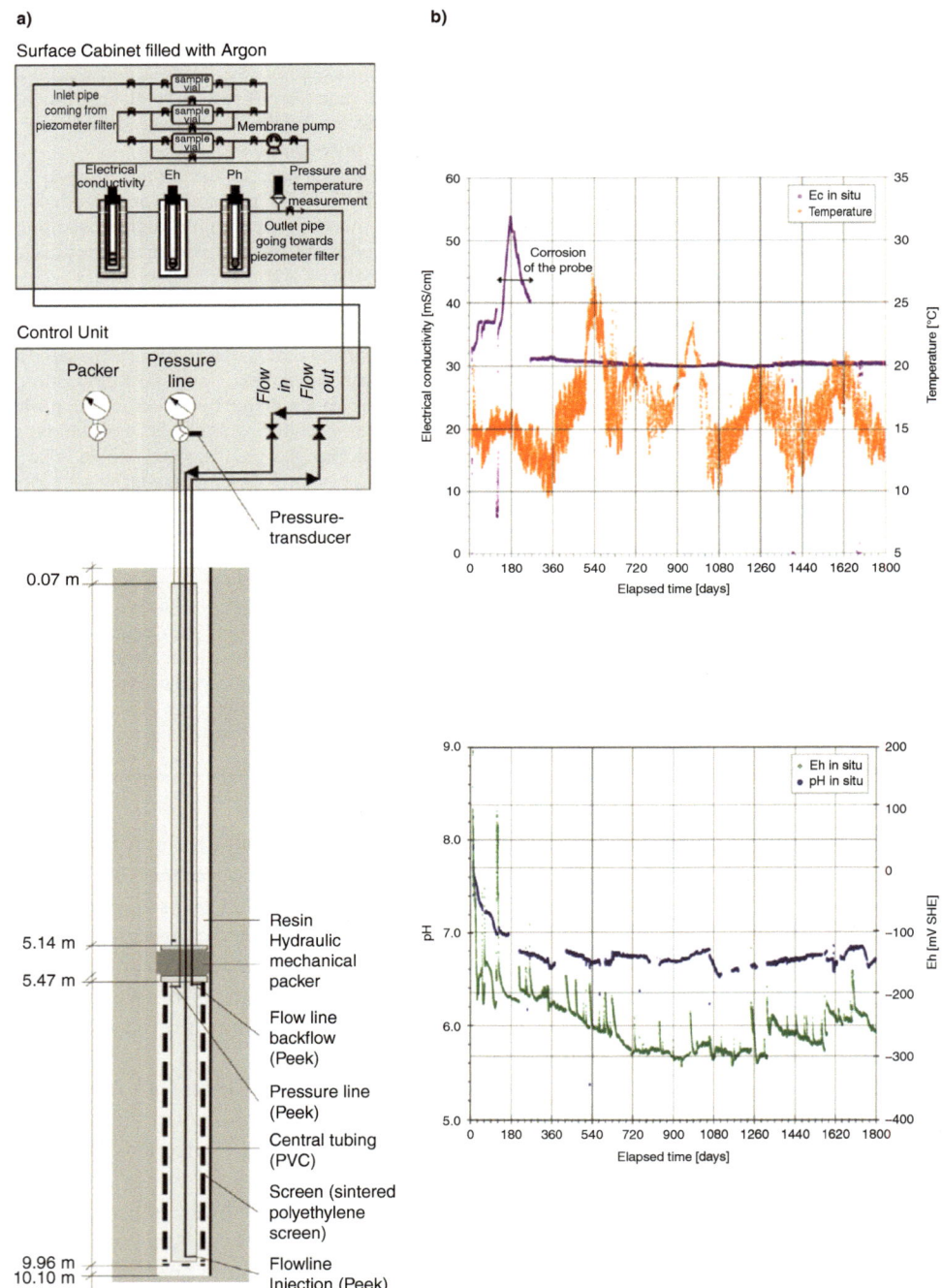

Figure 10 Geochemical monitoring a) principle and b) electrical conductivity and temperature, pH and Eh histograms (De Cannière *et al.*, 2011; Wersin *et al.*, 2011). For location: see Figure 2b.

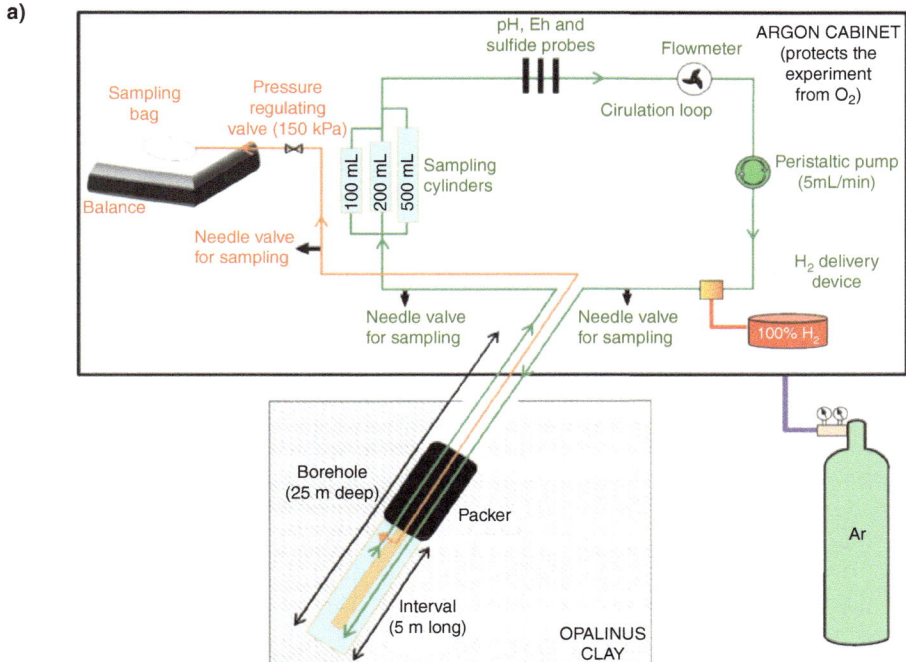

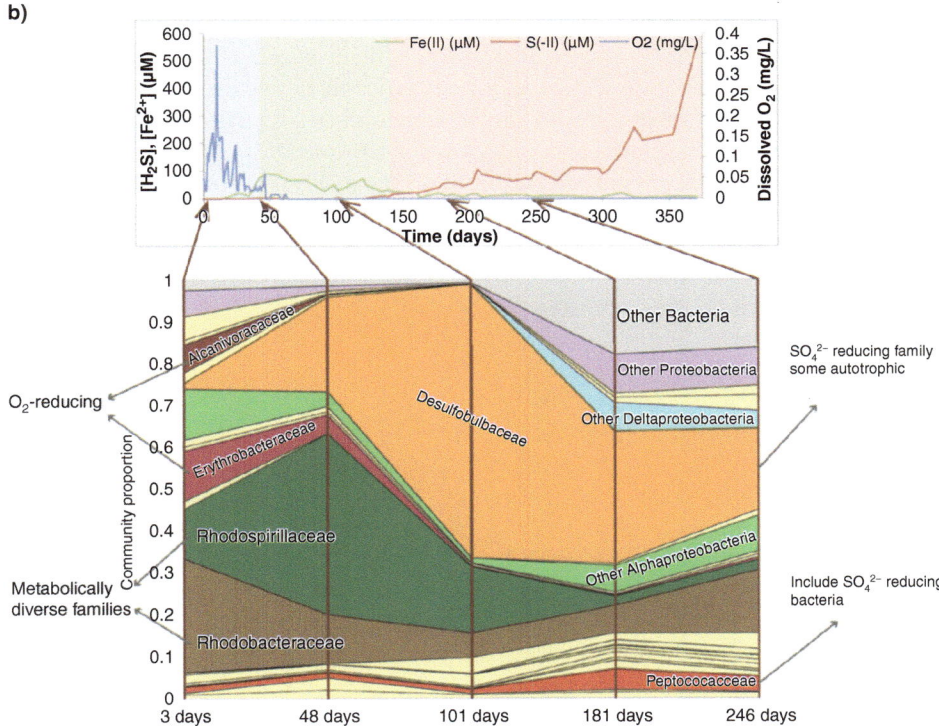

Figure 11 Microbial activity experiment a) surface and downhole equipment (Lettry *et al.*, 2012), and b) observed microbial species under different conditions (Bagnoud, 2015). For location see Figure 2b.

sampling and hydrogen injection. A peristaltic pump is used as driving force for the circulation. To avoid oxygen entering the circuit, components of the surface module are placed within a Plexiglas cabinet filled with argon. In order to keep the pressure within the circuit less than 3 bar, a hydraulic flow line of the packed-off borehole section is connected to a pressure regulation valve. This valve is connected to a tedlar bag, which is continuously weight-monitored in order to observe the outflow rate. Hydrogen injection into the circuit is realized by means of a semi-permeable membrane that is permeable for gas but impermeable for water. The dosage device contains two flow-through chambers, one for pore water and one for gas, separated by a semi-permeable membrane. The hydrogen dosage rate can be adjusted by pressure increase or decrease within the reservoir tank. Caution is needed to keep hydrogen concentrations less than 4 vol%; higher concentration could lead to self-ignition and explosion. In order to avoid or to minimize chemical interactions between pore water, downhole sampling interval, and surface equipment, we used only inert materials such as PVC, PEEK and polyamide. We sampled pore water at the needle valves under most sterile conditions and geochemically and analyzed the samples for their microbial content.

The results of these analyses are nicely outlined in Bagnoud (2015). He identified chemical reactions together with the bacterial species involved (Figure 11b). During the first 48 days, no hydrogen was added. Reduction of oxygen and iron could clearly be identified, together with the associated bacterial species. After 48 days, injection of hydrogen began. After 100 days, the pore water sulfates started to get reduced to sulfides. Sulfate-reducing bacteria became abundant as soon as hydrogen was added. DNA-analyses of different bacterial species led to the conclusion that most of these micro-organisms were brought artificially into the rock laboratory. However, at least two indigenous bacteria species could be identified (Poulain *et al.*, 2008), indicating that bacteria may survive in exotic environments such as deeply buried dense clay formations.

These findings have implications for the performance of a repository in the Opalinus Clay. On one hand, steel canister corrosion leads to the formation of hydrogen, and the sulfates in the pore water are reduced to sulfides such as H_2S, decreasing the hydrogen concentrations. Thus high hydrogen overpressures with subsequent processes like two-phase flow into the clay and mechanical dilation of bedding planes are not likely. On the other hand, H_2S dissolves easily into the pore water and will enhance corrosion once in contact with the steel canisters.

2.4 Heating experiments

A variety of in-situ and laboratory observations and results have been obtained for the thermal characterization of the Opalinus Clay. Two key in-situ heater tests were carried out in the Mont Terri rock laboratory. They consisted of equipping a vertical and a horizontal borehole with heating devices, the latter shown in Figure 12a. Observation boreholes arranged around the horizontal heater borehole were used to install temperature, pore water pressure, and deformation sensors at several locations. Heat was applied over a period of 15 months. The heater test field of the vertical borehole was fully dismantled after a recovery period.

Generally, the thermal expansion of a porous medium in saturated condition is governed mainly by thermal expansion of both pore water and solid minerals. The thermal volumetric expansion coefficient of pore water in the Opalinus Clay ($3.4 \cdot 10^{-4}\,K^{-1}$) is about

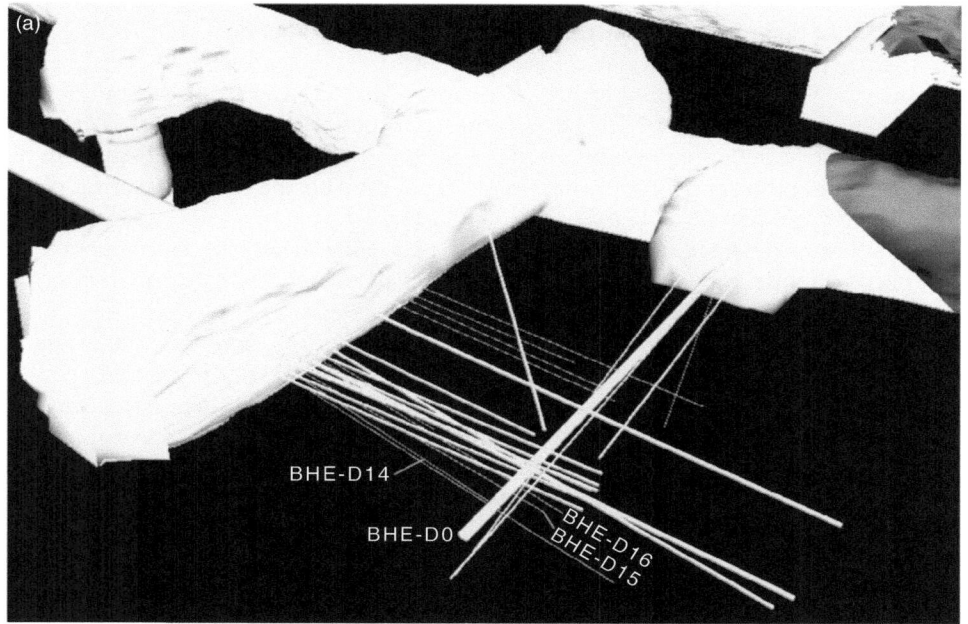

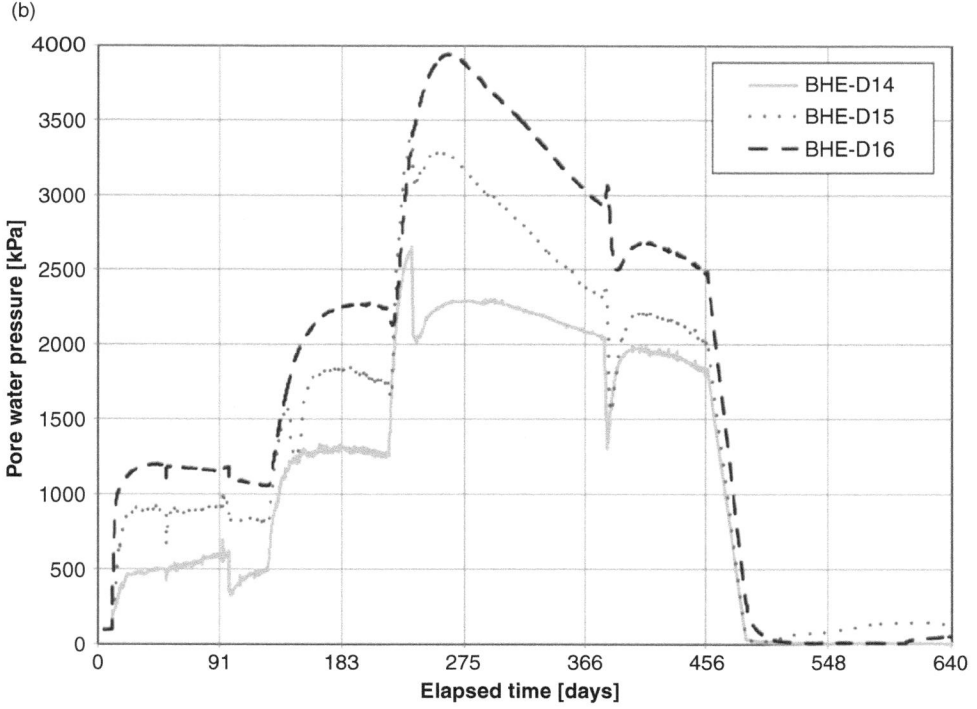

Figure 12 Heater experiment a) test array and borehole net of the horizontal heater experiment with heater borehole BHE-D0, and b) pore water pressure monitoring of three boreholes before, during two heating phases, and after heating. For location see Figure 2b.

two orders of magnitude higher than that of solid clay grains ($4.5 \cdot 10^{-6}$ K^{-1}). This is also the reason why a pore water pressure increase during heating is observed, which is shown in Figure 12b for the three boreholes BHE-D14, 15 and 16 outlined in Figure 12a (Wileveau, 2005). The first heating with a power supply of 650 W caused a rapid increase in the pore pressure up to 2.4 MPa at a distance of 0.8–1.4 m from the heater. The maximum heating power of 1950 W during the second heating phase resulted in peak pressures between 2.6–4.0 MPa. During the subsequent cooling phase, pressure dropped slowly down to 1.8–2.5 MPa. These measurements were then used to model temperatures and pore water pressures and to estimate thermal parameters (Zhang *et al.*, 2007).

We carried out additional field investigations such as gas sampling, gas pressure surveys, geoelectrical, and seismic surveys before, during, and after the heating period. With respect to the possibility of constructing a repository for heat-generating waste in clay-rich formations, the heating of the Opalinus Clay to 90 °C should be considered as the upper limit since self-sealing properties are hampered due to conversion of mixed-layer smectite-illite to illite). The maximum temperature to which the Opalinus Clay at Mont Terri has been exposed in the geological past was 85±5 °C. As a rule of thumb, artificial temperatures obtained from heat-producing waste should not exceed the natural peak temperatures of the geological past. Observations on dismantled boreholes used in the heating experiments clearly showed that thermally induced macro-fractures were not formed at these temperatures.

Although determination of thermal conductivity depends on accuracy of sensor locations, the quality of measurement of electric power injected, and the environment and accuracy of the probes, the thermal in-situ properties for the shaly facies are consistent with a mean value of 2.1 Wm^{-1}K^{-1} parallel and 1.2 Wm^{-1}K^{-1} normal to bedding, respectively (Table 1). The best estimate for the heat capacity is 860 Jkg^{-1}K^{-1}.

2.5 Geophysical and geotechnical experiments

During the last 10 years, we have carried out a very large number of rock mechanical laboratory tests on differently drilled Opalinus Clay drillcores and several in-situ experiments with the main objective of understanding deformation mechanisms under different in-situ and laboratory conditions. A further important objective was to derive in-situ primary stresses and to follow stress redistributions and related rock deformations during excavation of differently directed galleries. Finally, an important goal was to obtain reliable and robust rock mechanical parameters from the Opalinus Clay. During the in-situ investigations, major emphasis was placed on geophysical methods: downhole seismic measurements, seismic and electric tomography experiments, geoelectrical resistivity measurements on tunnel surfaces and in boreholes, and acoustic emission measurements.

The deformation mechanism of Opalinus Clay is cataclastic flow, which is the dominant deformation mechanism at relatively low effective confining pressures, low temperatures, and high strain rates, all conditions that exist in the Mont Terri rock laboratory. Creep behavior (stationary or transient) could not be confirmed by in-situ experiments. Non-linear and time-dependent behavior can be explained and reliably modeled using a coupled hydro-mechanical approach. Of special interest is the shrinkage (desaturation) and swelling (saturation) of Opalinus Clay (Möri *et al.*, 2010). Desaturated Opalinus Clay shows a significantly higher strength. Thus, strong air ventilation may stabilize newly excavated galleries. Shrinkage cracks will be sealed by subsequent saturation and swelling.

Opalinus Clay behaves as a mechanically transverse isotropic material with five independent elastic parameters: Young's modulus normal to bedding (2800 MPa), Young's modulus parallel to bedding (7200 MPa), Poisson's ratio normal to bedding (0.33), Poisson's ratio parallel to bedding (0.24) and shear modulus (1200 MPa). The uniaxial compressive strength is around 7 MPa parallel and 11 MPa normal to bedding, provided that undrained conditions prevail. These values are valid for the shaly facies; the sandy facies results in different values as shown in Table 1.

We carried out three different types of experiments to determine the in-situ stresses: 1) over- and undercoring experiments, 2) borehole slotter experiments, and 3) hydraulic fracturing experiments. The in-situ stresses are not yet well constrained. The major principal stress axis σ_1 is subvertically oriented with a magnitude of 6–7.5 MPa. The intermediate and minimal principal stresses σ_2 and σ_3 are subhorizontally oriented with magnitudes of 4–5 and 2–3 MPa (NNW-SSE directed), respectively (Martin & Lanyon, 2004). Anisotropy-controlled breakouts occur mainly at locations where bedding planes are tangent to the gallery circumference.

Seismic and geoelectrical methods are well suited for determining heterogeneities in the Opalinus Clay, e.g. in the excavation damaged zone, but also in tectonic fault zones (Schuster & Alheid, 2002). Undeformed and saturated Opalinus Clay shows electrical resistivities between 8 and 16 Ωm, while desaturated and loosened Opalinus Clay has resistivities between 20 and 60~Ωm. Best estimates of seismic P-wave velocities of normally consolidated and undeformed shaly Opalinus Clay are 2600 m/s normal and 3400 m/s parallel to bedding, while Opalinus Clay from tectonically deformed zones and from the excavation damaged zone shows considerable lower velocities in the range of 1500–2500 m/s.

An overview of applied methods and equipment of geophysical and geotechnical experiments is shown in Table 5. In the following we present three selected examples of these tests. The first example deals with geoelectrical measurements, the second presents ultrasonic interval velocity measurements and the third shows a load-plate test.

Geoelectrical measurements are often used in claystones to identify changes in saturation. Figure 13a shows a demonstration experiment, where a canister was placed onto bentonite blocks in 2001. The space between the canister and the rock wall was filled with a granular bentonite and artificially saturated over several years (Mayor et al., 2007). Before backfilling, the test section, a longitudinal and circular array of electrodes was installed on the tunnel wall (Kruschwitz & Yaramanci, 2002). The different saturation steps of the bentonite and the rock wall were continuously monitored by geoelectrical measurements over more than 10 years. Figure 13b shows the spatial resistivity distribution just before dismantling in 2013 (Furche & Schuster, 2014). It can be seen that the concrete floor and the canister are highly resistive (order of 100 Ωm), as expected, whereas the saturated bentonite below and above the canister shows the lowest resistivity values (below 3 Ωm). This corresponds directly with the spatial distribution of water content (Figure 13c), which was measured on bentonite samples just after the dismantling of the granular backfill (Palacios et al., 2013). The positive correlation of resistivity measurements and water content is presented in Figure 13d. Thus the resistivity shown in Figure 13b provides a good pattern

Table 5 Methods and applied equipment of geophysical and geotechnical experiments.

Method	Equipment
Hyperspectral imaging	Thermal light spectra are induced by illuminating a tunnel face with standard spot lights. The light spectra are captured by a hyperspectral digital camera. A large number of contiguous spectral bands are recorded and analyzed subsequently with digital image analysis software. The spectral signature can be calibrated to display mineralogy.
Geoelectrical measurements at tunnel walls and in boreholes	Array of 4 grounded electrodes, with 2 inner current, and 2 outer potential electrodes. Different geometries are used such as Wenner- and Schlumberger arrays. Apparent resistivities are obtained by applying Ohm's law. An example of resistivity distribution is shown in Figure 13b.
Seismic (ultrasonic) interval velocity measurements	Array consisting of 1 ultrasonic borehole source with a piezoelectric transducer, and 3 piezoelectric receivers at distances of 10, 20 and 30 cm from the source. Transducers are pneumatically attached to the borehole wall. Measurements are shown in Figure 14.
Seismic tomography	String of geophones or hydrophones installed in a dry, resp. water-filled borehole. A seismic source, for instance a seismic sparker, is moved to well-defined positions in a neighboring parallel source borehole.
Seismic monitoring	Hydrostatic leveling system (frequency range: $1 \cdot 10^{-6}$ to $1 \cdot 10^{-2}$), broad-band seismometer ($1 \cdot 10^{-3}$ to 10 Hz), seismic navigating system array ($1 \cdot 10^{-1}$ to 100 Hz).
Acoustic emission measurements	Acoustic emitter (20 kHz to 55 kHz) and array of acoustic receivers (2 Hz to 60 kHz). Identification of P- and S-waves. An example is shown in Figure 21.
Deformation measurements – Tunnel convergences – Strain measurements in boreholes	– Convergences: Temperature resistant Invar wires are spanned across tunnel profiles. Relative length changes are measured with a distometer – Classical strain measurements: sliding micrometer, inclinometer, conventional multipoint extensometer, magnetic extensometer, CSIRO cells – Strains measured with fiber-optic sensors: distributed sensors with BOTDA cables and point sensors with FBGs. Problem of de-convolution between temperature and strain signal with common Brillouin interrogators. Hybrid systems (Neubrex[TM]) measure both Brillouin and Rayleigh-peak of the spectra enabling measurement of both temperature and strain.
Swelling-pressure measurements	– Pressure cell with different types of pressure transducers (piezo-resistive, vibrating wire, resistance and semi-conductive strain gages) – Load plate (see Figure 15), simulating swelling pressures of bentonite.
Stress measurements	In-situ stress tensor measured by stress-relief methods such as – borehole slotter (Interfels[TM]), undercoring, overcoring using 12 elements strain gauges (CISRO strain relief cells). With these methods deformation due to the stress release induced by cutting or drilling is measured in different directions. – Hydraulic fracturing method using a double packer system to pressurize intervals along a vertical borehole. Shut-in pressure and re-opening pressure define stress magnitudes, impression packer tests (imprints of discontinuities on soft rubber) or optical logs define stress orientation.

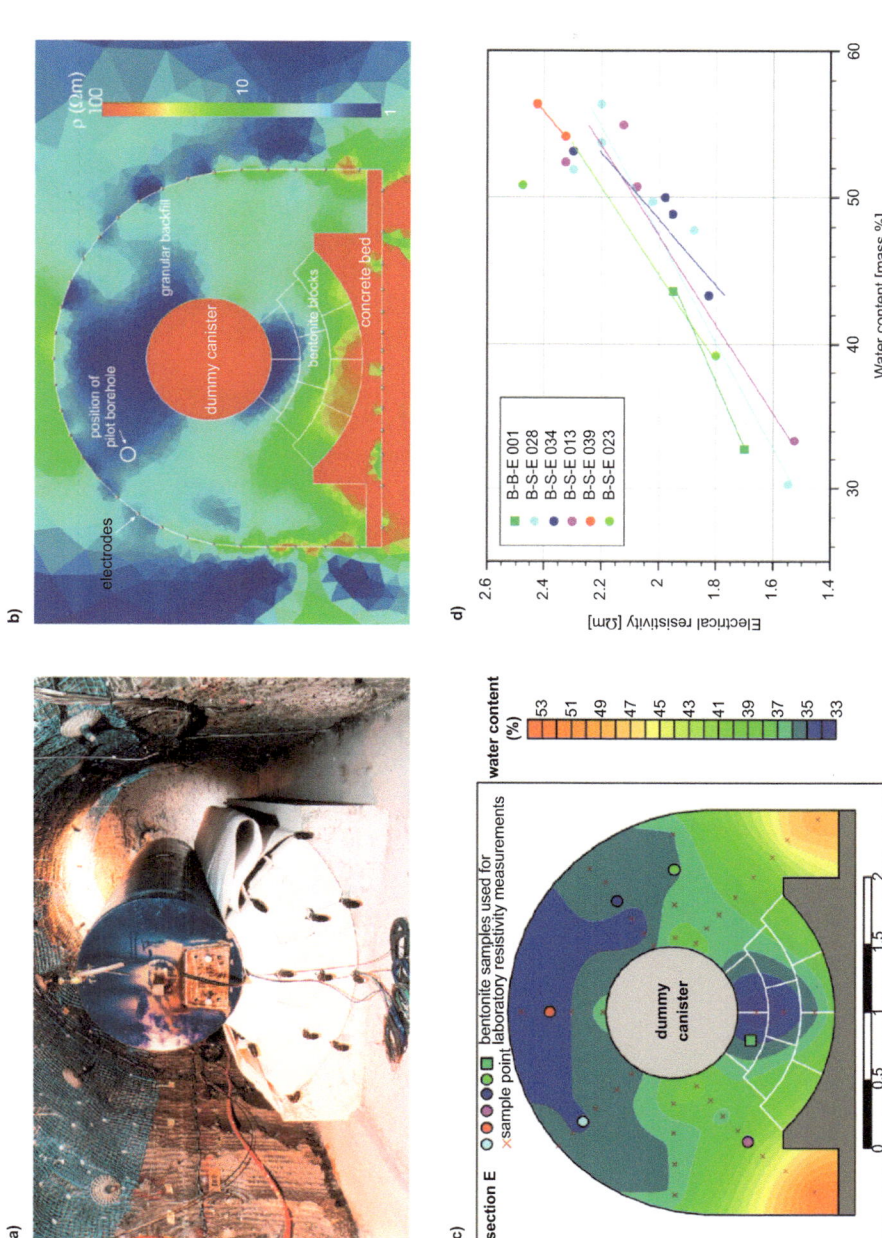

Figure 13 Demonstration experiment a) canister emplacement on bentonite block-bed b) electrical resistivity distributions after 10 years of saturation (Furche *et al.*, 2014) c) water content measurements after dismantling (Palacios *et al.*, 2013), and d) water content versus electrical resistivity. For location see Figure 2b.

of the spatial distribution of water content in a non-destructive way (without excavation and collection of samples).

Ultrasonic interval velocity measurements in boreholes are a powerful tool to detect small-scale rock heterogeneities. Figure 14a shows the ultrasonic traces and Figure 14b the P-wave velocities and normalized amplitudes derived from a 30 m-long horizontal borehole, which crosses a tectonic fault zone (Schuster, 2009). This fault zone is visible between 18 and 22 m and is characterized by clearly reduced P-wave velocities. But also small-scale heterogeneities can be detected, such as sandy intercalations on the centimeter scale with P-wave velocity peaks. Reduced velocities and amplitudes in the first 3 m reflect the excavation damaged zone of the tunnel. The rather pronounced velocity differences of three receivers in the undeformed sections can be explained by bedding anisotropy. The highest velocities were obtained with receiver 3 where the P-waves travelled parallel or slightly oblique to the bedding planes (see also Table 1). In the fault zone, this anisotropy is less developed, especially in the fault gouge and scaly clay zones. There the P-waves of the three receivers are almost identical.

We applied a load-plate test to simulate the swelling pressure of bentonite around a tunnel (Heitz *et al.*, 2003). Special emphasis was placed on the transmissivity changes in the EDZ. Hydrotests were carried out before the load plate was installed and also during the load-plate test. The test installations together with the borehole array are

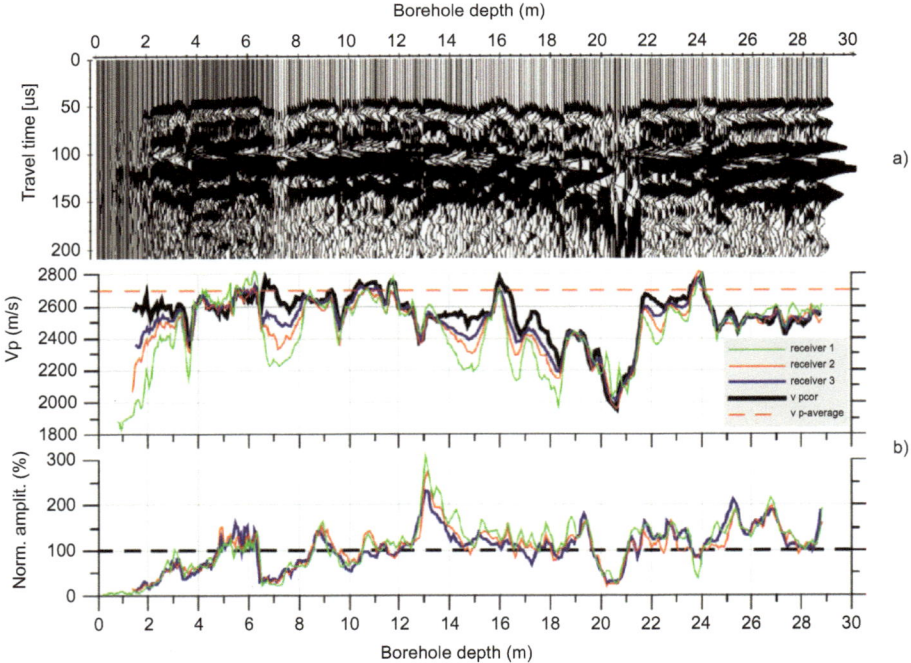

Figure 14 Ultrasonic interval-velocity measurements a) ultrasonic traces and b) P-wave velocity distribution and normalized amplitudes of the three receivers, which are attached at 120° angles to the borehole wall (Schuster, 2009). For location see Figure 2b.

shown in Figure 15a, and the transmissivity results are plotted in Figure 15b. During the first 850 days, without loading, transmissivity decreased from $4 \cdot 10^{-7}$ to $2 \cdot 10^{-9}$ m²/s. Then the load plate was installed from 850 to 1000 days, exerting an external stress stepping up from an initial 1 to 5 MPa, resulting in a further decrease of the transmissivity from $2 \cdot 10^{-9}$ to $4 \cdot 10^{-11}$ m²/s. As a result, the transmissivity decreased by 4 orders of magnitudes. The results of this test are interpreted as a follows: the EDZ fractures become sealed, first by swelling of the clay minerals on the fracture surfaces, and second by fracture closure due to mechanical stress exerted by bentonite swelling pressure. Both, swelling of the clay minerals in the EDZ fractures and mechanical fracture closure (swelling bentonite) contribute to the EDZ self-sealing. Thus transport of released radionuclides in the EDZ is governed by molecular diffusion and advection-dispersion can be neglected.

2.6 New equipment and materials

Since the initiation of the Mont Terri project, development of novel methodologies and new equipment has been a key issue. Materials science plays an important role in the development and application of surface and downhole equipment. Figure 16a presents a catalogue of applied materials and Figure 16b shows two examples where these materials were used for manufacturing downhole equipment (Fierz et al., 2008). Adequate materials are especially important in pore water sampling, in-situ diffusive equilibration and long-term diffusion experiments. For redox-sensitive measurements, polyamide or PEEK (poly-ether-ether-ketone) materials were used instead of stainless steel. However, polyamide is less inert than PEEK and the latter is therefore favored in diffusion and retention experiments. The PP-piezometer shown in Figure 7 contains hydraulically inflated polyurethane packers, which do not suffer from corrosion. Protective coating of surface and downhole equipment is often required to minimize interaction between experimental equipment surfaces and test fluids; a well-suited material is PFA (tetrafluoroethylene-perfluoro-copolymer). Recent material developments are related to ceramics, which are used for test interval filters and are applied in the field of geochemical evolution of pore water and dissolved gases. Ceramic filters do not serve as nutrients for bacteria, which may perturb the chemistry of pore water. Quite possibly, ceramics will also play an important role in the future for canisters containing radioactive high-level vitrified waste or spent uranium fuel: steel canisters could be replaced by ceramics, provided that their mechanical parameters (strength, elastic moduli) are comparable to those of steel. If such a material is found, processes such as anaerobic corrosion and hydrogen production would no longer pose problems.

3 MINE-BY TESTING

Several mine-by tests have been carried out in the Mont Terri rock laboratory (Martin & Lanyon, 2004; Martin et al., 2013). Major aims were characterization of excavation disturbed and damaged zones, estimation of its hydro-mechanical parameters, and determination of primary and secondary stress fields. The concept of mine-by testing consists of drilling an array of boreholes from an existing gallery. Downhole equipment

a)

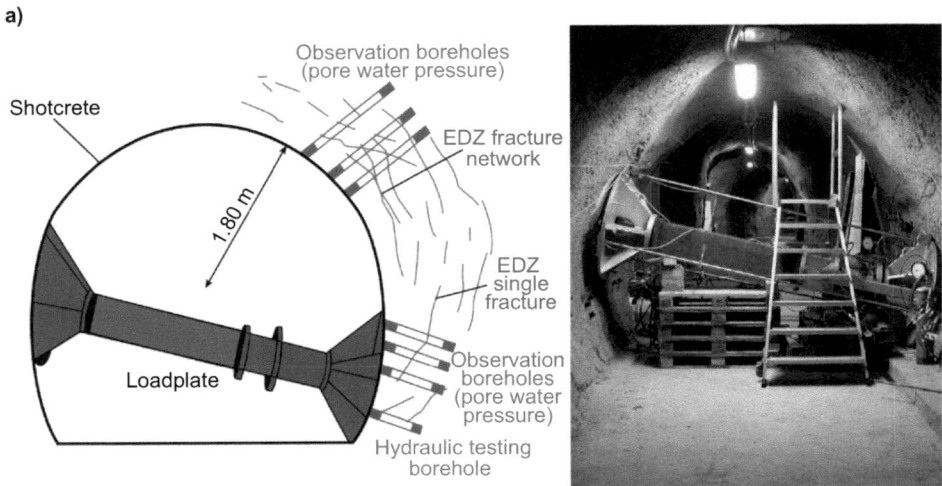

b)

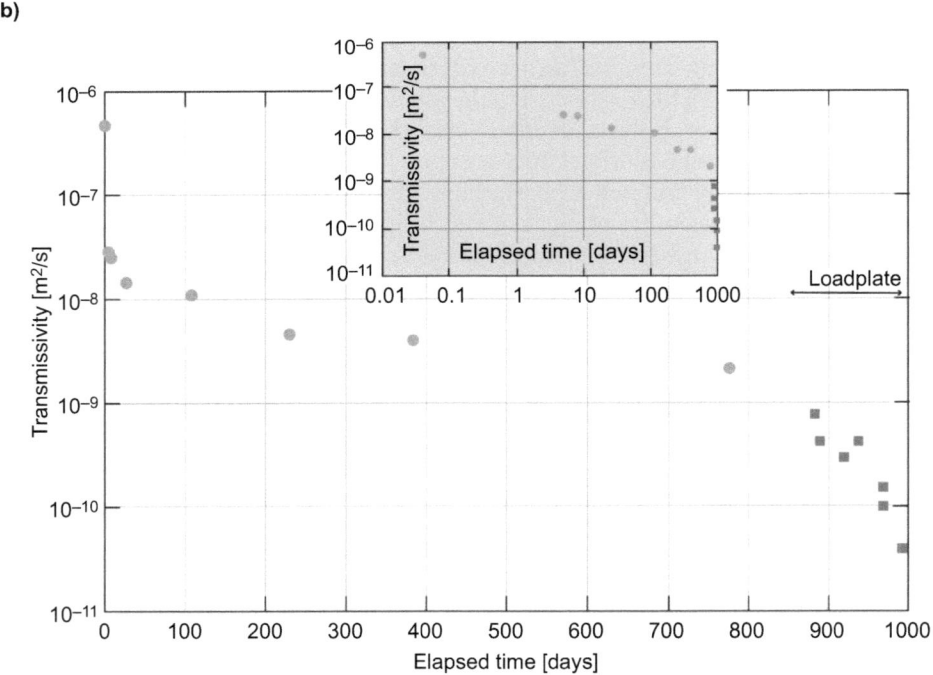

Figure 15 Load-plate experiment a) installation in the rock laboratory, and b) measurement results: from 0–850 days without load plate, indicated with dots (transmissivity T decreases by 2 orders of magnitudes), from 850–1000 days with load plate, indicated with squares (further decrease of T by 2 orders of magnitude). For location see Figure 2b.

	old → new				
Porewater sampling experiment	Stainless steel (packers, lines)	Stainless steel (packers)			
	Natural rubber (packer sleeve)	Natural rubber (packer sleeve)			
	Teflon (interval coating)	Teflon (interval coating)			
		PEEK (lines, fittings)			
		PVC (filter screen, interval)			
Porewater geochemistry experiment	Polyamide (lines)		PEEK (lines, fittings, flow through cells for electrodes)		
	Stainless steel (lines, electrode flow trough cells, circulation pump)		PVC (interval)		
	Polyurethane (packer)		Polyethylene (screen)		
	PVC (filter screen, interval)		Polyurethane (packer)		
Long-term diffusion experiment		Stainless steel (lines, flow through cells for electrodes)		PEEK (lines, fittings)	
		Polyurethane (packer)		PFA (interval and reservoir coating)	
				Teflon (screen, interval and sample vial coating)	
				Natural rubber (packer sleeve)	
Porewater and gas equilibrium experiment					PEEK (lines, fitting, electrode flow trough cells)
					PFA (interval packer)
					Ceramic (Al-Si-O, test interval screen)
					Teflon (sample vial coating)
					Polyurethane (packer)

b)

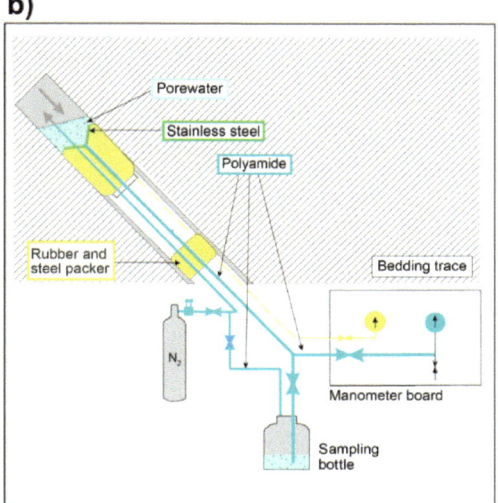

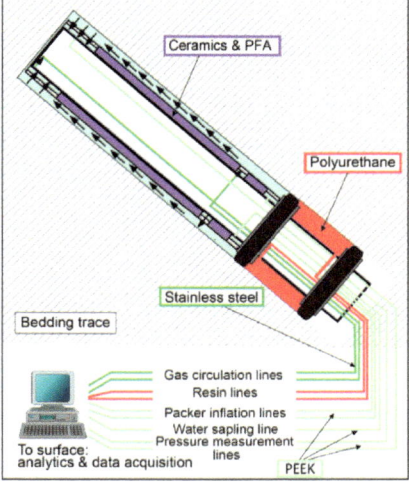

Figure 16 Material used for designing in-situ experiments a) material catalogue of different key experiments and b) schematic illustrations of a simple water sampling experiment and a complex gas injection experiment.

is installed in these boreholes to monitor deformations, pore pressure, acoustic emission and geophysical properties. Then a new gallery is excavated in the vicinity of these boreholes and the hydraulic and mechanical changes are carefully monitored in the boreholes. Hydro-mechanical modelling is used to simulate strains and pore water pressures, determination of hydro-mechanical parameters, and the stress field.

3.1 The Excavation Damaged Zone (EDZ)

We applied a variety of methods and in-situ experiments to understand the formation of the excavation damaged zone (EDZ) around underground openings in the Opalinus Clay. Key experiments included small-scale mapping and fracture-line counting of EDZ fractures on newly excavated surfaces, overcoring and fracture network analyses of resin-filled EDZ fractures, seismic single and cross-hole measurements, repeated pneumatic and hydraulic testing in order to record decreasing transmissivity in the EDZ fracture network with time, and mine-by testing to investigate hydro-mechanical coupled processes in the EDZ.

Figure 17 shows an example of an EDZ fracture network that formed due to stress redistribution during construction of Gallery 04 (Nussbaum *et al.*, 2004). The

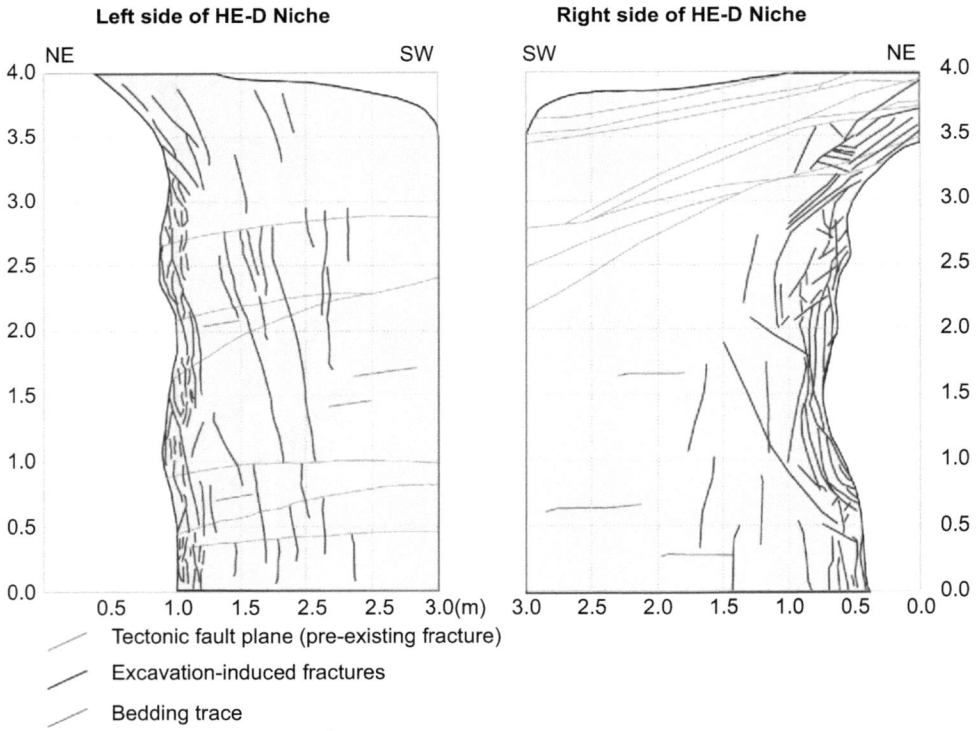

Figure 17 Example of a stress-induced excavation damaged zone: extensile fractures in a sidewall niche of the Gallery 04 (Nussbaum *et al.*, 2004). For location see Figure 2b.

technique consists of excavating a sidewall niche into the existing Gallery 04, followed by small-scale mapping of the fracture network in the wall of the Gallery 04. In the first 70 cm of the tunnel wall, an interconnected fracture network forms, which is then filled with tunnel air, thus supporting oxidation of pyrites. Deeper in the tunnel wall, extensile fractures become less frequent, and are isolated. These show no oxidation spots and are limited to a depth of one tunnel radius.

The redistribution of in-situ stresses during excavation is clearly one reason for the formation of EDZ fractures. On the other hand, the EDZ fractures are also governed by bedding anisotropy. Figure 19 shows an example of an EDZ around a borehole, which is mainly controlled by the bedding anisotropy and the bedding-parallel orientation of the borehole (Badertscher *et al.*, 2008). Borehole breakouts initiate where the bedding planes are tangent to the circumference of the borehole and they propagate into the rock in both directions, toward the ceiling and toward the floor. The resulting EDZ has the shape of a chimney, which extends several borehole diameters into the rock.

From these observations we conclude that two processes contribute to the formation of an EDZ: 1) a stress-induced EDZ caused by stress redistribution and increase of tangential stresses normal to the maximum principal stress direction. This results in an extensile fracture network as shown in Figure 17, and 2) an anisotropy-induced EDZ, which is observed along tunnels and boreholes oriented parallel to the bedding anisotropy where bedding planes become sheared and buckled until breakouts into the opening occur, as shown in Figure 18 (Kupferschmid *et al.*, 2015). In terms of stability and lining, this anisotropy-induced EDZ is much trickier to handle than the stress-induced one since its development is an ongoing process that still continues long after excavation is completed. In order to avoid tunnel deformation and breakouts from an anisotropy-induced EDZ, tunnels in claystones should not be drilled parallel to the bedding anisotropy but rather oblique or normal to it.

Pore water pressure plays an important role during creation of the EDZ. During excavation, we can assume an undrained response of the tunnel wall, since hydraulic conductivity is far too small and excavation times too short for a liquid-water discharge into the tunnel. Measured pore water over-pressures during excavations are mainly observed in the tunnel walls (Figures 19a and 7b), whereas normal or even under-pressures have been measured in the roof and bottom of the galleries. These low pore pressures in the bottom and ceiling may prevent bedding failure and bedding slip, until the pore pressure is recovered. On the other hand, overpressures in the tunnel wall may enhance bedding slip and reactivation of tectonic fractures. Furthermore, desaturation effects during ventilation of the tunnel have to be taken into account. Deformation such as extensile fracturing and shear deformation along bedding and fracture planes may be limited or even blocked if a strong ventilation, sufficient to maintain an unsaturated zone around the tunnel, is running. In this case uniaxial compressive strength immediately increases and this period should be used by the miners to install the lining.

Martin & Lanyon, 2004 developed a hydro-mechanical conceptual model that takes into account most of the above-mentioned observations and measurements such as pore water pressure changes at different sections around the tunnel, measured deformations on the tunnel walls and stress redistribution around the tunnels (Figure 19b). This model also illustrates the extent of the EDZ, which is on the order of 1 tunnel radius parallel to σ_1 and 0.5 tunnel radius parallel to σ_3. Also shown is the fracture

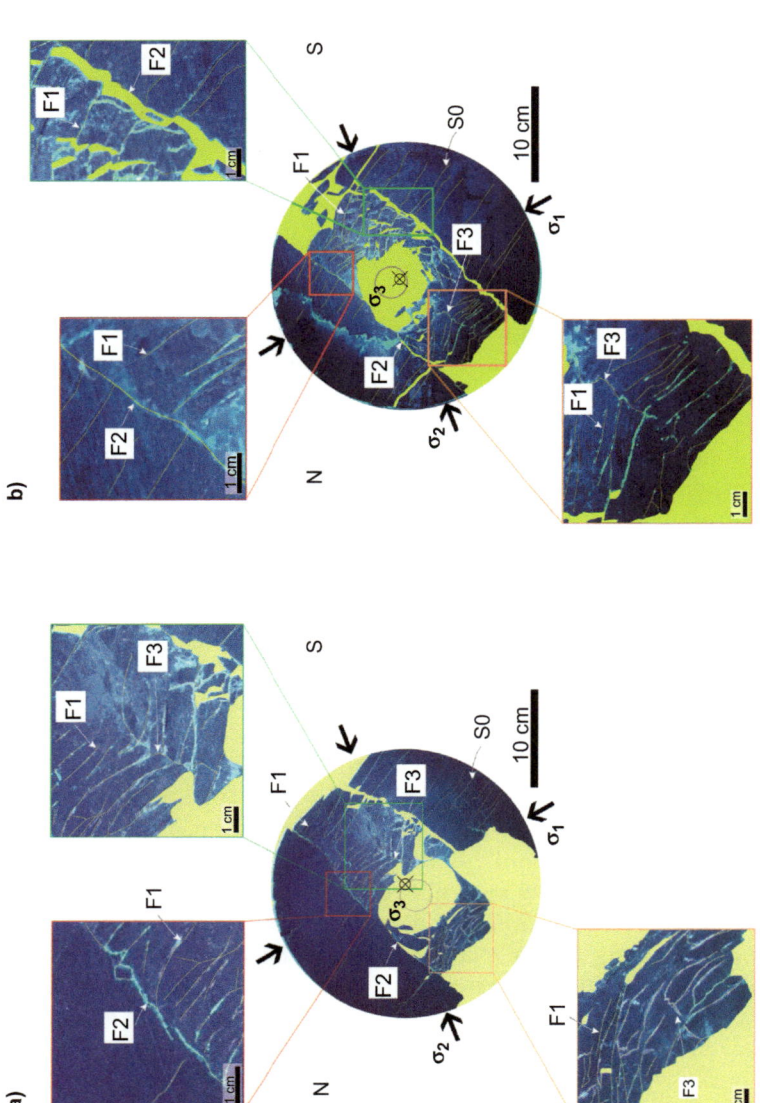

Figure 18 Example of an anisotropy-induced excavation damaged zone. Visualization of different fracture patterns by fluorescence-doped epoxy resin, which has been injected into a small diameter bedding-parallel borehole at the center, followed by overcoring and analyses under UV-light (Badertscher et al., 2008). Core disks from a) borehole BSE-3, depth of 7.92 m and b) depth of 8.59 m. S0: bedding plane. F1: bedding-parallel fractures. F2: bedding-normal fractures along the border of the central kink zone; F3: bedding-normal fractures at the center of the kink-zone. Indicated are the in-situ stress directions σ_1, σ_2 and σ_3. For location see Figure 2b.

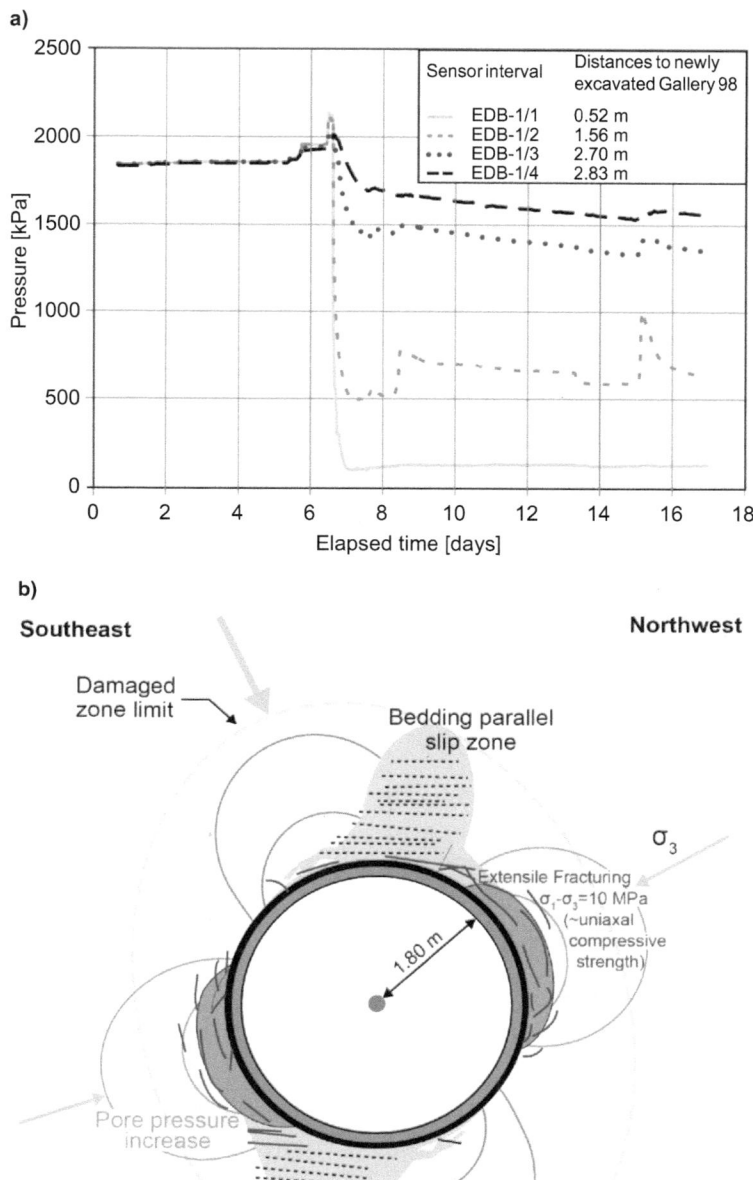

Figure 19 Pore-pressure observations in a mine-by experiment, a) pore-pressure changes before, during and after excavation of Gallery 98, measured in the horizontal borehole BED-B1 with 4 sensor intervals of different distances to Gallery 98. Note the pore-pressure increase at day 6 when the tunnel front passed the sensors (for location see Figure 2b), and b) hydro-mechanical model of the excavation damaged zone (Martin & Lanyon, 2004).

criterion: locations with deviatoric stresses σ1-σ3 ≥10 MPa are subjected to extensile fracturing.

Self-sealing of the fracture network was demonstrated by repeated hydrotesting carried out over more than 2 years in two different experiments. These show that the EDZ fracture transmissivity decreased by almost two orders of magnitude over a period of 3 years (see also Figure 15b). The relevant self-sealing process may be due to osmotic swelling of fracture walls, but other processes such as rock creep are also likely (Bossart *et al.*, 2004).

3.2 The EZ-G experiment

Geophysical surveys are useful methods for unraveling the development and evolution of the excavation damaged zone in time and space (Nussbaum & Bossart, 2014). The excavation of Gallery 08 provided the opportunity to study the junction between two galleries: an existing one represented by the end-face of Gallery 04 and a gallery under construction, called Gallery 08 (see location in Figure 2b). The rock-mass segment in between, called "EZG-08 segment," was excavated in July and August 2008. Before its excavation, this rock mass had been instrumented from the end-face Gallery 04 with several boreholes to conduct acoustic experiments and to monitor evolution of the EDZ of Gallery 08 during its excavation. The experiments consisted in two complementary measurements. The first was an active seismic method involving a controlled acoustic source. This method is applied to characterize elastic properties of a rock mass in terms of P-wave velocity. The second dealt with a passive seismic method based on the detection of acoustic emissions, *i.e.* micro-seismic events (MSEs). A MSE is a sudden release of elastic energy induced by microcracking in response to local stress variations, *e.g.* to stress redistribution around an excavated gallery.

The acoustic experiments consisted in three main experimental setups (Figure 20): (1) an acoustic source dedicated to the seismic survey experiments in order to probe the EZ-G08 segment by the use of a controlled acoustic signal inserted in the central BEZ-G5 borehole, (2) an array, A1, composed of 64 acoustic receivers located close to the acoustic source (not shown in Figure 20) and (3) a second array, A2, composed of 16 acoustic receivers installed in the four boreholes BEZ-G16 to BEZ-G19 to detect and record MSEs.

More than 56,000 events have been detected and recorded by the multichannel acoustic monitoring system (Le Gonidec *et al.*, 2012). Not all of these corresponded to natural MSEs induced by gallery excavation. For instance, some events are related to human activities such as a sledgehammer, which has an acoustic frequency of 3 kHz per impact. In order to avoid any confusion, only the MSEs monitored after excavation stopped (*i.e.* in the early morning of July 11) are considered for the analysis. In the A1 array configuration, the 16 acoustic receivers of one holding pole are aligned inside one borehole and the source is located at the end-face of Gallery 08, *i.e.* the source is nearly aligned with the receivers. With this alignment P- and S-wave velocities can be estimated as illustrated in Figure 21. The active acoustic experiment enables us to determine the P-wave velocity field that is required to locate the MSEs induced by the excavation process monitored by the acoustic array A2 (passive acoustic monitoring). Thanks to the efficiency of the location algorithm determined by the active acoustic survey, we could locate three groups of MSEs, *i.e.* the burst of hundreds of MSEs identified on July 11, the 71 MSEs on July 12, and the 16 MSEs on July 13 (Figure 22).

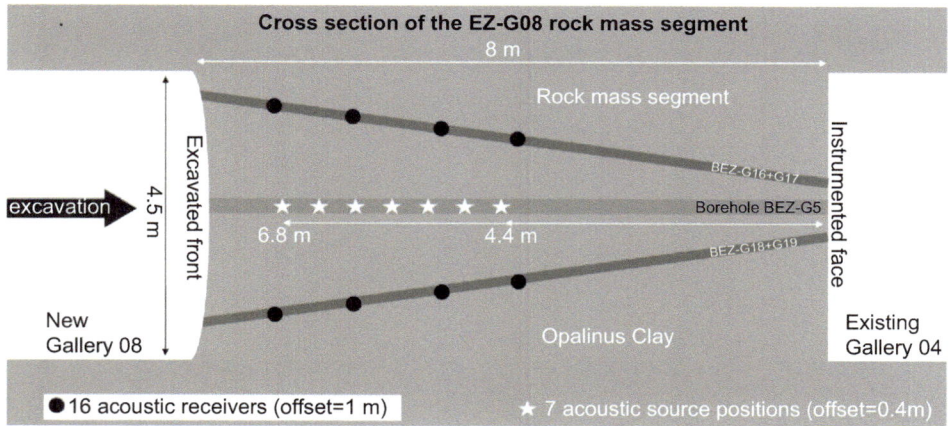

Figure 20 In July–August 2008, the excavation of the last rock-mass segment took place at the interface between Gallery 04 and 08 (for location see Figure 2b). The end-face of Gallery 04 was instrumented for acoustic experiments conducted to monitor the EZG-08 rock-mass segment. Instrumented Ga04 face with location of the boreholes: an acoustic source introduced in BEZ-G5 (black dot) and two arrays of acoustic receivers introduced in BEZ-G16-19 (black dots).

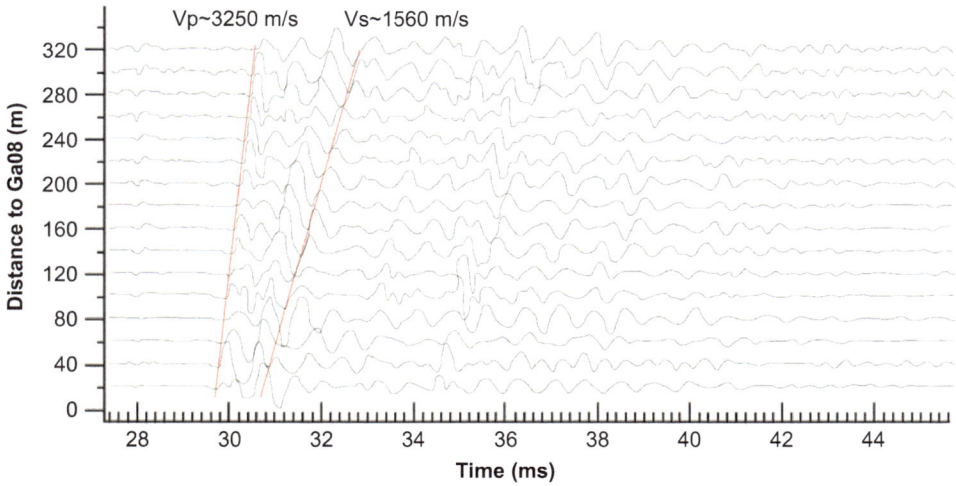

Figure 21 Identification of P- and S-waves and their velocities (red lines) from a recorded event induced at Gallery 08 front during the excavation procedure (modified from Le Gonidec *et al.*, 2012).

The MSEs located on July 11 are plotted in blue in Figure 22 where the events cluster on the right-hand side of Gallery 08. It is of prime importance to note that this location corresponds to the shaly facies sidewall and that no microseismic activity was observed in the sandy facies of the opposite sidewall. On July 12, the events located on the

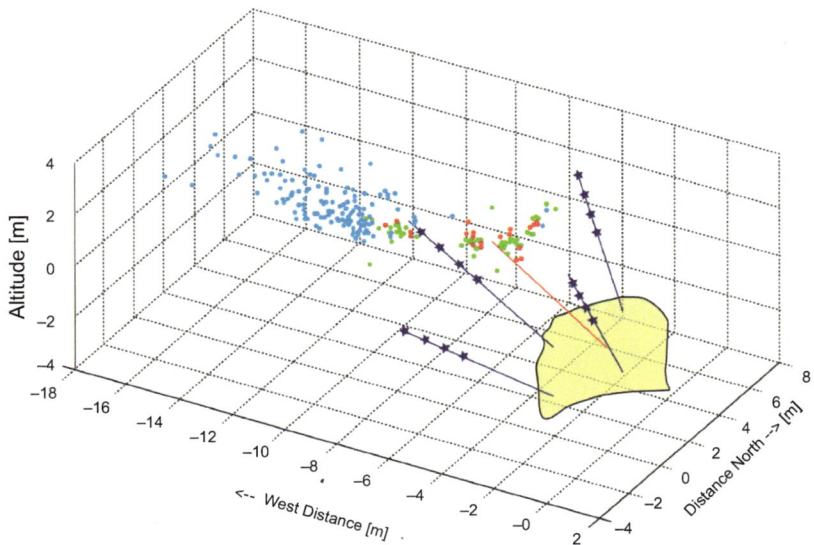

Figure 22 Spatial location of the events identified as micro-seismic events (MSEs) detected on July 11 2008 (blue), 12 July (green), and 13 July (red). The four dark-blue lines represent the boreholes with the 16 acoustic receivers (blue stars). The red line is the BEZ-G5 borehole containing the seismic source. The yellow surface indicates the face of Ga04 Gallery (modified from Le Gonidec *et al.*, 2014 and from Nussbaum & Bossart, 2015).

excavated front are plotted as green points in Figure 22. Interestingly, the MSEs detected one day later, on July 13, cluster in the same area. This suggests that both series of events located behind the tunnel face correspond to a similar failure mechanism (Le Gonidec *et al.*, 2014). To identify the failure mechanism we used an inversion algorithm based on the time-domain Moment Tensor (MT) implemented with the Insite software (Pettitt, 1998; Young *et al.*, 2000). In that method, both the first motion and associated amplitude of the P-waves are considered for the algorithm. The result of the MT inversion is plotted in the Hudson T-k plot and shown in Figure 23. In this diagram (T,k) = (0,1) corresponds to pure dilatation mechanism (ISO), (T,k) = (−1,0) to compensated linear vector dipole mechanism (CLVD), and (T,k) = (0,0) to pure double-couple (DC). The distribution of failure mechanisms is not random but clusters in the DC and positive CLVD domain (Le Gonidec *et al.*, 2014). CLVD mechanisms are found to be dominant in the sidewall of Gallery 08 in the shaly facies and may be assigned to extensional fracturing. DC mechanisms are restricted to behind the tunnel face and may be interpreted as activation of bedding planes and/or reactivation of bedding-parallel fault planes that are free to slip after excavation.

4 SUMMARY

A rock laboratory generally provides direct access to the geological environment under realistic repository conditions. The generic Mont Terri Rock Laboratory is purely for research purposes and no waste will ever be disposed of there. The

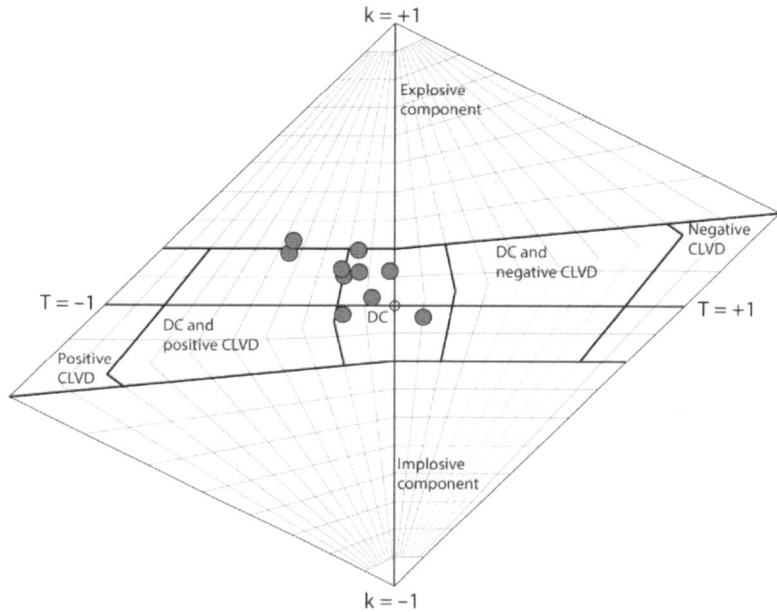

Figure 23 Hudson T-k plot of acoustic events showing damage mechanisms of microseismic events recorded on July 12, 2008 and July 13 after the excavation stopped (modified from Le Gonidec *et al.*, 2015). T- and k-axes represent shear and volumetric portion of the mechanism, respectively. The DC mechanism is defined by (T,k)=(0,0) and CLVD mechanism by (T,k) = (+1,0). The results of the analyses, based on few events associated to weak amplitude waveforms, clearly show that distribution of the damage mechanism is not random, and is characterized by a dominant DC mechanism.

experiments being carried out provide the basis for understanding the hydrogeological, rock mechanical and geochemical characteristics, and coupled processes that control the performance of natural (rock) and engineered barriers (canister, bentonite backfill) of a geological repository. Furthermore, rock laboratories are ideal locations for developing technologies required for the construction, operation, monitoring and closure of a repository. They are essential in providing information on the long-term performance of engineered barriers and monitoring systems. In addition, they increase public acceptance of geological disposal, especially when citizens critical of waste repositories can visit the underground facilities and discuss issues with scientists and engineers.

REFERENCES

Badertscher, N., Girardin, C. & Nussbaum, C. (2008) SE-H Experiment: EDZ structural analyses of resin impregnated sections from BSE-3 overcores, Internal Technical Note. Mont Terri Project, Bundesamt für Landestopografie swisstopo, 3084 Wabern, Switzerland.

Bagnoud, A. (2015) Microbial metabolism in the deep subsurface: Case study of Opalinus Clay. Ph.D thesis No 6727 (2015). Ecole Polytechnique fédérale de Lausanne, EPFL Lausanne.

Bossart, P., Trick, T., Meier, P.M. & Mayor, J.-C. (2004) Structural and hydrogeological characterization of the excavation-disturbed zone in the Opalinus Clay. Mont Terri Project, Switzerland. Applied Clay Science. [Online] 26 (1–4), 429–448. Available from: doi:10.1016/j.clay.2003.12.018.

Bossart, P. (2007) Overview of key experiments on repository characterization in the Mont Terri Rock Laboratory. Geological Society, London, Special Publications. [Online] 284, 35–40. Available from: doi:10.1144/SP284.3.

Bossart, P. & Thury, M. (2008) Mont Terri Rock Laboratory. Project, Programme 1996 to 2007 and Results, Report No. 3 of the Swiss Geological Survey. Mont Terri Project, Bundesamt für Landestopografie swisstopo, 3084 Wabern, Switzerland.

Caer, T., Maillot, B., Souloumiac, P., Leturmy, P., Frizon, de Lamotte, D. & Nussbaum, C. (2015) Mechanical validation of balanced cross-sections: the case of the Mont Terri anticline at the Jura front (NW Switzerland). Journal of Structural Geology. Available from: doi: 10.1016/j.jsg.2015.03.009.

De Cannière, P., Schwarzbauer, J., Höhener, P., Lorenz, G., Salah, S., Leupin, O.X. & Wersin, P. (2011) Biochemical processes in a clay formation in-situ experiment: Part C, organic contamination and leaching data. Applied Geochemistry 26, 967–979.

Fierz, T., Bossart, P. & Volckaert, G. (2008) New equipment and materials. In: Bossart, P. & Thury, M. (eds.), Mont Terri Rock laboratory. Project, Programme 1996 to 2007 and Results. Swiss Geological Survey Report No. 3. Bundesamt für Landestopografie swisstopo, 3084 Wabern, Switzerland.

Furche, M. & Schuster, K. (2014) Engineered Barrier Emplacement Experiment in Opalinus Clay: "EB" experiment, Geoelectrical monitoring of dismantling operation, Technical Report TR 2014-04. Mont Terri Project, Bundesamt für Landestopografie swisstopo, 3084 Wabern, Switzerland.

Gimmi, T., Leupin, O.X., Eikenberg, J., Glaus, M.A., Van Loon, L.R., Waber, H.N., Wersin, P., Wang, H.A., Grolimund, D., Borca, C.N., Dewonck, S. & Wittebroodt, C. (2014) Anisotropic diffusion at the field scale in a 4-year multi-tracer diffusion and retention experiment – I: Insights from the experimental data. Geochimica et Cosmochimica Acta 125, 373–393.

Heitz, D., Trick, T. & Bühler, Ch. (2003) Selfrac (SE) Experiment: Long term plate load experiment, Internal Technical Note. Mont Terri Project, Bundesamt für Landestopografie swisstopo, 3084 Wabern, Switzerland.

Jaeggi, D., Bossart, P., & Wymann, L. (2014) Kompilation der lithologischen Variabilität und Eigenschaften des Opalinus-Ton im Felslabor Mont-Terri, Beurteilung der sicherheitsrelevanten Gesteinsparameter aus dem Felslabor und ihre Übertragbarkeit auf die Standortgebiete, Expert Reüprt for EMSI, 09-08, 66 pp. Bundesamt für Landestopografie swisstopo, 3084 Wabern, Switzerland.

Kruschwitz, S. & Yaramanci, U. (2002) Engineered Barrier (EB) Experiment: EDZ Geophysical Characterisation. Detection and Characterisation of the Disturbed Rock Zone in Claystone with complex valued geoelectrics, Mont Terri Rock Laboratory, Technical Report TR 2002-01. Mont Terri Project, Bundesamt für Landestopografie swisstopo, 3084 Wabern, Switzerland.

Kupferschmied, N., Wild, K.M., Amann, F., Nussbaum, C., Jaeggi, D. & Badertscher, N. (2015) Time-dependent fracture formation around a borehole in a clay shale. International Journal of Rock Mechanics & Mining Sciences. [Online] 77, 105-114. Available from: doi:10.1016/j.ijrmms.2015.03.027.

Le Gonidec, Y., Schubnel, A., Wassermann, J., Gibert, D., Nussbaum, C., Kergosien, B., Sarout, J., Maineult, A. & Guéguen, Y. (2012) Field-scale acoustic investigation of a damaged anisotropic shale during a gallery excavation. International Journal of Rock Mechanics & Mining Sciences 51, 136–148.

Le Gonidec, Y., Sarout, J., Wassermann, J. & Nussbaum, C. (2014) Damage initiation and propagation assessed from stress-induced microseismic events during a mine-by test in the Opalinus Clay. Geophysical Journal International 198, 126–139.

Lettry, Y., Rösli, U. & Fierz, T. (2012) MA (Microbial activity in Opalinus Clay) experiment: Installation of surface equipment for use of H2 in the BRC-3 bioreactor borehole, Internal Technical Note. Mont Terri Project, Bundesamt für Landestopografie swisstopo, 3084 Wabern, Switzerland.

Marschall, P., Delay, J., Mayor, J.-C., Shao, H., Wieczorek, K. & Zhang, C.-L. (2008) Gas Migration Experiments. In: Bossart, P. & Thury, M. (eds.), Mont Terri Rock Laboratory. Project, Programme 1996 to 2007 and Results. Swiss Geological Survey Report No. 3. Bundesamt für Landestopografie swisstopo, 3084 Wabern, Switzerland.

Martin, C.C. & Lanyon, G.W. (2004) Excavation Disturbed Zone (EDZ) in Clay Shale: Mont Terri, Technical Report TR 2001-01. Mont Terri Project, Bundesamt für Landestopografie swisstopo, 3084 Wabern, Switzerland.

Martin, C.D., Macciotta, R., Elwood, D., Lan, H. & Vietor, T. (2013) Evaluation of the Mont Terri Mine-By response: Interpretation of results and observations. University of Alberta, Canada. Technical Report TR 2009-02. Mont Terri Project, Bundesamt für Landestopografie swisstopo, 3084 Wabern, Switzerland.

Mayor, J.C., Garcia-Siñeriz, J.L., Alonso, E., Alheid, H.-J. & Blümling, P. (2007) Engineered barrier emplacement experiment in Opalinus Clay for the disposal of radioactive waste in underground repositories. In: Bossart, P. & Nussbaum, C. (eds.), Mont Terri Project – Heater Experiment, Engineered Barriers Emplacement and Ventilation Tests, Swiss Geological Survey, Bern, Report No. 1. Bundesamt für Landestopografie swisstopo, 3084 Wabern, Switzerland.

Mazurek, M., Alt-Epping, P., Bath, A., Gimmi, T., Waber, H.N., Buschaert, S., De Cannière, P., De Craen, M., Gautschi, A., Savoye, S., Vinsot, A., Wemaere, I. & Wouters, L. (2011) Natural tracer profiles across argillaceous formations. Applied Geochemistry 26 (7), 1035–1064.

Möri, A., Bossart, P., Matray, J.-M., Frank, E., Fatmi, H. & Ababou, R. (2010) Mont Terri Project, Cyclic deformations in the Opalinus Clay. In: Clay I Natural and Engineered Barriers for Radioactive Waste Confinement: Proceedings of the Meeting "Nantes 2010", Nantes, France, pp. 103–124.

Nagra (2002): Opalinus Clay Project. Demonstration of feasibility of disposal for spent fuel, vitrified high-level waste and long-lived intermediate-level waste. Summary Overview. Swiss National Cooperative for the Disposal of Radioactive Waste, 5430 Wettingen, Switzerland.

Nussbaum, C., Graf, A. & Bossart, P. (2004) HE-D Experiment (THM behaviour of host rock): Structural and geological mapping of the HE-D niche: results and discussion. Internal Technical Note. Mont Terri Project, Bundesamt für Landestopografie swisstopo, 3084 Wabern, Switzerland.

Nussbaum, C., Bossart, P., Amann, F. & Aubourg, C. (2011) Analysis of tectonic structures and excavation induced fractures in the Opalinus Clay, Mont Terri underground rock laboratory (Switzerland). Swiss Journal of Geosciences. [Online] 104 (2), 187-210. Available from: doi:10.1007/s00015-011-0070-4.

Nussbaum, C. & Bossart, P. (2014) Mont Terri Rock Laboratory – Geophysical investigation of the Excavation Damaged Zone during a mine-by experiment. Ber. Landesgeol. 5, Wabern, swisstopo.

Palacios, B., Rey, M., Garcia-Siñeriz, J.L., Villar, M.V., Mayor, C. & Velasco, M. (2013) Engineered Barrier Emplacement Experiment in Opalinus Clay "EB" Experiment. As-built of dismantling operation. Long-term Performance of Engineered Barrier Systems PEBS. European Commission, 7th Euratom Framework Program for Nuclear Research and Training Activities (2007–2011). Report number: D2.1–4.

Palut, J.-M., Montarnal, P., Gautschi, A., Tevissen, E. & Lamoureux, M. (2003) Characterisation of HTO diffusion properties by an in-situ tracer experiment in Opalinus clay at Mont Terri. Special Issue of the Journal of Contaminant Hydrology 61, 203–218.

Pearson, F.J., Arcos, D., Bath, A., Boisson, J.-Y., Fernandez, A., Gaebler, H.-E., Gaucher, E., Gautschi, A., Griffault, L., Hernan, P., & Waber, H.N. (2003) Geochemistry of Water in the

Opalinus Clay Formation at the Mont Terri Rock Laboratory – Synthesis Report, Geological Report No. 5. Swiss National Hydrological and Geological Survey, Switzerland. Bundesamt für Landestopografie swisstopo, 3084 Wabern, Switzerland.

Pettitt, W.S. (1998) Acoustic emission source studies of microcracking in rock. Ph.D. Thesis, Keele University, UK.

Poulain, S., Sergeant, C., Simonoff, M., Le Marrec, C. & Altmann, S. (2008) Microbial investigations in Opalinus Clay, an argillaceous formation under evaluation as a potential host rock for a radioactive waste repository. Geomicrobiology Journal 25 (5), 240–249. Available from: doi: 10.1080/01490450802153314.

Schuster, K. & Alheid, H.-J. (2002) Engineered Barrier (EB) Experiment and Geophysical Characterisation of the Excavation Disturbed Zone (ED-C) Experiment: Seismic Investigation of the EDZ in the EB niche, Technical Report TR 2002-03. Mont Terri Project, Bundesamt für Landestopografie swisstopo, 3084 Wabern, Switzerland.

Schuster, K. (2009) Detection of borehole disturbed zones and small scale rock heterogeneities with geophysical methods. Conference paper EC-TIMODAZ-THERESA THMC, 29th September – 1st October 2009.

Li, S., Feng, X.-T & Li, Z. (2012) Evolution of fractures in the excavation damaged zone of a deeply buried tunnel during TBM construction. International Journal of Rock Mechanics and Mining Sciences 55 (10), 125–138.

Trick, T. (1999) EH Experiment: Constant Rate Injection/- Extraction. Gas Tests in Boreholes: BEH-1 to BEH-10 and Helium Dipole Tracer Test in Boreholes BEH-5/BEH-11 and BEH-6/BEH-11, Internal Technical Note. Mont Terri Project, Bundesamt für Landestopografie swisstopo, 3084 Wabern, Switzerland.

Volckert, G. & Fierz, T. (1999) The new piezometer. In: Thury, M. & Bossart, P. (eds), Mont Terri Rock Laboratory. Results of the Hydrogeological, Geochemical and Geotechnical Experiments performed in 1996 and 1997. Swiss National and Geological Survey (SNHGS), Geological Report No. 23. Bundesamt für Landestopografie swisstopo, 3084 Wabern, Switzerland.

Wersin, P., Baeyens, B., Bossart, P., Cartalade, A.,Dewonck, S., Eikenberg, J., Fierz, T., Fisch, H. R., Gimmi, T., Grolimund, D., Hernán, P., Möri, A., Savoye, S., Soler, J., van Dorp, F. & van Loon, L. (2006) Long-term Diffusion Experiment (DI-A): Diffusion of HTO, I⁻, ^{22}Na$^+$ and Cs$^+$: Field activities, data and modelling, Technical Report TR 2003-06. Mont Terri Project, Bundesamt für Landestopografie swisstopo, 3084 Wabern, Switzerland.

Wersin, P., Leupin, O.X., Mettler, S., Gaucher, E.C., Mäder, U., De Cannière, P., Vinsot, A., Gäbler, H.E., Kunimaro, T., Kiho, K. & Eichinger, L. (2011) Biogeochemical processes in a clay formation in-situ experiment: Part A – Overview, experimental design and water data of an experiment in the Opalinus Clay at the Mont Terri Underground Research Laboratory, Switzerland. Applied Geochemistry 26, 931–953.

Wileveau, Y. (2005) THM behavior of host rock: (HE-D experiment): Progress Report September 2003 – October 2004 Part 1, Technical Report TR 2005-03. Mont Terri Project, Bundesamt für Landestopografie swisstopo, 3084 Wabern, Switzerland.

Young, R.P. & Collins, D.S. (2001) Seismic studies of rock fracture at the Underground Research Laboratory, Canada. International Journal of Rock Mechanics and Mining Sciences 38 (6), 787–799.

Zhang, C.-L., Rothfuchs, T., Jockwer, N., Wieczorek, K., Dittrich, J., Müller, J., Hartwig, L. & Komischke, M. (2007) Thermal Effects on the Opalinus Clay, A Joint heating Experiment of ANDRA and GRS at the Mont Terri URL (HE-D project). Report number: 224. GRS Braunschweig, Germany.

Chapter 16

Thermal properties and experiment at Äspö HRL

Jan Sundberg

Division of Geoengineering, Department of Civil and Environmental Engineering, Chalmers University of Technology, Gothenburg, Sweden

Abstract: The underground hard rock laboratory at Äspö (Äspö HRL), situated north of Oskarshamn, Sweden, is where much of the research about the final repository for spent nuclear fuel is taking place. The Äspö HRL is a unique research facility situated 450 meters underground. One part of the investigations at Äspö HRL, and the adjacent Simpevarp and Laxemar areas, has focused on thermal transport properties in the rock. The thermal properties of the rock have large influence on the thermal design of an underground repository for spent fuel. Temperature requirements on the canister or the surrounding buffer influence the distances between canisters and tunnels. The relevant scale for the thermal processes is of importance. This chapter discusses influences on thermal transport properties in crystalline rock with respect of *e.g.* scale, mineral composition and anisotropy. Primarily these processes are discussed according to methods for determination of thermal conductivity, from laboratory measurements to large scale experiments. The chapter also discusses indirect methods that make it possible to calculate the thermal conductivity from mineral composition and from the density log, including the theoretical base. Results on thermal conductivity and heat capacity from common rock types are reported. The development of prediction methods in different scales are exemplified with large scale thermal response test and inverse modeling at the prototype repository. Finally the strategy is outlined for the site descriptive thermal modeling in the Swedish site investigation program for nuclear waste disposal.

1 INTRODUCTION

1.1 General

Thermal properties in the area of engineering geology have large influence on a number of different processes, *e.g.*:

- The thermal design of buried high voltage cables. Temperature requirements on the cable influence the design of the cable and distances between cables.
- The heat losses from a geothermal storage. Higher thermal conductivity gives larger thermal losses.
- The extraction rate from geothermal applications. High thermal conductivity means more effective heat exchange in the rock.
- The thermal dimensioning of underground repository for spent fuel. Temperature requirements on the canister or the surrounding buffer influence the thermal design of distances between canisters.

The latter bullet has relevance for Äspö and the adjacent Laxemar area, described in this chapter. The prototype repository is situated at Äspö HRL and Laxemar was one of the candidate areas for the final repository for spent nuclear fuel in Sweden. The heat generated by the spent nuclear fuel will increase the temperature of all components of the repository: barriers, tunnels, seals and the host rock itself. To ensure the long-term sealing capacity and the mechanical function of the bentonite buffer surrounding each individual canister, a maximum bentonite temperature is prescribed in the design premises. This requirement, which relates to the safety assessment, implies that the canisters cannot be deposited arbitrarily close to each other. Unnecessarily large distances between the canisters, on the other hand, will mean inefficient and costly use of the repository rock volume. In order to determine the minimum canister spacing required to meet the temperature criterion for all canister positions it is necessary to establish an adequate description of the site's rock thermal properties and their spatial variation at the relevant canister scale. In addition to being needed for the design and layout, this thermal site model will be important for predicting the thermo-mechanical evolution of the repository host rock at different scales.

1.2 Influence on thermal transport properties

The dominating thermal transport process in crystalline rock is thermal conduction. Forced convection or convection by gravitation may occur only in hydraulic conductive structures. The thermal conductivity of crystalline rock is mainly influenced by the following factors:

- Mineral composition
- Temperature
- Porosity and pressure
- Anisotropy and heterogeneity

Variations in the mineral distribution for a rock type results in differences in the thermal conductivity. Quartz has 3-4 times higher thermal conductivity than most other common minerals. Thus, the quartz content normally has a great influence on the total thermal conductivity. However, for rock types with low quartz content other minerals have a dominating effect. The thermal conductivity of some minerals, for example plagioclase, depends on the chemical composition of the mineral. Compared to thermal conductivity, the heat capacity of different minerals has a lower variation.

Studies of the temperature dependence of the thermal conductivity of common rocks normally shows a decrease in thermal conductivity and increase in heat capacity with increasing temperature. The porosity of crystalline rock is low, in general less than 1 %. The pressure dependency of thermal conductivity is also generally low, provided that the rock is water saturated (Walsh & Decker, 1966).

In anisotropic rocks, the thermal conductivity is different in different directions. The anisotropy factor is defined as the ratio of thermal conductivity of the high-conductive and low-conductive directions. Thus the anisotropy factor is always ≥1 (equal to 1 for an isotropic rock). There are different types of thermal anisotropy to consider. The first type is a structural anisotropy caused by foliation and lineation which occur within a rock type. The foliation and lineation imply a directional orientation of the minerals in the rock mass. A second type of anisotropy occurs in a larger scale and is a result of the spatial orientation of magmatic rock bodies, primarily subordinate rocks (*e.g.*

A   B   C   D

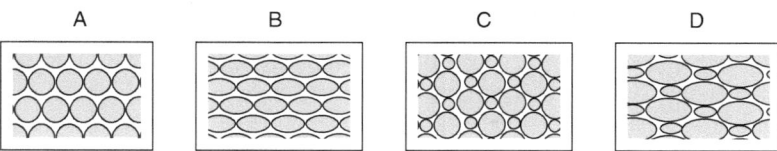

Figure I The concepts of heterogeneity and anisotropy: A) homogeneous isotropic material, B) homogeneous anisotropic material, C) heterogeneous isotropic material, and D) heterogeneous anisotropic material (see *e.g.* Norrman, 2004).

Amphibolites in a granite). Anisotropy may also be caused by heterogeneity within a rock type, *i.e.* by different spatial trends in the composition of a rock type in different directions.

Heterogeneity in thermal properties is an effect of the lithology in combination with heterogeneous mineral composition of individual rock types. Homogeneity is a question of scale. A homogenous rock as granite is of heterogeneous in the small mineral scale. The difference between anisotropy and heterogeneity is illustrated in Figure 1.

1.3 Definitions

Thermal conductivity and heat capacity is needed to describe the transient thermal transport process. The thermal conductivity, λ [W/(m·K)], describes the ability of a material to transport heat. The heat capacity denotes the capacity for a material to store thermal energy. The volumetric heat capacity, C [J/(m^3·K)], is the product of density, ρ, and specific heat capacity, c [J/(kg·K)]. The thermal diffusivity, κ [m^2/s], describes a material's ability to level temperature differences. It is defined as the ratio between thermal conductivity and volumetric heat capacity:

$$\kappa = \lambda/(\rho \cdot c) \tag{1}$$

The natural temperature field in the ground is a function of boundary conditions, internal heat production and thermal transport properties in the rock mass. In a crystalline rock mass, the thermal transport mainly results from conduction and locally, in fracture zones, of convection due to ground water movement.

1.4 Dominant rock type in the Äspö area

The bedrock in the Laxemar and Äspö area is dominated by porphyritic intrusive rock types that display a compositional variation between granite, granodiorite and quartz monzodiorite. In addition, the rock types are commonly intimately mixed and the contacts are gradational. Due to difficulties to distinguish between the different compositional varieties during the bedrock mapping at the surface and mapping of the drill cores without access to analytical data, they were collectively called Ävrö granite during the initial stages of the site investigation at Laxemar and Simpevarp. However, in connection with the final site modeling work at Laxemar (Wahlgren *et al.*, 2008), the Ävrö granite was subdivided in the two varieties Ävrö granodiorite and Ävrö quartz monzodiorite by the evaluation of results from modal and chemical analyses, as well as density data from the geophysical logging of the boreholes. This subdivision was of great importance, not the least for the rock mechanical modeling work and modeling of the thermal properties, since the mineralogical composition affects the rock mechanical and thermal properties of the rock types. The corresponding rock types at Äspö have traditionally been called Ävrö (Småland) granite and Äspö diorite since the pre-investigations and construction of the Äspö HRL. However, analytical data have shown that the rock type that has been documented as Äspö diorite is not a true diorite, but rather display a variation in composition between quartz monzodiorite and granodiorite (Rhén et al., 1997), i.e. similar to the situation at Laxemar.

1.5 Data on thermal properties

Data on thermal properties comes from both Äspö HRL and the adjacent Laxemar area (and to some extent from the adjacent Simevarp area) with similar rock types. In section 3 and 4 the data mainly come from the site investigations in Laxemar since the amount of thermal data is much larger. Specific data from Äspö HRL are described in section 5.1 and 5.2. The varieties in terminology between Äspö and Laxemar are described in section 1.4. It should be emphasized that the name "Ävrö granite" has not identical meaning in terminology used at Äspö HRL and the Laxemar site investigations.

2 USED METHODS

The methods used for determination of thermal conductivity can be subdivided in different ways. Here thermal methods are divided in:

- Laboratory methods
- Field methods
- Large scale methods
- Indirect methods – mineralogical composition and density logging

Focus for the methods is on determination of the thermal conductivity. If the method also allows for determination of the thermal diffusivity, the volumetric heat capacity can be calculated. Direct determination of heat capacity has only been made in laboratory scale.

2.1 Relevant scales

The thermal properties vary depending on the scale of observation. Heterogeneities exist at the whole spectrum of scales and the resulting variability must be handled. A common modeling approach is to use effective values to characterize the thermal conductivity at a particular scale. The effective value for a larger scale than the measurements represents can be approximated by calculations, i.e. upscaling (see e.g. Dagan 1979; Sundberg et al., 2005). The geometric mean is a good approximation of the effective thermal conductivity for a larger scale if the standard deviation is small (Dagan 1979; Sundberg et al., 2005). The mean of the distribution is generally affected only to a small extent by upscaling. However, the variance and standard deviation are usually reduced when the scale is increased. Upscaling of thermal conductivity is further discussed in section 2.4.1, 5.3.2 and e.g. Back & Sundberg (2007).

The different rock forming minerals exist at a micro- or millimeter scale. Thus, there is a large variation in thermal properties at this scale. If the rock is relatively homogeneous, variation of thermal conductivity at one scale is averaged out at a certain distance (a larger scale).

The *in situ* methods usually give a characteristic value for a larger volume compared to laboratory measurements. When laboratory measurements are used, upscaling may be necessary. However, this involves uncertainties and measurements at a significant scale for the actual problem are preferable.

2.2 Laboratory methods

2.2.1 TPS-method

There are a number of different types of laboratory methods to determine thermal properties. At Äspö HRL and the site investigation in Laxemar the recommended method is the TPS-method (Transient Plane Source). It is a transient method, unlike the common used dived bar method. The method uses a sensor element with an engraved pattern of a thin metal double spiral. The spiral is embedded in an insulation material and installed between two rock samples. The sensor both emits heat and measures the temperature. Both the thermal conductivity and diffusivity is possible to evaluate, and from these the volumetric heat capacity is calculated. Depending on desired measurement scale, different sensor sizes can be used. The TPS method is described by Gustafsson (1991).

The TPS-method also allows for measurement of thermal anisotropic conditions. However, such an evaluation demands that the sample is orientated due to the principal axes of the anisotropy and that the heat capacity is known and determined separately with an independent method.

2.2.2 Calorimetric method

In addition to the TPS-measurements, the calorimetric technique has been used in the site descriptive modeling in Laxemar to determine the heat capacity. The samples are placed in a temperature controlled water bath long enough to stabilize. The calorimeter is filled with water to a defined level and at nearly steady state conditions. The calorimeter is stirred and the temperature logged. The sample is quickly moved from the bath into the calorimeter. The temperature rise of water is recorded during the equalization process (Adl-Zarrabi, 2006).

2.3 Field and large scale methods

2.3.1 Multi-probe method

The multi probe method (Landström *et al.*, 1979) is a development of the well-known (single) probe method. The theory is originally based on the infinite line source theory. The typical scale for single- or multi-probe measurements is in the range of 0.2-1 m. *In situ* measurements with the multi-probe method have been performed at the prototype repository at Äspö HRL (Sundberg & Gabrielsson, 1999) and in the Laxemar area

(Mossmark & Sundberg, 2007). The latter measurements were made in larger scale, using a 2.4 m long heating probe.

Measurements with the multi-probe method can also be performed to analyze the thermal conductivity in different directions in anisotropic rocks. However, this demands simultaneously measurement of the temperature response in two directions, parallel and perpendicular to the anisotropy, and a more advanced evaluation technique (Mossmark & Sundberg, 2007).

2.3.2 Thermal response test

The method of thermal response tests has been suggested as a potential thermal characterization method for the site investigations (SKB, 2001). The method can in principle be described as a large-scale probe method that makes it possible to evaluate the mean of the thermal conductivity and, theoretically, the thermal diffusivity for the rock mass around a borehole, Figure 2. Primarily, an apparent thermal conductivity is determined with this method. The analysis assumes heat transfer through thermal conduction only but the measured actual heat transfer also includes possible convective heat transport, *e.g.* along the borehole. The method is described by (Gehlin, 2002) and has been tested and evaluated at Äspö HRL (Sundberg, 2002). The typical measurement scale for thermal response tests is in the range of 5–100 m.

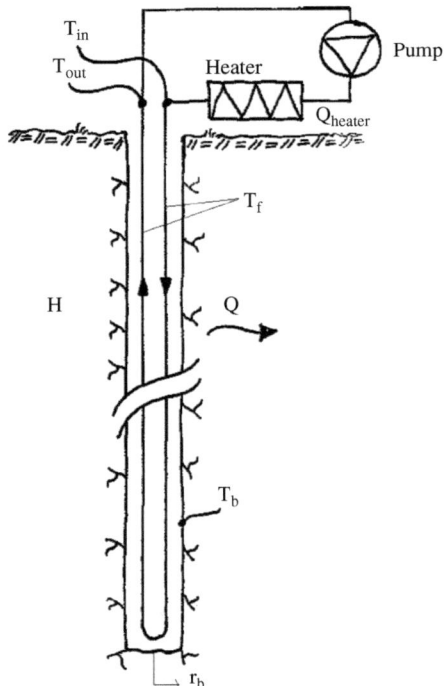

Figure 2 The principle of the thermal response test (Gehlin, 2002).

2.4 Indirect methods

2.4.1 Calculation from the mineralogical composition

The heat capacity of rock can be computed from volume integrations of the heat capacity for the minerals. The thermal conductivity of composite materials, such as rock, is much more complicated to calculate. In Sundberg (1988) an overview of different approaches to the subject is given.

For calculations of thermal conductivity from mineral compositions, the self-consistent approximation (SCA) of a 2-phase material was suggested by Bruggeman (1935). For hydraulic conductivity, this has later been redeveloped for n-phase materials (Dagan, 1979). Transformed to thermal conductivity (Sundberg, 1988), the method assumes each grain to be surrounded by a uniform medium with the effective thermal conductivity. In a n-phase material, the effective thermal conductivity, λ_e, can be estimated from the following expression by a number of iterations:

$$\lambda_e = \frac{1}{m} \left[\sum_{i=1}^{n} \frac{v_i}{(m-1) \cdot \lambda_e + \lambda_i} \right]$$

where m is the dimensionality of the problem, λ_i the thermal conductivity of a grain, v_i the associated volume fraction of the grain and n the number of phases. The accuracy of the method is depending on the accuracy of the thermal conductivities of the minerals, possible alterations and the limitations associated with the point-counting method used. For a log-normal distribution the more simply geometric mean is associated with transport in 2 dimensions (Dagan, 1979).

2.4.2 Calculation from density logging

Density measurements have been used as an indicator to distinguish between Smålands (Ävrö) granite (in the site investigations at Laxemar called Ävrö granodiorite) and Äspö diorite at Äspö HRL (Rhén et al., 1997). A relationship between density and thermal conductivity for investigated rock types was later observed by Sundberg (2002). Based on available measurements from Äspö HRL empirical relationships between density and thermal properties were derived in Sundberg (2003) and are shown in Figure 3. More recent data for Ävrö granite have led to a modified relationship between density and thermal conductivity (Sundberg et al., 2008), see section 3.2. The typical scale for density measurements by means of gamma-logging is approximately 0.2 m. Using the relationship it is possible to calculate the thermal properties from density loggings in boreholes, and investigate the spatial correlation structure. The theoretical base for the correlation between density and thermal properties have been analyzed in Sundberg et al. (2009) and is illustrated in Figure 9.

3 THERMAL CONDUCTIVITY OF COMMON ROCK TYPES

3.1 Measurements results

Summary statistics of thermal conductivity for each rock type are presented in Table 1. The statistics are mainly based on data from the Simpevarp and Laxemar subareas.

Table 1 Measured thermal conductivity (W/(m·K)) of some rock types using the TPS method. Data from the Laxemar and the Simpevarp subarea (Sundberg et al., 2008).

Rock name	Mean	St. dev	Max	Min	Number of samples	Mean density
Fine-grained granite	3.69	0.08	3.76	3.58	4	
Ävrö quartz monzodiorite	2.36	0.20	2.71	2.01	33	2756
Ävrö granodiorite	3.17	0.17	3.76	2.81	60	2677

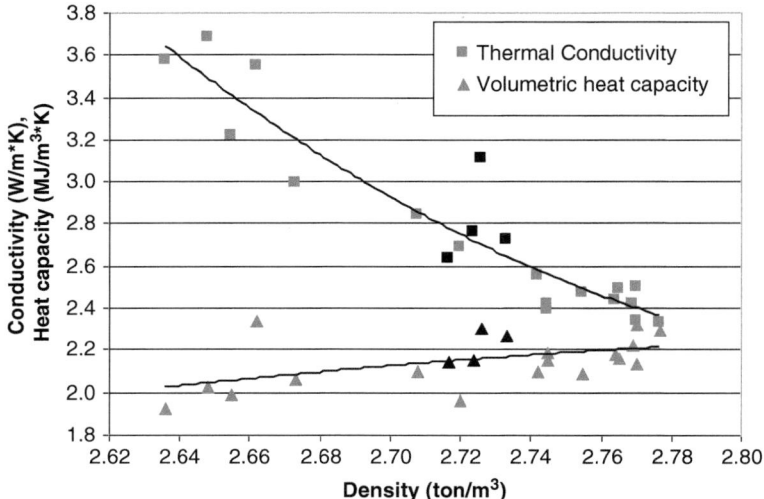

Figure 3 Estimated relationships between density and thermal properties of investigated rock types at Äspö HRL. Values of altered Äspö diorite in blue color (Sundberg, 2003).

The Ävrö granite has been recognized as bimodal with respect to thermal conductivity (Wrafter *et al.*, 2006). An investigation of the mineral composition indicates a broadly bimodal quartz content, which has resulted in the subdivision of Ävrö granite into two distinct rock types: Ävrö quartz monzodiorite and Ävrö granodiorite (Wahlgren *et al.*, 2008), see also section 1.4. The former is quartz poor (commonly < 15 %) while the latter is quartz rich (usually > 20 %). In the absence of modal analysis, a density of 2 710 kg/m³ has been identified as a suitable threshold value for distinguishing between the two rock types (Wahlgren *et al.*, 2008). The thermal conductivity for each type are summarized in Table 1 and a histogram of thermal conductivity values for both types is presented in Figure 4. The thermal conductivities of Ävrö quartz monzodiorite and Ävrö granodiorite are clearly differentiated from each other; the distributions show little or no overlap.

The thermal conductivities of Ävrö quartz monzodiorite and Ävrö granodiorite versus elevation are presented in Figure 5.

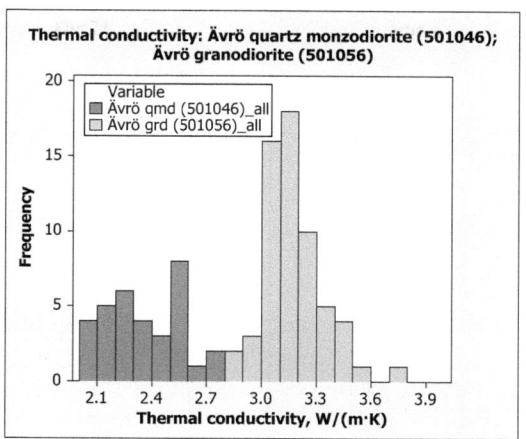

Figure 4 Distribution of thermal conductivity values for different varieties of Ävrö granite measured using the TPS method (Sundberg *et al.*, 2008).

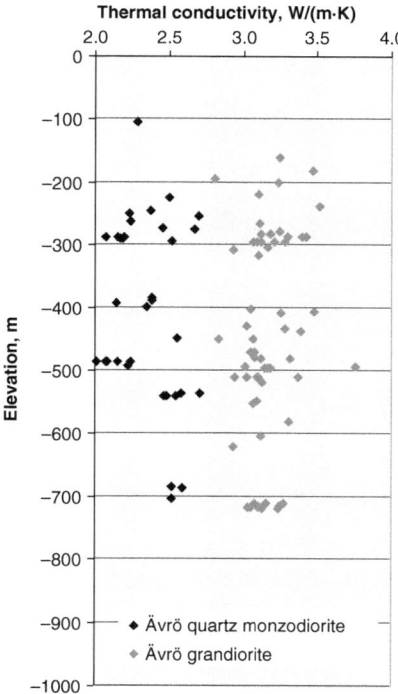

Figure 5 Thermal conductivity versus elevation for Ävrö granodiorite and Ävrö quartz monzodiorite using the TPS method. Samples from Laxemar and Ävrö boreholes (Sundberg *et al.*, 2008).

Table 2 Temperature dependence of thermal conductivity (per 100°C temperature increase) for some rock types. Mean of temperature dependence calculated by linear regression (Sundberg *et al.*, 2008).

Rock name	Sample location	Mean	St. dev	Number of samples
Ävrö quartz monzodiorite	Boreholes KLX04 and KA2599G01	−2.9 %	3.3 %	8
Ävrö granodiorite	boreholes KLX02 and KA2599G01	−6.8 %	3.6 %	5

3.1.1 Temperature dependence

The temperature dependence of thermal conductivity and heat capacity has been investigated by laboratory measurements, three to five different temperatures (mainly at 20, 50 and 80°C). In Table 2 the thermal dependence of thermal conductivity for the Ävrö quartz monzodiorite and the Ävrö granodiorite is summarized.

3.1.2 Influence of alteration

The degree of alteration has been classified as faint, weak, medium or strong, and is dominated by both oxidation and saussuritisation. The available data indicates that the thermal conductivity of rock, showing a weak or medium degree of alteration, is generally higher than that of fresh rock. For Ävrö granodiorite, altered samples indicate a 4 % higher thermal conductivity (Sundberg *et al.*, 2008).

3.1.3 Pressure dependence

The thermal conductivity is lower for stress-released samples compared to determinations at higher pressure (greater depths). The reason is assumed to be the closing of micro cracks. However, the pressure influence up to 50 MPa seems to be low if the samples are water saturated, approximately 1–2 % (Walsh & Decker, 1966). The pressure dependence after closing of fractures can be estimated to approximately 0.5-1 %/100 MPa, based on data presented in (Seipold & Hunges, 1998). All determinations of thermal conductivity have been made on water saturated samples. The pressure dependence has therefore been neglected.

3.1.4 Anisotropy

In Laxemar, a faint to weak foliation, which is not uniformly distributed over the area, is commonly present and affects all rock types (Wahlgren *et al.*, 2008). Measurements of anisotropy on core samples in laboratory have not been included in the thermal investigation programme. The anisotropy has instead been evaluated based on the field measurements with the multi probe method (Sundberg & Mossmark, 2007). Field measurements in Ävrö granite indicate that thermal conductivity parallel to the foliation plane are higher, by a factor of approximately 1.15, than conductivity perpendicular to the foliation. The spatial variability of this anisotropy is not known (Sundberg *et al.*, 2008).

3.2 Correlation with density

A relationship between density and measured (TPS) thermal conductivity for Ävrö granite in the Laxemar-Simpevarp area is well established (Sundberg, 2003; Sundberg *et al.*, 2005, 2009; Wrafter *et al.*, 2006). The observed relationships, see Figure 6, are consistent with the results of theoretical calculations based on mineral composition (Sundberg *et al.*, 2009).

Establishing relationships between density and thermal conductivity allows a more reliable use of borehole density data for analyzing the spatial distribution and correlation structure of thermal properties. This has been utilized in the site investigations in Laxemar (Sundberg *et al.*, 2008).

Using the relationship in Figure 6 it is also possible to calculate the thermal properties from density loggings in boreholes. An example of thermal conductivity versus depth modeled from density logging in a borehole at Äspö HRL is shown in Figure 7.

The relationship between density and thermal conductivity for all other investigated rock types is illustrated in Figure 8. Diorite-gabbro appears to display the reverse relationship between thermal conductivity and density as compared to Ävrö granite. Low density samples have low thermal conductivities (< 2.6 W/(m·K)), whereas high density samples have more variable, but generally higher, conductivities (up to 3.65 W/(m·K)). Furthermore, the marked variation in thermal conductivity displayed by samples with similar densities may be partly due to variable degrees of post-magmatic mineralogical changes as a result of, for example, hydrothermal alteration. The observed overall relationship between density and thermal conductivity for all igneous rock types

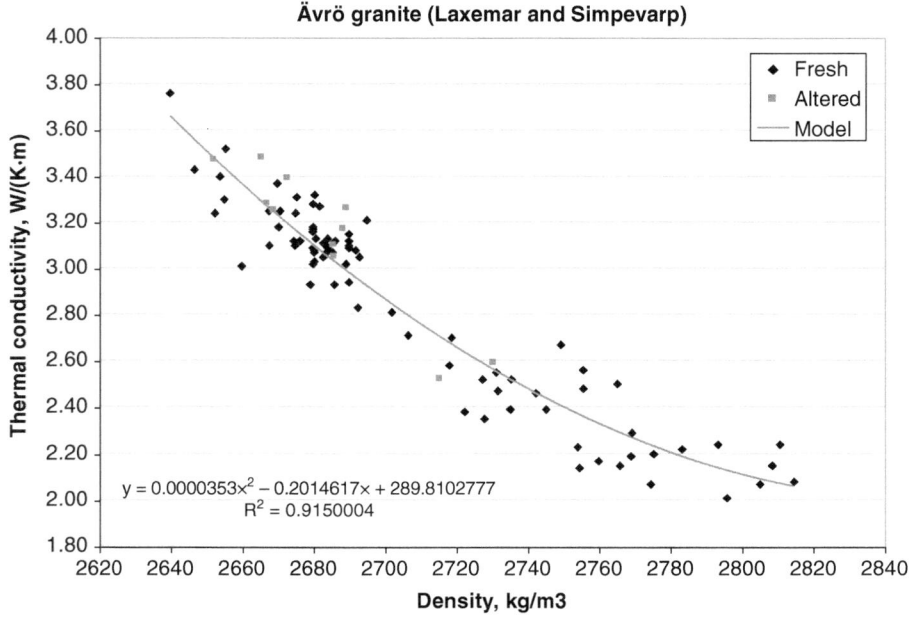

Figure 6 Relationships between density and thermal conductivity (TPS) for Ävrö granite. The equation is valid within the density interval 2 640 – 2 820 kg/m³ (Sundberg *et al.*, 2008).

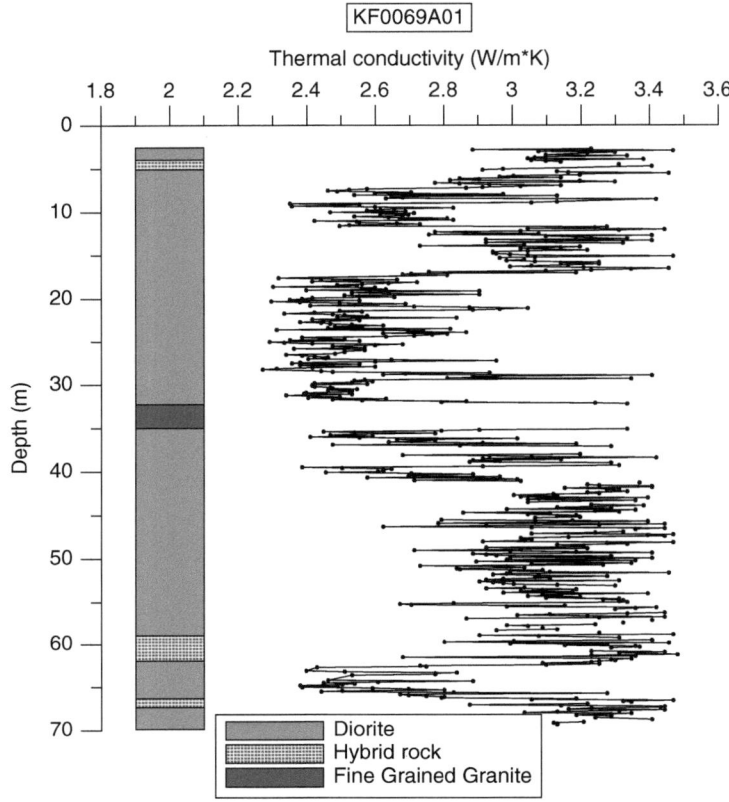

Figure 7 Calculated thermal conductivity (from density logging result) versus depth for borehole KF0069A01 at Äspö HRL (Sundberg, 2003).

together is consistent with the results of theoretical calculations presented in Sundberg *et al.* (2009) and in Figure 9. In the analysis, it was established that, for felsic rocks the thermal conductivity decreases with density, whereas for mafic rocks the opposite relationship applies. For example, the high-density mafic rocks (*e.g.* gabbros) contain abundant pyroxene and amphibole (and in some cases chlorite), minerals which, compared to the feldspars, have relatively high thermal conductivities.

It is reasonable to assume that there exists a corresponding relationship between density and heat capacity. Such a relationship can be seen in Figure 3 and Figure 11, but it is weaker than that for thermal conductivity vs. density.

3.3 Comparison of results with different methods

3.3.1 TPS-measurements versus calculation from modal analysis

For several of the drill cores in the Laxemar area from which samples have been taken for laboratory determination of thermal conductivity (TPS method), sampling for

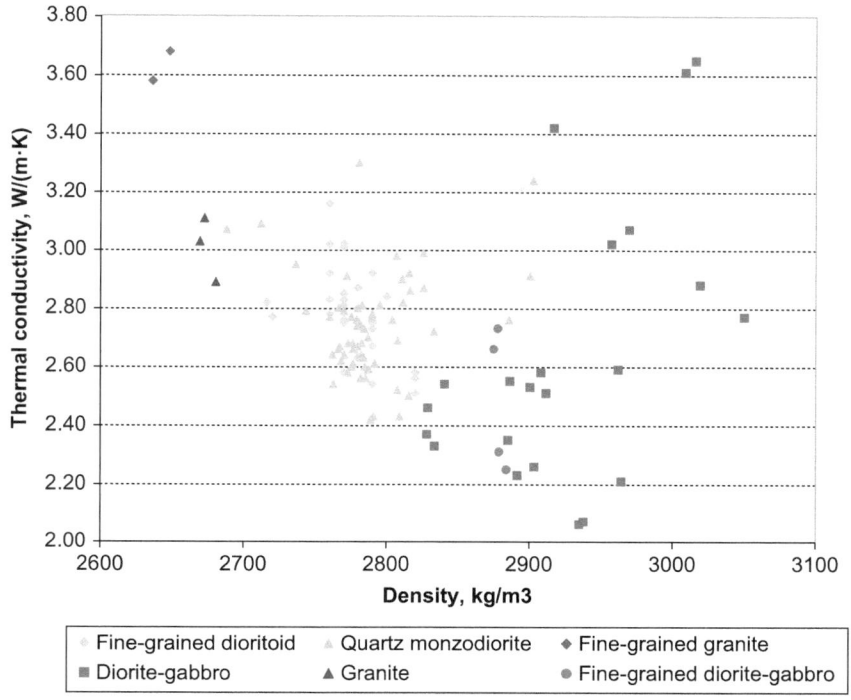

Figure 8 Relationships between density and thermal conductivity for rock types other than Ävrö granite (Sundberg *et al.*, 2008).

modal analysis and SCA calculations has also been carried out, see equation in section 2.4.1. The objective is to compare determinations from the different methods as well as to evaluate the validity of the SCA calculations. In Table 3 a comparison of TPS and SCA data is presented on a rock type basis. A comparison of individual samples is illustrated in Figure 10. It should be emphasized that the samples are not exactly the same, but come from adjacent sections of the borehole. Therefore, some of the observed differences are probably a result of the sampling.

The results indicate a quite a good agreement between the measured (TPS) and calculated thermal conductivity values for most rock types. An exception to this are the samples of Ävrö quartz monzodiorite with thermal conductivities less than 2.3 W/(m·K) as indicated by the TPS data. The SCA values of these four samples overestimate the thermal conductivity by on average 12 %. Possible explanations for this are (1): the degree of alteration assumed in the calculations of "fresh" samples may not have been affected by alteration to the same extent as other Ävrö quartz monzodiorite samples and may therefore not be representative for these samples; (2): the anorthite content of plagioclase influences the thermal conductivity and may be significantly higher in this type of Ävrö quartz monzodiorite than in more quartz rich varieties.

The comparison of SCA results with TPS data indicates that the SCA method yields quite good estimates of the mean thermal conductivities for the different rock types.

Table 3 Comparison of thermal conductivity of some rock types calculated from mineralogical compositions by the SCA method and measured with the TPS method.

Method	Ävrö quartz monzodiorite	Ävrö granodiorite	Diorite-gabbro
Calculated (SCA): Mean λ, (W/(m·K))	2.56	3.16	2.69
Measured (TPS): Mean λ, (W/(m·K))	2.48	3.17	2.72
Number of sample pairs	11	19	11
Diff. (SCA − TPS)/TPS	3.3 %	−0.3 %	−1.1 %

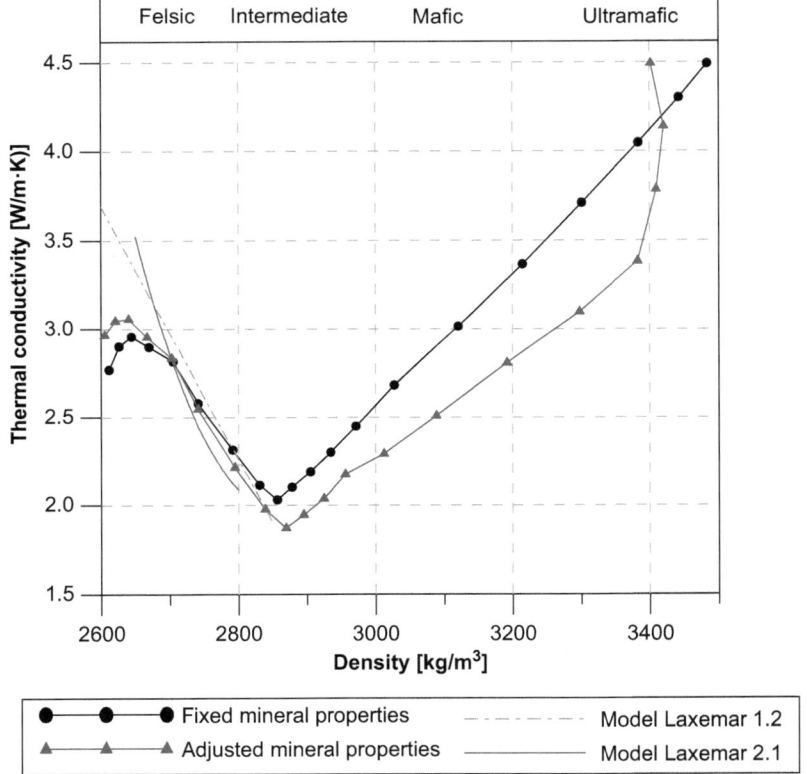

Figure 9 Thermal conductivity vs. density for synthetically defined data with different mineral compositions (Sundberg et al., 2009).

3.3.2 TPS versus divided bar

In Sundberg *et al.* (2003) a comparable study on 17 samples of measurements with the TPS method and the divided bar method were performed. The result indicated slightly higher thermal conductivity mean with the divided bar method than the TPS method in average 3.4 %. The difference between individual samples show larger variations, ranges between −8.9 % to +10.6 %.

Reasons for the differences are estimated mainly to be dependent on differences between the samples. The TPS measurements are performed using two pieces

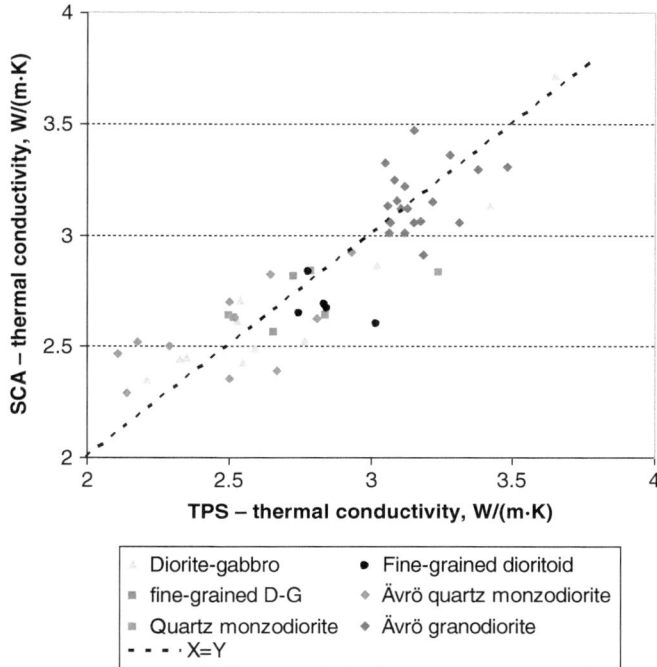

Figure 10 TPS versus SCA values for the "same" samples. The line through the data points represents TPS = SCA and is inserted to aid interpretation. (Sundberg et al., 2008).

(subsamples) of rock. Only one of these two subsamples were sent to the Geological Survey of Finland and measured using the divided bar method. Further, sample preparation for the divided bar method involved changes in the size of some of the samples (smaller size). The mean differences between the methods are for most samples within the margins of error reported by the measuring laboratories.

4 HEAT CAPACITY OF COMMON ROCK TYPES

4.1 Measurement results

Heat capacity has been determined indirectly from thermal conductivity and diffusivity measurements using the TPS (Transient Plane Source) method, and directly by calorimetric measurement. In Table 4 the results from heat capacity calculations from TPS measurements are summarized on a rock type basis. Ävrö granite has been divided into Ävrö quartz monzodiorite and Ävrö granodiorite. Comparison between indirect and direct method is made in section 4.4.

4.2 Temperature dependence

Temperature dependence of heat capacity have been investigated in the temperature interval 20 – 80°C. Results for main rock types are presented in Table 5, per 100°C temperature increase.

Table 4 Results of heat capacity (MJ/m³·K) determinations from TPS measurements on common rock types (Sundberg *et al.*, 2008).

Rock name	Ävrö quartz monzodiorite	Ävrö granodiorite	Fine-grained granite
Mean	2.23	2.20	2.04
St. dev.	0.17	0.16	0.08
N	33	60	4
Max	2.52	2.50	2.12
Min	1.73	1.81	1.93

Table 5 Temperature dependence of heat capacity (per 100°C temperature increase). The mean temperature dependence is estimated by linear regression (Sundberg *et al.*, 2008).

Rock name (name code) (sample location)	Mean	St. dev	Number of samples
Ävrö quartz monzodiorite (Boreholes KLX04 and KA2599G01)	26.0 %	7.04 %	8
Ävrö granodiorite (boreholes KLX02 and KA2599G01)	23.8 %	2.92 %	5

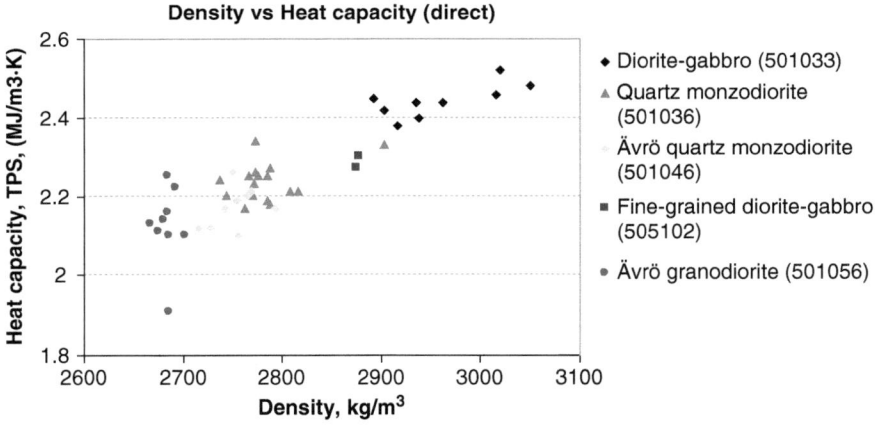

Figure 11 Density versus heat capacities (direct calorimetric measurements) for different rock types (Sundberg *et al.*, 2008).

4.3 Correlation with density

The relationship between thermal conductivity and density was described earlier. To investigate if a corresponding relationship between heat capacity and density exists, density was plotted against both indirectly and directly determined heat capacity values. The results show a wide range in indirectly determined heat capacity values within a restricted density range, see section 4.4. Figure 11 on the other hand shows a more consistent pattern of increasing heat capacity (direct measurements) with increasing density.

4.4 Comparisons of different methods

A comparison of direct (calorimetric) and indirect methods (calculation from diffusivity by the TPS-method) on the same samples is presented in Table 6 and Figure 12. Standard deviations are higher for the indirect determinations than for the direct measurement data and differences in the heat capacity values of up to c. 20 % are observed for individual rock samples. However, the average difference between the results of the two methods is less than 1 %, which indicates that the calculated values based on TPS determinations, although more uncertain, do not suffer from bias.

Table 6 Comparison between heat capacities ($MJ/m^3 \cdot K$) from TPS measurement and the calorimetric method for the same samples. Data from existing rock types in the Laxemar area (from Sundberg *et al.*, 2008).

Method	Rock name	Diorite-gabbro	Quartz monzodiorite	Ävrö quartz monzodiorite	Ävrö granodiorite	Fine-grained diorite-gabbro	All rock types
	Rock code	501033	501036	501046	501056	505102	
TPS	Mean	2.37	2.22	2.19	2.13	2.20	2.23
	St. dev.	0.22	0.10	0.13	0.17		0.164
	N	9	16	9	9	2	45
Calorimetric	Mean	2.44	2.24	2.17	2.12	2.29	2.24
	St. dev.	0.04	0.05	0.05	0.10		0.125
	N	9	16	9	9	2	45

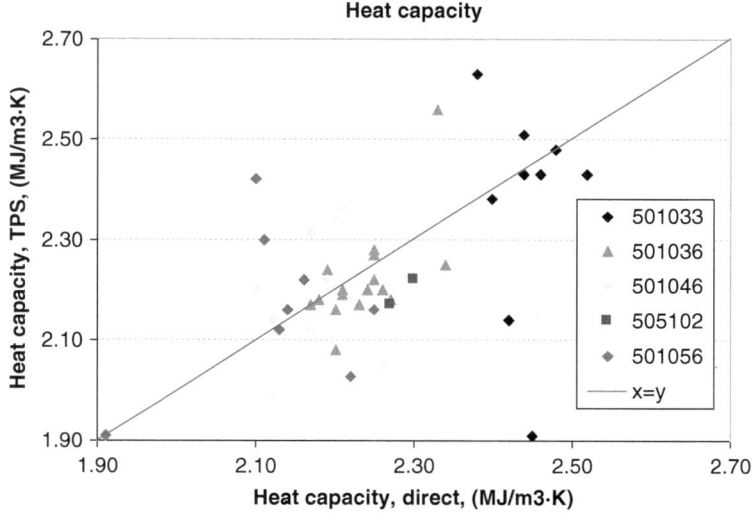

Figure 12 Comparison between heat capacities from TPS measurement and the direct calorimetric method. Data from existing rock types in the Laxemar area (Sundberg *et al.*, 2008).

5 DEVELOPMENT OF PREDICTION METHODS IN DIFFERENT SCALES

The thermal function of a repository can be studied in different scales, see section 5.3.1. Thermal data can be determined by measurement in a relevant scale or by modeling involving upscaling. In this section prediction methods are discussed and analyzed. Further, a strategy for thermal modeling in a relevant scale are described.

5.1 Large scale thermal response test

5.1.1 Introduction

A method to determine the thermal properties on a large scale in a borehole, thermal response test, has been tested at the Äspö Hard Rock Laboratory. The borehole, KA 2599 G01, was specially drilled for in-situ measurement of rock stress. The core has a total length of 128.3 m and is drilled vertically from the gallery in Äspö HRL at chainage 2599 m, see Figure 13. Parallel to the response test other methods have been used to determine the thermal properties of the rock mass in order to make a prediction of the result of the large scale thermal response test (Sundberg, 2002).

The objective with the study was:

- To investigate the possibility to predict the result of large scale measurement with other methods and to evaluate the results of the different methods and methodologies
- To investigate the influence on the prediction of the quality of the rock type mapping

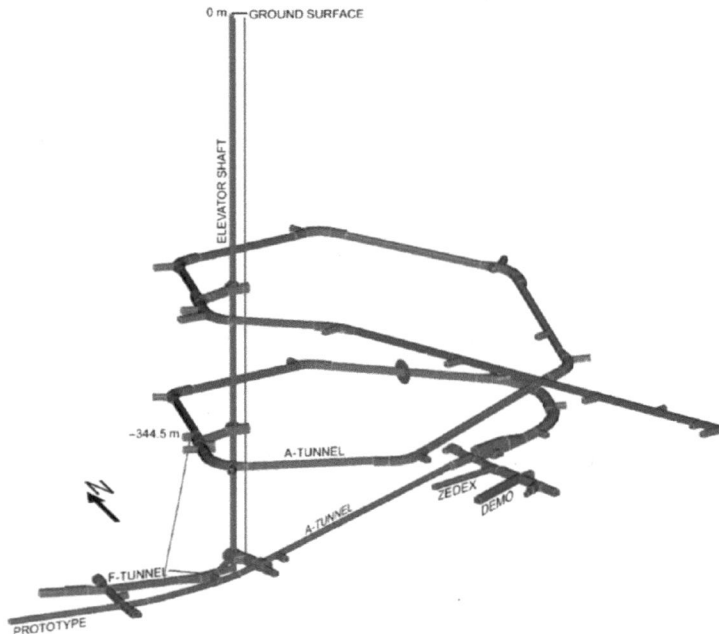

Figure 13 View of Äspö HRL from SSW. Borehole KA2599G01 is indicated (borehole top is located at level −344.5 m) (Sundberg, 2002).

- To compare the thermal properties of different rock types according to different methods and earlier results

5.1.2 Predictions

Laboratory measurements of 11 samples were made with the TPS-method. Calculations of the thermal conductivity were performed with the SCA-method, based on the mineralogical distribution of the samples. The drill core was originally mapped as Äspö diorite, Fine-grained granite, and Meta-basite. However, during the project a more detailed mapping was performed and it became clear that the term Äspö diorite involved a range of varieties including what has previously been mapped as Ävrö granite in the Äspö area and hybrid rocks (mingled types) between Ävrö granite and Äspö diorite, see section 1.4.

The thermal conductivity for the entire borehole has been predicted by using different assumptions and methods, as follows

1. The laboratory TPS-*measurements* are representative for equal parts of the drill core (irrespective of actual rock type distribution).
2. The laboratory *measured* thermal properties of each rock type are representative for all parts of the drill core where the particular rock type has been mapped according to the **original** mapping.
3. The *measured* thermal properties of each rock type are representative for all parts of the drill core where the particular rock type has been mapped according to the **revised** mapping.
4. The *calculated* thermal properties of each rock type are representative for all parts of the drill core where the particular rock type has been mapped according to the revised mapping.

The results for thermal conductivity are shown in Table 7. It is interesting to observe that quite different assumptions regarding representation of the measured values give

Table 7 Prediction of thermal response test in KA2599G01 with different assumptions and methods. values corrected for temperature a difference (25 °C – 14 °C) are in brackets (Sundberg, 2002).

Assumpt. method	Distribution of rock types / thermal conductivity (W/(m·K))					
	Äspö Diorite	Altered Äspö Diorite	Ävrö Granite	Fine-grained granite	Mingled	Meta-Basite
1			Distribution irrelevant **2.96**			
2	85 %	–	–	10 %	–	5 %
	2.83	–	–	3.63	–	2.58
				2.89		
3	54.5 %	1 %	25 %	11 %	3.7 %	4.8 %
	2.56	3.11	3.24	3.63	2.90	2.58
				2.84 (2.85)		
4	54.5 %	1 %	25 %	11 %	3.7 %	4.8 %
	2.35	3.38	3.01	3.45	2.68	2.58
			2.64			

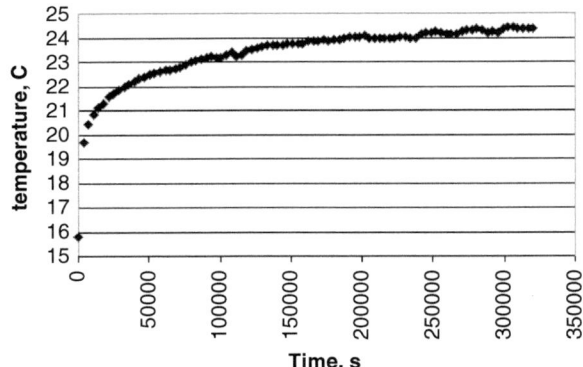

Figure 14 Time vs. temperature from thermal response test (Sundberg, 2002).

rather small differences in thermal conductivity (assumptions 1–3). The thermal conductivity varies between 2.64 and 2.96 W/(m·K) depending on assumed rock type distribution and method. The corresponding heat capacity varies within a smaller interval (2.08–2.11 MJ/(m³·K).

5.1.3 Evaluation of results from different methods

The thermal conductivity obtained from the thermal response test in borehole KA 2599 G01 has been estimated at 3.55 W/(m·K). The predicted thermal conductivity is in the interval 2.64 – 2.96 W/(m·K), depending on different assumptions and methods. Thus, a large difference in evaluated thermal conductivity exists between different methods, for the rock mass in borehole KA 2599 G01. The full-scale thermal response test resulted in a 25 % higher value compared to the mean thermal conductivity based on the different predictions.

With the thermal response tests an apparent or effective thermal conductivity is determined. The measured value is influenced by the specific natural conditions in the field. In the present case, large hydraulic pressure gradients exist, that probably induce convection, and increase the uncertainty of the measurements. This can also be seen as disturbances on the temperature response, see Figure 14. Conditions influencing measurements of samples in the laboratory are more easily controlled and observed. A more adequate determination of the thermal conductivity of the surrounding rock mass may be performed by filling the borehole with impermeable and thermal conductive material. The most reliable result, due to the discussion above, is the predicted thermal conductivity based on laboratory measurements and revised rock mapping, assumption No. 3.

5.2 Inverse modeling at the prototype repository

5.2.1 Methodology

The Prototype Repository, at the Äspö HRL (Hard Rock Laboratory), is a demonstration project for the deposition of spent nuclear fuel and provides a full-scale reference for

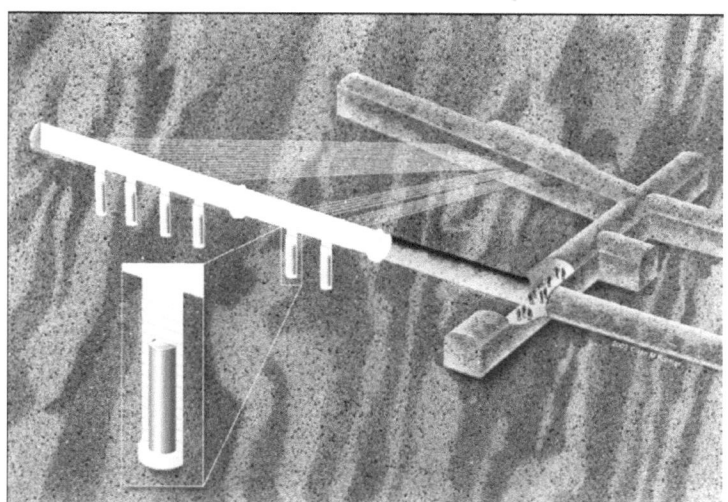

Figure 15 Schematic view of the layout of the Prototype Repository (not to scale) (SKB, 2005).

testing predictive models relating to a spent nuclear fuel repository, see recent description in Svemar *et al.* (2016). The layout involves six deposition holes, four in an inner section and two in an outer, each fitted with an electrically heated canister. The access tunnel is backfilled with a mixture of bentonite and crushed rock, see Figure 15. In 2001, the inner section was completed and monitoring of the heating process started. Temperature measurements in the rock mass are performed at 37 different points.

The measured thermal response in the surrounding rock was analyzed by inverse modeling of the thermal conductivity of the rock mass (Sundberg *et al.*, 2005; Sundberg & Hellström, 2009). A three-dimensional finite difference model of the prototype repository (canisters, buffers, tunnel, etc.) was used to calculate the transient temperature increase due to the heat generation in the canisters. The homogeneous rock thermal conductivity was varied until the best fit with measured data for each of the 37 temperature sensor points were obtained. The evaluation period for the fitting procedure was varied in order to study sensitivity to different time-scales and to avoid initial disturbances. Approximately 1.5 years of temperature data were available at the time for evaluation.

5.2.2 Prediction

The thermal properties were predicted based on both field and laboratory measurements. These predictions are verified by comparison with thermal conductivity values calculated through inverse modeling.

Prediction of the thermal properties has been made based on data from laboratory measurements on rock samples with the TPS method (Gustafsson, 1991; Sundberg & Gabrielsson, 1999; Sundberg, 2002) and by direct field measurements (Sundberg & Gabrielsson, 1999) with the multi probe method in the prototype tunnel. Different

Table 8 Results from different predictions of thermal properties (Sundberg *et al.*, 2005).

Prognosis	Population	Mean W/(m·K)	Std.dev W/(m·K)	Number of samples	Distribution
#1	Section 1 (inner), laboratory measurements	2.44	0.08	6	Lognormal
#2	Section 2 (outer), laboratory measurements	2.63	0.15	4	Lognormal
#3	Section 1+2, all Laboratory Measurements	2.52	0.15	10	Lognormal
#4	Section 1+2, field measurements	2.83	0.19	5	Lognormal
#5	Section 1+2, field and laboratory measurements. Combination of prognosis #3 and #4	2.62	0.22	15	Lognormal

prognosis models were established depending on where and how the samples were obtained, see Table 8.

Reasonable predictions may be based on all laboratory measurements or field measurements, see prognosis #3 and #4 in Table 8. The number of measurements is quite small and there is no reason to believe that the variability detected in different parts of the Prototype Repository would not be representative of the inner part where the temperature measurements have been conducted. The results for the field measurements are somewhat higher than the laboratory measurements. The field measurements are made 0.6 m below the tunnel floor and are possibly influenced by groundwater movements, partly induced by hydraulic gradient at 450 m depth toward the open tunnel.

However, the results from the temperature measurements at the prototype repository are also influenced by water movements, especially in the initial stage after installation of the backfill, and should consequently influence the back-calculated thermal conductivities from the inverse modeling. The prediction of the effective thermal conductivity, it could be argued, should therefore be based on field measurements only. However, the water movements are probably highest close to the tunnel, so that field measurements made near the tunnel floor probably overestimate the effective thermal conductivity in a scale relevant for the thermal impact on the canister. A combination of #3 (laboratory measurement-pure conduction) and #4 (field measurements-including a convective part) may be a reasonable prognosis (#5) of the effective thermal conductivity.

5.2.3 Inverse numerical modeling of the rock thermal conductivity

The material used in this study consists of data on mean canister power rates, rock temperatures at 37 temperature sensor locations for 525 days, initial temperatures and temperature on the canisters. In Figure 16, the heat sources are indicated as rectangles, and the sensors are indicated as circles. The thermal properties of the canister, buffer and tunnel backfill are given and held constant. Also the rock heat capacity is given. Ideally the heat capacity should be evaluated

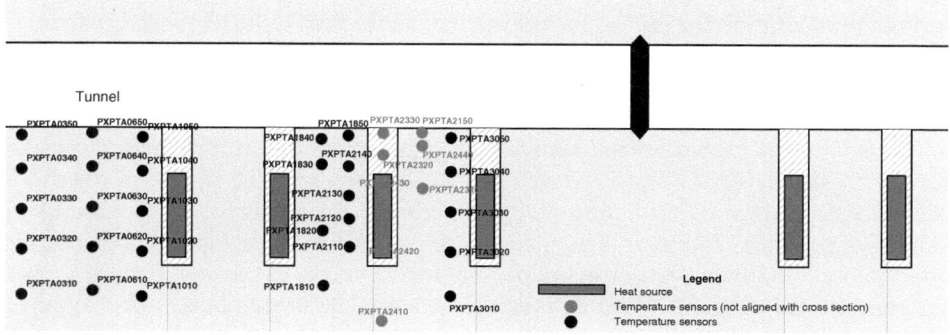

Figure 16 Location temperature sensors (Sundberg *et al.*, 2005). The inner part of the prototype tunnel is to the left. The numbering of canisters is from the left.

from the fitting procedure. However, the sensitivity is small in this pseudo steady-state process.

The study focuses on the thermal conditions in the rock mass. Thus, the conditions at the canister and in the bentonite buffer are not of primary interest in this study. Similar to the evaluation of thermal probe methods, it is assumed that these conditions have little influence on the evolution of rock temperatures after a certain initial time period. Most of the heat released during the initial period is absorbed by the canister and buffer. An initial period is therefore ignored during the parameter fitting procedure.

The water saturation of the bentonite buffers can be mentioned as an example of the complexity of the boundary conditions close to canister. Saturation of bentonite buffers around the canisters, which affects the thermal properties, is predicted to take 2-6 years if water is available.

The thermal properties of the different materials involved in the large-scale thermal process are assumed to be homogeneous. Thus, convection caused by the inflow to the backfilled tunnel is ignored.

This means that the thermal problem is linear. Different solutions to the heat conduction problem can then be superimposed on each other to form the complete temperature field. Here, this technique is used by superimposing two parts of the thermal response – the initial temperature field and the temperature increase due to the heat generation in the canisters. The initial temperature is assumed to be uniform, although it is apparent from the initial rock temperatures that there is also a certain superimposed thermal disturbance from activities in the deposition tunnel. Since this disturbance will decline with time, the influence on the evaluation will be reduced if a certain initial time period is omitted from the fitting procedure.

5.2.4 Fitting procedure

The measured temperature response is used to find the thermal conductivity that results in the best fit with the simulated thermal response. The thermal conductivity of the rock is set to a constant homogeneous value for each calculation of the temperature disturbance resulting from heat generation within each canister. The value of thermal conductivity is varied from 1.9 to 3.7 W/(m·K) with an increment of 0.1 W/(m·K). During the simulation for each value of thermal conductivity, the temperature increase at each temperature sensor location is calculated.

In the second step, the simulated values are interpolated in time to match the times at which the measured data were collected for each sensor. The third step involves comparison of the measured and simulated temperatures for each point and for each thermal conductivity during the chosen evaluation interval. The square of the difference between measured and simulated temperature for each measured time in the evaluation period is summed. This procedure is repeated for each value of thermal conductivity. Finally, we have the square sum for 19 values of thermal conductivity in the range from 1.9 to 3.7 W/(m·K) for each sensor. The thermal conductivity of the best match between measured and simulated values for a given sensor is found by minimizing a polynomial fit to the 19 values.

It should be emphasized that the fitted thermal conductivity values are calculated for each sensor point individually without regard to the thermal response of any other point. However, it is also possible to achieve an overall thermal conductivity by considering the best fit for all 37 sensor points. To obtain the overall thermal conductivity value the average square sum of the difference between measured and calculated temperatures for each of the 37 sensor points are summed up for each value of the thermal conductivity. The thermal conductivity of the best overall match between measured and simulated values taking all sensors into account is found by minimizing a polynomial fit to the 19 values.

5.2.5 Results

High thermal conductivity values are modeled in the inner parts of the tunnel, especially close to the tunnel surface, see Figure 17. These high values may be caused by water movements, which have been reported in the actual parts of the tunnel. Due to the limited period of heating, evaluated thermal conductivity close to the canister are probably the most reliable. The influence of a relative error in the measurements and

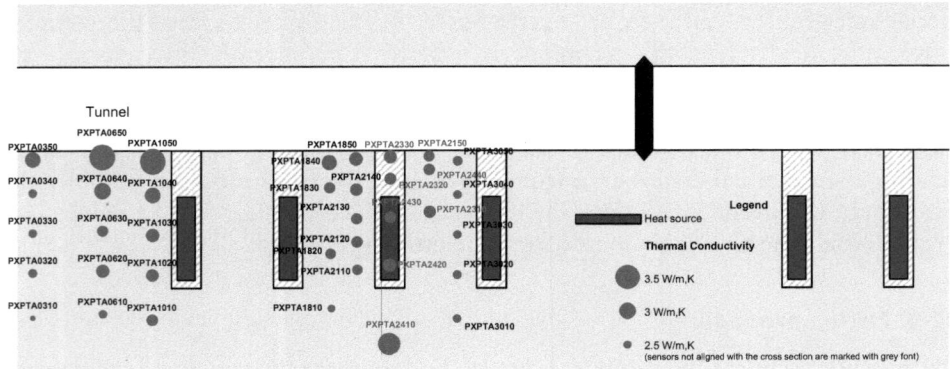

Figure 17 Thermal conductivity from simulation including curve fitting for 160 through 525 d (Sundberg & Hellström, 2009).

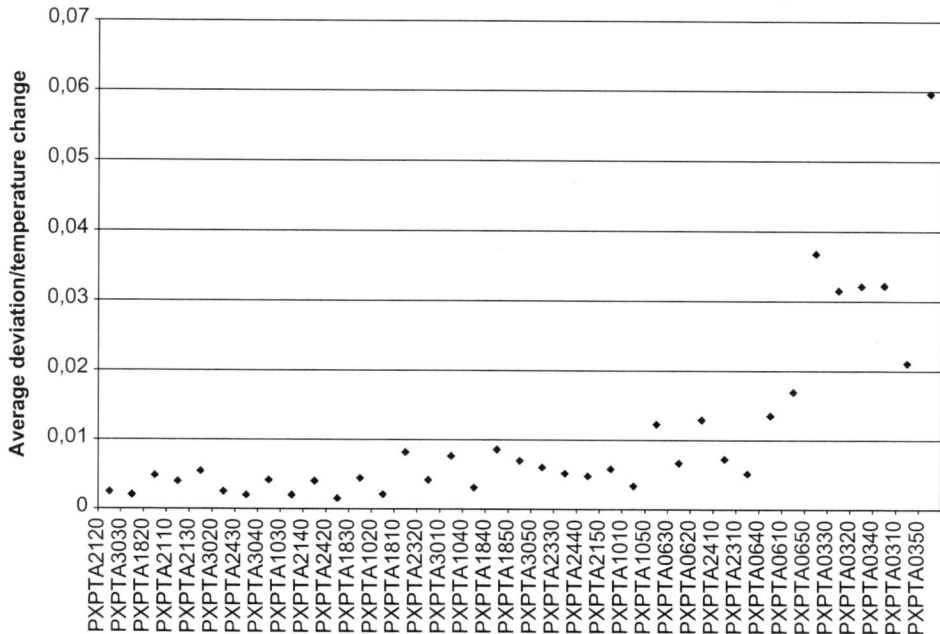

Figure 18 Average deviation between measured and simulated temperatures (°C) in relation to simulated temperature increase for each of the 37 simulated sensors. The period 160–365 days is used for fitting the measured and simulated response (Sundberg & Hellström, 2009). The sensors have been ordered according to magnitude of average simulated temperature increase. This reflects to some degree the temperature sensor's proximity to a canister. The influence of a relative error in the measurements and local deviations in the initial temperatures will be less for sensors close to canister.

local deviations in the initial temperatures will be less, see Figure 18. Consequently, sensors with largest deviation (>0.01) are usually not close to the canister or influenced by water movements.

5.2.6 Verification of predictive model

The measured thermal response has been used to estimate the effective thermal conductivity by finding the best fit with results obtained by a numerical simulation of the thermal process around the repository.

In Table 9 comparisons are made between different prognoses (Table 8) and inverse modeling results for all and individual sensors. We have argued above that a reasonable prediction of the effective thermal conductivity is a combination of laboratory (thermal conductivity) and field measurements (effective thermal conductivity) at the prototype repository (prognosis #5).

Prognosis #5 (Table 8) displays good agreement with back-calculated thermal conductivities from the inverse modeling and the above statement of a reasonable

Table 9 Modeled thermal conductivity (W/(m·K)) compared with prognosis (mean values). Most relevant values are in **bold** (judgment). Prognosis refers to Table 8 (Sundberg & Hellström, 2009).

Case	Modeled conductivity Best fit to		Prognosis	Comment
	all sensors	individual sensors		
Inverse modeling (160–525 days)	2.72	2.91 (N=37)		
Ditto but including possible global temp change −0.2°C/year	**2.65**	2.73 (N=37)		
Ditto but values above 3.4 excluded		**2.65** (N=34)		
Laboratory measurements section 1 +2 (prognosis #3)			**2.52**	"Best" prediction of thermal conductivity
Combination of prognosis #3 and #4 (prognosis #5)			**2.62**	"Best" prediction of effective thermal conductivity

prediction of the effective thermal conductivity seems to be verified. Values in bold are judged to be the most reliable and relevant.

5.2.7 Conclusions

There is good agreement between the prediction of thermal conductivity in the prototype repository and the result from inverse modeling based on 37 different temperature sensors. The rather low mean thermal conductivity value of around 2.6 W/(m·K) is verified. The estimated thermal conductivity gives an effective value that includes a small contribution from convection and is therefore probably overestimating the actual "pure" thermal conductivity.

There is a large discrepancy between the individual fit for some sensors and the overall best fit, see Figure 19. Sensors with the largest distance to canisters, or located near the tunnel floor or hydraulic structures, are overrepresented.

The accuracy of the evaluation would have been improved if the rock temperature had been allowed to equilibrate after the sealing of the tunnel and verified to be stable, before the heating of the canisters began.

The results of the inverse modeling indicate that data is influenced by a temperature drift in the rock mass and/or by water movements. These "errors" are probably most significant in early data. Evaluation of data from a longer time period would probably improve the inverse modeling and also allow more initial data to be omitted in order to improve accuracy. This would also enhance prediction of groundwater flow effects.

5.3 Strategy for thermal modeling in relevant scale

5.3.1 Scale of thermal processes in a repository

The thermal function of a repository for spent nuclear fuel can be studied at different scales. The following scales are believed to be relevant:

- 1–10 m for the thermal function of the canister (canister scale or local scale)
- 10–100 m for the thermal function of the tunnel (tunnel scale)
- 100–1000 m for the thermal function of the whole repository (repository scale)

The "global" temperature field around a repository mainly depends on the time-dependent generated heat, boundary conditions, initial temperature conditions and mean values of large-scale thermal transport properties. The thermal processes at this scale are quite slow and insensitive to local variations in the thermal properties. The demands for high accuracy in the thermal property distribution are lower compared to the local scale.

The local temperature field is of primary concern for the design of a repository. During the site investigations a design criterion was specified as the maximum temperature allowed in the bentonite buffer outside the canisters (SKB, 2006). A low thermal conductivity leads to larger distances between canisters than in the case of a high thermal conductivity. It is therefore of special interest to analyze the thermal impact on the canister if there is a variation in the thermal properties in the rock mass at the canister scale, 1-10 m, that will influence the canister temperature.

5.3.2 Strategy for thermal modeling

The strategy for thermal modeling in the site investigation in Laxemar is summarized below. A complete description is made in Back & Sundberg (2007), and further developed in Back et al. (2007) and Sundberg et al. (2008).

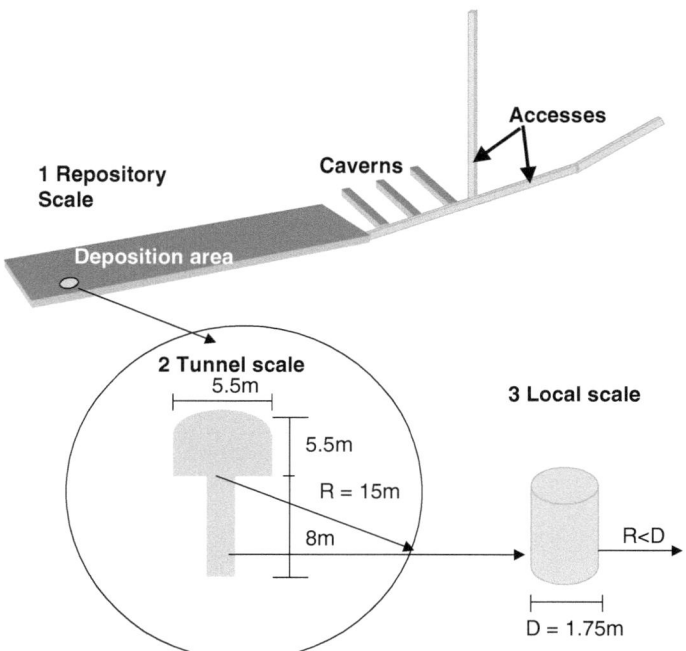

Figure 19 Illustration of the various scales of importance for rock mechanics considerations for siting and constructing a KBS-3 repository, from Andersson et al. (2002).

The spatial variability of the thermal properties of the rock mass is important for the canister spacing. The strategy for the thermal site descriptive modeling is to produce spatial statistical models of both lithology's and thermal properties and perform stochastic simulations to generate a spatial distribution of thermal properties that is representative of the modeled rock domain in canister scale. The properties must be determined and upscaled with such a degree of certainty and resolution that the temperature field around the repository can be described with sufficiently high degree of confidence.

The total variability within a rock domain thus depends on the lithology and the thermal properties of each rock type. Although the thermal conductivity of a single rock type may be close to normally distributed, the statistical distribution of thermal conductivity for the rock domain as a whole is far from normally distributed. Depending on their fraction of the total volume, the low-conducting rock types may determine the lower tail of the thermal conductivity distribution and influence the canister spacing.

The methodology involves a series of steps, see Figure 20, *e.g.* choice of simulation scale, upscaling in several steps, definition of Thermal Rock Classes (TRCs) within the rock domain, stochastic simulation of TRCs (lithology in the

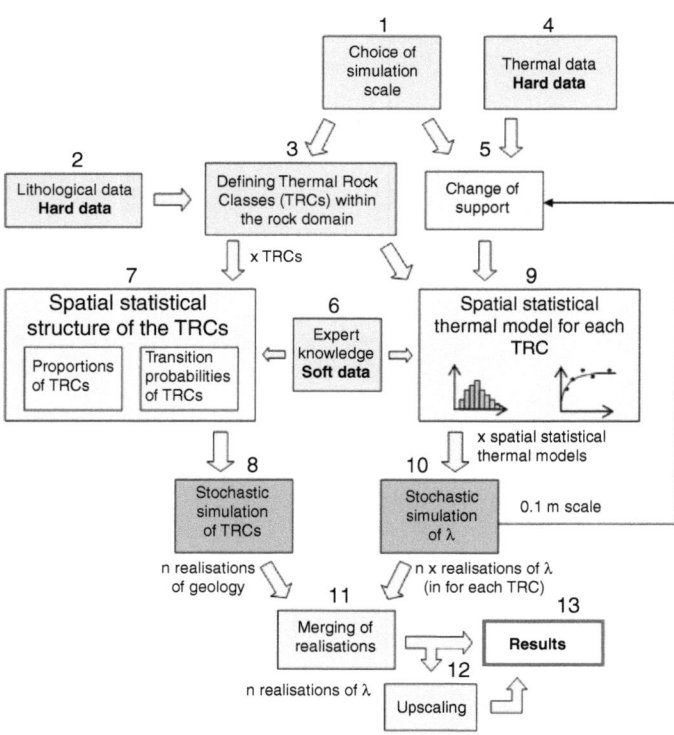

Figure 20 Schematic description of the procedure for thermal modeling (λ represents thermal conductivity) (Sundberg et al., 2008).

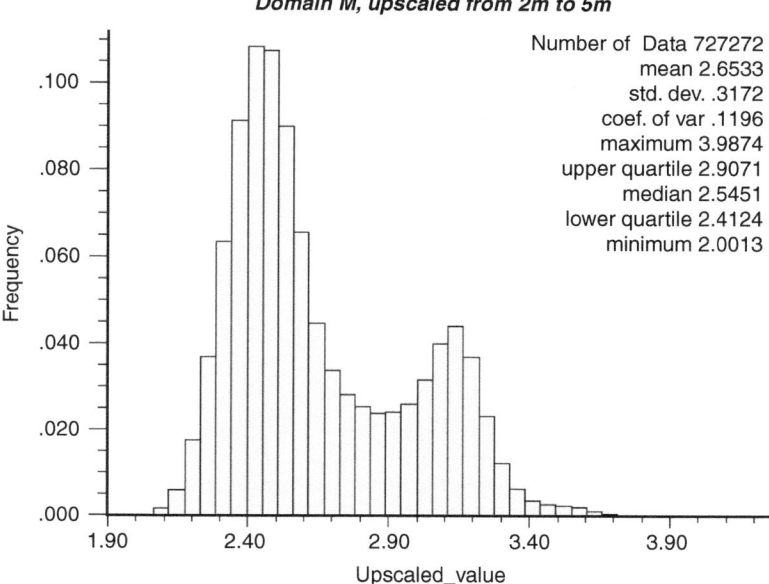

Figure 21 Histogram of thermal conductivity of domain RSMM01 in Laxemar simulated at the 2 m scale but upscaled to 5 m. The main result of the thermal modeling is the set of 1 000 realizations of thermal conductivity from the 2 m-simulations. The lower tail is a result of the low-conductive rocks, mainly Ävrö quartz monzodiorite and diorite-gabbro. Upscaling of the realizations to 5 m has the effect of smoothing the histogram (Sundberg *et al.*, 2008).

domain), stochastic simulation of thermal conductivity within each TRC, and finally merging of the realizations of TRCs (lithology) with the thermal conductivity realizations. The result is a set of 3D realizations describing the spatial variability of thermal conductivity within the rock domain. These realizations can thereafter be upscaled to a desired scale. The described methodology can also be used for other types of rock properties.

The distribution of thermal conductivities comes primarily from TPS-measurements on rock samples in laboratory. In order to describe the spatial structure of thermal conductivity a very large number of samples are needed. A correlation between thermal conductivity and density has been found and described above. The relationship can be explained by the mineral distribution for different rock types and the density and thermal conductivity for the minerals (Sundberg *et al.*, 2009). The relationship makes it possible to use density logging to estimate the spatial correlation structure within each rock type, instead of huge amount of thermal conductivity determinations.

The described strategy has been used in the thermal site models in both Laxemar and Forsmark. Some example results on rock domain level is showed in Figure 21 (rock domain RSMM01 includes the Ävrö granite and the distribution is bimodal).

6 SOME CONCLUSIONS

The Ävrö granite show a high variability and bimodal distribution in thermal conductivity, ranging from 2.01 W/(m·K) up to 3.76 W/(m·K). The sub division of the Ävrö granite into Ävrö quartz monzodiorite and Ävrö granodiorite have considerable lowered the variation respectively. The temperature dependence is quite low, especially for Ävrö quartz monzodiorite.

The found relationship between thermal conductivity and density is very useful to calculate the thermal conductivity from density logging and allows a more reliable use of borehole density data for analyzing the spatial distribution and correlation structure of thermal properties. The relationship shows that thermal conductivity decreases with density for felsic rocks, whereas for mafic rocks the opposite relationship applies. This can be explained by theoretical calculations on synthetic samples.

The comparison between TPS measurements and SCA calculations from the mineralogy indicates a good agreement for most rock types, in average. For individual samples the discrepancies are larger.

In large scale thermal conductivity tests it is important to have full control over the ground water situation in order to avoid disturbances that may influence the evaluated result. The thermal response test at Äspö HRL overestimated the thermal conductivity with 25% compared to the different predictions. The main reason is judged to be induced water movements. Quite different approach for the predictions resulted in small variations in thermal conductivity.

The inverse modeling of the thermal properties at the prototype repository at Äspö HRL are in good agreement with the prediction of thermal conductivity based on laboratory and field measurements. The rather low mean thermal conductivity value of around 2.6 W/(m·K) in the prototype repository is verified. The estimated thermal conductivity gives an effective value that includes a small contribution from convection and is therefore probably somewhat overestimating the actual thermal conductivity.

The thermal site descriptive model for Laxemar provides a spatial statistical description of the rock mass thermal conductivity in canister scale (Sundberg et al., 2008). The developed methodology employed for the thermal modeling, involving stochastic simulation of both lithologies and thermal conductivity, takes into account the spatial variability of thermal conductivity both within and between different rock types, and allows for upscaling to a relevant scale.

ACKNOWLEDGMENT

All investigations and research have been fully financed by SKB, Swedish Nuclear Fuel and Waste Management Company.

REFERENCES

Adl-Zarrabi B, 2006. Borehole KLX11A. Thermal properties of rocks using calorimeter and TPS method. Oskarshamn site investigation. SKB P-06-269. Svensk Kärnbränslehantering AB, Stockholm.

Andersson J, Christiansson R & Hudson J, 2002. Site Investigations. Strategy for Rock Mechanics Site Descriptive Model. SKB TR-02-01. Svensk Kärnbränslehantering AB, Stockholm.

Back P-E & Sundberg J, 2007. Thermal Site Descriptive Model. A Strategy for the Model Development during Site Investigations. Version 2.0. SKB R-07-42, Svensk Kärnbränslehantering AB, Stockholm.

Back P E, Wrafter J, Sundberg J & Rosén L, 2007. Thermal properties. Site descriptive modelling Forsmark – stage 2.2. SKB R-07-47, Svensk Kärnbränslehantering AB, Stockholm.

Bruggeman D A G, 1935. Annalen der Physik, 24, 636–679.

Dagan G, 1979. Models of groundwater flow in statistically homogeneous porous formations, Water Resources Research, 15(1), 47–63.

Gehlin S, 2002. Thermal response test. Method development and evaluation. Luleå University of Technology, 39. Doctoral thesis.

Gustafsson S, 1991. Transient plane source techniques for thermal conductivity and thermal diffusivity measurements of solid materials. Review of Scientific Instruments, 62, 797–804. American Institute of Physics, USA.

Landström O, Larson S-Å, Lind G & D Malmqvist, 1979. Heat Flow in Rock (In Swedish). Chalmers University of Technology, Department of Geology, Publ. B137, Göteborg.

Mossmark F & Sundberg J, 2007. Oskarshamn site investigation. Field Measurements of Thermal Properties. Multi Probe Measurements in Laxemar. SKB P-07-77, Svensk Kärnbränslehantering AB, Stockholm.

Norrman J, 2004. On Bayesian Decision Analysis for Evaluating Alternative Actions at Contaminated Sites. Dissertation Thesis, 2202. Chalmers University of Technology and University of Göteborg.

Rhén I, Gustafson G, Stanfors R & Wikberg P, 1997. Äspö HRL – Geosientific evaluation 1997/5. Models based on site characterisation 1986–1995. SKB TR 97-06. Svensk Kärnbränslehantering AB, Stockholm.

SKB, 2001. Site Investigations. Investigation methods and general execution programme. SKB TR-01-29. Svensk Kärnbränslehantering AB, Stockholm.

SKB 2005. Äspö Hard Rock Laboratory. Annual Report 2004. TR-05-10. Stockholm, Sweden: SKB Svensk Kärnbränslehantering AB, Stockholm.

SKB, 2006. Long-term safety for KBS-3 repositories at Forsmark and Laxemar – A first evaluation. Main report of the SR-Can project. SKB TR-06-09. Svensk Kärnbränslehantering AB, Stockholm.

Seipold U & Huenges E, 1998. Thermal properties of gneisses and amphibolites high pressure and high temperature investigations of KTB-rock samples. Tectonophysics 291, 173–178.

Sundberg J, 1988. Thermal Properties of Soils and Rocks. Publ. A 57, Dissertation. Geologiska institutionen, Chalmers University of Technology and University of Göteborg.

Sundberg J, 2002. Determination of thermal properties at Äspö HRL. Comparison and evaluation of methods and methodologies for borehole KA 2599 G01. SKB R-02-27. Svensk Kärnbränslehantering AB, Stockholm.

Sundberg J, 2003. Thermal properties at Äspö HRL. Analysis of distribution and scale factors. SKB R-03-17. Svensk Kärnbränslehantering AB, Stockholm.

Sundberg J, Back P-E & Hellström G, 2005. Scale dependence and estimation of rock thermal conductivity. Analysis of upscaling, inverse thermal modelling and value of information with the Äspö HRL prototype repository as an example. SKB R-05-82, Svensk Kärnbränslehantering AB, Stockholm.

Sundberg J, Back P-E, Ericsson L O & Wrafter J, 2009. A method for estimation of thermal conductivity and its spatial variability in igneous rocks from in situ density logging. International Journal of Rock Mechanics & Mining Sciences 46 (2009) 1023–1028.

Sundberg J & Gabrielsson A, 1999. Laboratory and field measurements of thermal properties of the rocks in the prototype repository at Äspö HRL. SKB IPR-99-17. Svensk Kärnbränslehantering AB, Stockholm.

Sundberg J & Hellström G, 2009. Inverse modelling of thermal conductivity from temperature measurements at the Prototype Repository, Äspö HRL. International Journal of Rock Mechanics & Mining Sciences 46, 1029–1041.

Sundberg J, Hälldahl L & Kukkonen I, 2003. Comparison of thermal properties measured by different methods. SKB R-03-18. Svensk Kärnbränslehantering AB, Stockholm.

Sundberg J, Wrafter J, Back P-E & Rosén L, 2008. Thermal properties Laxemar. Site descriptive modelling. SDM-Site Laxemar. SKB R-08-61. Svensk Kärnbränslehantering AB, Stockholm.

Svemar C, Johannesson L-E, Grahm P, Svensson D, Kristensson O, Lönnqvist M & Nilsson U, 2016. Prototype Repository. Opening and retrieval of outer section of Prototype Repository at Äspö Hard Rock Laboratory. Summary report. SKB TR-13-22. Svensk Kärnbränslehantering AB, Stockholm.

Wahlgren C-H, Curtis P, Hermanson J, Forssberg O, Öhman J, A Fox, P. La Pointe C-A Triumf, H Mattsson, H Thunehed & C Juhlin 2008. Geology Laxemar. Site descriptive modelling SDM-Site Laxemar. SKB R-08-54, Svensk Kärnbränslehantering AB, Stockholm.

Wrafter J, Sundberg J, Ländell M & Back P, 2006. Thermal modelling. Site descriptive modelling. Laxemar 2.1. SKB R-06-84, Svensk Kärnbränslehantering AB, Stockholm.

Walsh JB & Decker E R, 1966. Effect of pressure and saturating fluid on the thermal conductivity of compact rock. Journal of Geophysical Research, 71, 12.

Åkesson U, 2007. Boreholes KLX07A, KLX10, KLX05 and KLX12A. Extensometer measurements of the coefficient of thermal expansion of rock. SKB P-07-63, Svensk Kärnbränslehantering AB, Stockholm.

Chapter 17

Excavation response studies at AECL's underground research laboratory—1982 to 2010

Rodney S. Read
RSRead Consulting Inc., Okotoks, Alberta, Canada

Abstract: This chapter provides an overview of the major excavation response experiments conducted at AECL's Underground Research Laboratory (URL) between 1982 and 2010. Excavation response research was initially conducted as part of shaft construction. An excavation response test at the 240 Level of the URL involved a horizontal tunnel excavated through a subvertical water-bearing fracture. These precursor studies led to a series of experiments in the more highly stressed rock at the 420 Level of the URL to investigate the formation of rock damage around tunnels, and to assess the factors that influence the stability of underground excavations. These experiments included the Mine-by Experiment, the Heated Failure Tests, studies of borehole breakouts, the Excavation Stability Study, and the Tunnel Sealing Experiment. The excavation response experiments at the URL culminated in the Thermal-Mechanical Stability Study (TMSS), a comprehensive study to link characterization, numerical modeling, monitoring, and design of underground excavations. As part of decommissioning of AECL's URL, underground openings were flooded and the shaft was sealed. The facility was closed in 2010.

I INTRODUCTION

Understanding the rock mass response to the creation of an underground opening is important in designing civil structures (*e.g.*, transportation tunnels and hydroelectric powerhouses), developing underground mines, and drilling deep boreholes for petroleum production, gas storage, or other purposes. Rock mass response to excavation is particularly relevant in concepts for long-term geologic isolation of spent nuclear fuel and high-level radioactive waste.

Based on conceptual designs developed by Atomic Energy of Canada Limited (AECL), a deep geologic repository would comprise a series of underground openings including shafts and/or ramps, access tunnels, emplacement rooms, and other excavations required for underground operations and testing. Shafts, or ramps, and important access tunnels would remain open for the construction and operation stages of the repository, while emplacement rooms would be backfilled and sealed following emplacement of spent fuel. Temporary ground support would be used in the short-term to provide a safe working environment, but would likely be removed prior to filling of the tunnels, depending on restrictions regarding gas generating materials.

Excavation-induced damage (*i.e.*, increased crack density or crack volume) in the near-field rock mass is anticipated around repository excavations. This damage may increase connected permeability, and thereby increase potential for groundwater-borne transport of radionuclides away from the emplaced waste. Under certain in situ conditions, progressive damage development may create instability at the periphery of an underground opening.

In designing repository excavations, the key objectives therefore are to maintain stable rock mass conditions around each opening, and to minimize excavation damage. Over the repository's life cycle, the rock mass will be subjected to mechanical loads generated by stress redistribution around openings, support loads introduced by swelling buffer and backfill materials, and thermal loads generated by emplaced spent nuclear fuel. Other loads, such as those associated with glaciation, are also anticipated. To meet the design objectives, factors that may affect the rock mass response to excavation and to subsequent loading must be understood.

Over a period of almost 30 years, AECL conducted extensive research and development related to excavation response of the rock mass at its Underground Research Laboratory (URL). Investigations started with monitoring of the URL shaft construction, and progressed through several major in situ experiments focused on excavation damage development and progressive failure. The work conducted in granite at the URL is complemented by international research and development programs in similar rock types with different in situ conditions (i.e., fracture frequency, in situ stresses, and groundwater), and in other media (e.g., salt, argillite, tuff, and clay).

This chapter provides an overview of various excavation experiments and studies conducted at the URL from 1982 to its closure in 2010. The information is updated from an original publication (Read, 2004) that was part of a special issue of the International Journal of Rock Mechanics featuring AECL's URL. This chapter highlights the advances in our fundamental understanding of rock mechanics related to underground excavations, and in our means of designing stable underground openings with minimal excavation damage. The findings from this research are relevant to the design of not only nuclear fuel waste repositories, but also other underground excavations.

2 AECL'S UNDERGROUND RESEARCH LABORATORY

Prior to its closure and decommissioning, AECL's URL was located approximately 120 km NE of Winnipeg, Manitoba, Canada within the Lac du Bonnet granite batholith near the western edge of the Canadian Shield. As shown in Figure 1, the granite contains subvertical joint sets and several major low-dipping thrust faults (referred to as Fracture Zones) within the first few hundred meters of the surface at this site. The URL shaft intersected the most prominent of these thrust faults (Fracture Zone 2) at about 270 m below surface. Below Fracture Zone 2 and its splays, the granite is interrupted by a granodiorite dyke swarm, but is sparsely fractured and relatively unaltered.

During its operation, the URL consisted of four levels: two minor experimental/drilling stations at depths of 130 and 300 m (referred to as the 130 and 300 m Levels, respectively), and two major experimental levels at depths of 240 and 420 m (referred to as the 240 and 420 Levels, respectively). The levels were connected to ground surface by a 443-m-deep shaft. The shaft was rectangular (2.8 × 4.9 m) from surface to 255 m depth, and circular (4.6-m-diameter) from 255 to 443 m depth. A rectangular vent raise connected the 240 Level to ground surface. A circular bored vent raise was constructed to connect the 420 and 240 Level.

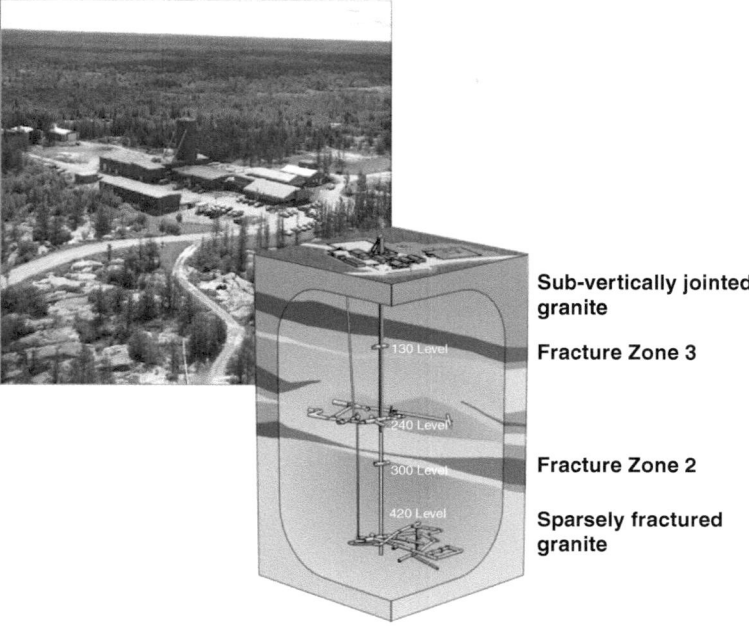

Figure 1   AECL's Underground Research Laboratory surface facilities, and generalized geology of the Lac du Bonnet batholith in the vicinity of the underground developments.

In situ stresses to the depth of Fracture Zone 2 are typical of other sites at similar depths in the Canadian Shield. Stress magnitudes and orientations (given as trend/plunge) at the 240 Level are $\sigma_1 = 26$ MPa (228°/08°), $\sigma_2 = 17$ MPa (132°/23°) and $\sigma_3 = 13$ MPa (335°/65°). The tangential stresses around openings at this level do not exceed the unconfined strength of intact rock, and the rock mass behaves essentially elastically outside of a blast-induced zone of damage.

Below Fracture Zone 2, the sub-horizontal principal in situ stresses are significantly higher. Stress magnitudes and orientations at the 420 Level are $\sigma_1 = 60$ MPa (145°/11°), $\sigma_2 = 45$ MPa (054°/08°) and $\sigma_3 = 11$ MPa (290°/77°), resulting in a maximum stress ratio of almost 6:1. These conditions tend to promote stress-induced damage development and progressive failure around underground openings, and are considered adverse in terms of excavation design and construction (Read *et al.*, 1998). In this environment, the rock mass response to excavation varies from linear-elastic further than about 1 m from a typical 3.5-m-diameter opening to non-linear/non-elastic with micro- and macro-scale fracturing near the opening.

3 EXCAVATION RESPONSE STUDIES

Incremental development of a rock mechanics, rock fracturing, and excavation stability knowledge base at the URL commenced with shaft sinking in 1982 (Martin & Simmons, 1993). Early results included data from instrument arrays and overcoring stress determinations in the URL shaft and at the 240 Level in moderately to sparsely fractured rock subjected to moderate in situ stresses. The knowledge base was expanded through in situ experiments at the highly stressed 420 Level, culminating in the Thermal-Mechanical Stability Study (TMSS). An overview of the major excavation experiments and studies conducted at AECL's URL, and a summary of pertinent results related to rock mass response to excavation, are presented in the following sections.

3.1 URL shaft studies

Seven excavation response tests using arrays of mechanical and hydrogeological instruments were conducted between 1982 and 1988 during shaft construction. These instrument arrays included triaxial strain cells, extensometers, convergence pins, and, in some cases, hydraulic borehole packers. An array of microseismic (MS) sensors supplemented the other instruments in the lower shaft. The upper rectangular and lower circular shafts were excavated using the drill-and-blast method. Responses were measured in relation to blast rounds following installation of instruments. Mapping of the shaft walls was carried out during excavation from a specially-designed platform. Particular attention was paid to mapping Fracture Zone 2 where it was intersected by the shaft, and the 300 Level shaft station. Core from each array of instrumentation boreholes was logged. In addition, Colorado School of Mines (CSM) dilatometer tests were conducted in 15-m-long boreholes around the upper section of the URL shaft in moderately fractured rock conditions to measure the deformation modulus of the rock mass.

The results of monitoring and characterization in the upper shaft in fractured granite showed variable mechanical responses dominated by subvertical joints. Continuum models used to analyze the results did not predict excavation responses very well; discontinuum models provided a better match between predicted and measured results assuming a linear increase in elastic modulus from 10 to 40 GPa from the shaft wall out to 1.5 m from the opening. Dilatometer tests measured an increase in modulus from about 20 GPa near the opening to 70 GPa about 9 m from the opening, and variability in modulus values near known and assumed subvertical fractures. Calculated hydraulic conductivity of subvertical fractures decreased as a result of excavation. At the 300 Level, tangential excavation-induced fractures on the micro- and macro-scale developed within 300 mm of the circular shaft perimeter (Martino & Chandler, 2004). Displacements within 2.5 m of the circular shaft wall in unjointed granite exceeded those predicted by linear elastic theory.

These early results suggested that rock mass response around drill-and-blast excavations can differ substantially from linear elastic assumptions, even in the absence of discrete jointing. The combination of instruments used in the shaft excavations simultaneously measured stress changes, displacement, and pore pressure responses. These types of instruments were used in later studies.

3.2 Room 209 excavation response test

The Room 209 Excavation Response Test (Simmons, 1992) was conducted in 1986 at the 240 Level to determine the hydraulic and mechanical response of a rock mass containing a narrow zone of *en echelon* permeable fractures. In addition, the test was used to estimate the mechanical and hydraulic properties of the rock mass, and to assess modeling capabilities.

A horizontal horse-shoe shaped tunnel (3.84-m-wide by 3.45-m-high) was excavated through the subvertical water-bearing fracture zone using a pilot-and-slash, drill-and-blast method. Displacement was measured using convergence arrays, three types of radial extensometers, and an ISETH sliding micrometer installed from a parallel tunnel. Stress changes were monitored using eight CSIRO Hollow Inclusion (HI) triaxial strain cells installed ahead and outside of the tunnel face. Hydraulic pressure was measured using pneumatic straddle packers installed across the fracture zone in ten boreholes. GEOKON vibrating-wire piezometers were used to monitor piezometric pressures in the fracture zone. Thermistors were installed with each of the extensometers, CSIRO cells, and convergence arrays. In addition, single and multi-step drawdown tests were performed in nine boreholes intersecting the fracture zone prior to tunnel excavation to estimate hydraulic transmissivity. CSM dilatometer tests measured rock mass modulus near the tunnel wall.

Displacement measurements showed a decrease in elastic modulus near the tunnel wall, suggesting confining-pressure-dependent behavior, and damage within 0.5 m of the wall. The rock mass modulus decreased by 10 to 20 GPa near the sidewall, and increased by 15 GPa near the crown, as a result of tunnel excavation. The zone in which the modulus was affected by excavation was limited to within 2 m from the tunnel perimeter. Dilatometer tests measured lower modulus values near the tunnel sidewall.

The calculated equivalent hydraulic aperture for the fracture zone ranged from 14 to 155 μm. Tunnel excavation reduced inflow from the fracture zone from 11 L/min flowing into boreholes prior to excavation to measured inflows to the pilot and slash excavations of 0.35 and 0.45 L/min, respectively. The measured transmissivity of the fracture zone decreased with the pilot tunnel excavation, then recovered slightly with the slash excavation. Transmissivity in two boreholes showed a permanent five-fold decrease following completion of the slash excavation.

Numerical modeling using three different models adequately predicted the mechanical response of the unfractured rock mass, but the hydraulic response was not simulated well by any of the models. These results indicated that the excavation response of even relatively simple fracture networks adjacent to underground openings can be complex, and difficult to simulate using the conceptual models for hydro-mechanical fracture response available at the time.

3.3 Mine-by experiment

The Mine-by Experiment (MBE) (Read & Martin, 1996; Martin & Read, 1996) was the first major excavation response experiment at the 420 Level. It was conducted between 1989 and 1995 to study the processes involved in excavation-induced damage development and progressive failure around an underground opening subjected to high differential stresses under ambient temperature conditions. The experiment involved a 3.5-m-diameter, 46-m-long test tunnel (Figure 2) excavated using a non-explosive hydraulic rock-splitting technique. The tunnel was advanced in 1-m rounds subparallel to the intermediate principal stress direction to maximize the stress ratio acting on the tunnel, and thereby promote rock mass failure.

An essential feature of the MBE was that monitoring instrumentation was installed and operating prior to excavation of the test tunnel. Instruments included extensometers, convergence arrays, triaxial strain cells, thermistors, and an acoustic

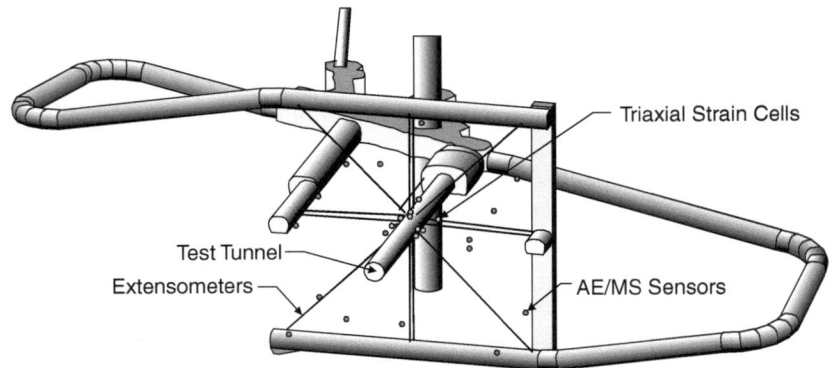

Figure 2 Layout of the Mine-by Experiment at the 420 Level.

emission/microseismic (AE/MS) monitoring system. By pre-installing the instruments, it was possible to monitor the complete excavation-induced response of the rock mass, including effects ahead of the advancing tunnel face. New instruments, such as the excavation damage extensometer (Ed-ex) (Thompson *et al.*, 1993), were also developed to take detailed measurements close to the tunnel wall during excavation.

During excavation of the MBE test tunnel, a multi-stage process of progressive brittle failure was observed. This process resulted in the development of v-shaped 'notches', typical of borehole breakouts under high stress conditions. These notches developed in the areas of compressive stress concentration in the crown and invert of the tunnel (Figure 3).

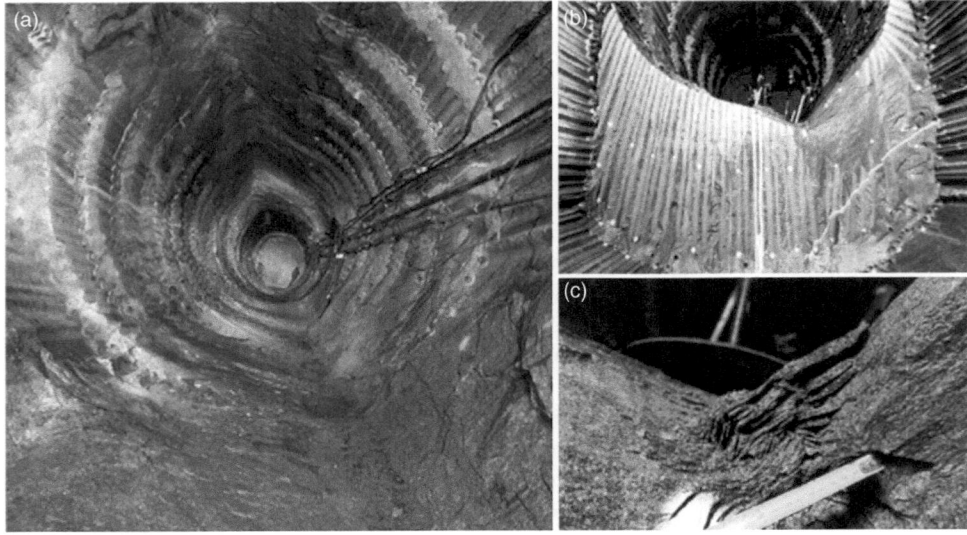

Figure 3 MBE test tunnel showing (a) final shape of v-shaped breakout notches, (b) cross-section through the notch in the tunnel invert, and (c) close-up of the notch tip in the tunnel invert.

Breakouts were more evident, and developed more readily, in areas of granite versus granodiorite lithology. The breakout notches were found to be in a state of meta-stable equilibrium, and were extremely sensitive to minor changes in boundary conditions such as confinement, stress changes from adjacent excavations, humidity, and temperature.

The progressive failure process and a typical slab generated during the process are shown in Figure 4. The sequence of events leading to failure was similar to that observed in laboratory compression tests on block samples containing a central hole.

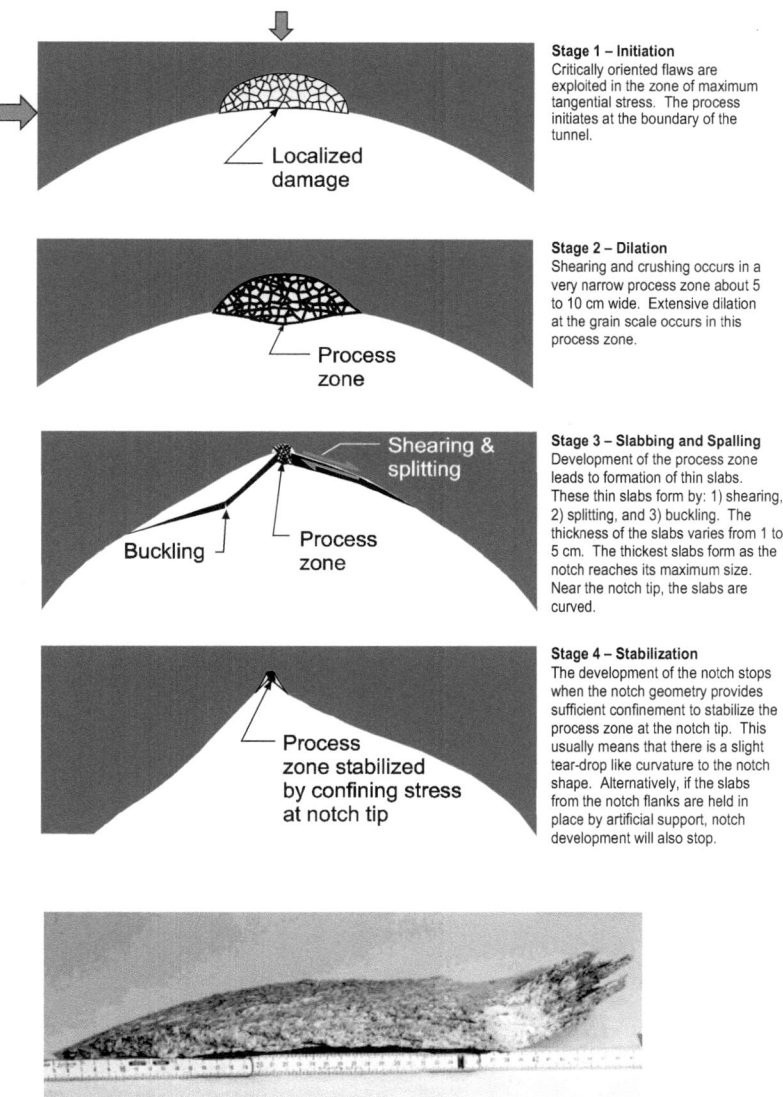

Stage 1 – Initiation
Critically oriented flaws are exploited in the zone of maximum tangential stress. The process initiates at the boundary of the tunnel.

Localized damage

Stage 2 – Dilation
Shearing and crushing occurs in a very narrow process zone about 5 to 10 cm wide. Extensive dilation at the grain scale occurs in this process zone.

Process zone

Stage 3 – Slabbing and Spalling
Development of the process zone leads to formation of thin slabs. These thin slabs form by: 1) shearing, 2) splitting, and 3) buckling. The thickness of the slabs varies from 1 to 5 cm. The thickest slabs form as the notch reaches its maximum size. Near the notch tip, the slabs are curved.

Shearing & splitting

Buckling

Process zone

Stage 4 – Stabilization
The development of the notch stops when the notch geometry provides sufficient confinement to stabilize the process zone at the notch tip. This usually means that there is a slight tear-drop like curvature to the notch shape. Alternatively, if the slabs from the notch flanks are held in place by artificial support, notch development will also stop.

Process zone stabilized by confining stress at notch tip

Figure 4 Progressive failure process in the MBE test tunnel.

However, the compressive stress at the point of failure on the tunnel wall in situ was only about 50 to 60% of the rock strength determined from uniaxial compression tests (Martin *et al.*, 1997).

The difference between the laboratory strength of 200 MPa and the back-calculated in situ strength of 120 MPa at the tunnel periphery was attributed, in part, to complex three-dimensional stress changes that occurred during excavation of the tunnel, especially in the vicinity of the advancing face (Read & Martin, 1996). It was hypothesized that as the tunnel advances, the magnitudes and orientations of the principal stresses in parts of this region change significantly, causing micro-scale damage which locally weakens (or pre-conditions) the rock mass. This process can lead to failure localization in the form of breakouts. These three-dimensional stress effects are not simulated in standard laboratory tests.

Another factor contributing to the discrepancy between the laboratory and in situ strength was the loading conditions associated with the standard unconfined compression test. Results from a series of modified laboratory tests demonstrated that the long-term strength of the rock corresponded with the point of volumetric strain reversal on a typical stress-strain curve (Martin, 1993; Lau & Chandler, 2004). The stress level associated with this point was termed the crack damage stress σ_{cd}. Long-term triaxial loading tests conducted in later studies confirmed that σ_{cd} represents a lower bound on strength under both unconfined and confined conditions.

Based largely on results from AE/MS monitoring (Figure 5), the MBE showed that rock damage occurred near the advancing face of the tunnel. Microseismic events in the

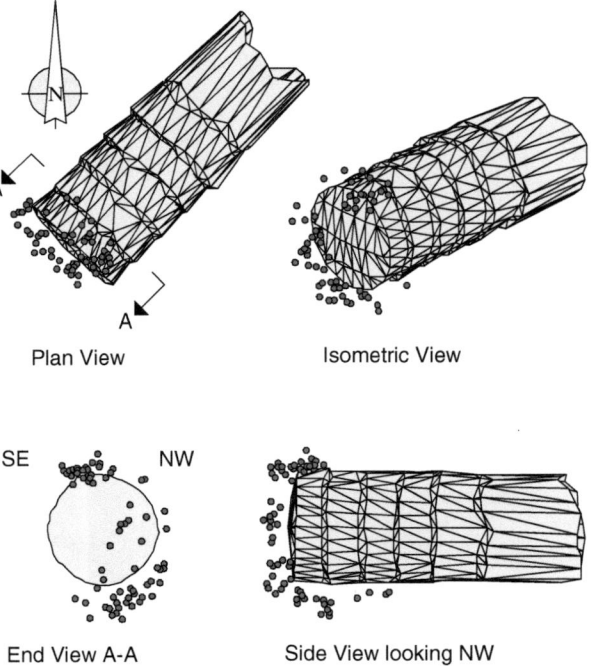

Plan View   Isometric View

End View A-A   Side View looking NW

Figure 5 Microseismic monitoring results from the MBE test tunnel showing microseismicity ahead of the tunnel face in regions where breakouts eventually develop as the tunnel advances.

50 Hz to 10 kHz frequency range occurred up to 0.6 m ahead of the tunnel face, and clustered in regions of compressive stress concentration where breakouts eventually developed as the tunnel was advanced. Where microseismicity was evident, the maximum deviatoric stress (σ_3–σ_3) exceeded a lower threshold of 70 MPa. This threshold was consistent with crack initiation stress σ_{ci} measured in laboratory tests on Lac du Bonnet granite.

Studies were conducted to establish the extent and characteristics of the EDZ around the MBE test tunnel (Read, 1996). Results from underground characterization, geophysics surveys, a focused AE study in the tunnel sidewall, and numerical modeling indicated that excavation damage was most evident in areas of compressive stress concentration in the tunnel crown and invert, and in areas of tensile stress concentration (or unloading) in the sidewalls. These results also showed that the EDZ associated with the 3.5-m-diameter MBE test tunnel was limited to within about 1 m from the original tunnel perimeter, and was three-dimensional, extending some distance ahead of the tunnel face (Figure 6).

The EDZ characteristics were found to vary with position around the tunnel (Figure 7), and were controlled or influenced by several factors. These factors included the nature of the stress concentration (*i.e.*, compressive versus tensile); mineralogy, grain size distribution, and fabric of the various lithologic units; in situ stress state (particularly the σ_1/σ_3 ratio) and the orientation of the principal stresses relative to the tunnel axis; excavation method and sequence; and internal confinement created by excavation debris and ballast on the tunnel invert.

A staged numerical modeling approach was adopted for the MBE to assess the state-of-the-art in numerical models, and the ability of available models to predict the in situ rock mass behavior. It was recognized early in the experiment that none of the numerical models available at the time was capable of simulating the complex processes involved in the progressive failure of brittle rock. Consequently, instead of comparing forward predictions with measured results, the emphasis of the numerical modeling component of the experiment was to improve our fundamental understanding of rock mass failure. It was reasoned that, by advancing understanding of processes involved in progressive failure, predictive numerical models could then be developed.

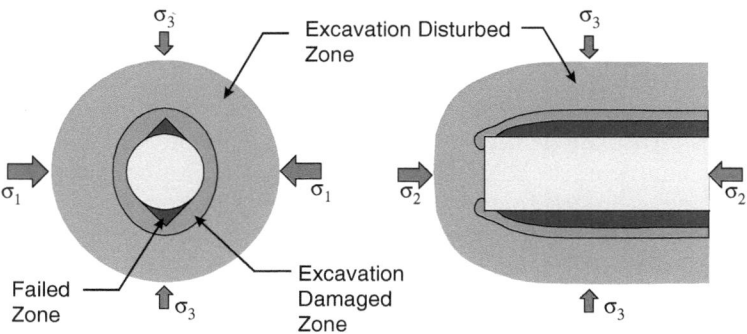

Figure 6 Three-dimensional EDZ around the MBE test tunnel. Stress orientations are generalized as horizontal and vertical to illustrate the various zones.

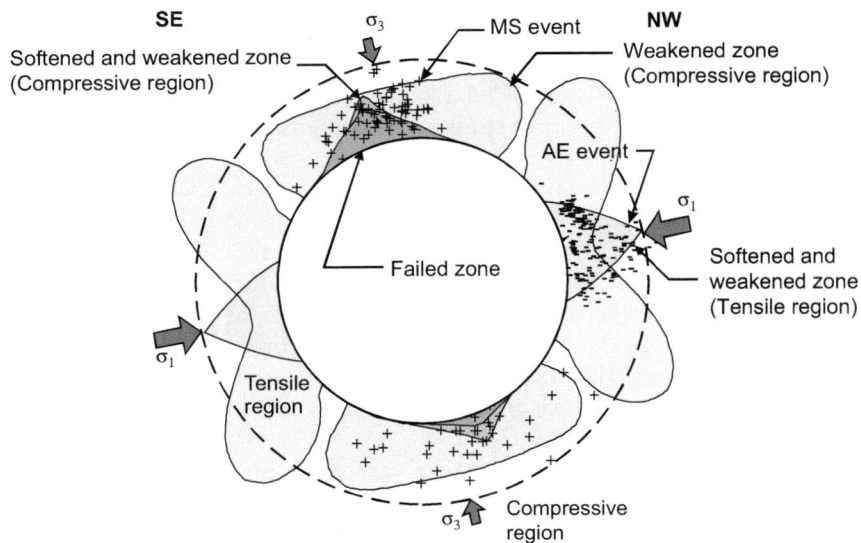

Figure 7 EDZ characteristics around the MBE test tunnel. Characteristics are position-dependent relative to the tunnel periphery and stress directions.

Numerical modeling for the MBE involved a variety of methods (Read & Martin, 1996). In assessing these methods, linear elastic models were shown to be useful for back analyzing monitoring results from the MBE test tunnel if observed changes in tunnel geometry were taken into account. Results showed that, with the exception of regions of failed rock near the surface of the tunnel, the rock mass could be modeled as a linear elastic continuum (Read, 1994). However, neither linear nor non-linear continuum analyses adequately predicted beforehand the extent of damage around the test tunnel, or the shape of the damaged and failed zones. As part of this study, in situ stresses and material properties were established through back analysis of measured displacements and strains from radial extensometers and convergence arrays (Read, 1994).

A comparison of different continuum and discontinuum codes available at the time of the MBE identified several shortcomings. Continuum codes were found to require complex constitutive relations to account for damage to the rock mass, and did not capture the observed transition from continuum to discontinuum behavior (*i.e.*, large dilation leading to buckling and slabbing). Discontinuum codes were also of limited use because simulation results were dependent on the network of discontinuities assumed in each analysis. Distinct element codes such as the Particle Flow Code (PFC) were considered for simulating the continuum-discontinuum transition, but were difficult to calibrate, and were limited by the state of computing technology as to the level of discretization that could be incorporated into simulations. Nonetheless, this latter type of model was considered the best candidate to simulate progressive brittle failure of the rock mass.

PFC was initially introduced to the MBE as a means of calibrating a constitutive 'lattice' model implemented in the Fast Lagrangian Analysis of Continua (FLAC) code (Cundall *et al.*, 1996). Initial results from application of PFC suggested that this code could be refined to directly model the complex failure process associated with the MBE. These conclusions led to a multi-phase modeling program (Read & Chandler, 2002; Potyondy & Cundall, 2004) involving the refinement of PFC input routines to assess tunnel stability and excavation damage.

3.4 Borehole breakout studies

In conjunction with the MBE, borehole breakouts were studied extensively in situ at the URL (Martin *et al.*, 1994). In a series of boreholes with diameters ranging from 150 to 1240 mm, observed borehole breakouts did not form diametrically opposite one another. None of the boreholes were drilled parallel to a principal stress direction. The same observation was true of breakouts in the 3.5-m-diameter MBE test tunnel. This phenomenon is contrary to the idea that breakouts initiate at the points of maximum tangential stress concentration around a circular opening because, in an elastic medium subjected to an anisotropic stress field, these points are diametrically opposite one another on the borehole wall.

Analysis of the MBE showed that three-dimensional stress history effects can result in non-uniform pre-conditioning of the rock mass near the tunnel periphery, and can lead to asymmetric breakout development (Read *et al.*, 1995). In the analysis, it was shown that the distribution of maximum deviatoric stress ($\sigma_1 - \sigma_3$) ahead of the face is significantly affected by the orientation of the tunnel relative to the principal stress directions (Figure 8). In conditions where the excavation is oriented parallel to a principal stress direction, the maximum deviatoric stress ahead of the face is distributed symmetrically with respect to the tunnel axis. However, for situations where the tunnel axis is not parallel to a principal stress direction, such as the MBE, the distribution of the maximum deviatoric stress ahead of the face becomes asymmetric with respect to the projected tunnel axis.

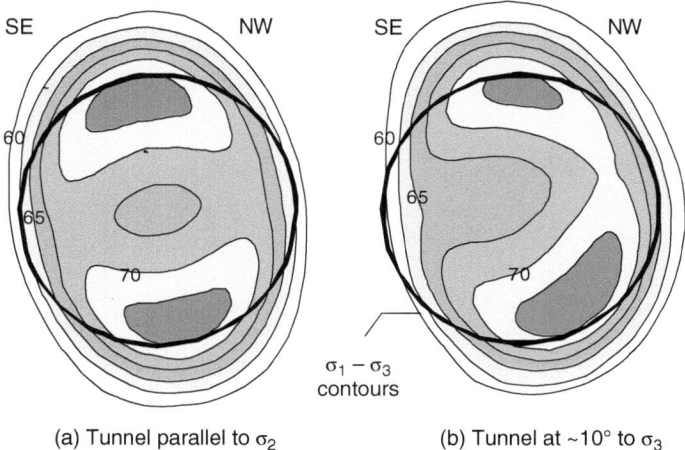

(a) Tunnel parallel to σ_2 (b) Tunnel at ~10° to σ_3

Figure 8 Pattern of maximum deviatoric stress in a plane 0.6 m ahead of a 3.5-m-diameter tunnel (a) parallel to the intermediate principal stress direction and (b) at 10° to the intermediate principal stress direction. Far-field stresses are those at the 420 Level.

A conclusion from the MBE was that the combination of high deviatoric stresses and rotation of principal stresses ahead of the tunnel face generated local weakened zones that were eventually exposed at the tunnel perimeter as excavation advanced. These zones were distributed asymmetrically with respect to the tunnel axis because the tunnel was not aligned with a principal stress direction. Recorded microseismic events ahead of the tunnel face provided evidence that weakening due to cracking occurred in the region of this calculated asymmetric stress distribution (Figure 8). Consequently, even though the points of maximum tangential stress at the tunnel wall resulting from tunnel advance were distributed symmetrically with respect to the tunnel axis, breakouts developed asymmetrically, initiating in these regions of reduced strength.

A similar effect was observed in a 1.24 m diameter vertical borehole drilled at the 420 Level. As shown in Figure 9, the breakout pattern was asymmetric in both orientation and depth. Stresses 0.1 diameters ahead of the borehole face at the projected borehole perimeter were computed using a 3D boundary element program; the maximum deviatoric stress is plotted versus azimuth in the lower part of Figure 9. The magnitudes of the maximum deviatoric stress concentrations were compared with the crack-initiation threshold determined from AE monitoring. The numerical model results showed a pattern of maximum deviatoric stress ahead of the face that was consistent with the observed pattern of breakouts. The azimuths of maximum stress concentrations coincided closely with the observed azimuths of the asymmetric breakouts (Figure 10a). Also, the breakout at azimuth 048° was predicted to be the deeper of the two breakouts, as was observed in situ.

To confirm the influence of stress orientation on borehole breakouts, a 600-mm-diameter borehole was drilled parallel to the σ_3 direction at the 420 Level (Figure 10b). The asymmetry in the borehole breakouts was virtually eliminated compared to the adjacent vertical 1.24-m-diameter borehole, both in terms of the difference in breakout azimuths, and in breakout depths (Read, 1994). The orientations and magnitudes of the calculated maximum deviatoric stress concentrations ahead of the borehole face, which in this case were symmetric about the borehole axis, were consistent with the field observations in this inclined borehole. The study supported the hypothesis that the stresses ahead of the tunnel face influenced the characteristics of breakouts at the borehole perimeter, suggesting that cracking and rock failure is a three-dimensional issue.

3.5 Heated failure tests

The rock mass response to thermal loading is important in situations where temperature conditions vary significantly from ambient conditions (*e.g.*, steam-assisted petroleum recovery processes, geothermal applications, and nuclear waste repositories). The underground emplacement of nuclear fuel waste in the Canadian concept, for example, is expected to cause an increase in temperature of up to 85°C within the rock around a repository, depending on design. Under these elevated temperature conditions, the rock mass will experience increased rock stresses and pore pressures, which may contribute to the development of excavation damage and progressive failure. The elevated temperature may also induce changes in the material properties, including strength characteristics of the rock mass.

In order to investigate the effects of thermal loading on excavation damage development and progressive failure, the Heated Failure Tests (HFT) (Read & Martino, 1996a, b;

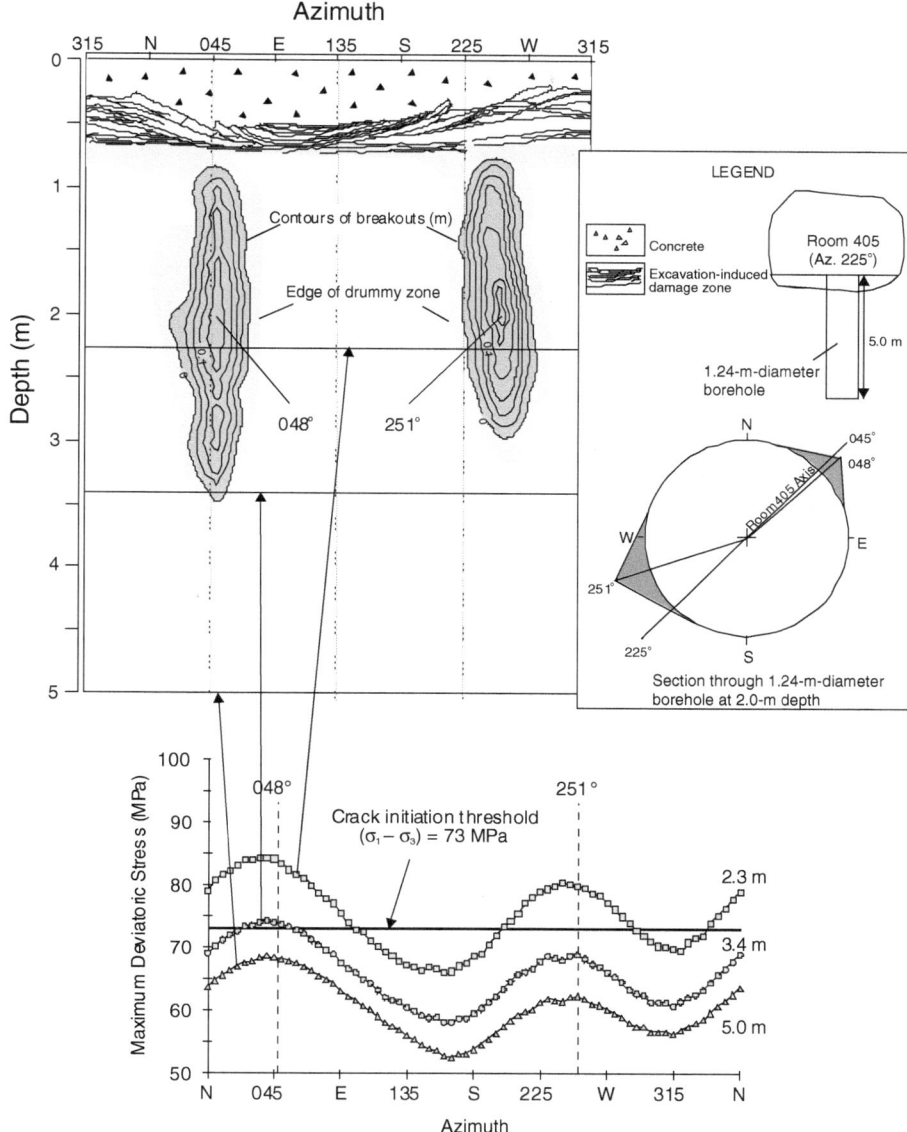

Figure 9 Perimeter map and cross-section showing borehole breakouts in a 1.24-m-diameter borehole at the 420 Level (top); maximum deviatoric stress calculated ahead of the borehole face at three depths (bottom). Breakouts extend to different depths on opposite sides of the bore-hole, and are offset from diametrically opposed positions by 23°.

Martino & Read, 1996) were conducted between 1993 and 1996 in the same area as the MBE at the 420 Level (Figure 11). These tests constituted a multi-stage in situ thermal experiment comprising four investigations designed to assess the effects of drilling/heating sequence (*i.e.*, loading path), borehole interaction, and internal confining pressure on the

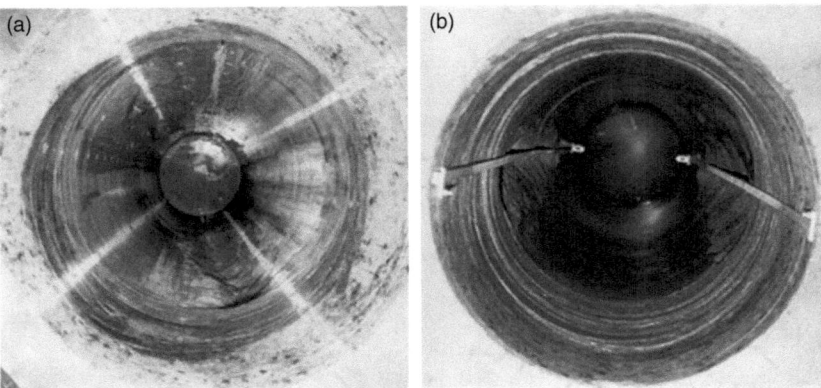

Figure 10 Borehole breakouts: (a) asymmetric breakouts in a vertical 1.24 m diameter borehole and (b) symmetric breakout notches developed in a 600-mm-diameter borehole aligned parallel to the minimum principal stress direction at the 420 Level.

Figure 11 The Heated Failure Tests at the 420 Level: installation of instrumentation borehole casing in invert of widened MBE test tunnel (left), and completed experiment construction (right).

progressive development of excavation damage around underground openings (Read *et al.*, 1997a). The various stages of the experiment incorporated 600-mm-diameter vertical boreholes subjected to different drilling/heating sequences and confining conditions.

The strategy adopted for the HFT was to monitor the development of excavation damage, characterize the extent of damage in situ, then use thermal-mechanical modelling to assess the relation between damage development and near-field in situ stresses. Monitoring was conducted using an AE system, extensometers, convergence arrays, piezometers, and thermistors/thermocouples. Characterization activities included geological mapping and photography/videotaping of lithology, induced fractures, and breakouts in each of five observation boreholes. Numerical modelling was carried out using linear elastic models to correlate stresses with breakouts, and with AE response, in order to assess the relation of progressive failure and excavation damage development to thermal-mechanical loading history.

Results from the four stages of testing showed that AE activity correlated well with observed damage development. AE monitoring delineated the extent, and captured the temporal and spatial history, of rock mass damage, including microcracking (or 'preconditioning') ahead of the base of the borehole associated with face advance, and macroscopic failure manifested as breakout development. AE monitoring results showed that the crack initiation stress established from the MBE (*i.e.*, $\sigma_{ci} \approx 70$ MPa) roughly correlated with stresses back calculated at AE locations ahead of the advancing borehole face recorded in Stage 1 of the tests. AE events along the borehole wall were associated with breakouts, which occurred at higher deviatoric stress levels than microcracking ahead of the borehole face.

The AE monitoring results illustrated that breakouts initiated through localized damage. Once initiated, the combination of stress localization due to the developing breakout geometry and increasing tangential stress due to either borehole advance or heating caused the breakouts to extend into relatively undamaged rock. The breakout terminus (Figure 12) was taken to be the point of equilibrium between failed and non-failed rock. The calculated stress at this location would, in theory, represent the in situ strength of the rock mass. The results of the experiment indicated that damage development, as indicated by AE activity, occurred primarily during periods of drilling, heating, and, to a lesser extent, cooling, and tended to decrease during periods of constant temperature (Figure 13).

The terminus of breakouts observed in Stage 1 of the HFT, where heating commenced after completion of drilling, were consistent with the thermal-mechanical tangential stresses calculated around the borehole, and with previous measurements of long-term rock strength. Geological variability in this borehole influenced development of breakouts, with granodiorite dykes less susceptible to breakouts than granite. Granodiorite dykes in this setting tend to be finer-grained and more equigranular than the host granite. Breakouts were limited to within 45 mm radially of the borehole periphery (*i.e.*, 15% of the borehole radius).

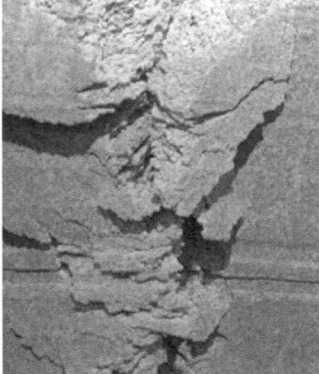

Figure 12 Breakouts developed in Stage 1 of the HFT: view from top of borehole HFT1 (left), and close-up of breakout terminus showing three-dimensional nature of the failure process (right).

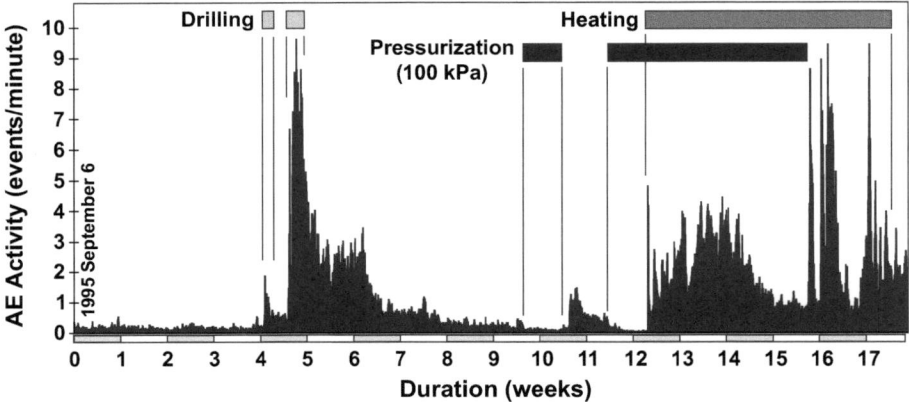

Figure 13 AE activity in relation to drilling, heating and pressurization the 600-mm-diameter borehole in HFT Stage 4.

Results from Stage 2 of testing, where the drilling/heating sequence was reversed, indicated more substantive weakening of the rock mass ahead of the advancing face in the regions eventually exposed at the borehole wall. This increased damage was believed to be caused by the combined thermal-mechanical load resulting in high deviatoric stresses generated near the advancing borehole face. Breakouts extended further vertically up and down along the borehole wall than those observed in Stage 1. These breakouts also extended up to 60 mm radially beyond the original borehole surface (*i.e.*, 20% of the borehole radius).

In Stage 3, the interaction of two adjacent boreholes caused breakouts between the holes that were more extensive than those in Stage 1, and resulted in up to a 40% reduction in the thickness of the 400-mm-thick rock web between the holes. However, even in this case, breakouts on the web-side of the boreholes were limited to within 95 mm of the borehole wall (*i.e.*, 32% of the borehole radius). Breakouts on the opposite sides of the boreholes were of similar radial extent as those observed in Stage 1 (*i.e.*, ~45 mm).

In Stage 4, the influence of a small (~100 kPa) internal confining pressure on the development of thermally-induced excavation damage was assessed. Prior to heating the surrounding rock, a vinyl liner was installed in the 600-mm-diameter borehole and a nominal 100 kPa of internal pressure was provided by a 10 m static head of water. The rate of observed AE activity showed that the confining pressure acted to inhibit the development of rock mass damage before heating (Figure 13). Pressurizing the lined borehole to approximately 100 kPa, or simply filling it with water, likewise reduced AE activity during heating. Removal of the confining pressure led to an increase in AE activity both before and during heating. The suppression of both AE activity and the associated damage demonstrated the highly sensitive nature of the breakout process to confining pressure, and reinforced the idea that the breakout is in a state of meta-stable equilibrium.

The findings from the HFT indicated that the extent of excavation damage was dependent mainly on the magnitudes of the radial and tangential boundary stresses at the borehole periphery, but could also be influenced by the thermal-mechanical loading sequence (*i.e.*, stress loading path). The dominant thermal effect in each of the four stages appeared to be the increased tangential stress at the borehole periphery. Thermally-induced pore pressure changes, although relatively minor by comparison (*i.e.*, about 1 MPa at 100°C) in this experiment, may have also contributed to the AE activity recorded and to changes in ultrasonic velocity in the experiment area.

Using the tangential boundary stress calculated at the upper and lower terminus of the breakout in each observation borehole as a measure of the in situ unconfined compressive strength of the rock mass, finer-grained granodiorite in this alignment was consistently stronger and less susceptible to pre-conditioning damage than the medium-grained granite. The in situ strength at the borehole periphery ranged from about 120 to 185 MPa for granite, and from about 170 to 210 MPa for granodiorite, for the different stages of testing. The lowest strength was associated with a borehole drilled into heated, and hence more highly stressed, granite, indicating significant pre-conditioning damage during drilling.

3.6 Excavation stability study

Based on the results of the previous experiments and associated studies, the Excavation Stability Study (ESS) (Read & Chandler, 1996, 1997) was undertaken between 1995 and 1997 to evaluate stability and the extent of excavation damage in underground openings as a function of tunnel geometry and orientation, geology, and excavation method. The ESS consisted of a series of drill-and-blast excavations at the 420 Level with different near-field stress distributions and stress histories, excavated in both granite and granodiorite lithology. Different near-field stress conditions were achieved by varying tunnel geometry and orientation. As part of the study, results from two circular drill-and-blast openings were compared to those from the mechanically-excavated MBE test tunnel (see Figure 3).

Under the adverse stress conditions at the 420 Level and a practical limitation on room aspect ratio of approximately 2.2:1, it was shown that for certain conditions, openings of an 'ovaloid' shape (Greenspan, 1944) produced lower peak tangential stresses and less stress localization than elliptical openings of the same aspect ratio (Figure 14). These conditions required that the major cross-sectional axis of the ovaloid

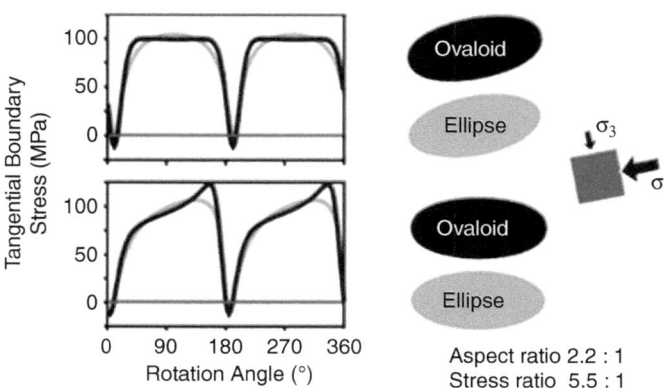

Figure 14 Comparison of elliptical and ovaloid tunnels of the same aspect ratio under in situ stress conditions at the 420 Level. Boundary stresses are calculated values.

tunnel was aligned with the maximum principal stress σ_1, which at this level plunges at 11° from horizontal. For non-circular tunnels with a horizontal major axis, elliptical openings at the 420 Level were shown to have a lower peak tangential boundary stress than ovaloid openings with the same aspect ratio. These results illustrate the importance of thorough characterization of the in situ stress tensor in excavation design, particularly if non-traditional tunnel shapes are planned.

The generalized ovaloid geometry is described by two parametric equations:

$$x = p \cos(\beta) + r \cos(3\beta) \tag{1}$$
$$y = q \sin(\beta) - r \sin(3\beta) \tag{2}$$

For sub-optimum room aspect ratios (*i.e.*, the room aspect ratio is less than the maximum-to-minimum stress ratio in the cross-sectional plane), the parameter r can be optimized to minimize the peak tangential stress, and to reduce stress localization, around the opening.

In total, eight ovaloid openings with different aspect ratios and/or inclination relative to the maximum principal stress direction, and two circular openings, were excavated for the experiment. The experiment arrangement and excavation cross-sections used in the ESS, and their associated peak compressive stress concentrations, are shown in Figure 15. Nine of these openings were excavated parallel to the intermediate principal stress direction (similar to the MBE tunnel); the other (M4) was aligned parallel to the maximum principal stress direction.

Observations in the ten tunnel segments showed that instability leading to breakout formation occurred only in those segments that were excavated in granite and had a calculated peak boundary stress exceeding 120 MPa. Tunnels that were optimized to the stress conditions (*i.e.*, with the major cross-sectional axis parallel to σ_1) were found to be inherently more stable than those openings with a horizontal major axis and peak compressive stress concentration localized at two points on the boundary. A comparison of segments U1 and M1 (Figure 16), both with an aspect ratio of 2.2:1 but different cross-sectional alignments relative to σ_1, illustrates the difference in performance. Progressive failure involving large-scale slabbing occurred in segment M1, but not in

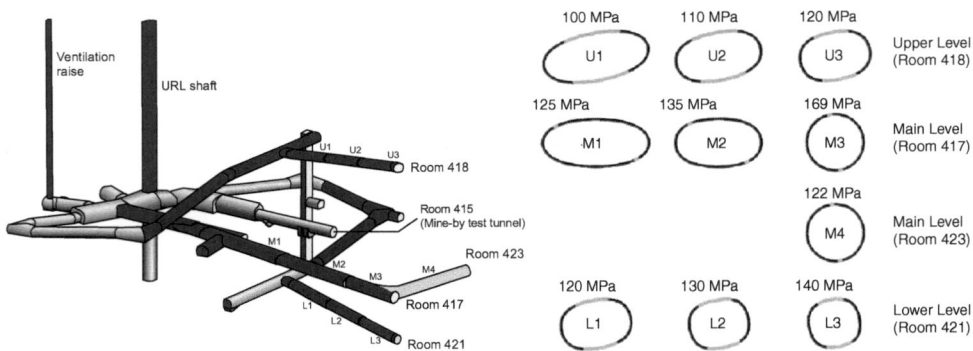

Figure 15 Arrangement of the ESS at the 420 Level and cross-sections of the 10 tunnels segments showing peak compressive boundary stress magnitude and distribution.

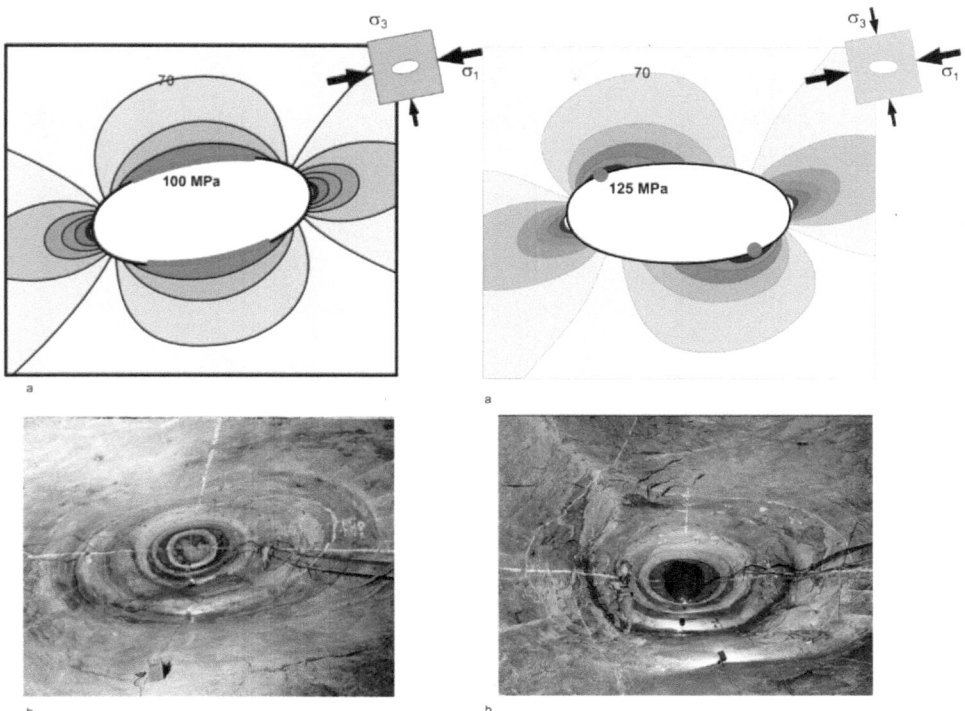

Figure 16 Tunnel segment UI (left) with a uniformly-distributed boundary stress of 100 MPa over the crown and invert of the tunnel, and tunnel segment MI with a localized peak boundary stress of 125 MPa in the crown and invert of the tunnel.

U1. The difference in tunnel stability in these two tunnel segments was related primarily to the difference in stress magnitude and stress distribution around the two tunnels.

Segment U3 (Figure 17), with a calculated boundary stress of 120 MPa, was also stable despite being excavated in the weakest rock type (coarse-grained granite) at the 420 Level, and exhibiting marked geological variability. By achieving a uniformly-distributed boundary stress, blast-damaged rock that spalled initially from the tunnel crown tended to smooth the tunnel profile without the localization of damage that could potentially lead to breakout formation. This spalling process did not change the trajectory of the maximum principal stress around the opening, and effectively reduced the thickness of damaged rock around the tunnel.

In addition to initial observations related to tunnel stability (Read & Chandler, 1996), characterization and monitoring activities conducted for the ESS provided further insight into the relation between in situ stress, damage, and acoustic velocity of the rock mass (Read *et al.*, 1997b). Microseismic monitoring confirmed that damage development (characterized by induced microseismicity) was more prevalent for openings with higher compressive boundary stresses, particularly those with stress concentrations localized at single points on the tunnel periphery, and for those tunnels excavated in granite versus granodiorite. Convergence measurements confirmed that

Figure 17 Stable tunnel segment U3.

most of the ESS excavations, with the exception of those that experienced macroscopic progressive failure, displayed linear elastic displacement behavior. Openings that experienced progressive failure were located in granite lithology, and had localized stress concentrations at the tunnel periphery. Deformation in these openings exceeded the predicted elastic response, and exhibited a time-dependent component.

The characteristics of excavation damage inferred from MVP measurements were found to correlate with damage expected in regions of high deviatoric stress, and reduced confining stress (Read *et al.*, 1997b). This suggests that damage development occurs in both highly compressed regions and those that have experienced unloading, confirming observations from the MBE.

The outer extent of damage, as indicated by borehole velocity measurements, was limited to less than 0.5 m from the tunnel wall in each of the ESS openings, with the exception of the circular tunnel segment M3, which had evidence of damage in the sidewall up to 1.5 m from the tunnel wall. Although a greater amount of damage was apparent in segment M1 (Figure 16), velocity measurements for this tunnel segment were conducted in boreholes in granodiorite where damage was less severe. This influence of subtle changes in rock characteristics on stability at AECL's URL is discussed in more detail by Everitt & Lajtai (2004).

Compared to the mechanically-excavated MBE test tunnel, the breakout in the crown of the parallel circular drill-and-blast tunnel in the ESS was of similar shape and extent. However, damage in the tensile sidewall region of the drill-and-blast tunnel included discrete radial fracturing, and extended further into the rock mass than in the MBE test tunnel.

The ESS confirmed that it is possible to construct large stable underground openings with only limited excavation damage in the adverse stress conditions at the 420 Level. Excavation geometries were selected based on comprehensive knowledge of the in situ stresses and a two-dimensional elastic analysis. For the most part, this analysis method was sufficient to design tunnels with damage in the crown and invert limited to a depth of about 20 cm in the same stress conditions that produced the large breakout notches around the MBE test tunnel. However, the ESS also highlighted the influence of geological variability on rock mass stability. There is a reasonably high degree of uncertainty associated with the two-dimensional, homogeneous, isotropic, linear elastic approach to excavation design. This uncertainty can be reduced by accounting for the three-dimensional stress paths leading to rock failure around excavations (Read *et al.*, 1998), and by characterizing the geological variability and the potential influence of rock fabric on excavation stability (Everitt & Lajtai, 2004).

3.7 Tunnel sealing experiment

Unlike other excavation experiments, the Tunnel Sealing Experiment (TSX) (Chandler *et al.*, 1998) was undertaken in 1995 to develop sealing technologies and to demonstrate their effectiveness in minimizing flow along a full-scale emplacement room under ambient and elevated temperature conditions. However, the effects of tunnel geometry, tunnel orientation, and excavation method on rock strength, failure mechanisms and damage zone development were also investigated. The nature of the EDZ in granite and its relationship to hydraulic properties was of particular interest in this case.

The rock engineering design objectives in the TSX were to minimize, within limits dictated by practical construction considerations, the effects of access and instrumentation excavations on the test chamber; and to achieve a stable test chamber with limited excavation damage in the rock surrounding the opening, typical of conditions expected in repository rooms. Effectiveness of cut-off keys in bulkhead design, and numerical modelling advances, were also investigated.

The experiment involved excavating access tunnels and a 30-m-long, 3.5-m-high elliptical test chamber (aspect ratio 1.25:1). Access excavations for the experiment were completed in 1997 using a full-face drill-and-blast method. To assess the likely performance of the main test chamber and bulkhead keys, a short test tunnel (Room 419) was excavated parallel to, and with the same aspect ratio as, the test chamber. These excavations provided additional observations of the effects of excavation shape, room orientation, and excavation method on excavation damage development. The test chamber was excavated between January and March 1997 using a full-face drill-and-blast technique, and optimized blast patterns/sequencing developed in the access excavations. The cross-sectional dimensions of the test chamber were large enough to be representative of a repository room, allowing sealing technologies to be tested at full-scale.

The sealing system design incorporated a 2.3-m-long clay bulkhead, and a 3-m-long concrete bulkhead, installed 12 m apart to seal the test chamber (Figure 18). The segment of the chamber between the two bulkheads was filled with permeable sand, and seal performance was tested by pressurizing the sand-filled chamber between the bulkheads with water. Based on experience gained in other experiments at the 420 Level of the URL, the excavation damage around the test chamber was expected to be limited to a thin annulus, typical of conditions expected in a repository room. Experience from the MBE and numerical modelling suggested that, by keying the tunnel bulkheads into the rock mass, the potential for connected permeability along the tunnel in a continuous damage zone would be reduced (Martin *et al.*, 1996).

A sensitivity study of various shapes and sizes of keyed slots (Dzik & Read, 1997) showed that annular keys of rectangular or triangular (with one vertical face) cross-

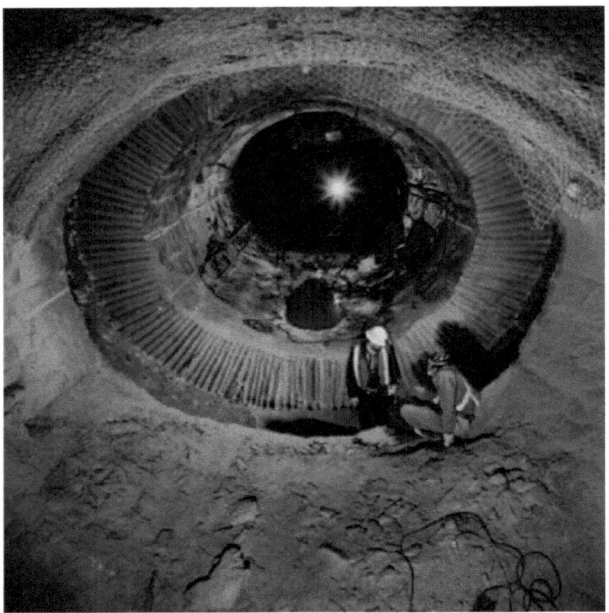

Figure 18 Test chamber for the Tunnel Sealing Experiment, showing the clay-bulkhead key (rectangular cross-section).

section cut off connected damage along the tunnel without creating significant additional stress-induced damage. To minimize the induced damage associated with the excavation of such keys, a method incorporating line-drilling and hydraulic rock splitters was used in lieu of blasting, similar to the technique used in the MBE. The rectangular clay-bulkhead key (Figure 18) was excavated between June and August 1997. The triangular concrete-bulkhead key (with the shape resembling a bathtub stopper) was completed between September and November 1997.

Rock instruments, including a microseismic (MS) array, thermistors, and hydrogeological instruments, were installed in the experiment area prior to excavation of the test chamber. Other hydrogeological instruments and boreholes for geophysical monitoring were installed around the seal locations following excavation of the test chamber. In addition, a smaller-scale acoustic emission (AE) array was installed around the clay bulkhead location to provide high frequency monitoring of acoustic events and velocity changes associated with damage development and changes in rock properties. The AE array was installed prior to excavation of the bulkhead keys to provide a complete record of AE activity associated with bulkhead performance.

Characterization of the rock mass surrounding the excavation using a variety of techniques indicated that most of the rock damage was limited to within 0.5 m of the tunnel surface (Martino & Chandler, 2004). Borehole velocity measurements, permeability measurements and observations of visible fracturing indicated an inner zone of severe velocity reduction within about 0.2 m of the TSX tunnel wall and an outer, less severe, damaged zone between 0.2 and 0.5 m. The inner damaged zone included visible fracturing oriented subparallel to the excavation surface, resulting from the drill-and-blast excavation method. In the sidewalls of the tunnel, where the high in situ stresses caused by excavation result in extensional strain, permeability and MVP measurements indicated microcrack damage as far as 1 m from the tunnel surface. Repeated ultrasonic velocity surveys through the damaged zone showed a decrease in both P- and S-wave velocity with time over a period of one year after excavation, supporting the use of time-dependent damage development models in numerical simulations.

Results from the TSX showed that the tunnel design was successful in limiting the extent of excavation damage around the main test chamber. The hydrogeological instruments installed prior to the test chamber excavation indicated pore pressure increase in the region of increased compressive stress in the tunnel crown, and a decrease in pore pressure in the sidewall area where the compressive stress was reduced. These results were consistent with expected behavior.

The TSX provided perhaps the most comprehensive monitoring and characterization associated with any of the experiments at the URL. The integration of data from these different techniques and instruments provided a unique understanding of the nature of excavation-induced damage.

4 THERMAL MECHANICAL STABILITY STUDY

The Thermal-Mechanical Stability Study (TMSS) (Read & Chandler, 2002) was under-taken between 1996 and 2001 to refine tools and capabilities for characterization, monitoring, and numerical modelling, and to develop an integrated system (or 'tool-box') for engineering design of repository excavations. Such a rock engineering toolbox would ultimately be used to assess stability of repository excavations under loading conditions expected over a repository's lifetime.

The tools identified or developed during the TMSS were aimed at quantifying expected change in material properties induced by stress changes in the rock adjacent to excavated openings. The project involved parallel development, followed by systematic integration, of tools and capabilities related to numerical modelling and complementary analytical approaches, AE monitoring, in situ characterization, and laboratory testing.

4.1 Numerical modeling

Significant advances were made in the use of discrete element codes and other models to investigate excavation damage development and excavation stability. The TMSS demonstrated that the PFC modeling approach has the following advantages when compared to a conventional continuum approach:

- Damage and its evolution are explicitly represented in the model as broken bonds; no empirical relations are needed to define damage or to quantify its effect on material behavior.
- Localized microcracks form and coalesce into macroscopic fractures automatically without the need for re-meshing or grid reformulation.
- Complex non-linear behaviors, such as hysteresis, dependence of strength on confining stress, dilatancy and evolution of material anisotropy arise as emergent features, given simple behavior at the particle level. There is no need to develop constitutive laws to represent these effects.

- Secondary phenomena, such as acoustic emission, occur in the PFC model without additional assumptions. In general, the model is believed to reproduce qualitatively all of the mechanical mechanisms and phenomena that occur in rock, although adjustments and modifications may be necessary to achieve quantitative matches in particular cases.

Tests conducted as part of the TMSS demonstrated that the PFC model for rock (Potyondy & Cundall, 2001, 2004) was capable of successfully tracking the development of damage and the progressive failure process observed in situ around various excavations at the URL. The comparison tests illustrated that minimal damage and no breakouts were predicted if the normal and shear strengths of the PFC bonds between particles were selected to reproduce only the laboratory-derived unconfined compressive strength of the rock.

Further analysis showed that observed damage and breakout development could be replicated by either reducing the bond strengths by 30 to 40%, or by activating a stress corrosion algorithm. The stress-corrosion algorithm simulates subcritical crack growth by introducing a time-dependent strength reduction for the bonds between particles when the local tensile stress exceeds a specified threshold. In PFC, tension will develop between some particles even when the overall stresses are compressive. These micro-tensions are dependent on both the stresses applied to the model boundaries, and on the redistribution of interparticle forces caused by the progressive development of bond breakages (which are analogous to microcracking).

A key finding of the PFC modelling was that model results would match observations if the process within the model was time-dependent, and directly related to the magnitude of local micro-tensions between particles. For unconfined conditions at the periphery of tunnels, the simulated inter-particle tensile forces orthogonal to the free surface increased as the tangential boundary stress increased. For tangential boundary stress distributions that are characterized by a localized peak compressive stress, the PFC model would predict damage localization and non-uniform weakening of the rock mass if a micro-activation stress threshold (*i.e.*, the limit above which stress corrosion becomes active) is exceeded. Depending on the applied stress, this damage development process would lead to localized dilation, and large-scale progressive slabbing of material from the tunnel periphery (Figure 19).

With developments in the application of continuum damage mechanics, modeling of important aspects of the brittle failure process came nearly full circle. PFC was initially used to develop a constitutive model for rock damage for input into a continuum code to simulate the MBE. More recently, a continuum damage formulation has been implemented to capture many of the important characteristics of the rock damage process as observed in the URL experiments (Mitaim & Detournay, 2004). One of the outcomes of the TMSS is the recognition that the incorporation of dilation is a critical element in models of the brittle failure process. This work will eventually result in fully functional T-M-H continuum modeling tools to complement PFC (Read & Chandler, 2002).

Thermoporoelastic modeling of the rock surrounding repository excavations provided another tool for assessing the effects of heating a low permeability rock mass. The fundamental theory of thermoporoelasticity is not new, but its application to repository analyses was an important advancement of technology brought about through the

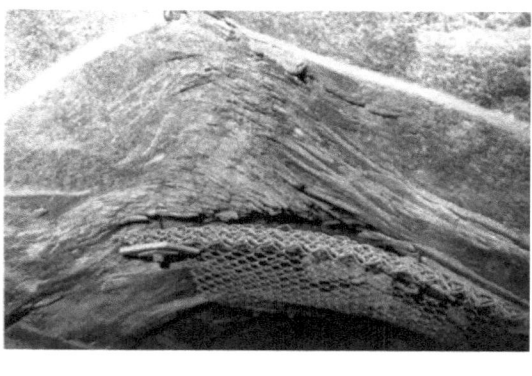

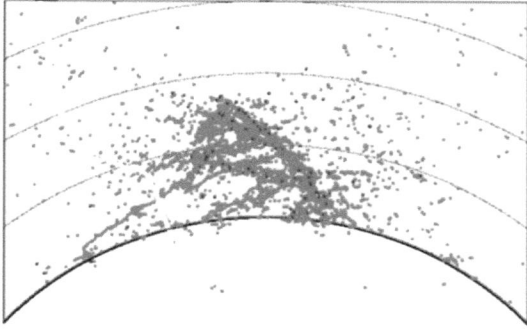

Figure 19 Photograph of slabs developed in the breakout region of the MBE test tunnel (top) and simulated damage in the PFC model of the MBE test tunnel after activating stress corrosion for two months (bottom).

TMSS. In particular, the laboratory and in situ characterization of thermoporoelastic parameters was an important contribution to coupled T-H-M analysis of rock (Berchenko *et al.*, 2004; Detournay *et al.*, 2004; Lau & Chandler, 2004). Thermally-induced propagation of fractures was examined using fundamental fracture mechanics theories, providing another tool for assessing the stability of existing fractures and faults under combined thermal and hydraulic loads (Berchenko *et al.*, 1997).

4.2 Monitoring and instrumentation

The TMSS focused on the advancement of AE/MS technology as a means of remotely monitoring underground excavations. It is likely that AE/MS technology will play an important role in the various phases involved in repository design, construction, operation, and post-closure monitoring. This includes site characterization and performance monitoring at a wide range of monitoring scales. The advantage of AE/MS technology is in the non-invasive and remote monitoring capabilities of these systems, and in combining active and passive monitoring methods. These techniques also have applications for investigating the behavior of man-made engineered barriers such as concrete seals. AE/MS technology currently provides one of the few methods available to validate numerical models used to predict the behavior of repository excavations.

Developments in AE/MS technology and application to monitoring URL underground excavations are discussed by Young & Collins (1999).

In the course of the TMSS, techniques to model and predict AE/MS activity were explored. The successful use of PFC to conduct dynamic modeling and analysis of AE events was a major step forward in understanding the underlying micromechanics of damage development and rock fracture (Young et al., 2004), and in linking numerical models directly to measurable damage characteristics. This technique is described by Hazzard & Young (2004).

The monitoring and instrumentation tools and capabilities that were added to the repository excavation design toolbox during the TMSS include state-of-the-art AE and MS monitoring tools including hardware, software, and methodologies for in situ monitoring of different rock volumes; processing algorithms; a cross-correlation technique; dynamic modeling capabilities for AE events; data management capabilities including internet access to data; sophisticated visualization capabilities; quality assurance procedures; and a systematic approach to applying AE/MS technology to analyze AE, MS, and velocity data to assess excavation damage.

4.3 Characterization

The characterization tools and techniques employed in the TSX tunnel demonstrated that no single characterization method can provide a complete understanding of excavation damage. However, by using multiple techniques ranging from remote sensing to invasive borehole-based tomographic surveys, a systematic approach to damage characterization was successful in identifying important features associated with excavation damage (Martino & Chandler, 2004). Integrated application of AE and velocity surveys using a damage model based on these characterization data proved effective in defining the nature of damage in different locations around the tunnel. Numerical modeling conducted in conjunction with these activities provided a possible tool to differentiate between stress-related effects and those related to excavation damage. Specialized laboratory testing provided data for calibration of models for the short- and long-term responses of granite and granodiorite at ambient and elevated temperature (Lau & Chandler, 2004). The data from these tests also complemented the in situ characterization studies.

4.4 Integration and design

The TMSS established a systematic excavation design approach that integrates characterization, monitoring, and numerical modeling tools and capabilities. The collective set of engineering tools can be applied to both back analysis and forward prediction of short- and long-term rock mass responses, and can be used to predict associated changes in material properties.

The excavation design approach developed from the TMSS includes an integrated sequence of steps related to in situ characterization, laboratory testing, numerical modeling, in situ monitoring, and in situ calibration tests (Read & Chandler, 2002). In this approach, preliminary repository excavation designs can be established using two-dimensional linear elastic modeling to compare the calculated peak compressive stress on the tunnel perimeter with the long-term strength of the various rock types

(Read & Martino, 2002). A fractured rock mass may require a discrete element simulation using codes such as UDEC. Detailed analyses using PFC are then conducted on the selected excavation design to assess the short- and long-term behavior under a range of expected boundary conditions.

Integration tools and capabilities that were added to the repository excavation design toolbox during the TMSS include strategies for data management, analysis, and comparison; technical coordination; project integration; in situ validation tests; and a systematic approach for integrating results from numerical modeling, characterization, and monitoring activities to assess damage development and stability of repository excavations.

4.5 Outstanding issues

While there were still technological gaps and limitations of the tools and capabilities available for repository excavation design, the advances in the TMSS were significant. The results of the TMSS provided convincing evidence that the basic physics of rock damage and fracturing are better understood than they were at the completion of the MBE, and that the tools and capabilities needed to design stable repository excavations with minimal excavation damage were either within the toolbox, or reasonably within reach.

Tools and capabilities have continued to advance as researchers and practitioners build on the experience from the TMSS. For example, as computing technology has advanced (*e.g.*, Potyondy, 2014), it is feasible to conduct full-scale tunnel simulations using particle discretization in two- and three-dimensions. This provides the means to address some of the outstanding issues from the TMSS such as the role of 3D stress effects near the advancing tunnel face in pre-conditioning the rock mass. The effect of geological heterogeneity and variability can also be addressed as tools continue to advance, and the use of multiple shear wave sensors in conjunction with AE/MS monitoring becomes standard practice for anisotropic analysis.

The collective work from the TMSS has resulted in the development of practical engineering tools and the advancement of our understanding of rock mechanics, the application of this understanding to investigate implications of a deep geologic repository in other rock types (*e.g.*, Read, 2008a, b), and has spawned extensive research by various universities and international agencies.

5 SHAFT SEALING AND FACILITY DECOMMISSIONING

As part of the Nuclear Legacy Liability Program (NLLP) funded by Natural Resources Canada (NRCan), a program was undertaken to decommission facilities that are no longer part of AECL's mandate or operations. Included in these facilities was AECL's URL. A decision was taken in 2003 to discontinue operation of AECL's URL and ultimately to decommission and permanently close the underground portion of this facility. The facility was permanently closed in 2010.

In addition to removal of underground appurtenances and flooding of underground openings, part of this decommissioning work involved the installation of seals at the intersection of the access shaft and ventilation shaft to limit the potential for mixing of deeper saline and shallower, less saline groundwater. The Enhanced Sealing Project

(ESP), a joint international project involving agencies from Canada, Finland, Sweden and France, was implemented to design, construct and monitor the evolution of a full-scale repository-type shaft seal. This project was aided by the well-characterized nature of the URL site, lessons learned from the precursor rock mass response studies, and the otherwise undisturbed rock mass in which the URL was constructed.

Full-scale shaft plugs of the types that might be used in various international repositories were installed where the shaft and vent raise intersect Fracture Zone 2 at about 270 m depth. The plug in the main shaft at the URL consists of a 6-m long vertical section of in situ compacted backfill bounded by two 3-m-thick concrete segments keyed into the rock wall of the shaft. The backfill is a 60-40% aggregate-bentonite clay mixture, mixed and compacted using conventional engineering techniques and equipment. The ventilation shaft plug was constructed using pre-compacted blocks that were field fit to the opening.

The ESP design included a suite of sensors installed in the shaft seal components to allow monitoring of the evolution of temperature, concrete strains, pore water pressure, total stress and water uptake by the clay component (Dixon et al., 2009, 2012; Martino et al., 2011).

Initial curing of the concrete was complete after three years, and the structures reached temperature equilibrium. As of 2012, hydration of the clay component was progressing with saturation achieved in the perimeter regions, and swelling pressures developing. Monitoring showed that groundwater table recovery has begun, and that there is no open hydraulic connection between the top and bottom of the seal (Holowick et al., 2011; Dixon et al., 2012), with a hydraulic head difference of more than 20 m maintained across the plug.

6 DISCUSSION

The results of the experiments described in the previous sections have implications in terms of the Canadian waste isolation concept, and in designing tunnels for other applications. In each of the experiments involving horizontal excavations under ambient conditions, the in situ strength of the rock mass at the tunnel periphery was significantly less than that expected based on laboratory tests. The discrepancy is attributable to the difference in accumulated damage between laboratory samples and the in situ rock mass associated with different stress histories and loading paths. The stress path experienced by the rock mass is a function of the far-field in situ stress conditions, tunnel geometry and orientation, excavation method, and position with respect to the tunnel centreline. At AECL's URL, three-dimensional stress effects around, and ahead of, the advancing tunnel face could reduce the strength of the rock mass near the tunnel periphery to 50% of the laboratory-derived unconfined compressive strength. In each of the experiments described, AE and/or MS monitoring was shown to be an effective tool in identifying damage development and assessing the likelihood of strength reduction around underground excavations.

For circular and other tunnel geometries that result in the peak tangential stress being localized at a point on the tunnel periphery, progressive failure in the form of breakouts occurred once the tangential stress exceeded the in situ strength at the tunnel wall. This process, once initiated, is meta-stable, and highly sensitive to subtle changes in boundary conditions, including confining pressure, thermally-induced stresses, and moisture. By applying design criteria that account for the stress history and three-dimensional stress path effects, it is possible to design openings that are stable under adverse stress conditions, thus avoiding conditions of meta-stable equilibrium. The effects of thermal-mechanical stress history can be taken into account in the same way.

Studies at AECL's URL have shown that excavation damage can increase the connected permeability of the near-field rock mass. The extent of damage was found to be a function of near-field stresses, stress history, excavation method, and geology. For typical horizontal openings at the 420 Level of the URL, damage (including breakouts) was limited to within about 0.5 to 1 m of the tunnel wall both in regions of increased compression, and tension (or reduced hydrostatic or mean stress). In bored vertical openings, the findings suggest that excavation damage will be limited to within a small distance (*i.e.*, less than 33% of the hole radius) from the opening, even when multiple openings are spaced close enough to interact. By employing refined excavation techniques, and accounting for near-field stress effects in excavation design, excavation damage around openings can be minimized.

Bulkhead keys were shown through numerical modeling to be effective in mitigating the effects of excavation damage. Experience from the TSX suggests that construction of bulkhead keys of both rectangular and triangular shape is feasible. As shown in the HFT, the use of sealing materials that apply even a small active confining pressure to the rock mass will reduce the likelihood of instability and will inhibit damage development, or continued growth of an existing excavation damage zone. Similar effects were observed from gravitational loading of the tunnel invert by rock debris during excavation of the MBE test tunnel, and by the sand fill in the TSX tunnel. This understanding was applied in the ESP shaft seal design.

Comparison of the peak tangential boundary stress for different excavations at the 420 Level suggests that slight variations in geology can also influence the susceptibility of the rock mass to damage and strength reduction. Granite at this level is generally coarsely crystalline with an inequigranular structure, whereas granodiorite is finely crystalline with an equigranular structure. The finer and more uniform crystal structure of granodiorite makes it less susceptible to damage development in particular tunnel orientations (Read & Martin, 1996). A key component in successfully constructing stable openings is therefore the thorough characterization of in situ conditions in the excavation area, such as lithologic variation, anisotropy, and in situ stress magnitudes and directions.

Based on the experimental work conducted at the URL, the following key factors affecting excavation damage and stability have been identified:

1. <u>In situ stress</u> – The magnitude of in situ stresses and stress ratios relative to the strength of the rock mass, and the orientation of the excavation relative to the principal stress directions, strongly influence the development of excavation damage.

2. <u>Geologic variability</u> – Subtle variations in lithologic composition, grain size distribution, rock fabric, and micro-structure can affect the macroscopic strength characteristics of rock around underground openings, and can result in variable and/or anisotropic mechanical properties.

3. <u>Excavation method</u> – The effects of excavation method were assessed by comparing a circular tunnel excavated by drill-and-blast in the ESS with the

mechanically-excavated MBE test tunnel. The extent of the failed zone in the compressive regions of the crown and invert of the two tunnels was similar, but the drill-and-blast tunnel showed more damage in the tensile sidewall region, including some evidence of discrete tensile cracking.

4. Tunnel geometry and orientation – The ESS and the TSX tunnels showed that excavation damage can be reduced through selection of tunnel geometries and orientations that reduce the near-field compressive stress concentrations, and avoid tensile regimes in the sidewalls. Results suggest that the optimum tunnel shape is one that avoids stress localization around the tunnel periphery (*i.e.,* creates conditions approaching a uniform boundary stress).

5. Adjacent excavations – In high horizontal stress environments, the development of excavations close to an existing opening can alter the near-field stresses enough to affect the extent and severity of excavation damage.

6. Operational practices – The use of excessive scaling (*i.e.,* removal of loose rock) from a failed region has been shown in the MBE to increase the level of damage development ongoing in regions of high compressive stress concentrations. Removal of tunnel supports during decommissioning may reinitiate instability, and requires careful execution to ensure worker safety. Decommissioning of AECL's URL has demonstrated that this can be achieved safely.

7. Thermal and humidity effects – Excavation damage is exacerbated by fluctuations in temperature and humidity in the ventilation air in underground openings. Thermally-induced stresses add to the compressive stress concentrations in the near-field, and humidity increases the rate of development of induced fracturing. Seasonal and daily fluctuations in these conditions are common in mining environments, and even small changes in environmental conditions can cause minor rock instability. These effects can be controlled during construction by ventilation and temperature control.

8. Rock mass quality – The quality of the rock mass will determine, to some extent, the size and nature of excavation damage (Read, 1997). Naturally-fractured rock masses may have excavation-induced deformations that localize along existing fractures rather than creating new fractures. Sparsely-fractured rock masses are expected to be more highly stressed owing to a greater rock mass stiffness, and hence, a greater capacity to carry lithologic stresses in the absence of stress-relief resulting from shear displacements along fractures.

9. Confining pressure – Application of a small confining pressure, either as ballast or as pressure on against an excavation surface, was shown to be effective in controlling damage development. This provides evidence that active seals with swelling materials, such as that in the ESP, will be effective in reducing excavation damage.

10. Chemical and biological effects – Natural chemical infilling and the development of biological growth in fractures in the damaged rock may alter transmissivity, either by plugging existing fractures, or by enhancing permeability through dissolution or biological action. Calcium carbonate leaching from bulkheads constructed using standard concrete mixtures and other structures (*e.g.,* grout, cement used in rock bolting or other support) may have similar plugging effects, and may change the pH of the disposal environment. Gas generation from corroding metal fixtures or rock bolts is also a consideration.

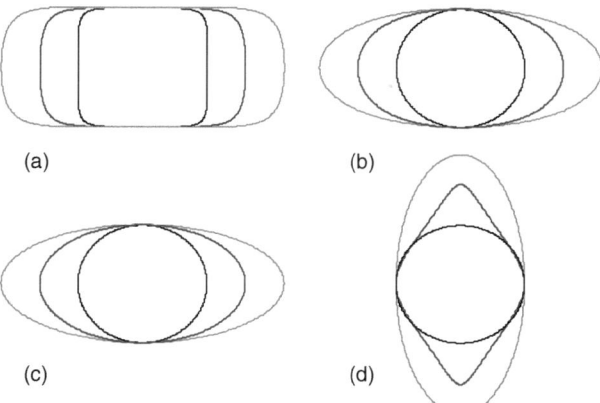

Figure 20 Options for tunnel geometry at the 420 Level of the URL: (a) rectangular, (b) oval, (c) elliptical, and (d) elliptical or notched shape to create controlled damage.

The design of excavations in sparsely fractured brittle rock must take into account these many factors. Thorough characterization of geological and geotechnical conditions of the rock mass for a proposed excavation is necessary to rank the various contributing factors, and to provide input data for numerical analyses. This is particularly true for designs aimed at avoiding progressive damage development and tunnel instability.

The TMSS provided insight into the micro-mechanics of damage development and progressive failure. This insight resulted in a detailed design approach for repository excavations. As part of this approach, preliminary design involves scoping analyses using relatively simple tools to identify tunnel geometries (Figure 20) that meet specific design requirements, such as tunnel size, shape, or orientation. Scoping analyses based on conditions at AECL's URL (Read & Martino, 2002) have shown that subtle differences between oval and elliptical geometries impact the magnitude and distribution of tangential boundary stresses. Likewise, misalignment of the tunnel cross-sectional axes with respect to the principal stress directions has the potential to affect the boundary stress distribution for oval, elliptical, and rectangular geometries. Fractured or layered rock conditions add complexity in terms of the behavior of jointed rock around openings, and the potential for block or wedge failure, and may require more sophisticated scoping analysis tools.

7 CONCLUSIONS

The findings from in situ experiments conducted at the URL since 1982 illustrate that excavation damage and stability are affected by a number of factors including the near-field stress history, geological variability, excavation method, tunnel geometry and confining pressure. The studies confirm that it is possible to construct large stable underground openings, with only limited excavation damage, under adverse in situ stress conditions. Keyed bulkhead seals were also shown to be an important consideration in mitigating the effects of excavation damage on connected permeability along the tunnel. These findings are relevant in terms of defining characterization methods and specifications for use in siting a disposal vault, and in designing openings at depth in rock masses typified by the Canadian Shield.

The findings from the TMSS and other rock mechanics experiments at the URL have advanced our understanding of excavation damage development and progressive failure in brittle rock. In addition, these experiments have led to improvements in available tools and capabilities for repository excavation design. The design approach developed as part of the TMSS is an integrated sequence of characterization, monitoring, modeling, and testing activities. The integration of the technologies and associated methodologies from the TMSS and from previous experiments provides the basis (*i.e.*, the design tools and capabilities) for predicting and back analyzing rock mass behavior around underground excavations under different boundary conditions. These tools and capabilities, together with associated calibration and scoping studies, will be required early in a repository siting program to establish preliminary repository designs, and to assess the significance of various rock mass characteristics on repository excavation design. These same tools and capabilities also have potential application outside the nuclear waste disposal sector.

ACKNOWLEDGMENTS

The author acknowledges the outstanding work by many individuals at AECL's Underground Research Laboratory, including AECL staff, contractors to AECL, and international participants from other agencies. Thanks also to Elsevier Limited for permission to reuse material originally published in the International Journal of Rock Mechanics and Mining Sciences in December 2004 as part of a special issue on AECL's URL.

REFERENCES

Berchenko, I., Detournay, E. & Chandler, N.A. (1997) Propagating natural hydraulic fractures. *Int. J. Rock Mech. & Min. Sci.*, 34(3/4), paper #63.

Berchenko, I., Detournay, E., Chandler, N. & Martino, J. (2004) An in-situ thermo-hydraulic experiment in a saturated granite I: Design and results. *Int. J. Rock Mech. & Min. Sci.*, 41(8), 1377–1394.

Chandler, N., Dixon, D., Gray, M., Hara, K., Cournut, A. & Tillerson, J. (1998) The Tunnel Sealing Experiment: An in situ demonstration of technologies for vault sealing. In: *Proc. 19th Annual Conference of the Canadian Nuclear Society*, Toronto, 1998. Canadian Nuclear Society.

Cundall, P.A., Potyondy, D.O. & Lee, C.A. (1996) Micromechanics-based models for fracture and breakout around the Mine-by Experiment tunnel. In:*Designing the Excavation Disturbed Zone for a Nuclear Repository in Hard Rock Proceedings, EDZ Workshop, Winnipeg, 1996.* Canadian Nuclear Society. pp 113–122.

Detournay, E., Senjuntichai, T. & Berchenko, I. (2004) An in-situ thermo-hydraulic experiment in a saturated granite II: Analysis and parameter estimation. *Int. J. Rock Mech. & Min. Sci.*, 41(8), 1395–1412.

Dixon, D.A., Martino, J.B. & Onagi, D.P. (2009) *Enhanced Sealing Project (ESP): Design, construction and instrumentation plan.* Nuclear Waste Management Organization (NWMO). APM Report, APM-REP-01601-0001, October 2009. Toronto, Canada.

Dixon, D.A., Priyanto, D.G. & Martino, J.B. (2012) *Enhanced Sealing Project (ESP): Project status and data report for period ending 31 December 2011.* Nuclear Waste Management Organization (NWMO). APM Report, APM-REP-01601-0005, July 2012. Toronto, Canada.

Dzik, E.J. & Read, R.S. (1997) *Scoping analysis of bulkhead keys for the Tunnel Sealing Experiment.* Atomic Energy of Canada Limited Report TSX-06.

Everitt, R.A. (2001) *The influence of rock fabric on excavation damage in the Lac du Bonnet granite.* Ph.D. Thesis, University of Manitoba, Winnipeg, Manitoba, Canada.

Everitt, R.A. & Lajtai, E.Z. (2004) The influence of rock fabric on excavation damage in the Lac du Bonnet granite. *Int. J. Rock Mech. & Min. Sci.*, 41(8), 1277–1304.

Greenspan, M. (1944) Effect of a small hole on the stresses in a uniformly loaded plate. *Q. Appl. Math.*, 2(1), 60–71.

Hazzard, J.F. & Young, R.P. (2004) Dynamic modelling of induced seismicity. *Int. J. Rock Mech. & Min. Sci.*, 41(8), 1365–1376.

Holowick, B., Dixon, D.A. & Martino, J.B. (2011) *Enhanced Sealing Project (ESP): Project status and data report for period ending 31 December 2010.* Nuclear Waste Management Organization (NWMO). APM-REP-01601-0004. Toronto, Canada.

Lau, J.S.O. & Chandler, N.A. (2004) Innovative laboratory testing. *Int. J. Rock Mech. & Min. Sci.*, 41(8), 1427–1446.

Martin, C.D. (1993) *The strength of massive Lac du Bonnet granite around underground openings.* Ph.D. Thesis, University of Manitoba, Winnipeg.

Martin, C.D., Dzik, E.J. & Read, R.S. (1996) Designing an effective excavation damaged zone cut-off in high stress environments. In:*Designing the Excavation Disturbed Zone for a Nuclear Repository in Hard Rock Proceedings, EDZ Workshop, Winnipeg, 1996.* Canadian Nuclear Society. pp 155–164.

Martin, C.D., Martino, J.B. & Dzik, E.J. (1994) Comparison of borehole breakouts from laboratory and field tests. In: *Rock Mechanics in Petroleum Engineering Proc. EUROCK'94, Delft, 1994.* Balkema: Rotterdam. pp 183–190.

Martin, C.D. & Read, R.S. (1996) AECL's Mine-by Experiment: a test tunnel in brittle rock. In: *Proc. 2nd North American Rock Mech. Symp., Montreal, 1996.* Balkema: Rotterdam. pp 13–24.

Martin, C.D., Read, R.S. & Martino, J.B. (1997) Observations of brittle failure around a circular test tunnel. *Int. J. Rock Mech. & Min. Sci.*, 34(7), 1065–1073.

Martin, C.D. & Simmons, G. R. (1993) The Atomic Energy of Canada Limited Underground Research Laboratory: An overview of geomechanics characterization. *Comprehensive Rock Engineering,* J.A. Hudson, ed., Vol. 3, Pergamon Press: Oxford, 915–950.

Martino, J.B. & Chandler, N.A. (2004) The existence of the EDZ at the URL. *Int. J. Rock Mech. & Min. Sci.*, 41(8), 1413–1426.

Martino, J.B. & Read, R.S. (1996) An overview of AECL's Heated Failure Tests. *ISRM Newsjournal,* 4(1), 24–31.

Martino, J.B., Dixon, D.A., Holowick, B.E. & Kim, C.-S. (2011) *Enhanced Sealing Project (ESP): Seal construction and instrumentation report.* Nuclear Waste Management Organization (NWMO). APM-REP-01601-0003. Toronto, Canada.

Mitaim, S. & Detournay, E. (2004) Damage around a cylindrical underground opening in hard rock. *Int. J. Rock Mech. & Min. Sci.*, 41(8), 1447–1457.

Potyondy, D.O. (2014) The bonded-particle model as a tool for rock mechanics research and application: Current trends and future directions. *Geosys. Eng.*, 17(6), 342–369.

Potyondy, D.O. & Cundall, P.A. (2001) *The PFC model for rock: predicting rock mass damage at the Underground Research Laboratory.* Ontario Power Generation, Nuclear Waste Management Division Report No. 06819-REP-01200-10061-R00.

Potyondy, D.O. & Cundall, P.A. (2004) A bonded-particle model for crystalline rock. *Int. J. Rock Mech. & Min. Sci.*, 41(8), 1329–1364.

Read, R.S. (1994) *Interpreting excavation-induced displacements around a tunnel in highly stressed granite.* Ph.D. Thesis, Department of Civil and Geological Engineering, University of Manitoba, Winnipeg.

Read, R.S. (1996) Characterizing excavation damage in highly-stressed granite at AECL's Underground Research Laboratory. In: *Proceedings of the International Conference on Deep Geological Disposal of Radioactive Waste EDZ Workshop, Winnipeg, September 20, 1996,* J.B. Martino & C.D. Martin, eds. Canadian Nuclear Society. pp 35–46.

Read, R.S. (1997) *Effect of rock mass quality on selection of a waste emplacement option.* Ontario Power Generation, Nuclear Waste Management Division Report No. 06819-REP-01240-0003 R00.

Read, R.S. (2004) 20 years of excavation response studies at AECL's Underground Research Laboratory. *Int. J. Rock Mech. & Min. Sci.,* 41(8), 1251–1276.

Read, R.S. (2008a) *The role of rock engineering in developing a Deep Geological Repository in sedimentary rock.* Nuclear Waste Management Organization. Technical Report NWMO TR-2008-16, December 2008.

Read, R.S. (2008b) *Developing a reasoned argument that no large-scale fracturing or faulting will be induced in the host rock by a Deep Geological Repository.* Nuclear Waste Management Organization. Technical Report NWMO TR-2008-14, December 2008.

Read, R.S., Chandler, N.A. & Dzik, E.J. (1998) In situ strength criteria for tunnel design in highly-stressed rock masses. *Int. J. Rock Mech. & Min. Sci.,* 35(3), 261–278.

Read, R.S. & Chandler, N.A. (1996) *AECL's Excavation Stability Study – Summary of observations.* Atomic Energy of Canada Limited Report, AECL-11582, COG-96-193.

Read, R.S. & Chandler, N.A. (1997) Minimizing excavation damage through tunnel design in adverse stress conditions. In: *Proc. 23rd General Assembly – Int. Tunnel. Assoc., World Tunnel Congress `97, Vienna, 1997.* Balkema: Rotterdam. pp 23–28.

Read, R.S. & Chandler, N.A. (2002) *An approach to excavation design for a nuclear fuel waste repository – The Thermal-Mechanical Stability Study final report.* Ontario Power Generation Report No: 06819-REP-01200-10086-R00.

Read, R.S. & Martin, C.D. (1996) *Technical summary of AECL's Mine-by Experiment. Phase 1 excavation response.* Atomic Energy of Canada Limited Report, AECL-11311.

Read, R.S., Martin, C.D. & Dzik, E.J. (1995) Asymmetric borehole breakouts at the URL. In: *Proc. 35th US Rock Mech Symp., Lake Tahoe, 1995.* Balkema: Rotterdam. pp 879–884.

Read, R.S. & Martino, J.B. (1996a) Effect of thermal stresses on progressive rock failure at AECL's Underground Research Laboratory. In: *Proc. Int. Conf. Deep Geol. Disposal Radioactive Waste, Winnipeg, 1996.* Canadian Nuclear Society. pp 743–753.

Read, R.S. & Martino, J.B. (1996b) In situ thermal testing at AECL's Underground Research Laboratory. In: *Proc. 2nd North American Rock Mech. Symp., Montréal, 1996,* Aubertin, H. & Mitri eds. Balkema: Rotterdam. pp 1487–1494.

Read, R.S. & Martino, J.B. (2002) To arch or not to arch – The role of tunnel design in controlling excavation damage development. In:*Proc. of the EDZ Workshop, NARMS-TAC 2002, July 6, 2002, Toronto, Canada.*

Read, R.S., Martino, J.B., Dzik, E.J. & Chandler, N.A. (1997b) *AECL's Excavation Stability Study - Analysis and interpretation of results.* Ontario Hydro, Nuclear Waste Management Division Report No. 06819-REP-01200-0028 R00.

Read, R.S., Martino, J.B., Dzik, E.J., Oliver, S., Falls, S. & Young, R.P. (1997a) *Analysis and interpretation of AECL's Heated Failure Tests.* Ontario Hydro, Nuclear Waste Management Division Report No. 06819-REP-01200-0070-R00.

Simmons, G.R. (1992) *The Underground Research Laboratory Room 209 Excavation Response Test – A summary report.* Atomic Energy of Canada Limited Report AECL-10564, COG-92-56.

Thompson, P.M., Martino, J.B. & Spinney, M.H. (1993) Detailed measurements of deformation in the excavation disturbed zone. *Int. J. Rock Mech. Min. Sci. & Geomech. Abstr.*, 30(7), 1511–1514.

Young, R.P. & Collins, D.S. (1999) Monitoring an experimental tunnel seal in granite using acoustic emission and ultrasonic velocity. In: *Rock Mechanics for Industry, Proc. 37th U.S. Symp. on Rock Mechanics, Vail, Colorado*, Amedai, Kranz, S. & Smealie, eds. Balkema: Rotterdam. pp 869–876.

Young, R.P., Collins, D.S. Reyes, J.M. & Baker, C. (2004) Quantification and interpretation of acoustic emission and microseismicity at the Underground Research Laboratory, Canada. *Int. J. Rock Mech. & Min. Sci.*, 41(8), 1317–1328.

Chapter 18

URL and rock mechanics in Finland

Erik Johansson
Saanio & Riekkola Oy, Helsinki, Finland

1 BACKGROUND

1.1 History of rock mechanics in Finland

Rock engineering in Finland has a long history with the first underground mining going as far back as the sixteenth century (Rönkä & Ritola, 1997). Civil engineering work began with the construction of railway tunnels – excavation of the first one began in 1896. Construction of the infrastructure began in the 1920s with the excavation of the main water tunnels. Good description of the current rock engineering projects in Finland is found in Särkkä & Aho (2011).

Rock mechanics as a science can be considered to start in Finland at the turn of the 1950s and 1960s. Underground construction was also initiated in the early 1960s in the Helsinki region with the requirement by Finnish legislation to construct more civil defense rock shelters for citizens.

The birth of the Finnish rock mechanics community took place in early 1967 when a sub-committee under the Finnish Mining and Metallurgical Society was established. The first annual Finnish Rock Mechanics Symposium was held in 1967. The Finnish Rock Mechanics Symposia have been held annually since then. The Finnish Rock Mechanics Society was established in 1971 with 51 members and later that same year it was accepted as a member of ISRM (International Society for Rock Mechanics).

One of the first pioneer rock mechanic projects was the Tytyri limestone mine that was presented in our first Rock Mechanics Symposium in 1967 (Finnish Mining and Metallurgical Society, 1967). Since 1964 several stress measurements mainly using Hast's overcoring method had been conducted in the mine. Understanding the stress state there markedly changed the shapes and volumes of stopes and pillars in the mine resulting in larger open rooms and more effective mining. This was first time horizontal in situ stresses were utilized in the mine planning.

1.2 Basic data to describe rock mechanics conditions in Finland (geology, rock properties, rock stress)

The Precambrian bedrock of Finland forms the core of the Fennoscandian Shield (Figure 1). Areas of similar hard bedrock areas are also found in different regions of the world for instance in Africa, Australia, Canada, China and South America (see *e.g.* Rönkä & Ritola, 1997).

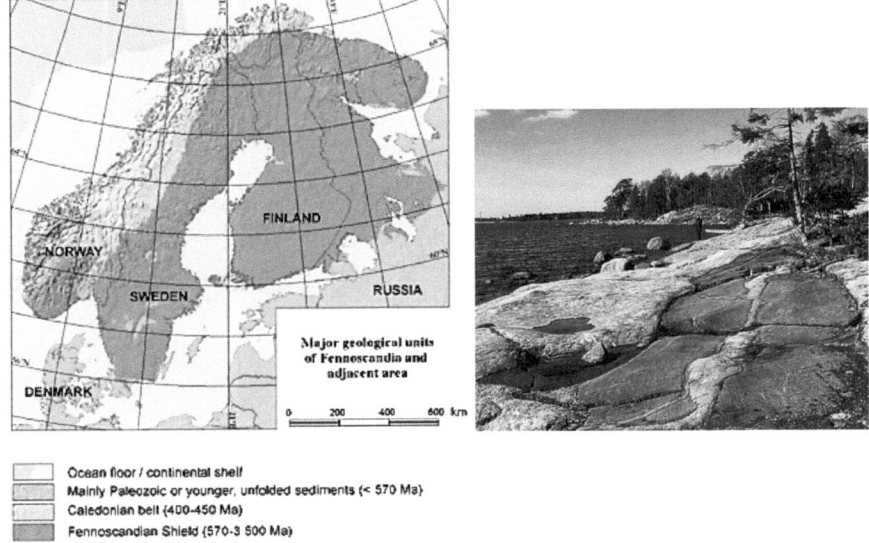

Ocean floor / continental shelf
Mainly Paleozoic or younger, unfolded sediments (< 570 Ma)
Caledonian belt (400-450 Ma)
Fennoscandian Shield (570-3 500 Ma)

Figure 1 Precambrian bedrock of Finland (Rönkä & Ritola, 1997).

The Finnish bedrock is divided into an Archaean complex (3.50–2.50 Ga) in the north and east and an Early Proterozoic (1.92–1.77 Ga) Svecofennian domain which dominates in the central and southern Finland. The Middle Proterozoic (1.65–1.54 Ga) rapakivi granites of southern Finland represent the predominant post-orogenic rock type. The old Finnish bedrock is mainly composed of granites, gneisses, migmatites and schistose rocks.

Only little sedimentary rock is found in continental Finland. Most of the layers have been eroded, thus exposing the bedrock. The Satakunta (sandstone) and Muhos (mostly siltstone, shales) formations in impact structures represent the two largest sedimentary rock areas in continental Finland.

The present features of the Finnish landscape are primarily the result of the erosion that took place during the numerous ice ages of the last hundred thousand years. Soil cover is then typically thin due to the glaciation.

The bedrock in Finland is sometimes intensively jointed near the surface, which is subjected to the effects of glaciation and tectonic movements as well as climatic and other erosive forces. The fracture, brittle deformation or weakness zones are also typical features in the Finnish bedrock, which present certain challenges for the rock engineering.

The solid, crystalline bedrock is typically very strong. The average uniaxial compressive strengths are around 150 ± 20 MPa as shown in Figure 2 (Johansson *et al.*, 1996). Tensile strengths are around 5–10 MPa indicating the Finnish rocks to be quite brittle. The rocks are also stiff with Young's modulus being typically between 50-70 GPa.

A typical feature of the Finnish bedrock is also high horizontal stresses due to the plate tectonics. In Fennoscandia, the orientation of the major principal stress is attributed to an E-W compression from the mid-Atlantic ridge push and a N-S compression from the Alpine margin, resulting in a roughly NW-SE orientation of the major

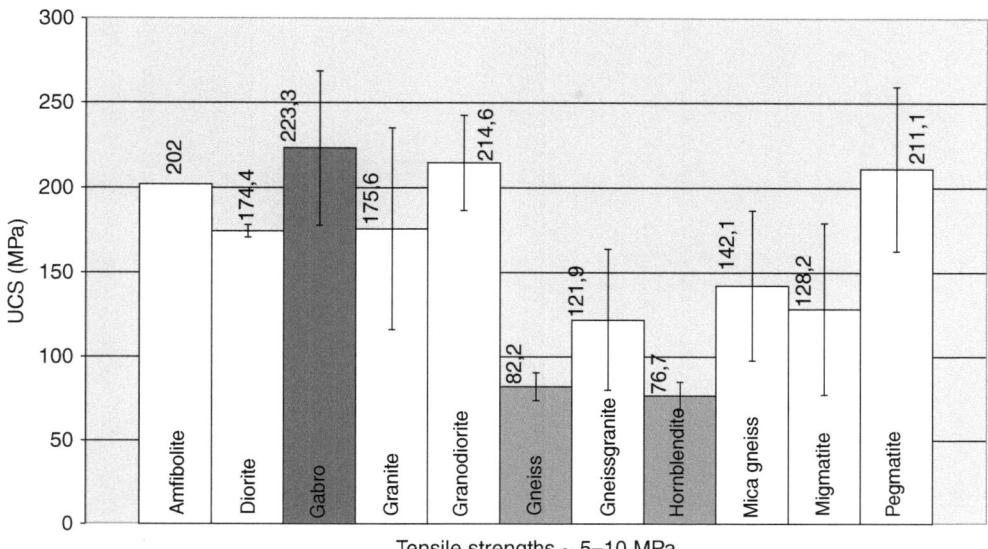

Figure 2 Peak strength (UCS) of some typical Finnish rock types (Johansson *et al.*, 1996).

principal stress (Heidbach *et al.*, 2008). A thrust faulting stress regime is typically present in Finland, *i.e.* the horizontal stresses are larger than the vertical stress, $\sigma_H > \sigma_h > \sigma_v$. The magnitudes of horizontal stresses are increasing with depth as shown in Figure 3 (Tolppanen & Johansson, 1996) and since the rock properties do not change with depth high stresses in relation to rock strength may cause instability problems or damage for underground facilities. On the other hand the horizontal stresses can also be utilized in stabilizing large rock rooms like caverns.

2 ROCK MECHANICS PRACTICE IN CIVIL ENGINEERING PROJECTS

Rock mechanics design in Finland follows the methodologies commonly used worldwide. They are based either on analytical, computational, empirical or observational methods or on the combination of those. Analytical methods are only suitable for simple geometries and for the complex situations and excavation geometries the computational (numerical) methods are commonly used. The empirical methods *i.e.* the rock mass classifications (mostly Q-system) are nowadays very commonly used to support the numerical methods in underground projects in Finland.

Also, some technical ordinances, guidelines or regulations exist in Finland such as the Civil Defense Ordinance that defines for instance the thickness of the rock cover and the amount of rock support (rock bolts and shotcrete). Currently also, the Eurocodes have been started to utilize but they do not explicitly state how to design rock spaces, but they define the minimum requirements on how to design structures. However, work is still to be done to revise the Eurocode 7 or EC7 to become a key design standard for

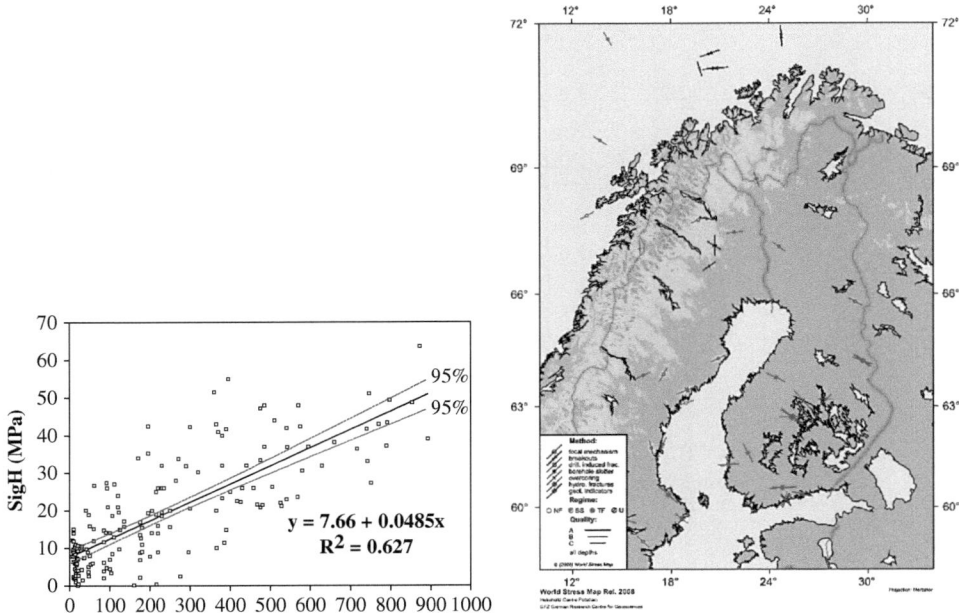

Figure 3 Magnitude (181 measurement sites) and orientation of maximum horizontal stress in Finland (Tolppanen & Johansson 1996).

geotechnical engineering. The next version of EC7 will be written by 2018, and will then be published in 2018 for adoption in 2020.

Rock performance monitoring is typically performed during the project execution. The objective is to compare the measured results with the predicted ones to ensure the mechanical stability of the excavated rooms. Monitoring is typically done during the excavation and construction phases but also sometimes depending on the project demands during the operation phase to ensure the long term stability. The two examples below will demonstrate the use of the rock mechanics design in the civil engineering projects.

2.1 Underground library cavern

2.1.1 General

Since the number of publications had continued to grow, the University of Helsinki has constructed an underground library extension (book archive) in the central Helsinki. An underground rock cavern was recognized as being the only solution which would overcome future storage problems.

The volume of the rock cavern is approximately 80,000 m³ and it has a length of 180 m, a span of 21 m and a height of 17 m (Figure 4). The basement floor lies 23 m below sea level. A four-story prefabricated building with a volume of 10,000 m³ was constructed inside the cavern. Four stairs and elevators lead to the surface. At the same time nearby Kluuvi parking caverns were excavated.

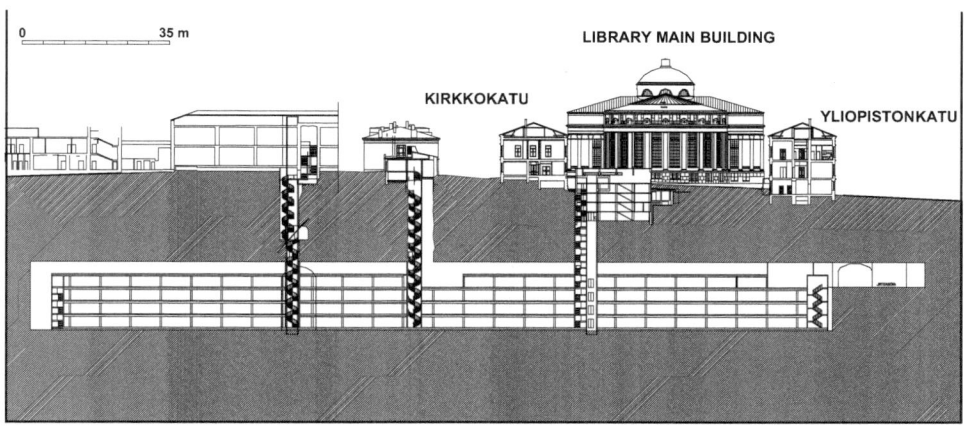

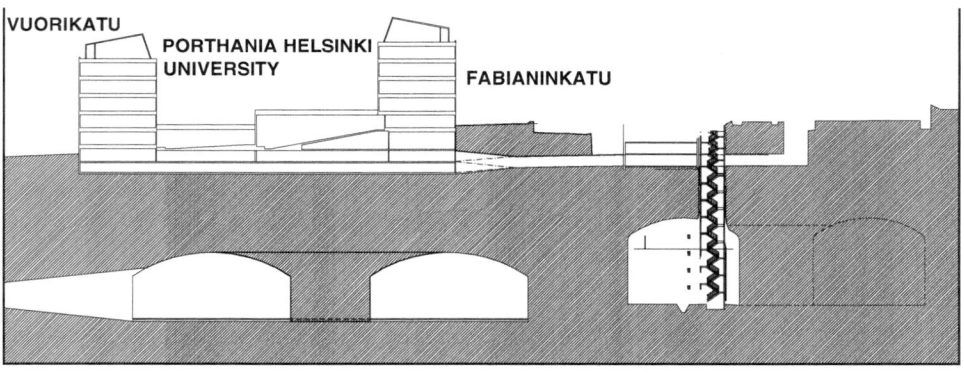

Figure 4 Library cavern under the library main building (top) and close to the Kluuvi underground parking caverns (below) (Johansson *et al.*, 2000).

2.1.2 Site conditions

The rock mass at site was migmatic gneiss-granite, which is typical rock type in the Helsinki area. Based on the borehole investigations the average fracture density was 2.8-4.3 pcs/m and the average RQD (Rock Quality Designation) value was 85 - 96% (*i.e.* good). A small number of narrow zones of weakness were encountered in one borehole. Rock quality according to the Q-classes varied from very poor in fracture zones to fair-good in the rock mass. Hydraulic testing indicated that the rock mass was tight and had a low hydraulic conductivity. Water inflow over the total cavern was estimated to be a maximum of approximately 7 l/minute, so, no pre-grouting was considered necessary.

2.1.3 Stability analyses

Rock mechanics stability analyses were performed using discontinuum finite element code UDEC (Johansson *et al.*, 2000). The parameters employed in the numerical

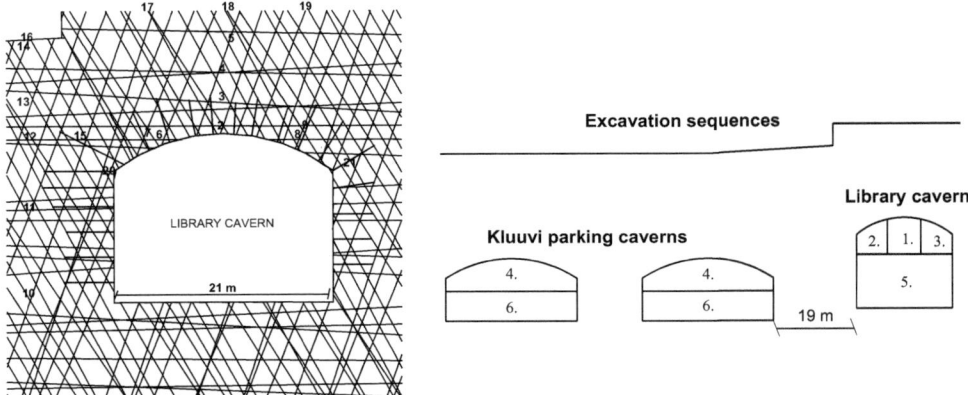

Figure 5 UDEC model for the library cavern (left) and one realization of excavation sequences (right) (Johansson *et al.*, 2000).

analyses were chosen on the basis of both the field investigations and the rock mechanics information obtained from existing underground facilities in the vicinity. The UDEC-model included jointing, support structures (rock bolts) and monitoring points (Figure 5). Special attention was also paid to effects of the excavation of the nearby Kluuvi parking facilities. The overall rock mechanics conditions (in-situ stress, rock strength, joint properties) were assumed to be normal. The maximum horizontal stress of 5 MPa at the cavern level was used in the analysis.

The results obtained from the numerical analyses indicated that the library cavern remain stable in all excavation cases. The *in-situ* stresses acting at the site were found favorable enough to stabilize the roof of the cavern.

2.1.4 Excavation

Excavation of the library cavern was carried out in sequences using the conventional drill-and-blast technique. Systematic rock bolting was used and galvanized rebar rock bolts 4–7 metres long were installed at a spacing of 1.6–2.5 metres. The roof and the walls were shotcreted to a thickness of 60–150 mm using fibre-reinforced shotcrete (wet mix). In the northern part of the cavern the rock quality was worse than predicted, which resulted in additional rock support being required. In those rock conditions (very poor) 20 meter-long Ischebeck-anchors were used. No other problems were encountered during the excavation work. The cavern was also found to be dry as predicted with a total measured water leakage of approximately 5-7 l/minute.

2.1.5 Rock monitoring

From a rock mechanics point of view, the library cavern was considered to be exceptional since the two large caverns of the Kluuvi parking facility were to be excavated at about the same time and the rock pillar separating the two projects was only 19 metres wide. All three caverns were also situated relatively close to the surface and had old

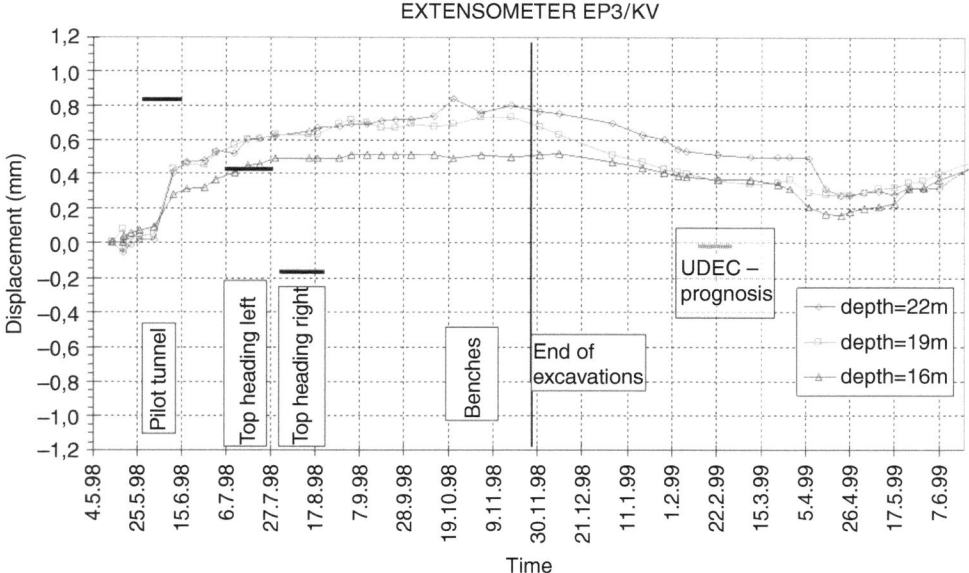

Figure 6 Measured and calculated displacements in one of the roof extensometers EP3 (Johansson et al., 2000).

historic buildings with wooden foundations situated on it. For all these reasons, extensive monitoring of the rock was carried out during the excavation and construction works to confirm that the underground facilities would remain stable. The monitoring programme consisted of rock displacement measurements (extensometers, convergence measurements and precision leveling), precision leveling of the above-ground buildings, rock quality control and groundwater monitoring.

The extensometers that had been installed prior to excavation indicated only very small displacements in the cavern roof, less than 1 mm (Figure 6). Upwards displacements in the top-heading phase indicated stabilizing horizontal stresses, which is very typical behavior in such projects in Finland. The displacements in the rock pillar between the library cavern and the Kluuvi parking facility were also very small. The measured behavior of the rock mass corresponded well with the predicted behavior, especially the final displacements. The displacements were within the anticipated displacements. The results indicated that the stability of the library cavern remained good, as well as the stability of the two parking cavers (Hakala et al., 1999). Precision leveling measurements of the above-ground buildings were carried out once a week and no significant displacements were found. Also, no significant changes were noted in the level of the groundwater table.

2.2 Viikinmäki sewage treatment plant

2.2.1 General

The city of Helsinki constructed the Viikinmäki wastewater treatment plant (Figure 7) at a central underground location to replace seven separate old above-ground plants. The plant

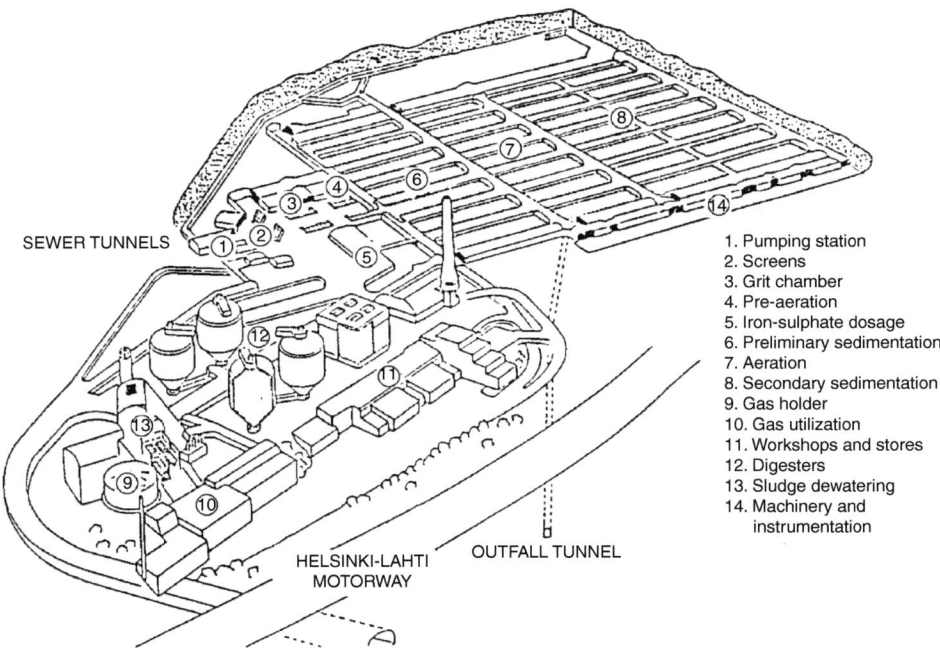

Figure 7 The Viikinmäki underground sewage treatment plant in Helsinki (Johansson & Kuula, 1996).

serves around 800 000 inhabitants and it was taken into operation in 1994. It was later in 2004 extended in order to boost nitrogen removal. The treated, purified water is conveyed in a 16 km long outlet tunnel to the sea. The 1.2 million cubic meters of total excavated volume made it at that time one of the largest civil underground projects in Europe. Currently, above the underground plant there is residential area of 3 500 inhabitants.

The seven main underground processing caverns are 17 m to 19 m wide and 10 m to 15 m high and are separated by 10 m to 12 m rock pillars.

2.2.2 Site conditions

Viikinmäki was typical hilly topographic site that is found in southern Finland with its soil layers and hard, crystalline bedrock. The area consists predominantly of granite and mica gneiss combined to form migmatite. Rock quality is uniformly good with weakness zones clearly linked to dips in the surface. Rock mass is typically fractured and some caverns were located in highly fractured and weathered rock (Q values from 0.01 to 4.0 and the joint spacing from 0.01 to 0.1 m). Rock cover was also in some locations only a few meters and serious consideration was even given to construct portions of the caverns in open excavations.

2.2.3 Stability analyses

It was decided early in the design process that the caverns should be placed above the sea level to limit the possibilities for sea water flooding so the caverns were located as close to the surface as possible. Due to the extreme conditions this was first time in Finland that stability analyses were required by the authorities. It was also the first time that numerical methods were extensively used to assist the design process of an underground project (Johansson *et al.*, 1988; Johansson & Lorig, 1990).

Preliminary modelling was performed for several major cross-sections using 2D-continuum code FLAC. Primary purpose of the analyses was to identify potential areas where stability problems might be encountered, so support was not included. Following this more detailed analyses were carried out with FLAC and weakness zones and with discontinuum code UDEC where fracturing was also included (Figure 8). In case where rock mass behavior was unacceptable, the analyses were repeated and adequacy of various rock support options (rock bolts, cable bolts, shotcrete) were evaluated. Finally, calculated support forces were compared to allowable forces. In Viikinmäki case, support was judged to be adequate if support forces were less than 40% of the allowable (Johansson *et al.*, 1988). Potential for surface loading from the planned above-ground buildings were estimated and recommendations were given to the city planning, *e.g.* no buildings were suggested in the areas of the weakness zones and how much can surface excavations for buildings be allowed.

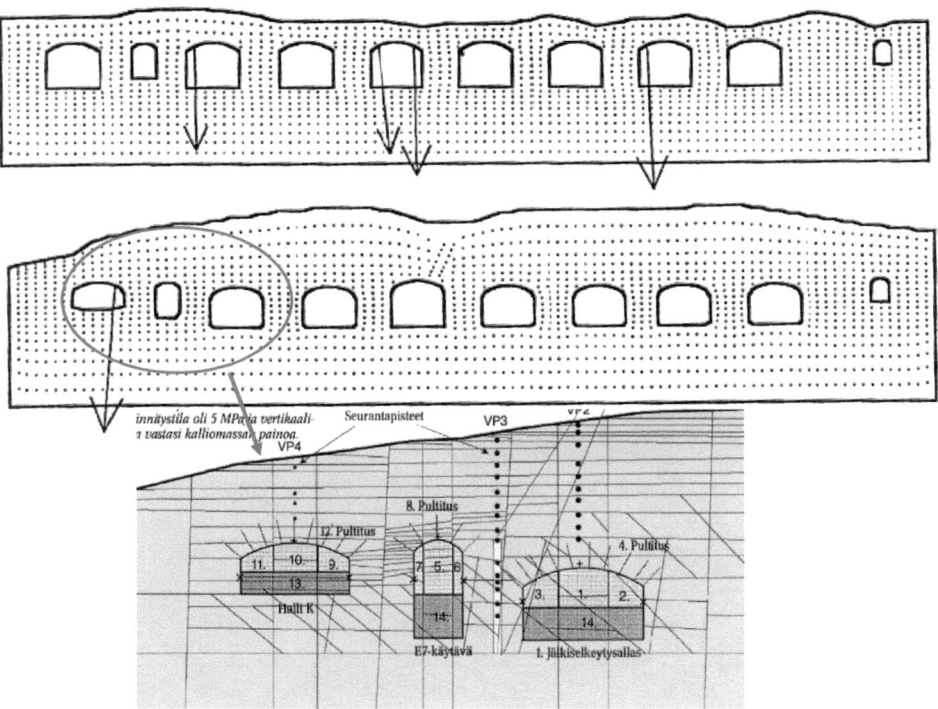

Figure 8 Preliminary Viikinmäki cavern cross-section analyses (top) and example of the more detailed analyses (bottom) with fracturing and support (Johansson *et al.*, 1988).

2.2.4 Rock monitoring

Rock monitoring was performed to control the excavation process (Johansson & Lorig, 1990). The instrumentation was primarily used in the areas where instability might have occurred. The objectives of the rock monitoring were to control the roof areas with low rock cover, pillar behavior, verify the modeling work and overall stability control. The instrumentation consisted of extensometers installed from surface and in caverns, strain gages in pillars, load cells in rock bolts, pressure cells in shotcrete and convergence measurements in some cross-sections.

Monitoring results indicated that the predicted rock behavior based on the stability analyses were reasonably close to the measured behavior, although there were slight differences in the magnitude of the displacements (Figure 9). Monitoring also revealed that the assumption of the zero horizontal stress in highly-fractured areas was overly conservative. This led to a better roof stability that was then taken account in the further analyses.

2.2.5 Back analyses

Three 3D numerical models were later used to model the deformation behavior of a rock mass and also, to confirm the description of geological structures and the validity of input parameters (Figure 10) (Johansson & Kuula, 1996). The geological features: joints, fracture zones and rock types were determined to the models with a close co-operation of a project geologist. Two of the models consisted of weakness zones where

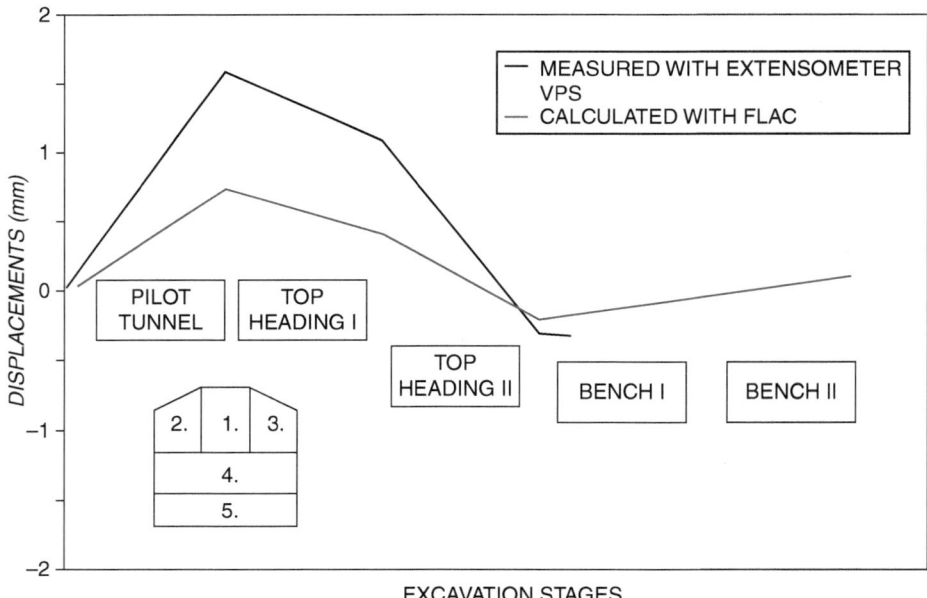

Figure 9 Comparison of the measured and predicted displacements in the roof of aeration cavern VII (Johansson & Lorig, 1990).

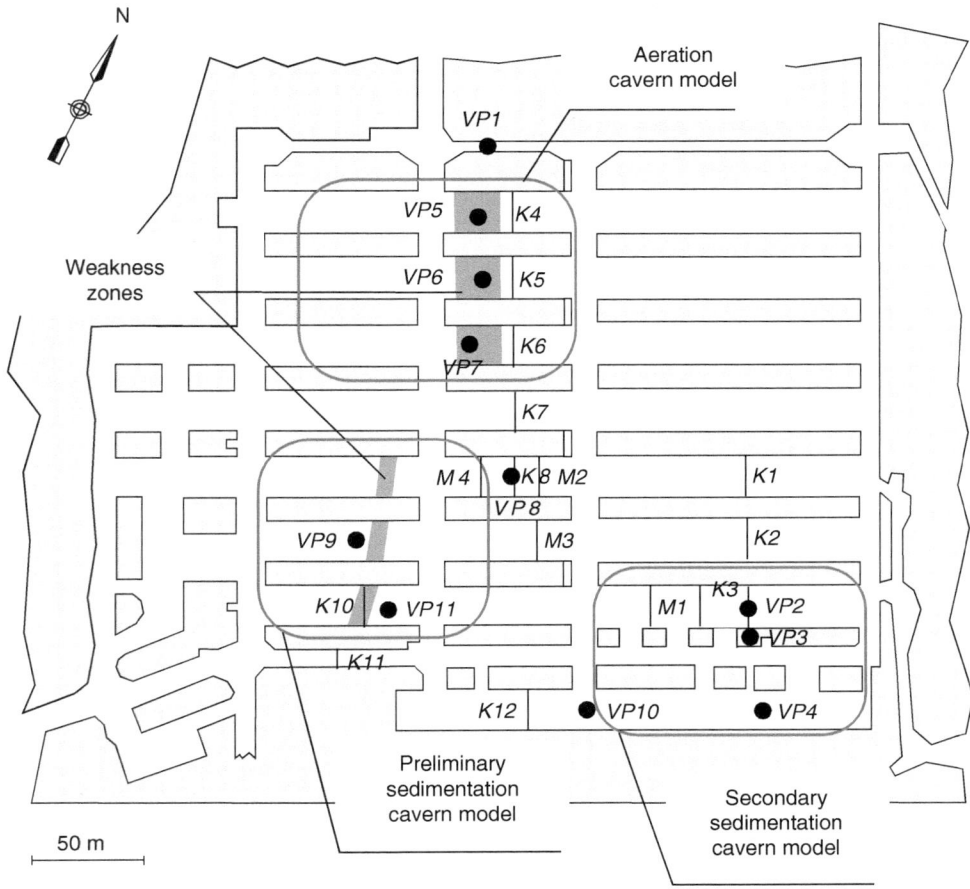

Figure 10 3DEC-models of Viikinmäki sewage treatment plant and locations of rock monitoring instrumentation marked as VP, M and K (Johansson & Kuula, 1996).

the strength of the rock mass was remarkable lower than that of the massive, intact rock mass.

The analyses confirmed the field observations that the rock mass around the excavated rooms was stable. The modeled results and the measured data were generally close to each other. Measured displacements were generally small in most monitoring points. However, in the pillar extensometer VP3, larger displacements from 2 to 7 mm, depending on the joint parameters, were observed in jointed rock mass area. The corresponding displacement was in the model only about 0.1 mm. This was caused by a larger rock block movement that was also seen as cracking in the shotcrete layer.

In the extensometer VP7, the measured and calculated displacements were identical in the fracture zone (Figure 11). The negative values in Figure 11 correspond to the deformation inwards in the cavern. The overall results of the performed three dimensional back-analysis calculations showed a good agreement between the calculated and

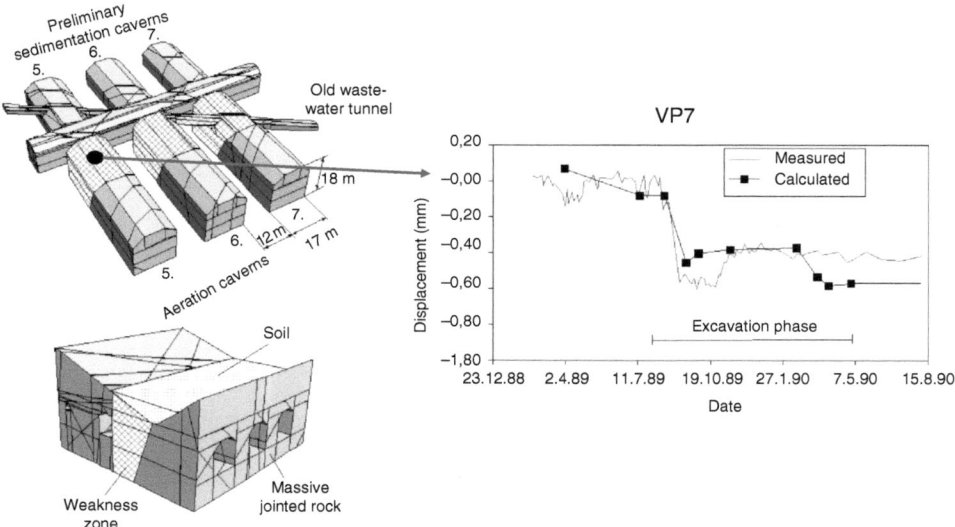

Figure 11 Excellent match between the measured and back-calculated deformations of the extensometer VP7 (Johansson & Kuula, 1996).

the measured results. This was basically true with respect to the input parameters of in situ stress state, the intact rock and the weakness zones, whereas the simple elastic-plastic joint constitutive law seemed to give somewhat misleading results. Thus, there are needs to define in more detail the rock joint behavior at laboratory and to use a non-linear joint model to understand the jointed rock mass behavior accurately enough. The results also showed the importance of the extensive rock pre-characterization program for understanding the multi-structured rock mass behavior during the excavation sequences in large underground projects.

3 ROCK MECHANICS IN NUCLEAR WASTE MANAGEMENT PROJECTS

3.1 Repositories for operating waste

There are two underground repositories (VLJ repository) for the final disposal of operating nuclear waste in Finland. They are both located at nuclear power plant sites in Olkiluoto, Eurajoki, western Finland and in Hästholmen, Loviisa, southern Finland. The operation of the low- and medium-level reactor waste repository at Olkiluoto started in 1992 and at Hästholmen in 1997. Both repositories are placed in hard, crystalline bedrock and their layout is based on the local geological conditions. The long-term safety of the repositories is ensured by surrounding the wastes with multiple barriers. The most effective barrier is the bedrock and accordingly the bedrock conditions have to be confirmed for disposal purposes. Rock mechanics stability analyses were performed in the design phase followed by extensive rock monitoring that started already during the repository construction phase and still continued. Later, these facilities will be enlarged to accommodate the decommissioning waste.

3.1.1 Olkiluoto repository

Olkiluoto silo-type VLJ repository has been constructed in a tonalitic formation which is less fractured than surrounding mica gneiss formation. Excavation of the final repository started in 1988, and the disposal of waste began in 1992. The waste packages are emplaced in the two silos, which are located at the depth between 60 and 100 m below the ground level (Figure 12). The rock silos are 24 m in diameter and 34 m high. The total excavated volume of bedrock is about 90,000 m³.

Prior to the excavations, extensive rock mechanics analyses were made to ensure the stability of the repository (Johansson, 1999). The analyses were performed both with the continuum finite-difference code FLAC and with the distinct-element codes UDEC and 3DEC. The input parameters for numerical calculations were based mainly on the field investigations and the laboratory test results. One of the key issues in the analyses were to determine the width of the pillar between large silos and to analyze the effects of sequential excavations of the crane hall to rock mass behavior. Analyses were also performed to predict the long-term effects of the rock mass responses during the operation phase due to the operational temperature increase.

Since the construction of the repository in the end of 1980's, the bedrock has been monitored systematically. The rock monitoring programme mostly concerns the measurements of rock stability, hydrogeology and groundwater chemistry and the instruments are mostly automatic and read daily by the datalogger. The bedrock stability in Olkiluoto has also been monitored with the microseismic monitoring since 2002 and the GPS measurements since 1995.

The results indicate that the rock conditions in the bedrock have been very stable during the last 25 years and that the environmental impact has been very small. The measured displacements in the extensometers have been small (< 0.7 mm) and they are

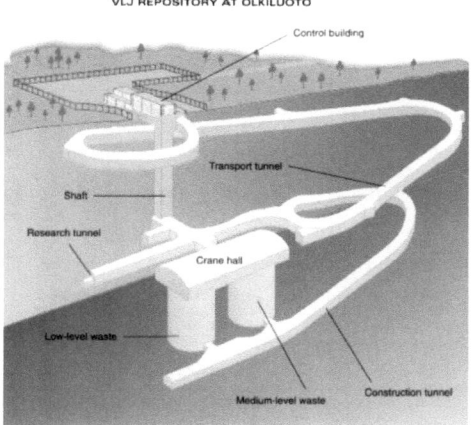
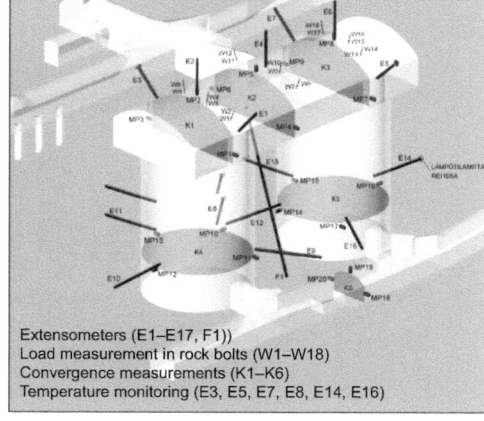

Figure 12 Layout of the Olkiluoto VLJ repository (left) and the rock monitoring system (right) (Figure: Teollisuuden Voima Oy).

due to the operational temperatures of the rock caverns as predicted (Öhberg *et al.*, 2011).

3.1.2 Hästholmen repository

Hästholmen tunnel-type VLJ repository has been constructed in a Precambrian rapa-kivi granite. The repository is located in an intact rock mass between the two upper-most sub-horizontal fractured zones in a stagnant brackish groundwater regime. The repository construction was started in 1993, and the disposal of low-level waste began in 1997. The repository includes two tunnels for maintenance waste (third one is under construction) and one cavern for solidified waste between depth levels -110-120 m (Figure 13). The interior of the solidified waste cavern was finalized in 2005-2008 and the disposal operation there should start soon. The total excavated rock volume is some 126,000 m^3.

Prior to the excavations, rock mechanics stability analyses were performed with the FLAC code (Johansson, 1999). The input parameters for numerical calculations were based on field investigation and the laboratory test results. Except for the normal stability analyses of the three waste caverns, special attention was paid to the behavior of one major fracture zone that cut the access tunnel. The fracture zone was assumed to cut the tunnel in different locations. Several support options were also analyzed. Relatively large leakages occurred in that zone during excavation. The analyses showed that sequential excavation and support method was necessary to stabilize the tunnel. The final support structure was based on rock bolts and shotcrete aided by pre-bolting. The actual excavation through the fracture zone succeeded well, and no stability problems occurred.

In the stability analyses of the waste caverns, several options were analyzed with respect to the magnitude and orientation of in-situ stresses, since some disagreement existed in the results of overcoring and hydraulic fracturing methods. However, later measurements and feed-back analyses confirmed the right option.

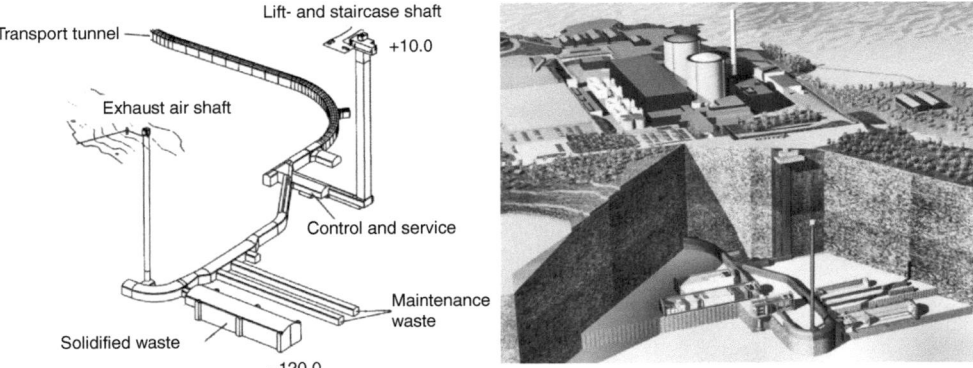

Figure 13 Layout of the Hästholmen VLJ repository, rooms for decommissioning waste shows on right figure (Figure: Fortum Oy).

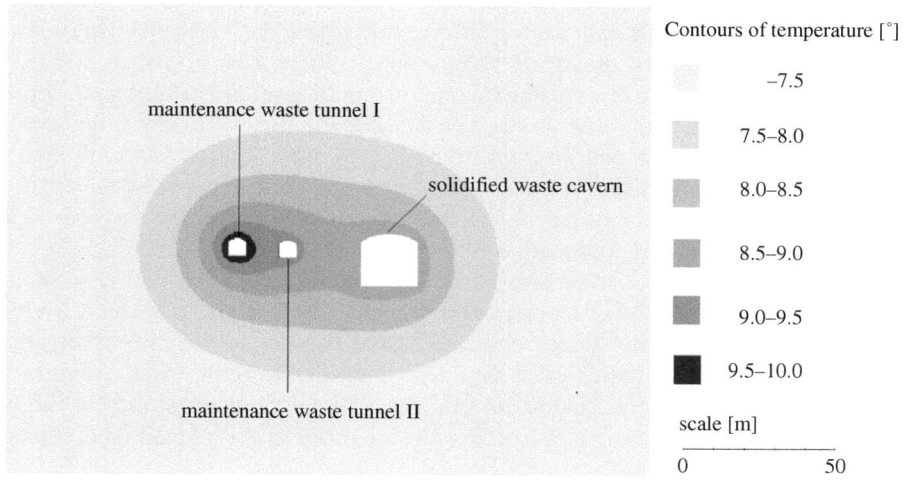

Figure 14 Calculated (prognosis) temperature distribution around the rooms of Hästholmen repository after ten years of operation (Johansson, 1999).

The monitoring programme of the bedrock of the Hästholmen VLJ repository is based on continuous (automatic) and repeated (1-12 times per year) manual measurements. The automated system consists of 14 extensometers, 7 load gauges, 9 thermal gauges and one fissurometer. Rock mechanics monitoring began right after the excavations in late 1996 in the VLJ repository, and the readings are taken daily by a datalogger. After fifteen years, the displacements around the caverns have been small, as expected (generally < 0.1 mm), indicating stable rock mechanics conditions in the repository (Öhberg *et al.*, 2011). The fissurometer that monitors movements in one particular open fracture in the tunnel roof has indicated a small 0.15 mm fracture opening.

In addition to the mechanical stability analyses, thermomechanical analyses were performed to evaluate the long-term effects of the operational temperatures on rock displacements and, especially, on rock monitoring points (Figure 14).

Also, excavation induced seismicity was monitored during the construction of the transport tunnel and the repository itself. The events revealed structures of fractured rock within 50 meters from the excavation. The events close to the transport tunnel were related to horizontal fracture/weakness zones. The events induced by the excavation of the repository seem to be associated with right-lateral strike-slip movement in the set of vertical fractures running in the direction SW-NE (Öhberg *et al.*, 2011).

3.2 Site investigations for HLW repository

The development of the siting programme for Finnish nuclear fuel waste disposal began in 1979. The work progressed from regional studies, to identification of investigation areas (time period 1983–1986, desk studies), to selection of five sites for preliminary characterization (time period 1987–1992) and to detailed characterization of four sites (time period 1993-2000) (Posiva, 2003). These investigation and research phases ended in 2001 when Olkiluoto site was selected by the Finnish Parliament for further site investigations. This phase is called site confirmation phase with an objective to launch an underground rock characterization programme in three stages (Posiva, 2003). The first stage consisted of surface-based investigations before construction of the first underground access. During this phase, the baseline conditions of the Olkiluoto site were established. An improved description of the potential target rock volumes was provided and the basis was set for the choice of the access locations of an underground facility. The summary of the different site characterization phases (surface based field investigations) before going underground are shown in Figure 15.

In the second stage of the underground characterization programme, Posiva started to construct an underground characterization facility called the 'ONKALO'. Methods and equipment were further developed and tested for the investigations at greater depth. In the third stage, the actual characterization of the target rock volumes was commenced. The goal of this stage was to determine the final suitability of the rock volume for repository purposes and to define the locations of the first deposition panels.

3.3 ONKALO underground rock characterization facility

Except the extensive surface based investigations described above, Posiva Oy started in 2004 the construction of an underground characterization facility. The ONKALO has now been completely excavated to the anticipated repository depth i.e. −430−450 m. The research conducted in the ONKALO gives further information on the bedrock (rock mechanics) and groundwater conditions of the final disposal site, as well as on the impact of the construction. The ONKALO has provided an excellent opportunity to investigate the rock at tunnel scale, to conduct in-situ testing in rock, to develop excavation and final disposal techniques in realistic conditions (Johansson et al., 2015a). The ONKALO has aided in collecting the data needed, supported by a Preliminary Safety Assessment, for the application of the construction license that was submitted in the end of 2012. The Finnish Regulator STUK concluded in its statement in February, 2015 that the criteria set forth in the Nuclear Energy Act are fulfilled and the final disposal facility can be built to be safe. According to STUK, both STUK's own experts and other Finnish and international experts were used for the review of the construction license application. The statement supports Posiva's research findings that the final disposal of spent nuclear fuel can be carried out in a safe manner in Olkiluoto, in the municipality of Eurajoki. The project can now be continued with detailed engineering on the basis of STUK's statement, and the development areas presented by STUK will be incorporated in the forward plans. The target is to begin disposal operations in 2022. According to current plans, the final disposal would end in 2112 and the repository would be sealed up by 2120.

3.3.1 ONKALO layout

The ONKALO consists of the access tunnel (~5 km) and three shafts that have been excavated to anticipated repository level i.e. around −430 m (Figure 16). The access ramp has been excavated by using the D&B method to a depth level of about −450 m. Three shafts (one personnel shaft Ø4.5 m and two ventilation shafts Ø3.5 m) have been raise bored to the depth of 450 m. Technical facilities are located at the depth of −437 m. In the ONKALO, one of the main focuses of investigations is also currently in the

SITE IDENTIFICATION SURVEY (DESK STUDIES) (1983–1986)

- Also Lavia test hole for borehole investigation methodology development

PRELIMINARY SITE INVESTIGATIONS at five sites (1987–1992)

- Geological mappings, drilling of deep boreholes KR1-KR6 and shallow boreholes
- Geophysical studies (airborne, ground survey, borehole logging)
- Network of multi-level piezometers
- Installation of multi-packer systems into deep boreholes KR1-KR5
- Monitoring of groundwater/hydraulic heads in shallow and deep boreholes (start)
- Sampling of groundwater and rain water from the surrounding area, from wells, piezometers, deep boreholes KR1-KR5
- Measurements of hydraulic conductivity of deep boreholes KR1-KR6
- Rock stress measurements in borehole KR1 at depth level of 470–900 m
- Rock mechanics laboratory tests from deep boreholes KR1-KR3, KR5
- Rock mechanics field tests (point load) from deep boreholes
- Thermal property laboratory tests from boreholes KR2, KR3 and KR5

DETAILED SITE INVESTIGATIONS/ PHASE I (1993–1996) and PHASE II (1997-2000) at four sites

- Geological mappings including research trenches TK1 and TK2, drilling of deep boreholes KR7-KR12 and extension drilling of KR2, KR4, KR6 and KR7
- Regional geological studies (lineament interpretation, gravimetric survey, mapping)
- Geophysical studies (ground survey, borehole logging, acoustic-seismic study of the seabed)
- Groundwater sampling from deep boreholes including pressurised water sampling
- Measurements of hydraulic conductivity of deep boreholes, long-term pumping tests
- Installation of shallow groundwater observation tubes
- Ecological studies related to EIA, EIA/nature survey
- Rock stress measurements in boreholes KR2, KR4 and KR10 at depth level of 300–800 m
- Extensive rock mechanics laboratory tests from deep borehole KR10 and few from KR2, KR4
- Rock mechanics field tests (point load) from deep boreholes
- Thermal property laboratory tests from boreholes KR1, KR2, KR4, KR9, KR11
- Monitoring of the deformation of bedrock with GPS network (start)

SITE CONFIRMATION (PRE-ONKALO) PHASE, Olkiluoto site (2001–2004)

- Geological mapping of research trench TK3, drilling of deep boreholes KR13-KR28
- Regional geological studies (mapping)
- Geophysical studies (ground survey, borehole logging)
- Water sampling from the surrounding area and rain water, from groundwater tubes, shallow boreholes and deep boreholes including pressurised groundwater sampling
- Measurements of hydraulic conductivity of deep boreholes
- Vegetation and forest inventories, Ground frost measurements
- Extension of groundwater monitoring network (observation tubes), hydraulic conductivity measurements in shallow boreholes
- Rock stress measurements in borehole KR24 at depth level of 290–390 m, Kaiser Effect study in borehole KR14
- Rock mechanics anisotropic laboratory testing from deep boreholes KR12 and KR14
- Rock mechanics field tests (point load) from deep boreholes
- Microseismic monitoring network (start)

Figure 15 Different surface based site characterization phases before the ONKALO. The rock mechanics and thermal investigations are in light gray (modified from Posiva, 2003).

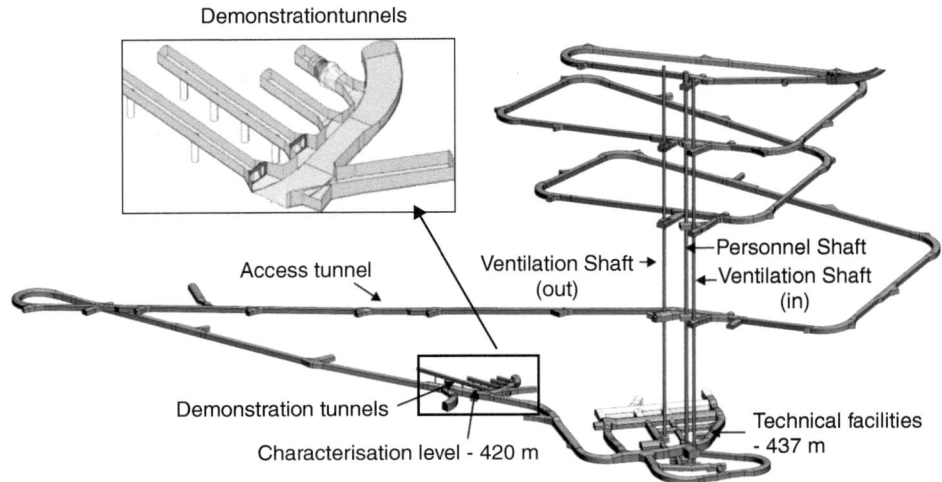

Figure 16 The layout of the ONKALO underground rock characterization facility (Figure: Posiva Oy).

demonstration tunnels at a depth of -420 m, where for example technical demonstrations and full scale tests are carried out and the methodology for locating suitable rock volumes is demonstrated.

The geological conditions in the ONKALO are characterized by crystalline bedrock, dominated by migmatitic, foliated gneiss. Massive, coarse-grained pegmatitic granites also occur as dykes in the area. The dip direction and dip of the overall foliation is estimated to be about 160°/40°. Based on observations from the access tunnel, horizontal or sub-horizontal, south-east-dipping fractures dominate; some sub-vertical fracturing also exists. The Olkiluoto rock mass is also characterized by the brittle deformation or fracture zones (Figure 17). The tunnel mapping during the ONKALO construction has shown that, after tunnel chainage 1300 m (depth ~130 m), the average rock mass quality has in general been good or very good.

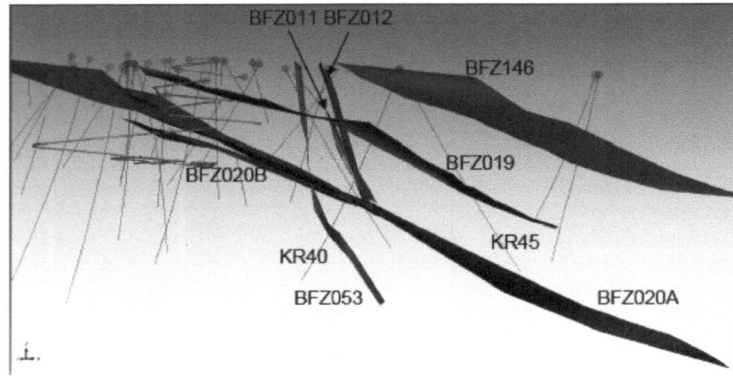

Figure 17 Main brittle fracture zones (BFZ) in dark gray and outline of the ONKALO facility seen left in light grey (Posiva, 2013).

3.3.2 Rock mechanics

Rock stress and rock strength are one of the most important parameters describing the rock mechanics conditions and predicting possible rock damage or spalling for the repository located at depth. Except the strength and deformation properties of the intact rock, also the thermal properties affect the rock's stability, the extent of rock damage or spalling in particular, and the potential for dissipating the heat produced in the spent nuclear fuel. These properties depend essentially on the intact rock's mineral composition and structure. The rock mechanics and thermal property models are thus strongly linked to the site lithology.

A thrust faulting stress regime is present in Olkiluoto, *i.e.* the horizontal stresses are larger than the vertical stress, $\sigma_H > \sigma_h > \sigma_v$. Also, the principal stresses are oriented horizontally and vertically, respectively. The recent LVDT-cell stress measurement results from the ONKALO (see Section 3.3.3.2) indicate that rock stress is affected by the major fracture zones (BFZs) and scatter exists in the results, especially near the surface. Below BFZ020 (-345 m to -408 m) the results are, however, quite well pronounced: the major in situ stress component is almost horizontal and the mean trend is 144° (±25°), which is roughly the regional NW-SE orientation of the major principal stress typically found in Scandinavia (see Section 1.2). The intermediate stress component is also horizontal and the minor is almost vertical. The mean magnitudes and standard deviations of the horizontal and vertical components for the depth range from 345 m to 400 m are: σ_H = 28.8 MPa ± 4.7 MPa, σ_h = 19.9 MPa ± 3.3 MPa and σ_V = 13.3 MPa ± 4.0 MPa (Posiva, 2013). The stress magnitudes are very close to the mean values measured in Finland (see Section 1.2).

Based on the fact that the Olkiluoto rock types are very heterogeneous and that the type and degree of foliation change rapidly, all gneisses can be considered in this context as one rock domain (type) with only the pegmatitic granite being considered a separate type – observed as several meter thick layers with their own parameter values (lower tensile strength). In spite of the high variation in rock strength properties due to the rock heterogeneity, the distributions of intact rock parameter values are reasonably well known. The spatial distribution of rock strength is not, however, determined by rock type, alteration or by ductile domains. Based on the laboratory tests the peak uniaxial strength for gneissic rocks is on average ~110 MPa with 50% confidence limits of 92 MPa and 121 MPa (Posiva, 2013). The strength values represent typical properties of migmatitic or gneissic rocks in Finland (see Figure 2 for comparison). Tests on clearly-foliated gneissic samples however showed that the uniaxial compressive strength, crack damage and tensile strength may vary depending on the loading orientation. Same study also indicated a mean anisotropy factor of 1.4 for the Young's modulus (Hakala *et al.*, 2005).

3.3.3 In situ testing

3.3.3.1 Rock strength

The in situ rock mass strength cannot be established from drillhole scale core samples because they are insufficiently large; instead, a larger scale deposition hole experiment is required. Previously, similar tests have been conducted at the URL (Canada) and

Äspö HRL (Sweden) but due to different geology adopting such information to Olkiluoto geology is at least questionable. Thus, an in situ spalling experiment called POSE (Posiva's Olkiluoto Spalling Experiment) was required whereby rock damage is induced by artificially increasing the pre-existing rock stress.

The in situ spalling experiment was carried out in the investigation niche location off the ONKALO ramp at about the -345 m depth level, tunnel chainage 3620 m, to determine the rock mass strength in representative rock conditions (Johansson *et al.*, 2015a). To obtain favorable stress conditions below the tunnel floor, the 4.5 m wide and 5.0 m high, originally EDZ niche was first reshaped to be 9 m wide and 7 m high. The niche, as well the expansion of the niche, has been excavated using careful blasting procedures to minimize the possible EDZ around the niche. The POSE plan consisted of two in situ tests; POSE Pillar Test where two near full-scale deposition holes, Ø1.52 m (compared to 1.75 m for actual deposition holes) and 7.2 m deep were drilled and leaving a 0.9 m pillar between the holes, and POSE Single Hole test with similar dimensions (Figure 18). The stresses around the holes were increased by additional heating to reach the stress level to generate the rock damage.

Three POSE holes were drilled during the summer 2010 using a full-face boring machine. The heating of the Pillar test took place in 2011 (Johansson *et al.*, 2014) and the heating period of Single hole test was in early 2013 (Valli *et al.*, 2014). Both test also included different investigations before and after the tests, monitoring during the test

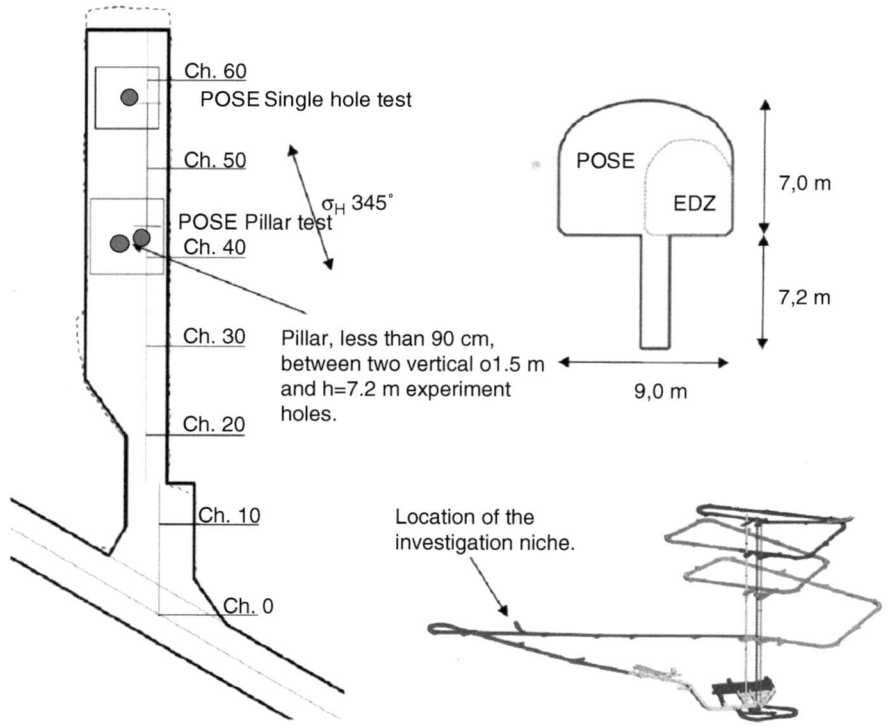

Figure 18 POSE in situ experiments in the POSE/EDZ investigation niche (Johansson *et al.*, 2015a).

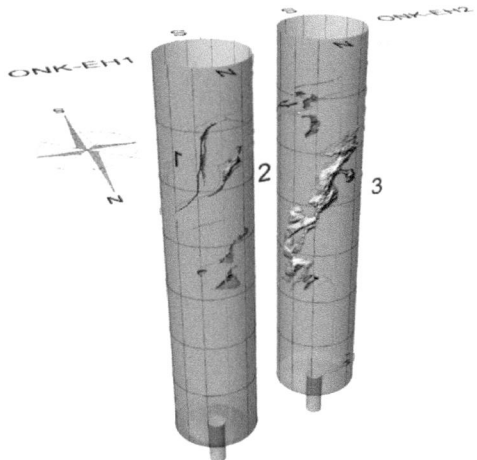

Figure 19 Observed damages in the POSE pillar test after the heating. Numbers 1–3 refer to fractures (red) that was formed before the heating (Johansson *et al.*, 2014).

executions and numerical predictions to estimate the rock spalling/damage potential using two different approaches *i.e.* fracture mechanics (Siren, 2011) and traditional continuum thermomechanics (Hakala & Valli, 2013, 2014).

The POSE Pillar test showed that the damage that was observed after boring the two holes was quite small *i.e.* two sub-vertical fractures were formed in the wall of one hole (EH1) and one sub-vertical fracture in the wall of the second hole (EH2) (Figure 19). Further damage occurred due to heating, but the damage was not located on the pillar side. The major damage was well localized and controlled by the foliation and rock type contacts which were known to be weak. Most damage seemed to have occurred on the S- and N-side walls. No major spalling type damage was observed except in some very limited areas in the pegmatite-granite. During the test, no significant rock pieces fell out and all the major damage was mostly related to fracture shearing/opening type damage. Later on the experiment holes were scaled, but only very minor pieces could be removed. The depths of the scaled (damage) areas were less than 100 mm (Figure 19). Hydraulic testing in the damaged areas showed the flows to be relatively small. Estimated distances from the damage to the hole wall varied from 20 to 180 mm with average value of 118 mm (Johansson *et al.*, 2014).

The Single hole test (hole EH3) was located almost completely in pegmatitic granite. Four fractures near the top of the hole were mapped after boring EH3, and a tensile failure located at the contact between mica-rich gneiss and pegmatitic granite was observed 18 months after boring, prior to the experiment. The heating phase did not show any clear damage and no spalling was occurred. However, acoustic emission and ultrasonic monitoring results pointed to events (damage) located in the immediate vicinity of a band of foliated gneiss, distributed in a N–S trend, although this could not be corroborated visually (Figure 20). Water loss measurements conducted in the hole indicated that the extent of damage was constrained to the first 100–200 mm of the hole wall (Valli *et al.*, 2014).

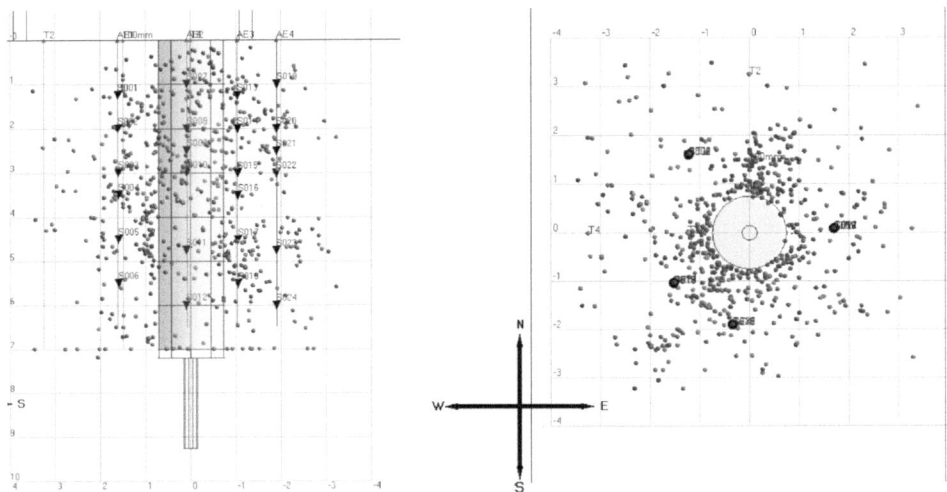

Figure 20 Located acoustic emission (AE) events during the monitoring period along the POSE Single hole test (Valli *et al.*, 2014).

Both POSE tests indicated that spalling type damage, rather it can be described as surface damage may not be a factor at Olkiluoto; failure in Olkiluoto rock conditions (anisotropic, heterogeneous) may be governed more by the relationship of the in situ stress state, the local geology and weakest rocks strength components. The depth of the damage zone around the holes was at maximum 200 mm in comparison to 60–120 mm prior to heating. The POSE in situ experiments resulted in the conclusion that the damage in Olkiluoto tends to be structurally controlled, the onset of the fracture initiation is around 40 MPa and the rock mass strength is around 90 MPa (Siren *et al.*, 2015).

3.3.3.2 Rock stress

Another important parameter to understand the rock mass behavior at depth is the in situ rock stress. Prior to the ONKALO stresses were measured at ground surface mostly with the overcoring and hydraulic fracturing methods. Measurements in very deep drillholes are not always robust and technical problems also occurred. The ONKALO made it possible to conduct more robust stress measurements *e.g.* with short drillholes. Due to some technical problems that were encountered with the traditional overcoring measurements, Posiva has developed a new, more reliable and accurate stress measuring device based on the LVDT (Linear Voltage Differential Transducer) cell and on the overcoring procedure (Hakala *et al.*, 2013). The in situ state of stress is solved by numerical inversion using the LVDT results of at least three optimally placed measurement locations around the excavation profile (Figure 21). The interpretation assumes a continuous, homogeneous, isotropic and linearly elastic material response (CHILE), but known transverse anisotropy or orthotropy material responses could also be applied. In high stress conditions at depth, overcoring may become vulnerable for

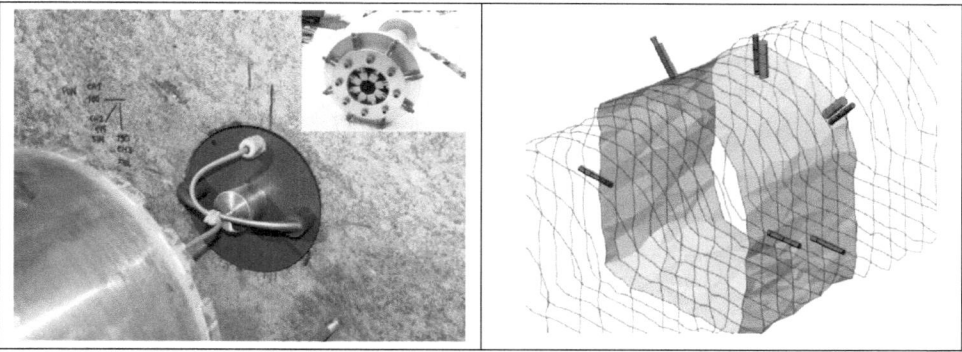

Figure 21 LVDT cell stress measurement method. The cell in left is inserted in a drillhole and then the drillhole is 'overcored' by a larger drillbit, thus releasing the rock stresses which change the LVDT readings. Typical measurement layout is shown right. (Posiva, 2013; Hakala *et al.*, 2013).

ring discing. To overcome this, a partial stress release method termed 'sidecoring' was developed. This is possible because the applied inverse best fit in situ stress solution scheme does not require total stress release, *i.e.* as opposed to analytical solutions, numerical 3D simulations can be undertaken for the geometry before and after sidecoring. LVDT cell stress measurements have been conducted at several locations in the ONKALO (shafts, ONKALO ramp, POSE investigation niche).

3.3.3.3 Thermal properties

Thermal properties are also needed to understand the thermally induced stresses and to design the canister spacing in the repository. Thermal properties of rock are also scale dependent as the rock strength, so the ONKALO gives good opportunity to study this, as well. Rock temperature monitoring in three drillholes with a length of 5–10 m has been carried out in one of the investigation niches (Suppala *et al.*, 2013). The objective was to test the methodology to estimate the thermal properties of gneissic rock from long term temperature monitoring. The results from one year monitoring indicated the median diffusivity parallel to the average foliation from the drillhole depth of 2 m to be $1.86 \cdot 10^{-6}$ m^2/s and between quartiles $1.82 \cdot 10^{-6}$ and $1.88 \cdot 10^{-6}$ m^2/s. The estimated diffusivities perpendicular to the average foliation vary between $1.47 \cdot 10^{-6}$ and $1.59 \cdot 10^{-6}$ m^2/s. This type of in situ upscaling test was considered useful to estimate the rock thermal diffusivity, which is important parameter for the thermal dimensioning of the repository.

3.3.3.4 Summary

The ONKALO facility has played an important role in the Finnish repository development programme. The ONKALO has enabled direct in situ measurements, experiments and testing of different investigation tools underground. In addition to the in situ tests, all the observations made during the ONKALO excavations and constructions have also provided valuable information for the rock mass property characterization to build up more reliable site descriptive models.

Several separate campaigns to measure the rock responses have been conducted at the different locations and different depths in the ONKALO underground rock characterization facility during its excavation works in 2007 – 2014 (Johansson *et al.*, 2015b). The rock response measurements have included microseismic and acoustic emission monitoring, rock displacement measurements with convergence pins and extensometers, strain gauge measurements and video camera measurements. The objectives of these campaigns were to better understand the rock properties (deformation-strength properties and *in situ* stress) and to develop a predictive capability for the design purposes and improve and refine the ability to predict the mechanical conditions ahead of excavation. Most of the rock response measurements have included predictions made beforehand as part of the ONKALO Prediction-Outcome campaign set in the early phase of the ONKALO construction.

The ongoing and future tests will confirm the elements of the final disposal concept and will demonstrate the design, construction and host rock suitability assessment process to be utilized during the final disposal. All that information is essential for the repository design and the safety analyses. The testing in the ONKALO still continues several years and more in situ testing is under planning to better understand *e.g.* rock mass properties.

REFERENCES

Finnish Mining and Metallurgical Society. (1967) Finnish Rock Mechanics Symposium. (in Finnish)

Hakala, M., Syrjänen, P., Roinisto, J., Salmelainen, J. & Aartolahti J.-V. (1999) Rock mechanical design of the Kluuvi underground car park/civil defence shelter. In G. Vouille & P. Berest (eds.), Proc. 9th International Congress on Rock Mechanics, Paris. Rotterdam, Balkema.

Hakala, M., Kuula, H. & Hudson, J. A. (2005) Strength and strain anisotropy of Olkiluoto mica gneiss. Working Report 2005-61. Posiva Oy, Olkiluoto.

Hakala, M., Siren, T., Christiansson, R. & Martin, D. (2013) In situ stress measurement with the new LVDT-cell method – Method description and verification. Posiva Report 2012-43. Posiva Oy, Olkiluoto.

Hakala, M. & Valli, J. (2013) ONKALO POSE Experiment – Phase 3: 3DEC prediction. Working Report 2012-58. Posiva Oy, Olkiluoto.

Hakala, M. & Valli, J. (2014) ONKALO POSE Experiment – Phase 1&2: 3D Thermo-mechanics prediction. Working Report 2012-68. Posiva Oy, Olkiluoto.

Heidbach, O., Tingay, M., Barth, A., Reinecker, J., Kurfeß, D. & Müller, B. (2008) The World Stress Map database release 2008. doi:10.1594/GFZ.WSM.Rel2008.

Johansson, E., Riekkola R. & Lorig, L. (1988) Design analysis of multiple parallel caverns using explicit finite difference methods. In P. Cundall *et al.* (eds.), Key questions in rock mechanics. Proc. 29th U.S. Rock Mechanics Symposium. Rotterdam, Balkema.

Johansson, E. & Lorig, L. (1990) Use of numerical models in design and excavation monitoring of multiple parallel caverns. Int. Symp. on Unique Underground Structures. Denver, Colorado School of Mines.

Johansson, E & Kuula, H. (1996) Three dimensional back-analysis calculations of Viikinmäki underground sewage treatment plant in Helsinki. Proc. of 8th International Congress on Rock Mechanics, Tokio. Rotterdam 1995, Balkema.

Johansson, E., Kuula, H. & Kajanen, J. (1996) Modelling of rock structure and determination of rock parameters. RIL K177-1996. Helsinki, RIL Finnish Association of Civil Engineers. (in Finnish)

Johansson. E. (1999) Use of FLAC in two underground repository projects for low and medium level radioactive waste in Finland. Proc. of International FLAC Symposium, Minneapolis. Rotterdam 1999, Balkema.

Johansson, E., Riekkola, R. & Kalliomäki, M. (2000) Rock engineering of an underground library for Helsinki University. Proc. of NGM-2000 Symposium. Finnish Geotechnical Society, Helsinki.

Johansson, E., Siren, T., Hakala, M. & Kantia, P. (2014) ONKALO POSE experiment – Phase 1 & 2: Execution and monitoring. Posiva Working Report 2012–60. Posiva Oy, Olkiluoto.

Johansson, E., Siren, T. & Kemppainen, K. (2015a) ONKALO – Underground rock characterization facility for in-situ testing for nuclear waste disposal. 13th International Congress on Rock Mechanics. Montreal, Canada.

Johansson, E., Siren, T. Hakala, M., Saari, J. & Malm, M. (2015b) Measured rock responses in the ONKALO – Underground rock characterization facility. Posiva Working Report 2015-20. Posiva Oy, Olkiluoto.

Öhberg, A, Johansson, E., Anttila, P. & Saari, J. (2011) Two decades of rock monitoring experiences at the two underground repositories for operating waste in Finland. World Tunnel Congress WTC2011, Helsinki, May.

Posiva. (2003) Baseline conditions at Olkiluoto. POSIVA Report 2003-02. Posiva Oy, Olkiluoto.

Posiva. (2013) Olkiluoto Site Description 2011. Report POSIVA 2011-02. Posiva Oy, Olkiluoto

Rönkä, K. & Ritola, J. (eds.) (1997) The fourth wave of rock construction. MTR-FTA Finnish Tunneling Association. Porvoo.

Särkkä, P. & Aho, J. (eds.) (2011) Rock – Sound of countless opportunities. MTR-FTA Finnish Tunneling Association. Keuruu.

Siren, T. (2011) Fracture mechanics prediction for Posiva's Olkiluoto spalling Experiment (POSE). Working Report 2011-23. Posiva Oy, Olkiluoto.

Siren, T., Hakala, M., Valli, J., Kantia, P., Hudson, J. & Johansson, E. (2015) In situ strength and failure mechanisms of migmatitic gneiss and pegmatitic granite at the nuclear waste disposal site in Olkiluoto, Western Finland. Int. J. of Rock Mechanics and Min. Sci. Volume 79, pp. 135–148.

Suppala, I., Kukkonen, I., Korpisalo, A. & Koskinen, T. (2013) Thermal diffusivity of a rock mass estimated from drillhole temperature monitoring in the ONKALO. Working Report 2013–35, Posiva Oy, Olkiluoto.

Tolppanen, P. & Johansson, E. (1996) Rock stress measurements in Finland 1961–1994. RIL and TEKES, Helsinki. (in Finnish)

Valli, J., Hakala, M., Wanne, T., Kantia, P. & Siren, T. (2014) ONKALO POSE Experiment – Phase 3: Execution and Monitoring. Working Report 2013-41. Posiva Oy, Olkiluoto.

Chapter 19

The Meuse/Haute-Marne underground research laboratory: Mechanical behavior of the Callovo-Oxfordian claystone

G. Armand[1], F. Bumbieler[1], N. Conil[1], S. Cararreto[1], R. de La Vaissière[1], A. Noiret[1], D. Seyedi[2], J. Talandier[2], M.N. Vu[2] & J. Zghondi[1]

[1]Meuse/Haute-Marne Underground Research Laboratory, National Radioactive Waste Management Agency (Andra), Bure, France
[2]National Radioactive Waste Management Agency (Andra) R&D, Chatenay-Malabry, France

Abstract: The Meuse/Haute-Marne Underground Research Laboratory (URL) is located in the eastern boundary of the Paris Basin. The URL started to be built in 2000 (shaft sinking operations) in the framework of Andra's (French national radioactive waste management agency) research program aimed at demonstrating the feasibility of a reversible deep geological disposal of high-level and intermediate-level long-lived radioactive waste (HLW, IL-LLW). Its underground drifts are used to study the Callovo-Oxfordian claystone layer between 420 m and 550 m in depth. Scientific and technological goals and demonstration experiments have been conducted together through a step by step approach. This chapter presents some of the field observations. The various configurations implemented in the field give insights of the influence of construction method on the excavation damaged zone, the hydromechanical behavior of drift and progressive loading of the support, which are key issues for the design and safety studies.

I INTRODUCTION

In the context of radioactive waste disposal, an underground research laboratory (URL) is a facility in which experiments are conducted so as to establish and to be able to demonstrate the feasibility of constructing and operating a radioactive waste disposal facility within a geological formation (NEA, 2001a, b). Twenty-six URLs were set up in 10 countries between 1965 and 2006 (Delay *et al.*, 2014). They are located in a range of geological formations: argillaceous sedimentary rocks, magmatic rocks, evaporites and volcanic tuff. Two main categories can be stand out (Blechschmidt & Vomvoris, 2010): 'methodological laboratories' and 'site-specific laboratories'.

Experiments in URLs meet two sets of needs: (a) characterization, that is, acquiring knowledge of the geological, hydro-geological, geochemical, structural and mechanical properties of the host rock and of its response to perturbations; and (b) construction and operation, that is, developing equipment to acquire know-how about the construction of all the components of a disposal facility up to its closure, and the emplacement and/or retrieval of the waste.

In France, Andra is in charge of long-term management of radioactive waste produced in France. It is an industrial and commercial public body established by the December 30, 1991 Waste Act. Its role was completed by the June 28, 2006 Planning Act concerning the sustainable management of radioactive materials and waste. One of its main missions is to study and design solutions for the sustainable management of radioactive waste for which there is not yet any specific disposal facilities, as a reversible deep geological disposal of high-level and intermediate-level long-lived radioactive waste (HL, IL-LL projects).

Clay formations in their natural state exhibit very favorable confining conditions for repository of radioactive waste because they generally have a very low hydraulic conductivity, small molecular diffusion and significant retention capacity for radionuclides. That is why Andra started the study the Callovo-Oxfordian claystone as a possible host rock for radioactive waste repository and build an underground research laboratory in Bure since 2000.

This chapter describes the approach and the strategy implemented in the Meuse Haute Marne URL to demonstrate and optimize repository components. An insight is made on the work performed in rock mechanics on testing and optimization of techniques for the construction of drifts (excavation/support) and building of HL waste disposal vaults (steel lined micro-tunnel of 0.7 m of diameter) with focus on the excavation induced damage, and short and long term deformation on the opening (convergence, strain, pore pressure change, strain/stress in the support). Thermo-Hydro-Mechanical (THM) behavior, seal and gas transfer studies, which are also important processes to design the final repository are considered in the program carried out at the Meuse Haute Marne URL, but are not presented in this section.

2 THE SITE

2.1 The Callovo-Oxfordian claystone (COx)

2.1.1 Location of the site

From a geological perspective, Eastern France aroused interest because it is part of the largest sedimentary basin (Paris Basin) in France. Historically known to be stable, it presented the advantage to be highly studied due to various oil drilling operation during the second half of the 20th century. The Callovo-Oxfordian claystone was identified as a potential host rock for a deep geological radioactive waste disposal. A favorable location to implement an underground research laboratory was found in the sector south of the Meuse and North of the Haute Marne department (Delay et al., 2007a). The major characteristics are:

–   Very favorable rock characteristics due to presence of clay minerals ensuring very low permeability and good retention capacity
–   Located between 400–600 deep; deep enough not to be affected by predictable geological phenomena and shallow enough to avoid encountering insurmountable difficulties during construction (between 420 m and 550 m in depth at the URL, figure 1)
–   Over 100 m thick (130 m thick at the URL location) and covering a large area.

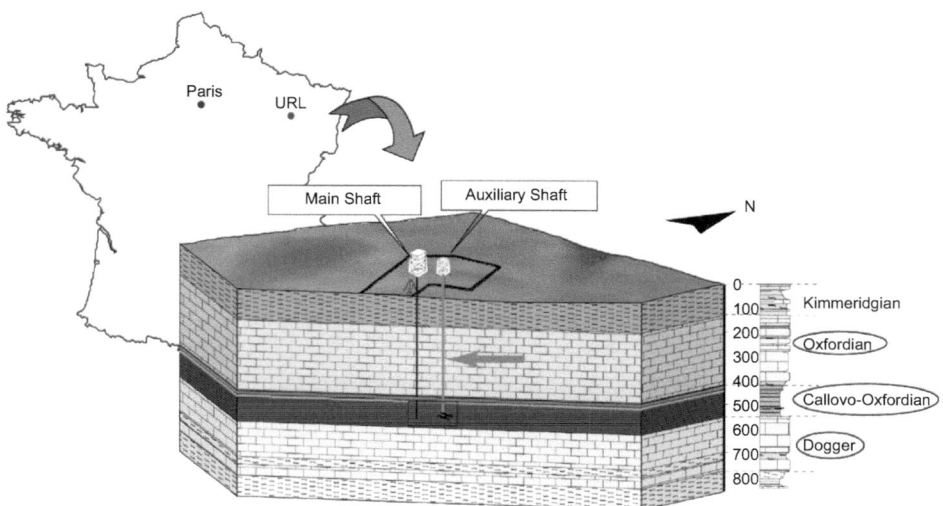

Figure 1 Location of the underground research laboratory in Meuse/Haute-Marne in the Paris Basin.

Geological and geophysical investigations showed the homogeneity of the Callovo-oxfordian claystone on and around the URL site and the absence of faults or even minor fracturing (Vigneron *et al.*, 2004). The Callovo-Oxfordian claystone is overlain and underlain by poorly permeable carbonate formations.

2.1.2 Mineralogy and physical properties of the layer

Sediments of the Callovo–Oxfordian unit consist of a dominant clay fraction associated with carbonate, quartz with minor feldspars, and accessory minerals (Lerouge *et al.*, 2011). At the MHM URL main level (−490 m depth), the clay fraction is high (40–60%), and clay minerals consist of illite, ordered illite/smectite mixed layers, kaolinite, tri-octahedral, iron-rich chlorite, and minor biotite (Lerouge *et al.*, 2011).

The Callovian–Oxfordian clay-rich rock porosity lies between 14% and 20% at the MHM URL site and is close to 18% at the URL main level (Yven *et al.*, 2007), and natural water content ranges between 5% and 8%. Due to a very small mean pore diameter (about 0.02 μm), the claystone has a low permeability (5×10^{-20} to 5×10^{-21} m^2).

2.1.3 Mechanical properties of the layer

Different laboratory tests have been conducted on core samples to obtain the hydro-mechanical properties of the claystone (Armand *et al.*, 2016). The very low level of permeability makes the measurement of pore pressure complex during triaxial tests. Research is on-going to get better data on that subject. The main features of the short-term mechanical behavior observed on the samples of claystone under triaxial tests can be summarized as follows: a linear behavior under low deviatoric stress; the loss of linearity of stress-lateral strain curves begins approximately at 50 % of the peak value of the deviatoric stress, which can be associated with damage. Under low confining pressures, a brittle failure of the samples is observed and corresponds to the formation of a shear band inclined with respect to the sample axis. There is a strong dependence of the mechanical behavior on the confining pressure, marked by a transition from a brittle toward a ductile behavior. It has been shown that a failure criterion based on the generalized Hoek and Brown criterion is well adapted to represent shear strength of the Callovo-Oxfordian claystone (Souley *et al.*, 2011).

Long term behavior of the claystone has been studied through creep tests. Zhang *et al.* (2012) showed that the creep behavior under an increasing load is characterized by two phases (under multi-step uniaxial loads over 6 years.). First a transient phase with decreasing rates governed by strain hardening is observed. This phase is followed by a second one with an asymptotically approached constant rate after strain recovery. The creep test under dropped load contrary evolves firstly backwards with negative rates and then inversely returns with time to a steady-state creep at a positive rate. The creep rates are in the order of magnitude of $1 \cdot 10\text{-}11$ to $6 \cdot 10\text{-}11$ s-1. Obviously, the creep behavior is dependent on the loading path. Similar creep rates have been measured by other authors (Gasc-Barbier *et al.*, 2004). Zhang *et al.* (2010) also pointed out that, in uniaxial creep tests, the creep rate varies very slowly and linearly with stress at low stresses (below 13–15 MPa). Above that, the creep rate increase deviates from the linearity. The acceleration of the creep rate seems to be linked with the damage onset and increase.

Table 1 represents the Callovo-Oxfordian claystone mechanical characteristics at the main level of URL. Sedimentation has led to a slightly anisotropic behavior of the COx. From compressive and shear wave measurements on cubic samples, the anisotropy ratio of the dynamic Young Modulus is found to be around 1.3. From triaxial tests, this ratio can reach 2 for some samples. All mechanical parameter are dependent of the mineralogy which varies with depth.

In order to provide a physical interpretation of the variation of the mechanical properties of Callovo-Oxfordian claystone with mineral composition, Abouchakra Guery (2007) developed three linear homogenization schemes considering the COx claystone as a three-phase material composed of a clay matrix and inclusions of quartz and calcite. It is shown that, unlike the dilute scheme and the self-consistent scheme, the Mori-Tanaka model describes the in situ experimental data well. The model was used to back calculate the Young modulus in deep boreholes from the measured mineralogy. Figure 2 shows the very good agreement between the measurement and the prediction of the homogenization model and emphasizes the evolution of the mechanical parameters as function of depth due to mineralogy variation.

2.2 In situ stress state and pore pressure

At the Meuse/Haute-Marne URL, the anisotropic stress state of the claystone was determined and discussed by Wileveau *et al.* (2007). The combination of sleeve and hydraulic fracturing and imaging method in vertical, inclined and horizontal boreholes has yielded to the determination of the complete stress state in the Callovo Oxfordian and the over and underlying limestone. The shafts sinking has also provided an opportunity to compare these results with the strain measurements obtained within the same depth range. Wileveau *et al.* (2007) shows a reasonable agreement between methodologies for estimating the major stress magnitude and stated that the horizontal stress anisotropy (σ_H/σ_h) shows significantly lower values within the strata (at the main level of the URL) than the over and underlying limestone.

Table 1 Callovo-Oxfordian claystone parameters at the main level of MHM URL.

Bulk specific gravity ρ(g/cm^3)	Porosity, n (%)	Young's modulus (MPa)		Poisson's ratio, ν	Uniaxial compressive strength, UCS (MPa)	Hoek-Brown criteria			Intrinsic permeability, k (m^2)	Water content, w(%)
		$E_\perp$	$E_{//}/E_\perp$			S	m	σ_c (MPa)		
2.39	18±2	4000 ± 1470	1.2–2	0.29±0.05	21±6.8	0.43	2.5	33.5	5×10^{-20}–5×10^{-21}	7.2±1.4

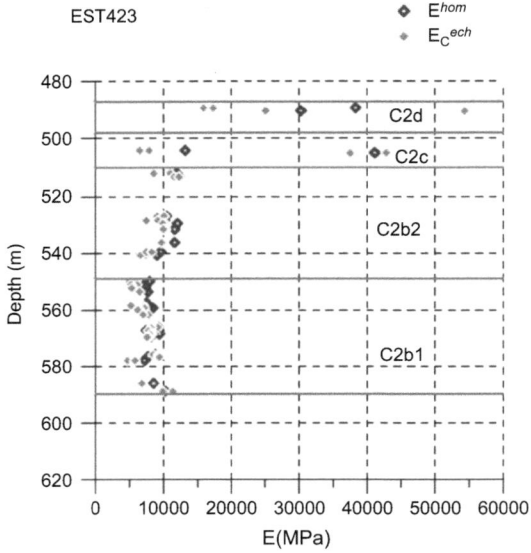

Figure 2 Young modulus versus depth in borehole EST423 (at about 12 km from the URL): comparison of measured cyclic Young modulus and estimated one from homogenization model developed by Abouchakra Guery (2007).

In the COx layer, the major stress (σ_H) is horizontally oriented at NE155°. The vertical stress (σ_v) is nearly equal to the horizontal minor one (σ_h):

$$\sigma_v = \rho g Z \tag{1}$$

$$\sigma_H = \sigma_v \tag{2}$$

where Z is the depth, ρ is the density and g is the gravity.

The ratio σ_H/σ_h is at maximum 1.3 and varies with depth and the rheological characteristics of the respective layers.

By opposition to the surrounding carbonated formations, the hydraulic head in the Callovo-Oxfordian is not uniform (Distinguin & Lavenchy, 2007) as it has been observed in other low permeable clay formations. A regular increase in the freshwater head is observed between the Oxfordian limestone, with a head of 305 m ASL, and the Callovo-Oxfordian, with a maximum head close to 350 m ASL. The maximum head within the Callovo-Oxfordian seems to take place between 430 and 475 m BGL. Pore-pressure at the main level (–490 m) is around 4.7 MPa.

3 DEMONSTRATION AND SCIENTIFIC PROGRAM AT THE MEUSE HAUTE MARNE URL

3.1 General objectives of the research

Delay *et al.* (2014) point out that over time, the work done in URLs has evolved both in type and in importance. In the early days, some 25–30 years ago, the objective was more to identify the main scientific and technical issues and to develop methods to cater for them. The priorities then were (a) to define programs relating to the study of confinement by the medium or by engineered barriers, (b) to develop experimental equipment and methods, (c) to conduct elementary technical feasibility studies, and (d) to collect fundamental geological data. Since the early 2000s and even more recently, most of the URLs have turned toward the creation of demonstrators. Demonstration experiments are supposed to represent, test and optimize potential disposal systems and repository components.

3.1.1 Phase: 2000–2005

In the 1990s a geological investigation covering over a hundred square kilometers was carried out involving geophysical survey and drilling of boreholes in eastern France near Bure village. The surface reconnaissance program helped to locate the site of the Meuse Haute-Marne URL, but also provided an impressive wealth of data covering a wide range of geoscience like hydrogeology (Distinguin & Lavanchy, 2007), structural geology (Vigneron *et al.* 2004), mineralogy (Gaucher *et al.*, 2004), geochemistry (Vinsot *et al.*, 2008) and rock mechanics (Lebon & Ghoreychi, 2000).

Since 2000, Andra began to build the URL by shafts sinking to study the Callovo Oxfordian claystone lying between 420 m and 550 m below ground. The main objective of the first research phase (2000 to 2005) was to characterize the confining properties of the clay through in situ hydrogeological tests, chemical measurements and diffusion experiments and to demonstrate that the construction and operation of a geological repository would not introduce pathways for radionuclides migration (Delay *et al.*, 2007b). The regional and local knowledge acquired through the geological survey campaigns, as well as the first results obtained in the URL, are presented in Andra's "Dossier 2005" which helped assess the progress of research at the end of the 15-year period of research work prescribed by the Waste Act of 30 December 1991 (Andra, 2005).

3.1.2 Phase: Post 2005

The ongoing research program (started in 2006) following the "Technology Readiness Level" Scale (TRL) is more dedicated to technologies improvement and demonstration issues of different disposal systems, even if characterization studies are still ongoing. The main research and development themes of the laboratory can be structured as follows:

1. Verification of the design of the disposal facility, development of construction methods and optimization of the design of the future structures of the disposal facility.
2. Verification of the ability to seal the vaults drifts and shafts, development of methods for the sealing and filling of the drifts and shafts.
3. Confirmation of the low impact of perturbations caused by the disposal facility, assessment of the behavior of the perturbed argillaceous rocks and materials at the interfaces.

4. Confirmation of the confinement abilities of the Callovo-Oxfordian formation, characterization of the argillaceous rocks.
5. Assessment of the conditions of transfer of radionuclides into the biosphere, observation of the hydrogeological context and the environment over time.
6. Development of observation and monitoring methods for the reversible management of the disposal facility.

3.2 Main components of the design of a possible disposal facility in France

The 2006 planning Act gives a schedule to build and operate an industrial site for geological disposal in France called Cigéo, which includes assembling all elements necessary for a license application to be filed in 2018. Pending authorization, the repository construction should begin in 2021 and be ready for commissioning in 2025, starting with a pilot zone to be operated over some 10 years. The Cigéo underground installations will be progressively built, operated and finally closed over a roughly 150 year period.

Figure 3 shows a possible architecture of the Industrial Centre for Geological Disposal, Cigéo. Surface facilities will be split in 2 parts with the nuclear activities (taking charge of the primary waste packages (rail terminal or others), checking the packages, re-packing them into disposal packages) and the excavations activities (workshops, cement plant, storage, drainage basins, excavation muck "dumping area", etc.). A ramp will be used to transport waste packages to the main level of the disposal. Shafts are mainly used for all the activities related to excavation/support works, ventilation and worker transportation. The main disposal area will be split in two parts related to the type of waste IL-LL or HL-LL and the type of excavation used to emplace the waste canister (respectively tunnel of 9 to 11 m in diameter for IL-LL waste and micro tunnel of 0.7 m in diameter for HL-LL waste).

The waste package disposal facilities and the transfer and emplacement processes are designed with the aim of simplifying waste package retrieval operations which may be decided, using, if possible, similar means to the ones used for emplacement. As a result, clearances for handling purposes that must be durably maintained are provided between the package and the cell walls.

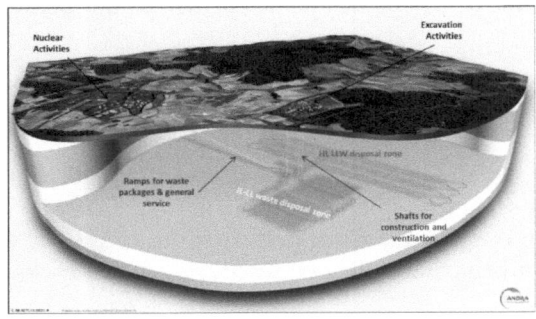

Figure 3 Possible architecture of the Industrial Centre for Geological Disposal Cigéo.

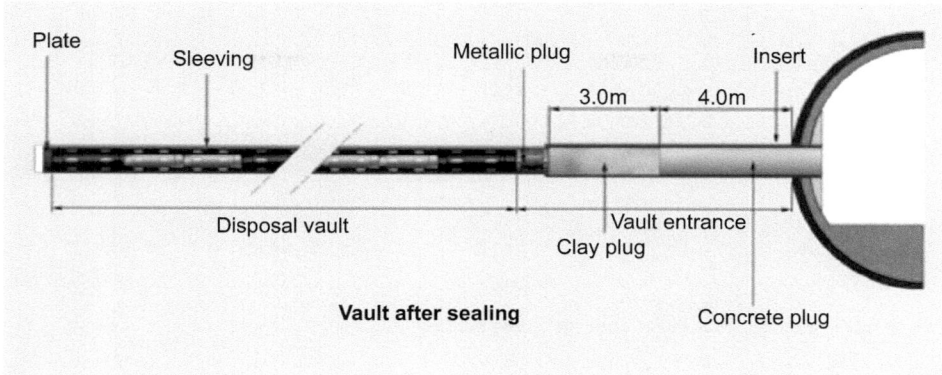

Figure 4 Concept 2009 of disposal cell for HL-LL activity waste.

3.2.1 Concept of disposal cell for HL-LL activity waste

Vitrified HL Waste Stainless Canister (primary waste) will be placed in thick steel overpacks to prevent glass leaching during the thermal phase. The overpacks will be stored in dead-end, horizontal micro-tunnels with an excavated (drilled) diameter of approximately 0.7m (Figure 4). In the Dossier 2009, the length of the benchmark HL-LL cell had been limited to 40m, but it was then extended to 80m in the Cigéo pilot phase. In the present stage of design, it comprises a body part, for packages disposal, and a head part for cell closure. They are favorably aligned with respect to the stress field. To prevent against rock deformation and allow the potential retrieval of waste containers during the reversibility period, both body and head parts have a non-alloy steel casing? The casing in the body part has a diameter slightly smaller than the one in the head part called "insert". That means it can slide in the insert. Thus, the effects of the thrust produced by its dilation, due to heat generated by the exothermic packages, are absorbed without consequence for the cell head.

The cell base is closed off by a "base plate", also made of non-alloy steel. A metal radiation-protection plug separates the cell head from the body part. For cell closure, the insert is partly backfilled with a swelling-clay plug and then sealed with a concrete plug to provide additional safety.

3.2.2 Concept of disposal vault for IL-LL activity waste

Before emplacement, IL-LW will be grouped into precast concrete rectangular robust containers. The concrete containers will be stacked (trolley stack technique, pre stacking technique...) in large diameter horizontal tunnels (between 9 to 11 m). The vaults of IL-LW are comparable to drifts and lined with thick concrete lining to limit long term deformations. Ventilation of IL-LW repository cells has to be maintained as long as they are not closed, that is why vaults are parallel and connected to an air outtake drift. The forecast length of the disposal vault is some 400 m to 500m.

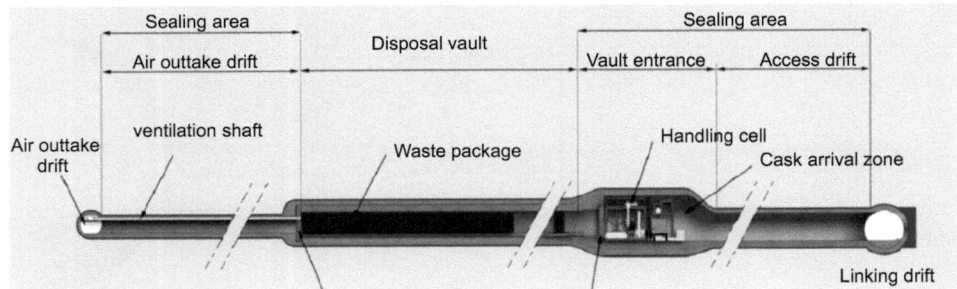

Figure 5 ILL Waste disposal vaults concept: cross section.

3.3 The Meuse/Haute Marne URL layout

Figure 6 shows the present and forecast drifts and shafts network at the Meuse Haute URL. Two shafts provide access to two levels of drifts at 445 and 490 m depth. The shafts with the connecting drift at –490m (between the two shafts) ensure also ventilation and security. Depth –445 m is in the upper part of the COx layer where the carbonate content is higher. The niche excavated at this depth was used to perform the first in situ experiments in the COx in the URL. Among those experiments, the first geomechanical experiment (REP experiment) was carried out during the excavation of the main access shaft (see section 4.2.3).

Depth –490m corresponds to the middle of the COx layer, at the URL, which is the most representative of the potential location of the waste repository. At the main level of the URL (Figure 6), the orientation of the experimental drifts has been determined

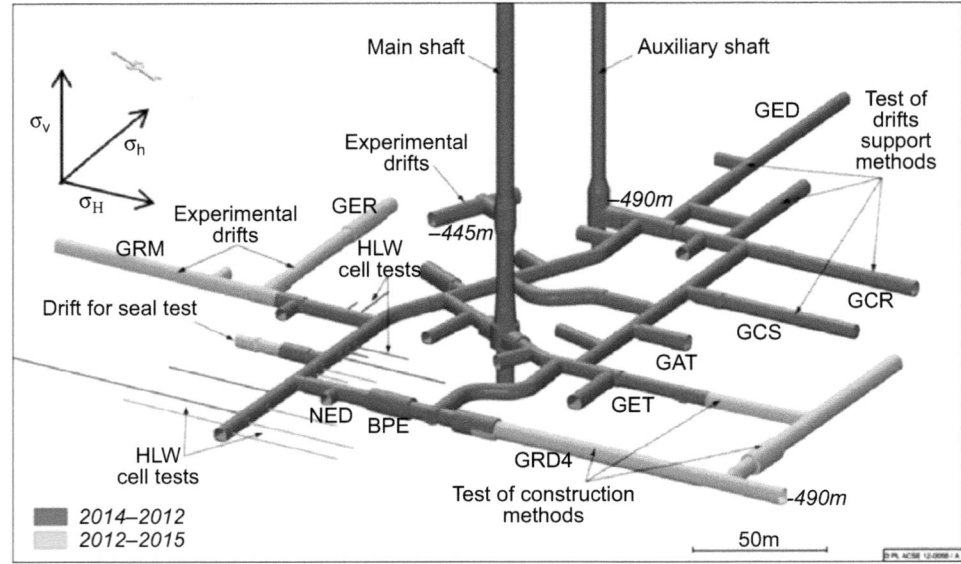

Figure 6 Meuse / Haute-Marne URL drifts network, (dark grey: already excavated, ligth grey: to be excavated).

according to the orientation of in situ stress field which also corresponds to the major orientation of the drifts in the repository. The excavation worksite in the host layer is a scientific experimentation in itself to characterize the impacts of digging, to understand the hydro-mechanical (HM) behavior of the claystone and to study the excavation damaged zone (EDZ). Most of the drifts have been excavated for a specific scientific and technological purpose, in order to emplace experiments or as a research tool when the Hydro mechanical behavior is studied. Remaining drifts are technical drifts necessary for operation (electrical power, equipment storage) or safety of the URL. Even in such drifts a minimum geomechanical survey has been performed during the excavation work (convergence measurements, geological mapping of the front face).

3.4 Experimental strategy in rock mechanics

Compared to conventional civil engineering works, Cigéo project has specific features like being an underground nuclear facility involving non-common rules for underground works; a long operational life time; size and geometry (complex drifts network and length of drift). In this context, rock mechanics experiments performed in the URL permit to achieve a good understanding of rock and structure interaction which will give important input to design (and optimize design) of different elements of the repository like support, and input data for the performance assessment calculation (mainly EDZ extent and transmissivity properties).

The horizontal underground structures dedicated to waste disposal fall under two types according to their sizes: the micro-tunnel (diameter less than 1 m) and drift (diameter from 4 to 11 m). Regarding the "drift" type structures, the primary aim of the tests and technological demonstrators is to develop a construction method and to study the interactions between the argillaceous rocks (intact and damaged) and the support and lining in space and time. Regarding the structures of the "micro-tunnel" or "large diameter borehole" type (disposal vaults of HL waste or spent fuel), the objective of developing a construction method must also be associated with the objective of enhancing the knowledge of the rock behavior in the vicinity of the structures.

The understanding of drift behavior is associated with the design of the IL-LL Waste disposal vaults and access drifts in the COx claystone. A huge program of experiments is ongoing to characterize the response of the rock to different drift construction methods. In situ experimental strategy for the study of drift behavior is based on:

- Use the underground laboratory for observing real behavior of drifts under different conditions (depth, size, geometry, ventilation, temperature): each drift excavation in the URL is a rock mechanics experiment in itself
- Sequencing of drift construction to highlight the role of support/excavation method on the HM behavior (support loading, EDZ)
- Start with "soft" support drift to study the behavior of COx, skip to rigid support emplaced after a delay of several months and then emplace rigid support just after the excavation of the gallery
- Using parallel drifts to compare the different behavior (Figure 6)
- Use mine by experiments (borehole emplaced before the excavation) to study the HM behavior (at short and long terms)

In the repository concept IL-LL cells are oriented along the major horizontal stress. That is why, in the laboratory, at the beginning the most of the drift were oriented along the major stress (Figure 6, drifts GET, GCS, GCR, GRD, GRM). After 2015, drifts GER and GVA have been excavated in the other direction. Those data could be compared to the ones acquired in GED drift and will provide the same level of knowledge for drift parallel to the minor horizontal stress.

In the shafts and the drift at −445 m a drill and blast method was used with a step of 3 m. At the main level (at −490 m) this excavation method could not be used due to the amount of induced fractures.

Figure 7 illustrates the excavation methods used at the main level of the URL: hydraulic hammer and road header. A full face tunnel boring machine (TBM) which could be used for access drift in the repository could not be easily used in the URL due to the small size of the shaft and the low capacity (in size and load) of the freight elevator. But a road header under shield has tested in 2013 with segments emplacement technique (Figure 7 c,d). Before 2008, most of the drifts have a horse-shoe shape cross section (17 m² area, with r ≈ 2.3 m) excavated with a hydraulic hammer. Since 2008, a counter vault has been used to reduce uplift of the foot plate (drift GED). The spans of the excavation are mainly 1 m long and are immediately covered with 3–5 cm of fibered shotcrete. Excavated zone was supported immediately by bolts, sliding arches and 10 cm thick layer of shotcrete was set in place. Martin *et al.* (2010) presented the

Figure 7 Excavation methods: (a) hydraulic hammer mounted on an electric device of the "Brokk" type, (b) Road header, (c) road header under a shield in the assembly building, (d) front face during the excavation with the road header under a shield.

efficiency of the different support elements – at the front – under different digging configurations during the laboratory construction.

Other excavation techniques using a road header has been tested with the same support and with other types of supports (GET, GCS, and GCR). In drift GCS (Figure 8-a-b), the support is ensured by an 18 cm thick fibre reinforced shotcrete

Figure 8 Example of different drifts support: (a) GCS drift emplacement of the yieldable concrete wedge, (b) GCS drift support with yieldable concrete wedge and shotcrete, (c) GCR with concrete lining grouted in place (thickness ≈ 27 cm), (d) BPE drift with 0.45 m of shotcrete projected in four layers, (e) segment emplacement under the shield (f) View of the GRD drift.

shell, interrupted by 12 yieldable concrete wedges (hiDCon®); completed by a crown of 12 HA25 radial bolts of 3 m length every meter. Drift GCR (Figure 8-c) was built in a similar way as drift GCS and left "unsupported" for seven months. A concrete lining of a 27 cm thick was casted in place at this time. Two types of concretes were used, namely a C60/75 and a C37/40 to obtain different stiffness values and strengths. A complete description of the work can be found in Bonnet-Eymard *et al.* (2011).

For BPE experiment (Figure 8-d), performed in GRD drift, excavation was performed with the hydraulic hammer and support is made of 4-layer fibre reinforced shotcrete (11 to 12 cm thick) emplaced after the four successive excavation steps. A thick lining of 45 cm is emplaced at a distance of less than one diameter (d = 6.2 m) from the excavation front.

In GRD4 drift (Figure 8e-f), road header under shield with segments emplacement technique was used. The lining is reinforced concrete segment of 45 cm thick. This method needs injection of backfill material behind the segments. Two types of material have been tested: a classical mortar used with TBM and a compressible mortar with polystyrene ball. The comparison between the two sections will help to understand the interface behavior and its role in the loading of the support. The main issue of this drift is to follow the stress/strain evolution (at short and long term) in the segments which will give insights to design the real support of the access drifts of Cigéo.

The behavior of HL-LL activity waste package cell over repository lifetime is complex due to the fact that lot of THM-C coupled phenomena occur at different time and with different rate. Demonstration of the working requires a good understanding of those coupled phenomena but also needs to overcome technological barrier. For example, the concept cannot be validated as a whole without demonstrating the geotechnical capacity to excavate and emplace the casing of the disposal cell.

A step by step approach was used from scientific and technological point of view in order to reach a scale 6 in the Technology Readiness Levels scale for the HL-LL activity waste cells, meaning testing of a prototype cell that is near the desired configuration in terms of performance (in our case without any radioactive waste). Step by step approach with demonstration and scientific objectives consists of different simple experiments at different scales, at borehole scale (diameter from 100 to 250 mm) and real scale with micro-tunnel of 0.7 m diameter. The first tests focused on one phenomenon or one demonstration issue. The complexity of the experiments increases with time in order to evaluate different aspects of the concept, optimize it and finally reach a demonstration experiment.

4 MAIN OBSERVATIONS

4.1 Characterization of the excavation damaged zone around the opening

4.1.1 Methods of characterization

Armand *et al.* (2014) recall the common measuring methods used to assess the EDZ extension around drifts and shaft and discuss how to determine the EDZ at the Meuse/Haute-Marne URL taking into account Tsang *et al.* (2005) definition. The main methods used are:

- Structural analysis of the core samples and 3D analysis to characterize the types of fractures and to define the fractures density according to depth,
- Geological survey of the drift face and sidewalls,
- Seismic measurements: interval velocity measurements to study the evolution of P and S wave velocities according to the distance from the drift wall in single or cross hole (Schuster *et al.*, 2001) or seismic refraction,
- Permeability measurements carried out through gas and/or hydraulic tests to assess the EDZ hydraulic conductivity with time and the degree of saturation of COx claystone (Bossart *et al.* 2002), and
- Overcoring of resin-filled EDZ fractures (Bossart *et al.*, 2002), a method to efficiently visualize and measure fractures' aperture.

At the URL, detailed geological survey with a 3D scan was conducted every meter in the first three meters of a new excavation and then at 5 meter intervals.

Since 2005, core samples from nearly 400 horizontal, vertical and inclined boreholes drilled in the URL drifts were investigated. More than 4000 excavation induced fractures were studied and no fault or natural fracture was identified in the cores. Some fractures induced by drilling were identified and were not taken into account during the analysis. Cores are reconstituted and oriented based on position in the borehole and the nearly horizontal stratification in the drifts and used to determine strike and dip of the fractures. The observation of the fracture face assess to classify the fractures according to their specific typology (shear, extension fractures).

Hydraulic pore pressure and permeability measurements at this formation requires equipment with long term stability, resistant to high pressure and allowing the fastest possible return in pressure. Often these measurements need to be started at the drift construction phase and continued for a longer period. This stresses that the equipment must be simple, robust and reliable. For this, Mini-Multi-Packer-System (designed by Solexperts AG, Fierz *et al.*, 2007) has been selected to define up to six test intervals optimally spread along the borehole. The test interval lengths range between 10 to 20 cm and are isolated by hydraulically inflated packers. The volume of the chamber is reduced in order to ensure rapid buildup and stabilization of the pressure. The rigid packers limit the compressibility of the system. Two stainless steel lines allow continuous monitoring of pressure and the running of pneumatic or hydraulic tests depend on the saturation degree in the interval. In the following sections of this chapter, mainly hydraulic test results are discussed.

In order to investigate the hydraulic properties of the rock around the opening, several test campaigns have been carried out in dedicated boreholes filled with synthetic water having the Callovo-Oxfordian formation fluid composition to prevent clay degradation as well as osmotic fluxes. Hydraulic permeability measurements were carried out using pulse test (instantaneous increase or decrease of the pressure) or constant head/rate injection test. The general model theory and application for hydraulic permeability test was described by Bourdarot (1996) and Horne (1997). Pickens *et al.* (1987) provided an application of the method for a radial composite flow, noting that the model assumes a continuous porous medium. Andra defined a strategy for the interpretation of hydraulic tests described in Baechler *et al.* (2011).

4.1.2 During shaft excavation

In the shaft at different depth, measurements were performed, mainly through seismic methods. For example a velocity survey has been emplaced from the niche -445 m to follow the damage induced during the shaft sinking between -465 m to -475m (Balland *et al.*, 2009). The excavation of the main shaft in the COx claystone of the Meuse/ Haute-Marne URL significantly disrupted the velocity field. The velocity field deteriorated from 10 m ahead of the shaft front with a maximum disruption during the shaft crossing. This disturbance decreased progressively up to 1.6 m from the shaft wall. The shaft crossing creates an oscillation of the main velocity directions with a significant velocity drop initially and then a differed velocity rebound. Finally, the order of magnitude of the velocity changes showed that deconfinement and damage to the rock were limited. Figure 9 shows the evolution of the velocity variations versus the radial position at different step of the shaft sinking (1) the initial state corresponds to the average of the velocity for each ray obtained with ten measurements (shaft front at a depth of –451 m); (2) the intermediate state corresponds to an average obtained for each ray over ten measurement points immediately after the blast hole 6 (shaft front at a depth of –468 m); and (3) the final state based on an average, for each ray, of the last ten measurements of the acquisition (front at a depth of –480 m). The radial position selected for a ray is the radial distance of the median point between the emitter and the receiver. The velocity of compression waves decreases during shaft sinking. The closer the measurement point is to the shaft wall the more significant is the decrease. The maximum reduction is in the order of 70 m/s, and this variation remains low (less than 2%) compared with the initial velocity ranging from 2,900 to 3,500 m/s. Two specific zones have been clearly identified, as follows:

- zone A, near the shaft wall (0 to 1.6 m) where the variations in the velocity of *P* waves are in the order of 50 m/s, with a stronger variation around the – 466 m level;
- zone B (>1.6 m) seems slightly affected by the variations in the velocity of P waves.

It confirms that the level of damage, at this depth, is low and that the fractures extent remains small. These results are coherent with other measurements performed in the vicinity of the shaft (permeability and deformation measurements and other seismic methods). The Figure 9-b shows the model of the extent of micro-cracks around the shaft in the upper part of the COx layer.

4.1.3 At -445 m (top of the COx layer)

The Figure 10 shows a front in the niche -445 m. The excavation was performed by a drill and blast method and the support was made of rock bolt and sliding arches. Nearly no induced fracture was observed on the front and the side wall, except fractures around blast drilled holes. Geological survey on drilled boreholes emphasized a scattered presence of fracture (Figure 10-b). In most of the boreholes no induced fracture is observed and in some boreholes up to three fractures in the first 0.3 m from the wall, mainly at the floor (horse shoe section), are observed. The low number of fractures is in accordance with the nearly elastic behavior observed in the convergence measurements.

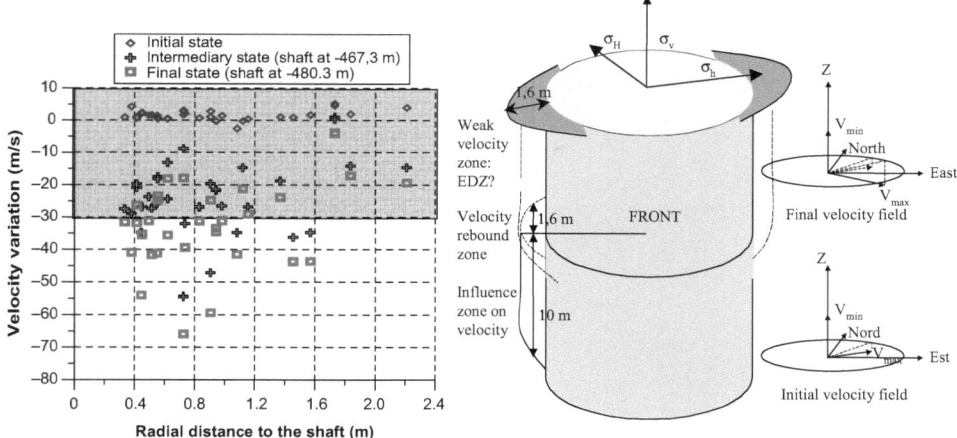

Figure 9 P-wave velocity analysis: (a) Velocity change according to shaft wall distance for the initial, intermediate and final states, (b) Interpretative figure of the velocity changes during the mine-by test (Balland *et al.*, 2009).

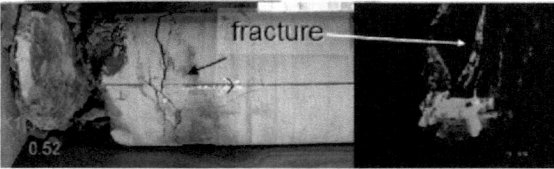

Figure 10 View of the front face during drift excavation at −445 m (b) resin injected borehole showing no fracture at the side wall and fracture near the wall at the floor.

4.1.4 At the main level of the laboratory (middle of the COx layer)

Number of fractures has been observed at the front and side (Figure 11-a) of all the new openings. The most significant information about induced fractures was obtained from drilling performed on the walls of the drifts parallel to major stress. In GET drift, 12 horizontal boreholes were performed in a narrow space. Optical imaging of these boreholes was carried out systematically to identify the dip and orientation of the fracture plans. Based on the data obtained, a 3D representation of the fracture plan in the GET drift was prepared (Figure 11-b). Two different zones of excavation induced fracturing can be clearly distinguished. The first zone near the drift consists of tensile and shear fractures. The orientation of both types of fractures is very heterogeneous

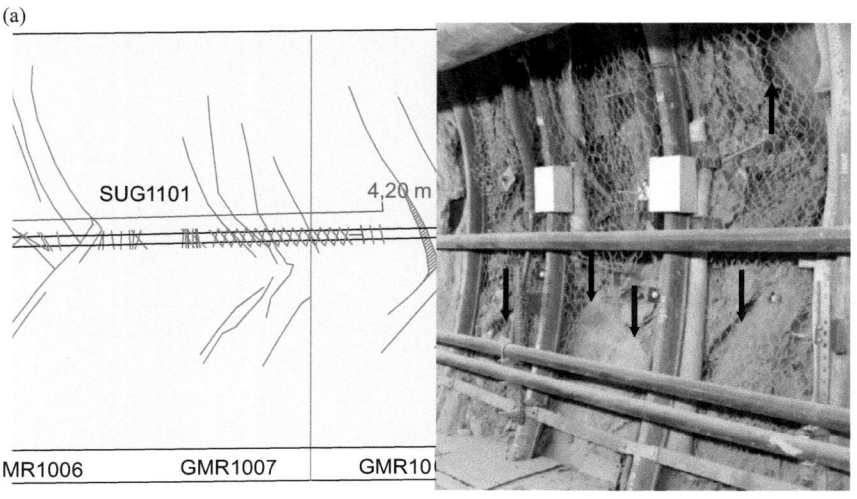

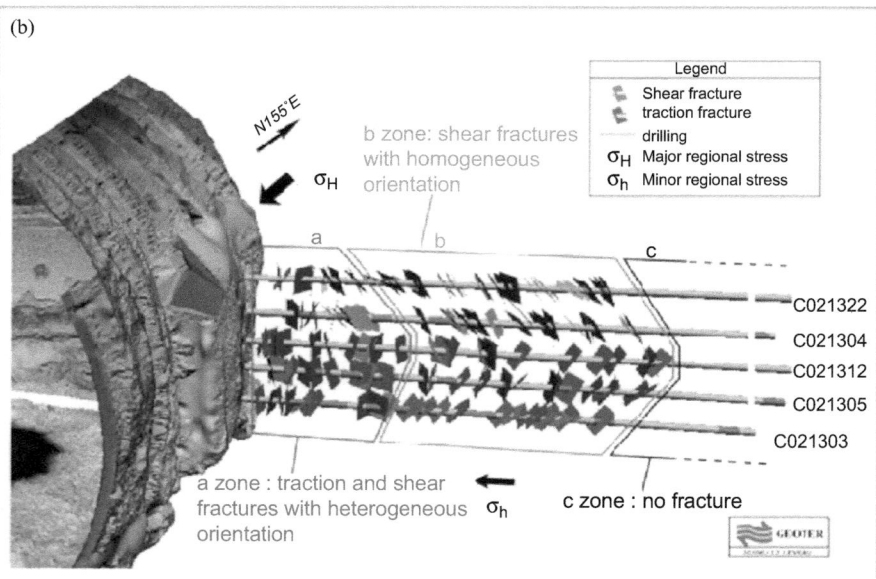

Figure 11 Excavation induced fracture pattern: (a) Chevron fractures pattern observed at gallery scale in relation with the face progression (left) and damaged zone like schistosity in the notch of chevron fracture (right), (b) 3D visualization of the fractures network in GET drift (CDZ boreholes).

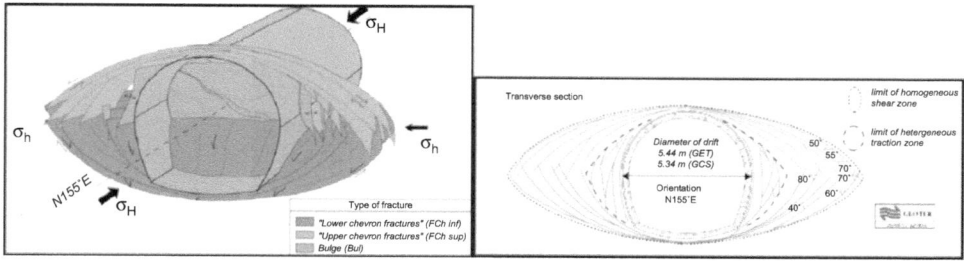

Figure 12 Conceptual model of the induced fracture network around a drift parallel to the major horizontal stress.

with high dip fractures. The second zone, farther from the drift consists of shear fractures only. Their orientation is more homogeneous with low dip fractures. The data showed that most of the excavation induced fractures (75%), mainly at the front, appear in mode II (shear fracture) and 25% are in mode I at the main level of the URL.

As a synthesis and interpretation of drifts geological survey, and impregnation data, a model of digging induced fracture network around drifts excavated parallel to major horizontal stress is proposed in Figure 12. Shear fractures are more widespread and expand farther in the rock. Extensional fractures are located near the wall and with a more heterogeneous dip and strike. This model tries to exhibit the geometry (strike and dip) of different type of fractures in different cross sections. It can be notice that the heterogeneity of the network is not well represented in the model due to the fact that the extent fluctuation along the digging axis is not enough identified and is related to the step of excavation. But the general understanding of the geometry of the fracture network is considered in the model. Figure 11a shows the front of a drift excavated parallel to σ_h, at 3.5 m from the wall. The footprint of the chevron fractures is well identifiable and confirms the reliability of the proposed model.

4.1.4.1 Impact of the orientation of the drift versus in situ stress state

Around drifts parallel to the minor horizontal stress, the same type of fracture has been identified than around drift parallel to σ_H at the front and on the core, but geometry (strike, dip, extent) differs:

- The shear chevron fractures are initiated ahead of the excavation face during work. They form symmetrically to the horizontal plane crossing the gallery axis. The fracture dip with respect to the horizontal plane is around 45°. Extension of the chevron zone ahead of the excavation face is about 1 drift diameter (about 4 m). The shear fractures are not impregnated with resin at the ceiling and roof.
- Shear vertical and oblique fractures arrest on the chevron fractures. They are oriented at a low angle (10° to 30°) with respect to the wall. This fracture system

is identified only in drill cores from the sidewall. It shows sub horizontal lineation (strike slip kinematics). The shear fractures are not impregnated with resin; *i.e.*, fractures are closed in *in-situ* conditions.

– Unloading tensile fractures: this fracture system is identified all around the drift (sidewall, vault and floor). Most fractures are parallel to drift wall, partially connected together by shear plays. This fracture system is comparable to the onion skin fracture identified at Mont Terri URL (Bossart *et al.*, 2002), but they are limited to a maximum depth of 0.5 – 0.7 m, whatever the direction considered, and there are very few of them (1 to 2). Most fractures initiated in extension mode are impregnated with resin.

Figure 13 presents the proposed model of excavation induced fractures for drift parallel to σ_h. It can be noticed that most of the fractures represented in the 3D view are shear fractures, because they expand more than unloading fracture. Near the wall the dip and strike of both family fractures are nearby and are difficult to distinguish in a 3D view. The description of the network is less significant due to less available data especially at ± 45°.

Table 2 gives the extent of the extensional and of shear fractures and emphasizes the effect of the in situ stress state on the induced fracture networks.

4.1.4.2 Impact of the size of the opening

Note that the shape of the excavated fracture networks around openings parallel to σ_H seems to be similar even at smaller scale; for drilling of 5 cm diameter and micro tunnel of 0.7 m diameter (Figure 14). All performed excavations (drifts,

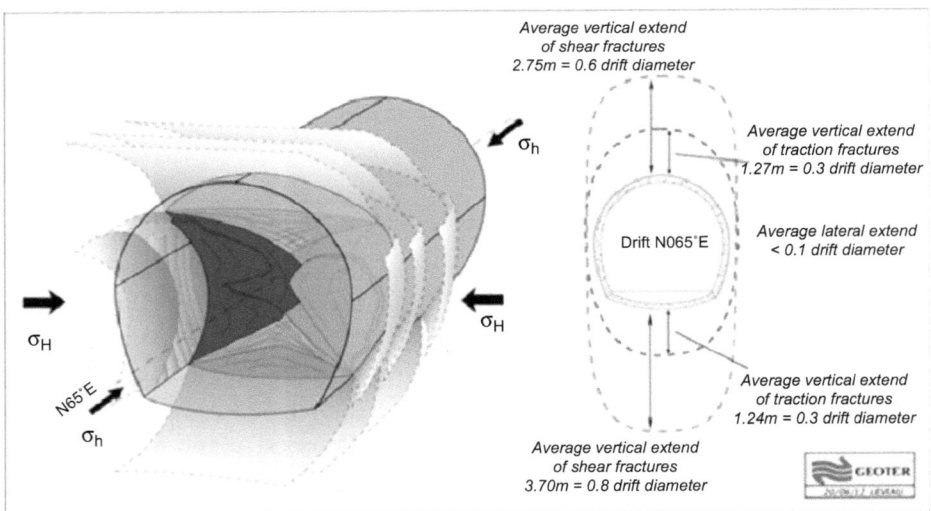

Figure 13 Conceptual model of the induced fracture networks around a drift parallel to the horizontal minor stress (Armand et al., 2014).

Table 2 Extent of excavation induced fracture zone (ratio of drift diameter from the wall).

Drift Orientation / location		Extent in diameter (×D)							
		Extensional fractures			Shear fractures			EDZ	EdZ
		Average	Min.	Max.	Average	Min.	Max.	Max.	Max.
N65//σ$_h$	Ceiling	0.3	0.2	0.4	0.6	0.5	0.8	0.4	0.8
	Wall	0.1	0.1	0.2	–	–	–	0.2	0.6
	Floor	0.4	0.2	0.5	0.8	0.8	1.1	0.5	1.1
N155 //σ$_H$	Ceiling	0.1	–	0.15	–	–	–	0.15	0.5
	Wall	0.2	0.01	0.4	0.8	0.7	1.0	0.4	1.0
	Floor	0.1	–	0.15	–	–	–	0.15	0.5

'–' means no extent farther the extensional fractures network (Armand *et al.*, 2014)

drillings, micro-tunnel) show roughly identical induced fractures network (type of fractures, geometry, extent). The lateral extent of the shear fractures zone for smaller opening is almost equal to 1 diameter of the opening, which seems to show that there is no significant scale effect from a borehole of 5 cm diameter to a drift of 5 m diameter. Excavation induced fractures density depends of the size of the opening.

4.1.4.3 Hydro Mechanical properties of the Excavation Induced Fractures

Figure 15 presents hydraulic conductivities in 8 and 9 boreholes, up to 6 m in depth to the wall, respectively in drift GCS (parallel to the horizontal major stress) and GED (parallel to the horizontal minor stress). Note that the first chamber in borehole SDZ1243 has not been tested to water, but with air (pneumatic test). Sensitivity analysis has shown that the uncertainty on hydraulic conductivity is between half and one order of magnitude. The highest hydraulic conductivity is observed in chambers, where fractures cross the borehole and are not representative of the

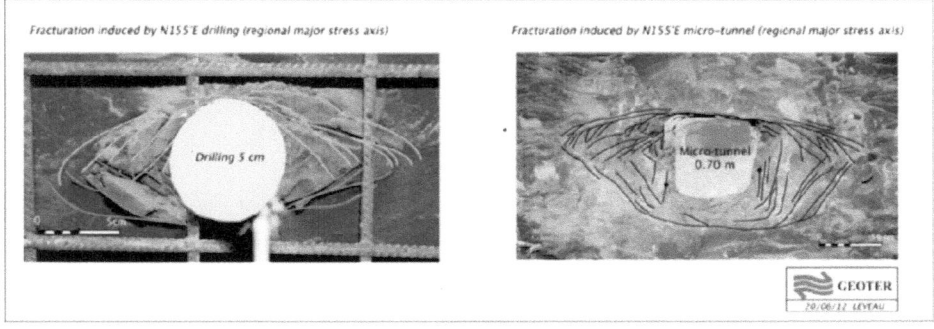

Figure 14 Induced fracture network around opening parallel to σ$_H$: (a) around a borehole of 5 cm diameter. (b) Around a micro tunnel of 0.7 m diameter (Armand *et al.*, 2014).

continuous rock matrix, but of the fractures transmissivity. This is confirmed by hydraulic tests performed on rock samples taken between fractures, which exhibit(s) hydraulic conductivity lower than 1×10^{-12} m/s.

In Figure 15 the extent of extensional fractures and purely shear fractures is also plotted for each single borehole. For all boreholes, conductivity decreases with depth from the wall, and the highest conductivity is found in the extensional fractures zone. In the extensional fractures zone, hydraulic conductivity varies from 2.0×10^{-6} m/s (permeability to gas) to 4×10^{-12} m/s (water permeability). In the area with only shear fractures, an overall slight decrease of conductivity is observed from 6×10^{-12} m/s to 1×10^{-12} m/s. In some boreholes, like OHZ1731 and OHZ1735, the conductivity is not always perfectly reduced with depth to the wall in the zone with only shear fractures, which can be due to the fact that in this zone some chambers cross fractures and others

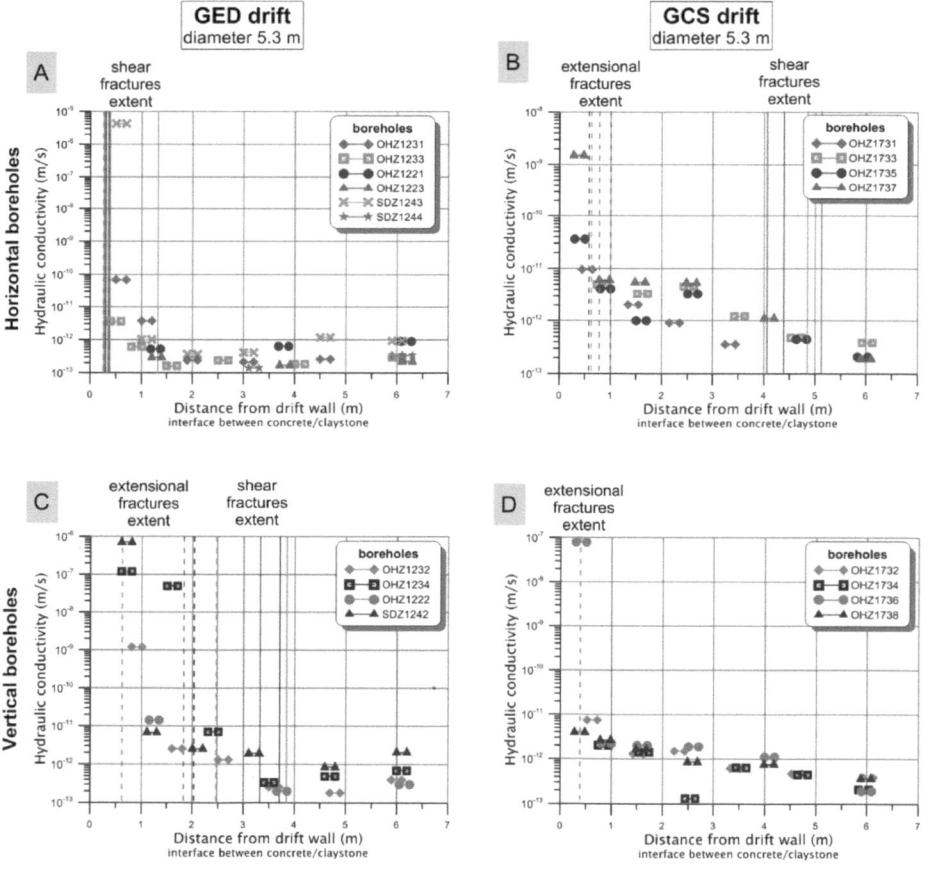

Figure 15 Initial hydraulic conductivity measured in GED (perpendicular to σ_H) and GCS (parallel to σ_H) drifts: (a) and (b) horizontal boreholes at the wall (c) and (d) vertical boreholes at floor and ceiling (modified from Armand *et al.*, 2014).

not or less. Even without any induced fractures, a slight decrease of conductivity is observed for example from 7×10^{-12} m/s to 2×10^{-13} m/s from 0.5 m to 6 m to the wall in drift parallel to σ_H.

From a general point of view, measurements in drift parallel to σ_h exhibit the same behavior to the one in drift parallel to σ_H:

(i)   the highest hydraulic conductivities are observed where extensional fractures are located, never elsewhere,

(ii)   farther deep in the rock hydraulic conductivity decrease from 10^{-10} m/s to 10^{-12} m/s in some meters

(iii) the location of high hydraulic conductivity differs with drift orientation because induced fracture networks depend on the drift orientation versus the in situ state of stress

(iv)   having in mind the uncertainty on the hydraulic conductivities back analysis, it could be considered that at one diameter to the wall the hydraulic conductivity is nearly equal to the conductivity of the virgin rock.

Then, discussion about EDZ around the drift has to be tackled taking into account Tsang *et al.* (2005) definition of EDZ (from the performance assessment point of view) to determine accurate values which are measurable in situ. In 2003 Bauer *et al.* reported conclusions of the performance assessment on the Meuse /Haute-Marne URL:

–   considering permeability of sealing systems lower than, or equal to 10^{-10} m.s^{-1}, and a damaged zone with a permeability no higher than 2 orders of magnitude less than the sound rock permeability, the geological barrier is the main radionuclide transfer pathway;

–   if the seals do not cut off the fractured zone around the openings and if the latter is continuous with a permeability higher than or equal to 10^{-9} m.s^{-1}, EDZ provokes a short circuit of the geological barrier.

The analysis of permeability presented in the previous section exhibits that permeability to water lower or equal to 10^{-10} m.s^{-1} is observed in chambers crossing fractures more or less connected to the wall. Taking into statement of Bauer *et al.* (2003), a permeability of 10^{-10} m.s^{-1} could be considered as the upper permeability limit of the EDZ for the Meuse / Haute-Marne URL. When permeability measurements are compared to the structural analysis, it can be observed that all chambers in which estimation of permeability is lower or equal to 10^{-10} m.s^{-1} is crossed by or nearby fractures which are located in the zone where extensional fractures have been observed. On the other hand, all the fractures located in the zone with extensional fractures do not necessarily exhibit high transmissivity lying to average permeability lower than 10^{-10} m.s^{-1}. Then, the extent of tensile fracture zone could be considered as an upper bound of the EDZ extension at the main level of the Meuse / Haute-Marne URL. The EDZ is mainly the connected fractures network with shear and extensional fractures. It's also proved that in this claystone we have to distinguish the mechanical damage and the EDZ, if the definition of Tsang *et al.* (2005) is used. In fact the propagation of shear fractures is larger than the extent of EDZ. In our case, the EDZ is exactly what Lanyon (2011) described as HDZ (Highly Damaged Zone defined as a zone where macro-scale fracturing or spalling may occur).

Farther in the rock, permeability decreases slowly to reach permeability of intact rock at around a diameter from the wall and defines the EdZ (Excavation disturbed Zone). The EdZ area is formed by the end of the shear fractures which not affect a lot the average permeability, like on the side wall of drift parallel to σ_H. The EdZ contains also micro-cracks and/or porosity change which implies slight change in permeability like at the ceiling of drift parallel to σ_H. Table 2 summarizes the extent of EDZ and EdZ as function of the drift orientation.

The discrepancy between permeability to water and permeability to gas is linked with the self sealing processes which reduce the hydraulic conductivity of the fractured zones. A meter-scale experiment (a few square meters in area) entitled "Compression of the Damaged Zone" (CDZ) has been implemented in the main level of the MHM URL to study the evolution of the EDZ hydrogeological properties (conductivity and specific storage) of the Callovo-Oxfordian claystone under mechanical compression and artificial hydration (de La Vaissière et al., 2015). The objective is to study the influence of mechanical loading and unloading conditions and the influence of hydration on selected properties of the EDZ. Mechanical compression is applied to the drift wall by a hydraulic loading device consisting of a 1 m² circular steel plate (Figure 16-a). Plate displacements and corresponding applied forces and strain in the rock (through extensometer) are monitored. Multi-packer systems for hydraulic measurements and performance testing (Fierz et al., 2007) were

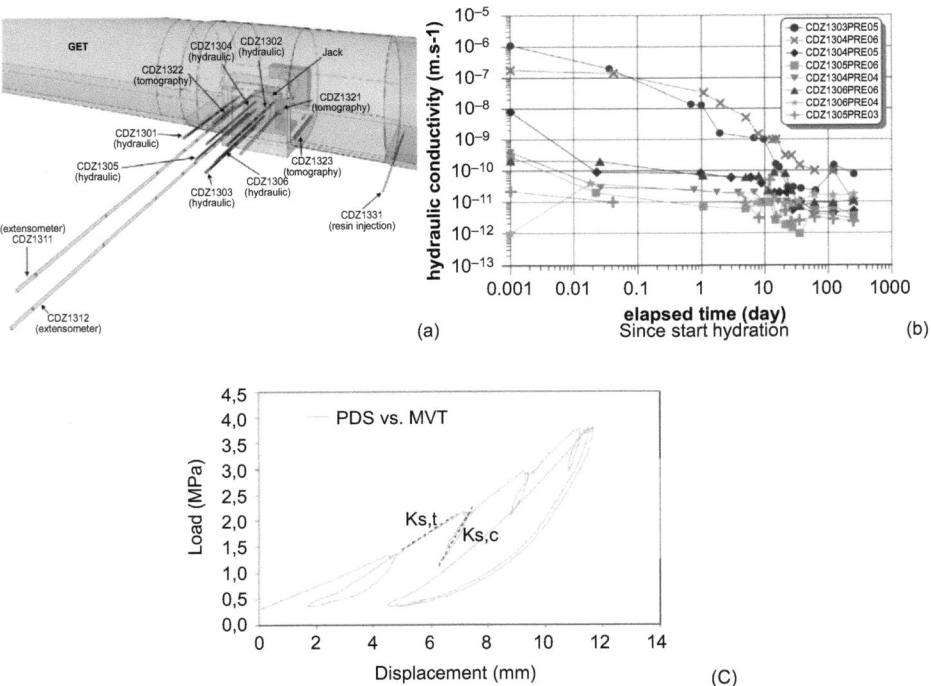

Figure 16 CDZ experiment: (a) experimental layout, (b) evolution of the hydraulic conductivity since the begging of hydration for the repetitive hydraulic tests (de La Vaissière et al., 2015), (c) stress/displacement curve during the loading test.

installed in six boreholes *i.e.* three boreholes located behind the loading plate and three boreholes located in the immediate vicinity.

Firstly, a loading cycle applied on a drift wall was performed to simulate the compression effect from bentonite swelling in a repository drift (bentonite is a clay material to be used to seal drifts and shafts for repository closure purpose). Gas tests (permeability tests with nitrogen and tracer tests with helium) were conducted during the first phase of the experiment. De La Vaisière *et al.* (2015) showed that the fracture network within the EDZ was initially interconnected and opened for gas flow (particularly along the drift) and then progressively closed with the increasing mechanical stress applied on the drift wall. Moreover, the evolution of the EDZ after unloading indicated a self-sealing process. Secondly, the remaining fracture network was resaturated to demonstrate the ability to self-seal of the COx claystone without mechanical loading by conducting from 11 to 15 repetitive hydraulic tests with monitoring of the hydraulic parameters. During this hydration process, the EDZ effective transmissivity dropped due to the swelling of the clay minerals near the fracture network (Figure 16-b). The hydraulic conductivity evolution was relatively fast during the first few days. Low conductivities ranging at 10^{-10} m/s were observed after four months. Cross-hole tests showed the disappearance of the preferential network interconnectivity pathway along the drift axis (major pathway for performance assessment purposes, due to its potential to act as a seal bypass), but heterogeneous conductivity effects persisted. Some uncertainty remains on this parameter due to volumetric strain during the sealing of the fractures. These in-situ measurements confirm the results of previous in-situ tests on plastic clay (at the Mol URL) and claystone (at the Mont Terri URL), as well as other tests performed on various samples (Davy *et al.*, 2007; Zhang, 2011).

The mechanical loading cycles on the wall show irreversible strain and an increase of stiffness with the increase of normal stress, which is a classical evolution for fractured media. Using a simple Boussinesq solution and strain measured under the plate, Young's modulus has been calculated (Figure 16-c). It emphasizes low modulus at the wall (up to 0.5 m) which increases with depth in the rock mass.

4.2 Hydro mechanical behavior of drift during and after excavation

4.2.1 Measurement methods

Mine-by-test is the best way to achieve reliable hydro-mechanical data around a drift to give insight in time-dependent and hydro-mechanical behavior of the EDZ. Such test has been implemented in various rocks in several countries, such as in Lac du Bonnet granite (Read & Martin, 1996), ED-B experiment in Opalinus claystone (Martin *et al.*, 2002) or CLIPEX experiment in indurate Boom Clay (Bernier *et al.*, 2002). At the main level of the Meuse/Haute-Marne URL, locations of the experimental drifts have been chosen, when it was possible, in order to perform mine-by experiments. The major issue was to provide deformation and pore pressure measurements in near and far field of the new drift before excavation works in order to record the HM impact of the digging. Pore pressure measurements have been instrumented at least two months before drift excavation, to let pore pressure builds up and stabilizes.

During the excavation others measurements are installed, mainly convergence measurements and extensometers.

After completion of the drift, an important drilling campaign has been performed in order to characterize the induced fracture networks with geological survey of the drill core and following the HM behavior in the near field with multi-packer system to measure pore pressure and permeability, and with velocity survey to try to pick up damage evolution if there is an evolution.

4.2.2 During shaft excavation

Figure 17 illustrates convergence measurements at different depth in the shaft as a function of the orientation to the north. Up to -468 m the convergence is a sinusoidal function of the orientation versus the major/minor horizontal stress. The larger amplitude are observed in the direction of the major horizontal stress, confirming the anisotropic in situ stress field and the nearly elastic behavior of the claystone up to that depth. Deeper in the layer, convergences in the direction of the minor horizontal stress become the larger ones, showing that plasticity/damage in this direction appears signed of a more elastoplastic behavior. The change in behavior is not due to the small change in vertical stress, but to the mineralogy change with an increase of the clay

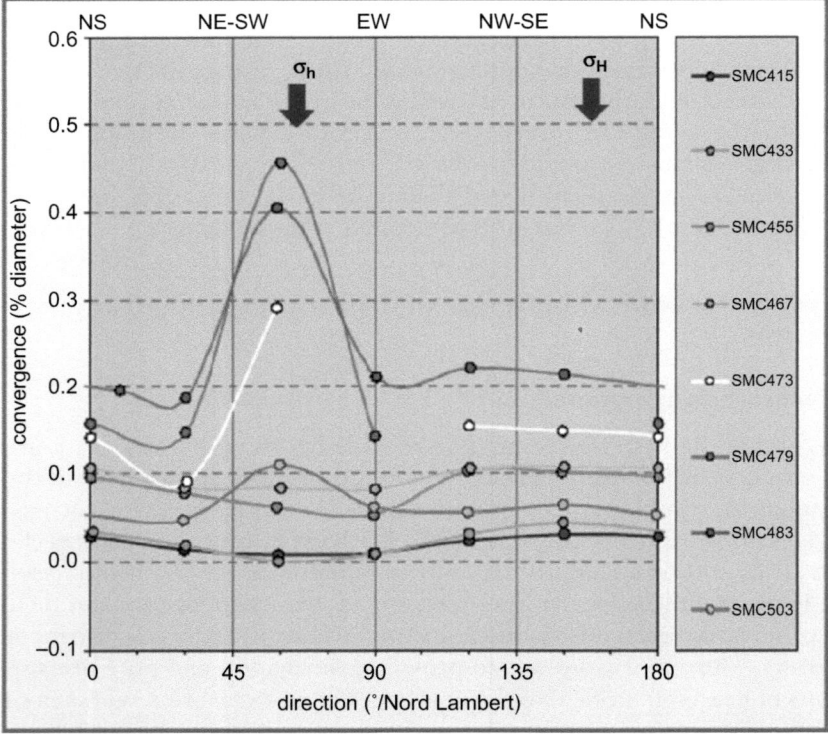

Figure 17 Convergence measurements at different depth in the COx layer (between 415 m to 503 m, when the front face is at 10 m below the section).

content (*i.e* a decrease of UCS from around 30 MPa at –445 m to 21 MPa at the main level).

4.2.3 At –445 m (top of the COx layer)

Figure 18 shows the evolution of pore pressure during the sinking of the access shaft in borehole REP2101 (oriented in the direction of the major horizontal stress) and REP2102 (oriented in the direction of the minor horizontal stress). The distances between the measurement sensors and the wall range from 1.8 to 5 m. Before sinking resumed, pore pressures were almost stable at 3.4 to 3.9 MPa. When sinking resumed, pressure-measurement chambers were located at least 8.5 m below the floor. Successive blasts induce an instantaneous variation in the pore pressure.

In borehole REP2101, the first blast generates a small reduction in pressure, followed by a new stabilization. This variation has an amplitude of 0.01 MPa for the farthest pressure-measurement chamber from working face and 0.09 MPa for the nearest. The amplitude of those jumps increases in proportion with the proximity of the working face with the chamber level. The most significant reduction is observed when shaft sinking passes the level of the measurement chamber and it increases when the wall is close to the chamber. It varies between 0.29 MPa in Chamber No. 1 at 4.85 m from the wall and 0.69 MPa in Chamber No. 5 at 1.7 m from the wall. After the passage of the level of every chamber, they all show a decrease in pore pressure. Pressures in the chambers tend to stabilize in August 2005 when sinking operations stopped at a depth of 483.36 m. Pressure distribution depends on the distance from the shaft, and pressures range from atmospheric pressure for the chamber located at 1.7 m from the wall to 2.0 MPa for the farthest chamber from the shaft (at 4.85 m from the wall).

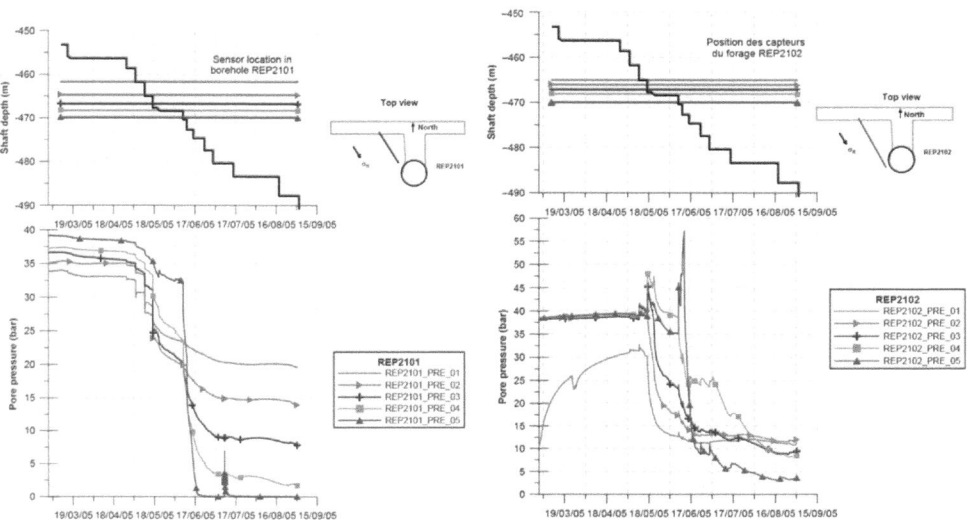

Figure 18 Evolution of pore pressures (a) in borehole REP2101 (// σ_h), (b) in borehole REP2012 (//σ_H), (Armand & Su, 2006).

In borehole REP2102, high overpressures are observed in chambers. Those overpressures increase in proportion to the distance of the measurement chambers from the shaft wall. They may vary from 0.06 MPa in Chamber No. 1 at 2.25 m from the wall to 1.6 MPa in Chamber No. 5 at 1.09 m from the wall. As soon as the working face passes the chamber level, the pore pressure begins to decrease.

The evolution of the measured pore pressures is therefore consistent with the poroelastic approach (analytical solution and numerical modeling): the highest overpressures occur in the closest measurement chambers to the direction of the minor horizontal stress and the strongest underpressures occur in the closest measurement chambers to the major horizontal stress, confirming the sound operation of the experimental system. The measurements of the evolution of mechanical and hydromechanical parameters confirm that:

– measured deformations remain low. COx claystone reacts almost elastically at this depth. Irreversible low-amplitude deformations are only observed at a distance of 1 to 1.5 m from the shaft wall;
– evolution of the pore pressure depends on the advance of the excavation work, the radial distance from the wall and the orientation of the measurement chamber in relation to the anisotropy of the *in-situ* stress field;
– the location of excavation-induced overpressures and underpressures is consistent with the stress concentrations generated by the sinking of a shaft within an anisotropic stress field, thus indicating a strong hydromechanical coupling (especially mechanical on hydraulic);
– Different measurement methods and approaches show a slight mechanical disturbance around the shaft. The mechanical disturbance generates a slight variation in both permeability (less than one order of magnitude) and the velocity of compression waves (reduction of less than 2% of the initial velocity).

4.2.4 At the main level of the laboratory (middle of the COx layer)

4.2.4.1 Convergence and deformation of drifts

Orientation of the drift, regarding to the horizontal major stress, plays an important role on the measured convergence (Armand *et al.*, 2014). Anisotropic convergences are

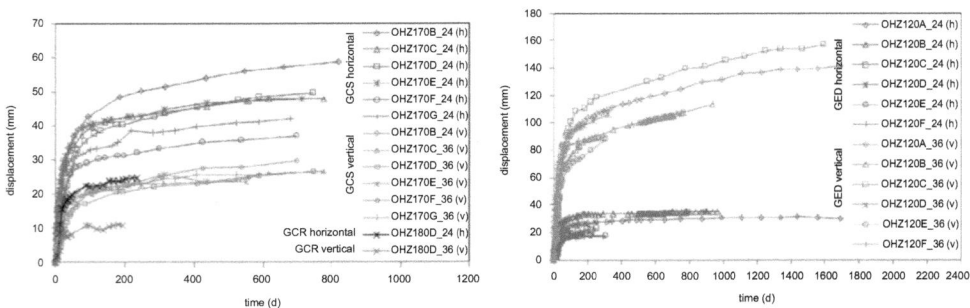

Figure 19 Convergence measurement in drift parallel: (left) to the horizontal major stress (drift GCS/GCR) and (right) to the minor horizontal stress (drift GED) (Armand *et al.*, 2013).

observed in all drifts. Figure 19 shows the convergence measurements in drift parallel to the major horizontal stress (drift GCS/GCR) and to the minor horizontal stress (drift GED).

In GCS drift, the horizontal convergences are perceptibly higher than the vertical ones ($C_h/C_v \approx 2$), while the in situ stress state is nearly isotropic. However, evolution of the horizontal and vertical convergences in time is similar. It can be seen that more than 80% of the convergence has been reached during the first 100 days. The convergence rates decrease with time and reach velocities lower that 5×10^{-11} s^{-1}. In GCR drift without yieldable wedges, the general trend of the curve and the convergence ratio are similar (Figure 19-a), but the amplitude of convergence is smaller due to the fact that the shotcrete support without wedges is stiffer than the one with yieldable wedges. It also means that the stress in the shotcrete lining increases more when yieldable wedges are not used, which is observed in the displacement/load measurements. The average convergence amplitudes are divided by 2.

In GED drift ($// \sigma_h$), the vertical convergence is higher than the horizontal one, which is the opposite of the observations in drifts parallel to the major horizontal stress. It is worth to note that the in situ horizontal stress is about 1.3 times the vertical one. This fact can be considered as the main reason of the anisotropic damage pattern observed around GED drift. The convergence ratio (C_h/C_v) is about 0.25. High convergence rates are observed during the first three months. A decrease of convergence rates is measured during the 3 years of observation.

Guayacán-Carrillo et al. (2016) demonstrate that convergence data of two drifts (GCS and GED) could be analyzed through the identification of the principal axes of deformation considering an elliptic deformation shape. The evolution in time of the two axes of the ellipse is interpreted as the convergence of the drift wall along the two principal directions of deformation, and fitted using the convergence law proposed by Sulem et al. (1987). The convergence law has five parameters, namely; T: related to the time-dependent properties of the ground, X: related to the distance of influence of the face and the extent of the decompressed zone around the drift, m: related to the ratio between the time-dependent convergence and the instantaneous convergence and C∞x: instantaneous convergence. These parameters have been evaluated for different drifts and for different monitoring sections in each drift. The model is validated by simulating the closure of a different drift (GCR) with the three parameters X, m, and T obtained in GCS and by only fitting C∞x. It is also shown that by monitoring convergence for about 40 days, the model can be applied for reliable predictions of the long term convergence evolution. The results of the analysis (Guayacán-Carrillo et al., 2016), for all the monitored sections of two drifts excavated in perpendicular directions, show that constant values for the parameters X and m can be assumed. Parameter T is constant along a drift and takes two different values.

In all the drifts, the major convergence is measured where the fracture zone is located. The convergence has directly an impact on the loading and deformation of the support. As the loading is anisotropic the deformation is also anisotropic with extension and compression zone depending of the location in the structure. The evolution of measured strain and stress in the casted concrete in GCR drift or in the concrete segment in GRD drift confirms such anisotropic stress field in the structure which has to be considered in the design and optimization of the support for the repository.

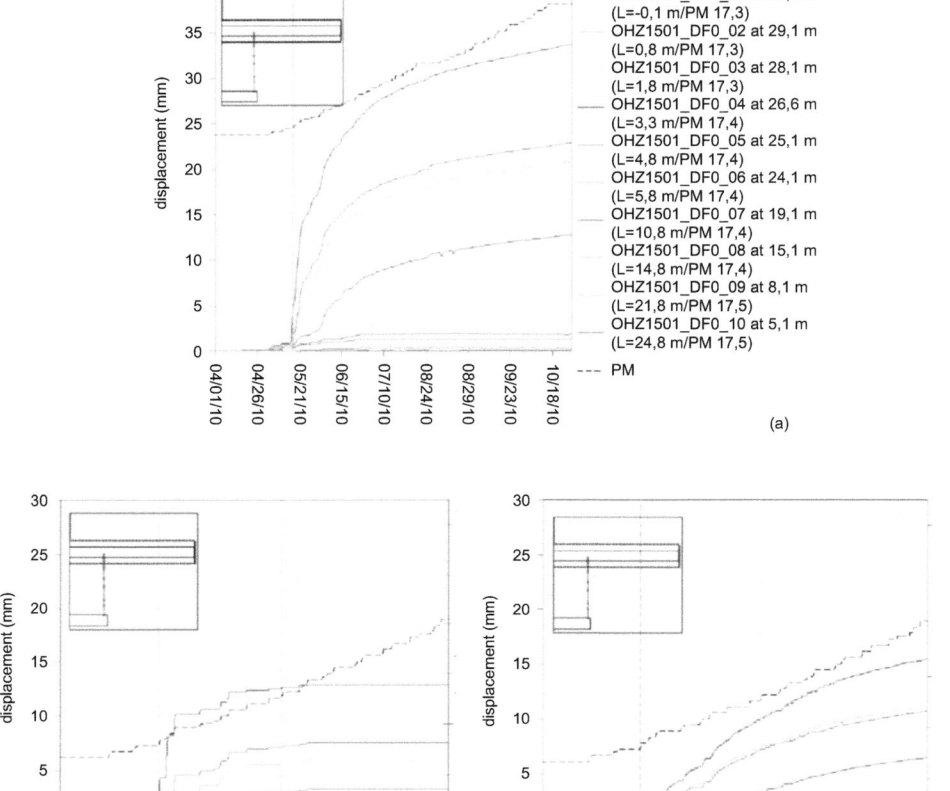

Figure 20 Radial displacement of drift GCS (//σ_H): (a) at different distance (d) from the wall (blue dash line is the date at which the front crosses the extensometer), (b) instantaneous response (elastoplastic behavior), (c) differed response. (L is the distance from the wall), (Armand et al., 2013).

To better understand deformation during the excavation work, an extensometer has been drilled from the GAT to the GCS drift ending close to the side wall of the GCS. The main advantage of this measurement compared to the convergence measurements or the classically installed extensometers is that it records displacement before the front face reaches the section. The black dashed line in Figure 20 shows the excavation steps. The extensometer starts to records radial displacement when the front is two spans (2.4 m ≈1 radius) ahead the section (Figure 20-a). Beyond 7 steps of excavation the displacement doesn't show any jump, meaning that the long-term deformation behavior is predominant (versus the elastoplastic one). Radial deformation measured in drift GCS has been analyzed in order to separate instantaneous and differed response to digging considering that the elastoplastic displacement is induced instantaneously during each excavation steps (Armand *et al.*, 2013). Between each excavation step, convergence increases due to differed effects. In this framework, each drift displacement curve can be seen as a set of "jumps" due to elastoplastic deformations and "slightly increasing curves" between excavation steps due to differed deformations. Two radial displacement curves are plotted for GCS drift in this manner (Figure 20-b-c). As it can be seen, no more "instantaneous elastoplastic deformation" is measured when the front reaches approximately 2 diameters. At this time, elastoplastic deformation (Figure 20-b) is slightly larger than the differed one (Figure 20-c). Elastoplastic deformation represents 40%–45% of the total deformation at 90 days, and around 30% at 900 days. Beyond 5.8 m deep in the rock, differed deformation is very small. The contribution of the far field is about 5% of the total elastoplastic deformation (it will be around 10% with an elastic behavior), 4% of the differed deformation at 90 days and less than 1% at 900 days.

4.2.4.2 Loading of supports

The following section illustrates the strain/stress measurement in different supports and emphasizes the relationship between strain/stresses in the support within the convergence of the rock and the excavation induced fracture zone. Figure 21 presents a comparison of deformations measured with extensometer (from the drift wall and 3 meters deep in the rock) in three drifts, with three types of support. 0° indicates horizontal extensometer at the wall, ± 90° indicates vertical upward and downward extensometer at the roof and floor respectively. The general trends are the same; the increase of the stiffness of the shotcrete lining by increasing thickness or removing yieldable wedge does not change the overall behavior, but deformations amplitude is a little reduced. However, the convergence ratio increases from 2.5 in GCS to 3.5 in BPE and to 6 in GCR.

During the excavation, deformation is the smallest in the GCR drift, meaning that at short term the 20 cm shotcrete lining is stiffer than shotcrete lining of 45 cm projected in four layers of 10 cm during two weeks. After the shotcrete maturation (around 28 days) the rate of deformation in BPE is two times lower the ones measured in GCS and

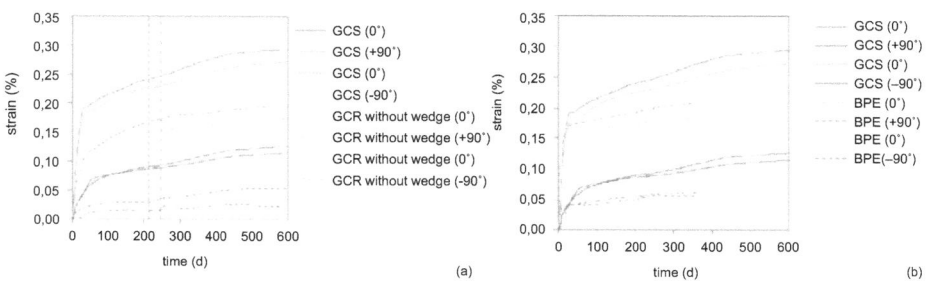

Figure 21 Comparison of deformation (measured from the wall up to 3m in rock) in drift parallel to the major horizontal stress: (a) in GCS (with yieldable concrete wedge) and GCR (without yieldable wedge (two black dashed lines show when each step of the lining is grouted)), (b) in GCS and BPE (45 cm of shotcrete put in four step).

GCR. This result illustrates the effect of the lining support stiffness on rock deformations. After 28 days, the similar rate of deformation in GCR compared to GCS could be explained by the fact that the load is more important in GCR support and shotcrete started to be damaged and stiffness of the support reduced. In GCR (Figure 21-a), small deformations variation occurs just after the grouting of the concrete lining, which has to be related to local stress state change (load of the counter vault lining) and thermal effect due to concrete curing. Installing the concrete lining decreases deformation rate in the rock, but cannot reduce it to zero. The shrinkage of the concrete allows deformation of the rock. In the three drifts, deformation rates are lower than 10^{-11} s-1, one year after excavation.

In GRD4 drift excavated with a road header under a shield with segments emplacement technique, strains and stresses in the concrete lining, as well as the compression of the backfill mortar are followed. Four rings (8 segments/ring - the key segment is not instrumented) have been equipped in Stradal's manufactory. Three total pressure cells and eight strain gages were installed on the reinforcement of each segment of the 4 monitored rings. Deformations and stresses of the instrumented lining are more significant on the classical backfill zone. Figure 22 shows an example of the segments behavior; it compares the total deformations (without correction) of two V4 segments one on the classical mortar (A22) and one on the compressible mortar (A62) (Zghondi *et al.*, 2015).

Figure 23 presents the corrected deformations (from the concrete shrinkage) and stresses evolution on the lining section circumference; the results show an anisotropic loading signature with opposite deformations and stresses between intrados and extrados. The load is located on the side wall and is coherent with the higher horizontal convergences observed in drift parallel to the major stress (see section 4.2.4.1).

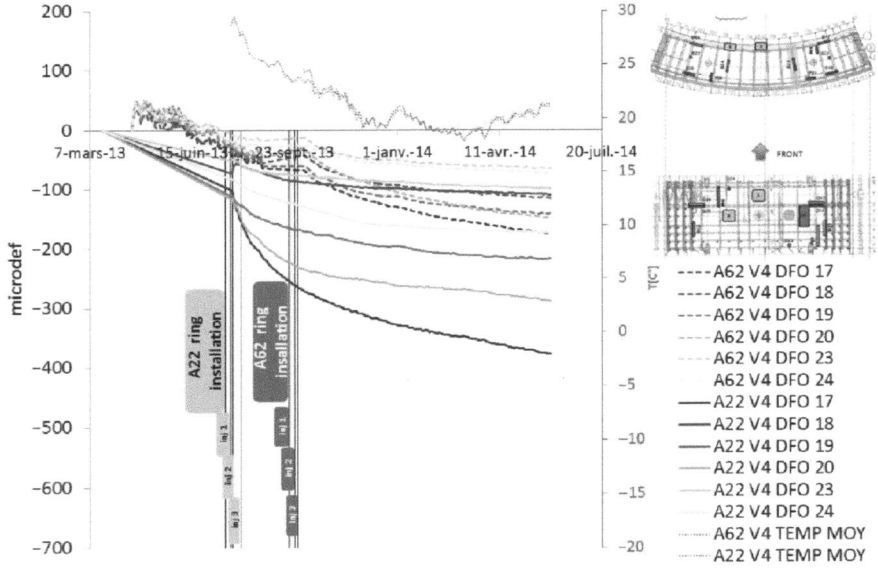

Figure 22 Comparison of deformations in the V4 segment lining for the two different backfill grout zones: A22 (classical grout zone) and A62 (compressible grout zone) (Zghondi *et al.*, 2015).

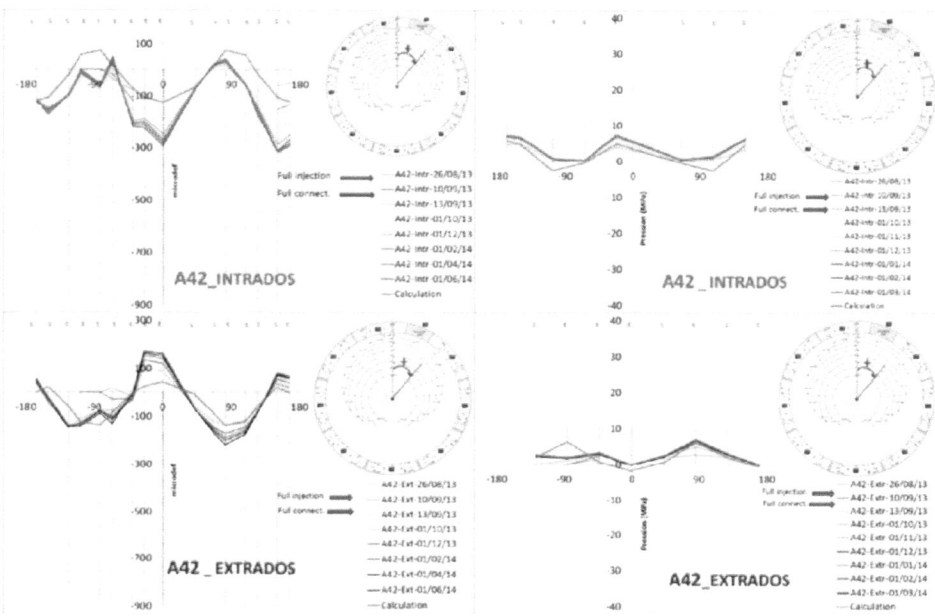

Figure 23 Evolution of the circumferential strains (microdef.) and stresses (MPa) on the extrados and intrados of the A42 (classical mortar backfill zone) ring (right: orthoradial strain measured with vibrating wire, left: orthoradial stress measured with total pressure cells (Zghondi *et al.,* 2015).

Anisotropic loading is observed from early age just after the first grout injection. Loading and deformations on the two mortar zones have similar shape but they are more significant on the classical mortar zone. This behavior confirms the impact of the behavior of the mortar on the loading of the segments. Compressible mortar decreases the orthoradial stress in concrete segments.

Reduced and full scale *in-situ* experiments (in the framework of the HL-LL activity waste program) consisting of equipping boreholes parallel to the major horizontal stress (σ_H) with instrumented steel tubing, have been performed to analyze the mechanisms involved in the casing/rock interface (Bumbieler *et al.,* 2015). The main characteristics of the short term mechanical load applied by the rock have been determined from local strain and convergence measurements. As for the drift in the same direction regarding the horizontal stress, the performed measurements exhibit a strongly anisotropic load with maximum tensile stress measured in the horizontal symmetry plane.

4.2.4.3 Pore pressure change during and after excavation

Two examples of hydro-mechanical response to the excavation of openings parallel to the major horizontal stress of different diameters (from 0.15 m and 5 m) show a strong coupling from the mechanical behavior on the hydraulic response (Figure 24). Figure 24-a and –b display the pore pressure change during the excavation of GCS drift. On the side-wall axis during the GCS drift digging:

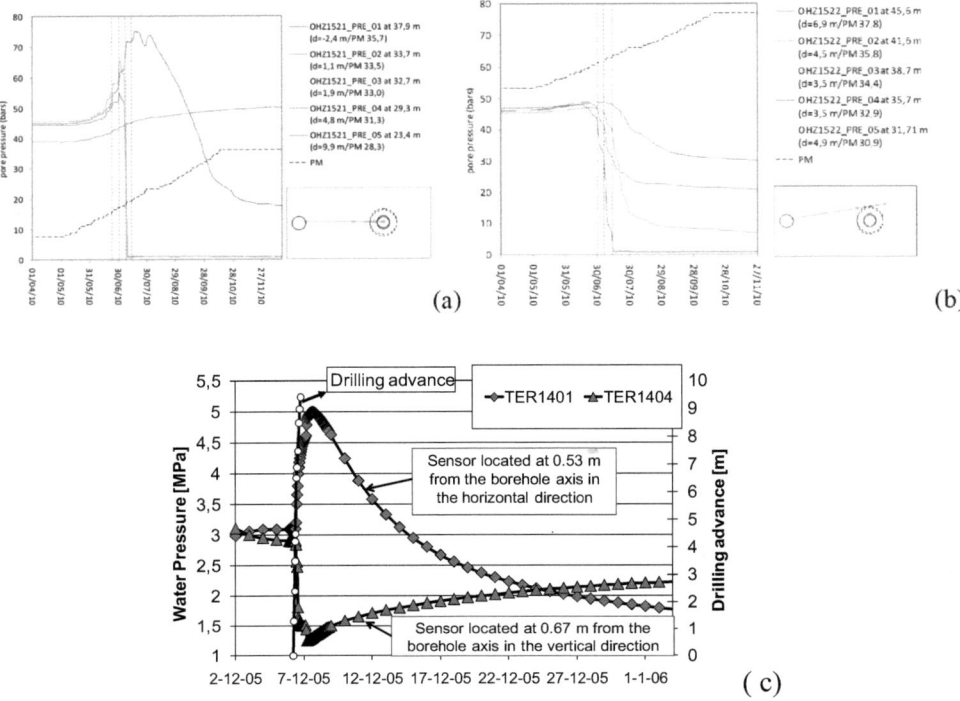

Figure 24 Pore pressure change during excavation of an opening parallel to the major stress (a) on the horizontal axis of drift GCS, (b) on the vertical axis of drift GCS and (c) drilling of a 150 mm-diameter borehole.

- Pore pressures first increased in all intervals since 18/05/2010, when the front reached 16.9 m. The distance of influence on the pressure field is about 20 m;
- For the interval ahead the face, pore pressure increases and reaches a peak when the face is at the 4.8 m the interval. Then pressure drops to 0 at 2.7 m probably due to the fact that the face hits the borehole;
- For the nearest intervals (up to 1.9 m), the peak of pore pressure was observed ahead the face. After that, pore pressure dropped and when the face reached the plane of the measurement interval, a drastic drop was observed down to the atmospheric pressure;
- Farther in the rock, an overpressure is generated during the excavation of the drift. It begins to dissipate after the excavation and finally the pressures become stable at 4.8 m from the wall. At 9.9 m from the wall the drop of pressure is not observed during the 6 first months after the excavation work, the pore pressure continues to increase.

Above the GCS wall during the excavation, the figure shows:

- Pore pressures first slightly increased in all intervals up to 2 to 3 bars. This increase is small compared to over pressure observed on the side wall (1 order of magnitude difference with the horizontal plane). The distance of influence of the face on the pressure field is about 20 m;

- The drop of pressure is observed two to three step of excavation before the front face cross the measurement section;
- After each step of excavation, a drop of pressure is observed.
- At 3.5 m from the drift wall, the pore pressure drops to the atmospheric pressure.

The results of two pore pressure sensors in a cross-section at 7 m from the bore-hole mouth are presented in Figure 24-c : one is located at 0.53 m from the borehole axis in the horizontal direction and the second one is located at 0.67 m from the borehole axis in the vertical direction. In the horizontal direction a very large increase of pore pressure is observed and a somewhat smaller reduction of pore pressure is obtained in the vertical direction, consistent with the fact that it is slightly farther away from the borehole. In both cases the water pore pressure response correlates closely with the drilling advance and a significant increase/decrease of pore water pressure is observed when the drilling tool passes the sensor section. Afterwards, the excess pore water pressure dissipates. The dissipation seems to occur more rapidly in the horizontal direction corresponding to the smaller distance to the draining neighboring borehole and also, most likely, to the higher permeability in the horizontal direction. The overall pore pressure evolution is similar to the one observed on the drift GCS of 5 m diameter, even if the overpressure is larger with respect to the distance (d/D) to the wall.

Even if the state of stress is nearly isotropic around the drift the pore pressure evolution is rather different at the side wall in the bedding plane than at the roof/floor of the drift. Significant over pressures are observed at the side wall far in the field. The mechanisms behind the pore water pressure response around an underground opening are twofold. The first type of mechanisms can be associated with nearly undrained behavior, due to the very low permeability of the COx claystone, and the related pore water pressure changes ranged here are induced by the stress redistribution triggered by the creation of the tunnel wall causing a reorientation of the principal stresses and influenced by the initial stress anisotropy. This means in the short term the pore pressure changes are due to volumetric deformation of the rock mass. The presence of over pressure deep in in the field (> 1 diameter) in a section with isotropic stress state emphasizes the role of the anisotropy in stiffness (the material is more stiff in the direction of bedding, *i.e.* horizontal). These pore water pressure changes are closely linked to the mechanical behavior of the rock and to the damage zone around the opening even beyond of this zone. Wild *et al.* (2015) demonstrated that major pore pressure drops in the tunnel near field can, in theory, be explained by both a pure elastic response in an isotropic or anisotropic rock mass characterized by an anisotropic stress state, or by dilatancy which accompanies failure.

The second type of mechanisms is related to drainage of excess pore water pressure relative to a state governed by the atmospheric water pressure condition at the gallery wall and the water flow law. The overpressure generated during the excavation is propagated in the field (increase of pressure are observed farther than 2 diameters) and also to the drift.

It was noted that the shape of the induced fracture network around openings parallel to σ_H seems to be similar even at the smaller scale *i.e.* for a 5 cm diameter

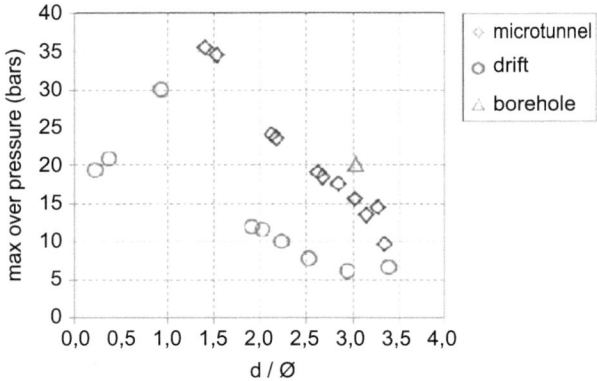

Figure 25 Maximum over-pressure measured at the wall side of different diameter opening :drift of 5 m diameter, micro tunnel of 0.7 m diameter and borehole of 0.15 m diameter (d: distance to the wall, φ: diameter of the drift) (Armand *et al.,* 2015).

borehole and micro tunnel of 0.7 m diameter (see section 4.1.4.2). All excavations (drifts, drillings, micro-tunnel) induce roughly identical fractures network (type of fractures, geometry, extent) with a lateral extent of the shear fractures zone around 1 diameter of the opening. This is consistent with the fact that the pore pressure evolution is rather similar for the different opening sizes, except in terms of magnitude. Figure 25 shows the maximum over pressure measured as a function of the distance to the wall side for different diameter opening (drift of 5 m diameter, micro-tunnel of 0.7 m diameter and borehole of 0.15 m diameter). The peak of over pore pressure is a decreasing function with distance to the wall, except in the area where the induced fractures appear. As the pressure response is strongly coupled to the volumetric strain, one could think that the larger opening will create larger over pressure. This is not the case due to the fact that the rate of excavation differs a lot from 5×R/h, 1×R/h to 0.006×R/h (with R the radius of the excavation opening), respectively for the borehole, the micro-tunnel and the drift. Another aspect is that there is a support in the drift which reduces the unloading of the radial stress, but this is not significant compared to the rate of excavation. With low rate of excavation, pore pressure releases and over-pressures are lower.

REFERENCES

Abou-Chakra Guéry A. (2007), Contributions à la modélisation micromécanique du comportement non linéaire de l'argilite du Callovo-Oxfordien, PhD thesis Université de Lille I.

Andra (2005), Dossier 2005 Argile. Evaluation de la faisabilité du stockage géologique en formation argileuse profonde. Rapport de synthèse, Juin 2005, Andra, France, www.andra.fr.

Armand, G. & Su, K. (2006), Hydromechanical coupling phenomena observed during a shaft sinking experiment in a deep argillaceous rock, In Proceeding of the Geoproc 2006 Conference, Nanjing, May.

Armand, G., Noiret, A., Zghondi, J. & Seyedi, D.M. (2013), Short- and long-term behaviors of drifts in the Callovo-Oxfordian claystone at the Meuse/Haute-Marne Underground Research Laboratory, Journal of Rock Mechanics and Geotechnical Engineering, 5, 221–230.

Armand, G., Leveau, F., Nussbaum, C., de La Vaissiere, R., Noiret, A., Jaeggi, D., Landrein, P. & Righini, C. (2014), Geometry and properties of the excavation-induced fractures at the Meuse/Haute-Marne URL Drifts, Rock Mechanic and Rock Engineering, 47 (1), 21–41.

Armand, G., Noiret, A., Morel, J. & Seyedi, D. (2015), Pore pressure change during the excavation of deep tunnels in the Callovo Oxfordian claystone, ISRM Congress 2015 Proceedings – International Symposium on Rock Mechanics – ISBN: 978-1-926872-25-4.

Armand, G., Conil, N., Talandier, J. & Seyedi, D.M. (2016) Fundamental aspects of the hydromechanical behaviour of Callovo-Oxfordian claystone: From experimental studies to model calibration and validation, Computers and Geotechnics, in press, http://dx.doi.org/10.1016/j.compgeo.2016.06.003

Balland, C., Morel, J., Armand, G. & Pettitt, W. (2009), Ultrasonic velocity survey in Callovo-Oxfordian argillaceous rock during shaft excavation, International Journal of Rock Mechanics & Mining Sciences, 46, 69–79.

Bauer, C., Pépin, G. & Lebon, P. (2003), EDZ in the performance assessment of the Meuse/Haute-Marne site: Conceptual Model used and Questions addressed to the Research, In Proceedings of the EC Cluster Conference on Impact of the EDZ on the Performance of Radioactive waste Geological Disposal, Luxembourg, November 3–5, 2003, European Commission Report EUR21028EN.

Bernier, F., Li, X.L., Verstricht, J., Barnichon, J.D., Labiouse, V., Bastiaens, W., Palut, J.M., Ben Slimane, J.K., Ghoreychi, M., Gaombalet, J., Huertas, F., Galera, J.M., Merrien, K., Elorza, F.J. & Davies, C. (2002), CLIPEX. Report EUR 20619, Commission of the European Communities, Luxembourg.

Blechschmidt, I. & Vomvoris, S. (2010). Underground research facilities and rock laboratories for the development of geological disposal concepts and repository systems. In: Ahn, J. & Apted, M.J. (eds), Geological Repository Systems for Safe Disposal of Spent Nuclear Fuels and Radioactive Waste, Woodhead Publishing, Cambridge, 82–118, Chapter 4.

Bonnet-Eymard, T., Ceccaldi, F. & Richard, L. (2011), Extension of the Andra underground laboratory: Methods and equipment used for dry, dust-free works, in World Tunnel Congress 2011, Helsinki.

Bossart, P., Meier, P.M., Moeri, A., Trick, T. & Mayor, J.-C. (2002), Geological and hydraulic characterization of the excavation disturbed zone in the Opalinus Clay of the Mont Terri, Engineering Geology, 66, 19–38.

Bourdarot, G. (1996), Essais de puits: méthodes d'interprétation. Éditions Technip; Institut français du pétrole, Paris, 350.

Boussinesq, J. (1885). Application des potentiels à l'étude de l'équilibre et du mouvement des solides élastiques. Gauthiers-Villars, Paris.

Bumbieler, F., Necib, S., Morel, J., Crusset, D. & Armand, G. (2015), Mechanical and SCC behavior of an API5L steel casing with the context of deep geological repositories for radioactive waste, Proceedings of the 2015 ASME Pressure Vessels & Piping Conference PVP2015, July 19–23, 2015, Boston, Massachusetts.

De La Vaissière, R., Armand, G. & Talandier, J. (2015), Gas and water flow in an excavation-induced fracture network around an underground drift: A case study for a radioactive waste repository in clay rock, Journal of Hydrology, 521, 141–156.

Delay, J., Rebours, H., Vinsot, A. & Robin, P. (2007a), Scientific investigation in deep wells for nuclear waste disposal study at the Meuse Haute Marne underground research laboratory, Northeastern France, Physics and Chemistry of the Earth, 32 (1–7), 42–57.

Delay, J., Vinsot, A., Krieguer, J.-M., Rebours, H. & Armand, G. (2007b), Making of the underground scientific experimental programme at the Meuse/Haute-Marne underground research laboratory, northeastern France, Physics and Chemistry of the Earth, 32 (1/7), 2–18.

Delay, J., Lebon, P. & Rebours, H. (2010), Meuse/Haute-Marne centre: Next steps towards a deep disposal facility, Journal of Rock Mechanics and Geotechnical Engineering, 2 (1), 52–70.

Delay, J., Bossart, P., Ling, L.X., Blechschmidt, I., Ohlsson, M, Vinsot, A., Nussbaum, C. & Maes, N. (2014), Three decades of underground research laboratories: What have we learned?, Geological Society, London, Special Publications, first published March 5, 2014. doi:10.1144/SP400.1

Distinguin, M. & Lavanchy, J. M. (2007). Determination of hydraulic properties of the Callovo-Oxfordian argillite at the bure site: Synthesis of the results obtained in deep boreholes using several in situ investigation techniques, Physics and Chemistry of the Earth, Parts A/B/C, 32 (1–7), 379–392.

Fierz, T., Piedevache, M., Delay, J., Armand, G. & Morel, J. (2007), Specialized instrumentation for hydromechanical measurements for hydromechanical measurements in deep argillaceous rock, FMGM 2007, Seventh International Symposium on Field Measurements in Geomechanics.

Gasc-Barbier, M., Chanchole, S. & Bérest, P. (2004), Creep behaviour of Bure clayey rock, Applied Clay Science, 26 (20), 449–458.

Guayacan-Carrillo, L.-M., Sulem, J., Seyedi, D.M., Ghabezloo, S., Noiret A. & Armand, G. (2016), Analysis of long-term anisotropic convergence in drifts excavated in Callovo-Oxfordian Claystone, Rock Mechanics and Rock Engineering, 49, 97–114. doi:10.1007/s00603-015-0737-7

Horne, R.N. (1997), Modern well test analysis. A computer-aided approach, Petroway, Inc., Palo Alto, 257.

Lebon, P. & Ghoreychi, M. (2000), Frenchunderground research laboratory of Meuse/Haute Marne THM aspect of argilite formation, Eurock 2000 Symposium, Aachen, Germany.

Lerouge, C., Grangeon, S., Gaucher, E.C., Tournassat, C., Agrinier, P., Guerrot, C., Widory, D., Fléhoc, C., Wille, G., Ramboz, C., Vinsot, A. & Buschaert, S. (2011), Mineralogical and isotopic record of biotic and abiotic diagenesis of the Callovian–Oxfordian clayey formation of Bure (France), Geochimica et Cosmochimica Acta, 75, 2633–2663.

Martin, C.D., Lanyon, G.W., Blümling, P. & Mayor J.C. (2002), The excavation disturbed zone around a test tunnel in the Opalinus Clay. In: R. Hammah, W. Baden, J. Curran & M. Telesnicki (eds.), Proceedings of the 5th North American Rock Mechanics Symposium and the 17th Tunnelling Association of Canada Conference: NARMS/TAC 2002, , 1581–1588, University of Toronto Press, Toronto.

Martin, F., Laviguerie, R. & Armand, G. (2010), Geotechnical feedback of the new galleries excavation at the ANDRA underground research laboratory – Bure (France), Eurock 2010, Lausanne.

NEA (2001a), Going underground for testing, characterization and demonstration (A Technical Position Paper). NEA/RWM (2001)6/rev1.

NEA (2001b), The role of underground laboratories in nuclear waste disposal programmes, Radioactive Waste Management, NEA 3142.

Pickens, J.F., Grisak, G.E., Avis, J.D., Belanger, D.W. & Thury, M. (1987), Analysis and interpretation of borehole hydraulic tests in deep boreholes: principles, model development, and applications, Water Resources Research, 23 (7), 1341–1375.

Read, R.S. & Martin, C.D. (1996), Technical summary of AECL's Mine-by Experiment, AECL-11311, AECL research (Series), Whiteshell Laboratories in Pinawa Man, 169p.

Schuster, K., Alheid, H.-J. & Böddener, D. (2001), Seismic Investigation of the EDZ in Opalinus Clay, Engineering Geology, 61, 189–197.

Souley, M., Armand, G., Su, K. & Ghoreychi, M. (2011), Modelling of the viscoplastic behaviour including damage for deep argillaceous rocks, Physics and Chemistry of the Earth, 32, 2–18.

Sulem, J., Panet, M. & Guenot, A. (1987), Closure analysis in deep tunnels, International Journal of Rock Mechanics and Mining Sciences & Geomechanics Abstracts, 24 (3), 145–154. doi:10.1016/0148-9062(87)90522-5

Vigneron, G., Delay, J., Distinguin, M., Lebon, P. & trouiller, A. (2004), Apport des investigations multi echelles pour la construction d'un modèles conceptual des plateformes carbonattées de l'Oxfordieen moyen et supérieur de l'Est du Bassin de Paris, Avancées et pespectives : hommage à Claude Mégnien, Paris, 16–17 Novembre 2004.

Vinsot, A., Mettler, S. & Wechner, S. (2008), In-situ characterization of the Callovo-Oxfordian pore water composition, Physics and Chemistry of the Earth, 33, S75–S86.

Wild, K.M., Amann, F. & Martin, C.D. (2015), Some fundamental hydro mechanical processes relevant for undertsanding the pore pressure response around excavation in low permeable clay rocks, ISRM Congress 2015 Proceedings – International Symposium on Rock Mechanics – ISBN: 978-1-926872-25-4

Wileveau, Y., Cornet, F.H., Desroches, J. & Blumling, P. (2007), Complete in situ stress determination in an argillite sedimentary formation, Physics and Chemistry of the Earth, 36, 1949–1959.

Yven, B., Sammartino, S., Geraud, Y., Homand, F. & Villieras, F. (2007), Mineralogy, texture and porosity of Callovo-Oxfordian argillites of the Meuse/Haute-Marne region (eastern Paris Basin), Mémoires de la Société géologique de France, 178, 73–90.

Zghondi, J., Carraretto, S., Noiret, A. & Armand, G. (2015), Monitoring and behavior of an instrumented concrete lining segment of a TBM excavation experiment at the Meuse Haute-Marne Underground Research Laboratory (France).

Zhang, C.L., Czaikowski, O. & Rothfuchs, T. (2010), Thermo-Hydro-Mechanical behavior of the Callovo-Oxfordian clay rock, GRS – 266.

Zhang, C.-L. (2011), Experimental evidence for self-sealing of fractures in claystone, Physics and Chemistry of the Earth, Parts A/B/C, 36 (17–18), 1972–1980.

The five-volume set *Rock Mechanics and Engineering* consists
of the following volumes

Volume 1: Principles
ISBN: 978-1-138-02759-6 (Hardback)
ISBN: 978-1-315-36426-1 (eBook)

Volume 2: Laboratory and Field Testing
ISBN: 978-1-138-02760-2 (Hardback)
ISBN: 978-1-315-36425-4 (eBook)

Volume 3: Analysis, Modeling and Design
ISBN: 978-1-138-02761-9 (Hardback)
ISBN: 978-1-315-36424-7 (eBook)

Volume 4: Excavation, Support and Monitoring
ISBN: 978-1-138-02762-6 (Hardback)
ISBN: 978-1-315-36423-0 (eBook)

Volume 5: Surface and Underground Projects
ISBN: 978-1-138-02763-3 (Hardback)
ISBN: 978-1-315-36422-3 (eBook)